Palm Oil

Production, Processing, Characterization, and Uses

Palm Oil

Production, Processing, Characterization, and Uses

Editors
Oi-Ming Lai
Chin-Ping Tan
Casimir C. Akoh

AOCS PRESS

AOCS Mission Statement
AOCS advances the science and technology of oils, fats, surfactants and related materials, enriching the lives of people everywhere.

ISBN: 978-0-12-810230-5

Library of Congress Cataloging-in-Publication Data
Palm oil : production, processing, characterization, and uses / editors, Oi-Ming Lai, Chin-Ping Tan, Casimir C. Akoh.
p. cm.
1. Palm oil. I. Lai, Oi-Ming. II. Tan, Chin-Ping. III. Akoh, Casimir C., 1955-
TP684.P3P357 2012
664'.36--dc23

2011044440

15 14 13 5 4 3 2

The paper used in this book is acid-free and falls within the guidelines established to ensure permanence and durability.

Contents

Preface

Palm oil is now the world's largest vegetable oil by volume. With proven nutritional benefits in addition to the presence of nutraceutical components, such as tocotrienols and carotenoids, palm oil is consumed in over 150 countries worldwide. It plays a pivotal role in the socioeconomic development of Asian, Latin American, and African regions. In 2010, global production of oils and fats stood at 172.1 million tonnes. Palm oil combined with palm kernel oil was the largest contributor to these, accounting for 51.6 million tonnes, or 30% of the total output. Of the 65 million tonnes of oils and fats entering global trade, palm oil and palm kernel oil accounted for close to 61% of the total volume, or 39.6 million tonnes, with Malaysia having 45% of the market share, dominating the international trade in palm oil.

The success of the palm oil industry in Malaysia, Indonesia, and elsewhere has never been easy and has only gotten harder in recent years. Its success has often been made more difficult by mistrust and misconceptions about the oil, its composition, and its nutritional benefits. Thus, with this critical issue in mind, this book has been assembled with the hope of being an authoritative, comprehensive, and highly informative compilation of recent advances and issues in various important areas of palm oil industry—from the production, processing, and characterization to the utilization of palm oil and its components.

The topics herein deal with both the production and processing of palm oil and kernels, and more importantly, the editors attempted to provide an international perspective to this area by drawing experts from around the world, especially Malaysia, Brazil, and Nigeria. Thus, this book has amassed diverse topics from these experts in specific areas from across the globe. This book begins with a chapter on the history of oil palm and subsequently three chapters related to oil palm breeding, oil palm genomics, tissue culture and genetic engineering of oil palm. The book also includes several topics related to postharvest activities, pests, and diseases of oil palm. Information on production of palm oil and palm kernel oil; physical, chemical, and polymorphic properties of palm oil and its components; as well as the measurement and maintenance of palm oil quality are included in the book, which may be of interest to researchers, scholars, technologists, and food manufacturers. To understand the importance of this plant-based oil, the variety of applications of palm and kernel oil, and their fractions in food, nutritional and oleochemical products are included. The potential use of palm oil as an alternative to *trans* fats in foods is also highlighted. There are also special chapters dedicated to the use of palm biomass in various wood-based products and as sources of bioenergy and biofuels. We have also included in-depth accounts of three current issues on oil palm/palm oil that are garnering attention and becoming globally relevant, namely waste and environmental management, trade and traceability, together with other issues relating to sustainable development of the palm oil industry. All the chapters in this book are presented in a stand-alone manner without the need to make references to other chapters in the book, thus allowing each of the contributors to provide comprehensive reviews of their field of expertise.

The editors have made every effort to select contributors who are internationally recognized as experts in their field to develop the chapters and cover all relevant topics in this book. To all the authors, we thank them for their outstanding efforts and timely submission of their chapters. The editors would also like to thank the editorial staff of AOCS Press for their helpful suggestions toward improving and coordinating the timely publication of this book. We hope, you the readers, will find this book a valuable read and a great addition and source of reference on the current developments in palm oil/oil palm industry.

Oi-Ming Lai
Chin-Ping Tan
Casimir C. Akoh

1

A Brief History of the Oil Palm

Ian E. Henson
7 Richmond Dale, Clifton, Bristol BS8 2UB, UK

Introduction

The African oil palm (*Elaeis guineensis* Jacq.) has been utilized by mankind as a source of oil and other products for thousands of years. In the last 50 years or so there has been a phenomenal expansion in its cultivation throughout the tropics, such that palm oil is now a major commodity of world trade, and the oil palm is a leading source of vegetable oil. This chapter traces the history of the African oil palm and its relatives in terms of its origins, evolution, distribution, and utilization; reviews the growth of the palm oil industry; and examines the progress made in enhancing production through selective breeding, improved cultivation practices and exploitation of optimum environments that together have resulted in progressive and substantial yield increases in this most productive of all oil-bearing crops.

History is often largely a matter of conjecture to the extent that it led Henry Ford (1916) to describe it as "more or less bunk". Certainly there is a need for careful interpretation of past records and available evidence, but this often generates opposing views and ideas. In the case of the African oil palm, *Elaeis guineensis,* and its relatives, while there is much documented evidence concerning its development as a major crop, there are nevertheless a number of uncertainties regarding its origins and distribution.

Perhaps the most comprehensive account of oil palm history is found in the classic work, *The Oil Palm.* It is now in its 4th edition, the first three of which are by Hartley and appeared in 1967, 1977, and 1988; and the fourth, by Corley & Tinker, was published in 2003. In these volumes one finds detailed reviews of the origins of the African and American oil palms, their spread, and the development of the oil palm industry in various parts of the tropics. No attempt will be made here to emulate these exhaustive and seminal treatments of the subject, which the reader needing further information is advised to consult. Rather, an attempt is made to summarize the more prominent events and developments in the history of the oil palm and highlight points of uncertainty that remain such as those concerning origins, taxonomic relationships, and geographical dispersal.

Taxonomy, Origins and Distribution of Oil Palms

Taxonomy and Classification

Three species of palm are currently accepted as belonging to the genus *Elaeis,* which is one of a number of genera within the sub-family *Arecoideae* of the family *Arecaceae* (formerly known as *Palmaceae*) (Box 1.1). The major oil palm of commerce, the African oil palm (*E. guineensis* Jacq.) (Fig. 1.1), was formally named as such by Jacquin in 1763 based on specimens collected in Martinique. The naming and classification of the species has been retained since that time, despite numerous synonyms having been proposed (Schultes, 1990). Its origin on the Guinea coast of West Africa had been attested to as early as the late 16th century (Lobelius, 1570, 1576, 1581).

Box 1.1. Scientific classification.

Kingdom: *Plantae*
Family: *Arecaceae*
Subfamily: *Arecoideae*
Tribe: *Cocoseae*
Subtribe: *Elaeidinae*
Genus: *Elaeis*

Source: GRIN (2011).

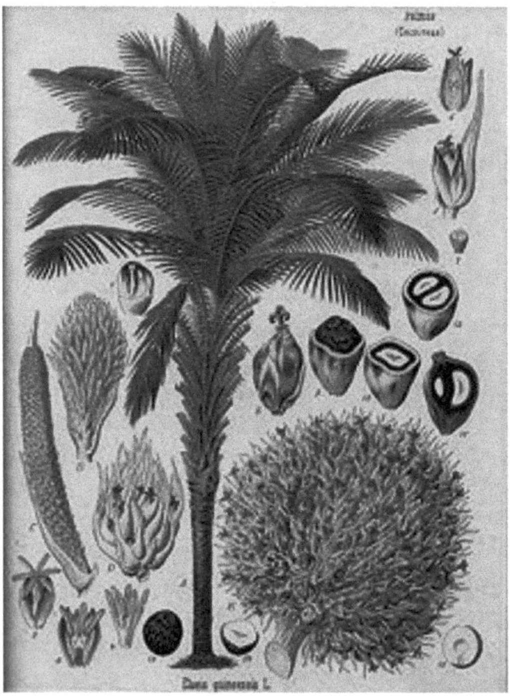

Fig. 1.1. The African oil palm (*Elaeis guineensis*, Jacq.). Source: Brandt et al. (1887).

The classification of the remaining species of *Elaeis* has proved more contentious. The American oil palm, now known as *E. oleifera* (HBK), Cortes has been considered variously as being either (1) a form of *E. guineensis*, (2) a separate species within the genus *Elaeis*, or (3) a member of the genus *Corozo* [Described by Bailey (1940) as being "transferred" from the earlier *Alfonsia oleifera* of Humbolt, Bonpland, and Kunth (HBK) (Hartley, 1977)]. Its formal description by Cortes as *E. oleifera* dates from 1897.

E. oleifera was initially known as *E. melanococca* and considered to be a form of *E. guineensis* (Corley, 1976). However, there are several distinct differences in morphology between the African and American oil palm that argue against this, notably the rate of trunk growth and the structure of the fronds and inflorescences.

Although *E. oleifera* was initially placed in the genus *Corozo,* due to its close affinities to *E. guineensis* as shown by its readiness to hybridize with this species and produce fertile offspring (Hardon & Tan, 1969), its placement in a separate genus was considered unwarranted. Hardon (1969) observed that the American oil palm appears to be the more primitive of the two, as indicated by the dominance in *E. guineensis* × *oleifera* hybrids of several characters that originate from the latter species.

The third species, *E. odora* Traill, was previously known as *Barcella odora,* but its minor morphological differences from those of the other *Elaeis* species (Corley, 1976), together with affinities revealed by molecular marker analysis (Corley & Tinker, 2003, citing work of Barcelos et al., 1999), are not thought sufficient to warrant it being placed in a separate genus.

Of the three species, the African oil palm is the only one cultivated on a wide scale. *E. oleifera* finds use for producing *E. guineensis* × *oleifera* hybrids, while *E. odora,* which grows wild in South America, has not yet been domesticated. Another proposed member of the genus was *E. madagascariensis* Becc, now generally thought to be a form of *E. guineensis* (Hartley, 1977). This chapter deals mainly with *E. guineensis* but reference is made to *E. oleifera* where appropriate.

Intra-Specific Variation in *E. guineensis*

Within *E. guineensis,* several genetically distinct forms are recognized based on the structure and coloration of the fruits. The major structural character is the thickness of the shell or endocarp that surrounds the seed or kernel (Fig. 1.2). Three main forms are generally recognized [i.e., *dura* (thick-shelled), *tenera* (medium or thin-shelled), and *pisifera*, (shell-less)]. There is also a fourth type, termed *macrocarya* (Cobley, 1963), with a very thick "stony" endocarp that is taken to represent an extreme form of the *dura* but which otherwise "has no genetic significance" (Hartley, 1977). Shell thickness is controlled by a single gene, as shown by segregation of the progeny of *tenera* selfs to give *dura, tenera,* and *pisifera* offspring in the ratio 1:2:1. It is the crosses between *dura* and *pisifera* that result in *tenera* hybrids that are the basis of modern plantations (Box 1.2).

Fig. 1.2. A ripe fruit bunch on a young *tenera* oil palm. Photograph by author.

Box 1.2. Did you know?

- The original description of the African oil palm by Jacquin in 1763 was based on material collected in Martinique to which he attributed a West African origin.
- It was extremely fortunate and purely by chance that the four *dura* palms imported into Indonesia in 1848 that formed the basis of the industry in Southeast Asia proved to be so uniform and potentially productive.

Other distinctions occur in the external coloration of the fruits. There are three main forms: *nigrescens*, being the most common with violet to black unripe fruits turning at maturity to reddish orange with a brown cap (*rubro-nigrescens*) or paler orange with a black cap (*rutilo-nigrescens*); *virescens*, with unripe green fruits and light reddish orange ripe fruits with a small greenish tip; and *albescens,* with deep green unripe fruits becoming pale yellow or ivory with apical green or black cap as they ripen. The color of the *albescens* is due to the virtual absence of carotene in the mesocarp. Both this and the *virescens* are relatively rare. The *virescens* is of interest as it is very easy for harvesters to identify ripe bunches from a distance without the need to check for the presence of loose fruit. As might be expected, the three forms differ in their chemical composition, such as carotene and anthocyanin contents. As with shell thickness, inheritance of fruit coloration is under mono-genetic control. "Mantled" fruits with supplementary carpels surrounding the main fruit are another variation. They occur naturally but were also found in high frequencies in early batches of tissue-cultured palms, resulting in low fertility and inferior oil yield and fruit quality. This contrasts with their previous inclusion in breeding programs aimed at increasing mesocarp content (Hartley, 1977).

In addition to fruit types, forms of *E. guineensis* with distinct vegetative characters are recognized, such as the *idolatrica* palm with fused leaflets (Hartley, 1977) and the dumpy (short-stemmed) palm with slow vertical growth.

Geographical Origins and Distribution

There has been divergence of opinion as to whether the African oil palm originated in Africa or in the Americas (Corley, 1976; Corley & Tinker, 2003; Hartley, 1977). While evidence supports both possibilities, the majority opinion favors West Africa as the centre from which *E. guineensis* has been dispersed, largely as a result of human activity. The main evidence supporting this, detailed by Rees (1965), Hartley (1967, 1977, 1988), Corley (1976), Gerritsma & Wessel (1997), Corley & Tinker (2003), and others, includes the following:

- The presence of fossil pollen similar or identical to that of *E. guineensis* in Miocene strata in Nigeria dating back 15 million years BP.
- The finding of oil that may have been palm oil in an Egyptian tomb dating back to 5000 BP.
- The presence of oil palm pollen in early deposits in the Congo dating from 24000 BP and in the Cameroons from 2730 BP.
- The discovery of fossil seeds and shells of oil palm in Uganda, the latter dating from 5000 BP.
- Reports by early European explorers and traders of the existence of palm oil and oil palms along the coast of West Africa, beginning with journeys by the Portuguese in 1435 and thereafter, that preceded the discovery of mainland South America by Colombus in 1498. Palm groves were observed in West Africa by Portuguese mariners in 1506 at about which time trade in palm oil commenced by the Portuguese. This and earlier reports make it highly unlikely for the palm to have been introduced into West Africa from South America, although, as mentioned by Hartley (1988), introduction in pre-Colombian times cannot be entirely ruled out.
- Vernacular terms for the African oil palm in South America are of African derivation rather than relating to indigenous American languages.
- Rees (1965) cites negative evidence opposing an American origin. Thus, there is no mention of palm oil or palm wine in early reports of the palms present in Brazil, nor is oil palm recorded as being among plants of American origin, such as the pineapple, that were introduced into West Africa.

Evidence in support of an American origin includes the following:

- There are considerably more palm species in America than are found in Africa.
- The majority of other species in the sub-family *Cocoideae* (now *Arecoideae*) are of American ancestry.
- The presence of wild groves of *E. guineensis* in Brazil is consistent with an American origin.

- Palm oil in Africa is largely eaten with cassava, which was introduced from Brazil.
- The dominance in the hybrids of characters inherited from the American oil palm *E. oleifera* over those of *E. guineensis* suggests the former to be the more primitive species, which favors the Americas as being the original home of the genus.

From the above, it seems the arguments for an American origin, while not entirely invalid, are less strong than those supporting the African case. Rees (1965), from a detailed study of historical records, opined that there was conclusive evidence for the presence in West Africa of *E. guineensis* at the time of the early Portuguese voyages and a clear description of palm oil there before the discovery of the new world.

A third possibility is that both species evolved in a common environment prior to the separation of South America and Africa by continental drift some 98 million years ago. But if this were the case, then one might expect to find native *E. oleifera* populations in tropical Africa.

The present day distribution of the African oil palm is now very wide, largely as a result of human dispersal. Within the African continent it occurs from 16°N in Senegal to 15°S in Angola (Fig. 1.3). It is also found off the East coast of the continent on the islands of Pemba, Zanzibar, and Madagascar and as far south as 21° on the

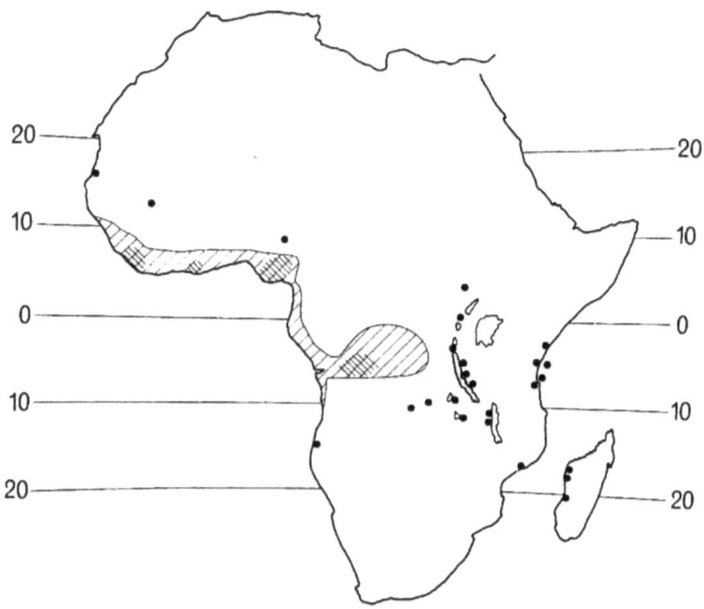

Fig. 1.3. Major locations of *Elaeis guineensis* in Africa. The main areas where the palm is ubiquitous are hatched, with the densest populations occurring in the double hatched areas. Black dots represent isolated populations. Reproduced with permission from Hartley (1977).

West coast of the latter island (Hartley, 1977). However, the main zone in which the palm is concentrated lies between 3°N and 7°S, a range demarcated by the 1600 mm isohyet (Gerritsma & Wessel, 1997). Indeed, rainfall appears to be the main climatic factor affecting distribution (Blach-Overgaard et al., 2010).

E. oleifera is considered endemic to equatorial America where it is located in a broad band stretching from 12°N in Nicaragua to 10°S in Brazil (Meunier & Hardon, 1976), mainly to the east of the Andean range. It is found in the form of wild stands, although these have not been exploited to the same extent as the African groves of *E. guineensis*. There is some variation in fruit and vegetative characters depending on location, but this is generally less than found in *E. guineensis*. An exception is a distinctly compact form of *E. oleifera* found in Surinam (Meunier & Hardon, 1976; Rao et al., 1989).

Habitat

Since the work of Zeven (1967), most workers have regarded forest fringes, river banks, and swampy areas at low altitudes to be the natural habitat of the African oil palm. Both *E. guineensis* and *E. oleifera* are thought to compete poorly with rain forest trees due to the dense shade in the understory. However, there is evidence (Henson, 1991) that oil palms have an ability to adapt to low irradiation at early stages of growth when conditions of high radiation accompanied by low humidity tend to inhibit photosynthesis.

Although oil palms, particularly at the juvenile stage, cannot withstand permanent flooding, they thrive well in areas with fluctuating water tables with lateral flow, and such areas are often found to be amongst the highest yielding.

While the densest, semi-natural stands of oil palm or palm "groves" are generally found at 500 m above sea level or less in major river deltas such as the Niger and Congo, oil palms are known to survive at altitudes as high as 1700 m (Gerritsma & Wessel, 1997).

Development of the Oil Palm Industry

Palm Groves

The oil palm is still considered to be largely a "wild" crop. This is because in spite of its long period of exploitation by man, the time during which scientifically based selection and breeding has been practiced is relatively short in terms of generation number. Unlike annual crops, breeding progress in perennials is inevitably slow, needing, in the case of oil palm, at least 9 years per generation to properly evaluate breeding material and to undertake progeny testing to determine the merits of *dura* and *pisifera* parents used to produce the *tenera* offspring grown by the planter.

Prior to the adoption of planned breeding programs, any improvement in the palm was largely *ad hoc* and associated with the exploitation of semi-wild groves. The establishment of groves is thought to have been facilitated by an increasing scarcity of land as the human population increased, which prompted more frequent clearance of forest as part of the prevailing system of shifting agriculture. This led to conditions being created that aided survival of the relatively shade-intolerant and, compared with forest trees, slower growing palms. From a study of Nigerian groves, Zeven (1967) suggested that the intensification of land use led to a process of gradual "succession" in development of the groves involving several transitional phases. Details are given by Hartley (1977) and Gerritsma & Wessel (1997).

The bunch yields of the palm groves have always been very low (Corley & Tinker, 2003), being mainly a function of palm density (Fig. 1.4). Oil yields are very poor by comparison with those of modern plantations (Table 1-A), partly because the majority of grove palms are thick-shelled *duras* (Hartley, 1977) with a low oil-to-bunch ratio and, in addition, receive little fertilizer, are often too tall to harvest easily and at the correct stage of ripeness, and are often used as sources of sap for palm wine production, a practice which reduces the production of bunches.

Nevertheless the palm groves, especially in West Africa, have long played an important role in the economy of the region, first supplying purely local needs and then facilitating development of a valuable export trade (particularly in palm kernels) with European merchants from the 16th century onwards. By the 19th century this trade provided a more profitable and convenient alternative to the, by then illicit, slave trade. Further details concerning the development of this early trade and the factors

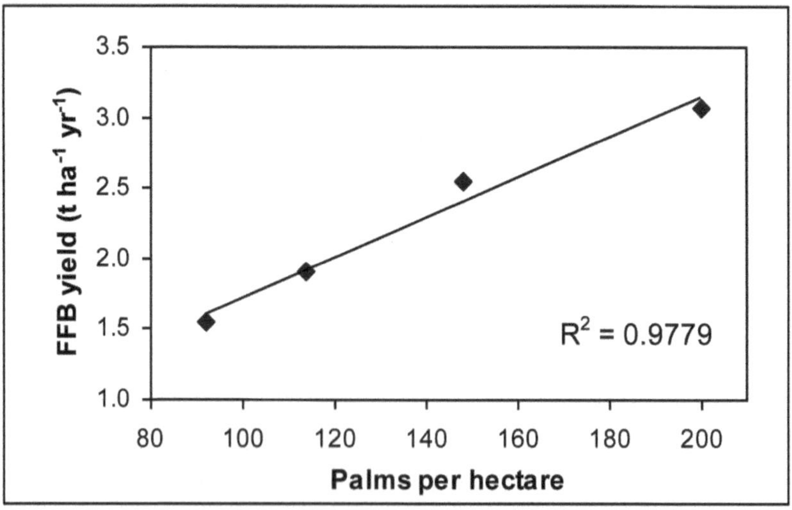

Fig. 1.4. Annual yield of fruit bunches in four types of oil palm grove in Nigeria (1949–1951). Grove types, in ascending order of yield, were farmland, groves near compounds, degraded groves, and dense groves. Data are from Corley & Tinker (2003), after Hartley (1988).

Table 1-A. Fresh Fruit Bunch (FFB) and Palm Oil Yield of Different Production Systems in Nigeria.

Production system	Area		FFB production	FFB yield	Palm oil yield	Oil/bunch ratio
	10^3 ha	% of total	10^3 tonnes	t ha^{-1} yr^{-1}	t ha^{-1} yr^{-1}	
Groves	2100	89.6	6300	3.0	0.30	10.00
Small-holdings	150	6.4	1200	8.0	1.00	12.50
Estates	95	4.1	950	10.0	1.89	18.95

Note. Modified from Table 1.2 of Corley & Tinker (2003) citing Omoti (pers. comm., 2000).

Box 1.3. Did you know?

- Palm oil from West African grove palms is favored by local domestic consumers because of its enhanced flavor associated with a high FFA content, while in contrast, large-scale producers aim to minimize FFA so as to prolong the stability of the oil and meet standards for export.
- In palm groves *dura* palms are favored over *teneras* for domestic use as cooking oil due to their more fluid properties.

Source: Poku (2002).

leading to its expansion are given by Hartley (1977), Gerritsma & Wessel (1997), and Corley & Tinker (2003).

Even today, the groves still dominate production in Nigeria (Table 1-A). Attempts to improve production of grove palms have not proved very successful. Part of the reason may be local consumers' preference of oil from *dura* rather than *tenera* palms for use in cooking, the former being regarded as having more flavor and being less prone to solidify (Poku, 2002) (Box 1.3).

Plantations

Attempts from the early 1800s to establish commercial-scale plantations in West Africa and to put palm oil production there on a firm commercial basis met with numerous obstacles that impeded development. There were impediments in terms of political instability and internal conflicts that hampered economic progress in general, as well as difficulties in obtaining land and providing infrastructure for transport and processing. This might explain why the main focus of development of oil palm as an "industrial" crop eventually shifted to Southeast Asia where the first plantations were established in Sumatra and Peninsular Malaysia (then Malaya) in the early part of the 20th century. This development was due to the exploitation of the "deli" *dura* palms that originated from four specimens planted in 1848 in the Botanic Gardens at Bogor in Java. The precise origin of the "Bogor" palms is not known for certain, but they are thought to have been taken from West Africa to the Botanic Gardens in Amsterdam before finding their way to Bogor. Two of them came directly to Java from Amsterdam while the other two arrived via the islands of Mauritius or Reunion in the Indian Ocean, probably via Amsterdam since all four palms proved very similar in terms of their growth and fruit characteristics. Seeds collected from the Bogor palms were planted as ornamental avenues on tobacco estates near Deli and elsewhere in Sumatra, and were found to be both uniform and productive, eventually providing material for the first commercial plantings in estates in Sumatra and Malaya, undertaken around 1911. Thus, the foundations were laid for what has now become an industry of major commercial importance dominated by Southeast Asian producers (Fig. 1.5).

Smallholdings

Oil palm does not easily lend itself to being a smallholder crop due to the need for long-term investment in land and planting material and access to mills to process the fruits. Unlike most smallholder crops, it is less suited for meeting immediate subsistence needs of the cultivator, being more in the nature of a "cash crop". Nevertheless, smallholdings of various kinds are responsible for a significant fraction of palm oil produced in Nigeria, Malaysia, Indonesia, and elsewhere (Tables 1-A and 1-B). A major factor in this has been the operation and provision of finance and infrastructure

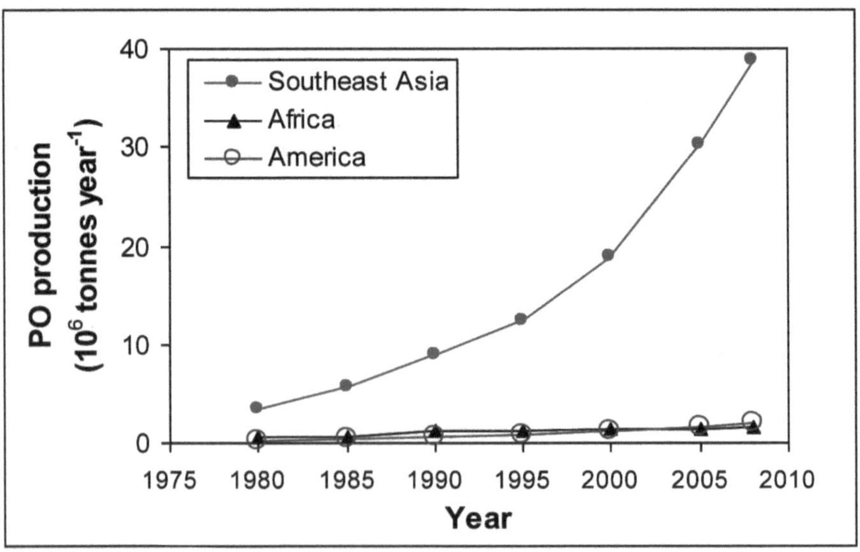

Fig. 1.5. Growth in palm oil production in the three regions of cultivation over 48 years. Sources: MPOB (2008), FEDEPALMA (2009).

Table 1-B. Distribution of Oil Palm Area Between Production Systems in Malaysia, Indonesia, and Papua New Guinea.

Production system	Malaysia (2007)		Indonesia (2006)		Papua New Guinea	
	10^3 ha	% of total	10^3 ha	% of total	10^3 ha	% of total
Smallholdings	470	10.9	2548	40.8	46.8	43.5
Private estates	2599	60.4	3023	48.3	60.9	56.5
Public schemes	1236	28.7	679	10.9	–	–

Sources: IPOB (2007); King et al. (2004); MPOB (2008).

by government agencies and by the establishment of "nuclear" estates whereby plantation companies provide processing and other facilities to support neighboring smallholders.

Uses of the Oil Palm

Through its long history, the oil palm has found many uses, and these continue to grow in line with technological developments. In Africa, the products of the oil palm initially served mainly to meet local needs and included the use of palm oil for cooking, as a vitamin source and for medicinal use, the "tapping" of palms to produce palm wine, the use of the fronds for thatching and fencing, and the harvesting of the edible palm "heart" or "cabbage" (Hartley, 1988). Because the palm oil produced by traditional methods was regarded as being of poor quality largely because of its high FFA content, the oil exported was initially used mainly for industrial purposes such as in the manufacture of soap and candles and as an antioxidant in tin plating (Harley, 1988). Its first use as an edible oil in Europe occurred in the late 1800s when it was used for margarine manufacture. Further use for edible purposes in Europe awaited improvements in oil quality that occurred with the introduction of mechanized mills and the regulation of standards, such as those introduced in 1949 by the Nigerian Oil Palm Produce Marketing Board (Berger, 2010). Later, the even higher quality of palm oil achieved by mills in Southeast Asia served to stimulate the trade from that region, thus contributing to its eventual dominance as a world source.

While the uses of palm oil have expanded considerably in recent years (e.g., MPOC & MPOB, 2007), some 77% of it is still used for food (USDA, 2008, cited by Sheil et al., 2009). Included in the expanding range of non-food uses is the production of bio-diesel and various oleochemicals. The combined demand for these products has encouraged the continued expansion of oil palm plantations.

Prior to the installation after WW2 of kernel crushing plants in West Africa, kernels were exported and the oil was extracted in the importing country (mainly the UK). With the installation of crushing plants, exports of kernel oil began and increased considerably, both from West Africa and Southeast Asia. Due to its different fatty acid composition, palm kernel oil has separate applications from palm oil and finds most uses in non-edible products such as detergents and cosmetics. In addition, the kernel "cake" or residue left after oil extraction finds use as animal feed. The non-food uses of palm oil and palm kernel oil have proliferated in recent years, as testified by other chapters in this volume.

Historical Trends in Yield Improvement

The yield of oil palm has increased remarkably from the low levels of production obtained with semi-domesticated palm groves in West Africa to that achieved today in the best managed plantations on the most favorable sites in Southeast Asia and elsewhere. Part of the increase is due to breeding and selection, part to the better environments in which the crop is now grown, and part to improvements in cultivation and management practices. The yield improvements are summarized in Table 1-C, the data in which is a mixture of results ranging from whole country averages to maximum yields obtained in experimental plots. Yield is normally strongly dependent on palm age and is thus affected by the number and actual years over which it is recorded, so the data presented below can only be a rough indication of progress. Even so, the improvement is remarkable.

Table 1-C includes estimates of *yield potential* and *potential yield*, which serve as benchmarks against which present yields and possible increases can be evaluated.

Yield potential (Evans & Fischer, 1999) is that realized by a cultivar growing in an environment to which it is adapted when water and nutrients are non-limiting and pests and diseases are absent. Breure (2003) calculated yield potential assuming levels of solar radiation to be adequate, as is generally the case in the more productive areas of Southeast Asia. Corley (2006) has termed this the *genetic yield potential* in contrast to *site yield potential* which allows for the effects of site-specific environmental constraints.

Potential yield (Corley, 2006; Evans & Fischer, 1999) is defined as that of a hypothetical genotype combining physiologically plausible (and presumably optimal) physiological attributes under optimal agronomic management. It must be admitted that these distinctions are rather subtle and easily confused. None of the definitions take account of whether yields represent a peak or an average over the life of the crop. Breure (2003) contested the estimated potential yield of Corley (1998) as being too high, although it pales into insignificance when compared to the much higher theoretical predictions of Murphy (2007).

Estimates of potential oil yield have progressively increased from 12 t ha^{-1} yr^{-1} to 18.5 t ha^{-1} yr^{-1} following renewed analysis (Corley et al., 1976; Corley, 1983a, 1983b, 1985, 1986, 1998, 2006). On the other hand, improvements in planting materials and other factors (Table 1-C) have reduced the gap between actual and theoretical yields. The yield potential estimated by Breure (2003) has actually been achieved,

Table 1-C. Improvements in Palm Oil Yield Over Time.

Production method and material		Palm oil yield ($t\ ha^{-1}\ yr^{-1}$)	References
Wild groves, Nigeria	National average	0.27	Ejemba (1989)
	Dense groves[a]	0.36	Gerritsma & Wessel (1997)
Plantations, Nigeria	–	1.80	Ejemba (1989)
Average national yields in 2008	Côte d'Ivoire	1.71	FEDEPALMA (2009), MPOB (2010)
	Ecuador	2.00	
	Nigeria	2.05	
	Thailand	2.56	
	Brazil	3.01	
	Costa Rica	3.42	
	Colombia	3.51	
	Indonesia	3.84	
	Malaysia[b]	4.08	
	Papua New Guinea	4.36	
	Sabah	4.85	
Malaysia, good estate yield	–	5.50	Corley (1983a, 1983b)
Malaysia, best estate yields	–	6.80	Goh et al. (2002)
Malaysia, best experimental yields	Best plots/years	8.60	Corley (1973) Corley et al. (1976)
	Best trees	10.00	Corley (1985; 1986)
	Best progenies	12.20	Rajanaidu et al. (1989)
	Best clone(s)	7.3–10.8	Basri & Arif (2009)
		12.56	Khaw et al. (1999)[c]
Maximum theoretical yields	Yield potential	10.6[d]	Breure (2003)
	Potential yield	18.5	Corley (2006)

Notes. [a]Calculated from mean FFB yield assuming a 9% extraction rate (Ejemba, 1989).
[b]Includes Sabah.
[c]Quoted by Ng et al. (2003). Mean of single clone from 4 to 9 years after planting.
[d]The average yield over the economic life of a planting after canopy closure (Breure, pers. comm.). Modified from Henson (1992, 2010).

at least under well-managed experimental conditions involving small populations of palms over limited time periods.

Variations in yield, such as those shown in Table 1-C, are a result of a combination of factors—genetic, environmental, and managerial. The main historical events that have led most to yield improvement compared with the grove palms of West Africa are considered to be the following:

- Introduction of the four *dura* seedlings of African origin to Bogor Botanic Gardens in Java (1848).
- Initial plantings of the Deli *dura* palms on a commercial scale in Sumatra and Malaya (1910–1911).
- The use of assisted pollination in Malaya, giving large yield increases (1922).
- The general improvement of the Deli *dura* for commercial planting achieved due to planned breeding and selection.
- Discovery of the monogenetic control of shell thickness leading to the widespread cultivation of *tenera* palms with a dramatic increase in the oil extraction rate (see below).
- The continued improvement of *tenera* material through conventional breeding and selection of *dura* female parents accompanied by testing of *pisifera* palms, firstly selecting for conventional traits (yield, mesocarp/fruit) and later for oil/mesocarp.
- The use of vegetative growth characters as selection criteria, facilitated by the development of growth measurements (Corley et al., 1971; Hardon et al., 1969).
- The introduction of the pollinating weevil into Southeast Asia from West Africa in 1981.

Arguably the most dramatic advance followed the discovery by Beirnaert & Vanderweyen (1941) of monogenetic control of shell thickness in the oil palm and the subsequent replacement of *dura* by *tenera* hybrids. A substantially higher oil yield is possible due to the thin-shelled character of the *tenera*, when compared with the thick-shelled *dura* (Table 1-D).

Breeding work in the Côte d'Ivoire over more than 40 years during which *tenera* replaced *dura* (Fig. 1.6) shows that improvements in both bunch composition and FFB yield contributed to increased oil yield. An 82% increase in palm oil yield resulted from a 20% increase in O/B (oil/bunch) and a 52% increase in FFB.

The inclusion of vegetative traits as selection criteria has been increasingly practiced in recent years. Direct comparisons of different generations of breeding material (Table 1-E) show that yield improvement has involved increases in bunch production with little change in vegetative growth, leading to an increase in the bunch index. This

Table 1-D. Comparison of *Dura* and *Tenera* Bunch Components and Yield.

Source	Material	Bunch yield (t ha^{-1} yr^{-1})	Fruit/ bunch (%)	Mesocarp/ fruit (%)	Oil/ mesocarp (%)	Oil/ bunch (%)	Palm oil yield (t ha^{-1} yr^{-1})
Corley & Lee (1992)[a]	*Dura*	21.6	–	–	–	17.2	3.72
	Tenera	22.2	–	–	–	22.9	5.08
Hardon (1995)[b]	*Dura*	27.0	65.0	61	50	19.8	5.2
	Tenera	27.0	62.0	78	50	24.1	6.6

Notes. [a]The data are means of seven experiments and 65 crosses. A planting density of 148 palms ha^{-1} is assumed.
[b]The data are for full sibs from a breeding trial.

Table 1-E. A Comparison of the Productivity of Different Generations of Deli *Dura* Palms Grown in Malaysia[a].

Material	Year of origin	VDMP	BDMP	TDMP	FFB yield	PO yield	O/B	BI	e*
					tonnes ha^{-1} $year^{-1}$		%	ratio	g Mj^{-1}
unselected	1878	15.15	9.15	24.30	16.3	3.09	17.9	0.38	1.00
F1	1927	15.25	11.31	26.56	20.1	3.68	17.2	0.43	1.13
F4	1969	14.74	14.24	28.98	24.8	4.95	18.4	0.49	1.26

Note. Results are reproduced or calculated from Lee et al. (1990). Those for "unselected" palms are means of two groups, Bogor and Tanjong Morawa palms, for which the year of origin is only approximate. F1 are from Elmina Estate and F4 from OPRS, Banting. VDMP is vegetative dry matter production calculated from petiole cross-section measurements taken in the 18th year and then adjusted, based on data from Breure (1988), van Kraalingen (1985), and Gerritsma et al. (1992), to match the age of palms when bunch yields were measured. BDMP and TDMP are bunch and total dry matter production, respectively. Palm oil (PO) yield was calculated from bunch (FFB) yields and oil/bunch ratios. BI is bunch index (BDMP/TDMP) and e* is the efficiency of conversion of radiation to dry matter (adjusted for oil content) calculated assuming an annual photosynthetically-active solar radiation (PAR) of 3.0 GJ m^{-2} and an extinction coefficient for PAR of 0.47. The Table is adapted from Henson (1998), and data are partly based on Lee et al. (1990). Similar analyses are presented by Corley & Lee (1992), Corley (2001), and Gerritsma & Wessel (1997).

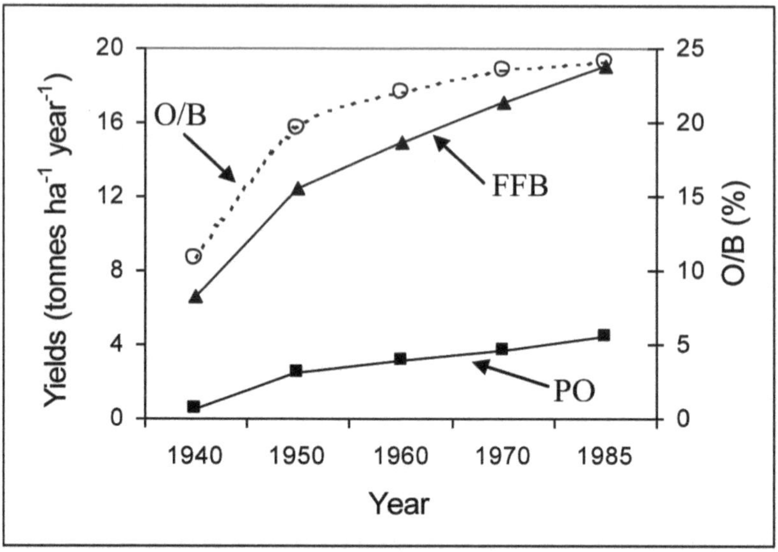

Fig. 1.6. Changes over time in FFB and palm oil (PO) yield and O/B of oil palm planting material produced in Côte d'Ivoire. From Henson (1998), based on data of Meunier (1989).

is comparable to the changes underlying yield increases in annual crops such as cereals (Rajanaidu & Zakri, 1988). Selection for vegetative characters such as reduced height and a compact canopy, allowing higher planting density without sacrificing yield per palm, shows much promise as a means to higher yield (Breure, 1985; Breure, 2010; Breure & Corley, 1983).

Yield improvement due to environment is most clearly demonstrated by comparing yields of the same materials grown in areas with differing rainfall patterns. The presence of regular dry periods in many areas of West Africa is associated with much lower yields than are obtained in Southeast Asia, even though the annual rainfall and solar radiation may not differ much. Thus, mean oil yield of the test cross L2T × D10D averaged 6.2 to 7.8 t ha^{-1} yr^{-1} in two trials in Indonesia but only 3 to 3.5 t ha^{-1} yr^{-1} when grown in the Côte d'Ivoire and Cameroon (Cochard et al., 1993). A further example is provided by Nouy et al. (1999), cited by Corley & Tinker (2003), in which the same cross was grown in Benin, the Côte d'Ivoire, and Indonesia. The importance of a regular water supply is confirmed by many experiments involving irrigation in dry areas (Corley, 1996; Palat et al., 2008).

A further factor in increasing yield in Southeast Asia has been the improvement in fruit set that followed the introduction to the region from West Africa in the early 1980s of the pollinating weevil, *Elaeidobius kamerunicus.* This increased average bunch weights, oil content, and oil yield (Corley & Tinker, 2003).

Table 1-F. Factors Leading to Palm Oil Yield Increases on a Malaysian Estate from 1951 to 1981.[a]

Factor	Palm oil yield ($t\ ha^{-1}\ yr^{-1}$)	Yield gain ($t\ ha^{-1}\ yr^{-1}$) due to: Breeding	Agronomy	Pollination	Processing
Initial condition	1.30				
Fertilizer use	2.50		1.20		
Improved *dura* palms	3.50	1.00			
Use of *tenera* palms	4.64	1.14			
Improved stand	4.78		0.14		
Water management	5.02		0.24		
Weevil introduction	5.08			0.06	
Improved extraction efficiency	5.43				0.35
Total gain	4.13	2.14	1.58	0.06	0.35
% of total	–	51.8	38.3	1.4	8.5

[a]Modified from Henson (1998) using data of Davidson (1993).

Other factors leading to increased yields over time, while not always easily quantified, include improved nutritional management aided by regular foliar and soil nutrient analyses, increased use of mill by-products as sources of nutrients and organic matter, and improvements in harvesting and milling technology. Using data provided by Davidson (1993), the contribution to yield improvement by breeding, agronomy, pollination, and fruit processing can be assessed for a single estate (Table 1-F).

Similar examples of progressive yield improvement, attributed either to improvements in management or in breeding, are presented by others (e.g., Goh et al., 1994a, 1994b; Jalani, 1998).

Palm Fruit Processing

The successful establishment and expansion of oil palm cultivation has been much dependent on the availability of suitable facilities for extracting oil and kernels from harvested fruit bunches. This was one of the factors that initially impeded development of the industry on a plantation scale in West Africa (Hartley, 1977). While traditional extraction methods co-evolved with the use of palm groves in West Africa

(Fig. 1.7) and are still practiced today, they are very labor intensive and result in poor recovery and poor quality of the products (Hartley, 1988). The methods entail expressing the mesocarp oil either by pounding ("soft" oil production) or treading ("hard" oil) bunches that have been split, sometimes boiled, and left to soften, often for several days. The expelled oil is separated from the resultant pulp or mash by skimming, usually repeated to increase recovery, and the nuts are then picked out and laid to dry before being individually hand cracked and sorted to obtain the kernels. These methods are recognized (Corley & Tinker, 2003) as being notoriously inefficient and to contribute to the high FFA content and low oil yield of grove palms.

However, while the traditional small-scale extraction methods used for bunches from grove palms produce an oil with a high FFA content that is unacceptable in oil for export, this is considered to add "bite" to the oil and enhance its flavor and is much preferred by domestic consumers (Poku, 2002).

Hand-operated presses (initially modified wine presses) were introduced around 1917, which improved recovery (Hartley, 1988). These were later augmented by auxiliary equipment to deal with processes such as sterilization, stripping, digestion, and clarification. The extraction of the kernels, at one time the main product of the West African groves entering international trade, was also improved with the advent of mechanical nutcrackers.

Although the first power-operated mills were constructed in Africa prior to 1914 and in Sumatra in 1919, it was not until the 1930s that the first industrial scale

Fig. 1.7. Artist's impression of the traditional "soft" palm oil extraction process used for generations in the palm groves of West Africa. Picture courtesy of N R Menon.

mechanical milling units appeared, while mills similar to those common today were only developed from the late 1960s onwards.

While mills are now an integral part of most large-scale plantations, small estates and smallholders remain reliant on sending fruits either to independent millers or to a plantation's mills. The latter arrangement has to some extent been formalized and encouraged by schemes whereby a "nucleus" estate serves the needs of adjacent, otherwise independent smallholders, both with regard to fruit processing and help with supply of fertilizers and other inputs and by providing management advice.

Progressive improvements in milling technology, while generally minor (Corley & Tinker, 2003), have nevertheless been partly responsible for increases in oil production (e.g., Table 1-F). The size of palm oil mills has tended to increase in proportion to the size of plantations and there are now mills handling up to 90 tonnes of FFB per hour with the consequent benefits of economies of scale.

With the increase in number and size of mills has come the problem of disposing of the by-products in ways that both minimize environmental pollution and make productive use of them. Materials such as fiber and shell are used as fuel to meet the energy needs of the mill while empty fruit bunches and palm oil mill effluent can be returned to the field as sources of nutrients and organic matter. A current challenge is to reduce the emission of methane, a potent greenhouse gas, from effluent retention ponds, and projects to do this are underway (e.g., Tantitham et al., 2009).

Recent Developments

Oil palm remains one of the most thoroughly studied of all tropical plantation crops and most producing countries and the larger plantation groups carry out some research and development aimed at improving the level and efficiency of production, as well as the quality and range of oil palm products. At the same time, there has been increasing recognition of the need to minimize any adverse effects that oil palm may have on the natural environment and on the safety, welfare, and lifestyle of people employed or otherwise affected by the industry.

Technology

In a look at the future challenges faced by the oil palm industry, Corley and colleagues in their 1976 review pointed to the increasing costs of labor that could be countered either by increased mechanization, increased yields, or increased product prices. Increased mechanization would not be achieved without costs, such as the need for greater capital investment, increased fuel use, and manpower training. Machinery might cause soil compaction that could result in reduced yields. While in-field fruit handling and fertilizer application have been successfully mechanized, harvesting is still a manual process and efficient mechanical harvesters have yet to be developed.

Increased mechanization means greater use of fossil fuels and thus greater greenhouse gas emission, so adding to current environmental concerns.

The yield improvements envisaged by Corley et al. (1976) have to some extent taken place, although for various reasons average country yields have tended to stagnate. Corley et al. (1976) were hopeful that clonal palms, just then being investigated, would provide a means to substantially improve yields. This initial optimism was subsequently dampened following the occurrence of floral abnormalities amongst clones, said to have delayed their successful adoption by about 10 years (Corley & Law, 1997). Now, however, quite large areas of clones have been planted (Basri & Arif, 2009), and their high yielding potential is beginning to be realized (Ng et al., 2003).

Product prices, while largely at the mercy of the market place and outside the control of the grower, might nevertheless be influenced by improvements in oil quality and expanding the applications of palm products. Examples of the latter are the recent use of palm oil for oleo-chemical and bio-diesel production, and there are other possibilities such as the commercial recovery of minor but high value constituents such as carotenes and vitamin E (Yusof & Lim, 1977).

A development not envisaged in 1976 has been the use of precision agricultural techniques facilitated by the increasingly accessible use of geographical information and remote sensing technology, which, together with computerized ground sensors, have the potential to increase efficiency of resource use in plantations, particularly with regard to fertilizer application.

Environment

The rapid expansion that has occurred in oil palm cultivation in the space of just a few decades has not unexpectedly been accompanied by several adverse impacts in terms of environmental and social disturbance. Key environmental effects include the loss of tropical rainforests due to land clearance and conversion to plantations (sometimes aided by burning leading to formation of haze), the destruction of wetland ecosystems following drainage, the pollution of waterways by eroded soil, the improper disposal of mill wastes, and the contamination of rivers due to leaching of nutrients and pesticides when applied improperly. Such expansion has reduced wildlife habitats, biological diversity and organic carbon stores, and contributed to the emission of CO_2 and other greenhouse gases that underpin climate change. In addition, social problems have arisen due to disputes over land ownership between plantation companies and native peoples. While various laws have been enacted in an effort to overcome such problems, lack of resources and political will have frequently led to their poor enforcement. As a result, palm oil and palm products have acquired an increasingly negative image. It is to be hoped that the situation will improve following the operation of various certification schemes that seek to promote "sustainable" production, such as those administered by some individual palm oil consumers and by the Roundtable for Sustainable Palm Oil.

Acknowledgments

I am most grateful to C. J. Breure, R. H. V. Corley, J-C. Jacquemard, V. Rao, N. R. Menon, and P. B. Tinker for many helpful comments and suggestions for improving the paper and for supplying useful references and information.

References

Arnott, G.W. The Malayan oil palm and the analysis of its products. Division of Agriculture Bulletin No. 113. Department of Agriculture, Federation of Malaya, 1963; pp. 33.

Barcelos, E.; Second, S.; Kahn, F.; Amblard, P.; Lebrum, P.; Seguin, M. Molecular markers applied to the analysis of genetic diversity and to the biogeography of *Elaeis* (Palmae). In Evolution. Variation and classification of palms. Memoirs 83 of the New York Botanical Garden; Henderson, A., Borchsenius, F., Eds.; New York Botanical Garden Press: New York, 1999; pp. 191–201.

Basri, M.W.; Arif, M.S. Cost of palm oil production. Planter 2009, 85, 375–394.

Beirnaert, A.; Vanderweyen, R. Contribution a l'etude genetique et biometrique des varieties d'*Elaeis guineenesis* Jacquin. Publ. Inst. Nat. Etude Agron. Congo Belge. Ser. Sci. 1941, 27, 1–101.

Berger, K.G. Palm oil – Early history. Origins in West Africa. Global Oils Fats Business Magazine 2010, 7(4), 43–48.

Blach-Overgaard, A.; Svenning, J-C.; Dransfield, J.; Greve, M.; Balslev, H. Determination of palm species distribution across Africa: the relative roles of climate, non-climatic environmental factors, and spatial constraints. Ecography 2010, 33, 380–391.

Blommendaal, H.N. Oil palm cultivation in the Dutch-East Indies. Tieenk Willink and Zn. Haarlem, 1937, 189 pp.

Brandt, W.; Gürke, M.; Köhler, F.E.; Pabst, G.; Schellenberg, G.; Vogtherr, M. Köhler's Medizinal-Pflanzen in naturgetreuen Abbildungen mit kurz ertäuterndem Texte: Atlas zur Pharmacopoea germanica, austriaca, belgica, danica, Helvetica, hungarica, rossica, suecica, Neerlandica, British pharmacopoeia, zum codex medicamentarius, sowie zur Pharmacopoeia of the United States of America. Pabst, G., Ed.; 1887, 161 pp.

Breure, C.J. Relevant factors associated with crown expansion in oil palm. Euphytica 1985, 34, 161–175.

Breure, C.J. The effect of different planting densities on yield trends in oil palm. Exp. Agric. 1988, 24, 37–52.

Breure, C.J. The search for yield in oil palm: Basic principles. In Oil Palm. Management for Large and Sustainable Yields. Fairhurst, T.H., Hardter, R., Eds.; Potash and Phosphate Institute, Potash and Phosphate Institute of Canada, and International Potash Institute, 2003; pp. 59–98.

Breure, C.J. Rate of leaf expansion: a criteria for identifying oil palm (*Elaeis guineensis* Jacq.) types suitable for planting at higher densities. NJAS–Wageningen J. Life Sci. 2010, 57, 141–147.

Breure, C.J.; Corley, R.H.V. Selection of oil palms for high density planting. Euphytica 1983, 32, 177–186.

Burkill, I.H. A dictionary of the economic products of the Malay Peninsula. 2nd ed. Ministry of Agriculture and Cooperatives, Malaysia, Kuala Lumpur.

Cobley, L.S. An Introduction to the Botany of Tropical Crops. 4th impression. Longmans: London, 1963; 357 pp.

Cochard, B.; Noiret, J.M.; Baudouin, L.; Flori, A.; Amblard, P. Second cycle reciprocal recurrent selection in oil palm *Elaeis guineenesis* Jacq. Results of Deli × La Me hybrid tests. Oleagineux 1993, 48, 441–451.

Corley, R.H.V. Effects of plant density on growth and yield of oil palm. Exp. Agric. 1973, 9, 169–180.

Corley, R.H.V. The Genus Elaeis. In Oil Palm Research. Corley, R.H.V., Hardon, J.J., Wood, B.J., Eds.; Elsevier: Amsterdam, 1976; pp. 3–5.

Corley, R.H.V. Potential productivity of tropical perennial crops. Exp. Agric. 1983a, 19, 217–237.

Corley, R.H.V. Oil palm and other tropical tree crops. In Symposium on Potential Productivity of Field Crops under Different Environments. International Rice Research Institute: The Philippines, 1983b; pp. 384–399.

Corley, R.H.V. Yield potentials of plantation crops. In Potassium in the agricultural systems of the humid tropics. Proceedings of 19th Colloquium International Potash Institute. International Potash Institute: Berne, 1985; pp. 61–80.

Corley, R.H.V. Yield potential of plantation crops. Better Crops International 1986, 2, 10–12.

Corley, R.H.V. Fifteen years experience with oil palm clones—a review of progress. In Proceedings of 1991 International Palm Oil Conference, Agriculture (Module 1). Palm Oil Research Institute of Malaysia: Kuala Lumpur, 1993; pp. 69–81.

Corley, R.H.V. Irrigation of oil palms—A review. J. Plantation Crops 1996, 24(Supplement), 45–52.

Corley, R.H.V. Oil palm. In Ecosystems of the World—19. Tree crop ecosystems. Last, F.T, Ed.; Elsevier Science: Amsterdam, 2001; pp. 299–320.

Corley, R.H.V. Potential yield of oil palm—an update. Paper presented at Symposium on Yield Potential in the Oil Palm. Rao, V., Henson, I.E., Rajanaidu, N., Eds.; International Society of Oil Palm Breeders: Phuket, Thailand, 2006.

Corley, R.H.V.; Law, I.H. The future for oil palm clones. In Proceedings of International Planters Conference May 1997. Incorporated Society of Planters: Kuala Lumpur, 1997; pp. 279–289.

Corley, R.H.V.; Lee, C.H. The physiological basis for genetic improvement of oil palm in Malaysia, Euphytica 1992, 60, 179–184.

Corley, R.H.V.; Tinker, P.B. The Oil Palm, 4th Edition. Blackwell Publishing: Oxford, 2003; 562 pp.

Corley, R.H.V.; Hardon, J.J.; Tang, Y. Analysis of growth of the oil palm (*Elaeis guineensis* Jacq.). I. Estimation of growth parameters and application to breeding. Euphytica 1971, 20, 307–315.

Corley, R.H.V.; Wood, B.S.; Hardon, J.J. Future developments in oil palm culture. In Oil Palm Research; Corley, R.H.V., Hardon, J.J., Wood, B.J., Eds.; Elsevier: Amsterdam, 1976; pp. 507–511.

Crowley, L. Oil palm genome sequencing may aid sustainability. Food Navigator-USA.com. 21 May 2008. http://www.foodnavigator-usa.com/Science-Nutrition (accessed October 2008).

Davidson, L. Management for efficient, cost-effective and productive oil palm plantations. In Proceedings of the 1991 International Palm Oil Conference. 1. Agriculture; Basiron, Y., Jalani, S., Chang, C.K., Cheah, S.C., Henson, I.E., Kamaruddin, N., Paranjothy, K., Rajanaidu, N., Mohd Tayeb, D., Eds.; Palm Oil Research Institute of Malaysia: Kuala Lumpur, 1993; pp. 153–167.

Ejemba, M.I. Oil palm development in Nigeria: past efforts and future prospects. Paper presented at: International Conference on Palms and Palm Products. Nigerian Institute for Oil Palm Research: Benin City, Nigeria, 21–25 November, 1989.

Evans, L.T.; Fischer, R.A. Yield potential; its definition, measurement and significance. Crop Sci. 1999, 39, 1544–1551.

FEDEPALMA. Statistical Yearbook. Federación Nacional de Cultivadores de Palma de Aceite, Bogotá, Colombia, 2009, 166 pp.

Ford, H. Chicago Tribune, 25th May 1916.

Friedel, C. Sur des matieres grasses trouvees dans les tombes egyptiennes d'Abydos. C R Academy of Science, Paris 1897, 124, 648–653.

Gerritsma, W.; Redshaw, M.J.; Soebagyo, F.X. Analysis of yield trends in two cultivar trials in Indonesia. In Proceedings of the 1990 ISOPB International Workshop on Yield Potential in the Oil Palm. Rao, V., Henson, I.E., Rajanaidu, N., Eds.; Palm Oil Research Institute of Malaysia: Kuala Lumpur, 1992; pp. 80–97.

Gerritsma, W.; Wessel, M. Oil palm: Domestication achieved? Netherlands J. Agric. Sci. 1997, 45, 463–475.

Goh, K.J.; Chew, P.S.; Teo, C.B. Maximising and maintaining oil palm yields on a commercial scale in Malaysia. In Proceedings of International Planters Conference, 24–26 October 1994. Incorporated Society of Planters: Kuala Lumpur, 1994a; pp. 121–141.

Goh, K.J.; Chew, P.S.; Teo, C.B. Commercial yield performances of oil palms in Sabah, Malaysia. The Planter 1994b, 70, 497–507.

Goh, K.J.; Gan, H.H.; Soh, A.C. Oil palm productivity: Commercial FFB yield analysis. Paper presented at: MPOA Seminar: R&D for competitive edge in the Malaysian oil palm industry. Malaysian Palm Oil Association: Kuala Lumpur, 2002.

GRIN; Germplasm Resources Information Network. United States Department of Agriculture. http://www.ars-grin.gov/cgi-bin/npgs. (accessed 2011).

Haddon, A.V.; Tong. Y.L. Oil palm selection and breeding: A progress report. Malayan Agric. J. 1950, 42, 124–156.

Hardon, J.J. Interspecific hybrids in the genus *Elaeis.* II. Vegetative growth and yield of F1 hybrids of *Elaeis guineensis* × *E. oleifera.* Euphytica 1969, 18, 380–388.

Hardon, J.J. Oil palm. *Elaeis guineensis* (Palmae). In Evolution of Crop Plants 2nd ed.; Smart, J., Simmonds, N.W., Eds.; Longmans: Harlow, UK, 1995; pp. 395–399.

Hardon, J.J.; Tan, G.Y. Interspecific hybrids in the genus Elaeis. I. Crossability, cytogenetics and fertility of F1 hybrids of *Elaeis guineensis* × *E. oleifera.* Euphytica 1969, 18, 372–379.

Hardon, J.J.; Williams, C.N.; Watson, I. Leaf area and yield in the oil palm in Malaya. Exp. Agric. 1969, 5, 25–32.

Hartley, C.W.S. The Oil Palm, 1st edition; Longman: London, 1967; 606 pp.

Hartley, C.W.S. The Oil Palm, 2nd edition; Longman: London, 1977; 806 pp.

Hartley, C.W.S. The Oil Palm, 3rd edition; Longman: London, 1988; 761 pp.

Henson, I.E. Adaptation to light environment by leaves if oil palm (*Elaeis guineensis*). PORIM Bulletin 1991, 22, 1–8.

Henson, I.E. Estimating potential productivity of oil palm. In Yield Potential in the Oil Palm (*Elaeis guineensis*); Rao, V., Henson, I.E., Rajanaidu, N., Eds.; Palm Oil Research Institute of Malaysia: Kuaha Lumpur, 1992; pp. 98–108.

Henson, I.E. The role of bunch composition in determining oil and kernel yields of oil palm. In Proceedings of 1996 International Conference on Oil and Kernel Production in Oil Palm—A Global Perspective. Palm Oil Research Institute of Malaysia: Kuala Lumpur, 1998; pp. 151–208.

Henson, I.E. Oil palm: Ecophysiology of growth and production. In Ecophysiology of Tropical Tree Crops; DaMatta, F.M., Ed.; Nova Science Publishers: New York, 2010; pp. 253–286.

Imam, M.M. Oil palm in Malaysia: chronology of events. PORIM Bulletin, Special issue, 1984, 5–8.

IPOB. Palm oil in numbers. Indonesian Palm Oil Board, Jakarta, Indonesia, 2007, 27 pp.

Jagoe, R.B. Deli oil palms and early introductions of *Elaeis guineensis* to Malaya. Malayan. Agric. J. 1952, 35, 3–10.

Jacauin, N.J. Selectarum Stirpium Americanum Historia. Vienna, 1763, 298 pp.

Jalani, B.S. Oil palm research and development in Malaysia: Progress and trends. The Planter 1998, 74, 73–92.

Jones, L. Propagation of clonal oil palm by tissue culture. The Planter 1974, 50, 374–381.

Khaw, C.H.; Ng, S.K.; Thong, K C.; Commercial production of clonal palms by tissue culture – prerequisites, constraints and issues. Paper presented at: International Palm Oil Congress. Agriculture Conference. Palm Oil Research Institute of Malaysia: Kuala Lumpur, 1999; pp. 37–43.

King, J.; Sipolo, S.; Taylor, C.; Fairhurst, T.H. Replanting of oil palm in Papua New Guinea: experience from Higaturu Oil Palms. Paper presented at: National Seminar 'Replant or perish', Ipoh Malaysia, 14–15 June 2004, 11 pp.

Kraalingen, D.W.G. Simulation of oil palm growth and yield. Thesis. Agricultural University, Wageningen, 1985, 106 pp.

Latiff, A. The biology of the genus Elaeis. In Advances in Oil Palm Research, Volume 1.; Basiron, Y., Jalani, B.S., Chan, K.W., Eds.; Malaysian Palm Oil Board: Kuala Lumpur, 2000, 19–38.

Lee, C.H.; Yong, Y.Y.; Donough, C.; Chiu, U.B. Selection progress in the Deli dura population. In Proceedings of Workshop on Progress of Oil Palm Breeding Populations. Palm Oil Research Institute of Malaysia: Kuala Lumpur, 1990, pp. 81–89.

Lobelius [Mathias de Lobel]. Plantarium seu Stirpium Historia. London, 1570.

Lobelius [Mathias de Lobel]. Plantarium seu Stirpium Historia. Antwerp, 1576.

Lobelius [Mathias de Lobel]. Kruydtboeck. Antwerp, 1581.

Meunier, J. Advances in oil palm breeding: progress and prospects. Paper presented at: International Conference on Palms and Palm Products. Nigerian Institute for Oil Palm Research, Benin City, 1989.

Meunier, J.; Hardon, J.J. Interspecific hybrids between *Elaeis guineensis* and *Elaeis oleifera.* In Oil Palm Research; Corley, R.H.V., Hardon, J.J., Wood, B.J., Eds.; Elsevier: Amsterdam, 1976; pp. 127–138.

MPOB. Malaysian Oil Palm Statistics 2007. 27th ed. Malaysian Palm Oil Board, Kuala Lumpur, 2008; 190 pp.

MPOB. http://econ.mpob.gov.my/economy/annual/stat2009/EID_statistics09.htm (accessed 2010).

MPOC; MPOB. Fact sheets. Malaysian Palm Oil. Malaysian Palm Oil Council, Kelana Jaya, Selangor, Malaysia and Malaysian Palm Oil Board, Bangi, Selangor, Malaysia, 2007; 63 pp.

Murphy, D.J. Future prospects for oil palm in the 21st century: biological and related challenges. Eur. J. Lipid Sci. Technol. 2007, 209, 296–306.

Ng, S.K.; Thong, K.B.; Khas, C.H.; Ooi, H.S.H.; Leng, K.Y.; Kayaroganam, P. Clonal oil palm: production, yield performance and nutritional requirements. In Oil Palm. Management for Large and Sustainable Yields. Fairhurst, T.H., Hardter, R., Eds.; Potash and Phosphate Institute, Potash and Phosphate Institute of Canada, and International Potash Institute, 2003, pp. 99–114.

Nouy, B.; Baudouin, L.; Djegui, N.; Omore, A. Oil palm under limiting water supply conditions. Plantations, recherché, développement 1999, 6, 31–45.

Palat, T.; Chayawat, N.; Clendon, J.H.; Corley, R.H.V. A review of 15 years of oil palm irrigation research in Southern Thailand. Paper presented at Indian National Conference on Oil Palm, Vijayawada, Andhra Pradesh, India, February 2008, 12 pp.

Paranjothy, K.; Rao, V. An introduction to the oil palm. PORIM Bulletin, Special issue November 1984, 9–18.

Poku, K. Small-scale palm oil processing in Africa. FAO Agricultural Services Bulletin, 148. Food and Agricultural Organisation, Rome, 2002, 56 pp.

Purseglove, J.W. Tropical crops. Monocotyledons. Longman Group: London, 1172; pp. 479–510.

Rajanaidu, N.; Rao, V.; Halim, A.; Ong, A.S.H. Genetic resources—new developments in oil palm breeding. Elaeis 1989, 1, 1–10.

Rajanaidu, N.; Zakri, A.H. Breeding for morphophysiological traits in crop plants. In Plant Breeding and Genetic Engineering; Zakri, A.H., Ed.; SABRAO, 1988, pp. 117–140.

Rao, V.; Chin, C.W.; Rajanaidu, N. Biology and performance of Surinam *Elaeis oleifera.* Elaeis 1989, 1, 109–118.

Rees, A.R. Evidence for the African origin of the oil palm. Principes 1965, 9, 30–36.

Schultes, R.E. Taxonomic, nomenclatural and ethnobotanic notes on *Elaeis.* Elaeis 1990, 2, 172–1187.

Sheil, D.; Casson, A.; Meijaard, E.; van Nordwijk, M.; Gaskell, J.; Sunderland-Groves, J.; Wertz, K.; Kanninen, M. The impacts and opportunities of oil palm in Southeast Asia: What do we know and what do we need to know? Occasional paper no. 51, CIFOR, Bogor, Indonesia, 2009, 67 pp.

Sloane, H. Catalogus Plantarum. London, 1696.

Syed, R.A. Studies on oil palm pollination by insects. Bull. Entomological Res. 1978, 69, 213–224.

Syed, R.A.; Law, I.H.; Corley, R.H.V. Insect pollination of oil palm: Introduction, establishment and pollinating efficiency of *Elaeidobius kamerunicus* in Malaysia. The Planter 1982, 58, 547–561.

Tantitham, S.; Khlaisombat, P.; Clendon, J.H.; Cambell-Board, M.; McIntosh, B. A review of three CDM biogas projects based on palm oil mill effluent in Southern Thailand. In Proceedings of Chemistry, Processing Technology and Bio-energy Conference. Malaysian Palm Oil Board: Kuala Lumpur, 2009; pp. 223–233.

USDA (United States Department of Agriculture). Foreign Agricultural Service. Commodity Intelligence Report, December, 2003.

USDA (United States Department of Agriculture). Palm oil: world supply and distribution. http://www.fas.usda.gov/psdonline/psdHome.aspx. (accessed 2008).

Yusof, B.; Lim, W.S. New technologies needed for the oil palm industry in the 21st century. In Plantation Management for the 21st Century. Vol. 1. Pushparajah, E., Ed.; Incorporated Society of Planters: Kuala Lumpur, 1997, pp. 37–45.

Zeven, A.C. On the origin of the oil palm. Grana Palynologica 1964, 5, 121–123.

Zeven, A.C. The semi-wild oil palm and its industry in Africa. Thesis. University of Wageningen, 1967, 178 pp.

Appendix

Chronology of Major Events in the History of the Oil Palm

Date	Event	Reference
120 million years BP	Emergence of the family *Palmae* (syn. *Arecaceae*) inferred from fossil remains in Cretaceous rocks (at a time when South America separated from Africa).	Latiff (2000) citing Purseglove (1972)
15 million years BP	Early presence of oil palm in Nigeria suggested by finding of fossil pollen similar or identical to that of *E. guineensis* in Miocene strata.	Corley (1976) citing Zeven (1964)
5000 years BP	Use of palm oil in antiquity suggested by samples of oil present in an Egyptian tomb, chemically similar to palm oil.	Corley (1976) and Berger (2010) citing Friedel (1897)
5000 years BP	Presence of broken palm kernels and tools for cracking them at Nigerian cave sites.	Berger (2010)
0900s	Probable introduction of *E. guineensis* into Madagascar.	Schultes (1990) citing Purseglove (1972)
1435–1600	Records by European explorers of the presence of the oil palm in West Africa.	Berger (2010), Corley & Tinker (2003), Hartley (1988), Rees (1965)
1500–1522	Trading in palm oil from West Africa by Portuguese merchants.	
1570–1696	Early descriptions of *E. guineensis*, its inflorescence, fruits and kernels.	Hartley (1977) citing Lobelius (1570, 1576, 1581)
1590	First record of palm oil export from West Africa to England.	Sheil et al. (2009)
1696	Record of importation of *E. guineensis* into Jamaica.	Hartley (1977) citing Sloane (1696)
1763	First formal description of *E. guineensis* by Jacquin.	Corley (1976), Corley & Tinker (2003), Hartley (1977, 1988)
1763	First description of *E. oleifera* by Jacquin.	Hartley (1977)
1790	Start of official records of palm oil imports into UK.	Berger (2010)
1807	Cultivation as commercial crop commenced in West Africa.	Hartley (1977)

Date	Event	Reference
1832	First export of palm kernels from West Africa to Europe.	Hartley (1977)
1848	Four palms planted in Bogor Botanic Gardens, Java, later used as source of first commercial plantings in Southeast Asia.	Gerritsma & Wessel (1997), Hardon (1995), Hartley (1977)
c. 1870	Introduction of oil palms to Singapore Botanic Garden.	Burkill (1966)
1870s	Start of margarine manufacture in Europe using palm oil.	Hartley (1977)
1875	Introduction of Bogor-derived palms into Deli, Sumatra.	Hartley (1977)
1875	Introduction of Bogor-derived palms into Malaya via Singapore.	Imam (1984) citing Arnott (1963)
1878	Trial planting of oil palm in Labuan, Sabah, using nuts obtained from Kew Gardens.	Berger (2010)
1897	Official description of *E. oleifera* by Cortes (Formerly considered a form of *E. guineensis*).	Hartley (1977), Latiff (2000)
1900s	Beginning of rapid rise in palm oil imports to UK due to industrial demand.	Hartley (1988), Berger (2010)
1903	Experimental trials of oil palm in Malaya.	Imam (1984) citing Jagoe (1952)
1909	First mechanized mills constructed in West Africa.	Corley & Tinker (2003), Hartley (1988)
1910	Estate scale planting of Deli palms commenced in Sumatra.	Gerritsma & Wessel (1997)
1911–1917	First commercial plantings of Deli palms in Malaya.	Hartley (1977)
1917	Introduction of hand operated presses to extract palm oil.	Hartley (1988)
1919	First mechanized mills established in Sumatra	Harley (1988)
1920s	First commercial planting of oil palm in Belgian Congo (Zaire).	Corley (1976)
1922	Assisted pollination used in Malaya found to give large yield increases.	Imam (1984)
1931	Palm oil mill established in Serdang, Malaysia.	Imam (1984)
1937	General introduction of mechanised oil palm mills.	Gerritsma & Wessel (1997) citing Blommendaal (1937)

Date	Event	Reference
1941	Discovery of monogenetic control of shell thickness leading to production and widespread cultivation of *tenera* palms.	Beirnaert & Vanderweyen (1941)
1952	Discovery of "dumpy" palm.	Jagoe (1952)
1956	Establishment in Malaya of Cooperative Breeding Scheme by Department of Agriculture involving four estates.	Imam (1984) citing Haddon & Tong (1959)
1959	Introduction of hydraulic hand presses to extract oil in West Africa.	Hartley (1988)
1969	MARDI (Malaysian Agricultural Research Institute) took over oil palm research from the Federal Department of Agriculture.	
1969–1971	Development of standard methods for assessing palm vegetative growth.	Corley et al. (1971), Hardon et al. (1969)
1973	Development of tissue culture techniques for production of clonal palms.	Corley (1993), Jones (1974)
1979	Establishment of PORIM (Palm Oil Research Institute of Malaysia) and transfer of oil palm research from MARDI.	Imam (1984)
1979	Confirmation of insect pollination of oil palm in West Africa.	Paranjothy & Rao (1984) citing Syed (1978)
1981	Introduction of the pollinating weevil, *Elaeidobius kamerunicus* from West Africa into Southeast Asia.	Syed et al. (1982)
1980s	Direct comparisons of the same planting materials in West Africa and Indonesia.	Gerritsma & Wessel (1997) citing Cochard et al. (1993)
2000	Amalgamation of PORIM and PORLA (Palm oil Licensing Authority) to form MPOB (Malaysian Palm Oil Board).	
2004	Formation of Roundtable on Sustainable Palm Oil (RSPO).	
2004	Oil palm replaces soybean as world's largest source of vegetable oil.	MPOB (2008)
2006	Indonesia overtakes Malaysia as world's leading producer of palm oil.	USDA (2007)
2008	First phase of sequencing the oil palm genome.	Crowley (2008) cited by Sheil et al. (2009)

2

Breeding and Genetics of the Oil Palm

Aik Chin Soh
The University of Nottingham Malaysia Campus, School of Bioscience, Jalan Broga, 43500 Semenyih, Selangor, Malaysia

Introduction

Origin, Distribution, and Historical Development

The African oil palm (*Elaeis quinensis*), which produces the palm oil for international trade, is endemic to tropical Africa, although its centers of origin diversity are concentrated in west (e.g., Côte d'Ivoire, Nigeria, Ghana, Cameroon) and central (e.g., Congo, Zaire, Angola) Africa. The American oil palm (*E. oleifera*) however is endemic to Latin America stretching from the Amazon to northern Mexico. Both occur as lowland semi-wild forest groves near river fringes and are often associated with native settlements or their migratory activities. Both species have been postulated to originate in Gondwanaland and diverged when the American and African continents drifted apart in prehistoric times (Zeven, 1965). The oil palm fruit forms an integral part of the native African and South American diet (fat, carbohydrate, protein, vitamins).

European (British, Dutch, French, German, Danish) colonial interest in palm oil to replace tallow in candle wax, soap, and margarine production resulted in the establishment of oil palm plantations in their colonies in southeast Asia, particularly Malaysia and Indonesia. The current large oil palm plantation industry owes its origin to four seedlings planted in Bogor Botanic Gardens in Indonesia that were derived apparently from the same fruit bunch in west Africa and obtained via Amsterdam and Mauritius/Reunion. The last surviving member of these palms has recently died. Further hybridizations and selections among the progenies of *these four Dura (D)* progenitors were distributed to the plantations in Deli province in Sumatra, Indonesia, and thence to Malaysia. The Deli Ds were the predominant commercial planting material for the early rapidly expanding plantation industry in Malaysia and Indonesia until the early 1960s (Rosenquist, 1986). In 1941, Beirnart and Vanderweyen elucidated the monogenic inheritance of the shell gene: the cross between the thick shell D parent palm and the shell-less (usually female sterile) *pisifera* (P) parent would give rise to 100% thin-shell *tenera* (T) palms, exhibiting incomplete dominance of the shell gene; the T × T, the T × P, and D × T crosses would give rise to segregating progenies in the classical Mendelian ratios of 1D: 2T: 1P for the first, 1T: 1P for the second, and 1D: 1T for the last (Figs. 2.1 & 2.2). With this discovery, the switch to the T hybrid as the commercial planting material was very rapid. This was achieved through the importation and planting of T seeds; the resultant P segregant palms provided the pollen for commercial D × P seed production. The Deli D since then has assumed the role of mother palm in commercial D × P hybrid production. The oil palm plantation expansion was spurred on by private plantation companies becoming interested in commercial hybrid seed production and investing in oil palm breeding. Mixed T or D × P hybrids have been the dominant commercial planting materials until today.

World Trade and Uses

The oil palm is an economically important crop to many third world countries in the humid tropics. It is the highest yielding and highly profitable oil crop and is relatively easy to grow by large plantations and small farmers alike. It is the leading export oil crop, providing a major source of external revenue to a number of countries, and its versatile uses spawn a host of domestic industries. Palm and kernel oil, with the production of about 51 million tons worth about $46 billion in 2009/2010, dominates the international edible oil trade (Oil World Annual, 2010). It is produced from about 13 million ha. of plantations. Indonesia accounted for 49% of the production, followed by Malaysia (32%), and others (Colombia, Côte d'Ivoire, Nigeria, Papua New Guinea), the remainder. The main markets are China, India, European Union, and Pakistan, with increasing interest shown by other developing countries.

The oil palm produces two types of oil from its fruits in international trade, mesocarp oil and kernel oil, known as crude palm oil (CPO), and palm kernel oil (PKO). Palm oil is the widest internationally traded vegetable oil. Palm oil is mainly used in food (80%) (e.g., cooking oil, margarines, vanaspati or vegetable ghee, shortenings) and 20% as oleochemicals, replacing those derived from mineral oil to feed the detergent, cosmetic, pharmaceutical/nutraceutical, plastic, and lubricant industries. Fig. 2.3 summarizes the palm oil fractionated products and their uses. With the recent hike in petroleum prices and the deadline for meeting the requirements of the Kyoto Protocol and subsequent initiatives in terms of "green" or renewable energy substitution, there has been a tremendous demand for palm oil as a source of biofuel (biodiesel). Also, with rising concern in consumer and environmental health, secondary and by-products from the palm oil industry have spawned new industries (e.g., vitamins A, E, and other antioxidant health supplements from the oil; animal feed and organic fertilizers from the kernel; and sludge cakes from the mill wastes) (Choo, 2000; Jalani et al., 1997).

Fig. 2.1. Commercial oil palm D × P hybrid seed production.

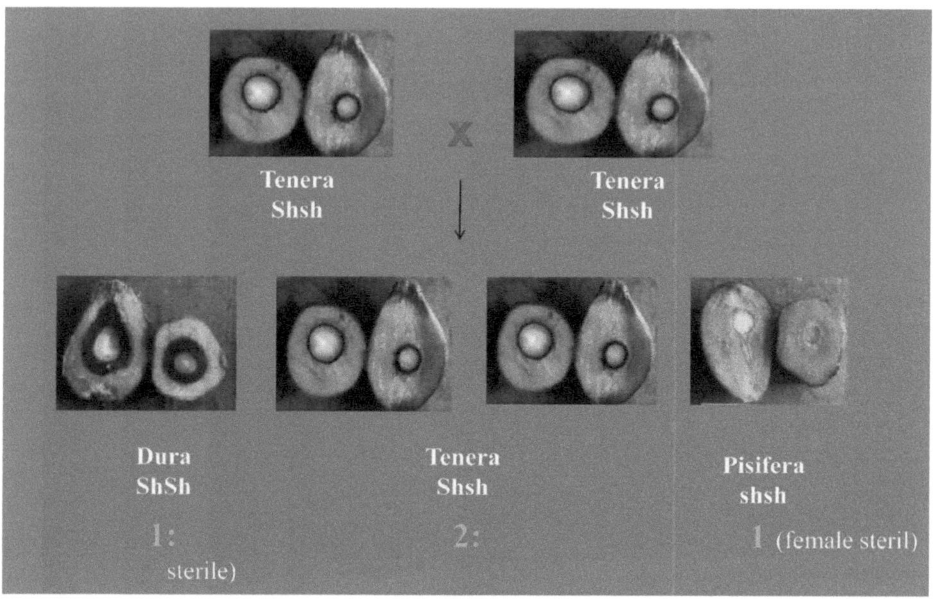

Fig. 2.2. Inheritance of fruit shell thickness. Source: Soh et al., 2009.

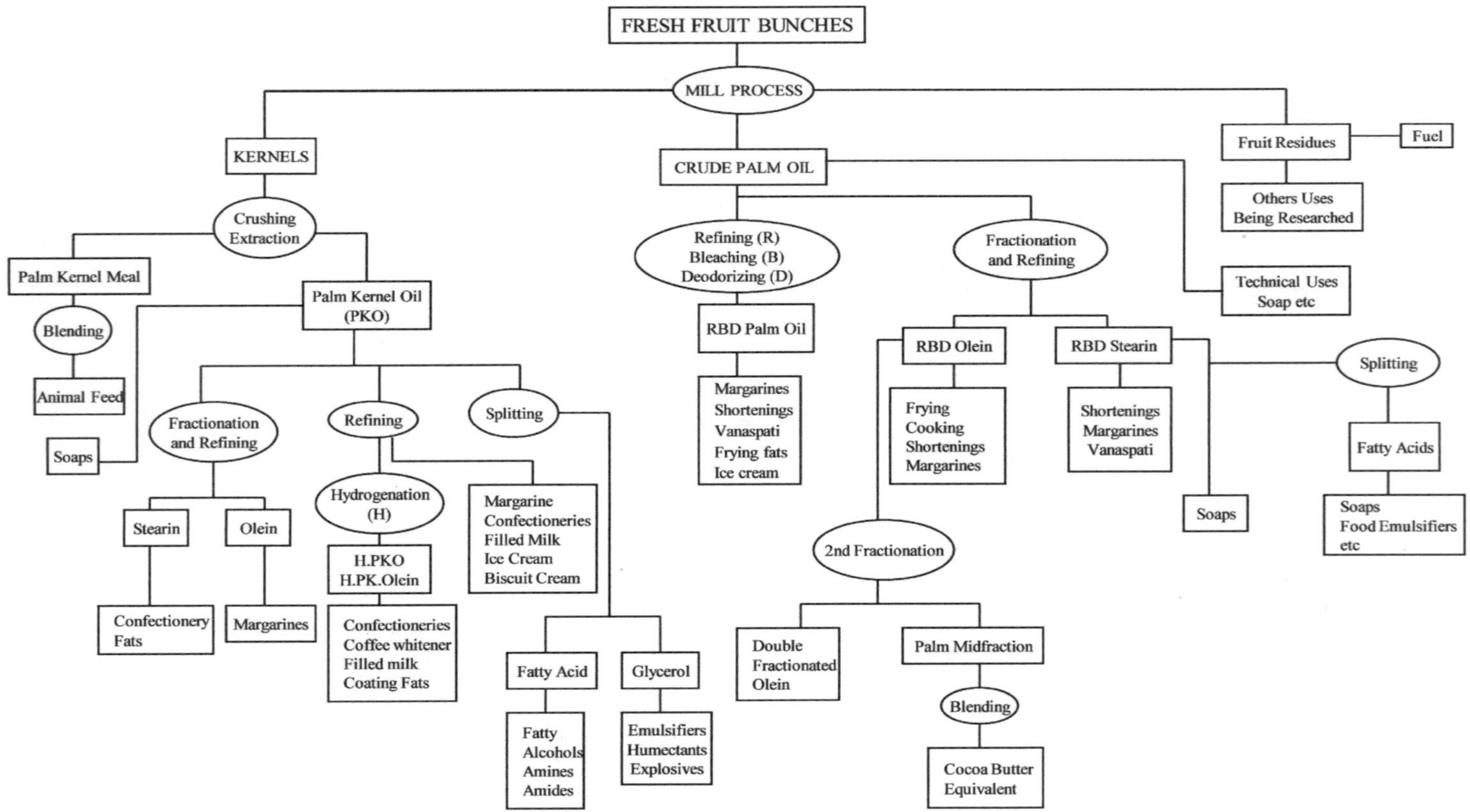

Fig. 2.3. Uses of palm oil. Source: Pantzaris, 2000.

As the oil palm is a very productive [world average of 3–5 t/ha/yr of palm oil compared to less than 1 t/ha/yr of others (Oil World Annual 2007)] and profitable crop with versatile uses from its oils, its areas are expected to increase further. The oil palm is seen as the main instrument for socio-economic development, particularly for the rural areas in Malaysia and Indonesia but also in other third world countries in the tropics. The palm oil industry provides employment to about 2 million people in Malaysia and close to 5 million people in Indonesia directly and indirectly (Darmosarkoro, 2010; Directorate of Estates, 2010; Malaysian Palm Oil Statistics, 2010).

Plant Variety Improvement

Current Breeding Objectives

Yield Potential

High oil yield remains the obvious primary objective of oil palm breeding improvement. Corley (1985) predicted an annual yield potential of 45 t/ha fresh fruit bunches (FFB) and 17 t/ha oil (mesocarp and kernel oil) based on theoretical considerations using factors adapted from other crops. The figure of 11–13 t/ha oil predicted by Breure (2003) based on the best achieved yield on experimental plots, which is the potential yield definition of Evans & Fischer (1999) (i.e., yield achieved under stress-free conditions), appears to be more realistic and acceptable. Fresh fruit bunch (FFB) yields close to 45 t/ha have been reported from commercial fields and experimental plots while palms having close to 40% oil to bunch (O/B) plus kernel oil to bunch (KO/B) have been reported in breeding materials (Rosenquist, 1990), although these palms did not have high FFB yields. It appears that current planting materials are capable of yielding close to the potential yield and that prospects of further improvement would be limited at least in the near term. Reported breeding improvements are likely to have been through adaptability to site limitations [i.e., moving the site yield potential (Tinker, 1984) towards the potential yield by circumventing the site limiting (biotic, abiotic) factors, rather than through improvement in genetic yield potential per se]. Nevertheless, proposals and attempts to breed for more efficient palms through higher harvest index (HI), more efficient light capture through optimum leaf area expansion rate, tolerance to high density planting, and reduced respiratory loss have been made (Breure, 1985, 1986, 2010; Breure & Corley, 1983). In cereals, yield potential improvement in wheat and rice have been achieved mainly through better HI by using the dwarfing gene, while in maize this has been attributed to mainly tolerance to high density planting (Evans & Fischer, 1999; Tollenaar & Wu, 1999). Semi-dwarf oil palm planting materials available earlier did not have improved HI (Fig. 2.4; Soh et al., 1981). Subsequent breeding improvement effort has given rise to

Fig. 2.4. Breeding of semi-dwarf dumpy. AVROS pisiferas.

progenies with smaller and more erect canopies with improved HI capable of higher density planting (Fig. 2.5; Soh, 2004).

Harvestable Yield

Harvestable or recoverable yield encompasses ease of harvesting, harvesting at optimal fruit ripeness and minimal loss of ripe fruit, and its oil content and quality. Dwarf palms and palms with longer bunch stalks would facilitate both manual and mechanical harvesting. Palms with fruits exhibiting distinct color change (e.g., virescens fruit), which changes from green to bright orange when ripe (Fig. 2.6), would facilitate good ripeness standard in harvesting while those with a non-abscising fruit habit would help minimize loose ripe fruit loss (Osborne et al., 1992; Rao, 1998). Low lipase in fruit would reduce loss in oil content, in oil quality from free fatty acid (FFA) production, and the need for frequent and regular harvesting rounds. Low lipase genotypes have been found in MPOB's Nigerian semi-wild accessions.

Oil Quality

The oils produced by the oil palm are very versatile in their food and industrial uses. Nevertheless, prompted by the industry, attempts have been made to improve the quality of palm oil such as improving the unsaturated oil content (i.e., high oleic acid)

Fig. 2.5. Breeding high yielding semi-dwarf palms. Source: Soh et al., 2009.

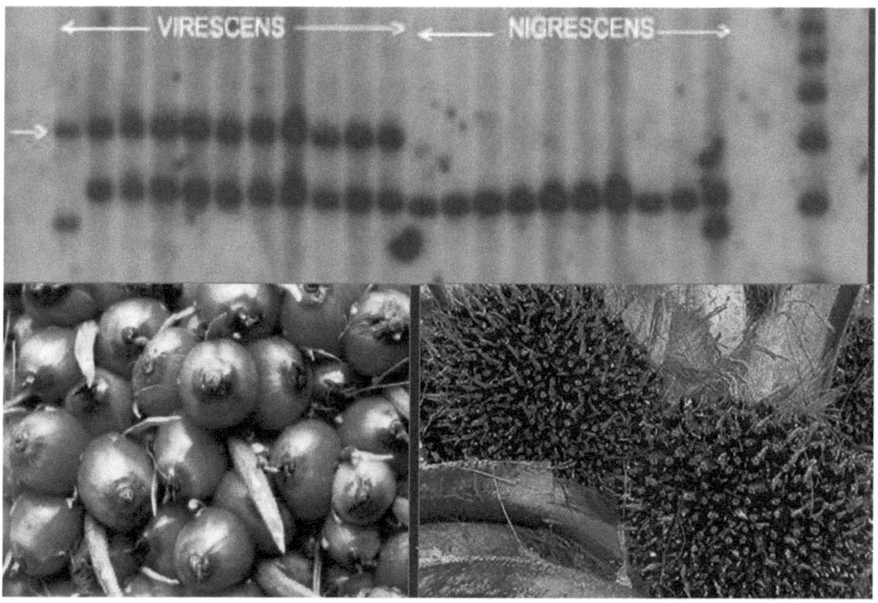

Fig. 2.6. Marker assisted selection for virescens fruits. Source: Soh, 2001.

to tap the cooking and salad oil market of the temperate countries and improving the beta carotene and tocotrienol contents for industrial vitamins A and E production, respectively, and also other antioxidant compounds. High oleic oil is also preferred for biodiesel production.

Biotic and Abiotic Stress Tolerance

Basal stem rot caused by *Ganoderma boninense* is the most important and worrisome disease of oil palm for urgent consideration in plant resistance breeding in southeast Asia (especially Malaysia and Indonesia), as chemical and cultural control methods have proved to be inefficient if not ineffective. The disease used to be confined to older palms planted on ex-coconut or oil palm land with a high water table. With continuous replanting (some into the third cycle) of susceptible and narrow genetic based genotypes, disease occurrence on younger replants have become more frequent and increasing in inland areas. Progress in resistance breeding has been hampered for a long time by the lack of a plant resistance source and an efficient screening technique. A partial resistance source has been found in a Zaire × Cameroon cross derived from Malaysian Palm Oil Board's or MPOB's semi-wild germplasm accessions, and its screening technique has been developed (Ariffin et al., 1995; Idris et al., 2004; Khairudin et al., 1993). The incorporation of this resistance into commercial varieties would take a longer period, as it is derived from a semi-wild genotype requiring 5–6 generations (30–40 years) of introgression or backcross breeding to advanced breeding parents. Centre Internationale pour les Reserche et Developement, or CIRAD, found partial resistance in their advanced breeding populations and has been collaborating with private plantation research laboratories in Indonesia to develop *Ganoderma* resistant commercial varieties (Breton et al., 2006; Durand Gasselin et al., 2005) and are expected to be available soon.

Breeding for tolerance to vascular wilt caused by *Fusarium oxysporum f. sp. elaedis* forms an integral part of the African programs (De Franqueville & Renard, 1990), and commercial materials produced possess some level of disease tolerance, as otherwise oil palm plantations would not be sustainable. The disease has been exported to Latin America. The disease does not occur in southeast Asia, although *F. oxysporum* does. Apparently, the Dumpy genotype is tolerant to this disease (Rosenquist et al., 1990).

Bud rot is a very serious disease in Latin America and has decimated plantations with *E. guineensis* palms. The causal agent has recently been attributed to *Phytophthora palmivora* (Martinez et al., 2009) instead of other non-pathogenic causes postulated earlier (Ariffin et al., 2000). The *E. oleifera* × *E. guineesis* (O × G) hybrids appear to be tolerant.

Oil palm plantations have in recent years expanded into more marginal areas (e.g., drier areas, deep peat, and breeding), for such adaptable varieties may have to be considered. Again, African commercial materials tend to be more drought tolerant as they have been bred under such conditions (Houssou et al., 1987). Normal commercial oil palms planted on sandy areas in Indonesia and Papua New Guinea exhibit magnesium (Mg) deficiency symptoms and breeding programs drawn up have tolerance to Mg deficiency as an important trait (Breure et al., 1986). Palms grown on peat have manifested some mineral stress symptoms. A commercial seed company in Sumatra has claimed that one of its commercial genotypes is particularly suitable for planting on peat.

Future Breeding Objectives

Future breeding objectives essentially would be related to sustainability and climate change issues: food security (quantity, quality), depleting arable land, environment degradation, natural resource depletion, high fuel cost and renewable energy, reduced farm labor, changing weather/climate pattern (subtle/drastic) [e.g., drought, floods, solar radiation, and peat and disease (P&D) stress affecting crop growth]. It would be impractical to breed (a long tedious effort) for all these traits, most of which could be better addressed by agronomic, engineering, and management approaches. Nevertheless, it would be useful to review whether current breeding priority traits would be able to address some of these concerns and where special breeding programs might be needed. Breeding for super yielding palms (F1 hybrids/clones, dwarf high density tolerant genotypes) with high quality micronutrients (e.g., oleic, carotene, tocotrienols, and other phytochemicals) would address the issues of food security, land and farm labor scarcity, renewable energy (oil and biomass from more trees per unit area), and buffer against the vagaries of biotic and abiotic environmental stresses for most situations. Breeding for quantitative basis of biotic (P&D) and abiotic (moisture, heat, mineral, radiation) stress tolerance would confer durable resistance for sustainable crop production and also alleviate the effects of short term stress. Genotype × environment (G × E) trials are mandatory as plantations continue to operate in diverse environments within and without national borders. Such trials would identify genotypes suitable for specific stress environments (i.e., specific adaptability) and others that generally perform well across all environments (i.e., general adaptability). Genotypes selected for high yield in good environments would perform similarly well in moderate environments except for extreme environmental conditions. Selection for the genotypes with specific adaptation would indirectly select for genotypes with good water use efficiency (WUE) and/or nutrient use efficiency (NUE) dispensing with special trials or provide leads to set up special trials if deemed mandatory.

Oleifera(O) × O and O × G materials where most of the novel and extreme traits are extant will feature prominently in future oil palm breeding programs as well as the use of biotechnological (e.g., in vitro culture) marker assisted selection (MAS) and other genomic tools and even transgenics as negative public opinions abate.

Genetics of Economic and Agronomic Traits

Most traits of economic and agronomic importance are usually quantitatively inherited, controlled by many genes (i.e., polygenic) and greatly influenced by the environment. Nevertheless, there are a number of useful qualitative traits (i.e., controlled by one or few genes). Also, as most oil palm breeding programs are commercially oriented, they cannot afford to undertake basic genetic studies. Thus such studies tend to be incidental, derived from breeding trials, and seldom meet the basic assumptions required for good genetic analyses.

Qualitative Traits

Although the presence of shell in the fruit is controlled by a gene exhibiting incomplete dominance, the thickness of the shell was purported to be modified by minor genes or under polygenic control (Van der Vossen, 1974). Okwuagwu (1988) and Okwuagwu & Okolo (1992, 1994) postulated maternal inheritance of kernel size with the involvement of a kernel-inhibiting factor. Fertile Ps do occur occasionally, and fertile P × fertile P crosses can produce highly fertile P progenies, suggesting perhaps fertility/sterility and absence of shell in the P are different genetic control (Wonki-Appiah, 1987). A number of other qualitative traits have also been reported from observations and genetic analyses to be simply inherited (Hartley, 1988), such as crown disease, leaf form (*idolatrica* = fused pinnae), fruit shape (mantled), and fruit color (*nigrescens* = black unripe, *virescens* = green unripe, albescence = whitish unripe; Fig. 2.6). Oil quality traits (e.g., high oleic, high stearic, low lipase) are also likely to be controlled by one or few major genes. Although the Dumpy trait appeared to be simply inherited, observations on segregating progenies did not confirm it (Soh et al., 1981). Population 12 from MPOB's Nigerian prospection has also a dwarf stature (Rajanaidu et al., 2000). Its inheritance has yet to be elucidated but is likely to have some major genes involved, likewise for long bunch stalk. The mode of inheritance for tolerance to *Fusarium* wilt has been postulated to be controlled by a few major genes (De Franqueville & Renard, 1990). The partial resistance trait in MPOB's Zaire × Cameroon and in CIRAD's advanced breeding parents are also likely to be controlled by a few major genes.

Quantitative Traits

Estimates of genetic variance and heritability have been sought by oil palm breeders to guide them for increased breeding efficiency. This involves information on the existence of sufficient genetic variability, the choice of the population to improve, and the adoption of appropriate breeding procedure to obtain the best selection response (Soh & Tan, 1983). Genetic experiments and mating designs commonly used to achieve these are North Carolina Model 1 (hierarchical) and Model 2 (factorial) mating designs and the parent-offspring method. Heritability estimates for various agronomic traits have been compiled by Soh & Tan (1983) and Hardon et al. (1985). With the use of data from breeding experiments derived from selected parents from very restricted populations, some of the assumptions such as random choice of parents, no correlation of genotypes at separate loci, and a definite reference population, necessary for valid quantitative genetic analyses could not be met and the biases incurred could not be estimated. These estimates merely serve as a guide for selection decisions, as any estimate of heritability is unique (i.e., dependent on the population and environment sampled and the estimation method used). For selection decisions or for prediction of selection response for a particular population and environment, specific heritability estimates obtained from the population concerned should be used. In lieu of reliable specific heritability estimates or if the responses are for different environments, average heritability values may be substituted.

Variety Groups and Genetic Resources

Variety Groups

Breeding Parents of Restricted Origin (BPRO)

Varieties in the strict sense do not exist in oil palm as the commercial materials are mixed hybrids from non-fully inbred (and sometimes out-crossed parents). They are inter-population hybrids with Deli commonly serving as the maternal parent source. The "variety" usually refers to the P parent source. As with Deli, P parental sources were usually derived from a few progenitors and hence have been referred as *breeding populations of restricted origins* (BPROs) (Rosenquist, 1986). Only a handful of BPROs are featured in commercial hybrid breeding programs.

Deli. The distribution of this BPRO to various countries followed by local selection led to the development of a number of subpopulations (e.g., Ulu Remis Deli, Elimina Deli, Serdang Deli in Malaysia; La Me Deli, Dabou Deli in Côte d'Ivoire. The Dumpy and Gunung Malayu are dwarf variants of the Deli).

AVROS. The ancestral palm was the "Djongo" or "Best" palm found at Zaire's Eala Botanical Garden. Seeds were brought and planted in Sungai Pancur, Sumatra by AVROS (Algemeene Vereniging van Rubberplantera ter Oostkust van Sumatra), the Dutch colonial plantation research institute, in1923. These gave rise to the famous ancestral palm SP540T. Subsequent crosses of this palm with others in Bangun gave rise to the AVROS BPRO and was imported into Malaysia and then distributed to other parts of the world (e.g., Thailand, Papua New Guinea, Colombia, Costa Rica). It is the main source of P parent palms in many D × P hybrid seed production programs. Vigorous palm growth and high oil yield from big fruits with thick mesocarp are characteristic features of this BPRO.

Yangambi. This BPRO originated from the breeding program of INEAC (Institut National pour l'Etude Agronomic du Congo) at Yangambi station using open-pollinated progenies of the Djongo palm and other Ts from Yawenda, Isangi, and N'gazi. This ancestral P source is also featured in a number of major breeding and seed production programs. Yangambi BPRO shares similar features as the AVROS. A short variant (16R) was found and further developed in this population.

La Me. This BPRO was bred from the 21 T palms, particularly the Bret 10 palm, derived from seeds collected from the wild groves in the Côte d'Ivoire by Institut de Recherches pour les Huiles et Oleagineux (IRHO), currently known as CIRAD. This BPRO, which has a smaller palm stature, produces many smaller bunches with small fruits and appears to be tolerant to less favorable growing environments forms the genetic base of the breeding and seed production programs in West Africa, Latin America, Thailand, and Indonesia advised by CIRAD.

Ekona. Breeding by Unilever Plantations Group using the wild palms found in the Ekona area of Cameroon gave rise to this BPRO. Through the Unilever plantations network, the Ekona BPRO, particularly progenies from the *Fusarium* wilt-resistant and high oil yielding palm CAM 2/2311, were distributed to Malaysia, Thailand, Costa Rica, and Thailand.

Calabar. The genetic base of the Nigerian Institute for Oil Palm Research (NIFOR) is much broader from collections made in Aba, Calabar, Ufuma, and Umuaibi. Progenies from its Calabar selections, particularly palm NF 32.3005, have been distributed to Ghana, Malaysia, Costa Rica, and subsequently Indonesia.

Derived and Recombinant BPROs. Second cycle breeding involving recombination or inter-breeding and introgression breeding have resulted in BPRO with mixed lineages or recombinant BPROs, such as URT (Ulu Remis T), Dumpy × AVROS, Dumpy × Yangambix AVROS, LaMe × AVROS, LaMe × Yangambi, LaMe × Dumpy AVROS (Soh, 2004).

Genetic Resources

Realizing the genetic improvement limitations and genetic vulnerability set by the very narrow genetic base of their existing oil palm breeding programs, beginning in the 1970s Malaysia (and subsequently Indonesia) made a series of expeditions to Africa and Latin America to prospect for new *E. guineensis* and *E. oleifera* germplasm for broadening the genetic base for breeding and for genetic conservation its *ex situ* genebank (Rajanaidu et al., 1979; Rajanaidu, 1990; Rajanaidu & Jalani, 1994; Rajanaidu et al., 2000). The first prospection in Nigeria was done systematically in collaboration with NIFOR and the then International Board of Genetic Resources with the objectives of genetic conservation and studying its population genetic structure (Obasola et al., 1983). Genetic studies done on the prospected progenies revealed that genetic variability was higher within families than between families and population. Subsequent collections in Angola, Cameroon, Gambia, Guinea, Madagascar, Sierra Leone, Tanzania, and Zaire were guided by this finding and also targeted towards accessions with prospective economically desirable traits. Prospections for *Oleifera* in Latin America were done in Brazil, Colombia, Costa Rica, Ecuador, Honduras, Panama, Peru, and Surinam and also included other oil bearing palms (*Bactris gossipaes* or pejibaye, *Jessenia-Oenocarpus, Orbignya martiana* or babassu) with unusual fatty acids and other properties. These prospections proved to be strategically very useful in that a number of extreme desirable traits, and traits lost/absent in existing breeding populations were discovered and are being introgressed.

Reproductive Biology and Plant Propagation

Hybrid Seed Production

Breeding System

The oil palm is a monecious bearing male and female inflorescences on the same palm occurring in different but overlapping cycles. Stress conditions favor male inflorescence production. *Pisifera* palms do produce female inflorescences, especially under good growing conditions, but they usually abort. Currently, natural cross pollination is effected efficiently by the pollinating weevil *Elaidobius kamerunicus* introduced from West Africa in the 1980s. Prior to the introduction, pollination was by wind and thrips (*Thrips hawaiiensis*), which were less efficient. Stringent, controlled pollination procedures had to be adopted. Anthesis and receptivity occur in the male and female inflorescences about 2 months after emergence. The pollen viability and the receptivity periods are about 3–5 days. The fertilized fruit ripens in about 5 months.

Controlled Pollination

The surface sterilized male inflorescence is isolated with an air permeable terelene/canvas/paper bag about 7–10 days prior to anthesis. The anthesized male inflorescens is harvested and air dried (38–39°C hot air or air conditioned chamber), the pollen shaken out, sieved and further dried in filter paper envelopes in a 38–39°C oven or dessicator until 6% moisture content, and stored in test or specimen tubes in a freezer. Such properly processed and stored pollen can retain its viability for 6 months to 1 year. Freeze drying is recommended for longer term storage. Isolation of the female inflorescence is done in exactly the same manner. Controlled pollination is effected when the female inflorescence is observed to be receptive, by puffing a 1:10 – 1:20 pollen : talcum powder mixture through a hole made in one or more plastic windows sown on to the isolation bag. The bag is removed and the fertilized fruits are left to ripen (in about 5 months). The oil palm seed is recalcitrant and requires heat treatment (at 37–38°C and 17–19% moisture content for D seed, 20–21% for T seed) for 40–60 days for germination at ambient conditions after rehydration (to 20–21% moisture for D and 27–28% for T seeds). Germination of very thin-shelled T and shell-less fertile P seeds is erratic and in vitro (embryo rescue) germination is recommended.

Oil Palm Seed Market

The world trade in oil palm seed is about 300 million seeds worth $250 million at around 80 cents per seed. The main supply and demand is in Indonesia with 133.8 million seeds (Purba, 2010), Malaysia with 109.5 million seeds (Review of the Malaysian Oil Palm Industry, 2010) and others the rest.

Tissue Culture Clonal Propagation

The actual process of production of oil palm clonal plants by tissue culture is the subject of a separate chapter. This section discusses the issues involved in the commercial propagation of oil palm by tissue culture.

Issues in Tissue Culture Clonal Propagation of Oil Palm

Commercial oil palm variety comprising a heterogeneous mixture of non-uniform hybrid progenies is genetically variable. Individual palms within it can yield considerably more than the field average in oil production. Reproducing these superior individual palms by conventional hybrid breeding would take more than 20 years. The original rationale of clonal propagation is to short circuit this process by capturing the elite individual palms as clones and mass propagate it for large scale commercial planting.

Despite the early success in developing the tissue culture clonal propagation technique (Hardon et al., 1987; Jones et al., 1974; Rabechault & Martin, 1976) and the subsequent expanded efforts in clonal propagation for trial and pilot field tests, large-scale commercial propagation and planting of proven elite clones did not take off until recently (Soh et al., 2003, 2011) owing to a number of critical issues: somaclonal variation, cloning efficiency, ortet (parent palm of clone) selection efficiency and feasibilities of recloning, and the liquid suspension culture system which needed resolution.

Cloning Efficiency

Applied Agricultural Resources Sdn. Bhd (AAR) pioneered the development of the commercial gel culture propagation method through research into protocol development and refinement to improve the cloning efficiency with minimal risk of the mantled fruit clonal abnormality and by cloning a package of ortets (Wong et al., 1997). This method, subsequently also adopted by other laboratories, is still an inefficient process due to the low percentage of palms amenable to mass propagation, lower plantlet production capability, and higher labor and space requirement. Commercial propagation is achieved by putting a large number of ortets into culture each time to circumvent the cloning inefficiency and to hedge on clonal abnormality risk. A continuous supply of superior ortets is thus needed and this compromises on the maximum yield potential of clones.

Ortet or individual palm selection is inefficient for oil yield improvement because of its poor heritability (Potier et al., 2006; Soh, 1986; Soh et al., 2003b). A clonal test is required to identify the outstanding clones. To exploit these outstanding clones as proven clones, recloning is needed. AAR demonstrated the feasibility of recloning with minimal clonal abnormality risk, contrary to earlier experiences. Unfortunately, recloning was still constrained by the inefficiency of the gel culture system. AAR subsequently demonstrated the feasibility of the liquid suspension culture system for both clones and reclones, again contrary to the high abnormality risk experienced by others (Soh et al., 2003a). The merits of the liquid system are its amenability to high capacity uniform plantlet production and to automation including the use of bioreactors. With these advances, the technology for commercial production and planting of superior proven clones appeared to be in place. Nevertheless, the ability to scale-up clonal production (with minimal somaclonal variation) to capture a significant proportion of the large hybrid seed market (>100 million) remains the biggest challenge (Soh et al., 2011).

Ortet Selection Efficiency and Cloning Improvement

CIRAD (Potier et al., 2006) in comparing 42 clone-ortet sets and 17 parent crosses from their second cycle reciprocal recurrent selection (RRS) program found that the mean oil yield (OY) of the clones although 7% better than the Control Cross mean (representing their previous generation D × P mean) was lower than the current generation crosses mean by 9%. The best clones were comparable in OY as the best crosses indicating that reclones would be only be as good as the reproduced best crosses. AAR made similar observations. This would imply that cloning ortets from advanced D × P populations is unlikely to result in substantial yield improvement as Soh (1986) predicted earlier. CIRAD also concurred with AAR (Soh, 1986; Soh et al., 2003b) on the low heritability for OY, which would imply the low likelihood of success in ortet selection from commercial fields as currently practiced by some oil palm tissue culture laboratories (Soh et al., 2011). This would question the advantage and relevance of commercial clonal plantings with their higher cost and attendant risks of somaclonal variation and pest, disease, and environmental stress vulnerabilities. Cloning ortets from the earlier or recombinant phases would be more efficient with the available wider genetic variability and higher heritability for yield and other desirable traits. The advantage of clones would be greatest in early commercial exploitation of new genetic materials from introgression breeding progenies derived from wide (intra and interspecific) crosses. The genetic base of the commercial crop would also be broadened to reduce its genetic vulnerability risk to biotic and abiotic stresses, ensuring crop sustainability. Lastly, adaptability trial tests are integral in any breeding and cultivar development program. The issues of commercial field plantings of clones in terms of their composition, planting configuration, and the agro-management practices, including G × E interactions, need to be resolved in order to fully exploit the advantages of clones or otherwise be further disadvantaged against hybrid seeds (Soh et al., 2003, 2011). In 2010, commercial oil palm clonal plantlet production in the world was just about 5 million dominated by about 10 tissue culture laboratories in Malaysia.

Clonal Seeds

Instead of cloning T ortets, one or both of the D and P parents of the best D × P crosses can be cloned for the production of semi-clonal and bi-clonal hybrid seeds, respectively. Bi-clonal seeds will reproduce the performance of the best crosses while semi-clonal seeds will reproduce the mean performance of the crosses from one best combining parent (D or P). Risk of mantling and other somaclonal variation are minimal, and conventional seed production is cheap and efficient. Furthermore, some clonal hybrids reproduced from advanced hybrid crosses are capable of yielding as well as the best clones (Potier et al., 2006; Soh et al., 2011). Semi-clonal seeds are now available for commercial sale by some companies (e.g., AAR, United Plantations) and are expected to capture a substantial portion of the hybrid seed market compared to clones.

Breeding and Selection Methods

Population Improvement and Hybrid Seed Breeding

The oil palm, being a cross-pollinated perennial tree crop, has adapted breeding methodologies developed in maize (e.g., recurrent selection, hybrid production) and animal breeding, such as sire (P) testing, index and best linear unbiased prediction (BLUP) selection, and the close temporal linkage of commercial hybrid production with each cycle of breeding. The major breeding programs adopted either the modified recurrent selection method (MRS) or the modified reciprocal recurrent

(RRS) method. The former is practiced by programs associated with the Unilever Group, while the latter is adopted by countries in West Africa and Indonesia under CIRAD's advice.

In MRS (Fig. 2.7), Deli D parents selected as commercial hybrid seed production mother palms and for further breeding are based on family and individual palm performances, or FIPS in short (Rosenquist, 1990). FIPS is also practiced in T selection and in T × T/P crosses for further P breeding. *Pisifera* (female sterile) parent selection as the male parent for commercial hybrid seed production, however, is based initially on its T sib performance in the T × T family followed by a D × P progeny test. This is done by crossing each P with a sample of the selected D female parents in a nested (NCM 1) mating design (i.e., top cross or sire testing). The variability within the commercial hybrid mix depends on the genetic diversity and inbreeding level of the parents used.

The main advantages of this scheme are faster turn-over of recombinant crosses, less space needed without extensive progeny-tests, and the genetic variability within the commercial hybrid mix would buffer against environmental and P&D stress changes. This scheme exploits only *general combining ability* (GCA) but not *specific combining ability* (SCA or hybrid vigour) of the parents (Soh, 1999).

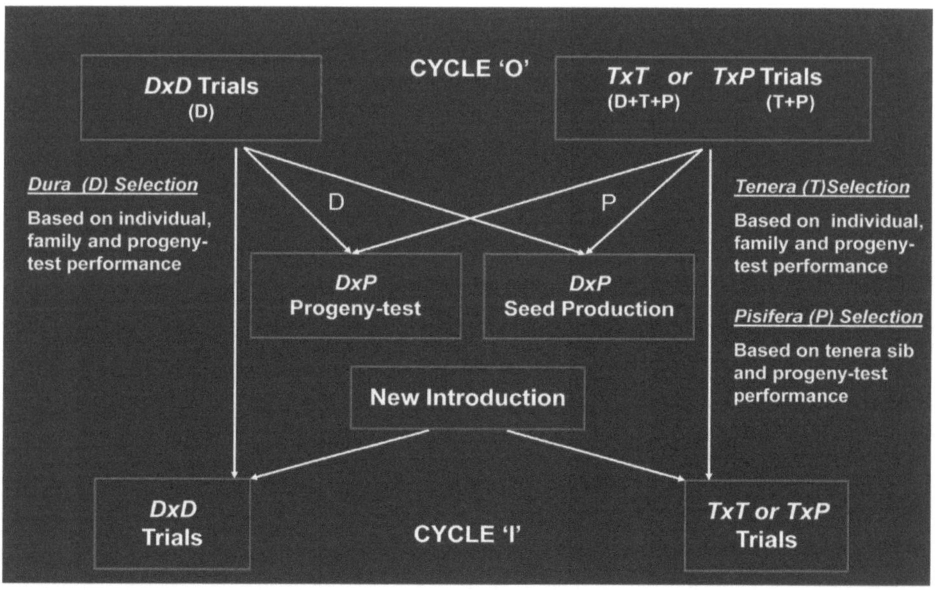

Fig. 2.7. Modified recurrent selection (MRS) in oil palm.

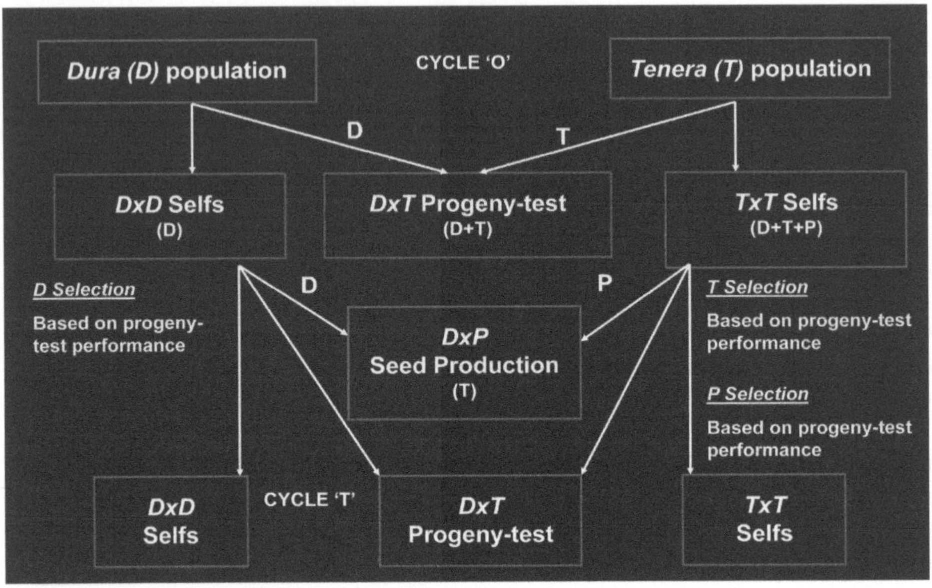

Fig. 2.8. Reciprocal recurrent selection (RRS) in oil palm.

In RRS (Fig. 2.8), the parents selected for further breeding and commercial hybrid seed production are based on progeny tests. Instead of D × P progeny test, D × T progeny test of the selected D and T parents from the D × D and T × T families is conducted (Cochard et al., 1993; Meunier & Gascon, 1972). To save time, selfs and sibs of the parents undergoing progeny testing are made and planted simultaneously as the progeny test crosses. The best hybrid crosses identified from the D × T progeny test are readily reproduced as commercial D × P hybrids using the Ds from the selfs of the D parents and the Ps from the selfs from the T parents. This scheme exploits both GCA and SCA. The scheme encompasses a *within hybrid improvement* phase to further refine a particular hybrid combination and a *recombination* phase involving outcrosses to maintain genetic variability for longer term genetic improvement. The former would eventually lead to the production of true *F1 Hybrid* or *Single-Cross Hybrid* varieties from fully homozygous parents. The latter constitutes population improvement.

Backcross Breeding

The backcross breeding method (Fig. 2.9) is usually used in introgression work in which a desirable trait from a relatively unimproved genotype is incorporated into an recurrent improved host genotype. The Dumpy AVROS (Fig. 2.4) was bred in

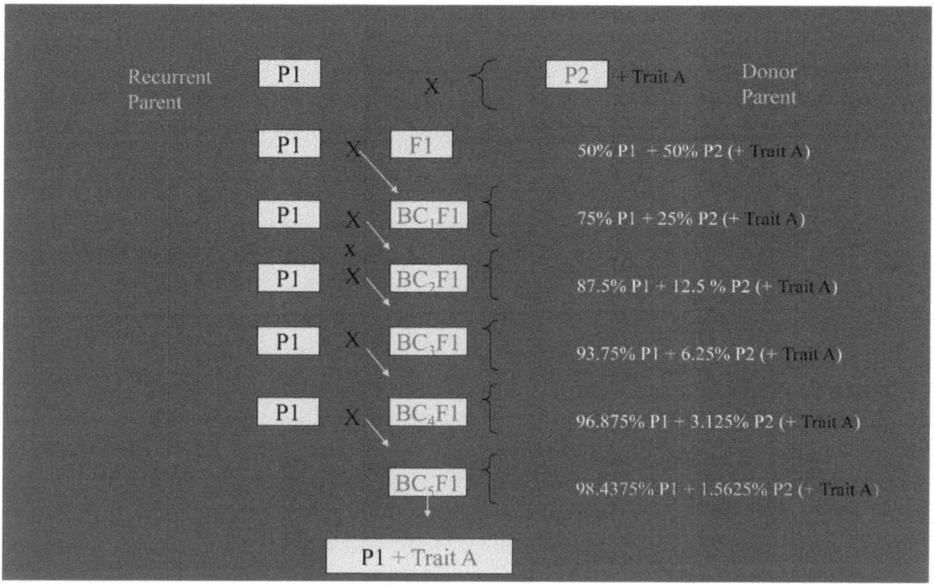

Fig. 2.9. Backcross breeding of dominant gene trait.

this manner (slightly modified due to the perennial nature of the crop), with the Dumpy as the donor parent for the dwarfing gene(s) and AVROS as the recurrent host (Soh et al., 1981). A similar approach is being adopted in the introgression of desirable traits (oil quality traits, dwarfness, disease resistance) from the relatively wild or unimproved recent *Guineensis* and *Oleifera* introductions to the advanced breeding populations.

Recombinant Inbred Breeding

Following an earlier proposal based on the dominance theory of heterosis or hybrid vigor than overdominance and epistasis (Pooni et al., 1989), programs to develop recombinant inbred varieties have been attempted. In this approach, there is no necessity to separate the breeding parents into heterotic groups as in MRS/RRS and a superior recombinant genotype can be developed by single seed descent into inbred varieties. Theoretically, one could eventually achieve an inbred D variety superior to the T hybrid that could then be propagated indefinitely by selfing. Most programs, however, would rather pursue D and T recombinant inbred line development to produce second/advance cycle hybrids as in maize breeding (Bernado, 2002) with the option to produce commercial hybrid at each progressive cycle.

Index and BLUP Selection

Tree crop breeding like cattle breeding has to contend seriously with space and time constraints and hence, the critical choice would be a few highly selected parents to go into the next cycle of breeding and also into hybrid seed production due to their close linkage. This is also true in the selection of ortets for clonal propagation. Ascertaining the breeding values of the parents and the combination of multiple traits to select are thus important selection techniques. The selection index method for multiple trait selection and for incorporating plot and family information were found to be useful for selecting ortets and could be extended to breeding parents, which are usually selected based on GCA for individual traits (Soh & Chow, 1989, 1993; Soh et al., 1994). Selection index is limited by the inability to obtain accurate estimates of genetic variance and heritability (White & Hodge, 1989). The BLUP technique, developed in cattle breeding, which can integrate unbalanced or "messy" data from mating and experimental designs and even production data, have been used to obtain parent breeding values in oil palm (Soh, 1994; Wong & Bernado, 2008). BLUP has also been used to predict hybrid performances in oil palm (Purba et al., 2000) as in other crops, such as maize (Bernado, 1995, 1996) and sugarcane (Chang & Milligan, 1992a, 1992b).

New Techniques in Breeding and Selection

Breeding for Clonal Propagation

In tree crop breeding where vegetative propagation is possible, maximum genetic segregation is attempted by crossing unrelated trees with complementary desirable traits, followed by identifying the superior segregant progeny tree or individual genotype and fixing it through cycles of field testing and selection (Kawano et al., 1998). With the ability to clonally propagate oil palm via tissue culture, it is tempting to adopt this approach. To do so would require repeated cycles of cloning (from clones), the feasibility of which is still uncertain. Hence, clonal propagation remains as an adjunct to breeding for hybrid seed production in oil palm.

Dihaploid Breeding

Dihaploids shortcut the tedious inbreeding/selfing process (>5 cycles) to obtain homozygous inbred parents for the production of F1 hybrids. This is particularly relevant to oil palm with its long generation cycle and proneness to inbreeding depression. Dihaploids can be obtained naturally and by microspore culture, inducer genes, and wide hybridization. Recently, the Sumatra Bioscience/London Sumatra group submitted a patent application for their natural haploid/dihaploid screening technique using flow cytometry, molecular markers and karyotyping, and oil palm F1 hybrid production, although presumably still at the R&D stage (Nelson et al., 2009). Natural haploids are rare and need to be doubled up by, for example, colchicine treatment to give dihaploids. Most will be weak seedlings and may not survive due to inbreeding depression. The survivors will need to be progeny tested and a large program would be required as the parents would have not undergone sequential selection as in the conventional inbreeding approach. If advanced breeding parents (e.g., advanced Deli and AVROS) are used, the expected improvement would not be substantial and a large testing program would be needed to pick this out. Also, some breeding programs have achieved reasonably advanced inbred parents by conventional means to produce near true F1 hybrids. Nevertheless, as in cloning, dihaploid breeding would be useful in the early cycles of a breeding program. MPOB is researching on the anther/microspore culture approach to obtain haploids and dihaploids.

Genomics and Molecular Breeding

The recent announcements of breakthroughs in obtaining the oil palm genome sequence map heralded a new chapter in oil palm genetic improvement (Basri, 2009). These breakthroughs will spur the further development of oil palm genomics and genomic/molecular tools. Genomics research would reveal the following:

- DNA structure of a gene, how it functions and interacts with other genes to produce a trait;
- *Homology* (similarity) and *synteny* (arrangement) of genes across species;
- Derived tools such as linkage maps, gene discovery (*candidate genes*), and efficient markers that would allow new genes or alleles to be discovered and transformed into breeding populations to broaden their genetic base for further breeding; and
- *Marker-assisted selection* (MAS) which saves effort, time, and space and can be more efficient than field *phenotyping* in conventional breeding.

As the oil palm is an out-bred perennial crop, the application of genomic knowledge and tools in its breeding is hampered by its long generation cycle, large space requirement for field testing, and consequently small population sizes and paucity of diverse uniform experimental lines (e.g., recombinant inbred/near isogenic/dihaploid/mutant/tranformed lines) (Soh, 2011). Reconciliation of the different draft genome maps (made with different analytical approaches and genotypes with separate proprietary interest) into a coherent map would be a challenging and tedious effort. Mapping genes even for only the desirable traits on the oil palm genome map would also be a protracted process as experienced in *Arabidopsis* (which has a generation cycle time only in terms of weeks!). The molecular genetic control in some of the plants' metabolic pathways is still unknown or very complex. Model species (e.g., *Arabidopsis* and rice) can provide a model to better understand these physiological processes and their genetic control besides serving as surrogates for testing the expression of oil palm *transgenes*.

MPOB has developed a number of putative transgenics, such as high oleic and high *polyhydroxy butyrate* (PHB) bioplastics (Ghulam et al., 2005). These need to be subjected to the necessary laboratory tests to exclude partial and multiple copies of the transgene, as most of the transformants were obtained via *biolistics* (instead of the preferred *Agrobacterium* transformation method). They also need to be field-tested for stability in inheritance and expression of the trait, and for biosafety. Malaysia has gazetted its Biosafety Act in 2007. In MAS, MPOB has also developed a number of putative tight markers, such as shell gene and virescens gene, and identified quantitative trait loci (QTL) or polygenes for a number of other traits including yield (Rajinder & Cheah, 2005; Rajinder et al., 2009). As these were developed from mapping populations or restricted breeding populations of small sizes, they need to be validated with other breeding populations available in the private industry and in different environments as MAS, especially for quantitative traits, has to contend with QTL × host genotype interaction, linkage, epistasis, QTL × environment interaction, inaccurate phenotyping, and false positive linkage issues (Collard et al., 2005; Xu & Crouch, 2009). Because of these limitations, there is a trend in pursuing the genome-wide MAS approach (combining MAS with phenotyping) for quantitative traits which can instead reduce time, space, and effort to produce cultivars, taking the lead from maize and animal breeding (Bernardo & Yu, 2007; Wong & Bernardo, 2008).

Conclusions

The frequent natural calamities of late, such as severe droughts and heat-waves, snow storms and typhoons, earthquakes, tsunamis and floods (some attributed to climate change), and man-made crises (political/civil strife, energy and financial) have driven home the message on the urgency to address food, nutritional, energy, environmental and economic securities in the developing countries and the world at large. The oil palm and palm oil has contributed to addressing some of these issues in some countries; notably, Malaysia, Indonesia, and other tropical countries are emulating this approach. The per capita consumption of the dietary fats and oil and fuel energy are still low in the poorer developing countries while in the new emerging economies (e.g., China, India, southeast Asia, Latin America) with their large populations and rising affluence, they are expected to escalate. The oil palm, being the highest yielding oil crop and its versatile oil uses besides food (e.g., biofuel, health, cosmetic, and other oleochemical products originally derived from petrochemicals) will play an increasingly important role, especially with new countries venturing into planting oil palm. The challenge is to be able to produce more palm oil sustainably in the future scenario of depleting suitable land, energy, environment, and, in certain acute situations, human resources. In terms of breeding, it is in the development of more efficient palms with the *ideotype* traits: high biomass production from high density planting of dwarf genotypes with high HI, NUE, WUE and added value products and easily harvestable trees with light mechanical tools (to reduce high energy use and soil compaction). Concerted efforts in broadening the oil palm genetic base and the development of improved biotechnogical/genomic tools to assist breeding will contribute greatly towards meeting this challenge.

References

Ariffin, D. Major diseases of oil palm. In Advances in Oil Palm Research; Basiron, Y., Jalani, B.S., Chan K.W., Eds. Malaysian Palm Oil Board: Kuala Lumpur, 2000; pp. 596–622.

Ariffin, D.A.; Idris, S.; Marzuki, A. Development of a technique to screen oil palm seedlings for resistance to Ganoderma. In Proc. 1995 PORIM National Oil Palm Conference on Technology in Plantation: The Way Forward. Palm Oil Res. Inst. Malaysia: Kuala Lumpur, 1995; pp. 132–141.

Basri, M.W. Sequencing the oil palm genome. In (Unedit.). Proc. of Agriculture, Biotechnology, & Sustainability Conference Vol. 1. PIPOC 2009. Malaysian Palm Oil Board: Kuala Lumpur, 2009; p 3.

Beirnart, A.; Vanderweyen, T. Contribution a l'etude genetique et biometrique des varieties *d'Elaeis guineensis* Jacq. Publ. INEAC, Serie Scientifique 1941, 27.

Bernado, R. Genetic models for predicting single-cross performance in unbalanced yield trial data. Crop Sci. 1995, 35, 141–147.

Bernado, R. Best linear unbiased prediction of maize single-cross performance. Crop Sci. 1996, 36, 50–56.

Bernado, R. Breeding for quantitative traits. Stemma Press: Woodbury, Minnesota, 2002.

Bernado, R.; Yu, J. Prospects for genome-wide selection for quantitative traits in maize. Crop Sci. 2007, 47, 1082–1090.

Breton, F.; Hasan, Y.; Lubis, Z.; de Franqueville, H. Characterization of parameters for the development of an early screening test for basal stem rot tolerance in oil palm progenies. J. Oil Palm Res. 2006 (Special Issue), 24–36.

Breure, C.J. Relevant factors associated with crown expansion in oil palm (*Elaeis guineensis* Jacq.). Euphytica 1985, 34, 161–175.

Breure, C.J. Parent selection for yield and bunch index in the oil palm in West New Britain. Euphytica 1986, 35, 65–72.

Breure, C.J. The search for yield in oil palm: basic principles. In The Oil Palm. Management for Large Sustainable Yields; Fairhurst, T., Hardter, R., Eds.; Potash & Phosphate Institute/Potash Institute of Canada/International Potash Institute, 2003; pp. 59–98.

Breure, C.J. Rate of leaf expansion: A criterion for identifying oil palm (*Elaeis guineensis* Jacq.) types for planting at high densities. NJAS-Wageningen J. Life Sci. 2010, doi:101016/j.jnas.2010.03.001. M.S.

Breure, C.J.; Corley, R.H.V. Selection of palms for high density planting. Euphytica 1983, 32, 177–186.

Breure, C.J.; Rosenquist, E.A.; Konimor, J.; Powell, M.S. Oil palm introductions to Papua New Guinea and the formulation of selection methods at Dami Oil Palm Research Station. In International Workshop on Oil Palm Germplasm and Utilization; Soh, A.C., Rajanaidu, N., Nasir, M., Eds.; Palm Oil Res. Inst. Malaysia: Kuala Lumpur, 1986; pp. 189–193.

Chang, Y.S.; Milligan, S.B. Estimating the potential of sugar cane families to produce elite genotypes using univariate cross prediction methods. Theor. Appl. Genet. 1992a, 84, 662–671.

Chang, Y.S.; Milligan, S.B. Estimating the potential of sugar cane families to produce elite genotypes using bivariate prediction methods. Theor. Appl. Genet. 1992b, 84, 633–639.

Cochard, B.; Noiret J.M.; Baudouin, L.; Flori, A.; Amblard, P. Second cycle reciprocal recurrent selection in oil palm (*Elaeis guineensis* Jacq.). Results of Deli × La Me hybrid tests. Oleagineux 1993, 48, 441–452.

Collard, B.C.Y.; Jahufer, M.Z.Z; Brouwer, J.B.; Pang, E.C.K. An introduction to markers, quantitative trait loci (QTL) mapping and marker-assisted selection for crop improvement: The basic concepts. Euphytica 2005, 142, 169–196.

Choo, Y.M. Specialty products. In Advances in Oil Palm Research; Basiron, Y., Jalani, B.S., Chan, K.W., Eds.; Malaysian Palm Oil Board: Kuala Lumpur, 2000; pp. 1036–1060.

Corley, R.H.V. Yield potentials of plantation crops. In Potassium in the Agricultural Systems of the Humid Tropics. Int. Potash Inst. Bern.: Switzerland, 1985, pp. 61–80.

Darmosarkoro, W. Facing climate change issue in oil palm industry in Indonesia. PLE. 2.1 Proc. (pre-edited) International Oil Palm Congress 2010: Yogyakarta, Indonesia, 2010.

De Franqueville, H.: Renard. J.L. Improvement of oil palm vascular wilt tolerance. Results and development of the disease at the R. Michaux plantation. Oleagineux 1990, 45, 399–405.

Directorate of Estates. Tree Crop Estate Statistics (Oil Palm). Secretariat of Directorate General of Estates Jakarta, 2010.

Durand-Gasselin, T.; Asmady, H.; Flori, A.; Jacquemard, A.C.; Hayun, Z.; Breton, F.; de Franquevillue, H. Possible sources of genetic resistance in oil palm (*Elaeis guineensis* Jacq.) to basal stem rot caused by *Ganoderma boninense*—prospects of future breeding. Mycopathologia 2005, 159, 93–100.

Evans, L.T.; Fischer, R.A. Yield potential: its definition, measurement, and significance. Crop Sci. 1999, 39, 1544–1551.

Ghulam, K.A.P.; Othman, A.P.; Yunus, M.A.M.; Ramli, A.M.; Sambanthanmurthi, R.; Bohari, B.; Shaipulah, N.F.M.; Yuen, L.H.; Ayub, N.H.; Hashim, A.A.; et al. Value addition of oil palm through genetic engineering. In Proc. Conf. Biotechnology of Plantation Commodities. Malaysian Palm Oil Board: Kuala Lumpur, 2005; pp. 198–210.

Hardon, J.J.; Corley, R.H.V.; Lee, C.H. Breeding and selecting the oil palm. In Improving Vegetatively Propagated Crops; Abbot, A.J., Atkin, R.K., Eds.; Academic Press: London, 1987; pp. 63–81.

Hardon, J.J.; Rao, V.; Rajanaidu, N. A review of oil palm breeding. In Progress in Plant Breeding; Rusell, E.G., Ed.; Butterworths: London, 1985; pp. 139–163.

Hartley, C.W.S. The oil palm (*Elaeis guineensis* Jacq.). Longman Scientific and Technical Publication. Wiley: New York, 1988.

Houssou, M.; Meunier, J.; Daniel., C. Breeding oil palm (*Elaeis guineensis* Jacq.) for drought tolerance. Preliminary results in Benin. In Proc. 1987 Int. Oil Palm/Palm Oil Conferences. Progress and Prospects; Hassan, H.A.H, Chew, P.S., Wood, B.J., Pushparajah, E., Eds.; Conference I–Agriculture, Palm Oil Res. Inst. Malaysia: Kuala Lumpur, 1987; pp. 23–26.

Idris, A.S.; Kushairi, A.; Ismail, S.; Ariffin, D. Selection for partial resistance in oil palm to Ganoderma basal stem rot. J. Oil Palm Res. 2004, 16, 12–18.

Jalani, B.S.; Cheah, S.C.; Rajanaidu, N.; Darus, A. Improvement of oil palm through breeding and biotechnology. J. Am. Oil Chem. Soc. 1997, 47, 1451–1455.

Jones, L.H. Propagation of clonal oil palm by tissue culture. Oil Palm News 1974, 17, 1–8.

Kawano, K.; Narintaraporn, K.; Nariataraporn, P.; Sarakarn, S.; Limsila, A.; Limsila, J.; Suparhan, D.; Sarawat, V.; Watananonta, W. Yield improvement in multistage breeding program for cassava. Crop Sci. 1998, 38, 325–332.

Khairudin, H.; Lim, T.K.; Abdul Rahman, A.R. Pathogenicity of *Ganoderma boninense* Pat. on oil palm seedlings Proc. 1991 PORIM Int. Palm Oil Conference-Agriculture (Module 1). Palm Oil Res. Inst. Malaysia: Kuala Lumpur, 1993; pp. 418–423.

Malaysian Palm Oil Statistics 2010–30th Edition. Malaysian Palm Oil Board: Kelana Jaya, Selangor, Malaysia, 2011, 194 pp.

Martinez, G.; Sarria, G.A.; Torres, G.A.; Varon, F.; Romero, H.M.; Sanz, J.I. Advances in the research of *Phytopthora palmivora*, the causal agent of bud rot of oil palm in Colombia. In Proc. of Agriculture, Biotechnology, & Sustainability Conference Vol. 1. PIPOC. Malaysian Palm Oil Board: Kuala Lumpur, 2009; pp. 875–890.

Meunier, J.; Gascon, J.P. General scheme for oil palm improvement at IRHO. Oleagineux 1972, 27, 1–12.

Nelson, S.P.C.; Wilkinson, M.J.; Dunwell, J.M.; Forster, B.P.; Wening, S.; Sitorus, A.; Croxford, A.; Ford, C.; Calgari, P.D.S. Breeding for high productivity lines via haploid technology. In Proc. of Agriculture, Biotechnology, & Sustainability Conference Vol. 1. PIPOC 2009. Malaysian Palm Oil Board: Kuala Lumpur, 2009; pp. 203–225.

Obasola, C.O.; Arasu, N.T.; Rajanaidu, N. Collection of oil palm genetic material in Nigeria. I. Method of collection. MARDI Report, 1983.

Oil World Data Base. 2007 ISTA. Mielke GmbH. Germany.

Oil World Data Base. 2010 ISTA. Mielke GmbH. Germany.

Okwuagu, C.O. The genetic basis for the low correlation between Dura and Tenera full-sib shell to fruit ratio in the oil palm (*Elaeis guineensis* Jacq.). In Proc. 1987 Int. Oil Palm Conference–Agriculture; Hassan, A.H., Chew, P.S. Wood, B.J., Pushparajah, E., Eds.; Palm Oil Res. Inst. Malaysia: Kuala Lumpur, 1988; pp. 143–154.

Okwuagu, C.O.; Okolo, E.C. Maternal inheritance of kernel size in the oil palm (*Elaeis guineensis* Jacq.). Elaeis 1992, 4, 72–73.

Okwuagu, C.O.; Okolo, E.C. Genetic control of polymorphisn for kernel to fruit ratio in the oil palm (*Elaeis guineensis* Jacq.). Elaeis 1994, 6, 75.

Osborne, D.J.; Henderson, J.; Corley, R.H.V. Controlling fruit-shedding in the oil palm. Endeavour 1992, 16, 173–177.

Pantzaris, T.P. Pocketbook of palm oil uses. Palm Oil Res. Inst. Malaysia: Kuala Lumpur, 2000.

Pooni, H.S.; Cornish, M.A.; Kearsey, M.J.; Lawrence, M.J. The production of superior lines and second cycle hybrids by inbreeding and selection. Elaeis 1989, 1, 17–30.

Purba, A.R. Update on oil palm planting materials producers in Indonesia. Paper presented at "2nd Oil Palm Summit: Sustainable Yields, The Way Forward", Centre for Management Technology, Bali, Indonesia. 9–10 December 2010.

Purba, A.R.; Flori, A.; Baudouin, L.; Hamon, S. Prediction of oil palm (*Elaeis guineensis* Jacq.) agronomic performances using best linear unbiased predictor (BLUP). Theor. Appl. Genet. 2000, 102, 787–792.

Rabechault, H.; Martin, J.P. Multiplication vegetative du palmier a huile (*Elaeis guineensis* Jacq.) a l'aide de culures de tissues foliares. C.R. Acad. Sci. Paris, Ser. D 1976, 283, 1735–1737.

Rajanaidu, N. PORIM Oil Palm Genebank. Palm Oil Res. Inst. Malaysia: Kuala Lumpur. 1990.

Rajanaidu, N.; Arasu, N.T.; Obasola, C.O. Collection of oil palm (*Elaeis guineensis* Jacq.) genetic material in Nigeria II. Phenotypic variation of natural populations. MARDI Research Bul. 1979, 7, 1–27.

Rajanaidu, N.; Jalani, B.S. Oil palm genetic resources—collection, evaluation, utilization and conservation. Presented at Colloquium on Oil Palm Genetic Resources. Palm Oil Res. Inst. Malaysia: Kuala Lumpur, 13 Sept. 1994.

Rajanaidu, N.; Kushairi, A.; Rafii, M.; Mohd, D.; Maizura, I.; Jalani, B.S. Oil palm breeding and genetic resources. In Advances in Oil Palm Research; Basiron, Y., Jalani, B.S., Chan, K.S., Eds.; Malaysian Palm Oil Board: Kuala Lumpur, 2000, pp. 171–227.

Rajinder, S.; Cheah, S.C. Potential application of marker assisted selection (MAS) in oil palm. MPOB Bulletin, 2005, 51, 1–9.

Rajinder, S.; Ong Abdullah, M.; Ooi, L.C.I.; Ooi, S.E.; Low, E.T.; Seng, T.Y.; Sharifah, S.R.S.A.; Ho, C.L. Molecular markers in oil pal breeding and tissue culture. In Proc. of Agriculture, Biotechnology, & Sustainability Conference Vol. 1 PIPOC, Malaysian Palm Oil Board: Kuala Lumpur, 2009; pp. 226–243.

Rao, V. Ripening in the virescens oil palm. In Oil and Kernel Production in Oil Palm–A global perspective; Rajanaidu, N., Henson, I.E., Jalani, B.S., Eds.; Proc. Int. Conference. Malaysian Palm Oil Board: Kuala Lumpur, 1998; p. 226.

Review of the Malaysian Oil Palm Industry 2010. Malaysian Palm Oil Board: Kelana Jaya, Selangor, Malaysia, 2011; p. 95.

Rosenquist, E.A. The genetic base of oil palm breeding populations. In Int. Workshop on Oil Palm Germplasm and Utilization; Soh, A.C., Rajanaidu, N., Nasir, M., Eds.; Palm Oil Res. Inst. Malaysia: Kuala Lumpur, 1986; pp. 16–27.

Rosenquist, E.A.; Corley, R.H.V.; De Greef, W. Improvement of tenera populations using germplasm from breeding programs in Cameroon and Zaire. In Proc. Workshop Progress of Oil Palm Breeding Populations. Palm Oil Res. Inst. Malaysia: Kuala Lumpur, 1990; pp. 37–69.

Soh, A.C. Expected yield increase with selected oil palm clones from current DxP seedling materials and its implications on clonal propagation, breeding and ortet selection. Oleagineux 1986, 41, 51–56.

Soh, A.C. Ranking parents by best linear unbiased prediction (BLUP) breeding values in oil palm. Euphytica 1994, 76, 13–21.

Soh, A.C. Breeding plans and selection methods in oil palm. In Proc. Symp. The Science of Oil Palm Breeding; Rajanaidu, N., Jalani, B.S., Eds.; Palm Oil Res. Inst. Malaysia: Kuala Lumpur, 1999; pp. 65–95.

Soh, A.C. Selecting the ideal oil palm: What you see is not necessarily what you get! J. Oil Palm Res. 2004, 16, 121–128.

Soh, A.C. Genomics and plant breeding. J. Oil Palm Res. 2011, 23, 1019–1028.

Soh, A.C.; Chow, C.S. Index selection in oil palm for cloning. In Iyama, S., Takeda, T., Eds.; Proc. 6th Int. Congr. SABRAO: Tsukuba, Japan, 1989; pp. 713–716.

Soh, A.C.; Chow, C.S. Index selection utilizing plot and family information in oil palm. Elaeis 1993, 5, 27–37.

Soh, A.C.; Chow, S.C.; Iyama, S.; Yamada, Y. Candidate traits for index selection in choice of oil palm ortets for clonal propagation. Euphytica 1994, 76, 23–32.

Soh, A.C.; Gan, H.H.; Wong, G.; Hor, T.Y.; Tan, C.C. Estimates of within family genetic variability for clonal selection in oil palm. Euphytica 2003b, 133, 147–163.

Soh, A.C.; Tan, S.T. Estimation of genetic variance, variance and combining ability in oil palm breeding. In Proc. 4th Int. SABRAO Congress Crop Improvement Research; Yap, T.C., Graham, K.M., Eds.; The Incorporated Society of Planters: Kuala Lumpur, 1983; pp. 379–388.

Soh, A.C.; Vanialingam, T.; Taniputra, P.; Pamin, K. Derivatives of the Dumpy palm. Some experimental results. The Planter 1981, 57, 227–239.

Soh, A.C.; Wong, C.K.; Ho, Y.W.; Choong, W.C. In Oil Palm, Handbook of Plant Breeding, Oil Crops; Vollmann, J., Rajcan, I., Eds.; Springer: New York, 2009; pp. 333–367.

Soh, A.C.; Wong, G.; Hor, T.Y.; Tan C.C.; Chew, P.S. Oil palm genetic improvement. Plant Breed. Rev. 2003a, 22, 165–219.

Soh, A.C.; Wong, G.; Tan, C.C.; Chew, P.S.; Chong, S.P.; Ho, Y.W.; Wong, C.K.; Choo, C.N.; Nor Azura, H.; Kumar, K. Commercial-scale propagation and planting of elite oil palm clones: research and development towards realization. J. Oil Palm Res. 2011, 23, 936–952.

Soh, A.C.; Yong, Y.Y.; Ho, Y.W.; Rajanaidu, N. Commercial potential of oil palm clones: early results of their performance in several locations. In Rao, V., Henson, I.E., Rajanaidu, N., Eds.; Recent Developments in Oil Palm Tissue Culture and Biotechnology. Int. Society for Oil Palm Breeders: Selangor, Malaysia, 1995; pp. 134–144.

Tinker, P.B. Site specific yield potentials in relation to fertilizer use. Proc. 8th Colloquium on Nutrient Balances and Fertilizer Needs in Temperate Agriculture. Int. Potash Inst.: Gardone, Riviera, Italy, 1984.

Tollenaar, M.; Wu, J. Yield improvement in temperate maize is attributable to greater stress tolerance. Crop Sci. 1999, 39, 1597–1604.

Van Der Vossen, H.A.M. Towards more efficient selection for oil yield in the oil palm (*Elaeis guineensis* Jacq.). Thesis. Univ. Wageningen, The Netherlands, 1974.

White, T.L.; Hodge, G.R. Predicting breeding values with applications in forest tree improvement. Kluwer Academic: Dordrecht, 1989.

Wong, C.K.; Bernado, R. Genomewide selection in oil palm: increasing selection gain per unit time and cost with small populations. Theor. Appl. Genet. 2008, 116, 815–824.

Wong, G.; Tan, C.C.; Soh, A.C. Large scale propagation of oil palm clones: Experience to date. Acta Hort. 1997, 447, 649–658.

Wonki-Appiah, J. B. Genetic control of fertility in the oil palm (*Elaies quineensis* Jacq.). Euphytica 1987, 36, 505–511.

Xu, Y.; Crouch, J. Marker-assisted selection in plant breeding: From publications to practice. Crop Sci. 2009, 48, 391–407.

Zeven, A.C. The origin of the oil palm. J. Nigerian Inst. Oil Palm Research 1965, 4, 218–225.

3

Oil Palm Genomics

Meilina Ong Abdullah, Maria Madon, Eng Ti Leslie Low, Maizura Ithnin, and Rajinder Singh
Advanced Biotechnology and Breeding Centre, Malaysian Palm Oil Board, No. 6 Persiaran Institusi, Bandar Baru Bangi, 43000 Kajang, Selangor, Malaysia

Introduction

During the era of pre-high throughput assays and when $1000 genome was unheard of, the molecular analysis of plants often focused on the single gene level. Recent technological advances have changed this norm, enabling the analysis of genome organization, expression, and interaction of organisms. Genomics generally refers to the study of the way genes and genetic information is organized within the genome, which subsequently determines their biological functionality. Plant genomics is reversing the previous paradigm of identifying genes behind biological functions and instead focuses on finding biological functions behind genes (Campos-De Quiroz, 2002). This aids in closing the gap between phenotype and genotype.

The oil palm belongs to the genus *Elaeis* which contains two species, *E. guineensis* and *E. oleifera*. It is regarded as the most efficient oil bearing crop in the world compared to other crops of the same genre (e.g., soybean, rapeseed, and sunflower) in terms of land use. Producing oil yield at an average of 3.74 tonne/ha/year, on land 10 times smaller than the requirement for soybean (Oil World, 2007) and with a productive life cycle of 25–30 years, make the oil palm a lucrative agricultural crop. However, of late the oil yield has reached stagnation and there is an urgent need for improvement. Hence, this chapter focuses on the current trend of leveraging genomics in the production of new and improved planting materials. The first section deals with laying the foundation for oil palm genomics research, which subsequently leads to the second section that concentrates on developing genomic platforms. These genomic platforms form the resource towards achieving the target of producing superior oil palms.

Building the Foundation for Oil Palm Genomics Research

Karyobiology

Karyobiology, or the study of cell nuclei and the structure and function of oil palm chromosomes, has been initiated intensively since the 1990s by Malaysian Palm Oil Board (MPOB). Amongst the tools developed are classical cytogenetics method to analyze mitotic and meiotic processes, techniques such as fluorescent *in situ* hybridization to locate the specific sequences on chromosomes, genomic *in situ* hybridization to differentiate parental chromosomes in *E. oleifera* × *E. guineensis* (O × G) hybrids, and finally the use of flow cytometry technique to estimate oil palm genome size and ploidy level. These methods were developed since they can assist in understanding the oil palm genome structure.

Cytogenetics

Classical cytogenetic tools such as squashing of samples, for example root tips, calli, or anthers, on glass slides followed by staining with fluorescent or non-fluorescent stains assist in the analysis of mitotic (in root tips or calli) or meiotic (pollen mother cells or microspores in anthers) processes. These tools would allow the determination of a normal or aberrant process undergone by the cells. Using metaphase chromosome spreads and image analysis measurement software, karyotypes of plant species can be obtained. Fig. 3.1 shows the karyotypes of *E. guineensis* and *E. oleifera*, the two

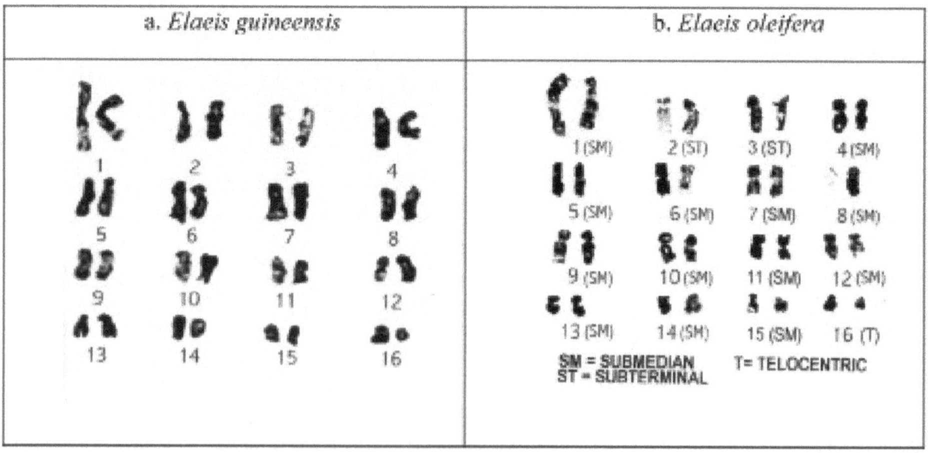

Fig. 3.1. Shows karyotypes of (a) *Elaeis guineensis* and (b) *E. oleifera*. Source: Madon et al., 1998.

species of oil palm that belong to the family Palmae, containing over 225 genera and 2600 species. Madon et al. (1998) reported the chromosome numbers of *E. guineensis* and *E. oleifera* to be 2n = 32 and found that in both species the chromosomes can be divided into three groups on the basis of length: Group I of only Chromosome 1 (the longest), Group II of Chromosomes 2–9 (medium length) and Group III of Chromosomes 10–16 (medium/short).

Cytogenetical analysis of meristematic cells in the root tips is able to illustrate whether the mitotic processes are normal or abnormal. For example, normal activities of mitotic cells such as interphase, prophase, metaphase, anaphase, and telophase can be observed without any chromosomal anomalies, such as, chromosome bridging or lagging. The cytogenetic analysis performed on the anthers containing pollen mother cells (PMC) is able to indicate whether meiosis occur normally or vice versa. The normal progression of meiosis in PMC produces four potential microspores that eventually become pollens, while abnormal meiotic activities produce microspores with unequal amount of DNA, and these pollens have the tendency to be sterile.

Fluorescence In Situ Hybridization (FISH)

FISH uses DNA probes that are labeled with differently colored fluorescent tags to visualize one or more specific regions of the genome. The technique involves denaturing both chromosomes on a slide and hybridizing them to a biotin, digoxigenin, or fluorophore labeled probe. After hybridization, post washes and incubation with suitable fluorescing conjugates such as FITC (fluorescein isothiocyanate)–avidin or rhodamine-anti-DIG, the sites of hybridization can be observed using an epifluorescence microscope. FISH technology enables the location of DNA sequences to be identified on oil palm chromosomes such as ribosomal DNA, transgenes, and specific DNA markers. In so doing, identification of oil palm individual chromosome pairs can be done.

FISH technique has been used to detect multicopy genes such as 5S and 18S–25S ribosomal DNA on oil palm chromosomes. These genes, which are highly conserved across the plant species, can be used as landmark probes for genome mapping and to give further information about genome structure and evolution of the oil palm. FISH of the 18S–25S rDNA from *Linum usitatissimum* (flax) and 5S rDNA from *Triticum aestivum* (wheat) hybridized to *Elaeis* chromosomes demonstrated that there is homology between the genes in these plant species. In *E. guineensis,* FISH with 18S–25S from flax were located on the satellite DNAs of an acrocentric chromosome pair and these regions are called nucleolar organizer regions (NOR) and the pair of chromosomes are called NOR chromosomes. The 5S rDNA is situated on the proximal arms of two medium length chromosome pairs (Group II). The occurrence of two classes of ribosomal genes on different chromosomes is common in other plant species as well. Locating low copy sequences via FISH technique is a challenging task and is being developed. Molecular markers such as RFLP, AFLP, SSR, SNP, and ISSR have been used to construct the linkage map of the oil palm. The individual oil palm chromosome pairs are difficult to distinguish, hence recently ISSR and RFLP markers that mapped together on the same linkage group have been identified in the effort to establish chromosome-specific markers. Currently, a modified FISH technique using ISSR and RFLP probes in simultaneous double labeling FISH experiments is being developed to assist in the oil palm genome mapping program.

Genomic In Situ Hybridization (GISH)

GISH uses total genomic DNA (consisting of the entire DNA complement of a plant) as probe for hybridization to chromosome spreads. This method permits discrimination of the genomic origin of the chromosomes in hybrid plants and allotetraploid species. Hybrid breeding between the *E. oleifera* and *E. guineensis* species is aimed at introgressing the genes for high oil unsaturation and slow height increment into the commercially important *E. guineensis*. The application of GISH in oil palm hybrid breeding is described in more detail in a later section.

Flow Cytometry

Flow cytometry (FCM) is a rapid and convenient technique that allows accurate determination of the nuclear DNA content or genome size and ploidy level in plants (Dolezel, 1991). The analysis is based on the use of propidium iodide (DNA intercalating fluorochrome) and analysis of the relative fluorescence intensity emitted by the stained nuclei (Dolezel et al., 1994). To determine the nuclear DNA content in absolute units, the fluorescent intensity of nuclei with unknown DNA content is compared with the fluorescence intensity of nuclei isolated from a species with known nuclear genome size (Dolezel et al., 1994).

Estimation of Oil Palm Nuclear DNA Content

The knowledge of DNA content in absolute units (genome size) is important in many areas of research ranging from evolutionary studies to genome mapping (Galbraith et al., 1997). The genomic DNA content of *E. guineensis* (*tenera*) as reported by Rival et al. (1997) using flow cytometry was $2C = 3.786 \pm 0.125$ pg. However, Jones et al. (1982) reported $2C = 2.4 \pm 0.4$ pg using the microdensitometric method on root tips of oil palm regenerants. Bennett & Smith (1991) and Röser et al. (1997) reported the genome size for *E. guineensis* as $4C = 4.00$ and 4.8 pg, respectively, also using the microdensitometric method. A study was done by Madon et al. (2008) to establish a standard protocol in order to estimate the genome size and to observe any variation in the intra- and interspecific hybrids of the oil palm by using a laser-sourced flow cytometer with *Glycine max* cv. Polanka ($2C = 2.5$ pg) (soyabean) as the external standard, as well as to identify the most suitable oil palm frond and lysis buffer. From this study, it was found that leaflets from Frond -1 and LBO1 lysis buffer supplemented with 50 mg

Propidium Iodide and RNaseA, followed by 48 hours incubation at 4°C are the most suitable conditions to obtain histogram peaks with 1–3% CV. Madon et al. (2008) found that the DNA contents in Frond -1 were generally higher and more consistent and the average 2C DNA content for *E. guineensis* was found to be 3.86 ± 0.26 pg, and F_1OxG hybrids has 2C = 4.16 ± 0.32 pg.

Ploidy Analysis of Induced Oil Palm Polyploids

Ploidy manipulation has been found to be a valuable tool in the genetic improvement of many plants and ploidy induction has been carried out for a variety of reasons. For example, in citrus, tetraploid (4n) parents were produced to create seedless triploids (3n) by crossing 4n and 2n parents (Wu & Mooney, 2002). By doubling the chromosome numbers using anti-mitotic chemical such as colchicine, variants generated may possess useful characteristics, and by doubling the gene products polyploids also provide a wider germplasm base for breeding studies (Thao et al., 2003). For oil palm, ploidy manipulation has been done using anti-mitotic chemicals such as colchicine and oryzalin in order to increase the genetic variability in oil palm (Madon et al., 2005). These chemicals act by binding to the tubulin dimers preventing the formation of microtubules. Consequently, spindle fibers failed to form during cell division, hence chromosomes arrest at the metaphase stage and the cell doubles its chromosome number. For further details on materials and methods, please refer to the article by Madon et al., (2005). The polyploids were shorter and their leaves thicker and greener than those of the controls. FCM analysis was used to screen for ploidy level in the treated seedlings. In triploids the chromosome numbers were 3n = 48 and tetraploid, 4n = 64. This study has demonstrated that in oil palm, the anti-mitotic chemicals, oryzalin and colchicine can induce change in the ploidy level. However, the stability of the changes needs to be monitored along with the other morphological traits. Also, in this study the most efficient method to analyze for the ploidy level was flow cytometry as it was more accurate and convenient than the conventional methods of chromosome counting and measuring stomatal length and density.

Unraveling the Oil Palm DNA

In general biological terms, DNA is the genetic material that is the building blocks of a genome. It contains the total gene complementation that includes both the coding (that is the genes) and the non-coding regions of an organism's genetic makeup. The study that encompasses the investigation and understanding of genomes is called genomics. It does not only study individual genes but treats the genome in its entirety as a unit of investigation. Recently, the field of study has been divided into structural and functional genomics. Therefore, in conducting genomics research, it is just as important to understand the basic structure of the organism's genome (karyotype through cytogenetics research) as well as its regulatory elements, non-coding, and genic regions.

Studies on the oil palm regulatory regions (controls) and genes provide important information on key biological processes, such as general plant development, flowering, and fruit development. Hence, generation of sequence information of the expressed genes and genome is necessary. In fact, the annotated sequences from other organisms can provide clues as to the identity of the oil palm genes and determine their possible role in the biological process studied. With this in mind, a number of research studies have exploited the sequence information of other organisms to identify specific oil palm genes, such as a 7S globulin (Aberlenc-Bertossi et al., 2008; Morcillo et al., 2001), Myb-related genes (Teoh et al., 2003) and AP2/EREBP transcription factor (Morcillo et al., 2007). This has allowed researchers to identify the specific conditions or developmental stage where these genes accumulate and provide useful leads for future studies. For example, Morcillo et al. (2001) was able to identify a 7S globulin gene (GLO7A) from a collection of 641 oil palm sequences by comparing the oil palm sequences with sequences in GenBank. This allowed the authors to correlate the expression of the GLO7A with the accumulation of 7S globulin in oil palm zygotic and somatic embryos in a previous publication (Morcillo et al., 1998) and determine that the expression of GLO7A is influenced by abscisic acid and sucrose (Morcillo et al., 2001). Since then, further studies have shown that GLO7A and two cysteine proteases were expressed during somatic embryogenesis and the subsequent germinative growth stage (Aberlenc-Bertossi et al., 2008).

The advantage of using prior knowledge to drive research in oil palm is even more apparent when working with well characterized gene families. This is seen in the study of MADS box genes in oil palm (Adam et al., 2006; Adam et al., 2007; Alwee et al., 2006). MADS box genes are a family of transcriptional factors that control floral organ identity (Cseke & Podila, 2004; Okamura et al., 1993). Adam et al. (2006) and Alwee et al. (2006) designed primers based on the available MADS box genes and were able to identify and classify 13 and 14 oil palm MADS box genes, respectively, into different subfamilies. Further studies of the 12 MADS box genes identified by Adam et al. (2006) showed that oil palm uses the generic dicot ABCDE model to control the determination and evolution of its flower form (Adam et al., 2007). This research provides a good foundation for more detail studies in oil palm flowering and floral abnormalities (mantled) caused by tissue culture.

Nevertheless, there are still a lot of gaps in the knowledge base of many of the biological processes in plants. Each organism would also have genes or processes that are unique to the organisms. Therefore, just using information from other organisms might not be able to fulfill some of the needs of oil palm genomics research. Global gene sequencing methods would need to be carried out to identify "new" oil palm genes. Initially, one of the techniques used was differential display. Singh & Cheah (2000) and San & Shah (2005) were able to identify five flower-specific and four kernel-specific genes, respectively. Tregear et al. (2002) also used differential display to identify a defensin gene (EGAD1) that was detected in inflorescence tissue and have the potential to be a marker for epigenetic somaclonal variation events. However, the rate of new gene discovery was slow. Subsequently, as technologies progressed, more high-throughput methods were developed.

One of these techniques is the expressed sequence tag (EST) method. The ESTs, first reported by Adam et al. (1991), have been effective in determining a higher number of genes from different developmental tissues and stages. The interesting part of the method described by Adam et al. (1991) is that an expressed gene could be identified by matching only 150 to 400 bases of the gene to sequences in the public domain. This provided a good opportunity for researchers to use the method for novel gene discovery (Albert et al., 2005; Pavy et al., 2005; Ralph et al., 2008; Verdun et al., 1998) and digital gene expression studies (Fields, 1994; Tittarelli et al., 2009; Vizoso et al., 2009).

For oil palm, the EST approach was especially useful, as very few oil palm sequences were available in GenBank a decade ago. Since then, the EST approach has been used to identify genes expressed in the different tissues/stages related to tissue culture (Chan et al., 2010; Ho et al., 2007; Lin et al., 2009; Low et al., 2008; Roowi et al., 2010), flowering (Beulé et al., 2011; Ho et al., 2007; Jouannic et al., 2005), shoot apices and apical meristem, zygotic embryos (Ho et al., 2007; Jouannic et al., 2005), fruit development (Nurniwalis et al., 2008; Xu et al., 2011), and roots (Ho et al., 2007). To improve the chances of identifying low abundant genes or genes expressed in specific developmental stages, normalized (Chan et al., 2010) or substractive hybridization cDNA libraries (Beulé et al., 2011; Xu et al., 2011) were also constructed. The abundance of some of these EST populations were analysed to deduce their relative frequencies in the tissues studied. This allowed the identification of differentially expressed genes for further studies (Ho et al., 2007; Lin et al., 2009; Low et al., 2008).

The EST collection also serves as an important resource for molecular markers, such as simple sequence repeats (SSR) (Low et al., 2008; Singh et al., 2008b; Ting et al., 2010) and single nucleotide polymorphism (SNP) (Riju et al., 2007). The gene clones that were used to generate the EST sequences were also used as restriction fragment length polymorphism (RFLP) markers (Maizura et al., 2006; Singh et al., 2008a). These markers have since been used for genetic diversity studies (Maizura et al., 2006; Singh et al., 2008b; Ting et al., 2010) and quantitative trait loci (QTL) analysis (Singh et al., 2009; Ting et al., 2006). The EST sequences had also been mined for miRNAs, where Nasaruddin et al. (2007) identified miRNAs that could regulate auxin response, floral development, and basal transcription.

The gene clones that were used to generate the EST sequences are also useful. The gene clones can be used to identify specific genes, such as cysteine proteinase (Aberlenc-Bertossi et al., 2008), KNOX transcription factor (Jouannic et al., 2007), and methytransferases (Rival et al., 2008) for hypothesis driven or targeted gene research projects. The clones can also be printed on cDNA microarrays or macroarrays for gene expression studies. cDNA microarrays using ESTs have been used to study tissue culture amenity and abnormality (Low et al., 2006). Beulé et al. (2011) and Lin et al. (2009) also used ESTs as probes on macroarrays to study gene expression. Beulé et al. (2011) was able to find one F-box and one RING family gene that were down-regulated in mantled inflorescence while Lin et al. (2009) identified transcripts that could play a role in auxin response and cellular polarization in somatic embryos.

Another method for high-throughput sequencing is the next generation sequencing (NGS) techniques. NGS is changing the way genomic research is being conducted. The technology has allowed the sequencing and assembly of all three fruit forms of *E. guineensis* and *E. oleifera*. Whole genome transcriptomics sequencing was able to provide detail profiles of genes during oil palm fruit development, maturation, and ripening. The authors also identified a transcription factor WRINKLED1 that was co-regulated with the fatty acid biosynthesis genes and could play a role in transcription regulation of oil synthesis in oil palm mesocarp tissue (Bourgis et al., 2011; Tranbarger et al., 2011). Both these developments of genome and transcriptome sequencing showed the potential of NGS techniques.

Genomic Platforms for Producing New and Improved Planting Materials

The oil palm has gone through at least two known cycles of yield improvements since its introduction as an oil seed crop in Malaysia. The first wave being the introduction of the hybrid *tenera* (D × P), which replaced the *dura* as commercial planting material. This is a demonstration of achieving an increase in oil yield of up to 30% by merely manipulating a single gene (Kushairi et al., 2006). However, the oil yield in Malaysia has been hovering around 3.5 and 3.9 t/ha/yr in the last two decades, and year 2010 was no different (MPOB, 2010). Having dropped to the number two spot in palm oil production, Malaysia is in need of a major leap in yield improvement. This is further compounded by the fact that agricultural land is becoming a rarity; therefore, increasing production by planting larger areas is no longer seen as an alternative.

Through years of breeding and selection, the industry has already witnessed individual palms yielding as high as 13.6 t/ha/yr (Sharma & Tan, 1999), which are close to the theoretical yield of 18.2 t/ha/yr (Corley, 1998). The best experimental plot has produced an average of 9.8 t/ha/yr of palm oil (Musa & Gurmit, 2008), with selected progenies able to achieve up to 12.2 t/ha/yr (Rajanaidu et al., 1990). Cloning these super palms may provide the industry with the much needed high-yielding planting materials to get it out of the stagnation rut. Hence, clones for commercial use are touted as the second wave of crop improvement for the oil palm.

Crop improvement is a continuous process and breeders have for years wielded tremendous effort in creating cultivars with appropriate complement of genes, often sourced from different wild populations (Brady & Provart, 2007). Identification of desirable genes in the oil palm, like other plants, has mainly been based on their manifestation as a characteristic that can be scored as a phenotype such as height, color, and yield. However, with the advent of DNA sequencing information, a more precise approach can be employed to plant genetics in the selection of agronomical important traits. Using molecular markers, introgression of favorable traits into elite germplasms can be accelerated.

Tools for Oil Palm Tissue Culture

Application in Oil Palm Tissue Culture

Vegetative propagation of the oil palm by tissue culture was first described in the 1970s. Since then, the commercial advantage of tissue culture materials over seedlings has been well established. The initial problems or setbacks associated with oil palm tissue culture (for example, flowering abnormalities) have also been minimized by adopting low phytohormone protocols and high culling rates in the laboratory and nursery. This has led to renewed interest in *in vitro* production of oil palm clones on a large scale. The availability of simple, highly robust and low cost quality control probes for fingerprinting ortets to verify identity, monitor line uniformity, and culture mix-ups will greatly enhance confidence in the commercial production of tissue culture materials.

Previously, the use of RFLP probes for quality control in the oil palm tissue culture process was described by Cheah et al. (1996). Although the process is effective, the drawbacks associated with using RFLP markers are that the analysis is expensive, time consuming, and laborious.

To overcome these shortcomings, microsatellites, or SSR probes, were developed for use in quality control for oil palm tissue culture (Singh et al., 2006). However, the SSR primers do not address the issue of clonal abnormalities arising from somaclonal variation. The SSR markers make it possible to monitor clonal fidelity during the tissue culture process. This is shown in Fig. 3.2, where the DNA from ramets of four clones were compared with their respective ortets' DNA. The results clearly demonstrate that there was no change in the SSR profiles of the ramets compared to those

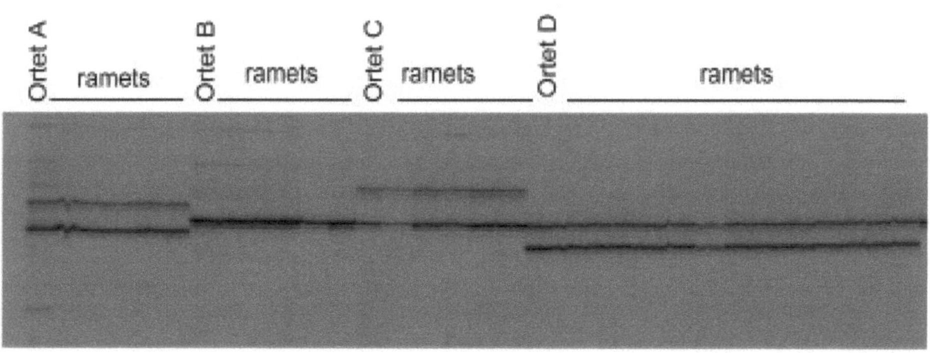

Fig. 3.2. DNA fingerprinting of tissue culture clones using SSR analysis. Source: Singh et al., 2006.

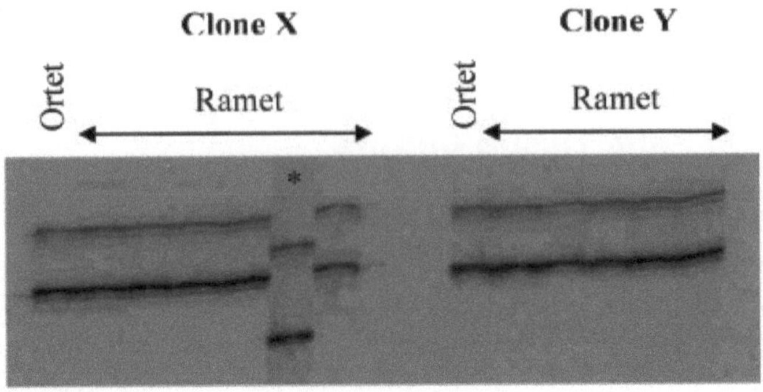

Fig. 3.3. Detecting culture mix-up. *denotes a sample that did not originate from ortet X. Source: Singh et al., 2006.

of their corresponding ortets. It is also clear that the SSR fingerprints can distinguish the four different ortets tested and making it possible to identify culture mix-up in the lab, as demonstrated in Fig. 3.3.

Biomarker Discovery for Abnormality

The scale up of ramet production in the early 1980s resulted in an increased frequency of abnormal clones following field transplanting. The observed abnormality included androgynous inflorescences, mantled, and parthenocarpic fruits. These abnormalities frequently result in bunch rotting with subsequently no harvestable fruit bunches. The frequency of abnormalities among clones increased from 30% in 1981 to 90% in 1983 (Meins et al., 1991). In 1986, information of abnormal clones were made public which resulted in diminished interest in clonal propagation of oil palm and commercial ramet production halted (Corley et al., 1986). Since then various programs have been initiated to address the abnormality issue in the oil palm.

Through these efforts significant advances in understanding the basic biology in the areas of floral development, karyotypic analysis, and somaclonal variation were achieved. While diagnostic markers for the mantled trait, in which the phenotype involves a modification of the stamens and the staminodes (vestigial stamens) of female and male flowers, respectively, into pseudocarpel structures (Morcillo et al., 2006) is still not yet available technologies to discover such markers, as well as candidate genes, have been established and exploited. Most important, it is now apparent that epigenetic mechanisms play a major role in mantling and other unstable variation arising from tissue culture. This was indicated by the occurrence of reversions of abnormal palms in the field over a period of years (Durand-Gasselin et al., 1990), and through conventional genetic crossings conducted by Rao & Donough (1990), this trait was shown to behave in a non-Mendelian manner. Earlier attempts that employed techniques such as flow cytometry, random amplified polymorphic DNA (RAPD), or the classical amplified fragment length polymorphisms (AFLP) analysis failed to yield any detectable differences between mantled and normal palms (Matthes et al., 2001; Rival et al., 1997, 1998). However, when methylation sensitive or related technologies were utilized, the methylation level of the mantled genome appears to be altered (Jaligot et al., 2002; Jaligot et al., 2004; Matthes et al., 2001). This subsequently became the basis for further research.

The mantled phenotype suggests that the homeotic MADS box genes that regulate the identity of the flower whorls may contribute to this effect. Hence, a set of MADS box genes were isolated and characterized from the oil palm (Alwee et al., 2006; Auyong, 2006). MADS box directed differential display (van der Linden et al., 2002) has been shown to be a powerful technique for isolation of MADS gene families. Genes that fall in the A class (EgMADS2, -3 -10, and -14), B class (EgMADS16), and C class (EgMADS12) of the ABC model were isolated in a single experiment. In addition, the E class or SEP genes (EgMADS5 and -8), which play important roles in regulation of the B and C functions, had also been isolated (van der Linden et al., 2005). This method has also generated two potential markers: MM77 (could be linked to reversion) and MM78 (unpublished). The latter was subsequently developed into a PCR based assay to enable a more rapid and simpler screening technique for utilization in checking for conformity amongst tissue culture derived materials (Fig. 3.4). MM78 was found to be exclusive for abnormal ramets, however, its success as a biomarker for conformity lies in its robustness when confronting large scale analysis. Other noteworthy candidate genes are EGAD1, EGCEL1, and EgM39A (Morcillo et al., 2006; Tregear et al., 2002) that exhibit higher expression in abnormal material when compared against their normal counterpart.

Following along the epigenetic track, it is now evident that the previously silent transposable elements may play a major role in reshaping the genome, or more precisely, the epigenome of the somaclonal variant (Rodriguez-Enriquez et al., 2011). Earlier investigations in the oil palm suggest that the mantled phenotype was not caused by major rearrangements of transposable elements but more by changes in the methylation pattern of other genomic components (Kubis et al., 2003). However, we reckon it is too early to draw a conclusion on the function of transposable elements in mantling, as formerly genetic characterization was done on single elements. Now that the oil palm genome has been sequenced, genome-wide analysis of large populations of transposable-elements can be applied (Feschotte et al., 2002). In addition, it has been reported that the regulatory system governing silencing of the transposable elements is a multipartite epigenetic code that involves small RNAs, chromatin, and DNA methylation (Lisch, 2009). The amalgamation of these components makes the mechanism highly dynamic and it now appears that to get to the crux of somaclonal variation of the oil palm, the epigenetic landscape needs to be resolved.

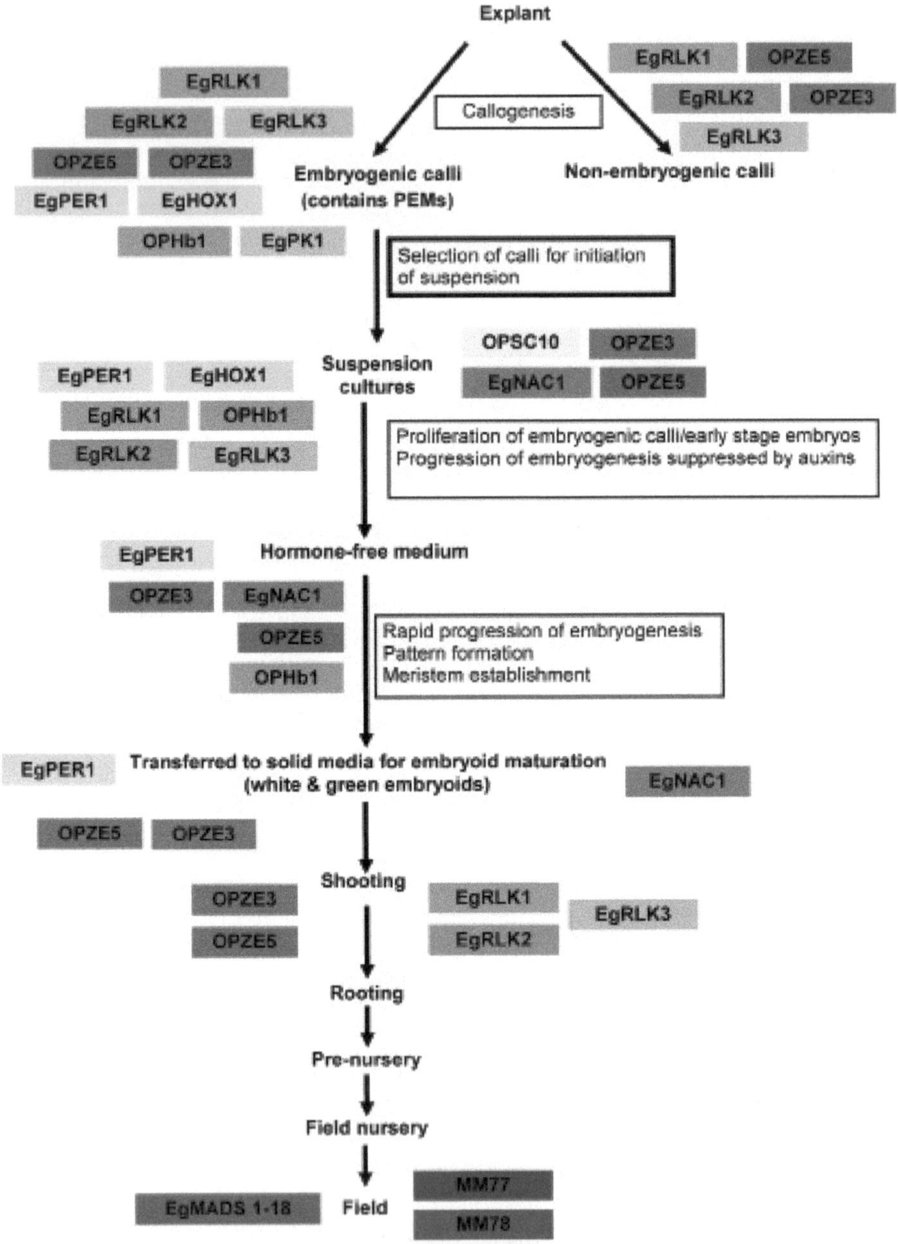

Fig. 3.4. Schematic representation of the stages whereby up-regulation in expression of potential markers for somatic embryogenesis and flowering, including potential markers for clonal conformity. PEMs: proembryogenic masses. Source: Ong-Abdullah & Ooi, 2007.

Biomarker Discovery for Embryogenesis

Much of the nineties were spent exploring various technologies in understanding the abnormality phenomenon as described briefly above. Although clonal propagation of elite oil palm offers a simple, rapid, and direct approach for yield improvement, it is often laborious as high number of clones and large explant sample sizes are required to produce sufficient population for a commercially viable production venture. As clonal improvement is 10% ahead of conventional breeding by about 5 years (Soh et al., 2003), it would be wise to work on improving the cloning process. Based on a rough estimation conducted in 2005, tissue culture laboratories would be operating at a loss of approximately RM80,000 per year just for cloning oil palm materials without any prior screening to gauge the embryogenic potential or the tissue culturability of the material (Ong-Abdullah, 2005). The obvious reason for this is that callogenesis and embryogenesis are random processes. The availability of suitable markers or group of markers at the molecular level that can aid in screening prolific and non-prolific material very early on or estimate the embryogenic potential would help in reducing production costs and eliminate the necessity to handle large culture volumes.

Earlier studies on somatic embryogenesis concentrated on developing methodologies for initiation and production of somatic embryos (Ahee et al., 1981; Jones, 1974; Pannetier et al., 1981). The manipulations mainly involved the composition of the exogenous phytohormones applied in the media as well as the introduction of tissues with better clonability to further improve the process. Histological analysis of somatic embryogenesis was carried out on leaf-derived callus detailing the emergence of callus and the subsequent formation of somatic embryos, with shoot and root apices (Schwendiman et al., 1988). Besides that, the occurrence of biochemical indicators that are useful in the prediction of embryogenic potential were also investigated (Turnham & Northcote, 1982).

As the field of molecular biology gained in importance, understanding molecular switches that occur in somatic cells that provoked them to becoming embryogenic and maintaining conformity as they develop into plants became mainstream research (Dudits et al., 1995). Biomarkers are molecular indicators that are highly predictive of biological processes. To date, several potential biomarkers for embryogenesis and abnormality have been developed and are presently being validated at the industrial scale. Prior to the discovery of these markers, the route taken was to study differential gene expression at various stages of development. The genome on its own is merely a source of information. However, in order for the system to function, it must be expressed. Traditionally, gene expressions were studied on a gene by gene basis, but with the invasion of large scale omics studies, the global approach is now the way to go. The study of the transcriptome became fashionable towards late 1990s (Velculescu et al., 1997), however, this approach needs the information support through the traditional hypothesis-driven research in order to answer interesting questions and are therefore synergistic to each other (Weinstein, 2001).

Based on expression studies, EgPER1, EgHOX1, EgPK1, and OPSC10 (Fig. 3.4) may target the callus stage in early somatic embryogenesis. Whether these proteins are involved in the induction of embryogenesis is not yet known. Functional studies have to be carried out to elucidate their functions during embryogenesis. OPHb1 most probably plays a part at a later stage of meristem establishment. As for the other genes, such as RLKs, Calmodulin (OPZE3), and QM (OPZE5) proteins, their expressions are more ubiquitous, indicating that they are possibly involved in more general mechanisms during somatic embryogenesis (Ong-Abdullah & Ooi, 2007). These molecular markers are useful to detect the early stages of oil palm tissue culture process such as embryogenic and non-embryogenic callus. However, they have not been thoroughly tested at stages earlier than the embryogenic calli or on a wider range of oil palm genotypes. Thus, the subsequent direction taken was to verify the utility of these biomarkers and eventually convert these markers into diagnostic tools for screening prolific tissues of oil palm. More biomarkers continue to be discovered for tissue culture amenity through a major tripartite collaborative effort between MPOB, local universities, and industries employing mostly DNA based techniques. Amongst some of the new discoveries are Eg707 (Thuc et al., 2010), EgSERK (a somatic embryogenesis receptor kinase; unpublished), EgHAD (a hydrolase; unpublished) and EgSAPK (a serine threonine protein kinase; unpublished). Eg707 and EgHAD seemed to be most promising as markers for early detection of tissue culture amenity.

Auxins have been the primary plant hormone used in oil palm tissue culture media. At some point, 2,4-D was blamed for the genetic variation observed in cultures (Machakova et al., 2008; Sogeke, 1998; Sogeke et al., 1999), and replacing 2,4-D with NAA was deemed to be a prudent move. However, to date, due to the stability of 2,4-D and its enhancement of culture friability (Street, 1979), it is still an ingredient in some tissue culture protocols albeit at low concentrations. In liquid culture, 2,4-D in combination with NAA was found effective for the growth of the embryogenic suspensions (Tarmizi, 2002). A preliminary study was conducted recently to investigate the response of selected plant hormone-responsive genes in explant cultures to exogenously applied auxins, NAA and 2,4-D, using real time quantitative PCR (Ooi et al., 2009). Different auxin concentrations were applied on explants from three clonal ortets (thus having an identical genetic background), and explant samples were collected at several time points for expression analysis. Through this experiment, EgIAA9 and EgIAA13, putative members of the Aux/IAA family, were found to be potentially viable as indicators for ortet response towards auxin and tissue culture amenity.

Aside from the hypothesis-driven approach, the hunt for potential biomarkers went global via microarrays. A customized oil palm cDNA array (Ooi et al., 2001) was built and several genes differentially expressed that merit further verification either for their high expression in EC and EMB (Chan et al., 2005) or that they were specifically enhanced in NEC (Low et al., 2006) were identified. It was also interesting to note that from a time course experiment, some of the genes that were found to be expressed after 7 days on callus induction media were a heat-shock protein, a pathogenesis-related protein, MAP kinase 5, and an early flowering protein (Sambanthamurthi et al., 2009).

Tools for Oil Palm Breeding

Molecular Cytogenetics

In Malaysia, *E. guineensis* is the crop species planted commercially since it produces high oil yield. However, *E. oleifera* produces more unsaturated oil compared to *E. guineensis* but has a lower oil yield. *E. oleifera* also has the advantage of slow height increment. In order to introgress the highly unsaturated oil trait and slow height increment into *E. guineensis*, oil palm breeders have crossed these two species to produce O×G hybrids. The O×G hybrids have the advantage of having more unsaturated oil and show a slower height increment compared to *E. guineensis*. However, the hybrids suffer from the disadvantage of poor oil yield. They are also vigorous in vegetative growth, making it difficult to harvest the fruit bunches. For these reasons, the O×G hybrids are backcrossed to the *guineensis* parent to improve on these characteristics. Therefore, it would be useful to breeders to be able to assess the genomic composition of the backcross progenies at the nursery stage and select only those with the desired genomic composition.

To facilitate the hybrid breeding program, the GISH technique is relevant. The GISH technique is able to distinguish both *oleifera* and *guineensis* chromosome sets in O×G hybrids (Madon et al., 1999). In O×G hybrids equal amounts of oleifera and guineensis chromosomes (16 each) are inherited. During meiosis in the O×G hybrids, each gamete receives different numbers of *E. oleifera* and *E. guineensis* chromosomes, while still maintaining the haploid number of $n = 16$. In a backcross of O×G to *E. guineensis* as the male parent, the O×G gametes fuse with pollens from *E. guineensis* ($n = 16$) during fertilization. Seeds derived from the backcross will therefore contain variable numbers of *oleifera* chromosomes due to the independent assortment of chromosomes in the hybrid. Therefore, the backcross individuals can have a genome with all the 32 chromosomes originating from *E. guineensis* ($2n = 32$) or a variable number of *E. oleifera* chromosomes to a maximum of 16. In interphase nuclei of O×G hybrids and their backcrosses, it can be observed that the chromosomes originating from the two species occupy discrete non-intermixed domains in the nuclei. In conclusion, with the development of oil palm genetic mapping program and the advancement of fluorescence and genomic *in situ* hybridization techniques, the future and speed of interspecific hybrid breeding program can be further enhanced. These tools will enable breeders to identify palms with interesting genes and on which genome (*oleifera* or *guineensis*) the gene/genes reside. Hence, the breeders can make informed decisions on how to proceed with the breeding program.

DNA Markers for Oil Palm Breeding

A wide repertoire of DNA markers have been developed for oil palm starting with RFLP in the 1990s (Cheah et al., 1993; Mayes et al., 1996), AFLP in late 1990s (Singh et al., 1999), SSR markers in the early and late 2000s (Billotte et al., 2001; Singh et al., 2008), and more recently SNP markers (Ooi et al., 2010). The early use of the markers in breeding was demonstrated in detecting pollen contamination in controlled crosses (Singh et al., 1999). Such application of markers in oil palm breeding is of interest, especially in maintaining the fidelity of controlled crosses and/or in conducting paternity tests. However, the more interesting application of DNA markers in oil palm breeding is in the association of markers with traits of interest for implementing marker assisted selection (MAS) programs.

The most widely used method to identify marker/trait association in oil palm has been through the construction of genetic maps of specific populations. The first genetic map for oil palm was constructed by Mayes et al. (1997) for a population derived from the selfing of a tenera palm. More importantly, Mayes et al. (1997) had successfully mapped the locus associated with shell thickness, an important monogenic trait of oil palm. The group had mapped a co-dominant RFLP marker about 8 cm from the shell locus. Subsequently a group in Brazil, using a combination of bulked segregation analysis (BSA) and genetic mapping, also reported a RAPD marker closely linked to the shell thickness locus (Moretzsohn et al., 2000). More recently Billotte et al. (2005) also reported the mapping of an AFLP marker close to the same locus. A marker linked to shell thickness is important in breeding programs. The marker could be used to separate the *dura, tenera,* and *pisifera* plants in the nursery; the advantage here being that they could be planted separately based on shell thickness. Currently there are ongoing attempts in several laboratories to clone the shell gene itself using the information from the genetic maps.

Work has also progressed in identifying markers linked to more complex traits in oil palm. Rance et al. (2001), expanding on the map generated by Mayes et al. (1997), identified QTLs linked to vegetative parameters and yield components of oil palm. QTLs linked to fatty acid composition (Singh et al., 2009) have also been reported. Although the research on associating markers with specific traits in oil palm is now progressing in several laboratories, it is important to take note of some issues that will affect finding marker/trait association especially for complex traits. Finding marker/trait association is only possible for phenotypes that can be scored or for which bioassays are available. Another important consideration is the cost and accuracy of phenotyping the traits concerned. The more difficult and costly a trait to phenotype will likely be more desirable for selection using DNA markers. Yield components in oil palm for example require substantial costs for accurate phenotyping as extensive field trials are required. However, due to the costs involved which is also linked to the perennial nature of oil palm, number of replicates and palms/replicate are often limited, reducing sensitivity of the QTL analyses. To overcome some of these constraints, especially with respect to number of palms in cross, Billotte et al. (2010) had reported identification of QTLs linked to vegetative parameters and yield components in a multi parent population design. The identification of QTLs in multi parental design (involving several crosses sharing a parent) was more fruitful compared to crosses involving single or two parents.

The successful mapping of QTLs and its application in oil palm breeding has to be evaluated carefully. In cases where QTLs are detected in crosses drawn from advanced breeding materials (elite lines), the alleles are most likely already present in the ongoing breeding programs (Langbridge & Chambers, 2004). This can be overcome to some extent by detecting alleles of interest in wild relatives or germplasm collections as demonstrated in relation to disease resistance genes incorporated from wild relatives into elite breeding lines of rice and barley. The development of techniques such as advanced backcross QTL analysis (Tanksley & Nelson, 1996) allows introgression of QTLs of important traits from wild relatives into advanced lines improving the traits concerned and also the level of diversity in the advanced lines (Langbridge & Chambers, 2004).

Exploiting Germplasm Collections for Producing New and Improved Varieties

Apart from the development and application of MAS in breeding, various molecular markers have also been employed to study the genetic diversity of MPOB germplasm collections. Field evaluation of these materials has indicated clearly that the collections are a meaningful source for oil palm improvement programs, which points to its significance for long term conservation. However, conserving the oil palm genetic materials in field genebank is an expensive endeavor due to the land requirement to maintain the collections. As such, establishment of core collections involving only populations that hold valuable genes and widest range of genetic variability is desirable. The field assessment of the palms provides direct scores for selecting such populations. Additionally, estimates of genetic variability using molecular markers allows for exclusion of duplicated and/or least variable populations, thus reducing the land area required and cost of maintenance.

Comprehensive screening of the MPOB oil palm germplasm collections using isozymes (Hayati et al., 2004), RFLPs (Maizura et al., 2006), and AFLPs (Kularatne et al., 2000) revealed measures of genetic diversity that has helped in identifying populations with unique and useful alleles as well as high genetic variability. High number of alleles per locus and levels of polymorphic loci were reported among populations from Nigeria (Hayati et al., 2004; Kularatne et al., 2000; Maizura et al., 2006) suggesting their value as core collection. Analysis of molecular data at the population level further specified the importance of preserving populations 12, 35, 36, 37, and 45 of the Nigerian collection. Four of these populations are from the eastern part of Nigeria (Rajanaidu, 1995), and they have also been found to exhibit good yield performance in the current MPOB ex situ living collection (Rajanaidu, 1994). Palms from populations 12 are of interest due to its slow height increment. Successful exploitation of palms from Population 12 would facilitate harvesting of fruits and extend the economic life span of oil palm. Besides the Nigerian populations, other populations that show high diversity include population 6 of Guinea, population 2 of Senegal, population 5 and 13 of Sierra Leone, and 22 and 29 of Cameroon (Hayati et al., 2004), which should be considered for conservation.

With regards to the allelic frequencies, emphasis should be given to populations that possess both common and rare alleles. This would safeguard the entire alleles present in the ex situ collections, irrespective of their phenotypic merits. The occurrence of rare alleles was reported among populations from Nigeria, Congo DR, Tanzania, Angola, Senegal, Sierra Leone, Madagascar, Ghana, and Guinea. Rare alleles are present at a frequency of less than 0.100 and have often been associated with adaptation in fluctuating environments. This is demonstrated by the Madagascar palms that grow very poorly in their land of origin due to low rainfall. Despite the low levels of polymorphic loci, the Madagascar palms exhibit rare alleles not detected in other populations (Hayati et al., 2004). The unfavorable growth environment resulted in relatively shorter palms that produce oils with high iodine value and linoleic acid content (Kushairi et al., 2002) and, as such, can be exploited to produce planting materials with slower height increment and higher levels of un-saturated palm oil.

Clustering analyses based on the genetic distance values computed among the assayed populations revealed clear separation of Madagascar populations from the rest of the genetic collections from Africa (Hayati et al., 2004; Kularatne et al., 2000). In general, the populations were grouped according to their country of origin, which suggests that choosing representative populations from each cluster would secure a greater range of diversity. The mean number of alleles scored per locus and polymorphic loci can also guide in selecting the appropriate populations.

Analysis of the germplasm collections from Africa also showed that the mean genetic differentiation among the populations (G_{ST}) was 0.143 (Maizura, 1999). This indicates that 85.7% of the total variation was due to the differences within palms in the populations. Similarly, Rajanaidu (1985) reported large component of variance within populations with regards to bunch characteristics. Therefore, sampling more palms within a few populations would be more effective in capturing the genetic variation in future conservation.

MPOB has also collected *Elaeis oleifera* germplasm materials from South America. Field and molecular characterization of the palms are being carried out currently (Teh, 2010). The same criteria described above will also be used to help create a core collection of *E. oleifera* palms. Preliminary evidence indicates that the *E. oleifera* revealed lower levels of genetic diversity compared to *E. guineensis*. Nevertheless, some of the genetic variability parameters computed, for instance, the mean number of alleles per locus and levels of polymorphism suggests that populations 3, 5, 8 and 12 of Panama, 6, 8 and 9 of Colombia, 2, 4, 14 and 16 of Costa Rica, and populations 2 and 3 of Honduras offer wider range of genetic variability. These populations also encompassed rare alleles and preserving those populations could safeguard the most number of alleles present in the germplasm collections.

Production of Doubled Haploids

Haploid plants, which contain only one set of chromosomes, have received increased interest recently due to their potential application in breeding programs. Haploids which are also hemizygous (one allele per locus) are mainly used for the rapid production of completely homozygous doubled haploid lines (DH) (even in self-incompatible species). As such, DH is completely homozygous diploids, produced without the time consuming task of selfing or sib-mating and selection over many generations. The advantage of the use of DH in plant breeding is the shorter breeding cycle and increased selection efficiency. Furthermore, DH is valuable in both applied and fundamental genetic studies. In particular, the homozygous nature of the lines and the potential for unlimited propagation has made DH populations highly desirable for molecular mapping and gene tagging studies.

Conclusions

As with other plant genomes, the oil palm has now been sequenced (Basri & OPGP, 2009), and with the plummeting price of sequencing as time goes by more genomes will follow suit. Most recently published and related to the oil palm is the date palm genome (Al-Dous et al., 2011). Sequencing a genome is just the tip of the iceberg; the real challenge comes from reaping the benefits from knowing the blueprint of life. Bhat (2011) aptly puts it that the investment in whole genome sequencing needs to be balanced with its intended downstream applications. Hence, the decision to sequence the oil palm genome was therefore a well thought out plan. MPOB is known to have the largest oil palm germplasm collection in the world (Basri et al., 2004). This is an ideal source for identifying new alleles essential for breeding. In addition, to fully exploit the genome information, having appropriate mapping populations to tag important agronomic traits is an advantage. Now that the underpinning issues are in place, MPOB and the industry as a whole is poised to move forward into a genomics guided breeding and epigenetics guided tissue culture era.

References

Aberlenc-Bertossi, F.; Chabrillange, N.; Duval, Y.; Tregear, J.W. Contrasting globulin and cysteine proteinase gene expression patterns reveal fundamental developmental differences between zygotic and somatic embryos of oil palm. Tree Physiol. 2008, 28, 1157–1167.

Adam, H.; Jouannic, S.; Morcillo, F.; Richaud, F.; Duval, Y.; Tregear, J.W. MADS box genes in oil palm (*Elaeis guineensis*): patterns in the evolution of the SQUAMOSA, DEFICIENS, GLOBOSA, AGAMOUS, and SEPALLATA subfamilies. J. Mol. Evol. 2006, 62, 15–31.

Adam, H.; Jouannic, S.; Orieux, Y.; Morcillo, F.; Richaud, F.; Duval, Y.; Tregear, J.W. Functional characterization of MADS box genes involved in the determination of oil palm flower structure. J. Exp. Bot. 2007, 58(6), 1245–1259.

Adam, M.D.; Kelley, J.M.; Gocayne, J.D.; Dubnick, M.; Polymeropoulus, M.H.; Xiao, H.; Merril, C.R.; Wu, A.; Olde, B.; Moreno, R.F. Complementary DNA sequencing: Expressed sequenced tags and human genome project. Science 1991, 252, 1651–1656.

Ahée, J.; Arthuis, P.; Cas, G.; Duval, Y.; Guénin, G.; Hanower, J.; Hanower, P.; Liévoux, D.; Lioret, C.; Malaurie, B. La multiplication vegetative in vitro de palmier à huile par embryogenèse somatique. Oléagineux 1981, 36, 113–118.

Albert, V.A.; Soltis, D.E.; Carlson, J.E.; Farmerie, W.G.; Wall, P.R.; Ilut, D.C.; Solow, T.M.; Mueller, L.A.; Landherr, L.L.; Hu, Y. Floral gene resources from basal angiosperms for comparative genomics research. BMC Plant Biol. 2005, 5, 5.

Al-Dous, E.K.; George, B.; Al-Mahmoud, M.E.; Al-Jaber, M.Y.; Wang, H.; Salameh, Y.M.; Al-Azwani, E.K.; Chaluvadi, S.; Pontaroli, A.C.; DeBarry, J. *De novo* genome sequencing and comparative genomics of date palm (*Phoenix dactylifera*). Nature Biotechnol. 2011, 29, 521–527.

Alwee, S.S.; Van der Linden, C.G.; Van der Schoot, J.; de Folter, S.; Angenent, G.C.; Cheah, S.C.; Smulders, M.J.M. Characterization of oil palm MADS box genes in relation to the mantled flower abnormality. Plant Cell Tissue Organ Cult. 2006, 85, 331–344.

Auyong, S.M.A. Amplified fragment length polymorphism-based detection of deoxyribonucleic acid methylation and the analysis of three MADS-box genes in oil palm (*Elaeis guineensis* Jacq.). Master of Science dissertation, March 2006.

Basri, M.W.; Abdullah, S.N.A.; Henson, I.E. Oil Palm—Achievements and potential. In New directions for a diverse planet. Proceedings of the 4th International Crop Science Congress: Brisbane Australia, 26 Sept–1 Oct.

Basri, M.W.; Oil Palm Genome Programme Sequencing the oil palm genome: the beginning. In: Proceedings of the 2009 PIPOC on Palm Oil—Balancing Ecologics with Economics; Kuala Lumpur, 2009.

Bennett, M.D.; Smith, J.B. Nuclear DNA amounts in angiosperms. Phil. Trans. R. Soc. London 1991, B334, 309–345.

Beulé, T.; Camps, C.; Debiesse, S.; Tranchant, C.; Dussert, S.; Sabau, X.; Jaligot, E.; Alwee, S.S.R.S.; Tregear, J.W. Transcriptome analysis reveals differentially expressed genes associated with the mantled homeotic flowering abnormality in oil palm (*Elaeis guineensis*) Tree Gene. Genomes 2011, 7, 169–182.

Bhat, S.R. Rationalizing investment and effort in whole genome sequencing for harvesting applied benefits. Curr. Sci. 2011, 100(11), 1633–1637.

Billotte, N.; Jourjon, M.F.; Marseillac, N.; Berger, A.; Flori, A.; Asmady, H.; Adon, B.; Singh, R.; Nouy, B.; Potier, F. QTL detection by multi-parent linkage mapping in oil palm (*Elaeis guineensis* Jacq.). Theor. Appl. Genet. 2010, 120, 1673–1687.

Billotte, N.; Marseillac, N.; Risterucci, A.M.; Adon, B.; Brottier, P.; Baurens, F.C.; Singh, R.; Herrán, A.; Asmady, H.; Billot, C. Microsatellite-based high density linkage map in oil palm (*Elaeis guineensis* Jacq.). Theor. Appl. Genet. 2005, 110, 754–765.

Billotte, N.; Risterucci, A.M.; Barcelos, E.; Noyer, J.L.; Amblard, P.; Baurens, F.C.; Development, characterisation, and across-taxa utility of oil palm (*Elaeis guineensis* Jacq.) microsatellite markers. Genome 2001, 44, 413–425.

Bourgis, F.; Kilaru, A.; Cao, X.; Ngando-Ebongue, G.F.; Drira, N.; Ohlrogge, J.B.; Arondel, V. Comparative transcriptome and metabolite analysis of oil palm and date palm mesocarp that differ dramatically in carbon partitioning. Proc. Natl. Acad. Sci., USA, 2011. PNAS 1106502108.

Brady, S.M.; Provart, N.J. Extreme breeding: Leveraging genomics for crop improvement. J. Sci. Food Agric. 2007, 87, 925–929.

Campos-de Quiroz, H. Plant genomics: an overview. Biol. Res. 2002, 35, 385–399.

Chan, P.L.; Low, E.T.L.; Boon, S.H.; Rozana, R.; Ooi, L.C.L.; Rajinder, S.; Cheah, S.C. Monitoring gene expression changes in different stages of oil palm tissue culture via cDNA microarray. In Proceedings of the 2005 Conference on Biotechnology of Plantation Commodities, MIECC, Mines Resort City, Selangor, 9–11 August 2005; pp. 445–453.

Chan, P.L.; Ma, L.S.; Low, E.T.L.; Elyana, M.S.; Ooi, L.C.L.; Cheah, S.C.; Singh, R. Identification of somatic embryogenesis-related genes from normalized embryoid cDNA library of oil palm. Electron. J. Biotechnol. 2010, 13(1), 14.

Cheah, S.C.; Ooi, L.C.L.; Rahimah, A.R. Quality control process for oil palm tissue culture using DNA probes. PORIM Information Series No. 40, Palm Oil Research Institute of Malaysia (PORIM), 2006.

Cheah, S.C.; Siti Nor Akmar, A.; Ooi, L.C.L.; Rahimah, A.R.; Maria, M. Detection of DNA variability in the oil palm using RFLP probes. In Proc. of the 1991 PORIM Int. Palm Oil Conf.: Agriculture (Module 1), PORIM, Malaysia, 1993; pp.144–150.

Corley, R.H.V. What is the upper limit to oil extraction ratio? In Proc. Oil and Kernel Production in Oil Palm—A Global Perspective. Palm Oil Res. Inst. Malaysia: Kuala Lumpur, 1998; pp. 256–269.

Corley, R.H.V.; Lee, C.H.; Law, I.H.; Wong, C.Y. Abnormal flower development in oil palm clones. The Planter 1986, 62, 233–240.

Cseke, L.J.; Podila, G.K. MADS-box genes in Dioecious Aspen II: A review of MADS box genes from trees and their potential in forest biotechnology. Physiol. Mol. Biol. Plants 2004, 10(1), 7–28.

Dolezel, J. Flow cytometric analysis of nuclear DNA content in higher plants. Phytochem. Anal. 1991, 2, 143–154.

Dolezel, J.; Dolezelova, M.; Novak, F.J. Flow cytometric estimation of nuclear DNA amount in diploid bananas (*Musa acuminata* and *M. balbisiana*). Biologia Plantarum 1994, 36, 351–357.

Dudits, D.; Gyorgyey, J.; Bogre, L.; Bako, L. Molecular Biology of Somatic Embryogenesis. In In vitro Embryogenesis in Plants. Current Plant Science and Biotechnology in Agriculture; Thorpe, T.A., Ed. Kluwer Academic Publishers: The Netherlands, 1995; pp. 267–308.

Durand-Gasselin, T.; Guen, V.L.; Konan, E.; Duval, Y. Oil palm (*Elaeis guineensis* Jacq.) plantations in Cote d'Ivoire obtained through in vitro culture: first results. Oleagineux 1990, 45, 1–11.

Feschotte, C.; Jiang, N.; Wessler, S.R. Plant transposable elements: where genetics meets genomics. Nature Rev. Gen. 2002, 3, 329–341.

Fields, C. Analysis of gene expression by tissue and developmental stage. Curr. Opin. Biotechnol. 1994, 5, 595–598.

Galbraith, D.W.; Lambert, G.M.; Macas, J.; Dolezel, J. Analysis of nuclear DNA content and ploidy in higher plants. Robinson, J.P., Eds. In Current protocols in cytometry. John Wiley & Sons: New York, 1997.

Hayati, A.; Wickneswari, R.; Maizura, I.; Rajanaidu, N. Genetic diversity of oil palm (*Elaeis guineensis* Jacq.) germplasm collections from Africa: implications for improvement and conservation of genetic resources. Theor. Appl. Genet. 2004, 108, 1274–1284.

Ho, C.L.; Kwan, Y.Y.; Choi, M.C.; Tee, S.S.; Ng, W.H.; Lim, K.A.; Lee, Y.P.; Ooi, S.E.; Lee, W.W.; Tee, J.M. Analysis and functional annotation of expressed sequence tags (ESTs) from multiple tissues of oil palm (*Elaeis guineensis* Jacq.). BMC Genom. 2007, 8, 381.

Jaligot, E.; Beule, T.; Baurens, F.C.; Billotte, N.; Rival, A. Search for methylation-sensitive amplification polymorphism associated with the mantled variant phenotype in oil palm (*Elaeis guineensis* Jacq.). Genome, 2004, 47, 224–228.

Jaligot, E.; Beule, T.; Rival, A. Methylation sensitive RFLPs: characterization of two oil palm markers showing somaclonal variation-associated polymorphism. Theor. Appl. Genet. 2002, 104, 1263–1269.

Jones, L.; Barfield, D.; Barrett, J.; Flook, A.; Pollock, K.; Robinson, P. Cytology of oil palm cultures and regenerants plants. Plant Tissue Culture 1982.

Jones, L.H. Propagation of clonal oil palms by tissue culture. Oil Palm News 1974, 17, 1–8.

Jouannic, S.; Argout, X.; Lechauve, F.; Fizames, C.; Borgel, A.; Morcillo, F.; Aberlenc-Bertossi, F.; Duval, Y.; Tregear, J. Analysis of expressed sequence tags from oil palm (Elaeis guineensis). FEBS Letters 2005, 579(12), 2709–2714.

Jouannic, S.; Collin, M.; Vidal, B.; Verdeil, J.L.; Tregear, J.W. A class I KNOX gene from the palm species *Elaeis guineensis* (Arecaceae) is associated with meristem function and a distinct mode of leaf dissection. New Phytol. 2007, 174, 551–568.

Kubis, S.E.; Castilho, A.M.M.F.; Vershinin, A.V.; Heslop-Harrison, J.S. Retroelements, transposons and methylation status in the genome of oil palm (*Elaeis guineensis*) and the relationship to somaclonal variation. Plant Mol. Biol. 2003, 52, 69–79.

Kularatne, R.S.; Shah, F.H.; Rajanaidu, N. Investigation of genetic diversity in African natural oil palm populations and Deli using AFLP markers. In Proceedings of the International Symposium on Oil Palm Genetic Resources and Their Utilisation, Malaysian Palm Oil Board: Bangi, Malaysia, 2000; pp. 497–555.

Kushairi, A.; Rajanaidu, N.; Jalani, B.S. Genetic variation of iodine value, carotene and tocotrienols in palm oil and its application in pharmaceuticals. Malaysian Oil Sci. Technol. 2002, 1(2), 92–105.

Kushairi, A.; Tarmizi, A.H.; Zamzuri, I.; Ong-Abdullah, M.; Rohani, O.; Samsul Kamal, R.; Ooi, S.E.; Ravigadevi, S.; Mohd Basri, W. Current status of oil palm tissue culture in Malaysia. Presented at the Clonal and Quality Replanting Material Workshop, 10 August 2006, Palace of Golden Horses, Seri Kembangan, Selangor.

Langbridge, B.; Chambers, K. The principles and application of molecular markers. In Biotechnology in Agriculture and Forestry; Nagata, T., Lorz, H., Widholm, J.M., Eds. Springer:Verlag, Berlin, Germany, 2004; pp. 3–22.

Lin, H.C.; Morcillo, F.; Dussert, S.; Tranchant-Dubreuil, C.; Tregear, J.W.; Tranbarger, T.J. Transcriptome analysis during somatic embryogenesis of the tropical monocot *Elaeis guineensis:* evidence for conserved gene functions in early development. Plant Mol. Biol. 2009, 70, 173–192.

Lisch, D. Epigenetic regulation of transposable elements in plants. Ann. Rev. Plant Biol. 2009, 60, 43–66.

Low, E.T.L.; Alias, H.; Boon, S.H.; Shariff, E.M.; Tan, C.Y.A.; Ooi, L.C.L.; Cheah, S.C.; Raha, A.R.; Wan, K.L.; Singh, R. Oil palm (*Elaeis guineensis* Jacq.) tissue culture ESTs: Identifying genes associated with callogenesis and embryogenesis. BMC Plant Biol. 2008, 8, 62.

Low, E.T.L.; Tan, J.S.; Chan, P.L.; Boon, S.H.; Wong, Y.L.; Rozana, R.; Ooi, L.C.-L.; Ma, L.S.; Ong-Abdullah, M.; Cheah, S.C. Developments towards the application of DNA chip technology in oil palm tissue culture. J. Oil Palm Res. 2006, Special Issue, 87–98.

Machakova, I.; Zazimalova, E.; George, E.F. Plant growth regulators I: introduction: auxins, their analogues and inhibitors. In Plant Propagation by Tissue Culture 3rd Edition Vol. 1; George, E.F., Hall, M.A., De Klerk, G.–J., Eds.; Springer: Netherlands, 2008; pp. 175–204.

Madon, M.; Clyde, M.M.; Cheah, S.C. Cytological analysis of *Elaeis guineensis* and *Elaeis oleifera* chromosomes. J. Oil Palm Res. 1998, 10(1), 68–91.

Madon, M.; Clyde, M.M.; Cheah, S.C. Application of genomic in situ hybridization (GISH) on *Elaeis* hybrids. J. Oil Palm Res. October 1999, Special Issue, 74–80.

Madon, M.; Clyde, M.M.; Hashim, H.; Mohd, Y.; Mat, H.; Saratha, S. Polyploidy induction of oil palm through colchicine and oryzalin treatments. J. Oil Palm Res. 2005, 17, 110–123.

Madon, M.; Phoon, L.Q.; Clyde, M.M.; Mohd Din, A. Application of flow cytometry for estimation of nuclear DNA content in *Elaeis*. J. Oil Palm Res. 2008, 20, 447–452.

Maizura, I. Genetic variability of oil palm (*Elaeis guineensis* Jacq.) germplasm collection using RFLP markers. PhD dissertation. Universiti Kebangsaan Malaysia, 1999.

Maizura, I.; Rajanaidu, N.; Zakri, A.H.; Cheah, S.C. Assessment of genetic diversity in oil palm (*Elaeis guineensis* Jacq.) using Restriction Fragment Length Polymorphism (RFLP). Genet. Resour. Crop Evol. 2006, 53, 187–195.

Matthes, M.; Singh, R.; Cheah, S.C.; Karp, A. Variation in oil palm (*Elaeis guineensis* Jacq.) tissue cultured derived regenerants revealed by AFLPs with methylation sensitive enzymes. Theor. Appl. Genet. 2001, 102, 971–979.

Mayes, S.; Jack, P.L.; Marshall, D.F.; Corley, R.H.V. Construction of a RFLP genetic linkage map for oil palm (*Elaeis guineensis* Jacq.). Genome 1997, 40, 116–122.

Mayes, S.; James, C.; Horner, S.F.; Jack, P.L.; Corley, R.H.V. The application of restriction fragment length polymorphism for the genetic fingerprinting of oil palm (*Elaeis guineensis*). Mol. Breed. 1996, 2, 175–180.

Meins, F.; Beversdorf, W.D.; Oono, K.; Hamzah, I. Final Report Oil Palm Tissue Culture Consultant Panel, 1991.

Mohd, B.W.; Abdullah, S.N.A.; Henson, I.E. Oil Palm–Achievements and Potential. In "New directions for a diverse planet". Proceedings of the 4th International Crop Science Congress, 26 Sep–1 Oct 2004, Brisbane, Australia.

Morcillo, F.; Aberlenc-Bertossi, F.; Hamon, S.; Duval, Y. Differential accumulation of storage protein, 7s globulins, during zygotic and somatic embryos development in oil palm *(Elaeis guineensis* Jacq.). Plant Physiol. Biochem. 1998, 36, 509–514.

Morcillo, F.; Gagneur, C.; Adam, H.; Richaud, F.; Singh, R.; Cheah, S.C.; Rival, A.; Duval, Y.; Tregear, J.W. Somaclonal variation in micropropagated oil palm. Characterization of two novel genes with enhanced expression in epigenetically abnormal cell lines and in response to auxin. Tree Physiol. 2006, 26, 585–594.

Morcillo, F.; Gallard, A.; Pillot, M.; Jouannic, S.; Aberlenc-Bertossi, F.; Collin, M.; Verdeil, J.L.; Tregear, J.W. *EgAP2-1*, an *AINTEGUMENTA-like* (*AIL*) gene expressed in meristematic and proliferating tissues of embryos in oil palm. Planta 2007, 226, 1353–1362.

Morcillo, F.; Hartmann, C.; Duval, Y.; Tregear,; J.W. Regulation of 7S globulin gene expression in zygotic and somatic embryos of oil palm Physiol. Plant 2001, 112, 233–243.

Moretzsohn, M.C.; Nunes, C.D.M.; Ferreira, M.L.; Grattapaglia, D. RAPD linkage mapping of the shell thickness locus in oil palm (*Elaeis guineensis* Jacq). Theor. Appl. Genet. 2000, 100, 63–70.

MPOB (2010) (http://econ.mpob.gov.my/economy/Overview%202010_final.pdf).

Musa, B.; Gurmit, S. Utilization of MPOB germplasm at United Plantations. In Proceedings of the 3rd Seminar on Performance of MPOB PS1 and PS2 Materials and Elite Germplasm MPOB; Bangi, Selangor, 2008.

Nasaruddin, N.M.; Harikrishna, K.; Othman, R.Y.; Hoon, L.S.; Harikrishna, J.A. Computational prediction of microRNAs from Oil Palm (*Elaeis guineensis* Jacq.) expressed sequence tags. AsPac J. Mol. Biol. Biotechnol. 2007, 15(3), 107–113.

Nurniwalis, A.W.; Suhaimi, N.; Akmar, S.N.A.; Aminah, S.; Manaf, M.A. Gene discovery via expressed sequence tags from the oil palm (*Elaeis guineensis* Jacq.) mesocarp. J. Oil Palm Res. Special Issue on Malaysia-MIT Biotechnology Partnership Programme (Oil Palm Metabolic Engineering) 2008, 2, 87–96.

Oil World. www.oilworld.biz/annual; (accessed 2007).

Okamuro, J.K.; den Boer, B.G.; Jofuku, K.D. Regulation of *Arabidopsis* flower development. Plant Cell 1993, 5, 1183–1193.

Ong, L.M. An examination of embryogenic and non-embryogenic cultures of oil palm (*Elaeis guineensis* Jacq.) Ph.D dissertation. Universiti Putra Malaysia, 2001.

Ong-Abdullah, M.; Ooi, S.E. Biomarkers: Finding a niche in oil palm tissue culture. Part 1—Laying the foundation. Oil Palm Bull. 2001, 53, 30–35.

Ong-Abdullah, M.; Ooi, S.E. Biomarkers: Finding a niche in oil palm tissue culture. Part 2—Targeting the transcriptome. Oil Palm Bull. 2007, 54, 68–74.

Ong-Abdullah, M.; Ooi, S.E.; Sharifah, S.R.S.A. A molecular insight to oil palm tissue culture. In Proceedings of the 2005 Conference on Biotechnology of Plantation Commodities. MINES Resort City, Selangor, 2005 pp. 46–63.

Ooi, C.L.; Cheah, S.C.; Rahimah, A.R.; Bresnahan, P. Preliminary microarray gene profiling and differential gene expression studies in oil palm. Poster presented at the MUGV meeting, Victoria, B.C., Canada. 12–14 August 2001.

Ooi, C.L.; Ong, P.W.; Ithnin, M.; Singh, R. Discovery of oil palm SNPs and their application in the characterization of oil palm genetic resources and breeding materials. In Proceedings of the National Biotechnology Seminar 2010, Kuala Lumpur, 2010.

Ooi, S.E.; Zamzuri, I.; Ang, C.L.; Mohd Azmir, M.; Ong-Abdullah, M. Response of hormone-responsive genes to 2,4-D vs. NAA treatment of explant cultures via gene expression analysis. In Proceedings of Agriculture, Biotechnology & Sustainability Conference; PIPOC 2009 International Palm Oil Congress, Vol. III. KLCC: Kuala Lumpur, 2009; pp. 1339.

Pannetier, C.; Arthuis, P.; Lievoux, D. Neoformation of young *Elaeis guineensis* plants from primary calluses obtained from leaf fragments cultured *in vitro*. Oleagineux 1981, 36, 119–122.

Pavy, N.; Paule, C.; Parsons, L.; Crow, J.A.; Morency, M.J.; Cooke, J.; Johnson, J.E.; Noumen, E.; Guillet-Claude, C.; Butterfield, Y. Generation, annotation, analysis and database integration of 16,500 white spruce EST clusters. BMC Genomics 2005, 6, 144.

Rajanaidu, N. The oil palm (*Elaeis guineensis*) collections in Africa. Proc. of Int. Workshop on Oil Palm Germplasm and Utilization, 1995, pp. 59-83.

Rajanaidu, N. PORIM oil palm genebank, collection, evaluation, utilization and conservation of oil palm genetic resources. In conjunction with the Release of Elite Oil Palm Planting Materials and Launching of Oil Palm Genebank. PORIM, Bangi, Selangor, 1994.

Rajanaidu, N.; Rao, V.; Halim, A.H.; Ong, S.H.; Genetic resources: New developments in oil palm breeding. Elaeis 1990, 1, 1–10.

Ralph, S.G.; Chun, H.J.E.; Kolosova, N.; Cooper, D.; Oddy, C.; Ritland, C.E.; Kirkpatrick, R.; Moore, R.; Barber, S.; Holt, R.A. A conifer genomics resource of 200,000 spruce (*Picea* spp.) ESTs and 6,464 high-quality, sequence-finished full-length cDNAs for Sitka spruce (*Picea sitchensis*). BMC Genomics 2008, 9, 484.

Rance, K.A.; Mayes, S.; Price, Z.; Jack, P.L.; Corley, R.H.V. Quantitative trait loci for yield components in oil palm (*Elaeis guineensis* Jacq.). Theor. Appl. Genet. 2001, 103, 1302–1310.

Rao, V.; Donough, C.R. Preliminary evidence of a genetic cause for the floral abnormalities in some oil palm ramets. Elaeis 1990, 2, 199–207.

Riju, A.; Chandrasekar, A.; Arunachalam, V. Mining for single nucleotide polymorphisms and insertions / deletions in expressed sequence tag libraries of oil palm. Bioinform. 2007, 2(4), 128–131.

Rival, A.; Bertrand, L.; Beule, T.; Combes, M.C.; Trouslot, P.; Lashermes, P. Suitability of RAPD analysis for the detection of somaclonal variants in oil palm. Plant Breeding 1998, 117, 73–76.

Rival, A.; Beule, T.; Barre, P.; Hamon, S.; Duval, Y.; Noirot, M. Comparative flow cytometric estimation of nuclear DNA content in oil palm (*Elaeis guineensis* Jacq.) tissue cultures and seed derived plants. Plant Cell. Report 1997, 16, 884–887.

Rival, A.; Jaligot, E.; Beule, T.; Finnegan, E.J. Isolation and expression analysis of genes encoding MET, CMT, and DRM methyltransferases in oil palm (*Elaeis guineensis* Jacq.) in relation to the 'mantled' somaclonal variation. J. Exper. Bot. 2008, 59(12), 3271–3281.

Rodriguez-Enriquez, J.; Dickinson, H.G.; Grant-Downton, R.T. MicroRNA misregulation: an overlooked factor generating somaclonal variation? Trends Plant Sci. 2011, 16(5), 242–248.

Roowi, S.H.; Ho, C.L.; Alwee, S.S.R.S.; Abdullah, M.O.; Napis, S. Isolation and characterization of differentially expressed transcripts from the suspension cells of oil palm (*Elaeis guineensis* Jacq.) in response to different concentration of auxins. Mol. Biotechnol. 2010, 46, 1–19.

Röser, M.; Johnson, A.T.; Hanson, L. Nuclear DNA amounts in palms (Areaceae). Botanica Acta, 1997, 110, 79–89.

Sambanthamurthi, R.; Rajinder, S.; Parveez, G.K.A.; Ong-Abdullah, M.; Kushairi, A. Opportunities for the oil palm *via* breeding and biotechnology. Breeding Plantation Tree Crops; Mohan Jain, S., Priyadarshan, P.M., Eds. Springer Science and Business Media, 2009; pp. 377–421.

San, C.T.; Shah, F.H. Differential gene expression and characterization of tissue-specific cDNA clones in oil palm using mRNA differential display. Mol. Biol. Rep. 2005, 32, 227–235.

Schwendiman, J.; Pannetier, C.; Michaux-Ferriere, N. Histology of somatic embryogenesis from leaf explants of the oil palm *Elaeis guineensis.* Annals Bot. 1988, 62, 43–52.

Sharma, M.; Tan, Y.P. Oil palm breeding programmes and the performance of D×P planting materials at United Plantations Berhad. In Proc. Seminar on Sourcing of Oil Palm Planting Materials for Local and Overseas Joint Venture; Rajanaidu, N., Jalani., B.S., Eds. Palm Oil Res. Inst. Malaysia: Kuala Lumpur, 1999; pp. 118–135.

Singh, R.; Cheah, S.C. Analysis of the inheritance of AFLP markers in an interspecific cross of oil palm using the pseudo-testcross strategy. J. Oil Palm Res. 1999 (Special Issue), 64–73.

Singh, R.; Cheah, S.C. Identification of flower specific proteins in the oil palm (*Elaeis guineensis*) by two-dimensional gel electrophoresis. J. Oil Palm Res. 2000, 12, 46–54.

Singh, R.; Noorhariza, M.Z.; Ting, N.C.; Rozana, R.; Tan, S.G.; Low, L.E.T.; Maizura, I.; Cheah, S.C. Exploiting an oil palm EST the development of gene-derived and their exploitation for assessment of genetic diversity. Biologia 2008b, 63, 1–9.

Singh, R.; Rahman, R.A.; Ooi, L.C.L.; Cheah, S.C. Microsatellite probes for fingerprinting oil palm clones. MPOB TT. MPOB Information Series (ISSN 1511-7871), 2006, No. 305.

Singh, R.; Tan, S.G.; Panandam, J.; Rahman, R.A.; Cheah, S.C. Identification of cDNA-RFLP Markers and their use for molecular mapping in oil palm (*Elaeis guineensis*). AsPac J. Mol. Biol. Biotechnol. 2008a, 16(3), 53–63.

Singh, R.; Tan, S.G.; Panandam, J.M.; Rahman, R.A.; Ooi, L.C.L.; Low, E.T.L.; Sharma, M.; Jansen, J.; Cheah, S.C. Mapping quantitative trait loci (QTLs) for fatty acid composition in an interspecific cross of oil palm. BMC Plant Biol. 2009, 9, 114.

Sogeke, A.K. Stages in the vegetative propagation of oil palm (*Elaeis guineensis* Jacq.) through tissue culture. J. Oil Palm Res. 1998, 10(2), 1–9.

Sogeke, A.K.; Odewale, J.O.; Exe, C.R.; Omamor, I.B. Vegetative Propagation of oil palm (*Elaeis guineensis* Jacq.). Initiation of callus from leaf explants. In Proceedings of the 1999 PORIM International Palm Oil Congress (Agriculture). 1999, pp. 536–541.

Soh, A.C.; Wong, G.; Hor, T.Y.; Tan, C.C.; Chew, P.S. Oil palm genetic improvement. Plant Breed. Rev. 2003, 22, 165–219.

Street, H.E. Embryogenesis and chemically induced organogenesis. In Plant Cell and Tissue Culture; Sharp, W.R., Larsen, P.O., Paddock, E.F., Raghavan, V., Eds., Ohio State Univ. Press: Columbus; pp. 123–154.

Tanksley, S.D.; Nelson, J.C. Advanced backcross QTL analysis: a method for the simultaneous discovery and transfer of valuable QTLs from unadapted germplasm into elite breeding lines. Theoret. Appl. Genet. 1996, 92, 191–203.

Tarmizi, A.H. Oil palm liquid culture–MPOB Protocol. MPOB Information Malaysian Palm Oil Board, 2002, Series No. 138.

Teh, C.K. Genetic diversity of Central and South American Wild Oil Palm (*Elaeis oleifera*) populations using microsatellite markers. Dissertation. Universiti Kebangsaan, Malaysia, 2010.

Teoh, W.C.; Cheah, S.C.; Harikrishna, K.; Tan, S.H. Isolation and characterization of *Myb-Related* genes from oil palm (*Elaeis guineensis* Jacq.) J. Plant Biol. 2003, 46(2), 95–104.

Thao, N.T.P.; Ureshino, K.; Miyajima, I.; Ozaki, Y.; Okubo, H. Induction of tetraploids in ornamental *Alocasia* through colchicine and oryzalin treatments. Plant Cell Tissue Organ Culture 2003, 72, 19–25.

Thuc, L.V.; Sarpan, N.; Ky, H.; Ooi, S.E.; Napis, S.; Ho, C.L.; Ong-Abdullah, M.; Chin, C.F.; Namasivayam, P. A novel transcript of oil palm (*Elaeis guineensis* Jacq.), Eg707, is specifically upregulated in tissues related to totipotency. Mol. Biotechnol. 2010, DOI: 10.1007/s12033-010-9356-4.

Ting, N.C.; Cheah, S.C.; Zamzuri, I.; Tan, S.G.; Faridah, Q.Z.; Maizura, I.; Singh, R. Statistical mapping of quantitative trait loci controlling time to first callusing in oil palm (*Elaeis guineensis* Jacq.) tissue culture. Pertanika J. Trop. Agric. Sci. 2006, 29, 35–45.

Ting, T.; Zaki, N.M.; Rosli, R.; Low, E.T.L.; Ithnin, M.; Cheah, S.C.; Tan, S.C.; Singh, R. SSR mining in oil palm EST database: application in oil palm germplasm diversity studies. J. Genet. 2010, 89(2), 135–145.

Tittarelli, A.; Santiago, M.; Morales, A.; Meisel, L.A.; Silva, H. Isolation and functional characterization of cold-regulated promoters, by digitally identifying peach fruit cold-induced genes from a large EST dataset. BMC Plant Biol. 2009, 9, 121.

Tranbarger, T.J.; Dussert, S.; Joet, T.; Argout, X.; Summo, M.; Champion, A.; Cros, D.; Omore, A.; Nouy, B.; Morcillo, F. Regulatory mechanisms underlying oil palm fruit mesocarp maturation, ripening, and functional specialization in lipid and carotenoid metabolism. Plant Physiol. 2011, 156, 564–584.

Tregear, J.W.; Morcillo, F.; Richaud, F.; Berger, A.; Singh, R.; Cheah, S.C.; Hartmann, C.; Rival, A.; Duval, Y. Characterization of a defensin gene expressed in oil palm inflorescences: induction during tissue culture and possible association with epigenetic somaclonal variation events. J. Exp. Bot. 2002, 53(373), 1387–1396.

Turnham, E.; Northcote, D.H. The use of acetyl-CoA carboxylase activity and changes in wall composition as measures of embryogenesis in tissue cultures of oil palm (*Elaeis guineensis*). Biochem. J. 1982, 208, 323–332.

Van der Linden, C.G.; Sharifah, S.A.; Angenent, G.; Cheah, S.C.; Smulders, M.J.M. Molecular characterization of flower development in oil palm in relation to the mantling abnormality. In Proceedings of the 2005 PIPOC on Technological Breakthroughs and Commercialization—The Way Forward. Kuala Lumpur, 2005; pp. 531–549.

Van der Linden, C.G.; Vosman, B.; Smulders, M.J.M. Cloning and characterization of four apple MADS box genes isolated from vegetative tissue. J. Exper. Botany 2002, 53, 1025–1036.

Velculescu, V.E.; Zhang, L.; Zhou, W.; Vogelstein, J.; Basrai, M.A.; Bassett, D.E.; Hieter, P.; Vogelstein, B.; Kinzler, K.W. Characterization of the yeast transcriptome. Cell 1997, 88(2), 243–251.

Verdun, R.E.; Di Paolo, N.; Urmenyi, T.P.; Rondinelli, E.; Frasch, A.C.; Sanchez, D.O. Gene discovery through expressed sequence tag sequencing in *Trypanosoma cruzi.* Infect. Immun. 1998, 66, 5393–5398.

Vizoso, P.; Meisel, L.A.; Tittarelli, A.; Latorre, M.; Saba, J.; Caroca, R.; Maldonado, J.; Cambiazo, V.; Campos-Vargas, R.; Gonzalez, M. Comparative EST transcript profiling of peach fruits under different post-harvest conditions reveals candidate genes associated with peach fruit quality. BMC Genomics 2009, 10, 423.

Weinstein, J.N. Searching for pharmacogenomic markers: the synergy between omic and hypothesis-driven research. Disease Markers 2001, 17(2), 77–88.

Wu, J.; Mooney, P. Autotetraploid tangor plant regeneration from *in vitro Citrus* somatic embryogenic callus treated with colchicine. Plant Cell, Tissue and Organ Culture 2002, 70, 99–104.

Xu, L.; Yuan, Y.; Zhang, L.; Wan, L.; Zheng, Y.; Zhou, P.; Li, D. Identification and characterization of differential gene expression in the mesocarp and kernel of oil palm nuts using suppression subtractive hybridization. Tree Genet. Genomes 2011, DOI 10.1007/s11295-011-0390-6.

4

Tissue Culture and Genetic Engineering of Oil Palm

Ghulam Kadir Ahmad Parveez, Omar A. Rasid, Ahmad Tarmizi Hashim, Zamzuri Ishak, Samsul Kamal Rosli, and Ravigadevi Sambanthamurthi
Advanced Biotechnology and Breeding Centre, Malaysian Palm Oil Board, No. 6, Persiaran Institusi, Bandar Baru Bangi, 43000 Kajang, Selangor, Malaysia

Introduction

The continuous improvement in oil palm planting materials thus far is due to conventional breeding approach. However, the number of crosses and backcrosses required is a major limitation to the application of this approach as the sole means for oil palm improvement. With a generation interval of 15 years (vs. 6 months for cereals), progress can be rather sedate. And with the oil palm being a large plant, large areas are required (Rajanaidu et al., 1990; Sambanthamurthi et al., 2000). There is also a limit to the genetic material that can be introgressed into any genome at one time, and the gene(s) would have to be from a related species.

Due to these limitations of conventional breeding, biotechnology, mainly tissue culture and genetic engineering, was chosen as an additional means to improve the yield and quality of this important monocot oil bearing tree. Using tissue culture method, regenerants (ramets) carrying the same genetic material as their originator (mother palm or ortet) could be theoretically produced. This allows a large scale duplication of highly productive palms not achieved via conventional breeding. Tissue culture of oil palm was initiated in the 1960s, and the first successful production of plantlets was reported in the 1970s (Jones, 1974; Rabéchault & Martin, 1976). The above success inspired Malaysian Palm Oil Board (MPOB) and many industry players to delve into oil palm tissue culture. Oil palm tissue culture, though successful, was not without its difficulties, especially due to somaclonal variation, namely abnormality in the regenerants (Corley et al., 1986). This somaclonal variation caused a furor among oil palm tissue culturists, and consequently, a number of laboratories slowed down their production efforts. However, after two decades of experimentations and experience, many laboratories have again embarked on large-scale cloning of oil palm (Kushairi et al., 2006). With the increase in world population, the demand for palm oil has also increased, and it is hoped that the planting of clonal palms would increase the oil yield without the need to increase the arable land area.

On the other hand, using genetic engineering, specific genes of interest can be directly introduced into a genome without any time spent in growing the plant then crossing it. Transgenic plants have already been produced for many agronomic traits, such as disease and herbicide resistance, in addition to novel products (Christou, 2004; Jaworski & Cahoon, 2003; Scheller & Conrad, 2005). Furthermore, genetic engineering also obviates a major impediment faced by breeders—the genes to be introduced are no longer limited to compatible species but can be from any living organism, such as bacteria, fungus, and animal. As genetic engineering only introduces specific gene(s) of interest, and in some cases it could be targeted to a desired organelle or genome location, this method is more precise compared to breeding where random genomic rearrangement and recombination could result in very high variation within progenies.

This chapter will focus on the work and progress made mainly by MPOB on their tissue culture and genetic engineering efforts on oil palm. While there is significant tissue culture research carried out in other places, such as Indonesia, Thailand, and Latin America, the research to genetically engineer oil palm was not widely known and published, making it difficult to elaborate on their research findings. Being in the European Union, scientists at CIRAD were not allowed to deal with genetic modification of oil palm, hence no progress has been made beyond transient expression. Genetic engineering in oil palm is relatively recent as it was only initiated at MPOB in the late 1980s. Major progress and breakthroughs were made in the 1990s and has become an impetus to venture further in modifying the oil palm for a better return. Decreasing saturated fatty acid content in palm oil has always been the main aim of genetic engineering of oil palm. Consequently, the target genes are those involved in the biosynthesis for producing more oleic acid in palm oil (Sambanthamurthi et al., 2000). Besides increasing the oleic acid content, MPOB also plans to produce the following traits in transgenic oil palm: high stearic acid palm oil; biodegradable plastics; pharmaceuticals and nutraceuticals, such as lycopene and palmitoleic acid; and industrial oils such as ricinoleic acid (Parveez, 2003; Parveez et al., 1999; Sambanthamurthi et al., 2000).

With all the above advantages of tissue culture and genetic engineering, application of these biotechnological approaches to oil palm will enhance the competitiveness and ensure sustainability of the industry (Basiron & Chan, 2005; Kushairi et al., 2006; Sambanthamurti et al., 2000). In the following section, progress made in the above research areas, mainly in Malaysia, will be elaborated.

Tissue Culture of Oil Palm

Tissue Culture Protocol

Since the first protocol for oil palm tissue culture was published in the 1970s, continuous improvements of the protocol have been made and the comprehensive protocol was reviewed by Rohani et al. (2000, 2003).

It is vital to propagate vegetatively outstanding palms or ortets from breeding trials that have extensive vegetative and reproductive information. Ortets are selected from field trials after at least four years' data on yield and bunch analysis, as stated in Malaysian Standard MS 2099:2008, and other desirable vegetative and physiological traits such as palm height, bunch index, oil characteristic, and so on. Usually, young leaves enclosed within a cylinder or cabbage of leaf petioles of the ortet's palm crown were taken down, cut into small pieces aseptically and cultured on callus initiation medium. Explant cultures are maintained for about a year by repeated subculture every three months. The palm age does not seem to have a strong influence on the callusing propensity (Wooi, 1995), whereas the rates of callusing were genotype dependent. The primary callus is selected and transferred to medium for embryoid initiation. The transfer is repeated every 3 months for a duration of about a year, giving the cells time for embryogenesis. A callus already forming embryoids will continue to produce more embryoids when subcultured further onto fresh medium (Guzman, 1995). Among the various types of embryoids, the nodular type is the most prolific. The embryoids expand and produce more compact clumpy, nodular structures called polyembryogenic (PE) cultures. The embryoid clumps are further subdivided into 1.5–2 cm diameters and subcultured repeatedly for longer periods. The PE cultures are maintained on nutrient medium and subcultured at two monthly intervals. Currently the subculture cycles of PE have been limited up to 15 (about 30 months). The cloning process can be summarized as in Table 4-A (Tarmizi et al., 2011).

Table 4-A. Stages of Oil Palm Tissue Culture Process.

No	Stages	Activities
0	Mother palm selection and preparation	a. Ortet selection b. Cabbage sampling
I	Establishing an aseptic culture	a. Explant inoculation for callus initiation b. Callus culture for embryoid initiation.
II	Multiplication of suitable propagules	a. Embryoid culture (in solid or liquid culture system) b. Polyembryoid (PE) multiplication
III	Formation of shoots and roots	a. Shoot elongation b. Root initiation
IV	Transfer to natural environment	a. Weaning

Fig. 4.1. A worker is performing routine culturing in an oil palm tissue culture laboratory.

Currently, there are 13 tissue culture laboratories in Malaysia with a production capacity of 2.44 million ramets annually (Fig. 4.1). This is enough for individual laboratories plantation's internal consumption only. However, for fulfilling the national demand, capacity of production has to be increased. Almost all laboratories have their own media sets and protocols for cloning and field trials (Tarmizi et al., 2011). It was reported that oil palm clones were mostly uniform and could increase the yield by a minimum of 20% as compared to conventional seed derived plants (Kushairi et al., 2006). Although somaclonal variation could not be ruled out, the level is manageable. As the government encourages replanting of palms older than 25 years old using better planting materials, laboratories are now confidently increasing their capability to help meet the demand.

Recent Progress in Tissue Culture of Oil Palm

Tissue culture is an efficient way to increase yield of plants. However, it is very costly and labor intensive. MPOB tissue culturists have innovated many systems and equipment to improve the tissue culture system for oil palm, which could be applied to other plants as well. There are a number of companies and organizations who have adopted some of these systems and equipment (Tarmizi et al., 2011). Some of the major inventions are briefly elaborated in the following section.

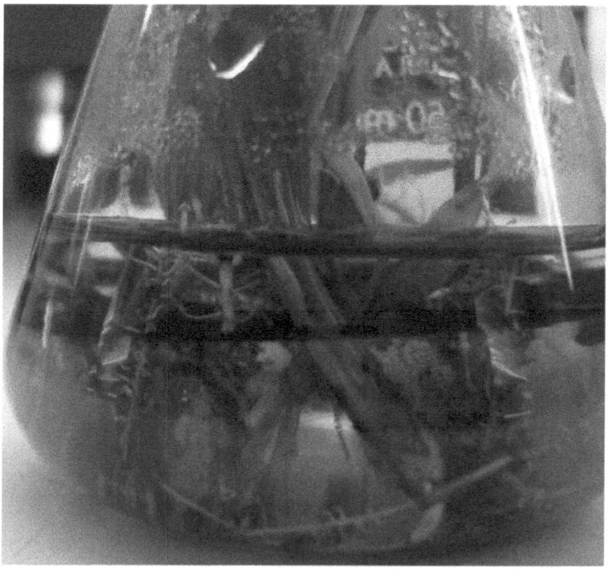

Fig. 4.2. Double layer technique for root maturation.

Fig. 4.3. Flameless sterilizer.

Double-layer Rooting Technique

An innovative method for topping up liquid medium in established cultures (Fig. 4.2) is obviating transplanting the individual shoots in other vessels for rooting or elongation (Zamzuri, 1998, 2001). This could increase productivity up to 17.5× in the rooting stage and reduce wages and cost of root culture medium by 80%.

Flameless Sterilizer

This is an electrical device that sterilizes glass containers (in the mouth-rim region) and surgical apparatus simultaneously (Zamzuri, 2002, 2009). This improved heater uses a circular heating coil (heated at 300°C) in the center where a culture vessel could be inserted upright from below for its mouth to be heated while the top of the device is a chamber that holds glass beads for sterilizing surgical apparatus, such as forceps and scalpels (Fig. 4.3). Among the advantages of this sterilizer are (1) less risk of fire due to no flame, (2) dual purpose device—sterilizes vessel mouth and surgical apparatus simultaneously, and (3) adjustable platform to accommodate vessels of different heights.

MPOB Oil Palm Liquid Culture

This is a basic protocol for liquid culture that includes selecting a suitable callus (friable type), media formulation, aggregate sieving, maturation induction, embryoid regeneration, and production of rooted plantlets in either solid or liquid medium (Tarmizi, 2002; Tarmizi et al., 2011).

MPOB Fast Transfer Technique (MoFaTT) in Liquid Culture

This is a rapid and convenient method for liquid medium replenishment during maintenance and maturation of oil palm suspension cultures (Fig. 4.4). The advantages of using MoFaTT are (1) it simplifies medium replenishment from 10 steps

Fig. 4.4. Normal liquid culture of oil palm (top) as compared to MoFaTT (bottom).

to only two steps, (2) no movement of cultures from culture room to laminar flow, (3) replenishment could be done onsite or on the shaker itself, (4) reduces risk of contamination due to less movement and handling, and (5) potential for automation (Tarmizi & Zaiton, 2004).

Two-in-One MPOB Simple Impeller (2-in-1 MoSLIM) in Liquid Culture

This economical system uses commonly available culture bottles in the laboratory for liquid culture instead of bioreactors or special commercial flasks (Tarmizi, 2002; Tarmizi et al., 2003). It provides simultaneous aeration and agitation (two-in-one) in the form of a single device for liquid culture propagation (Fig. 4.5). This new vessel could produce culture aggregates with fresh weight increment of two- to six-fold over a 30- to 40-day period.

Simple Impeller with Fast Transfer Technique (SLIM-FaTT) in Liquid Culture

The integration of MoSLIM and MoFaTT for synergistic effects resulted in the development of the SLIM-FaTT (Fig. 4.6) (Tarmizi et al., 2011). The advantages of this SLIM-FaTT technique are (1) it simplifies the media replenishment process from

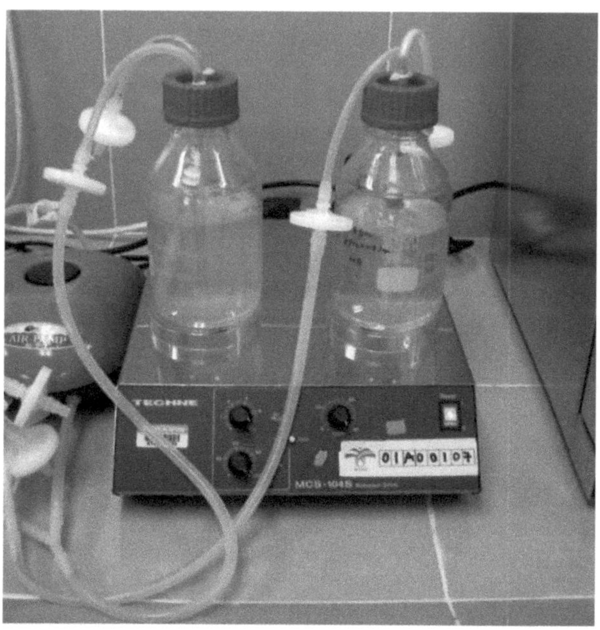

Fig. 4.5. Two-in-One MPOB Simple Impeller (2-in-1 MoSLIM).

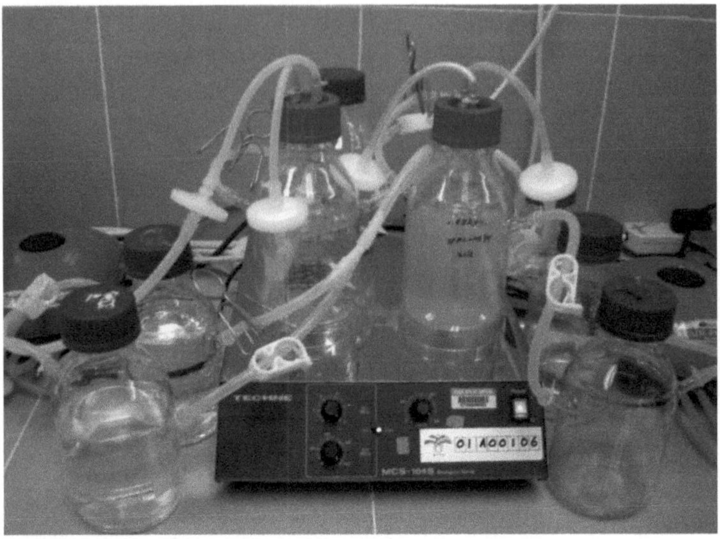

Fig. 4.6. Simple impeller with fast transfer technique (SLIM-FaTT).

eight steps to only one step, (2) it involved no culture movement from culture room to laminar flow, (3) replenishment of medium could be done in culture room itself which could further reduce risk of contamination, and (4) it has the potential for automation.

Multiplication of Oil Palm Liquid Cultures in Bioreactors

B-Braun 'Biostat® B' 2L Version 1.0. The use of this bioreactor in combination with a medium containing 1 mg litre-1 2,4-Dichlorophenoxyacetic acid (2,4-D) and 0.1 mg litre-1 α-Naphthaleneacetic acid (NAA) demonstrated an increase in culture fresh weights between 5- and 12-fold after 60 days for all the clones tested, in contrast to shake flasks (used as control), which only showed 2.5-fold weight increment (Tarmizi et al., 2011). All cultures proliferated in this bioreactor could be regenerated into whole plants when transferred onto solid media. For example, ramets for clone PL 90 were successfully established and transferred to the pre-nursery and field nursery. On the other hand, using another bioreactor, Biotron, fresh weight of two selected clones was recorded to increase only about five-fold after 60 days. As bioreactor systems were known to have the capability to save time and space, efforts are now made towards developing a semi-or fully automated oil palm clonal production system.

Sixfors™ Bioreactor (Infors AG, Bottmingen, Switzerland). This is one of the versatile and practical bioreactors that could allow simultaneous online monitoring and control of growth conditions, including temperature, agitation, and dissolved oxygen, in six independent reactor vessels. It is supported by analytical equipment, including a chromatographic system for analysis of the culture medium, secondary metabolites, biomass, and so on (Tarmizi et al., 2011). This bioreactor could help in providing a better understanding on the behavior of oil palm cultures with respect to the physical and chemical conditions. It was observed that after 60 days, 1.6-, 2.3-, 2.1-, 3.3-, and 3.5-fold of fresh weight were obtained for clones E132, E136, E146, E147, and E34, respectively. The main advantage of this bioreactor is that up to 6 clones or samples could be tested simultaneously or 6 conditions could be tested simultaneously for a same clone.

MPOB Modified Vessel (MoVess) for Liquid Tissue Culture System. This modified vessel was designed mainly to reduce cost and time during culture proliferation in bioreactors (Fig. 4.7) (Tarmizi et al., 2011). Using 3 clones in this vessel, it was found that increment in fresh weight of cultures were 5- to 35-fold after 30–60 days. The advantages of using MoVess system are (1) simple procedure for scaling up of culture volume and at more economical rate, (2) it uses a simpler inoculation method, (3) it

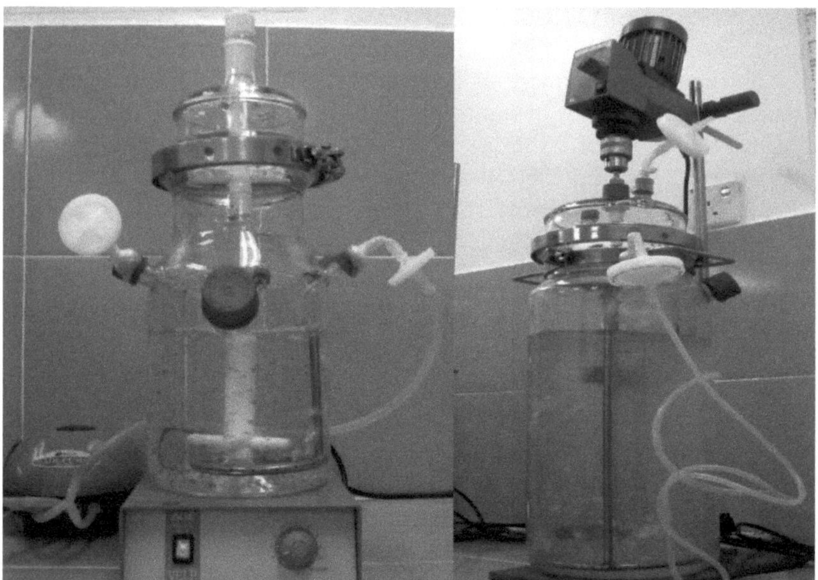

Fig. 4.7. MPOB modified vessel (MoVess), left and MPOB motorized vessel (MPOB-MotoVess), right.

could be applied to other crops or even animal cultures, and (4) it has the potential of system automation.

MPOB Motorized Vessel (MPOB-MotoVess) for Liquid Tissue Culture. Motovess is an improved version of MoVess (Tarmizi et al., 2007). This system omitted the use of magnetic stirrer as used in MoVess. The magnetic stirrer limits the culture volume between one and two liters, and the omission will allow handling of a larger culture volume of between two and nine liters (Fig. 4.7). MotoVess consists of a motor with a stand and a shaft with impeller for agitation, made of perforated stainless steel to also aerate the medium. It was recorded that two- to six-fold increment in fresh weight could be obtained after 40 days using MotoVess.

Oil Palm Tissue Culture in Other Countries

Indonesia

Indonesian Oil Palm Research Institute or IOPRI has a tissue culture laboratory located in Marihat, North Sumatra. The laboratory has operated since 1985 in collaboration with CIRAD-France. The clones produced have been planted since 1987, spread out in some areas such as North Sumatra, Riau, West Sumatra, Lampung, West Java, Aceh, West Borneo, and South Sulawesi with a total area of +2000 ha. In effort to commercialize IOPRI's Clone (known as Marihat Klon, MK), IOPRI has developed a system for quality control and efficiency improvement for mass production. The increase in the number of ramets was due to the increase in the number of ortets sampled, the degree of multiplication, the reduction in contamination, the rate of plantlet survival during acclimatization (more than 80%), and the low rate of clone abnormality (less than 5%). Abnormality can be traced using a database system. The system records all the history of the clones at every step. It makes it possible to trace and control clone production. MK has superior characters compared with plants from germinated seed. It has more homogenous vegetative, more precocious, and higher productivity at 20–25% (35–42 tons FFB/ha/year) than common D × P.

Latin America

Started in 1986, ASD's tissue culture research unit expanded into commercial scale clone production in 1999. The production of clones free of abnormalities and techniques to preserve ramets (in test tubes for over a year with a minimum growth) are among ASD's latest success. A very promising genetic material, exclusive of ASD, is the so-called "Compact". It shows very short leaves and reduced trunk growth rate as a result of an interspecies cross between the African (*Elaeis guineensis*) and the American (*E. oleifera*) oil palms. Compact ortets planted at 180 palms per hectare have a potential to produce between 12 and 19 t of oil/ha/yr, using current commercial extraction values.

Thailand

Univanich operates an Oil Palm Research Centre (OPRC), which is a leader in oil palm research in Thailand. The company's advanced palm breeding program produces high yielding hybrid seeds for its own replanting program and for sale to oil palm farmers, both locally and overseas. In 2006, Univanich OPRC commissioned Thailand's first laboratory for oil palm tissue culture. This project is to produce high yielding clonal palms in support of the Company's oil palm breeding program. Some of the clones are being planted and monitored for their performance in the field.

Genetic Engineering of Oil Palm

Since the initiation of oil palm genetic engineering program in the mid1980s, significant progress has been made in efforts to genetically engineer oil palm for value addition. In the following section, efforts in genetically manipulating oil palm is elaborated in 4 sections: (1) gene isolation, (2) promoter isolation, (3) transgenic oil palm development, and (4) production of value-added transgenic oil palm.

Gene Isolation

Ketoacyl ACP Synthase II (KAS II)

β-ketoacyl-ACP synthase II (KAS II) has been confirmed to be one of the enzymes responsible for the high accumulation of palmitic acid in palm oil and to have regulatory control over the flux of oil palm fatty acid biosynthesis. Specifically, KAS II is the rate-limiting enzyme for the formation of stearic, oleic, and other downstream fatty acids. This was evidenced by the negative correlation between KAS II activity and C16:0 and positive correlation with C18:1 and C18:2 levels, either individually or together (Sambanthamurthi et al., 2000; Sambanthamurthi et al., 1999; Umi Salamah & Sambanthamurthi, 1996). Another implication of this finding is that stearoyl-ACP desaturase, the enzyme that converts C18:0 to C18:1, is not rate-limiting. Thus, a major effort was made to isolate the KAS II gene so as to up-regulate it in oil palm to increase oleic acid at the expense of palmitic acid in its oil.

A full-length KAS II cDNA clone of *Elaeis guineensis* was isolated from oil palm mesocarp using PCR amplification (Umi Salamah et al., 2006; Umi Salamah et al., 2004). The 2.0 kb cDNA clone (Fig. 4.8) shared 80–97% identity to corresponding KAS II sequences in the Genbank database. It contained an ORF (open reading frame) of 1650 bp coding for 549 amino acid residues. The function of the isolated cDNA clones was later verified in two model systems, namely *E. coli* and *Arabidopsis thaliana* (Fig. 4.9). Over-expression of the gene in *E. coli* resulted in a significant build-up of oleic acid in the transformed cells. Down regulating the gene then significantly decreased both stearic and oleic acids with high accumulation of palmitic acid in the transformed plants (Umi Salamah et al., 2004).

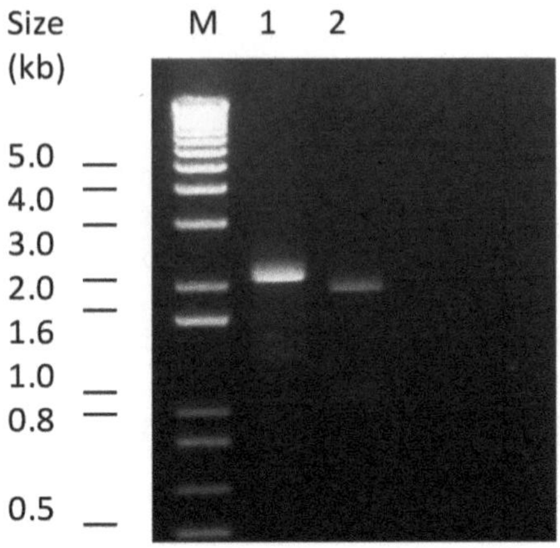

Fig. 4.8. Full-length KAS II gene of *E. guineensis*. PCR fragments (~2kb) were amplified using two different pairs of primers flanking the KAS II coding region. Note: M: 1 kb DNA ladder, lanes 1–2 amplified product for end-to-end PCR. Adapted from Umi Salamah et al., 2004.

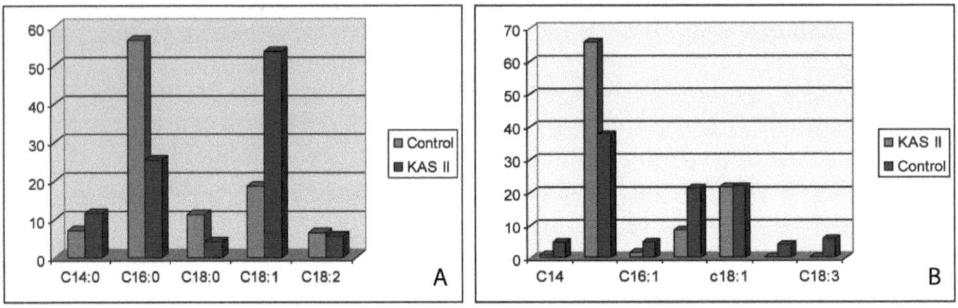

Fig. 4.9. A) Expression studies of KAS II in *E. coli*. Control is *E. coli* without oil palm KAS II. KAS II refers to *E. coli* transformed with oil palm KAS II. Gas chromatography was carried out to determine the fatty acid composition of *E. coli*. B) Analysis of fatty acids in antisense-KAS II transformed *Arabidopsis*. Control is *Arabidopsis* without oil palm KAS II and KAS II refers to *Arabidopsis* transformed with oil palm KAS II in antisense orientation. Gas chromatography was carried out to determine the fatty acid composition of *E. coli*. Adapted from Umi Salamah et al., 2004.

An attempt was also made to isolate KAS II cDNA from *E. oleifera* in view of the fact that the unsaturation level in its oil corresponded positively with KAS II activity (Umi Salamah et al., 2004). The effort led to amplification and cloning of the full length KAS II, the sequence of which was later shown to be highly identical to that of its counterpart from *E. guineensis*.

Palmitoyl-ACP Thioesterase

Biochemical studies on oil palm thioesterase indicate that there are at least two forms of thioesterase that catalyze the hydrolysis of acyl chains into free fatty acids. The enzyme encoded by the *FatA* gene is responsible for hydrolysis of unsaturated fatty acids, including oleic acid. In contrast, the *FatB* gene encodes an enzyme specific to saturated fatty acids, especially palmitoyl-ACP (Abrizah, 2001). This finding has an important implication on the attempt to modify the fatty acid composition of palm oil, especially to increase oleic acid. Since the two thioesterase genes can be independently manipulated, antisensing *FatB* can promote the flux of acyl chains for production of unsaturated fatty acids without affecting hydrolysis of the unsaturated fatty acids (Parveez et al., 2005; Parveez et al., 2003; Sambanthamurthi et al., 2000).

Full-length palmitoyl-ACP thioesterase cDNA clone was isolated and its specificity confirmed by over-expression study in *E. coli.* The expressed enzyme was able to hydrolyze medium-chain and long-chain acyl-CoAs. The gene was present in only single copy and was shown to be expressed at high levels in the leaves, flowers and 15-, 17- and 18-week mesocarp tissues (Abrizah et al., 2000). Functional analysis of the gene was performed by expressing it in *Arabidopsis* under the control of the CAMV 35S promoter. As expected, fatty acid analysis of the T3 seeds showed an increase or decrease in the palmitate (C16:0) content in the transgenic plants carrying the sense and antisense construct, respectively (Fig. 4.10) (Parveez et al., 2010a).

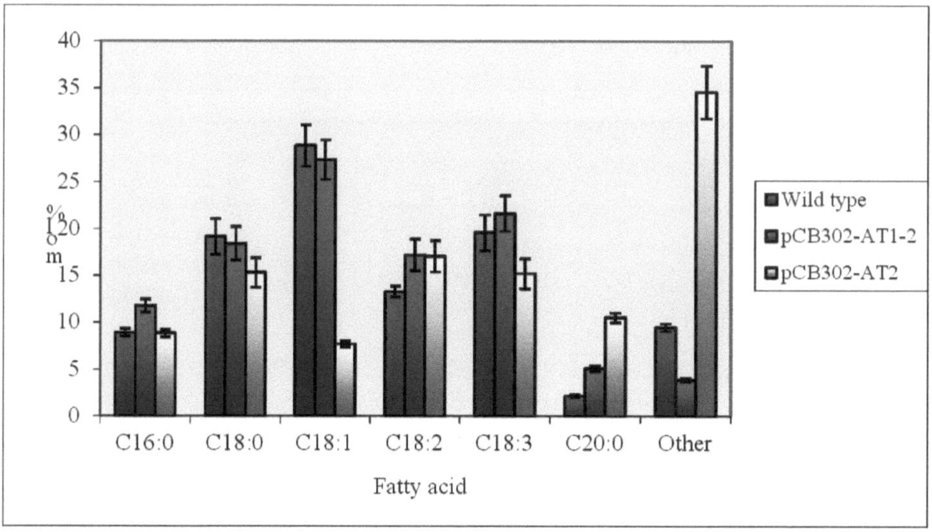

Fig. 4.10. Fatty acid composition of seed from *Arabidopsis thaliana* var Columbia of wild type and transgenic of pCB302-AT1 (sense) and pCB302-AT2 (antisense) carrying the thioesterase gene. Source: Parveez et al., 2010.

Stearoyl ACP Desaturases

Stearoyl-ACP desaturase is among the first genes isolated and characterized from oil palm fatty acid synthesis pathway (Shah et al., 2000; Siti Nor Akmar et al., 1999; Rasid & Shah, 1996). A cDNA clone from *E. oleifera* has also been reported (Siti Nor Akmar et al., 2004; Siti Nor Akmar et al., 2001). This enzyme is very active in oil palm as evidenced by its conversion of almost all stearoyl-ACP to oleoyl-ACP, explaining why palm oil contains <5% stearic acid but accumulates about 39% oleic acid (Sambanthamurthi et al., 1996a, 2000, 2002). The functions of the isolated cDNAs have been examined in transgenic *Arabidopsis* (Safiza et al., 2009a). Overexpression of the SAD1 cDNA sequence fused to *Rubisco* plastid targeting sequence in wild-type Columbia and *fab2* mutant *Arabidopsis* resulted in increased oleic acid and decreased saturated and polyunsaturated fatty acids in the oils of the T_3 homozygous lines. In comparison, the construct of SAD1 with its own plastid targeting sequence gave higher over-expression and complementation in the *Arabidopsis* wild-type Columbia and *fab2* mutant, respectively. However, even though the reduction in stearic acid in the oil of *fab2*-03a seeds was significant, the level remained much higher than in the oils of the wild-type Columbia plants (Figs. 4.11 and 4.12).

Oleoyl-CoA Desaturase

Desaturation of oleic to linoleic acid requires CoA as a cofactor and occurs in the cytosol. The reaction is catalyzed by oleoyl-CoA desaturase. There is a strong correlation between KAS II activity and oleic acid content in palm oil but an even stronger

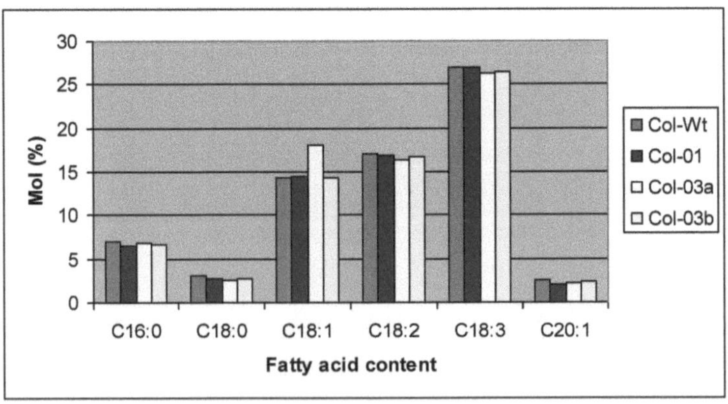

Fig. 4.11. Fatty acid composition of *A. thaliana* after over-expression of SAD1 in wild type *A. thaliana* var. Columbia (WT COL,) and transgenic plant transformed with mature oil palm SAD1 cDNA sequence fused to *Rubisco* plastid targeting sequence (COL-01 and COL-3b) and the precursor protein of SAD1 consisting of the plastid targeting sequence (COL-3a). Adapted from Safiza et al., 2009a.

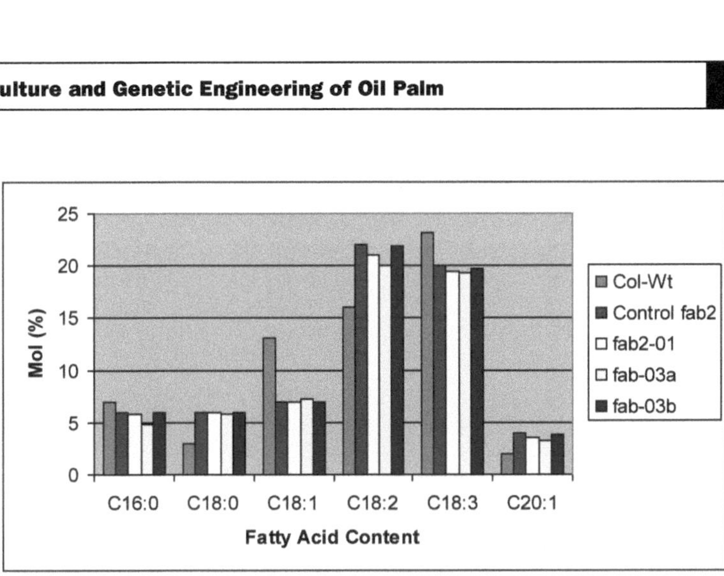

Fig. 4.12. Fatty acid composition of *A. thaliana* after complementation of SAD1 gene into *Arabidopsis* fab2 mutant and wild type (WT Columbia). The oil palm SAD1 cDNA sequence was also fused to *Rubisco* plastid targeting sequence (fab2-01 and fab2-3BCOL-3b) and the precursor protein of SAD1 consisting of the plastid targeting sequence (fab2-3a). Adapted from Safiza et al., 2009a.

correlation with linoleic (C18:2) (Sambanthamurthi et al., 1996a). Overexpressing KAS II and down-regulating palmitoyl-ACP thioesterase will not only increase oleic acid but also linoleic acid in the mesocarp oil. Therefore, for high oleic palm oil, down regulating oleoyl-CoA desaturase should also be considered to stop the overflow of oleic into linoleic acid (Sambanthamurthi et al., 1999, 2002).

A full length oleoyl-CoA desaturase clone has been obtained (Syahanim et al., 2007). The deduced amino acid sequence shares 67–75% identity with the putative Δ^{12} desaturase-like genes of other crops. Southern analysis indicated at least two copies of oleoyl-CoA desaturase genes in the oil palm genome. Northern analysis showed that it is highly expressed only in the later stages of mesocarp development in *E. guineensis* and *E. oleifera*. Low expression was detected at an early stage in the kernel and vegetative tissues.

Acetyl-CoA Carboxylase (ACCase)

ACCase is the enzyme that catalyzes the first step in fatty acid biosynthesis formation of malonyl-CoA from acetyl-CoA. Thus, it is believed to have a regulatory role in the biosynthesis. Its overexpression could increase the supply of the initial precursor and subsequently the oil content of oil palm fruits. Conversely, its down-regulation can divert acetyl-CoA for the synthesis of other high value products such as biodegradable plastics (PHB and PHBV). In oil palm, both the multi-subunit and multifunctional forms of ACCase are present. Isolation of partial length clones for multisubunit and multifunctional ACCase genes has been carried out (Sambanthamurthi et al., 2002)

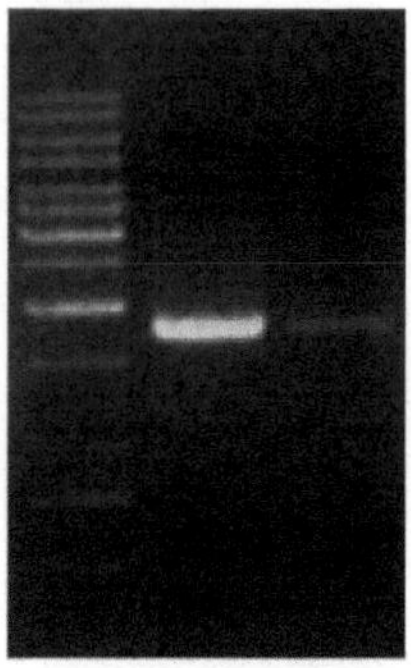

Fig. 4.13. PCR products for amplification of full-length biotin carboxylase by end-to-end PCR. Note: M: 1 kb DNA ladder, lanes 1–2 amplified products for end-to-end PCR. Adapted from Wan Saridah et al., 2008.

and followed by the full length cDNA coding for oil palm biotin carboxylase (Fig. 4.13) (Wan Saridah et al., 2008).

Transferases

Isolation and characterization of genes coding for proteins in the Kennedy pathway were also carried out. cDNA clones coding for lysophosphatidic acid acyltransferase (LPAAT) and glycerol 3-phosphate transferase (GPAT) have been isolated (Manaf et al., 2005). Functional study of LPAAT in transgenic *Arabidopsis* was carried out; however, the effect on the plant fatty acid content is still not known (Safiza et al., 2009b).

β-ketothiolase

β-ketothiolase is an important enzyme in isoprenoid metabolism and PHB biosynthesis. A full length cDNA coding for β-ketothiolase has been obtained through RT-PCR and RACE techniques. The clone contains an open reading frame that encodes a protein of 415 amino acids. The lack of peptide signal within the sequence suggests the protein to be cytosolic. The expression of the gene was indicated to be higher in ripe than young fruits (Fig. 4.14) (Teen et al., 2008). The gene was subsequently used in an intervention strategy by replacing the bacterial gene with it for the production of biodegradable plastic in oil palm.

Carotenogenic Genes

The potential health benefits of carotenoids as anticancer and antioxidant agents have been demonstrated. In particular, lycopene and β-carotene have been shown to reduce

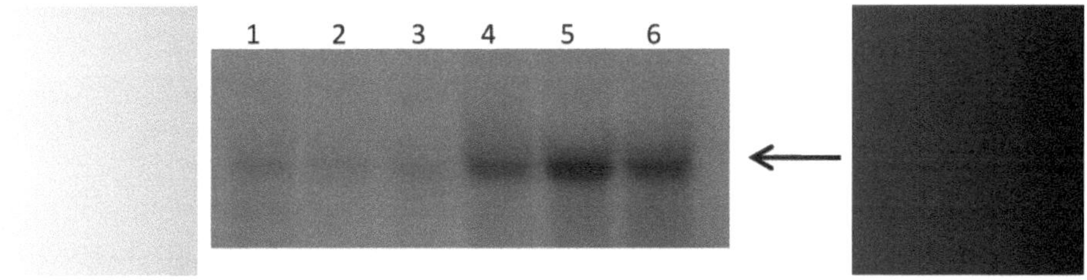

Fig. 4.14. Northern analysis for β-ketothiolase cDNA in oil palm with 2 weeks' exposure. A signal of ~1.5 kb (arrow) was detected in all samples. Lanes 1–6 represent 10 ug total RNA from 6, 8, 11, 13, 17, 20 WAA mesocarp, respectively. Adapted from Teen et al., 2008.

the risk of chronic coronary heart disease, certain cancers, and macular degradation. Crude palm oil is one of the richest natural resources of carotenoids; the current oil from commercial planting materials contains around 500–700 ppm carotene. This high content provides an opportunity for genetic manipulation, especially to increase lycopene at the expense of carotene as lycopene is a more potent antioxidant (Parveez et al., 2003; Sambanthamurthi et al., 2002).

For genetic modification purposes, to date, the cDNAs coding for phytoene synthase (Rasid et al., 2008; Wan Nur Syuhada, 2006), phytoene desaturase (Rasid et al., 2006), zeaxanthin epoxidase (Rasid et al., 2005, 2007a), lycopene ε-cyclase (Rasid et al., 2007b), and lycopene β-cyclase (Rasid et al., 2007b) have been isolated and characterized. The levels of these cDNA transcripts were shown to be correlated to the accumulation of carotenoids in developing fruits (Fig. 4.15). In

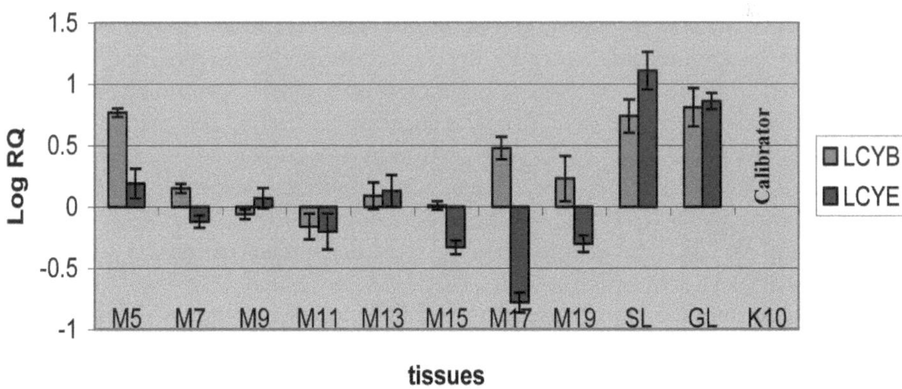

Fig. 4.15. Expressions of oil palm lycopene β-cyclase (*lcyb*) and lycopene ε-cyclase (*lcye)* genes in developing mesocarp tissues of different ages (M5-M19), Spear (SL) and green (GL) leaves relative to kernel (K10) obtained from real-time PCR analysis. The expression was in fold ($\log_{10}$ RQ).

addition, isolation of cDNAs coding for 1-deoxy-D-xylulose 5-phosphate synthase (GenBank Accession no. AY583783 and AY611205; Khemvong & Suvachittanont, 2005) and 1-deoxy-D-xylulose 5-phosphate reductase (GenBank Accession no. AY583783 and AY611205; Khemvong & Suvachittanont, 2005) have also been reported.

In addition to modification of carotenoids, there is also the possibility of further increasing the content of carotenoids. *E. oleifera* produces oil with higher carotenoids than *E. guineensis,* that is, up to 4,000 ppm (Choo et al., 1997; Mohd Din et al., 2004; Mohd Din et al., 2000). This may be due to a higher activity of *psy* in *E. oleifera.* Therefore, overexpressing the *psy* gene from *E. oleifera* may increase carotenoid content in *E. guineensis.* Currently, a full coding sequence of *E. oleifera psy* has been obtained (Rasid et al., 2009) and could be used to confirm the high activity of *E. oleifera psy* hypothesis.

Promoter Isolation

In addition to isolation and characterization of genes and enzymes in the oil palm fatty acid and carotenoid biosynthetic pathways, focus has also been given to isolation of regulatory sequences. The availability of promoter sequences is an absolute prerequisite for any genetic engineering work. Promoters with spatial and temporal specificities are particularly important for effective targeting of transgenes into specific tissues. Promoters with spatial specificity are important to ensure optimal timing of the transgene expression.

Mesocarp Specific Promoter

As the target of genetic engineering in palm oil, which is produced in the mesocarp, the genes to be engineered, either by up-regulation or down-regulation, will need to be targeted into the above tissue using a mesocarp specific promoter. Isolation of mesocarp-specific promoter (MT3-A) was successfully carried out (Siti Nor Akmar et al., 2002; Siti Nor Akmar et al., 2001). An ethylene-responsive element in reverse orientation (EREreverse) and two I boxes were identified at positions -317, -603, and -943, respectively (Siti Nor Akmar & Zubaidah, 2008). The ERE is similar to the ERE and AATTCAAA of the ripening-specific *E4* gene of tomato (Montgomery et al., 1993) and the ERE and ATTTCAAA associated with regulation of the carnation glutathione-S-transferase gene (*GST*) in senescing tissues. I-box was also found in the fruit-specific promoter of strawberry (Agius et al., 2005) and other light-regulated genes (Terzaghi & Cashmore, 1995). The specificity of the promoter was later confirmed using transient expression of *GUS* reporter and PHB genes in mesocarp tissues (Omidvar et al., 2008).

Root-specific Promoter

Using substractive hybridization, another metallothionein cDNA clone (MT3-B) weakly expressed in the mesocarp and root tissues was obtained (Siti Nor Akmar et al., 2002). It has no significant homology with the reported promoter sequence of *MT3-A*. *MT3-B* contains the putative metal-responsive and root-specific elements. The differences in the promoter sequences *MT3-A* and *MT3-B*, together with having different regulatory elements, may be responsible for the distinctly different expression profiles of the two closely related genes. The promoter demonstrated low GUS expression in transgenic *Arabidopsis* roots. Induction with different metal significantly increased the expression in transgenic *Arabidopsis* root (Zubaidah & Siti Nor Akmar, 2005).

Leaf-specific Promoter

There are certain value-added products, such as biodegradable plastics or gene for insect resistance, that will be targeted to the leaves and will hence require leaf-specific promoters. A leaf-specific gene, coding for light-harvesting chlorophyll A/B binding protein (LHCB) gene, was amplified using a RT-PCR approach. Its leaf-specificity was confirmed by Northern blot analysis. Higher expression was detected in the young and mature green leaves compared to yellowish spear leaves. No expression was detected in non-photosynthetic tissues. A genomic clone of 1.0 kb corresponding to LHCB promoter was amplified using the genome walking approach (Chan et al., 2008).

Constitutive Promoters

Besides tissue-specific promoters, constitutively expressing promoters were also isolated for expressing the marker genes or other genes that require constitutive expression. A number of putative EST clones were shown to be constitutively expressed in all oil palm tissues by reverse northern analysis (Masura & Parveez, 2007). Reverse Northern and Northern analyses were carried out to further confirm their constitutive nature. One of them was a clone that codes for the ubiquitin extension protein (UEP1). Based on sequence information of the clone, the promoter region was amplified using the genome walking approach. The promoter contains several motifs, suggesting its potential multiple roles in hormonal signaling pathways and physical and abiotic stresses. The constitutive promoter was verified by a transient expression study using *gusA* gene (Fig. 4.16) (Masura et al., 2010). Another oil palm constitutive promoter derived from the translationally controlled tumor protein (TCTP) gene was later isolated. The TCTP promoter showed constitutive expression of *gusA* reporter gene in monocot and dicot systems (Fig. 4.17) (Masura et al., 2011).

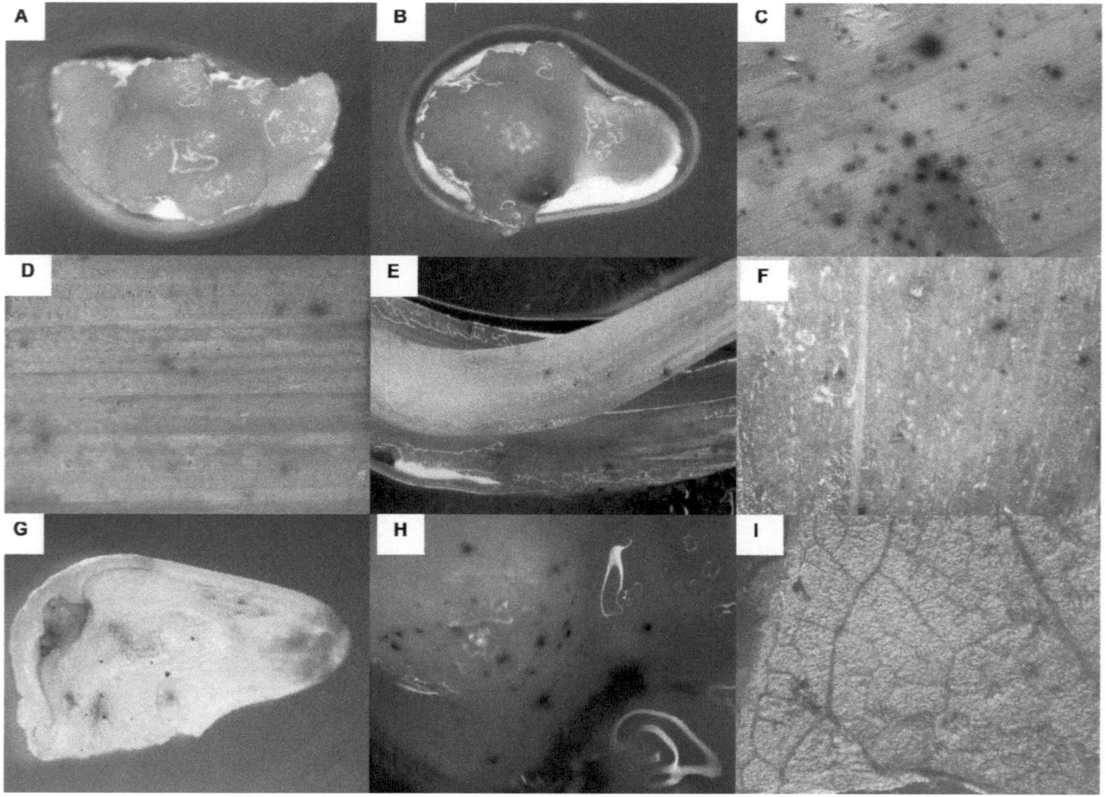

Fig. 4.16. Comparison of transient histochemical *gusA* gene expression in various oil palm tissues and tobacco leaf after bombardment with plasmid carrying UEP1 promoter. Embryogenic calli bombarded (A) without plasmid DNA and (B) with pUEP1 (UEP1). UEP also shows transient expression in (C) young leaves, (D) green leaf, (E) stem, (F) mesocarp, (G) immature embryo and (H) embryoids of oil palm, and (I) tobacco leaf. Source: Parveez et al., 2011.

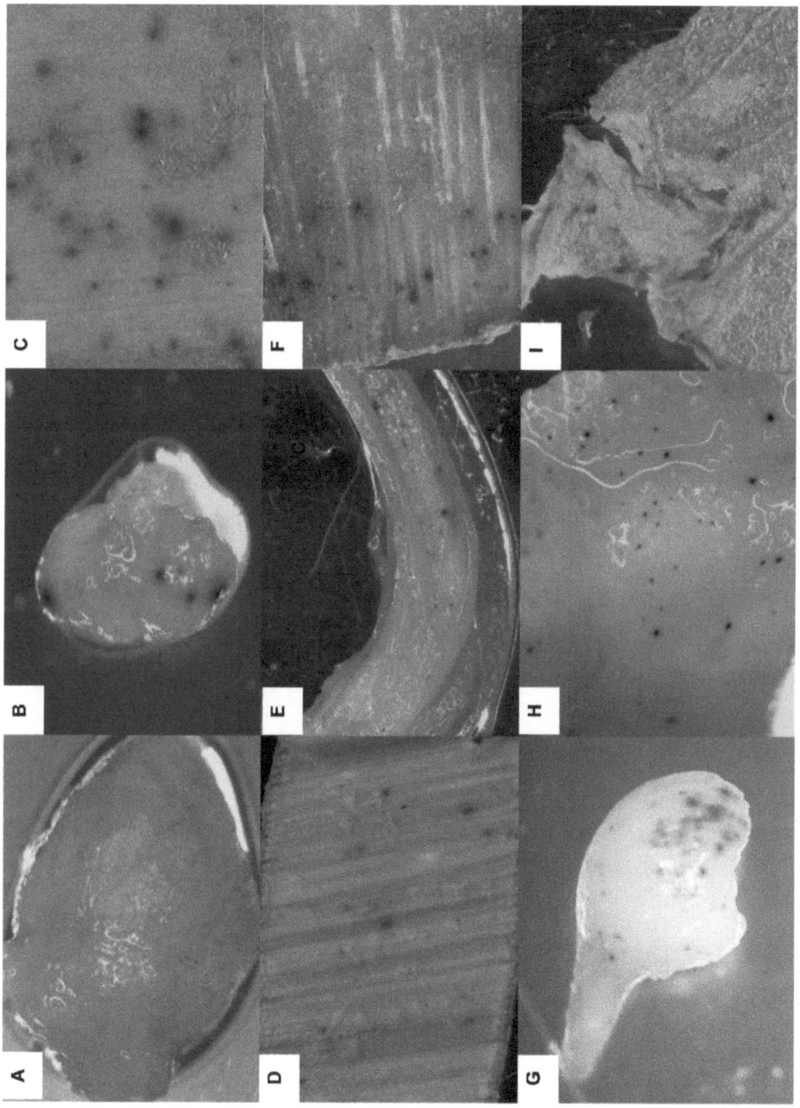

Fig. 4.17. Comparison of transient histochemical *gusA* gene expression in various oil palm tissues and tobacco leaf after bombardment with plasmid carrying TCTP promoter. Embryogenic calli bombarded (A) without plasmid DNA and (B) with pTCTP (TCTP). TCTP also shows transient expression in (C) young leaves, (D) green leaf, (E) stem, (F) mesocarp, (G) immature embryos and (H) embryoids of oil palm, and (I) tobacco leaf. Source: Parveez et al., 2011.

Transgenic Oil Palm Development

Biolistic-mediated Transformation

Biolistics-mediated transformation was initially chosen as the method to produce transgenic oil palm. The system was initiated by optimizing the biological and physical parameters affecting DNA delivery into target tissues and followed by identifying the best promoter and selection agent for producing transgenic oil palm (Fig. 4.18) (Chowdhury et al., 1997; Parveez et al., 1996, 1997, 1998). Using the optimized parameters, the bombarded embryogenic calli were selected on Basta until resistant embryogenic calli were obtained and regenerated into whole plants (Figs. 4.19 and 4.20). The transgenic status was confirmed by molecular and protein analyses (Fig. 4.21) (Parveez, 2000). This method was later routinely used to produce transgenic oil palm with genes of interest.

Agrobacterium*-mediated Transformation*

The *Agrobacterium tumefaciens*-mediated transformation was later established after optimizing a number of parameters (Dayang Izawati et al., 2009). In this experiment, *Agrobacterium* strain LBA4404 carrying a plasmid with Basta-resistant gene driven by

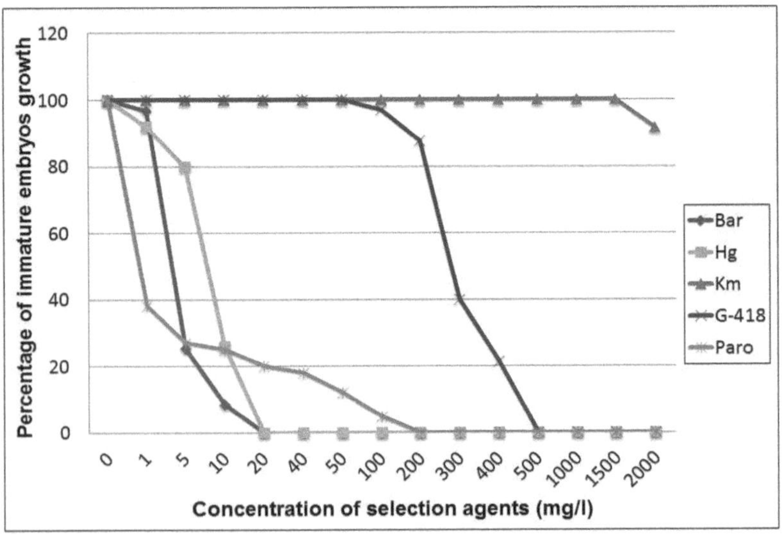

Fig. 4.18. Determination of optimum concentration of selection agents on oil palm immature embryos. Source: Parveez et al., 2007.

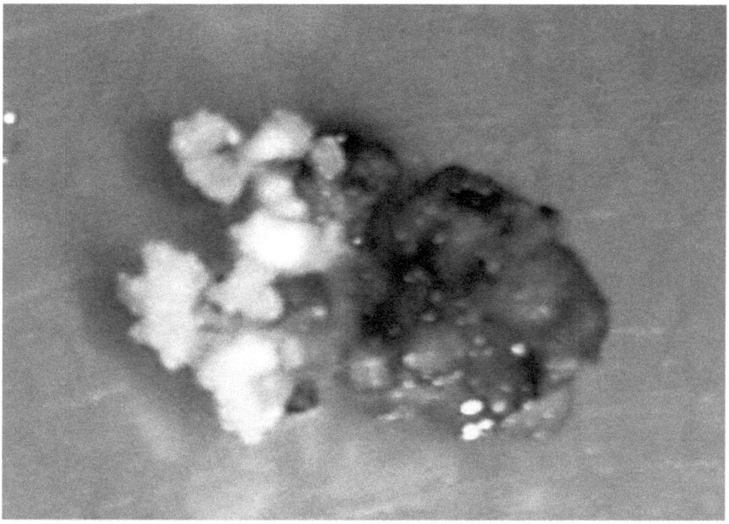

Fig. 4.19. Proliferation of Basta-resistant embryogenic calli on medium containing Basta. Source: Parveez et al., 2011.

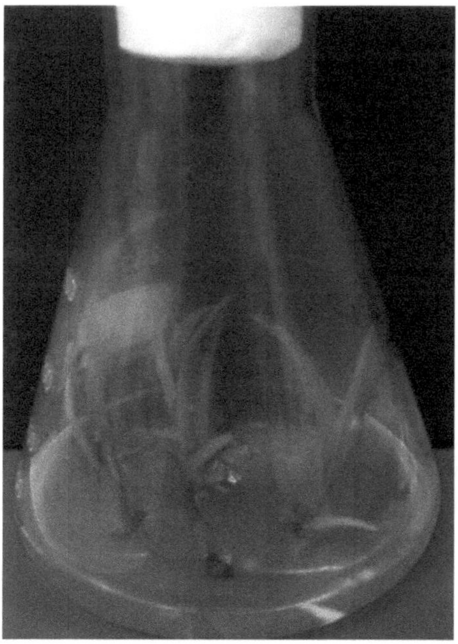

Fig. 4.20. Regeneration of transgenic oil palm from Basta-resistant embryogenic calli. Source: Parveez et al., 2011.

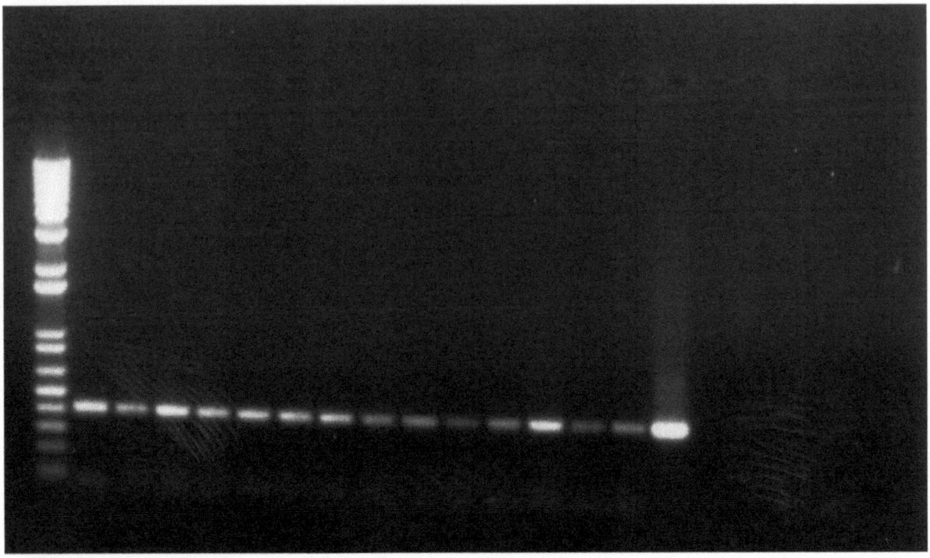

Fig. 4.21. Molecular analysis (PCR) to prove integration of transgenes in transgenic oil palm genome. Source: Parveez et al., 2011.

a maize ubiquitin promoter was transformed using the optimized conditions. Transgenic palms were eventually regenerated on Basta selection medium and followed by confirmation using molecular analyses (Fig. 4.22).

Containment Studies

The first batch of transgenic palms produced was grown to maturity on soil in huge polybags in the biosafety nursery at MPOB headquarters. However, due to the unnatural conditions in the screenhouse and polybags, their growth was retarded, taking six years before male flowering and a further two years before the first female flowers appeared. The female flowers were bagged before anthesis and after controlled pollination; the fruits were allowed to develop in bags. The biosafety screenhouse is covered with size 50 stainless steel mesh netting and fixed with a double door to prevent movement of insects and small animals during laboratory personnel entry/exit (Fig. 4.23). Entry is only for authorized personnel. All parts of the transgenic palms removed were collected and autoclaved prior to appropriate disposal. A new screenhouse with increased capacity was later constructed to plant the transgenic plants under near natural growth conditions (Fig. 4.24). The screenhouse has a sprinkler system for irrigation and the palms will be planted in a completely randomized design.

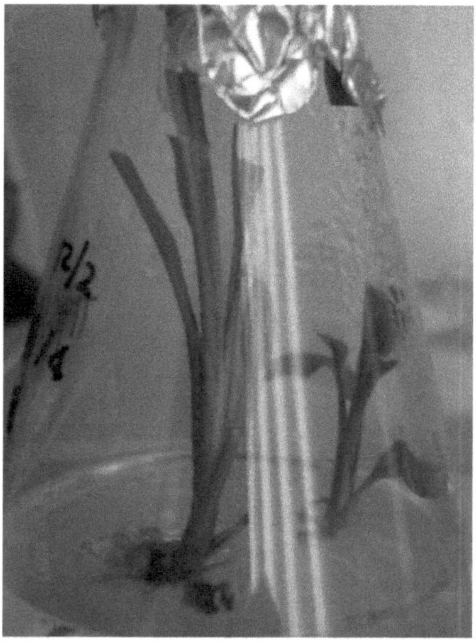

Fig. 4.22. Regeneration of transgenic oil palm after *Agrobacterium*-mediated transformation. Source: Dayang Izawati et al., 2009.

Fig. 4.23. Biosafety screenhouse with double doors at MPOB HQ. Source: Parveez et al., 2011.

Fig. 4.24. Newly constructed biosafety screenhouse at MPOB. Source: Parveez et al., 2011.

Production of Value-Added Transgenic Oil Palm

High Oleate Oil Palm

Oleic acid was targeted to be increased for use as feedstock for the oleochemical industry. In order to perform the above, MPOB first sought to understand why palm oil has a fatty acid composition of 44% palmitic and 39% oleic, using common plant fatty acid biosynthesis pathway (Fig. 4.25) as a guide. Based on the pathway, the following enzymes were studied: β-ketoacyl-ACP synthase II (KAS II), palmitoyl-ACP thioesterase, stearoyl-ACP desaturase, and oleoyl-CoA desaturase. Using the biochemical knowledge, the following was proposed: (1) KAS II be up-regulated, (2) palmitoyl-ACP thioesterase down-regulated, and (3) oleoyl-CoA desaturase may also need to be down-regulated avoiding oleic acid from overflowing to linoleic and linolineic acids (Sambanthamurthi et al., 2000). The cDNAs and genes for the enzymes were later isolated. Furthermore, in order to target genetic modification in the mesocarp (fruit), isolation of mesocarp-specific promoter was also carried out. More than one mesocarp-specific promoter may be needed if more than one gene is to be inserted to avoid homology-dependent gene silencing (Matzke et al., 1989).

It was demonstrated that by down-regulating oleoyl-CoA desaturase, oleic acid content of soybean oil was increased from 21.5 to 78.9% as a result of reduction in linoleic from 55 to 3% (Broglie et al., 1997). Similarly, in cottonseed oil, oleic

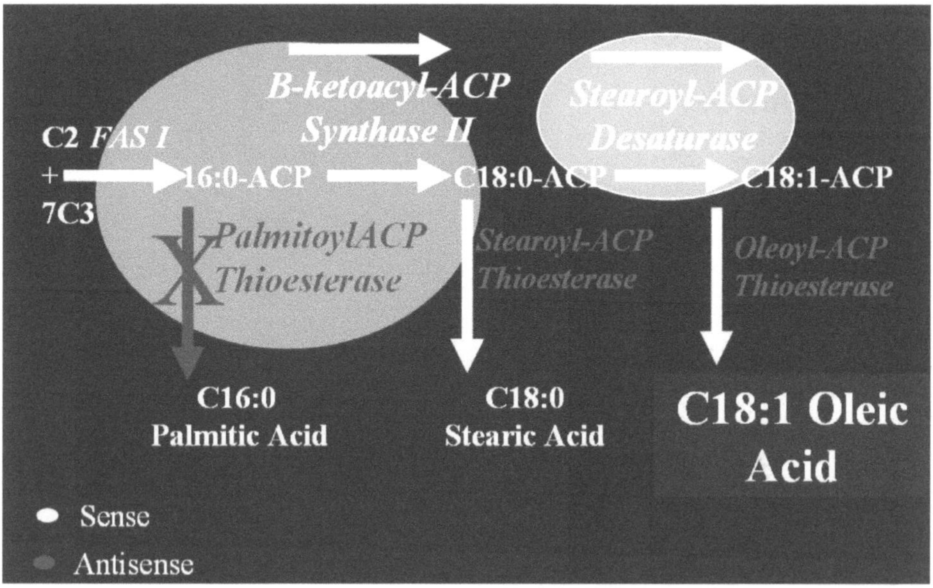

Fig. 4.25. Possible reactions involved in production of high oleic oil transgenic oil palm. Source: Parveez et al., 2011.

acid content was increased from 15 to 78% using the same gene (Liu et al., 2000). For soybean and cotton seed, oleoyl-CoA was down regulated because of their high linoleic acid content. However, for oil palm, as palmitic acid is the dominant fatty acid, down-regulation of palmitoyl-ACP thioesterase and up-regulation of KASII is essential. As oil palm stearoyl-ACP desaturase is very effective, up-regulation of the gene may not be needed (Sambanthamurthi et al., 1996a, 2002).

Four sets of transformation vectors were constructed based on the proposed strategy: (1) antisense palmitoyl-ACP thioesterase gene driven by CaMV35S promoter (Abrizah et al., 2000), (2) antisense palmitoyl-ACP thioesterase and sense KAS II genes driven by mesocarp-specific promoter (Masani & Parveez, 2008), (3) antisense palmitoyl-ACP thioesterase, sense KAS II and sense stearoyl-ACP desaturase genes driven by a mesocarp-specific promoter (Masani & Parveez, 2008), and (4) antisense palmitoyl-ACP thioesterase, antisense oleoyl-CoA desaturase, sense KAS II and sense stearoyl-ACP desaturase genes driven by a mesocarp-specific promoter (Fig. 4.26) (Abdul Masani, Pers. Comm., 2009). Constructs were bombarded into oil palm embryogenic cultures and a few hundred events were obtained after herbicide Basta selection. Putative transformed plantlets were transferred to soil in the

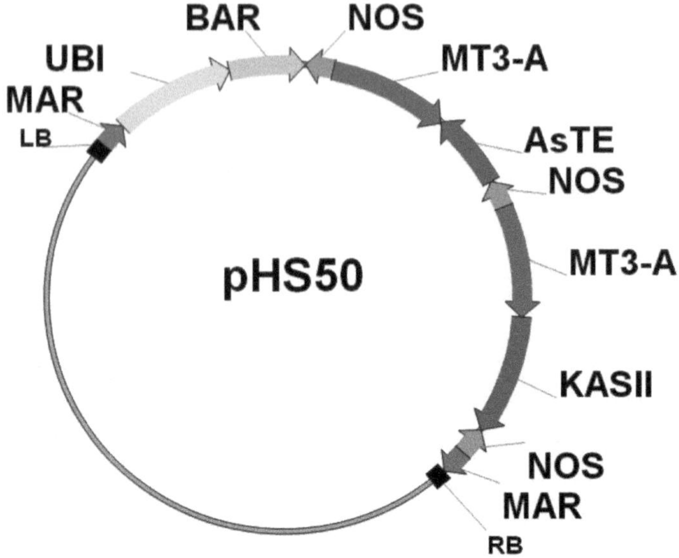

Fig. 4.26. Map of transformation vector carrying high oleic acid genes driven by a mesocarp specific promoter. Figure courtesy of Mr. Abdul Masani, MPOB.

fully-contained biosafety screenhouse (Fig. 4.27). Initial molecular analysis performed on these plantlets revealed that many plantlets were positive for the *bar* and palmitoyl-ACP thioesterase genes (Fig. 4.28) (Parveez et al., 2003a). The PCR positive cultures and plantlets were later subjected to fatty acid analysis for determining possible fatty acid composition changes at the early stage of the plant development.

High Stearate Oil Palm

The second trait of interest is increasing stearic acid content for novel applications, such as production of cocoa butter substitutes and personal care products like lotions, shaving cream, and rubbing oils (Parveez et al., 1999). From biochemical studies, it was reported that oil palm has an active stearoyl-ACP desaturase (Sambanthamurthi et al., 2000), so down-regulating it can reduce conversion of stearoyl-ACP to oleoyl-ACP, which will subsequently increase stearic acid by reducing its conversion to oleic acid (Fig. 4.25). Previous reports showed that down-regulating stearoyl-ACP desaturase in rape could increase stearic acid from 1.8 to 39.8% by weight, due to reduction in oleic acid (Knutzon et al., 1992). Using the same approach, stearic acid content in cottonseed oil was increased from 2 to 40% (Liu et al., 2000). In this study, based on biochemical observations, down-regulating stearoyl-ACP desaturase in the mesocarp was proposed. However, it was anticipated that this would result in a large pool of

Fig. 4.27. Transgenic plantlets in the fully-contained biosafety screenhouse. Source: Parveez et al., 2011.

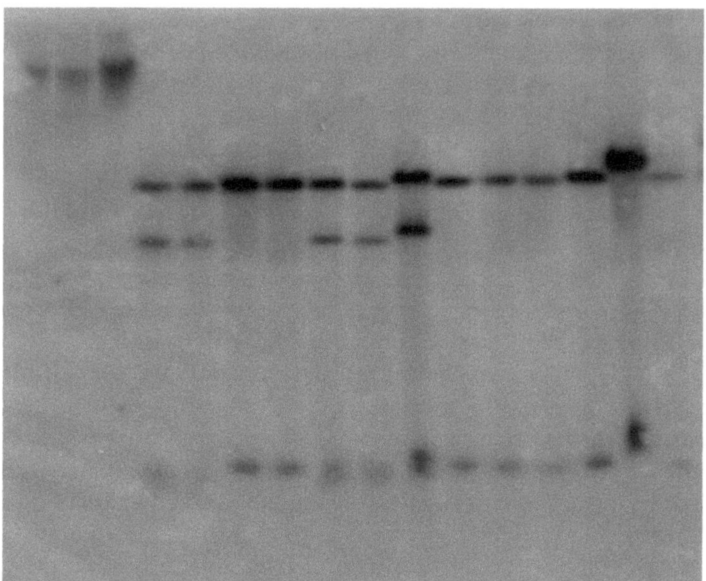

Fig. 4.28. Southern analysis of transgenic oil palm using palmitoyl-ACP thioesterase gene as probe. Source: Parveez et al., 2011.

stearoyl-ACP, which may interfere with the equilibrium of the pathway and result in accumulation of palmitic acid. Therefore, there may be need to manipulate KAS II and palmitoyl-ACP thioesterase as well. Down-regulating palmitoyl-ACP thioesterase and up-regulating KAS II may direct almost the entire stearoyl-ACP pool to stearic acid.

In this study, three transformation vectors were constructed carrying an antisense stearoyl-ACP desaturase gene driven by the ubiquitin, CaMV35S or mesocarp-specific promoters in addition to another construct carrying the above gene and antisense copy of the palmitoyl-ACP thioesterase gene driven by a mesocarp-specific promoter. All the constructs were bombarded into embryogenic cultures and so far resulted in a number of small plantlets obtained and are being grown in a biosafety screenhouse. It would be interesting to look at the fatty acid composition of the palms fruit once they matured. The morphology and growth of the transgenic plants will be analyzed later to study the effect of constitutively increasing stearic acid in the cells. Molecular analysis of the regenerated transgenic palms is being carried out.

Production of Biodegradable Thermoplastics

Polyhydroxybutyrate (PHB), which could be used to produce biodegradable plastic, is the most common polyhydroxyalkanoate (PHA) produced as storage material by bacteria under restricted growth conditions (Senior & Dawes, 1973). PHB is produced from acetyl-CoA by a sequence of three-enzymatic reactions catalysed by β-ketothiolase, acetoacetyl-CoA reductase and PHB synthase (Anderson & Dawes, 1990) (Fig. 4.29). PHA could be exploited to produce a wide range of environmentally friendly industrial polymers. PHB is a stiff and relatively brittle polymer (Holmes, 1988). Its melting point is only 10°C lower than its degrading point, making it a difficult substance to work with. On the other hand, another co-polymer, PHBV, is more crystalline with a larger difference between melting and degrading points.

Initially, large scale production of PHA by bacterial fermentation was carried out but it is not favorable due to high cost of production. Alternatively, production of PHA in plants was proposed as the cost is more reasonable (Poirier et al., 1995). As oil palm has higher acetyl-CoA pool, it is believed to be a viable system for PHA production. This research was carried out under the Malaysia-Massachusetts Institute of Technology (MIT) Biotechnology Partnership Programme, which ended in 2005 (Parveez et al., 1999).

Plant cells are known to contain acetyl-CoA pool in several organelles, such as cytosol, plastid, mitochondria, and peroxisome. Therefore, theoretically, PHA could be synthesized in any of those organelles (Poirier, 2002). Production of PHB in plants was first demonstrated in *Arabidopsis* cytoplasm. However, the amount was very low at 0.1% of the plant dry weight (Poirier et al., 1992). The PHB was produced by only transforming the reductase and synthase genes, as the thiolase gene is available endogenously. Similarly, later up to 0.3% dry weight of PHB was synthesized in the cytoplasm of transgenic cotton carrying the PHB genes driven by fibre-specific promoter

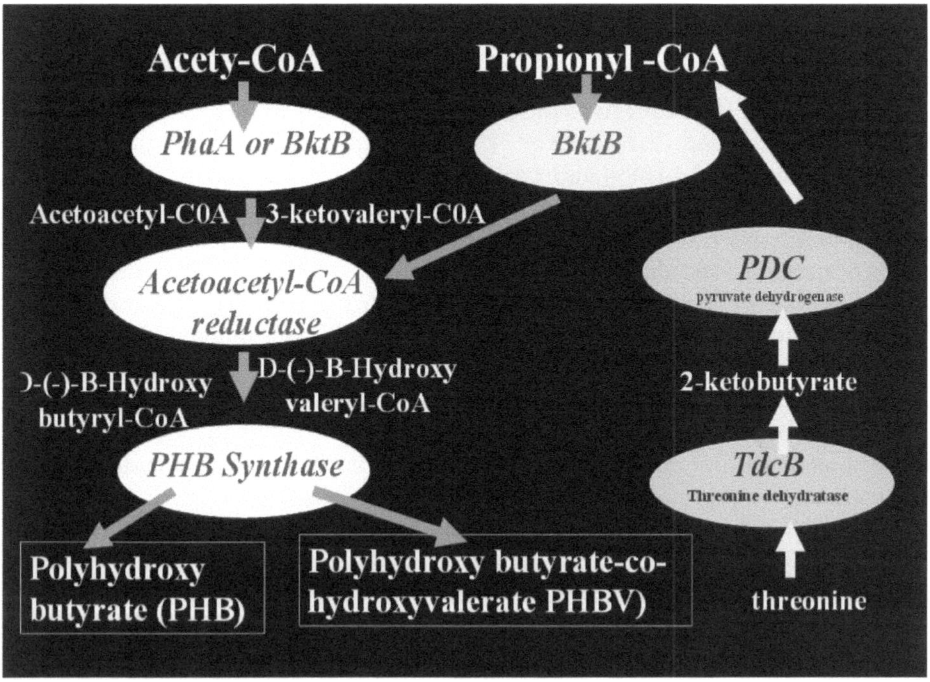

Fig. 4.29. Pathway for the biosynthesis of PHB and PHBV. Source: Parveez et al., 2011.

(John & Keller, 1996). The low yield in those reports was believed to be due to the limited supply of acetyl-CoA in cytoplasm (Nawrath et al., 1994a). This hypothesis was later proven; constructs carrying the three PHB genes were targeted to plastid of *Arabidopsis* where a higher amount of acetyl-CoA pool were available and resulted in the synthesis of up to 14% of the dry weight. The entire PHB granules were contained in the plastids (Nawrath et al., 1994b). However, the yield was still thought to be not optimum as the plants were obtained after crossing three transgenic events carrying different PHB genes. Differential expression levels and possible position effects of the three genes in the genome could have affected the synthesis. Therefore, in the following experiment all three genes were fused on a single plasmid, and PHB was synthesized up to 4% of their fresh weight. This level was postulated to be the maximum that *Arabidopsis* could synthesize (Bohmert et al., 2000).

Besides PHB, production of a copolymer PHBV [poly(3-hydroxybutyrate-co-3-hydroxyvalerate)] was demonstrated in rape and *Arabidopsis* (Slater et al., 1999). An additional enzyme is required for PHBV production, namely threonine dehydratase (*tdcB*) or threonine deaminase (*ilv*), to convert threonine into propionyl-CoA and finally into β-ketovaleryl-CoA (Fig. 4.29). Constitutive expression of *bktB, phbB, phbC,* and *ilv* genes into *Arabidopsis* plastid resulted in 1.6% of dry weight PHBV synthesized. In a separate experiment, using a seed-specific promoter and plastid targeted, up to 2.3% dry weight of PHBV was synthesized in rape.

Synthesis of PHB and PHBV constitutively or targeted to mesocarp and leaves of oil palm was proposed (Parveez et al., 1999). An oil palm transit peptide from the ACP gene (Rasid et al., 1999) and a tobacco RB7 matrix attachment region (MAR) (Matzke & Matzke, 1991) were also included in the constructs for plastid targeting and stabilizing the transgene expression, respectively. MAR is essential, as multiple copies of mesocarp-specific or leaf-specific promoters were used to drive three or four genes for PHB or PHBV, respectively, for minimizing homology dependent gene silencing (Matzke et al., 1989). For production of PHBV, a propionyl-CoA pool will be created using the threonine *dehydratase* (*tdcB*) gene from *E. coli* (Guillouet et al., 1999). In this study, mesocarp was targeted because this oil-synthesizing tissue is expected to have the most abundant acetyl-CoA pool. Alternatively, leaf was also targeted as it will not compete with oil synthesis; however, lower accumulation of polymers is to be expected due to acetyl-CoA availability.

A total of eight transformation vectors carrying the PHB and PHBV genes driven by the ubiquitin promoter alone (Willis et al., 2008), a combination of constitutive promoters (actin, ubiquitin and CaMV35S) (Masani et al., 2001, 2003), mesocarp-specific promoter (Masani et al., 2008), and leaf-specific promoters (Masani et al., 2009) were designed, constructed, and transformed into oil palm (Fig. 4.30). Currently, a few hundred putative transgenic plantlets have been produced and transferred to soil in the biosafety screenhouse. Initial PCR analysis revealed the presence of bar and PHBV/PHBV gene in some of the plantlets (Fig. 4.31). The PCR positive cultures and plantlets were later subjected to PHB/PHBV analysis via gas chromatography to examine the synthesis of PHB/PHBV (Parveez et al., 2008).

Modification of Carotenoid Content

A general schematic for synthesis of plant carotenoids is given in Fig. 4.32. Descriptions and reviews of the pathway are available in a number of publications and are referred to (Cunningham & Gantt, 1998; Frazer & Bramley, 2004; Hirschberg, 2001). Since phytoene sythase (PSY) catalyzes the first committed step in the plant carotenoid pathway and is believed to be rate limiting, it has been a preferred target for genetic manipulation, especially in the early work. Upregulating *psy* has been shown to increase total carotenoids in plants, including tomato (Frazer et al., 2002), *Brassica napus* (Shewmaker et al., 1999), and potato (Ducreux et al., 2005). For example, overexpression of the *E. uredovera* phytoene synthase gene in *Brassica napus* resulted in a 50-fold increase in its total carotenoid content, mainly in α-carotene, β-carotene, and phytoene. In addition to *psy*, several other carotenogenic genes have also been upregulated. Upregulation of phytoene desaturase (*pds*) in tomato resulted in 2- to 4-fold increases in β-carotene (Ronen et al., 2000). The transformants produced orange

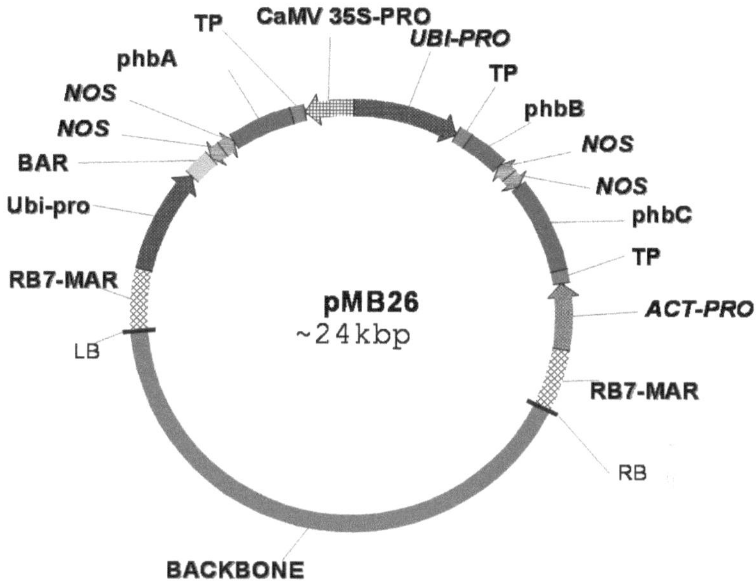

Fig. 4.30. Map of transformation vector for synthesis of biodegradable plastics in oil palm. PHB genes driven by combination of three constitutive promoters. Ubi-pro: maize ubiquitin promoter, CaMV 35S-PRO: cauliflower mosaic virus 35S promoter, ACT-PRO; rice actin promoter, TP: oil palm ACP transit peptide, LB: left border, RB: right border. Adapted from Masari et al., 2008.

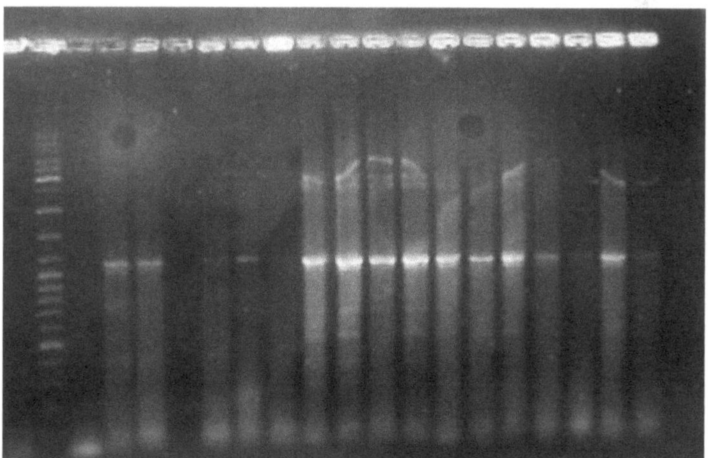

Fig. 4.31. PCR analysis demonstrating integration of *phaB* gene in transgenic oil palm samples. Source: Parveez et al., 2011.

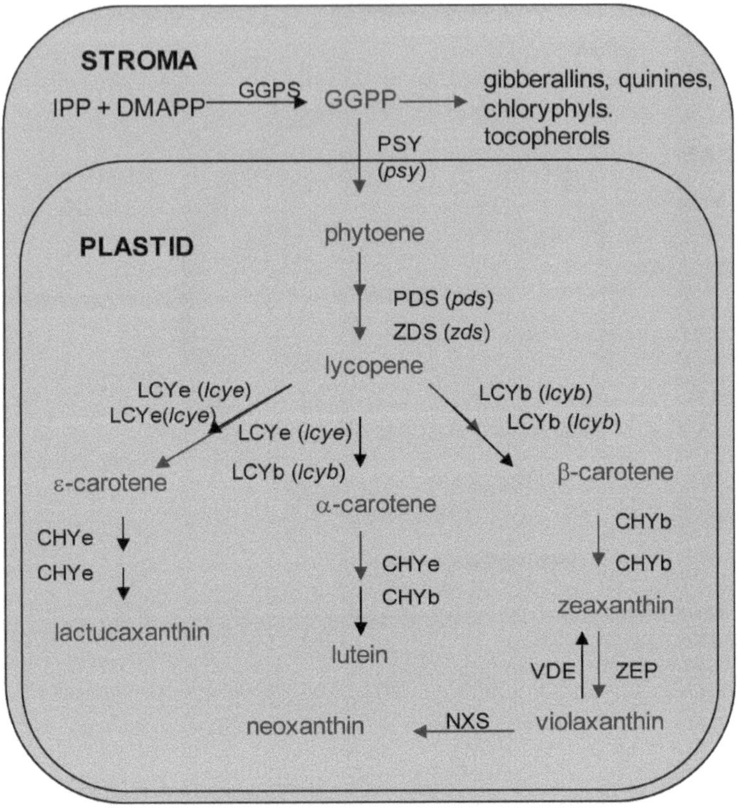

Fig. 4.32. Schematics of the carotenoid synthetic pathway in plants. The first committed synthesis is of phytoene from condensation of two GGPP molecules. Phytoene is then converted to lycopene through a series of desaturation steps. In most plants, lycopene will branch into α- and β-carotenes. However, in lettuce most of the lycopene is converted to ε-carotene and, ultimately, lactucaxanthin. PSY, phytoene synthase; PDS, phytoene desaturase; ZDS, zeta carotene desaturase; LCYb, lycopene β-cyclase; LCYe, lycopene ε-cyclase, CHYb, carotene β-hydroxylase; CHYe, carotene ε-hydroxylase; ZEP, zeaxanthin epoxidase; VDE, violaxanthin de-epoxidase; NXS, neoxanthin synthase. The gene for each enzyme is given in parentheses next to the enzyme. Adapted from Cunningham (2002).

fruits. Interestingly, upregulation of lycopene β-cyclase (*lcyb*) also resulted in a 5-fold increase in β-carotene in transgenic tomato (Romer et al., 2000) but with unchanged total carotenoids.

These results clearly indicate that the plant carotenoid pathway is amenable to genetic engineering. This obviously provides opportunities for improvement of the nutritional value of various food plants. It also opens the possibility for metabolic engineering of the compounds further downstream in the carotenoid pathway, for which naturally occurring mutants or genotypes are not available. Modification of the oil palm carotenoid content, especially to increase lycopene, is one of the main targets of the oil palm genetic engineering program (Parveez et al., 2003a; Parveez & Na'imatulapidah, 2008; Sambanthamurhti et al., 2002).

The carotenoids content in palm oil, which are predominantly α- and β-carotenes, clearly indicates that lycopene is mostly converted to its derivatives as the two lycopene cyclases are very active in oil palm fruits. Therefore, an obvious strategy to increase lycopene is to block its conversion to carotenes (Parveez et al., 2004; Sambanthamurtni et al., 2002). This can simply be done by silencing the two cyclase genes either by antisensing or RNAi.

In addition to modifying the carotenoid components in palm oil, there is also the possibility of increasing the total carotenoid content. The mesocarp oil of *Elaeis oleifera*, the South American oil palm, contains higher carotenoids than *E. guineensis* mesocarp oil (Choo et al., 1997; Mohd Din et al., 2004; Mohd Din et al., 2000). This could be due to its higher *psy* activity. Therefore, using *E. oleifera psy* may increase carotenoid accumulation in *E. guineensis* oil. In this study, a transformation construct containing the antisense copy of both oil palm cyclases driven by the MSP1 promoter was generated from two intermediate vectors. The resulting clone was designated pHL64 (Fig. 4.33). The construct was used to transform oil palm calli. To

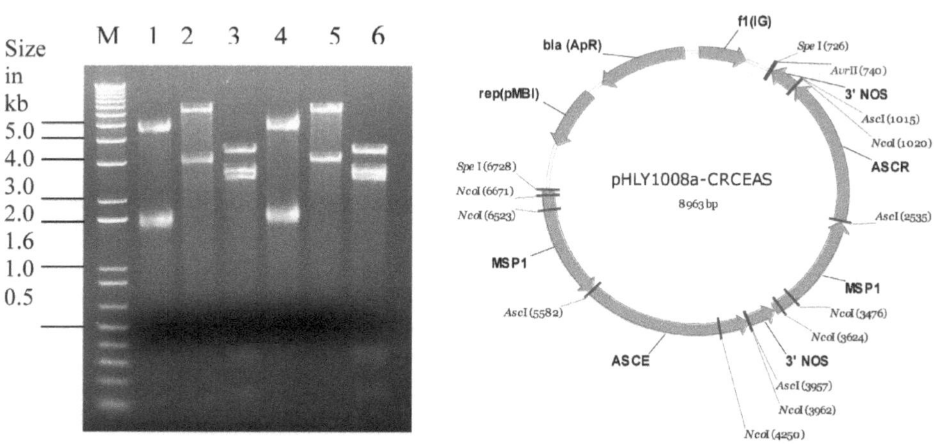

Fig. 4.33. Confirmation of clones (pHL64) containing antisense copy of both lycopene cyclases driven by MSP1 promoter. The clones were digested with Asc I, Spe I, and Nco I. Lanes 1–3 are pHL64G and lanes 4–6 pHL64H. ASCE is antisense lycopene ε-cyclase and ASCR antisense lycopene β-cyclase. Source: Parveez et al., 2011.

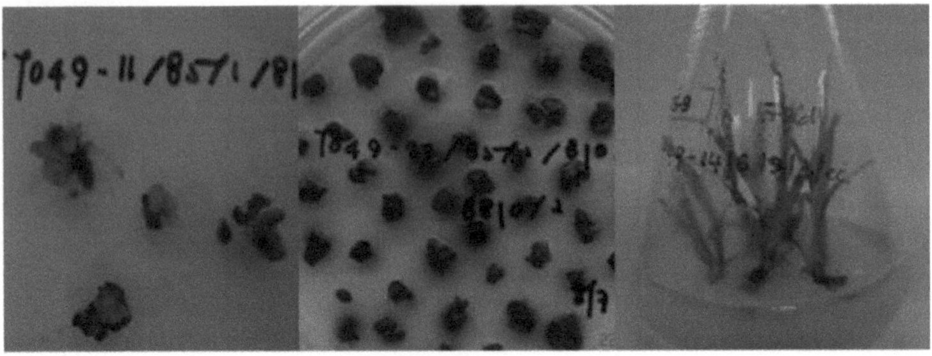

Fig. 4.34. Oil palm calli bombarded with pHL64 cultured on medium containing Basta. A number of resistant clumps were obtained from the bombarded tissues. A) proliferating resistant callus clumps, B) control non-resistant tissues, and C) shoots regenerated form resistant calli. Source: Parveez et al., 2011.

date, a few transformed plantlets were produced and are undergoing root induction (Fig. 4.34).

High Palmitoleate Oil Palm

Palmitoleic acid is an important fatty acid for pharmaceutical applications. It is postulated to have anti-thrombotic effects, which can help prevent stroke (Abraham et al., 1989). At present, it is mainly obtained from Macadamia oil (*Macadamia integrifolia*), which contains ~17% palmitoleic acid. A previous study by PORIM on oil palm protoplasts showed that *E. guineensis* protoplasts can synthesise up to 30% palmitoleic acid in their total lipids (Sambanthamurthi et al., 1996b). This illustrates the palms' inherent ability to produce high levels of this fatty acid.

Palmitoleic acid is produced by desaturation of palmitic acid. It is envisaged that Δ9-stearoyl-ACP desaturase, which acts mainly on stearic acid, can also use palmitic acid as substrate to produce palmitoleic acid. Several mutant stearoyl-ACP desaturases with increased specificity towards palmitoyl-ACP have been created and have been demonstrated to drive high levels of palmitoleic acid synthesis in *Arabidopsis* (Cahoon & Shanklin, 2000; Cahoon et al., 1997). For producing palmitoleic acid in oil palm, a mutant castor Δ9-stearoyl-ACP desaturase will be overexpressed into oil palm.

Seven transformation vectors carrying stearoyl-ACP desaturase driven by the CaMV35S promoter, a double mutant castor desaturase driven by the ubiquitin, CaMV35S and mesocarp-specific promoters, single mutant castor desaturase driven by the ubiquitin and mesocarp-specific promoters, and a single mutant desaturase together with antisense palmitoyl-ACP thioesterase driven by a mesocarp-specific promoter (Fig. 4.35) (Masani, personal communication, 2008) have been constructed.

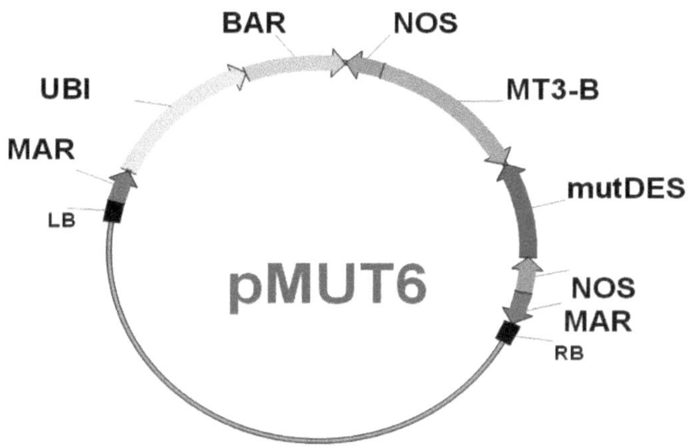

Fig. 4.35. Map of transformation vector carrying high palmitoleic acid gene driven by the mesocarp specific promoter. Figure courtesy of Mr. Abdul Masani, MPOB.

These constructs have been transformed into oil palm calli, and Basta-resistant embryogenic calli has already been obtained. Regeneration of resistant calli is being done with some transgenic plantlets already obtained for majority of the constructs used. Some of the plantlets are already transferred to soil in the biosafety screenhouse.

High Ricinoleate Oil Palm

Ricinoleic acid (D-12-hydroxyoctadec-*cis*-9-enoic acid) is an important industrial fatty acid. It is hydroxylated and a constituent of seed storage oil in at least 12 genera from 10 families of higher plants (Van de Loo et al., 1993). Ricinoleic acid can be used to produce cosmetics (~40% lipsticks), plastics (mainly nylon 11, nylon 10-10, nylon 6-10), surfactants, deodorants, plasticizers, diesel lubricants and coatings due to its effective drying properties (McKeon, 2009). The hydroxyl group in ricinoleate makes it a good lubricant and good/double effective as biodiesel. The agriculturally important source of ricinoleate is castor oil (*Ricinus communis* L.). In this plant, ricinoleic acid is restricted to seed triacylglycerides and constitutes 85–90% of the fatty acids (Atsmon, 1989).

Oleic acid is converted to ricinoleic acid by oleate 12-hydroxylase. The LFAH12 gene encoding for oleate 12-hydroxylase from *L. fendleri* has been isolated and characterized (Broun et al., 1998). Analysis of the fatty acid composition of oil produced by transgenic plants expressing the *L. fendleri* gene indicates that the enzyme exhibits both desaturase and hydroxylase activity. A cDNA encoding oleate 12-hydroxylase from castor bean was later isolated and transformed into *Arabidopsis thaliana* under control of the *Brassica napus* napin promoter. The resulting transgenic plants accumulate ricinoleic acid up to 17% of the seed fatty acids (Broun & Somerville, 1997). Further investigation indicated that co-transformation of castor oleate 12-hydroxylase with diacylglycerol acyltransferase 2 (DGAT2) could improve the ricinoleic acid content in transgenic *Arabidopsis* seeds (Burgal et al., 2008). Therefore, introduction of castor oleate 12-hydroxylase into oil palm should result in synthesis of ricinoleic acid.

Three transformation vectors carrying oleate 12-hydroxylase from castor bean driven by ubiquitin, CaMV35S and mesocarp-specific promoters were assembled (Fig. 4.36) (Masani, personal communication, 2008). Transformation of these constructs into oil palm cultures is being done, and Basta-resistant embryogenic calli are already obtained. Regeneration of resistant calli is being done with some transgenic plantlets already obtained. Some of the plantlets are already transferred to soil in the biosafety screenhouse (Fig. 4.37).

Future Trends in Tissue Culture and Genetic Engineering of Oil Palm

In striving to achieve an overall elevated productivity, one must first look at improvements that tackle core issues such as ortet selection. Members of the industry are urged to intensify their breeding programs for ortet selection and comply with Malaysian Standard (MS). Policies to enable cloning and recloning of elite palms should be implemented to encourage large scale production. The industry can also exploit the *in vitro* technology by cloning the parental palms to produce clonal seeds. This is a source of more affordable quality planting materials. The reasonably priced planting materials

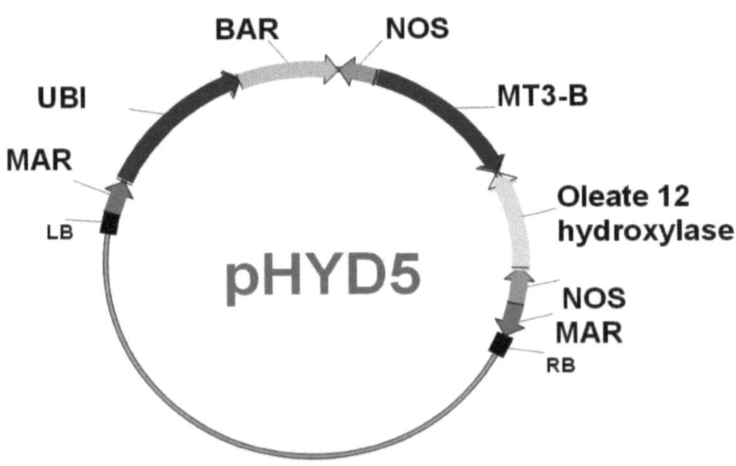

Fig. 4.36. Map of transformation vector carrying oleate 12 hydroxylase gene driven by the mesocarp specific promoter. Figure courtesy of Mr. Abdul Masani, MPOB.

Fig. 4.37. Transgenic oil palm plantlets carrying genes for ricinoleic acid in biosafety screen-house. Source: Parveez et al., 2011.

may encourage more participation from all sectors including smallholders leading to a more aggressive replanting program. The high demand for clonal materials would require labs to expand their production and encourages the setting up of new labs. To ensure successful implementation of the strategic plan towards increasing oil yield, full support from various sectors and the government is needed. Clonal planting material is expected to create the greater impact in future yield improvement. With tissue culture labs imposing higher than minimum selection standards, coupled with Good Agricultural Practices (GAP) in plantations, oil yield is expected to increase substantially from existing national average of about 4 tonnes t/ha/yr, towards achieving the national aspiration of Vision 35:25 (35 t/ha/yr fresh fruit bunch and 25% oil extraction rate).

For the genetic engineering research, the current production of transgenic oil palm is mainly derived from transformation using biolistics method, which has been known to routinely introduce multiple copies of the transgenes. This phenomenon may result in gene silencing. Therefore, optimization of the *Agrobacterium*-mediated transformation and subsequently using this developed method as the method of choice for transforming useful genes may be beneficial for producing transgenic oil palm with single or lower copy number which will be easily accepted by the regulators. Concern over the potential hazard of transgenic plants to the environment and human health has resulted in suggestions for creating antibiotic marker-free transgenic crops. Alternative methods have been proposed, such as (1) eliminate the marker genes, (2) replacing the marker genes, and (3) self-contain the transgenes. Eliminating the marker genes could be carried out either by transformation using two plasmids (Komari et al., 1996) or using established systems such as the Multi Auto Transformation Vector (Ebinuma et al., 1997). Alternatively, replacing the antibiotic marker genes with more friendly genes, such as *phosphomannose isomerase* gene (Joersbo et al., 1998) may be carried out. Finally, self-containment of the transgenes could be carried out using readily available transformation system such as chloroplast- or plastid-mediated transformation (Daniell et al., 1998). Additional concern about transgenic plants is the integration of the plasmid, or vector DNA, into the plant genome besides the gene of interest, as demonstrated in direct gene transfer (Kohli et al., 1999) and *Agrobacterium*-mediated transformation (Cluster et al., 1994). Furthermore, studies have shown that transforming rice with three minimal transgene cassettes (promoter, open reading frame, and terminator) demonstrated simple and low-copy-number integration events as compared to the use of supercoiled or linearized plasmids (Fu et al., 2000). The above are some of the approaches available for consideration and are being planned at MPOB. We hope these new approaches would help MPOB to proceed with producing clonal and transgenic plants that are suitable for commercialization.

Conclusions

The first part of this chapter has documented the innovations made in relation to tissue culture process to make it more efficient, time saving, and cheaper. Many innovations have been produced for adoption by the oil palm industry. This is to ensure effort towards large scale planting of clonal palms for increasing yield will be a reality. The second part of this chapter elaborated the biochemical studies on fatty acid biosynthetic enzymes, followed by isolation of the essential genes and promoters from oil palm for producing various products. This was followed by the construction of various transformation vectors carrying combinations of genes, promoters, and other genetic materials for transforming oil palm. Many transgenic oil palm plantlets carrying the above vectors have been produced and are currently growing in the MPOB screenhouses. Initial molecular analysis demonstrated the integration of the transgenes introduced. Further analysis is required to identify the useful and functional events for potential future commercialization.

These transgenic plants are currently planted in screenhouses due to genetically modified [living modified organism (LMO)] status and is in accordance to the Malaysian Biosafety Act 678 (2007). As their safety has not been fully assessed in the screenhouse, they are not allowed to be planted in the open field. The Biosafety Act requires any parties who are interested in releasing any LMO into the environment to apply for an approval from the National Biosafety Board (NBB). The NBB will later get the scientific advice on the safety aspect of the LMO from the Genetic Modification Advisory Committee (GMAC) on each application before making any decision whether to approve or not the release of an LMO. The plants are being grown for proof of concept to ensure they produced the expected or targeted products. Once the plant with the expected product is identified, it will be subjected to safety evaluation prior to field evaluation for amplifying the plants and followed by approval for commercial release for cultivation by farmers.

Transgenic oil palms are being produced for niche market, such as oleochemical industry, lubrication, and nutraceuticals. It is not targeted for used as commodity oil and the current palm oil produced are free from LMO source. It is expected that the LMO-free status of palm oil will remain for at least 15 years due to the long regeneration time as well as time required for yield testing and multiplying the plants for commercialization. The acceptance of LMO-derived products is slow due to no direct benefit to the consumer. However, if the LMO could produce useful consumer products at a higher capacity and at a lower price, it is expected that the acceptance will gradually increase. We hope by the time we are ready for commercialization, the consumer-benefiting products that we are targeting would be more acceptable. However, this remained to be seen.

References

Abraham, R.; Riemersma, R.A.; Wood, D.; Elton, R.; Oliver, M.F. Adipose fatty acid composition and the risk of serious ventricular arrhythmias in acute myocardial infarction. Am. J. Cardiol. 1989, 63, 269–272.

Abrizah, O. Isolation and characterisation of an acyl-acyl carrier protein (ACP) thioesterase from oil palm. PhD thesis submitted to the University of Bristol, UK, 2001.

Abrizah, O.; Lazarus, C.; Fraser, T.; Stobart, A.K. Cloning of palmitoyl-acyl carrier protein thioesterase from oil palm. Biochem. Soc. Trans. 2000, 28, 619–622.

Agius, F; Amaya, I; Botella, M.A.; Valpuesta, V. Functional analysis of homologous and heterologous promoters in strawberry fruits using transient expression. J. Exp. Bot. 2005, 56(409), 37–46.

Anderson, A.J.; Dawes, E.A. Occurrence, metabolism, metabolic role and industrial uses of bacterial polyhydroxylalkanoates. Microbiol. Rev. 1990, 54, 450–472.

ASD (http://www.asd-cr.com/paginas/english/tissue_culture.html)

Astmon, D. Castor. In Oil Crops of the World; Robbelen, G., Downey, K.R., Ashri, A., Eds; New York: McGraw-Hill, 1989; pp. 438–447.

Basiron, Y.; Chan, K.W. The role of research and development strategies in food safety and good agricultural, manufacturing and distribution practices in the Malaysian palm oil industry. Oil Palm Industry Econ. J. 2005, 5(1), 1–16.

Bohmert, K.; Balbo, I.; Kopka, J.; Mittendorf, V.; Nawrath, C.; Poirier, Y.; Tischendorf, G.; Trethewey, R.N.; Willmitzer, L. Transgenic *Arabidopsis* plants can accumulate polyhydroxybutyrate to up to 4% of their fresh weight. Planta 2000, 211, 841–845.

Broglie, R.; Kinney, Y.; Hitz, B.; Knowlton, S. Soyabeen oil with improved functional and nutritional properties. In Proceedings of the Biochemistry and Molecular Biology of Plant Fatty Acids and Glycerolipids Symposium. South Lake Tahoe, USA, 1997.

Broun, P.; Boddupalli, S.; Somerville, C.R. A bifunctional oleate 12-hydroxylase: desaturase from *Lesquerella fendleri.* Plant J. 1998, 13(2), 201–210.

Broun, P.; Somerville, C. Accumulation of ricinoleic, lesquerolic, and densipolic acids in seeds of transgenic *Arabidopsis* plants that express a fatty acyl hydroxylase cDNA from castor bean. Plant Physiol. 1987, 113(3), 933–942.

Burgal, J.; Shockey, J.; Lu, C.; Dyer, J.; Larson, T.; Graham, I.; Browse, J. Metabolic engineering of hydroxy fatty acid production in plants: RcDGAT2 drives dramatic increases in ricinoleate levels in seed oil. Plant Biotechnol. J. 2008, 6(8), 819–831.

Cahoon, E.B.; Lindqvist, Y.; Schneider, G.; Shanklin, J. Redesign of soluble fatty acid desaturases from plants for altered substrate specificity and double bond position. PNAS 1997, 94, 4872–4877.

Cahoon, E.B.; Shanklin, J. Substrate-dependent mutant complementation to select fatty acid desaturase variants for metabolic engineering of plants seed oils. PNAS 2000, 97, 12350–12355.

Chan, P. L.; Siti Nor Akmar, A.; Roohaida, O. Light-harvesting chlorophyll A/B binding protein (LHCB) promoter for targeting specific expression in oil palm leaves. J. Oil Palm Res. 2008, Special Issue on Malaysia-MIT Biotechnology Partnership Programme: Volume 2–Oil Palm Metabolic Engineering, 21–29.

Choo, Y.M.; Ma, A.N.; Yap, S.C. Carotenes, vitamin E and sterols in oils from *Elaeis giuneensis, Elaeis oleifera* and their hybrids. Palm Oil Dev. 1997, 27, 1–9.

Chowdhury, M.K.U.; Parveez, G.K.A.; Saleh, N.M. Evaluation of five promoters for use in transformation of oil palm. Plant Cell Rep. 1997, 16, 277–281.

Christou, P. Third generation transgenic plants and applications to developing country agriculture. Proc. Agr. Congress 4–7 October 2004, 36–58.

Cluster, P.D.; O'dell, M.; Metzlaff, M.; Flavell, R.B. Details of T-DNA structural organization from transgenic Petunia population exhibiting co-suppression. Plant Mol. Biol. 1994, 32, 1197–1203.

Corley, R.H.V.; Lee, C.H.; Law, I.H.; Wong, C.Y. Abnormal flower development in oil palm clones. The Planter 1986, 62, 233–240.

Cunningham, F.X. Regulation of carotenoid synthesis and accumulation in plants. Pure Appl. Chem. 2002, 74, 1409–1417.

Cunningham, F.X.; Gantt, E. Genes and enzymes of carotenoid biosynthesis in plants. Ann. Rev. Plant Physiol. Plant Mol. Biol. 1998, 49, 557–583.

Daniell, H.; Datta, R.; Varma, S.; Gray, S.; Lee, S.B. Containment of herbicide resistance through genetic engineering of the chloroplast genome. Nature Biotechnol. 1998, 16, 345–348.

Dayang Izawati, A.M.; Parveez, G.K.A.; Masani, M.Y.A. Transformation of oil palm using *Agrobacterium tumefaciens*. J. Oil Palm Res. 2009, 21, 643–652.

Debergh, P.; Maene, L.J. A scheme for commercial propagation of ornamental plants by tissue culture. Scientia Horticulturae 1981, 14, 335–345.

Ducreux, L.J.M.; Morris, A.L.; Hedley, P.E; Sheperd, T; Davies, H.V; Millam, S.; Taylor, M.A. Metabolic engineering of high carotenoid potato tubers containing enhanced levels β-carotene and lutein. J. Exp. Bot. 2005, 56, 81–89.

Ebinuma, H.; Sugita, K.; Matsunaga, E.; Yamakado, M. Selection of marker-free transgenic plants using the isopentenyl transferase gene. Proc. Natl. Acad. Sci. USA. 1997, 94, 2117–2121.

Franqueville (de), H.; Renard, J.L. "Bilan de l'amélioration du niveau de tolérance du palmier à huile à la fusariose. Evolution de la maladie sur la Plantation R. Michaux." Oléagineux 1990, 45(10), 399– 405.

Frazer, F.D.; Bramley, P.M. The biosynthesis and nutritional uses of carotenoids. Progress Lipid Res. 2004, 43, 228–265.

Frazer, F.D.; Romer, S.; Shipton, C.A; Mills, P.B.; Kiano, J.W.; Misawa, N.; Drake, R.G.; Schuch, W.; Bramley, P.M. Evaluation of transgenic tomato plants expressing an additional phytoene synthase in a fruit-specific manner. Proc. Natl. Acad. Sci. USA 2002, 99, 1092–1097.

Fu, X.; Duc, L.T.; Fontana, S.; Bong, B.B.; Tinjuangjun, P.; Sidhakar, D.; Twyman, R.M.; Christou, P.; Kohli, A. Linear transgene constructs lacking vector backbone sequences generate low-copy-number transgenic plants with simple integration patterns. Transgenic Res. 2000, 9, 11–19.

Guillouet, S.; Rodal, A.A.; An, G.H.; Lessard, P.A.; Sinskey, A.J. Expression of the *Escherichia coli* catabolic threonine dehydratase in *Corynebacterium glutamicum* and its effect on isoleucine production. Appl. Environ. Microbiol. 1999, 65, 3100–3107.

Guzman, N.M. Present status of clonal propagation of oil palm (*E. guineensis*) in Costa Rica by culture of immature inflorescences. ASD de Costa Rica, SA, 1995, 6 pp.

Hirschberg, J. Carotenoid biosynthesis in flowering plants. Curr. Opin. Plant Biol. 2001, 4, 210–218.

Holmes, P.A. Biologically produced (R)-3-hydroxyalkanoate polymers and co-polymers. In Development in Crystalline Polymers-2; Basset, D.C., Ed. Elsevier: London, 1988; pp. 1–65.

IOPRI (http://isopb.org/docs/P5_Taufiq%20Hidayat%20IOPRI_.pdf)

Jaworski, J.; Cahoon, E.B. Industrial oils from transgenic plants. Curr. Opin. Plant Biol. 2003, 6, 178–184.

Joersbo, M.; Donaldson, I.; Kreiberg, J.; Petersen, S.G.; Brunstedt, J.; Okkels, F.T. Analysis of mannose selection used for transformation of sugar beet. Mol. Breeding 1998, 19, 798–803.

John, M.E.; Keller, G. Metabolic pathway engineering in cotton. Biosynthesis of polyhydroxybutyrate in fiber cells. Proceedings National Academy of Science USA 1996, 93, 12768–12773.

Jones, L.H. Propagation of clonal palms by tissue culture. Oil Palm News 1974, 17, 1–8.

Khemvong, S.; Suvachittanont, W. Molecular cloning and expression of a cDNA encoding 1-deoxy-D-xylulose 5-phosphate synthase from oil palm *Elaeis guineensis* Jacq. Plant Sci. 2005, 169, 571–578.

Kohli, A.; Griffiths, S.; Palacios, N.; Twyman, R.M.; Vain, P.; Laurie, D.A.; Christou, P. Molecular characterization of transforming plasmid rearrangements in transgenic rice reveals a recombination hotspot in the CaMV 35S promoter and confirms the predominance of microhomology-mediated recombination. Plant J. 1999, 17, 591–601.

Komari, T.; Hiei, Y.; Saito, Y.; Murai, N.; Kumashiro, T. Vectors carrying two separate T-DNAs for co-transformation of higher plants mediated by *Agrobacterium tumefaciens* and segregation of transformants free from selection markers. Plant J. 1996, 10, 165–174.

Knutzon, D.S.; Thompson, G.A.; Radke, S.E.; Johnson, W.B.; Knauf, V.C.; Kridl, J.C. Modification of *Brassica* Seed Oil by Antisense Expression of a Stearoyl Acyl Carrier Protein Desaturase Gene. Proc. Natl. Acad. Sci. USA 1992, 89, 2624–2628.

Kushairi, A.; Tarmizi, A.H.; Zamzuri, I.; Ong-Abdullah, M.; Rohani, O.; Samsul Kamal, R.; Ooi, S.E.; Ravigadevi, S.; Mohd Basri, W. Current status of oil palm tissue culture in Malaysia. Clonal and Quality Replanting Material Workshop. 10 Aug 2006. Seri Kembangan.

Liu, Q.; Singh, S.; Green, A. Genetic modification of cottonseed oil using inverted repeat gene silencing techniques. Biochem. Soc. Transactions 2000, 28, 927–929.

Malaysian Standard MS 2099:2008 Oil palm clones for commercial planting: specification for ortet selection. Department of Standards Malaysia, 2008.

Manaf, M.A.A.; Abrizah,O.; Umi Salamah, R. Characterization of genes encoding key enzymes in oil synthesis in the oil palm. Proc Agr., Biotechnol. Sustainability Conference, PIPOC 2007. 583–606.

Masani, A.M.Y.; Ho, C.L.; Parveez, G.K.A. Construction of PHB gene expression vectors for the production of biodegradable plastics in transgenic oil palm. Proceedings of the 2001 MPOB International Oil Palm Congress: Kuala Lumpur, Malaysia, 2001; pp. 694–711.

Masani, A.M.Y.; Ho, C.L.; Parveez, G.K.A. Transformation of oil palm with PHB and PHBV gene constructs driven by combination of constitutive promoters. Proceedings of the 2003 MPOB International Palm Oil Congress, 2003, 803–812.

Masani, A.M.Y.; Ho, C.L.; Parveez, G.K.A. Construction of PHB and PHBV transformation vectors for bioplastics production in oil palm. J. Oil Palm Res, 2008, Special Issue on MMBPP, 2, 37–55.

Masani, A.M.Y.; Parveez, G.K.A. Development of transformation vectors for the production of high oleate transgenic oil palm. Electronic J. Biotechnol. [online]. 15 July 2008, 11(3). Available from: http://ejbiotechnology.info/content/vol11/issue3/full/11/index.html ISSN 0717-3458

Masani, A.M.Y.; Parveez, G.K.A.; Izawati, A.M.D.; Chan, P.L.; Siti Nor Akmar, A. Construction of PHB and PHBV multiple-gene vectors driven by an oil palm leaf-specific promoter. Plasmid. 2009, 62, 191–200.

Masura, S.S.; Parveez, G.K.A. Constitutively Expressing cDNA in Oil Palm. Proceeding of the PIPOC 2007 International Palm Oil Congress (Agriculture, Biotechnology & Sustainability), 2007; pp. 1063–1070.

Masura, S.S.; Parveez, G.K.A.; Ismanizan, I. Isolation and characterization of oil palm constitutive promoter derived from ubiquitin extension protein (*uep1*) gene. New Biotechnol. 2010, 27, 289–299.

Masura, S.S.; Parveez, G.K.A.; Leslie, L.E.T. Isolation and characterization of an oil palm constitutive promoter derived from translationally control tumor protein gene. Plant Physiol. Biochem. 2011, 49, 701–708.

Matzke, M.A.; Matzke, A.J.M. Differential inactivation and methylation of a transgene in plants by two suppressor loci containing homologous sequences. Plant Mol. Biol. 1991, 16, 821–830.

Matzke, M.A.; Primig, M.; Tranovsky, J.; Matzke, A.J.M. Reversible methylation and inactivation of marker genes in sequentially transformed tobacco plants. EMBO J. 1989, 7, 643–649.

Mohd Din, A.; Kushairi, A.A; Noh, A.; Isa, Z.A.; Junaidah, J.; Rajanaidu, N. Innovations in oil palm breeding and genetics. Book of Abstracts of AGRICONGRESS 2004, Selangor, Malaysia; pp. 448–449.

Mohd Din, A.; Rajanaidu, N.; Jalani, B.S. Performance of *Elaeis oleifera* from Panama, Costa Rica, Colombia and Honduras in Malaysia. J. Oil Palm Res. 2000, 12, 71–80.

Montgomery, J.; Goldman, S.; Deikman, J.; Margossian, L.; Fischer, R.L. Identification of an ethyleneresponsive region in the promoter of a fruit ripening gene. Proc. Natl. Acad. Sci. USA 1993, 90, 5939–5943.

Nawrath, C.; Poirier, Y.; Somerville, C.R. Plastid targeting of the enzymes required for the production of polyhydroxybutyrate in higher plants. In Biodegradable plastics and polymers; Doi, Y., Fukuda, K., Eds. Elsevier: Amsterdam, 1994a; pp. 136–149.

Nawrath, C.; Poirier, Y.; Somerville, C. Targeting of the polyhydrxybutyrate biosynthetic pathway to the plastids of *Arabidopsis thaliana* results in high levels of polymer accumulations. Proceedings National Academy of Science USA 1994b, 91, 12760–12764.

Omidvar, V.; Siti Nor Akmar, A.; Marziah, M.; Maheran, A.A. A transient assay to evaluate the expression of polyhydroxybutyrate genes by oil palm mesocarp-specific promoter. Plant Cell Rep. 2008, 27, 1451–1459.

Parveez, G.K.A. Production of transgenic oil palm (*Elaeis guineensis* Jacq.) using biolistic techniques. In Molecular Biology of Woody Plants Vol. 2; Jain, S.M., Minocha, S.C., Eds. Kluwer Academic Publishers: Netherlands, 2000; pp. 327–350.

Parveez, G.K.A. Novel products from transgenic oil palm. AgBiotechNet 2003, 5, 1–8.

Parveez, G.K.A.; Abrizah, O.; Abdul Masani, M.Y.; Umi Salamah, R.; Sambanthamurthi, R.; Bahariah, B.; Nur Fariza, M.S.; Yuen, L.H.; Nur Hanin, A.; Ahmad Tarmizi, H.; et al. Value addition of oil palm through genetic engineering. Proc. 2005 Conference Biotechnology Plantation Commodities 2005, 198–209.

Parveez, G.K.A.; Abrizah, O.; Masani, A.M.Y.; Omar, R.; Siti Nor Akmar, A.; Umi Salamah, R.; Ravigadevi, S.; Bahariah, B.; Yeun, L.H.; Nur Fariza, M.S. Production of transgenic oil palm carrying genes for modifying fatty acid composition of palm oil. Proceeding of 2004 Oils and Fats International Congress, 2004.

Parveez, G.K.A; Abrizah, O.; Nurhafizah, R.; Bahariah, B. Functional analysis of oil palm palmitoyl-acyl-ACP thioesterase (FatB) gene via down-regulation in a model plant: *Arabidopsis thaliana.* J. Oil Palm Res. 2010, 22, 803–813.

Parveez, G.K.A.; Abrizah O.; Tarmizi, A.H.; Zamzuri, I.; Kushairi, A.D.; Salmah, Y.; Bahariah, B.; Sabariah, K. Transformation of oil palm with antisense palmitoyl-acp thioesterase gene for increasing oleic acid content. Proceedings of the 2003 MPOB International Palm Oil Congress, 2003a; pp. 869–878.

Parveez, G.K.A.; Bahariah, B.; Nur Hanin, A.; Masani, A.M.Y.; Tarmizi, Ah.; Zamzuri, I.; Ahmad Kushairi, D.; York, G.; Jo, Y.B.; Sinskey, A.J. Transformation of PHB and PHBV genes driven by maize ubiquitin promoter into oil palm for the production of biodegradable plastics. J. Oil Palm Res. Special Issue on MMBPP 2008, 2, 76–86.

Parveez, G.K.A.; Chowdhury, M.K.U.; Saleh, N.M. Determination of minimal inhibitory concentration of selection agents for oil palm transformation. Asia-Pacific J. Mol. Biol. Biotechnol. 1996, 4, 219–228.

Parveez, G.K.A.; Chowdhury, M.K.U.; Saleh, N.M. Physical parameters affecting transient GUS gene expression in oil palm using the biolistics device. Industrial Crops and Product 1997, 6, 41–50.

Parveez, G.K.A.; Chowdhury, M.K.U.; Saleh, N.M. Biological parameters affecting transient GUS gene expression in oil palm embryogenic calli via microprojectile bombardment. Industrial Crops and Product 1998, 8, 17–27.

Parveez, G.K.A.; Masani, A.M.Y.; Salmah, Y.; Bahariah, B.; Sabariah, K.; York, G.; Jo, Y.B. Transfer of PHB and PHBV genes into oil palm for the production of biodegradable plastics. Proceedings of the 2003 MPOB International Palm Oil Congress, 2003b; pp. 859–868.

Parveez, G.K.A.; Na'imatulapidah, A.M. Factors affecting green fluorescence protein (GFP) gene expression in oil palm after microprojectile bombardment. J. Oil Palm Res. 2008, 20, 494–506.

Parveez, G.K.A.; Na'imatulapidah, A.M.; Alizah, Z.; Omar, A.R. Determination of minimal inhibitory concentration of selection agents for selecting transformed immature embryos of oil palm. Asia Pacific J. Mol. Biol. Biotechnol. 2007, 15(3), 33–146.

Parveez, G.K.A.; Omar, R.; Masani, A.M.Y.; Haliza, H.F.; Na'imatulapidah, A.M.; Kushairi, A.D.; Tarmizi, A.H.; Zamzuri, I. Transgenic oil palm: where are we?. In Murata, N., Ed. Advanced Research on Plant Lipids. Kluwer Academic Publishers: The Netherlands, 2003c; pp. 415–418.

Parveez, G.K.A.; Rasid, O.A.; Sambanthamurthi, R. Genetic engineering of oil palm. In Further Advances in Oil Palm Research (2000–2010); Wahid, M.B., Choo, Y.M., Chan, K.W.; MPOB: Kuala Lumpur, 2011; pp. 141–201.

Parveez, G.K.A.; Sambanthamurthi, R.; Siti Nor Akmar, A.; Rasid, O.; Masri, M.M.; Cheah, S.C. Production of transgenic oil palm—current success and future considerations. Proceeding PORIM International Palm Oil Congress, 1999; pp. 3–13.

Poirier, Y. Polyhydroxyalkanoate synthesis in plants as a tool for biotechnology and basic studies of lipid metabolism. Progress in Lipids Res. 2002, 41, 131–155.

Poirier, Y.; Dennis, D.E.; Klomparens, K.; Somerville, C. Polyhydroxybutyrate, a biodegradable thermoplastic, production in transgenic plants. Science 1992, 256, 520–523.

Poirier, Y.; Nawrath, C.; Somerville, C. Production of polyhydroxyalkanoates, a family of biodegradable plastics and elastomers, in bacteria and plants. Bio/Technology 1995, 13, 142–150.

Rabéchault, H.; Martin, J.P. Multiplication végétative du palmier á huile (*Elaeis guineensis* Jacq.) á l'aide de cultures de tissus foliares. C.R. Acad. Sci. Paris, 1976, 283, 1735–1737.

Rajanaidu, N.; Rao, V.; Halim, A.H.; Ong, S.H. Genetic resources: New developments in oil palm breeding. Elaeis 1990, 1, 1–10.

Rasid, O.A.; Cheah, S.C.; Arif, A.M. Isolation and sequencing of cdna clones coding for oil palm ACP. J. Oil Palm Res. (Special Issue) 1999, 88–95.

Rasid, O.A.; Nor Hanin, A.; Masura, S.S.; Abdul Masani, M.Y.; Singh, R.; Ho, C.L.; Suhaimi, N.; Parveez, G.K.A.; Sambanthamurthi, R. Lycopene: Genetic engineering of a potential nutraceutical. Proc Agriculture, Biotechnology and Sustainability Conference, PIPOC 2007, 2007b, 248–273.

Rasid, O.A.; Nor Hanin, A.; Masura, S.S.; Ho, C.L.; Suhaimi, N.; Sambanthamurthi, R. Regulation of oil palm (*E. guineensis*) phytoene desaturase in developing mesocarp tissues determined by real-time PCR. Proc. Agriculture Congress 2006, 41–47.

Rasid, O.A.; Nor Hanin, A.; Masura, S.S.; Ho, C.L.; Suhaimi, N.; Sambanthamurthi, S. Regulation of oil palm (*E. guineensis*) zeaxanthin epoxidase in developing mesocarp tissues determined by real time PCR. Proceedings 2007 Conference on Plantation Commodities, 2007a; p. 553.

Rasid, O.A.; Parveez, G.K.A.; Ho, C.L.; Sambanthamurthi, R.; Napis, S. Plant carotenoids: Molecular genetics and regulation. J. Oil Palm Res. 2007b, 21, 588–601.

Rasid, O.A.; Shah, F.H. The isolation, characterization and sequencing of cDNA clones coding for stearoyl-ACP desaturase gene from oil palm. Proceeding PORIM International Palm Oil Congress: Kuala Lumpur, Malaysia, 1996; pp. 575– 582.

Rasid, O.A.; Singh, R.; Sambanthamurthi, S.; Cheah, S.C. RT-PCR amplification and cloning of partial DNA sequence coding for oil palm (*E. guineensis*) zeaxanthin epoxidase. Proceedings 2005 Conference on Plantation Commodities, 2005; pp. 367–375.

Rasid, O.A.; Wan Nur Syuhada, W.S.; Nor Hanin, A.; Masura, S.S.; Ho, C.L.; Suhaimi, N.; Sambanthamurthi, R. RT-PCR amplification and cloning of partial DNA sequence coding for oil palm (*E. oleifera*) phytoene synthase. Asia Pac. J. Mol. Biol. Biotechnol. 2008, 16(1), 17–24.

Rohani, O.; Sharifah, S.A.; Mohd Rafi, Y.; Ong, M.; Tarmizi, A.H.; Zamzuri, I. Tissue culture of oil palm. Advances in Oil Palm Research; Yusof, Jalani, Chan, Eds., MPOB, 2000; pp. 238–283.

Rohani, O.; Zamzuri, I.; Tarmizi, A.H. Oil palm cloning: MPOB protocol. MPOB Technology, MPOB, Bangi 2003, 26..

Romer, S; Fraser, P.D.; Kiano, J.W.; Shipton, C.A.; Misawa, N.; Schuch, W.; Bramley, P. Elevation of the provitamin A content of transgenic tomato plants. Nature Biotechnol. 2000, 18, 666–669.

Ronen, G.; Carmel-Goren, L.; Zamir, D.; Hirschberg, J. An alternative pathway to β-carotene formation in plant chromoplasts discovered by map-based cloning of *Beta* and *old-gold* color mutations in tomato. Proc. Natl. Acad. Sci. USA, 2000, 97, 11102–11107.

Safiza, M.; Abrizah, O.; Siti Nor Akmar, A.; Mohamad Arif, A.M. Functional characterization of oil palm stearoyl-ACP desaturase by overexpression in wild-type *Arabidopsis* and complementation of *fab2* mutant. Proceeding of the PIPOC 2009 International Palm Oil Congress (Agriculture, Biotechnology & Sustainability), 2009a, 1229–1243.

Safiza, M.; Mohamad Arif, A.M.; Abrizah, O. Functional complementation study in *Arabidopsis thaliana* of oil palm lysophosphatidyl acyltransferases. Proceeding of the PIPOC 2009 International Palm Oil Congress (Agriculture, Biotechnology & Sustainability), 2009b, 1244–1253.

Sambanthamurthi, R.; Abrizah, O.; Ramli, U.S. Towards understanding the biochemical factors that affect oil composition and quality in the oil palm. In Proceedings ISOPB International Conference on Oil and Kernel Production in Oil Palm: Kuala Lumpur, Malaysia, 1996a.

Sambanthamurthi, R.; Abrizah, O.; Umi Salamah, R. Biochemical factors that control ol composition in the oil palm. J. Oil Palm Res. (Special Issue of MMBPP) 1999, 24–33.

Sambanthamurthi, R.; Parman, S.H.; Noor, M.R.M. Oil palm (*Elaeis guineensis*) protoplasts: isolation, culture and microcallus formation. Plant Cell, Tissue and Organ Culture 1996b, 46, 35–41.

Sambanthamurthi, R.; Parveez, G.K.A.; Cheah, S.C. Genetic engineering of the oil palm. In Advances in Oil Palm Research Vol I; Basiron, Y., Jalani, B.S., Chan, K.W., Eds. Malaysian Palm Oil Board: Kuala Lumpur, Malaysia, 2000; pp. 284–331.

Sambanthamurthi, R.; Siti Nor Akmar, A.; Parveez, G.K.A. Genetic manipulation of the oil palm: Challenges and future prospects. The Planter 2002, 78(919), 547–562.

Scheller, J.; Conrad, U. Plant-based material, protein and biodegradable plastic. Curr. Opin. Plant Biol. 2005, 8, 188–196.

Senior, P.J.; Dawes, E.A. The regulation of poly-β-hydroxybutyrate metabolism in *Azotobacter beijerinckii*. Biochem. J. 1973, 134, 225–238.

Shah, F.H.; Rasid, O.; Cha, T.S. Temporal regulation of two isoforms of cDNA clones encoding the Δ9 stearoyl-ACP desaturase from oil palm. Plant Sci. 2000, 152, 27–33.

Shewmaker, C.K.; Sheehy, J.; Daley, M.; Colburn, S.; Ke, D.Y. Seed-specific overexpression of phytoene synthase: increase in carotenoids and other metabolic effects. Plant J. 1999, 20, 402–412.

Siti Nor Akmar, A.; Cheah, S.C.; Abrizah, O.; Mohd Basri, W. Stearoyl-ACP desaturase genes from oil palm. MPOB Technol. Trans. 2004, no. 220.

Siti Nor Akmar, A.; Cheah, S.C.; Aminah, S.; Leslie, C.L.O.; Sambanthanmurthi, R.; Murphy, D.J. Characterization and regulation of oil palm (*Elaeis guineensis*) stearoyl-ACP desaturase genes. J. Oil Palm Res. (Special Issue of MMBPP), 1999, 1–17.

Siti Nor Akmar, A.; Cheah, S.C.; Murphy, D.J. Isolation and characterization of two divergent type 3 metallothionein from oil palm (*Elaeis guineensis*). Plant Physiol. Biochem. 2002, 40, 255–263.

Siti Nor Akmar, A.; Ravigadevi, S.; Ahmad Parveez, G.K. Genetic modification of oil palm for producing novel oils. Proceeding of the 2001 PIPOC International Palm Oil Congress (Agriculture), pp. 18–30.

Siti Nor Akmar, A.; Zubaidah, R. Mesocarp-specific metallothionein-like gene promoter for genetic engineering of oil palm. J. Oil Palm Res. Special Issue on Malaysia-MIT Biotechnology Partnership Programme: Volume 2–Oil Palm Metabolic Engineering 2008, 1–8.

Slater, S.; Mitsky, T.A.; Houmiel, K.L.; Hao, M.; Reiser, S.E.; Taylor, N.B.; Tran, M.; Valentin, H.E.; Rodriguez, D.J.; Stone, D.A.; et al. Metabolic engineering of *Arabidopsis* and *Brassica* for poly(3-hydroxybutyrate-co-3-hydroxyvalerate) copolymer production. Nature Biotechnol. 1999, 17, 1011–1016.

Syahanim, S.; Abrizah, O.; Siti Nor Akmar, A.; Mohamad Arif, A.M.; Ho, C.L. Cloning of an oleoyl-Coa desaturase from oil palm. Proceedings of the PIPOC 2007 International Palm Oil Congress (Agriculture, Biotechnology & Sustainability), KL. 1001–1009.

Tarmizi, A.H. Oil palm liquid culture–MPOB Protocol. MPOB Information Series, MPOB, Bangi 2002, no.138.

Tarmizi, A.H.; Zaiton, R MPOB fast transfer technique (MoFaTT) in liquid culture system. MPOB Information Series 2004, no. 261.

Tarmizi, A.H.; Zaiton, R.; Rosli, M.Y. MPOB Modified Vessel (MoVess) for liquid tissue culture system. MPOB Information Series 2007, no. 355.

Tarmizi, A.H.; Zamzuri, I.I.; Hashim, H. Oil palm tissue culture tracking system (OPTRACKS): Version 1. MPOB Information Series 2003, no. 185.

Tarmizi, A.H.; Zamzuri, I.; Meilina, O.A.; Ooi, S.E.; Samsul, K.R.; Chan, P.L.; Rohani, O. Forging ahead with clones In Further Advances in Oil Palm Research (2000–2010); Wahid, M.B., Choo, Y.M., Chan, K.W., Eds.; MPOB, 2011; pp. 102–140.

Teen, Y.J.; Masani, M.Y.; Parveez, G.K.A.; Sambanthamurthi, R. Activity studies, gene characterization and manipulation of β-ketothiolase of oil palm (*Elaeis guineensis* Jacq.) mesocarp. J. Oil Palm Res. Special Issue on Malaysia-MIT Biotechnology Partnership Programme: Volume 2–Oil Palm Metabolic Engineering 2008, 118–133.

Terzaghi, W.B.; Cashmore, A.R. Light-regulated transcription. Ann. Rev. Plant Physiol. Plant Mol. Biol. 1995, 46, 445–474.

Umi Salamah, R.; Ravigadevi, S.; Cheah, S.C.; Sharifah Shahrul Rabiah, S.A.; Omar, R.; Siti Nor Akmar, A.; Ahmad Parveez, G.K.; Mohd Arif, A.M.; Abrizah, O.; Abdul Masani, M.Y.; et al. Beta-ketoacyl ACP synthase II (KAS II). MPOB Information Series June 2004, no. 226.

Umi Salamah, R.; Ravigadevi, S.; Cheah, S.C.; Sharifah Sharul Rabiah, S.A.; Siti Nor Akmar, A.; Omar, R.; Ahmad Parveez, G.K.; Mohd Arif, A M; Abdul Masani, M.Y.; Dzulfazly, A.; et al. Isolation and characterization of β-ketoacyl ACP synthase (KAS) II gene in oil palm (*Elaeis guineensis* Jacq.). Abstract of 17th International Symposium of Plant Lipids. Michigan, USA, 16–21 July 2006.

Umi Salamah, R.; Sambanthamurthi, R. β-keto acyl ACP synthase II in oil palm (*Elaeis guineensis*) mesocarp. In Physiology, biochemistry and molecular biology of plant lipids; Williams, J.P., Khan, M.U., Lem, N.W., Eds. Kluwer Academic Publishers: The Netherlands, 1996; pp. 69–71.

UNIVANICH (http://uvan.listedcompany.com/corporate_profile.html)

Van De Loo, F.; Fox, B.; Somerville, C.R. Unusual fatty acids. In Plants Lipids; Moore, T., Ed. Boca Raton: CRC Press, 1993; pp. 91–126.

Wan Nur Syuhada, W.S. Isolation of cDNA cloe coding for oil palm pyhtoene synthase. B.Sc Thesis, University Malaya, Kuala Lumpur, Malaysia, 2006.

Wan Saridah, W.O.; Laura, B.W.; Chokyun, R.; Anthony, J.S.; Umi Salamah, R.; Abdul Masani, M.Y.; Ahmad Parveez, G.K.; Ravigadevi, S. Isolation and utilization of acetyl-CoA carboxylase from oil palm (*Elaeis guineensis*) mesocarp. J. Oil Palm Res. Special Issue on Malaysia-MIT Biotechnology Partnership Programme: Volume 2–Oil Palm Metabolic Engineering, 2008, 97–107.

Willis, L.B.; York, G.M.; Lessard, P.L.; Jo, Y.B.; Omar, R.; Mohammd Arif, A.M.; Park, C.; Sinskey, A.J. Reagents for generation and analysis of bioplastic producing plants. J. Oil Palm Res. Special Issue on MMBPP. 2008, 2, 56–68.

Wooi, K.C. Oil palm tissue culture—current practice and constraints. In Proc. of the 1993 ISOPB Intl. Symp. on Recent Developments in Oil Palm Tissue Culture and Biotechnology; Rao, V., Henson, I.E., Rajanaidu, N., Eds. Palm Oil Research Institute of Malaysia: Kuala Lumpur, 1995; pp. 565, 21–32.

Zamzuri, I. Efficient rooting of oil palm *in vitro* plantlets using double-layer technique. PORIM Bull. 1998, 36, 23–36.

Zamzuri, I. Double-layer technique in rooting of oil palm *in vitro* plantlets. MPOB Information Series, 2001, No. 99.

Zamzuri, I. Flameless sterilizer. MPOB Information Series TT, 2002, No.139.

Zamzuri, I. 2009. An electrical flameless steriliser. Patent granted MY-140341-A.

Zubaidah, R.; Siti Nor Akmar, A. The effects of metal ions on root-specific expression of the oil palm *MT3-B* gene promoter. Proceedings of the PIPOC 2005 International Palm Oil Congress, 1104–1110.

5

Ripening, Harvesting, and Transport of Oil Palm Bunches

Ian E. Henson
7 Richmond Dale, Clifton, Bristol BS8 2UB, UK

Introduction

The oil palm (*Elaeis guineensis* Jacq.) produces a continuous supply of fruit bunches in the axils of its fronds throughout its economic life commencing the second or third year after planting. In favorable environments bunches ripen throughout the year, although there are peaks and troughs in production. Bunches are harvested manually using a variety of methods and tools. Harvesting involves a compromise between the need to maximize oil yield and oil quality while minimizing cost of labor. The harvester must examine the maturing bunches on the palm and select for harvest only those deemed to be ripe, as indicated by a change in fruit color from black to bright orange or red or the shedding of fruits. The degree of ripeness at harvesting has a crucial impact on both the oil content of the mesocarp and on the number of fruits that become detached from the bunch. The detached or "loose" fruits eventually fall to the ground or are caught in the axils of lower fronds. Loose fruits are usually collected and combined with the cut bunches for delivery to the mill where palm oil and palm kernels are extracted. Collecting the loose fruits is important in order to maximize oil yield since they generally have higher oil contents than most of the attached fruits. The frequency of harvesting also has a large effect on oil recovery. Harvesting at too long of intervals results in bunches becoming over ripe with a higher percentage of loose fruits, some of which may escape collection, while the collected fruits may have a high free fatty acid content as a result of lipase activity following detachment. Conversely, harvesting too frequently may result in a higher percentage of unripe bunches with a lower oil content. In practice, a 10- to 14-day harvest interval is commonly adopted.

There has been considerable debate regarding the relationship between fruit detachment and oil synthesis. Some studies suggested oil synthesis ceases throughout the bunch at the start of fruit abscission. There is other evidence that oil synthesis continues well past this stage in later maturing inner fruits. However, it is generally agreed that oil synthesis ceases once the fruit is detached since it depends on a continued inflow of photosynthetic assimilates.

Harvesting and transport of bunches represents about 15% of the cost of fruit bunch production. Various methods are used to collect and transport bunches after harvest, depending largely on the scale of the operation and on the terrain. Thus, for small plantings or on steep terrain, wheel barrows are used to transport bunches from within the field to the roadside. In some countries draught animals such as buffaloes may be used for in-field transport. Bunches left at the roadside are loaded onto trucks or trailers, sometimes with the aid of nets and a crane. In some coastal estates railway wagons are used while in hilly areas overhead pulleys may be installed for moving bunches downhill to collection points. In large estates and on flat ground, this operation is usually mechanized. While progress has been made in mechanizing in-field collection of fruit bunches, mechanization of the harvesting progress, including collection of loose fruits, has yet to be effectively introduced.

Whatever system is used, it is of utmost importance to minimize the time between harvesting and the start of processing so as to reduce the production of free fatty acids (FFA) that takes place due to the action of endogenous and exogenous lipases.

The main harvest from the oil palm consists of fresh fruit bunches (FFB) from which are obtained a wide variety of products, the chief of which is crude palm oil (CPO) extracted from the mesocarp or flesh of the fruit and palm kernel oil (PKO) extracted from the palm kernel or seed within the fruit.

Harvesting is one of the most crucial operations in the management of any crop and in the case of oil palm it is one of the most labor-intensive components of production. This is a consequence of the manual nature of the harvesting process combined with the continuous production of FFB by the oil palm that constitutes the yield. Harvesting commences after a relatively brief period of immaturity lasting 2 to 3 years from the transfer of nursery palms to the field and continues up until the time when the palms are felled some 25 or more years later. It is the increasing difficulties of harvesting as the palms grow taller, as well as a decline in yields, that are major factors in determining the age at which replanting of the palms becomes necessary.

This chapter describes the changes occurring in bunches as they mature, the criteria used to determine when fruit bunches should be harvested, the tools and methods used in harvesting, and the subsequent collection and transport of FFB to the palm oil mill. It considers only the African oil palm, as there are few if any reports specific to the American oil palm, *E. oleifera*, or to its hybrids with *E. guineensis*. Previous reviews of oil palm harvesting can be found in books by Hartley (1967, 1977, 1988), Turner & Gillbanks (1974, 2003) and Corley & Tinker (2003), and in articles by Southworth (1976, 1977), Corley (2001) and Gillbanks (2003), while Ahmad Hitam & Ahmad Zamri (2000) review progress in the mechanization of various aspects of harvesting operations. The present chapter seeks to emphasize biological aspects involving bunch development, ripening, oil synthesis, and fruit abscission.

Bunch Development, Oil Synthesis, and Fruit Maturation

In favorable environments, (adequate solar radiation, sufficient but not excessive rainfall, and moderate temperatures) oil palms produce fruit bunches more or less continuously throughout the year, and so harvesting is also a continuous process. However, even in the most favorable and climatically constant locations, there are seasonal (i.e., intra-annual) variations in FFB production, that result in variations in harvesting load and FFB supply to the mills.

In oil palm, separate male and female inflorescences are produced in the axils of the fronds, usually in alternate phases, the timing and length of which varies with individual palms. Cross pollination is the norm, nowadays achieved largely through the agency of the West African weevil, *Elaeidobius kamerunicus*, which was introduced into Southeast Asian and other producing countries in the 1980s and which generally ensures a high rate of fruit set and the formation of large and compact bunches. Previous to this, older palms were pollinated by *Thrips hawaiiensis*, but in young palms and at all ages in areas where *T. hawaiiensis* was absent, hand pollination was necessary. Oil palm bunches take about 5 to 6 months from the time of pollination and fertilization to develop before they are ready for harvest. Changes in bunch composition during bunch development are shown in Fig. 5.1, while the salient events during bunch maturation are shown in Box 5.1.

Box 5.1. Major events during bunch maturation.

Days after pollination	Event
70	PKO synthesis starts
110	PO synthesis starts
130	PKO synthesis completes
140	Fruit changes colour
150	Fruit abscission starts
170	> 80% fruit are detached
180	PO synthesis completes and bunch decays

Based partly on data of Corley & Law (2001). Timings are approximate.

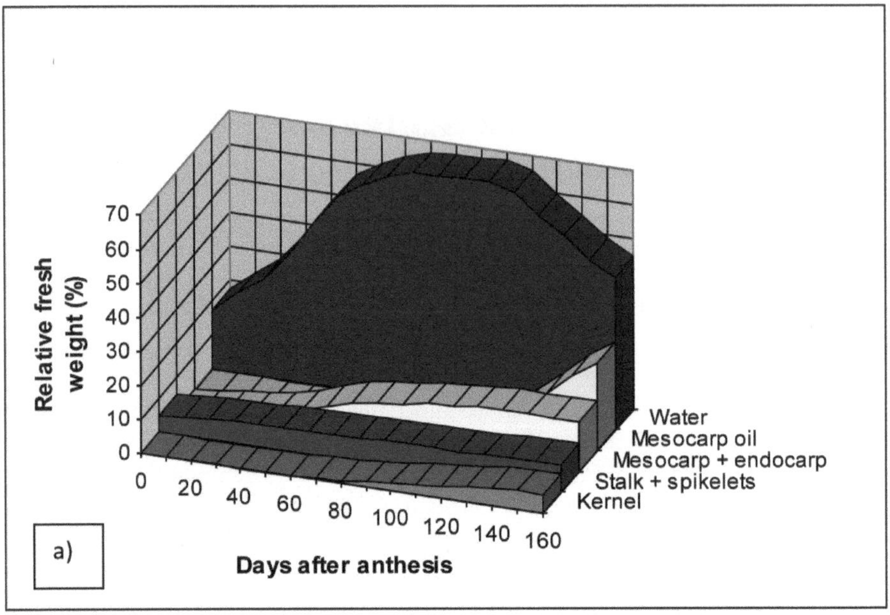

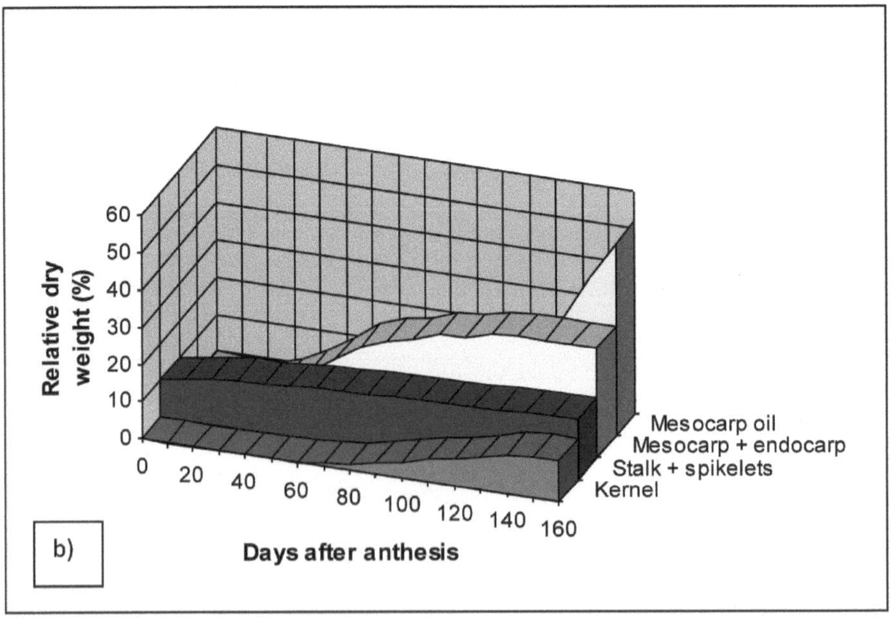

Fig. 5.1. Changes with time in major constituents of an oil palm bunch (*c.* 20 kg fresh weight) with (a) and without (b) the water component. Based on a graph from Corley (1986), redrawn from Henson (1994).

Oil accumulation in the mesocarp takes place in oil bodies termed oleosomes (Arumughan et al., 1990), starting first in cells near the fruit base (Aziz Ariffin, 1990). Oil synthesis depends on concurrent import of assimilates *via* the phloem, there being little evidence of synthesis from reserves stored within the fruit (Corley, 1986). The presence of chloroplasts (Aziz Ariffin et al., 1990) and hence, chlorophyll (Aziz Ariffin, 1985) in the mesocarp from 3 to 17 weeks after pollination suggests there may be some photosynthetic activity that could contribute to fruit growth and early lipid synthesis. However, this would be insignificant in the case of the majority of fruits which are shielded from light. Soluble carbohydrates accumulate transiently in small quantities from the fifth to ninth week after pollination (PORIM, 1986), but this precedes the major phase of oil synthesis. It appears, therefore, that fruits are unable to continue to synthesize oil once they are detached from the bunch.

The expansion of the oil bodies serves to displace water from the mesocarp so that oil effectively replaces water, while the non-oil dry matter content of the mesocarp remains relatively unchanged (Fig. 5.2). Water loss occurs by transpiration *via* stomates on the fruit surface or, particularly in the case of inner fruit, by reverse xylem flow (Jeje et al., 1978; PORIM, 1984). The stability of the non-oil dry matter content of the fruit provides a basis for evaluating oil production at different stages of fruit ripening.

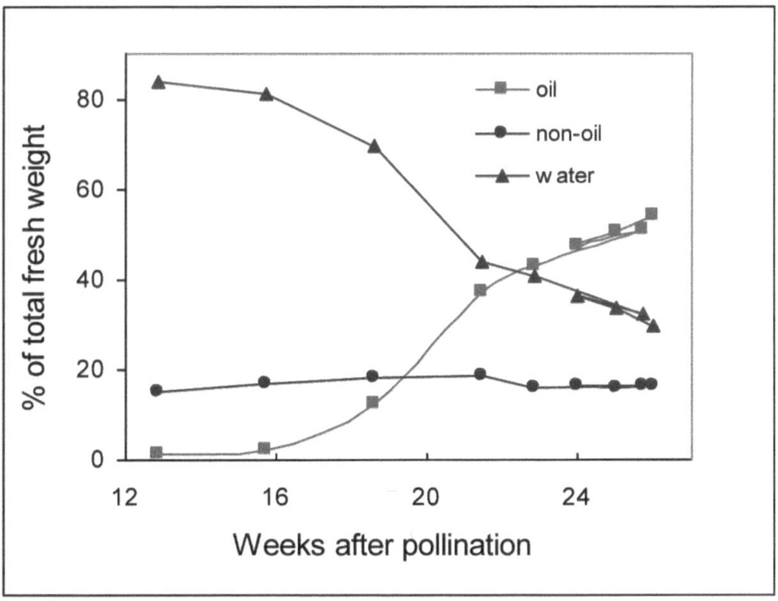

Fig. 5.2. Changes in mesocarp components during fruit maturation. Data are from PORIM (1987). Fruit abscission commenced 23 weeks after pollination.

As seen in Fig. 5.3, the formation of oil within the mesocarp takes place over a relatively short period during the last few weeks before maturity. An examination of 11 independent studies (Henson, 1998) showed the start of oil synthesis to vary from 15 to 19 weeks after pollination (WAP) with a mean of 17 weeks and to reach completion at between 20 and 24 weeks with a mean of 21 weeks. The data suggest that oil synthesis commences later in West Africa (19 WAP) than elsewhere (16 WAP), although differences in planting material, in addition to location, could also be responsible, with ripening in *dura* palms being later than in *tenera*. Palm age also affects the rate of oil synthesis, this being more rapid in younger palms (see below).

As discussed by Southworth (1976) and others, there are methodological problems in assessing changes in the amount of mesocarp oil over time. Three main alternative sampling methods are possible. The first involves sampling individual fruits at different times from the same bunch. This minimizes effects of variation between bunches but raises the possibility that removal of fruits at intervals might affect development of the ones remaining. It is also difficult to ensure that sampling within the bunch is random, particularly since only outer fruits (which mature first) are easily accessible and changes in inner fruits cannot be examined. The second method, which avoids these problems, involves sampling of whole bunches. This is a more costly approach which again

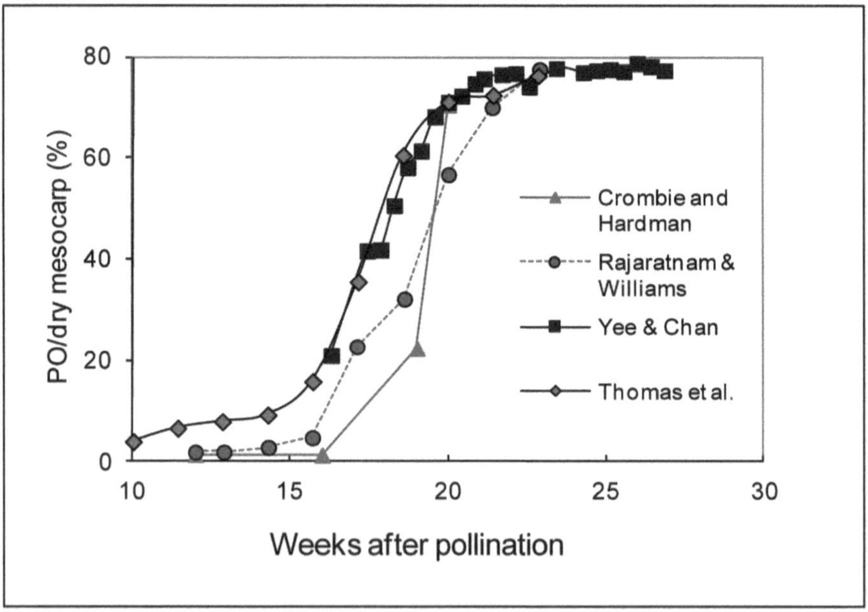

Fig. 5.3. Accumulation of palm oil in the mesocarp over time. Data of Crombie & Hardman (1958) are for *dura* palms in West Africa. Other data are from Malaysia. Those of Rajaratnam & Williams (1970) and Yee & Chan (1985) are for *tenera* palms while Thomas et al. (1971) used a mixture of 50% *dura* and 50% *tenera*.

requires care to avoid bias due to variation in bunch size. Southworth (1976) suggested that the latter problem could be avoided by selecting potential bunches for sampling prior to pollination. Finally, where facilities are available, the third approach can be taken of processed batches of FFB at the mill (Siregar, 1977; Wood et al., 1985b).

The second problem is how to express the changes in oil content. As mentioned already, use of oil/wet mesocarp is undesirable due to the large changes that occur in mesocarp water content (Fig. 5.2). Use of oil/dry mesocarp (Fig. 5.3) or oil/bunch are open to similar objections as oil is present in both numerator and denominator. The use of oil/non-oil (NO) dry matter, as suggested by Southworth (1976), does not seem to have been generally adopted, although it may be a better measure than the other ratios since %NO in the mesocarp remains fairly constant over time (Fig. 5.2). The oil/nut ratio is another possibility but may be unreliable if nut size varies. Differences both in sampling methods and in the basis for expressing oil content have probably contributed to the different conclusions concerning the point at which oil synthesis is complete.

The development of the kernels, from which the biochemically distinct kernel oil is extracted, precedes the formation of the oil in the mesocarp. Hence, kernel oil production is little affected by harvesting practices. In Malaysia it takes place from 12 to 17 WAP (Corley, 1986; Oo et al., 1985; Oo & Ong, 1989), but is completed a little later in West Africa (Fig. 5.4). Again, as with mesocarp oil, whether this is a result of

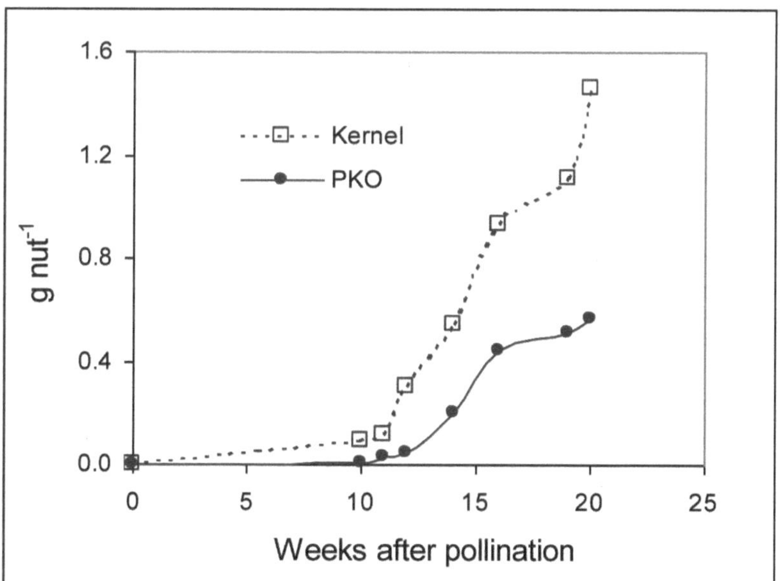

Fig. 5.4. Growth of the palm kernel and accumulation of palm kernel oil (PKO) in West Africa. Data are from Crombie (1956) as presented by Hartley (1977). Graph adapted from Henson (1998).

differences in climate, age, or planting material (the majority of African studies used *dura* palms while the material used in Malaysia was mostly *tenera*), is not certain. As pointed out by Corley & Law (2001), the role of the kernel oil is to serve as the source of assimilate for the young seedling so that biologically, maturity is reached when kernel oil is at a maximum, whereas from the perspective of the harvester (who seeks to maximize the mesocarp oil yield), maturity is reached later.

Ripening of the fruits is indicated by a color change which, for the most common fruit form *nigrescens*, involves a change from black to reddish orange (see Chapter 1). In the rarely occurring *virescens* palm, the unripe fruits are green, turning a light reddish-orange when mature. This trait serves as a conspicuous indicator of readiness for harvesting and thus attempts have been made to select for it in breeding programs, simply to aid harvesting (Donough et al., 1995; Rao, 1998).

As ripening progresses, the outer fruits separate from the bunch through a two-stage abscission process stimulated by endogenous ethylene production (Osborne & Henderson, 1993). The loose fruits either fall to the ground or are retained on the surface of the bunch, amongst epiphytes or in frond axils. Abscission is undesirable both because of the time needed to collect the loose fruits and because of the losses of oil and kernel that frequently result due to incomplete collection.

Oil synthesis and abscission are considered to be independent processes (Corley & Law, 2001), and hence inhibiting the abscission of "ripe" fruits by chemical or genetic manipulation (Chan et al., 1972; Osborne & Henderson, 1993), could provide a means of avoiding losses due to failure to collect all loose fruit. Non-shedding palms have been identified and efforts have been made in breeding programs to select for this character (Corley et al., 2006; Donough et al., 1995).

As mentioned before, loose fruits rapidly accumulate high levels of free fatty acid (FFA), presumably because they suffer damage that results either in release of endogenous lipase or favors the action of exogenous lipase of microbial origin. High FFA represents a decline in oil quality. A rise in FFA occurs at about the time of first fruit loosening (Chan & Lee, 1994; Siregar, 1977) and continues thereafter, while FFA levels in attached fruits remain low. The relationship between FFA and the number of loose fruits has been found to be approximately linear (Chan and Lee, 1994; Corley and Law, 2001; Siregar, 1977) (Fig. 5.5).

Harvesting Schedules and Criteria

Several factors determine the yield and quality of mesocarp oil obtained from the fruit bunches. These include the intervals between successive harvest "rounds", the level of ripeness at which bunches are cut, the extent of damage due to handling of the cut bunches and loose fruits, the completeness of loose fruit collection, and the time between harvest and arrival of bunches at the mill. The first two factors are somewhat interdependent and together define the harvesting schedule. The objective to

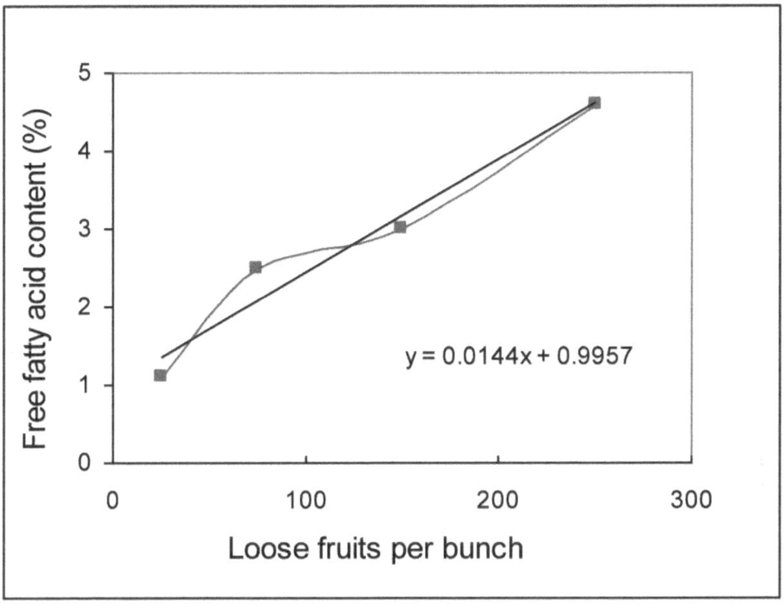

Fig. 5.5. Approximate relationship between free fatty acid contents of detached fruit and loose fruit number per bunch. Data are from Corley & Law (2001) and are means of three clones.

maximize oil yield and oil quality while minimizing the cost of harvesting inevitably requires some degree of compromise.

While the decision to harvest a bunch is based mainly on color change of the external fruits, it is generally supplemented by an assessment of the number of loose fruits, indicated either in the case of young palms by the number of empty "sockets" on the bunch or, for older taller palms, by the number of fruits on the ground (Box 5.2), assessed either before or after cutting the bunch. Needless to say the number on the ground will be less than the total number detached and the total will be greater after cutting than before, due to the impact of the bunch hitting the ground. Loose fruit number may be adjusted to take account of bunch size as larger bunches have more fruits to shed. While doing this is a more precise and logical approach, it requires a prior estimate of mean bunch weight. The bunch weight for the field recorded in the previous harvest round can be used.

Ng & Southworth (1973) found that the rate of fruit detachment differed with bunch size and with age of palm. For palms of the same age it occurred faster in small bunches than in large ones and was more rapid in five- and eight-year old palms than

Box 5.2. Elements of a harvesting system.

Frequency of harvesting (days)
Varies from 10 to 20 days depending on palm age ripeness standard crop load labour availability
Ripeness standard (loose fruit numbers)
Can be based on loose fruit numbers per bunch, per kg bunch weight or as a % of bunch weight, on the ground either before or after cutting. Varies from 1 loose fruit/bunch to 5–15 per bunch on ground before cutting or 2–10 fruits per kg.

Box 5.3. Influence of palm age on harvesting parameters.

Character	Palm age	
	young	old
Bunch size	small	large
Ripening rate	fast	slow
Fruit/bunch ratio	low	high
Harvest interval	short	long
Loose fruit numbers	few	many

in 11-year-old palms (Box 5.3). For small bunches on palms, 5 years old detachment was completed over an 11-day period whereas for older palms with larger bunches this may require 20 or more days (Gan et al., 1995).

Changes occur during fruit development in the fatty acid (FA) composition of mesocarp oil but these are not particularly useful for gauging ripeness, as the relative amounts of major FAs tend to reach their final values several weeks prior to fruit color change or the commencement of fruit abscission (Fig. 5.6).

There has been a shift in opinion over the years concerning the optimum stage at which to harvest. Dufrane & Berger (1957; reviewed by Hartley, 1988) found oil in the mesocarp of bunches weighing c. 5 kg to increase linearly with a rise in the number of loose fruits up to around 70 per bunch. Other early studies (e.g., Ng & Southworth, 1973; Southworth, 1976, 1977; Siregar, 1977) suggested that to maximize oil recovery requires a harvesting standard as high or higher, than five loose fruits per kg which, for a 20 kg bunch, amounts to 100 loose fruits. Siregar (1977) recommended that 25% of the outer fruits should be detached, amounting to one loose fruit per kg on the ground before cutting. In contrast, later work (Aziz Ariffin, 1985; Aziz Ariffin et al., 1993; Chan & Yee, 1985; Gan et al., 1995; Rajanaidu et al., 1985, 1988; Wood et al., 1985b) suggested that little further oil synthesis occurred in a bunch once fruit abscission had commenced, leading to a recommended harvesting standard of only

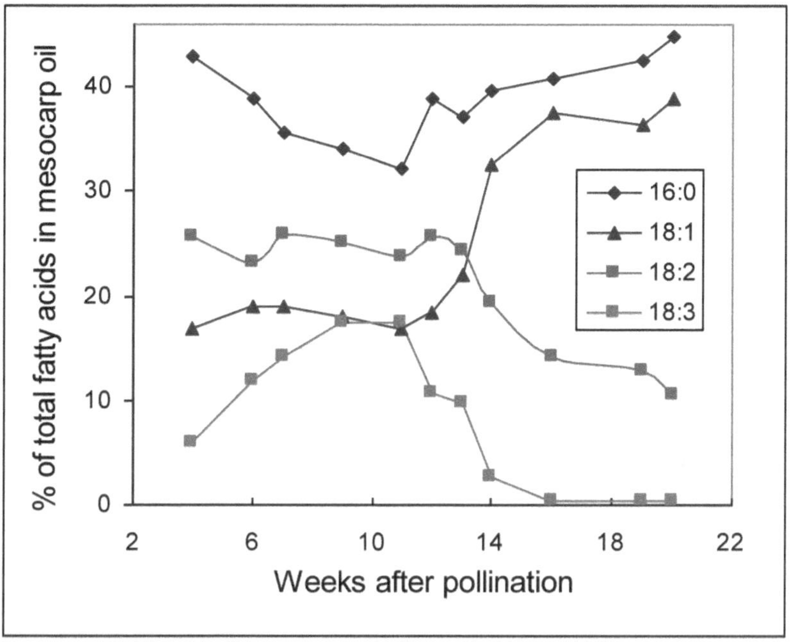

Fig. 5.6. Changes during fruit development in content of major fatty acids in the mesocarp. The four FAs shown, palmitic (16:0), oleic (18:1), linoleic (18:2), and linolenic (18:3), account for >91% of total FAs. Data are from Aziz Ariffin et al. (1993).

one loose fruit per bunch (Aziz Ariffin et al., 1993). This was deduced partly from observing similar color changes in response to staining the mesocarp of fruits taken from various positions along a spikelet. However, no checks were reported to relate staining intensity to oil content. In the early 1990s, declines in oil extraction rate (OER) in Malaysia stimulated much research into its causes (Ariffin Darus & Jalani, 1994). While a number of factors are known to affect OER, the adoption of the "one loose fruit" harvesting standard may have been at least partially responsible for the decline, as Gillbanks (2003) found that the fall in OER was reversed following re-introduction of a harvesting standard of "five loose fruit on the ground before harvest". Corley & Law (2001) and Rao et al. (2001) provided abundant evidence that oil continued to accumulate in bunches long after the first loose fruit was detected (Fig. 5.7), thus supporting the findings of other workers (e.g., Table 5-A). The problem with establishing the stage at which oil synthesis is complete is due partly to both the conventional method of bunch analysis (that results in an oil/wet bunch ratio), and the assessment of OER at the mill that also uses bunch fresh weight as the basis

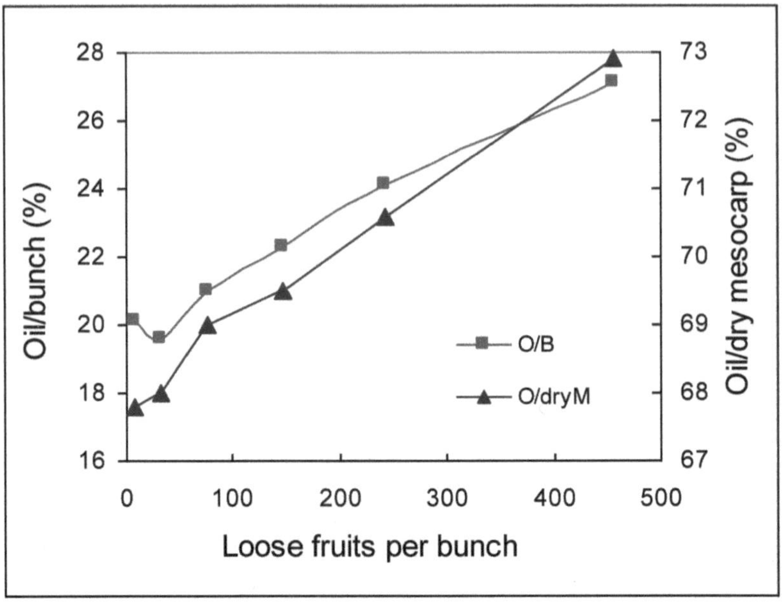

Fig. 5.7. Relationship between oil/bunch and oil/dry mesocarp and loose fruit number per bunch in oil palm clone 997. Data are from Corley & Law (2001).

Table 5-A. Changes in Bunch Components as Bunches Mature.[a]

Appearance of bunch	F/B	DM/F	O/DM	O/B	K/F	K/B
black yellow	61	30.4	73.3	13.59	14.1	8.6
black red	62	32.5	75.6	15.23	15.3	9.5
red	59	44.9	78.4	20.77	11.0	6.5
red with 31 loose fruit	58	48.8	78.7	22.28	8.6	5.0

[a]Data are from Wood et al. (1985a) and are for four bunches cut at the same time from a single palm. F = fruit; B = bunch; DM = dry mesocarp; O = mesocarp (palm) oil; K = kernel; PO = palm oil.

for expressing oil content. This means, as indicated in Fig. 5.2, that changes in % oil content in the mesocarp can occur simply as a result of changes in water content, and that changes in bunch fresh weight after harvest lead to changes in oil bunch and in OER.

Table 5-B. Composition of Fruits at Different Positions Within the Bunch.[a]

	Region of spikelet		
	Inner	Middle	Outer
Mean fruit wt (g)	5.78	7.44	10.11
WM/F	71.08	75.92	82.27
K/F	13.04	10.61	7.30
PO/WM	40.6	46.5	48.2
PO/F	28.9	35.3	39.7
PO (g/fruit)	1.67	2.63	4.01
PKO (g/fruit)	0.38	0.39	0.37

[a]Based on data of Yee et al. (1985) and adapted from Henson (1998). WM = wet mesocarp; PO = palm oil; PKO = palm kernel oil. PKO was assumed to equal half by weight of PK. Other abbreviations as in Table 5-A.

As discussed by Gillbanks (2003), the harvesting interval needs to be related to the harvesting standard to ensure bunches achieve but do not exceed the required level of ripeness between successive harvests. If bunches are harvested too early the inner fruits, which are the last to mature (Table 5-B), will be unripe. If bunches are harvested only after large numbers of fruits have become detached, then much time will be used in collecting the loose fruits, some will be missed and those collected will tend to have high FFA levels. Short intervals favor completeness of harvesting but may be more costly to implement unless the number of loose fruits is minimized so that loose fruit collection time can be reduced. If intervals are too long then loose fruit numbers increase (Fig. 5.8). Potential losses of mesocarp oil due to incomplete loose fruit collection are shown in Fig. 5.9. Usually, harvesting every 10 to 14 days will permit bunches to be harvested at an appropriate degree of ripeness using the standards presently favored.

The size of oil palm bunches increases with the age of the palms while their number decreases. Thus, old palms with larger but fewer bunches should be quicker to harvest. Loose fruit collection will also be aided due to reduced ground cover and maintenance of palm circles (weeded areas around the palm trunk). However, these advantages are offset by increasing height, which imposes a need for greater effort and skill by the harvester, both in locating as well as cutting, the ripe bunches. In addition, there may be greater loose fruit losses due to the greater impact of bunches hitting the ground.

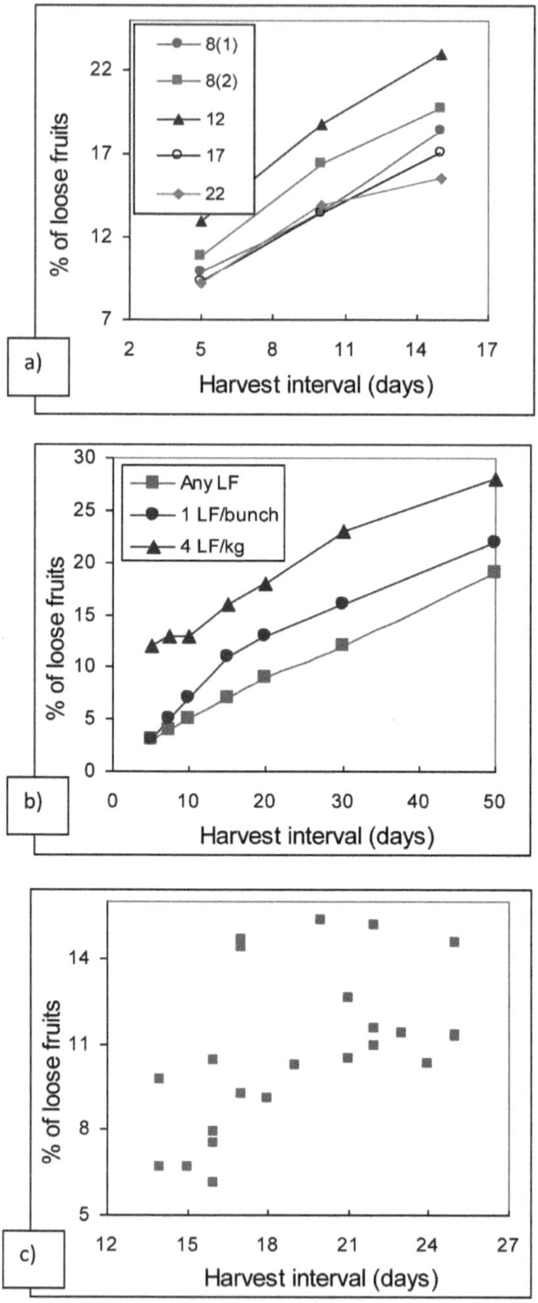

Fig. 5.8. Effects of harvest interval on amounts of loose fruit as a % of total bunch fresh weight. a) Data of Ng & Southworth (1973) for different ages of palm from 8 to 22 years and a ripening standard of one loose fruit per lb (2.205 kg). b) Data of Wood et al. (1985b) for three minimum ripeness standards (any loose fruits, one loose fruit per bunch, and four loose fruits per kg. c) Data of Sellan (2007) for different fields and months of harvest.

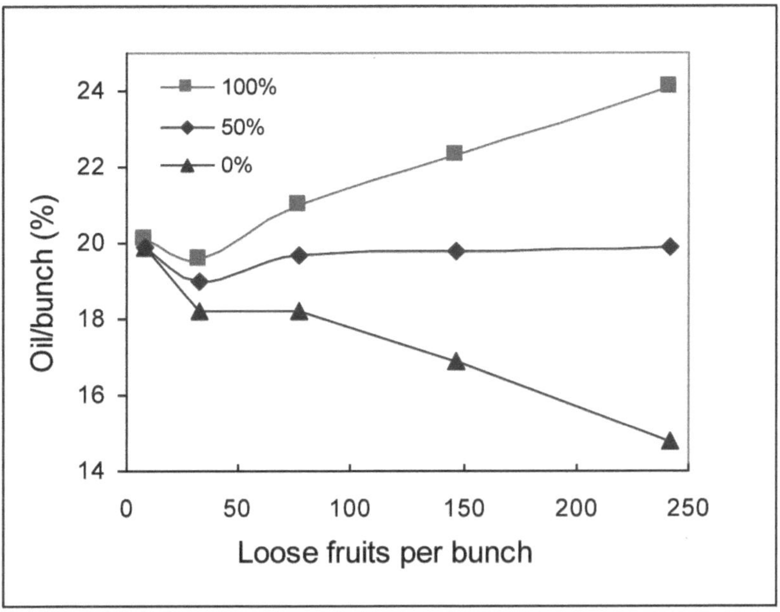

Fig. 5.9. Effects of three levels of loose fruit collection on oil/bunch ratio. Data are shown assuming 100%, 50%, and zero recovery of loose fruits and are for a single clone (#975). Note that bunch weights include the weight of loose fruits. Data are from Corley & Law (2001).

Harvesting Practice and Tools

In terms of labor requirement, harvesting represents a substantial proportion of the labor costs of the plantation. This is due both to the continuous nature of FFB production by the palms and to the difficulty in mechanizing a process that depends very much on skilled observations and decision making and entails relatively long travelling distances between palms and the need to transverse difficult terrain within the field.

Some improvements have been made in the design of harvesting tools and in methods used to handle and transport cut bunches within the field, but the basic need to locate individual bunches at the right stage for harvest and to maneuver a bulky machine within the plantation to access them poses a somewhat intractable problem for the engineer.

Bunches are presently harvested manually using a variety of tools to cut the bunch stalk; either a chisel for young palms, or a "pruning" hook attached to a long pole for palms too tall for use of a chisel (Fig. 5.10). Palms as tall as 15 m can be harvested using appropriate extensible poles. In the palm groves of West Africa (Harley, 1988), the cutter may climb the palm and chop off bunches using a machete, but this is very slow and inefficient, as well being hazardous.

Fig. 5.10. Harvesting oil palm bunches: a) using a chisel for young palms (note presence of loose fruits on ground), and b) using a pole with an attached pruning hook (shown in inset) for tall palms. In b) the "carrier" is collecting loose fruits into a bag. Note the large bunches from these old palms. Photographs courtesy of Malaysian Palm Oil Board Multimedia Unit.

The collection of loose fruits (Fig. 5.11) is usually a separate operation from the cutting of the bunch itself, and is very important for maximizing oil yield, partly because the loose fruits and outer fruits in general have the highest oil contents (Table 5-B).

Harvesting is usually a team effort with the bunch cutter being accompanied by a "carrier" whose job it is to collect loose fruits, remove excessive stalk from the bunches, and transport them to collection points or "platforms" by the road. To

Fig. 5.11. a) Harvested bunches and b) loose fruits from an over-ripe bunch, awaiting collection at the roadside. Photographs by the author.

gain access to the bunches of tall palms, the cutter also needs to remove adjacent fronds. Frond removal is avoided or minimized with young palms so as to maximize light interception and dry matter production. Cut fronds are stacked in piles usually located along alternate avenues or palm rows. In some cases bunches ready for harvest may be identified and labelled in advance (Mosquera et al., 2009), which eases the job of the cutter and avoids the harvesting of immature bunches, although, due to labor shortage, this practice is not common. The collection of loose fruits is aided by maintaining a weed-free "circle" round the base of each palm so that fruits do not get lost in vegetation.

Breeding for Efficient Harvesting

The majority of oil palm breeding programs include selection for minimum increase in the growth in height of the trunk so as to prolong the economic life of the palms, which is often limited by the practicalities of harvesting tall palms. Selection for minimum height increment may have other advantages such as less assimilates being used for trunk dry matter production and, thus, more being available for bunches.

Other characters have also been identified that should improve harvesting efficiency. These include the production of "non-shedding" palms, the use of the change in fruit color at ripening of the *virescens* palm, and the selection for bunches with long stalks that are easier to cut.

A number of non-shedding palms, the fruits of which show delayed abscission, were identified and the inheritance of delayed abscission was investigated (Corley et al., 2006). Such palms retain fruits on the bunch longer, which should lead to there being fewer losses and reduced need for loose fruit collection, with perhaps a delayed rise in FFA and a longer period of oil synthesis. The use of such variants, however, poses a problem for the harvester who uses fruit abscission as a sign of ripeness and readiness for harvest. Corley et al. (2006) suggested two ways to overcome this problem. The first is to use the black to red color change as the indicator of ripeness with a long harvesting interval (30 days or so) that allows greater opportunity for bunches to mature after color change. This may also increase the cutter's efficiency as more palms will have ripe bunches at the time of each visit and time spent walking between palms with ripe bunches would be reduced. On the other hand, the possibility that some bunches might not be seen and so remain un-harvested, could increase. The second proposal is to select for the *virescens* character which exhibits a more distinctive and easily recognized change in pigmentation than the common *nigrescens* palm, with fruits turning from green to orange during ripening. Combining this trait with delayed abscission would yield a double benefit, but this could prove quite a challenge.

The third objective of introducing long bunch stalks is intended to aid the cutting of bunches (Le Guen et al., 1990) and may be augmented by selection for other vegetative traits such as thinner petioles and changes in the angle of presentation of fronds. Presently, short stalks make it difficult to cut bunches on tall palms without removing adjacent fronds to gain access. Extra growth implied by long stalks could reduce dry matter allocation to the rest of the bunch, but this may be equally well compensated by retention of leaf area due to reduced frond pruning.

Additionally, it has been suggested (Corley, 2001) that efforts should be made to select palms that commence oil synthesis early, so that it is complete before fruit abscission starts.

Harvesting Costs and the Need for Mechanization

Although harvesting is a major and virtually continuous operation in oil palm plantations, it is, perhaps surprisingly, not the most costly item of production. Several studies (Table 5-C) show that the proportion of estate costs due to harvesting are around 15% of total costs, the actual figure being affected by differences in costs of other items and how the individual items are defined. A striking example of the former case is a reduction in the fractional costs of other items between 2007 and 2008 due to a rise in cost of fertilizers (Mohd Basri & Mohd Arif, 2009). In terms of the estate, harvesting is estimated to involve from 50 to as much as 85% of the workforce (Chew et al., 1996; Clendon, 1990; Jalani, 1998; LMC, 1999).

A time and motion study (Table 5-D) indicates that cutting bunches occupies only about 20% of the cutter's time, while about 60% of the carrier's time is spent collecting loose fruits. This work would be expected to increase with the interval between harvests. Harvesting costs tend to increase with age and thus with height of palm, as tall palms are more difficult to harvest (Table 5-E). Ling (2001) stated that harvesting costs in Sabah in 2000 varied from RM13 (~US$4) to RM30 (~US$10) per tonne FFB, depending on palm age.

While a fully mechanized system for harvesting bunches may not be impossible, the capital costs are likely to be very high. An alternative is to ease the effort and reduce the time needed to cut bunches manually. The Malaysian Palm Oil Board have

Table 5-C. Relative Costs (% of Total) of Field Operations in Oil Palm Estates.

Item	Wood (1995)	Ling (2001)	Mohd Noor et al. (2005)	Mohd Basri and Mohd Arif (2009)	
				2007	2008
field upkeep	13.0	18.3	12.0	10.7	8.1
replanting	14.7	–	–	–	–
fertilizers	22.8	23.2	22.0	32.0	52.1
harvesting	9.1	15.3	19.0	17.3	12.1
transport	1.3	9.3	19.0	16.9	12.4
general charges	39.1	33.9	28.0	23.1	15.2

Table 5-D. Relative Labor Requirements for Main Harvesting Operations Using Different Harvesting Intervals.[a]

		% of total time for each operator			
	Operation	10 days	12 days	14 days	16 days
Cutter	walking	45	30	23	16
	cutting fronds	21	24	31	32
	stacking fronds	15	27	26	29
	cutting bunches[b]	19	19	20	23
Carrier	walking	27	22	26	13
	trimming stalks	2	5	3	2
	loading bunches	6	7	7	16
	collecting loose fruit	56	56	51	71
	transport to road	9	10	13	8
Total time for both operators (minutes/palm)		3.47	3.13	4.23	4.88

[a]Data, from Gan et al. (1994), modified from Table 10.11 of Corley & Tinker (2003), are means of four harvesting intervals and are for 14-year-old palms using a minimum ripeness standard of one to ten loose fruits per bunch. [b]Includes time spent in adjusting length of harvest pole.

Table 5-E. Harvest Efficiency in Relation to Palm Height.[a]

Palm height (m)	Bunches harvested per manday
<3	200–500
3–6	75–125
6–12	50–75
>12	25–45

[a]Data are from Abdul Razak et al. (1998) and are based on a team of two workers.

introduced a mechanized cutter consisting of a vibrating sickle powered by a two-stroke engine that is reported to considerably increase harvesting productivity (Abdul Razak, et al., 2005; Mohd Basri & Mohd Arif, 2009).

Given the importance of maximizing oil yield by collecting loose fruits, mechanizing this process would be especially valuable (Corley & Law, 2001; Sellan, 2007). Prototype machines have been constructed (Ahmad Hitam & Ahmad Zamri, 2000), but they have yet to be generally adopted by the industry.

While mechanization of field operations reduces labor requirements, there may be some undesirable side effects such as increased soil compaction and emission of greenhouse gases associated with additional fuel consumption. A high level of machine maintenance and correct training of operators is vital if maximum benefits are to be obtained from introducing mechanized systems.

Handling and Transport of Bunches

Oil palm bunches can undergo quite rapid changes after harvesting, the least desirable and most common of which is the accumulation of FFA. This is promoted by damage to fruits, incurred as they fall to the ground and during subsequently handling both before reaching the mill and after at the mill unloading bay (Clegg, 1973; Ng & Southworth, 1985). To reduce FFA to a minimum, bunches need to reach the mill and be processed as soon as possible after cutting and to escape damage in the process. This requires an efficient handling system with as few transfer steps as possible. These include the collection of bunches within the field, their transport to the roadside, subsequent loading onto a lorry or other form of transport used to take them to the mill (railway wagon, barge, or overhead cable, where these are installed), followed by actual transport to the mill, which may be close by or at some distance from the field. In addition, speedy handling at the mill of the FFB received should be ensured.

Various methods to evacuate bunches from within the field have been used. Except in cases of very hilly terrain, a harvest path free of obstructions is provided between alternate palm rows, which allows for easy passage of workers and machinery. While mechanization of bunch cutting is presently some way off, more progress has been made in mechanizing bunch collection and transport within the field. Bunches may be moved using wheelbarrows, carts, or specialized mini-tractors. In some places draught animals such as buffaloes, bullocks, or mules are employed (Hartley, 1977; Wan, 1973). Details of alternative transport methods are reviewed and illustrated by Ahmad Hitam & Ahmad Zamri (2000). The efficiency of in-field transport is directly related to the degree of mechanization which needs to be balanced against the capital cost.

Bunches are usually taken to specific points or "platforms" along the road cleared of vegetation to await collection for onward dispatch to the mill. Here, the loading of bunches may be sped up by use of a net and crane system that again helps to minimize damage to fruits. Where the terrain allows it bunches can be pushed from a ramp onto waiting lorries parked below. Increasing cost and scarcity of labor has also prompted planters to consider how best to plan the layout of an estate and its road network so as to minimize distances travelled during harvesting and collection of bunches.

Conclusions

Harvesting of bunches is one of the most crucial aspects of oil palm estate management and its organization and execution has a large impact on the productivity and profitability of the enterprise. Despite much research there is still uncertainty concerning the optimum stage at which to harvest bunches in order to maximize oil yield while minimizing costs of bunch cutting and loose fruit collection. Two complementary developments that may facilitate such objectives in the future are the improvement and increased adoption of mechanization to harvest and handle bunches and the continued breeding and selection for traits that make harvesting easier and more effective. Given the considerable benefits to be expected, continued investment in further research and development in these areas is fully justified.

Acknowledgments

I am most grateful to Dr R.H.V. Corley and Mr. N.R. Menon for reviewing the manuscript, providing reference details, and offering valuable suggestions and advice.

References

Abdul Razak, J.; Desa Ahmad; Ahmad Hitam; Azmi Yahya; Johari Jamak. Force and energy requirements for cutting oil palm frond. J. Oil Palm Res. 1998, 10, 10–24.

Abdul Razak, J.; Ahmad Hitam; Abd Rahim Shuib; Johari Jamak; Malik Noor; Yosri Gono; Motorised mechanical cutter for efficient oil palm harvesting. In Proceedings of PIPOC 2005 International Palm Oil Congress (Agriculture, Biotechnology and Sustainability Conference). Malaysian Palm Oil Board: Kuala Lumpur, 2005; pp. 1189–1192.

Ahmad Hitam; Ahmad Zamri, M.Y. Mechanisation in oil palm plantations. In Advances in Oil Palm Research, Volume 1.; Basiron, Y., Jalani, B.S., Chan, K.W., Eds.; Malaysian Palm Oil Board: Kuala Lumpur, 2000; pp. 653–696.

Ariffin Darus; Jalani, B.S.; Eds. Proceedings of the National Seminar on Palm Oil Extraction Rate: Problems and Issues. Palm Oil Research Institute of Malaysia: Kuala Lumpur, 1994.

Arumughan, C.; Mohankumar, C.; Kaleysa, R. Histochemical localization of oil palm fruit lipase. In Proceedings of the 1989 International Palm Oil Development Conference. 11. Agriculture; Jalani, B.S., Zin Zawawi, Z., Paranjothy, K., Ariffin, D., Rajanaidu, N., Cheah, S.C., Mohd Basri, W., Henson, I.E., Mohd Tayeb, D., Eds.; Palm Oil Research Institute of Malaysia: Kuala Lumpur, 1990; pp. 92–100.

Aziz Ariffin, A. The biochemical aspects of ripeness standard. In Proceedings of the Symposium on Impact of the Pollinating Weevil on the Malaysian Oil Palm Industry. Palm Oil Research Institute of Malaysia: Kuala Lumpur, 1985; pp. 165–176.

Aziz Ariffin, A. A simple flotation technique to gauge ripeness of oil palm fruits and their maximum oil content. In Proceedings of the 1989 International Palm Oil Development Conference. 11. Agriculture; Jalani, B.S., Zin Zawawi, Z., Paranjothy, K., Ariffin, D., Rajanaidu, N., Cheah, S.C., Mohd Basri, W., Henson, I.E., Mohd Tayeb, D., Eds.; Palm Oil Research Institute of Malaysia: Kuala Lumpur, 1990; pp. 87–91.

Aziz Ariffin, A.; Rosnah, M.S.; Mohamadiah, B.; Wan Zailan, W.O. Morphological changes of the cellular component of the developing palm fruit (*tenera: Elaeis guineensis*). PORIM Bull. 1990, 21, 30–34.

Aziz Ariffin, A.; Rosnah, M.S.; Mohamadiah, B.; Wan Zailan, W.O.; Chen, K.W. Ripeness standard: any sign of loose fruit and with one loose fruit per bunch as the minimum standard. In Proceedings of 1991 PORIM International Oil Palm Conference—Agriculture. Basiron, Y., Jalani, B.S., Chang, K.C., Cheah, S.C., Henson, I.E., Kamaruddin, N., Paranjothy, K., Rajanaidu, N., Mohd Tayeb, D., Eds.; Palm Oil Research Institute of Malaysia: Kuala Lumpur, 1993; pp. 120–129.

Chan, K.W.; Corley, R.H.V.; Seth, A.K. Effects of growth regulators on fruit abscission in oil palm, *Elaeis guineensis*. Ann. Appl. Biol. 1972, 71, 243–249.

Chan, K.W.; Lee, K.H. OER: a concern facing oil palm industry. In Proceedings of the National Seminar on Palm Oil Extraction Rate: Problems and Issues; Ariffin Darus, Jalani, S., Eds.; Palm Oil Research Institute of Malaysia: Kuala Lumpur, 1994; pp. 1–16.

Chew, J.S.; Gan, L.T.; Chew, O.K.; Yeoh, O.T. Mechanically assisted in-field collection (MAIC)—Sime Darby's approach and experiences. In Proceedings of 1996 PORIM International Palm Oil Congress. Agriculture Conference.; Ariffin Darus, Mohd Basri, W., Rajanaidu, N., Mohd Tayeb, D., Paranjothy, K., Cheah, S.C., Chang, K.C., Ravigadevi, S., Eds.; Palm Oil Research Institute of Malaysia: Kuala Lumpur, 1996; pp. 280–293.

Clendon, J.H. The need for mechanisation on oil palm estates. Planter 1990, 66, 412–419. Quoted by Corley & Tinker (2003).

Clegg, A.J. An analysis of damage incurred by oil palm bunches during handling and transport. In Advances in Oil Palm Cultivation; Wastie, R.L., Earp, D.A., Eds.; Incorporated Society of Planters: Kuala Lumpur, 1973; pp. 421–431.

Corley, R.H.V. Oil palm. In CRC Handbook of Fruit Set and Development; Monselise, S.P., Ed.; CRC Press: Boca Raton, Fla., 1986; pp. 253–258.

Corley, R.H.V. Oil palm. In Ecosystems of the World–19. Tree crop ecosystems.; Last, F.T., Ed.; Elsevier Science: Amsterdam, 2001; pp. 299–320.

Corley, R.H.V.; Donough, C.R.; Teo, K.W.; Rao, V. Progress with non-shedding oil palms. Planter 2006, 82, 799–807.

Corley, R.H.V.; Law, I.H. Ripening, harvesting and oil extraction. Planter 2001, 77, 507–524.

Corley, R.H.V.; Tinker, P.B. The oil palm. 4th ed. Blackwell Publishing: Oxford, 2003.

Crombie, W.M. Fat metabolism in the West African oil palm (*Elaeis guineensis*). Part 1. Fatty acid formation in the maturing kernel. J. Exp. Bot. 1956, 7, 181–193.

Crombie, W.M.; Hardman, E.E. Fat metabolism in the West African oil palm (*Elaeis guineensis*). Part 3. Fatty acid formation in the maturing exocarp. J. Exp. Bot. 1958, 9, 247–253.

Donough, C.R.; Corley, R.H.V.; Law, I.H. "Non-shedding" variants of the oil palm. Int. Soc. Oil Palm Breeders News. 1995, 11, 10–11.

Dufrane, M.; Berger, J.L. Etude sur la recolte dans les palmeraies. Bull. Agric. Congo. Belge 1957, 48, 581–640.

Gan, L.T.; Ho, C.Y.; Chew, J.S.; Lam, K.S.; Kee, T.T. Recent developments in harvesting practices to maximise labour productivity, oil yields and profits. In Management for Enhanced Profitability in Plantations.; Chee, K.H., Ed.; Incorporated Society of Planters: Kuala Lumpur, 1994; pp. 315–336.

Gan, L.T.; Ho, C.Y.; Lam, K.S.; Chew, J.S. Optimum harvesting standards to maximise labour productivity and oil recovery. In Proceedings of the 1993 PORIM International Palm Oil Congress, Conference I, Agriculture.; Jalani, S., Ariffin, D., Rajanaidu, N., Mohd Tayeb, D., Paranjothy, K., Mohd Basri, W., Henson, I.E., Chan, C.K., Eds.; Palm Oil Research Institute of Malaysia: Kuala Lumpur, 1995; pp. 195–211.

Gillbanks, R.A. Standard agronomic procedures and practices. In Oil Palm. Management for large and Sustainable Yields.; Fairhurst, T., Hardter, R., Eds.; Potash & Phosphate Institute/ Potash Institute of Canada and International Potash Institute, 2003; pp. 115–150.

Hartley, C.W.S. The oil palm. 1st edition. Longmans: London, 1967; 606 pp.

Hartley, C.W.S. The oil palm. 2nd edition. Longmans: London, 1977; 806 pp.

Hartley, C.W.S. The oil palm. 3rd edition. Longmans: London, 1988; 761 pp.

Henson, I.E. Factors determining mesocarp oil to bunch ratio in the oil palm: a physiological perspective. In Proceedings of the National Seminar on Palm Oil Extraction Rate: Problems and Issues.; Ariffin Darus, Jalani, S., Eds.; Palm Oil Research Institute of Malaysia: Kuala Lumpur, 1994; pp. 27–35.

Henson, I.E. The role of bunch composition in determining oil and kernel yields of oil palm. In Proceedings of: International Conference on Oil and Kernel Production in Oil Palm—A Global Perspective. Rajanaidu, N., Henson, I.E., Jalani, B.S., Eds.; Palm Oil Research Institute of Malaysia: Kuala Lumpur, 1998; pp. 151–208.

Jalani, B.S. Research and development of oil palm towards the next millennium. In Proceedings of International Oil Palm Conference.; Angga Jatmika, et al., Eds.; Indonesian Palm Oil Research Institute: Medan, 1998; pp. 93–109. Quoted by Tinker (2000).

Jeje, A.A.; Odetola, A.O.; Zimmermann, M.H. Transpiration and oil accumulation rates for developing oil palm fruits *Elaeis guineensis* Jacq. Acta Botanica Neerlanda 1978, 27, 213–228.

LCM International. Technological change in the vegetable oil market: is palm oil being left behind?; Stringfellow, R., Ed.; LCM International: Oxford, 1999; pp. 1–6. Quoted by Tinker (2000).

Le Guen, V.; Ouattara, S.; Jacquemard, J.C. Oil palm selection with a view to easier harvesting. Initial results. Ol éagineux 1990, 45, 523–531.

Ling, A.H. Current cost structure of the palm oil industry in Sabah. In Proceedings of ISP National Seminar Strategic Directions for the Sustainability of the Oil Palm Industry.; Pushparajah, E., Ed.; Incorporated Society of Planters: Kuala Lumpur, 2001; pp. 1–11.

Mohd Basri, W.; Mohd Arif, S. Cost of palm oil production. Planter 2009, 85, 375–394.

Mohd Noor, M.; Mohd Arif, S.; Azman Ismail; Jusoh Latif. Palm oil cost of production in Malaysia: trend and performance. Planter 2005, 81, 745–762.

Mosquera, M.; Fontanilla, C.A.; Martínez; Sánchez, A.C.; Alarcón, W. Identifying oil palms with ripe bunches before harvesting, IRBBH: a strategy for increasing labor productivity. In Proceedings of International Palm Oil Conference, Agriculture, Biotechnology and Sustainability, Vol. III; Malaysian Palm Oil Board: Kuala Lumpur, 2009; pp. 993–1006.

Ng, K.T.; Southworth, A. Optimum time of harvesting oil palm fruit. In Advances in Oil Palm Cultivation; Wastie, R.L., Earp, D.A., Eds.; Incorporated Society of Planters: Kuala Lumpur, 1973; pp. 439–458.

Ng, S.B.; Southworth, A. Towards a better understanding of oil extraction ratio and free fatty acid. In Proceedings of the 10th Oil Palm Seminar; Guthrie, C., Foong, S.K., Siow, S.L., Eds.; Kumpulan Guthrie Sdn Bhd, 1985; pp. 120–131.

Oo, K.C.; Ong, A.S.D. Lipid synthesis in oil palm kernel. In Biological role of plant lipids.; Biacs, P.A., Gruiz, K., Kremma, T., Eds.; Acadamiai Kiado, Budapest and Plenum Publishing Corporation: New York, 1989; pp. 95–96.

Oo, K.C.; Teh, S.C.; Khor, H.T.; Ong, A.S.D. Fatty acid synthesis in the oil palm (*Elaeis guineensis*): Incorporation of acetate by tissue slices of the developing fruit. Lipids 1985, 20, 205–210.

Osborne, D.J.; Henderson, J. The mechanism of fruit abscission in the oil palm. In Proceedings of the 1991 International Palm Oil Conference. 1. Agriculture.; Basiron, Y., Jalani, B.S., Chang, K.C., Cheah, S.C., Henson, I.E., Kamaruddin, N., Paranjothy, K., Rajanaidu, N., Mohd Tayeb, D., Eds.; Palm Oil Research Institute of Malaysia: Kuala Lumpur, 1993; p. 470.

PORIM Annual Research Report for 1984. Vol. 1. Palm Oil Research Institute of Malaysia: Kuala Lumpur, 1984; pp. 117–119.

PORIM Annual Research Report for 1985. Vol. 2. Palm Oil Research Institute of Malaysia: Kuala Lumpur, 1986; pp. 447–453.

PORIM Annual Research Report for 1986. Vol. 1, part 2. Palm Oil Research Institute of Malaysia: Kuala Lumpur, 1987; pp. 339–341.

Rajanaidu, N.; Abdul Azis, A.; Wood, B.J.; Sarjit, S. Ripeness standards and harvesting criteria for oil palm bunches. In Proceedings of the 1987 International Oil Palm/Palm Oil Conferences. 1. Agriculture; Abdul Halim, H., Chew, P.C., Wood, B.J., Pushparajah, E., Eds.; Palm Oil Research Institute of Malaysia: Kuala Lumpur, 1988; pp. 224–230.

Rajanaidu, N.; Tan, Y.P.; Rao, V. Harvesting and estimation of oil in a bunch. In Proceedings of the Symposium on Impact of the Pollinating Weevil on the Malaysian Oil Palm Industry. Palm Oil Research Institute of Malaysia: Kuala Lumpur, 1985; pp. 177–186.

Rajaratnam, J.A.; Williams, C.N.; Ripeness and oil synthesis in tenera oil palm in Malaya. Planter 1970, 46, 339–341.

Rao, V. Ripening in the virescens oil palm. In Proceedings of International Conference on Oil and Kernel Production in Oil Palm—A Global Perspective. Rajanaidu, N., Henson, I.E., Jalani, B.S., Eds.; Palm Oil Research Institute of Malaysia: Kuala Lumpur, 1998.

Rao, V.; Gomez, M.; Chayawat, N.; Chia, C.C.; Lim, C.C. Some factors influencing oil/bunch, OER and KER. In Proceedings of 2001 International Palm Oil Congress—Agriculture. Malaysian Palm Oil Board: Kuala Lumpur, 2001; pp. 167–191.

Sellan, G. Arguments for separate oil palm loose fruit collection with special reference to a high yielding estate in Sabah. Planter 2007, 83, 181–191.

Siregar, I.M. Assessment of ripeness and crop quality control in oil palm. In International Developments in Oil Palm.; Earp, D.A., Newall, W., Eds.; Incorporated Society of Planters: Kuala Lumpur, 1977; pp. 711–725.

Southworth, A. Harvesting. In Oil Palm Research.; Corley, R.H.V., et al., Eds.; Elsevier Scientific: Amsterdam, 1976; pp. 469–477.

Southworth, A. Oil palm harvesting—a practical approach to the optimization of oil quantity and quality. In International Developments in Oil Palm; Earp, D.A., Newall, W., Eds.: Incorporated Society of Planters: Kuala Lumpur, 1977; pp. 726–739.

Thomas, R.L.; P'hang, S.; Mok, C.K.; Chan, K.W.; Easau, P.T.; Ng, S.C. Fruit ripening in the oil palm *Elaeis guineensis*. Ann. Bot. 1971, 35, 1219–1225.

Tinker, P.B. The future research requirements for the oil palm plantation. In Proceedings of International Planters Conference, 'Plantation tree crops in the new millennium; Pushparajah, E., Ed.; Incorporated Society of Planters: Kuala Lumpur, 2000; pp. 3–40.

Turner, P.D.; Gillbanks, R.A. Oil palm cultivation and management. 1st edition. Incorporated Society of Planters: Kuala Lumpur, 1974.

Turner, P.D.; Gillbanks, R.A. Oil palm cultivation and management. 2nd edition. Incorporated Society of Planters: Kuala Lumpur, 2003.

Wan, D. The use of buffaloes in oil palm fruit collection. In Advances in oil palm cultivation; Wastie, R.L., Earp, D.A., Eds.; Incorporated Society of Planters: Kuala Lumpur, 1973; pp. 432–437.

Wood, B.J. Research and development in the plantation industry. Planter 1995, 71, 171–188.

Wood, B.J.; Said Ismail; Loong, S.G.; Chew, S.C. A preliminary report on a long-term study of the effect of oil palm harvesting strategy on product recovery, including a comparison before and after weevil pollination. In Proceedings of the Symposium on Impact of the Pollinating Weevil on the Malaysian Oil Palm Industry. Palm Oil Research Institute of Malaysia: Kuala Lumpur, 1985a; pp. 187–219.

Wood, B.J.; Loong, S.G.; Said Ismail; Lee, M.H.; Quah, S.K. Mill recovery of palm oil from fresh fruit bunches (FFB) harvested to various "ripeness" standards. In Proceedings of the Symposium on Impact of the Pollinating Weevil on the Malaysian Oil Palm Industry. Palm Oil Research Institute of Malaysia: Kuala Lumpur, 1985b; pp. 255–266.

Yee, C.B.; Chan, K.W. Optimizing oil yield through harvesting standard. In Proceedings of the 10th Oil Palm Seminar, Guthrie, Chemara; Foong, S.K., Siow, C.L., Eds.; Kumpulan Guthrie Sdn Bhd, 1985; pp. 108–119.

Yee, C.B.; Lim, K.C.; Ong, E.C.; Chan, K.W. The effect of *E. kamerunicus* on bunch components. In Proceedings of the Symposium on Impact of the Pollinating Weevil on the Malaysian Oil Palm Industry. Palm Oil Research Institute of Malaysia: Kuala Lumpur, 1985; pp. 129–139.

6

Effect of Pests and Diseases on Oil Palm Yield

Gait Fee Chung
Crop Protectionist, Subang Jaya, Selangor, Malaysia

Introduction

In the cultivation of many crops in agriculture, whether cultivated in large plantations or small farms, pests and diseases prevent normal healthy growth and cause significant reduction of crop yield. Oil palm (*Elaeis guineensis* Jacq.) originated in tropical Africa and is currently planted in many countries. Indonesia has the largest planted area of 7.825 million ha, and second is 4.853 million ha in Malaysia (Ramesh, 2011). In many oil palm growing areas, oil palm is plagued by many pests and diseases.

A definition of "pest" is animals causing damage or annoyance to man, his animals, crops, or possessions. Hence, pests of oil palm include insects, mites, nematodes, rodents, birds, and other animals. A simple definition of a plant disease is any abnormal condition that damages a plant and reduces its productivity or usefulness to man. Diseases of oil palm affect different parts of the plants, namely roots, stem, fronds, inflorescence, and fresh fruit bunch (FFB). Oil palm diseases can be caused by fungi, bacteria, viruses, phytoplasma, nematodes, and parasitic plants. The damage of pests and diseases affect the growth of seedlings in nursery. In some cases, affected seedlings are killed if a control measure is not applied timely. Similarly, pests and diseases affect the healthy growth of immature planting, prolong the immaturity period, and even kill palms in the field. In mature plantings, pests and diseases also retard the healthy growth of palms, cause reduction in crop yield (fruit bunches production), and kill many palms (early replanting required).

Increased world population will demand more food production, which can only be achieved by increasing crop yields. The effective and efficient management of pests and diseases are very important to ensure healthy growth of oil palms at all stages and to maximize crop yield by achieving high oil yield per ha. In Malaysia, management of pests and diseases in oil palm incorporates integrated pest management approaches.

In this chapter, lists of selected pests and diseases are included. Selected pests (leaf eating caterpillars, leaf miner, long horn grasshopper, bunch moth, rhinoceros beetle, red palm weevil, cockchafer beetle, rat), selected diseases (basal stem rot, upper stem rot, vascular wilt, red ring disease, sudden wither, bunch rot), and their effects on oil palm yield are discussed briefly.

Pests of Oil Palms (Selected Examples)

Many insects, mites, nematodes, rodents, birds, and other animals damage oil palm in all the different growing areas of the world. Information on pests of oil palm in Southeast Asia, West Africa, and South America are given by Wood (1968, 1976a), Mariau (1976a), and Mariau (1976b), respectively. A report on vertebrate pests is given by Wood (1976b). Insect pests and vertebrate pests are reported by Basri & Norman (2000) and Chung (2000), respectively. Further reports on pests of oil palm are given in books by Basri et al. (2003), Corley & Tinker (2003), and Turner & Gillbanks (2003). Selected pests of oil palms in Central and South America are reported by Gomez et al. (2005), in Asia and Pacific regions are reported by Likhitekaraj et al. (2005), de Chenon et al. (2005), and Norman et al. (2005). Tables 6-A to 6-C list some important pests of oil palm in 3 regions.

Table 6-A. Some Pests of Oil Palm in Africa.

Common name	Scientific name	Order / Family	Reference
Leaf miners	*Coelaenomenodera lameensis (C. minuta, C. elaeidis)*	Coleoptera / Chrysomelidae	1, 2, 3
Rhinoceros beetle	*Oryctes boas, O. monoceros, O. owariensis*	Coleoptera / Scarabaiedae	1, 2, 3
Red palm weevil	*Rhynchophorus phoenicis*	Coleoptera / Curculionidae	1, 3
Leaf eating caterpillars	*Parasa pallida, P. viridissima*	Lepidoptera / Limacodidae	1, 3
Rats	*Dasymys incomtus, Lemniscomys* sp.	Rodentia / Muridae	1, 2
Cutting grass / cane rat	*Thryonomys swinderianus*	Rodentia	1, 4
Weaver bird	*Ploceus cuculatus*	Psittaciformes / Ploceidae	1, 2

Note. 1: Corley & Tinker, 2003; 2: Turner & Gillbanks, 2003; 3: Mariau, 1976a; 4: Wood, 1976b.

Table 6-B. Some Pests of Oil Palm in Central & South America.

Common name	Scientific name	Order / Family	Reference
Red palm weevil	*Rhynchophorus palmarum*	Coleoptera / Curculionidae	1, 5, 6
Leaf eating caterpillars	*Natada subpectinata*, *Sibine fusca*, *S. nesea*, *Darna metaleuca*	Lepidoptera / Limacodidae	1,3
Bunch and stem borer	*Cyparissius (Castnia) daedalus*	Lepidoptera / Castniidae	4
Root miners	*Sagalassa valida*	Lepidoptera / Glyphipterigidae	1, 3, 4
Bird (American black vulture)	*Coragyps atratus*	Cathartidae	1, 2

Note. 1: Corley & Tinker, 2003; 2: Turner & Gillbanks, 2003; 3: Mariau, 1976a; 4: Gomez et al., 2005a; 5: Chinchilla & Escobar, 2005; 6: Gomez et al., 2005b.

Leaf-eating Caterpillars

Leaf-eating caterpillars, which include bagworms, nettle caterpillars, hairy caterpillars, and other caterpillars, are serious pests of oil palm in Malaysia (Table 6-C). Leaf-eating caterpillars are always present in small numbers. They feed on oil palm leaves to complete their life cycles. The reproductive potential with many eggs and a life cycle of a few weeks to 3 months are important features of outbreaks. These caterpillars are normally under good control by natural enemies (predators, parasites, and diseases). Under normal natural balance conditions, no outbreak in oil palm is observed. Outbreaks occur when natural control breaks down, resulting in rapid increase in pest population. These caterpillars are classified as "occasional pests."

Brief Description of Pests

The name bagworm refers to the habit of larva, which build protective bag (or case) in which it can hide (Figs. 6.1 and 6.2). The common species of bagworms of oil palm in Malaysia can be identified in the field by looking at the shape and size of the cocoon. The three species of bagworms are *Pteroma pendula* (Fig. 6.3), *Metisa plana* (Fig. 6.4), and *Mahasena corbetti* (Fig. 6.5) (Lepidoptera: Psychidae). The size of the bag for *P. pendula* is about 1–1.5 cm, for *M. plana* is 2.5 cm, and for *M. corbetti* is 10 cm. Only the male moth is winged. The female moth is wingless and legless. The male moth on emergence seeks out the female for mating. Eggs are laid inside the bag.

Table 6-C. Some Pests of Oil Palm in Asia and Pacific Regions.

Common name	Scientific name	Order / Family	Reference
Bagworms	*Mahasena corbetti, Metisa plana, Pteroma pendula*	Lepidoptera / Psychidae	1, 2, 3, 4, 7, 9
Nettle caterpillars	*Setothosea asigna, Setora nitens, Darna trima, D. diducta*	Lepidoptera / Limacodidae	1, 2, 3, 7, 8, 9
Hairy caterpillars	*Amathusia phidippus*	Lepidoptera / Nymphalidae	2, 3
	Calliteara (Dasychira) horsfieldii	Lepidoptera / Lymantriidae	2, 3
Long horn grasshopper	*Segestes (Sexava) nubila, Segestidea defoliaria, Sexava coriacea*	Orthoptera / Tettigonidae	1, 2, 8
Bunch moth	*Tirathaba mundela*	Lepidoptera / Pyralidae	1, 2, 3, 7, 10
Rhinoceros beetle	*Oryctes rhinoceros*	Coleoptera / Scarabaiedae	1, 2, 3, 4, 7, 9
Red palm weevils	*Rhynchophorus schach*	Coleptera / Curculionidae	1, 2, 3
Cockchafer beetles	*Adoretus compressus, A. borneensis*	Coleoptera / Rutelidae	2, 3
Termites	*Coptotermes curvignathus*	Isoptera / Rhinotermitidae	2, 3, 10
Rats	*Rattus tiomanicus, R. r. diardii, R. argentiventer*	Rodentia / Muridae	1, 2, 3, 5, 6
Squirrel	*Callosciurus notatus*	Rodentia / Sciuridae	2, 3, 6
Porcupine	*Hystrix brachyura*	Rodentia / Hysticidae	2, 3, 6
Wild pigs	*Sus scrofa, S. barbata*	Artiodactyla / Suidae	2, 3, 6
Elephant	*Elephas maximus*	Proboscidea / Elephantidae	2, 3, 6
Birds (parrots / lorikeet / parakeets /	*Psittinus cyanurus, Lorriculus galgulus, Psittacula longicauda*	Psittaciformes / Psittacidae	1, 2, 3, 5, 6

Note. 1: Corley & Tinker, 2003; 2: Turner & Gillbanks, 2003; 3: Wood, 1968; 4: Wood, 1976a; 5: Wood, 1976b; 6: Chung, 2000; 7: Basri & Norman, 2000; 8: de Chenon et al., 2005; 9: Norman et al., 2005; 10: Lim, 2011.

Fig. 6.1. Large larva of *Metisa plana* feeding on upper leaf surface.

Fig. 6.2. Large larva of *Metisa plana* pulled out of bag.

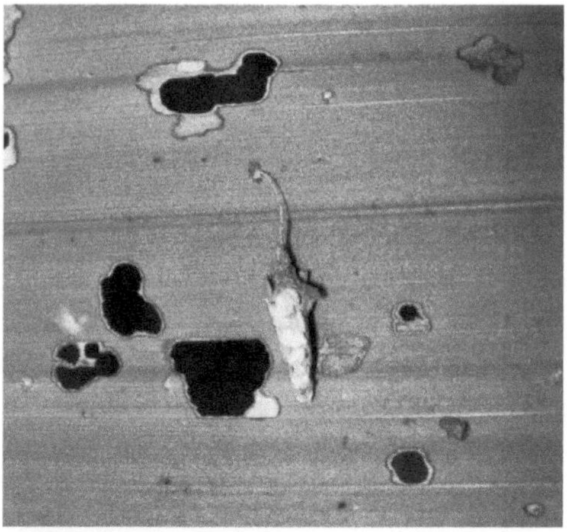

Fig. 6.3. Cocoon of *Pteroma pendula* (1–1.5 cm) hanging on thin thread.

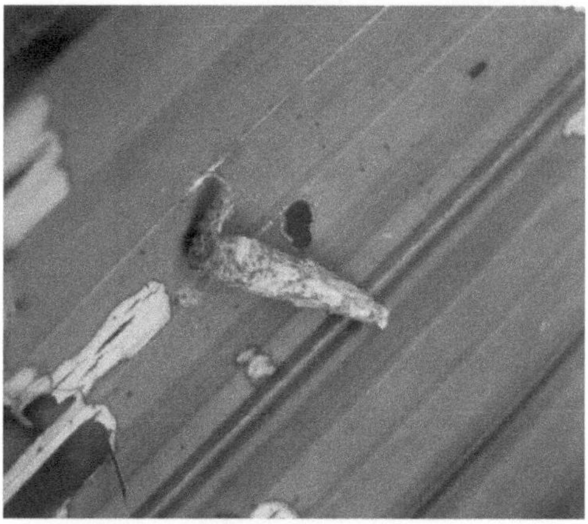

Fig. 6.4. Crook shaped cocoon of *Metsa plana* (2.5 cm).

Fig. 6.5. Cocoon of *Mahasena corbetti* (10 cm) with large pieces of leaves attached.

The common species of nettle caterpillars in Malaysia are *Darna trima* (Fig. 6.6), *Darna diducta* (Fig. 6.7), *Setora nitens* (Fig. 6.8), and *Setothosea asigna* (Lepidoptera: Limacodidae). The caterpillar of some species have tubercles with urticating hairs. Some caterpillars have bright warning colors. The sting can be quite painful. The caterpillar of *Darna* spp. can grow up to 1–2 cm long and the caterpillars of *Setora* and *Setothosea* up to 3–4 cm long. The smaller caterpillars pupate on the fronds. The larger species move down the trunk to pupate on frond bases or on the debris above ground. The adults are small, hairy moths and both male and female are winged.

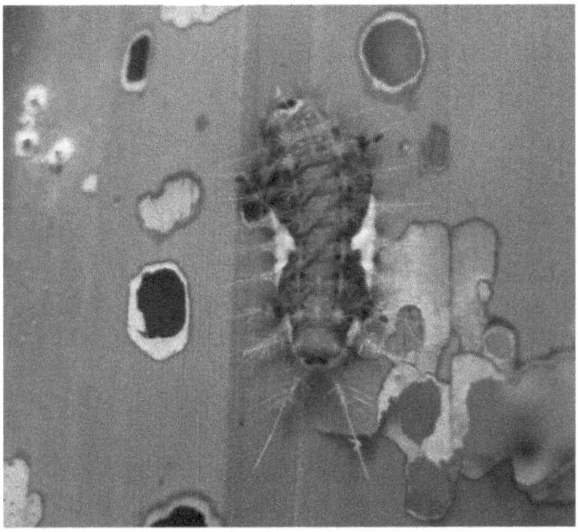

Fig. 6.6. Larva of *Darna trima* (1–1.5 cm).

Fig. 6.7. Larva of *Darna diducta* (2 cm).

Fig. 6.8. Larva of *Setora nitens* (3–4 cm).

Damage Symptoms

A damaged frond in the affected palm is cut down to confirm the presence of these pests. The early symptoms of bagworm damage are small feeding holes on fronds. Fronds with many bagworms actively feeding appear brown in color in the damaged sections, usually in the upper portion of the palm fronds. When many bagworms are feeding on the frond, the entire frond is damaged, eventually drying up completely. Fronds damaged by heavy bagworm infestation are brown in color (Fig. 6.9). The number of brown fronds increases until the whole canopy is completely brown. In the case of nettle caterpillars, larger species of bagworm, and hairy caterpillars, the damage symptoms appear as irregular feeding of the pinnae along the leaf margin and finally leaving only the bare mid-ribs (Fig. 6.10). Palms infested with caterpillars have increasingly damaged foliage, as the infestation increases until all the fronds are stripped bare.

Management Approaches

The management of leaf eating caterpillars in oil palm is a combination of ecological, biological, and chemical controls with emphasis on early detection of infestation and regular pest census. It has proven to be cost effective in solving many outbreaks in the recent past. The manipulation of the environment towards conservation and the

Fig. 6.9. Fronds damaged by bagworms (*Metisa plana*) are brown in color.

Fig. 6.10. Fronds damaged by nettle caterpillars leaving mid-ribs.

increase in the number of natural enemies would be practical to manage bagworm outbreaks (Ho, 2002). Planting of beneficial plants and field release of predatory bugs can help to prevent or reduce incidence of outbreaks and will contribute to the reduction in chemical usage. Two weekly censuses are carried out to monitor the pest population and damage on foliage in order to determine the need and timing of applying control measures. Spraying and application of natural disease-causing agents like fungi and viruses are also practiced by some larger plantations. Control of leaf-eating caterpillars using *Bt* products or selective/short residue insecticides are carried out when the pest counts exceeded action thresholds.

During outbreaks on tall palm, trunk injection of systemic insecticides is preferred because it has no adverse effect on the predators, parasites, pollinating weevils, and many non-target organisms (Chung, 1989).

Effect of Pest Damage on FFB Yield

If serious pest outbreak is not controlled, severe defoliation occurs and results in serious crop losses. Economic losses caused by defoliators have been estimated by Wood et al. (1973) through artificial defoliation trial to be as high as 40–50% loss of crop over a period of 2 years (after a single defoliation of 100%). Basri (1993) recorded up to 40% crop loss caused by an outbreak of *Metisa plana*. Liau & Ahmad (1995) reported that substantial yield loss (30–36% over 2 years) can be caused by defoliation in the immature period.

Leaf Miners

Leaf mining hispids, *Coelaenomenodera elaeidis* Mlk. (Coleoptera: Hispidae), is the most important oil palm pest in West Africa (Mariau, 1976a). After 1980, a second species *C. minuta* is recognized as the more important pest. A third species, *C. lameensis* is reported by Berti and Mariau (1999). The leaf miner (*Coelaenomenodera* spp.) is a serious oil palm pest of West Africa, regularly causing heavy defoliation in Côte d'Ivoire, Ghana, Benin, Nigeria, and western Cameroon.

Brief Description of Pests

Besides oil palm, this pale yellow beetle with reddish wing cases (4–5 mm long) is found on coconut and *Borassus* palms (Corley & Tinker, 2003). Mariau (1976a) reviewed the biology of this insect and its control. The life cycle is completed in a total of 94 days (egg stage 20 days, larval stage 44 days, pupal stage 12 days, adult to egg laying 18 days). The eggs are laid in a small cavity on the lower leaf surface. Up to 400 eggs are laid in 3–4 weeks. The larval development has 4 stages. The larvae can grow up to 7 mm and mine under the upper epidermis of the leaflets of palms of all ages. The pupae are found in the dead and dry tissues of the fronds.

Damage Symptoms

The larva mine the leaves causing serious damage to the fronds. The mining activity results in longitudinal galleries. In severe attack the greater part of the leaf tissue are destroyed. The adult beetles also are responsible for some amount of leaf destruction. The typical appearance of infested palms in the young fronds are green and the remaining older fronds are grey-brown and desiccated.

Management Approaches

The management of hispid leaf miners in oil palm is a combination of biological and chemical controls with emphasis on pest census. Various hymenopteran parasites have been attempted in biological control. The ant, *Crematogaster* sp., was reported to reduce hispid leaf miner attack, and collection for release in infected areas was carried out (Timti, 1991). Census of hispid leaf miner involves counting of adults and larvae on a frond in the lower canopy from between 25 and 30. Census is repeated at intervals depending on the pest population (Table 6-D).

When the census results record more than 20 larvae or more than 3 adults, treatment is required. Treatment is carried out by spraying or fogging of insecticide

Table 6-D. Census Frequency in Relation to Pest Counts.

Census frequency	Larvae count	Beetle count	Remarks
3 monthly	< 10	< 1	Continue census
Monthly	10–20	1–3	Continue census
Weekly	> 20	> 3	Treatment require

(Philippe, 1990a, 1990b). One or two rounds of chemical spraying are adequate to give control. Trunk injection of systemic insecticides has been recommended for the control of this pest (Phillipe & Diarrassouba, 1979).

Effect of Pest Damage on FFB Yield

The effect of severe defoliation on yield is similar to that caused by other leaf-eating pests. Philippe et al. (1979) estimated there was a 40% yield loss in the 2 years following an outbreak.

Long Horn Grasshoppers

The long horn grasshoppers are important pests of oil palm in some parts of Indonesia and Papua New Guinea (PNG). The species reported by de Chenon et al. (2005) are *Sexava coriacea*, *Sexava nubila*, and *Segestes decoratus* (Orthoptera: Tettigoniidae) in Indonesia. *Segestes decoratus*, *Segestidea defoliaria*, and *Segestidea novaeguinea* are species reported in PNG.

Brief Description of Pests

The brownish green grasshopper is 5–6 cm long with antennae reaching 20 cm long. The eggs are laid at night deep into the soil near palm bases. The life cycle lasts for 6 to 9 months.

Damage Symptoms

Both adults and nymphs are active at night and feed on the fronds. Initially, they feed on the lower and older fronds. Finally, the whole canopy is defoliated, completely leaving the midribs and rachis.

Management Approaches

The management of long horn grasshopper in oil palm is a combination of cultural, biological, and chemical controls, with emphasis on early detection of infestation and regular pest monitoring. Outbreaks are initially monitored and reported by the harvester, frond pruners, and field supervisors. After damage levels are assessed, pest counting is carried out and defoliation maps of affected areas are prepared to arrive at control recommendations. Five nymphs or adults per palm/ha is the critical level. The grasshoppers are active at night and it is difficult to assess. Therefore, light damage is considered critical level. The measures in cultural control are removal of natural host plants (wild bananas, *Heliconia* spp., sago palms, native palms), planting of suitable cover crops (exclusion of natural host plants), regular pruning of lower fronds, and hand picking of nymphs/adults for small young plantings. Preference for shade means that it is important that oil palms are kept well pruned and kept free of old fronds that tend to hang down from the central cluster.

Conservation of natural enemies and parasitoids remains the only biological control. Chemical control involves trunk injection of systemic insecticides. A second follow-up treatment is required because the incubation of eggs can last up to 3 months.

Effect of Pest Damage on FFB Yield

de Chenon et al. (2005) reported that 300 ha of oil palm has been destroyed in one night in PNG. After such defoliation, there was no production for 2 to 3 years. Long horn grasshoppers are the main insect pests of oil palm in PNG and constitute more than 80% of the reports of pest problems in oil palm in PNG each year (Page, W.Wt). The losses reported are more serious. Defoliation of upper fronds is particularly damaging. Moderate/severe defoliation reduced yield by 50% for up to 2 years. Full recovery may take up to 5 years.

Bunch Moth

The oil palm bunch moth, *Tirathaba mundella* Wlk. (Lepidoptera: Pyralidae), is a serious caterpillar pest of Malaysia and Indonesia. The bunch-infesting caterpillars is described by Wood (1968). It was once a common pest of newly mature young oil palms in the first generation planting during the 1960s and 1970s. Recently, with expansion of oil palm planting in the peat areas in East Malaysia and some areas in Indonesia, the bunch moth has again been reported as a serious pest, causing extensive damage on inflorescences and fruit bunches (Lim, 2011). The main cause could be the poor sanitation standards of young palms at maturity and partly due to difficulty in carrying out this operation in rather deep planting in peat areas.

Brief Description of Pests

The caterpillar of the oil palm bunch moth, *Tirathaba mundella,* is an important pest of fruit bunches. The caterpillars are light to dark brown in color and can grow to 4 cm (Fig. 6.11). The caterpillar lives inside the seeds. Pupation takes place in a cocoon of frass inside the bunch. The wing span of the adult moths is 2–2.5 cm. The larval period is about 2–3 weeks and the life cycle is about 1 month.

Fig. 6.11. Caterpillar of *Tirathaba mundella* feeding on spikelets of male inflorescence and fresh reddish granular frass.

Besides oil palm, *Tirathaba* spp. have been recorded attacking various palm species, such as *Areca catechu* (inflorescences), *Cocos nucifera* (flowers), *Nypa fruticans*, *Prestoea* sp., and *Pritchardia* sp. (Chung, 2010).

Damage Symptoms

All stages of inflorescences and fruit bunches are attacked by the caterpillars. The first sign of attack is the scoring and pitting of fruit surface (Fig. 6.12). Recent pest activity is indicated by the presence of fresh reddish granular frass over the surface of the infested inflorescence (Fig. 6.11) or fruit bunch (Fig. 6.12). Severely damaged fruit bunches are fully covered with frass. The damage caused by the boring of the caterpillars will lead to premature fall of fruits on young bunches, and on developing fruits the kernel is replaced with a hollow center. The feeding on mature fruits will form a characteristic ring around the style after the soft tissue is eaten.

Fig. 6.12. Caterpillar of *Tirathaba mundella* feeding on ripe fruits with scoring and pitting of fruit surface.

Management Approaches

The management of oil palm bunch moth in oil palm is a combination of cultural (sanitation), biological (spraying *Bt* product), and chemical (short residue insecticides) controls, with emphasis on early detection of infestation and regular pest monitoring. In new oil palm areas, rotten fruit bunches attract the moths. Sanitation to remove old male inflorescences and rotten fruit bunches must start early, especially in deep planted palms in peat areas. In estates with known history of infestation, sanitation is carried out early, even when no infestation occurs. In infested palms, sanitation is completed well before the spraying of insecticide commences. Spraying of *Bt* product at bi-weekly intervals is required to achieve good control. In yielding fields, it is important to practice regular complete harvesting at 10-day intervals. This operation will greatly reduce over-ripe missed bunches and rotten bunches.

Effect of Pest Damage on FFB Yield

In severe outbreak, all fruit bunches for all palms in the fields are heavily damaged, and crop loss of more than 50% have been reported (Lim, 2011).

Rhinoceros Beetle

Rhinoceros beetle, *Oryctes rhinoceros* L. (Coleoptera: Scarabaeidae), is an important pest of oil palm. *Oryctes rhinoceros* is distributed throughout Asia and the western Pacific. Besides Malaysia and Indonesia, the native range of this beetle include Bangladesh, Cambodia, China, Laos, Myanmar, Pakistan, Philippines, Sri Lanka, Thailand, and Vietnam (Invasive Species Specialist Group). It has been reported in PNG, Fiji, and other Pacific islands. In oil palm, attack by adults may reduce yield and kill palms if there are repeated attacks. They may provide entry points for lethal secondary attacks by the red palm weevils (*Rhyncophorus* spp.) or by other pathogens.

Brief Description of Pests

The adult beetles measures about 4 cm long and 2 cm wide (Fig. 6.13). Many predators, parasites, and disease-causing agents have been recorded to affect rhinoceros beetles (Wood, 1968), but none can exert adequate effective control in the prevailing

Fig. 6.13. Beetle of *Oryctes rhinoceros* with slightly curved horn.

oil palm agro-ecosystem, especially when suitable breeding materials are abundant. Thus, the rhinoceros beetle is classified as a key pest and is capable of causing serious damage in replants. The female beetles, having located a breeding site, lay the eggs into the rotting organic materials. In oil palm plantations, decaying palm logs, trunk chips, and heaps of empty fruit bunches are suitable breeding materials which are very attractive for egg laying and development of grubs. The grubs develop to maturity in the decaying materials after feeding voraciously on the organic matter, in dead standing coconut palms, decaying organic materials like compost and sawdust heaps, decaying pandanus trunk, heaps of decaying cocoa pod shells, heaps of cattle-dung, heaps of rotting paddy straw, farm yard manure, and floating logs with decaying tissues (ISSG). Pupation takes place in the breeding sites (e.g., the harder part of the decaying log or in the soil underneath, within a hard cocoon constructed from vegetative or soil matter). The adults emerge and feed, usually on palm crowns, before returning to suitable sites for egg laying. Life cycle of rhinoceros beetle in oil palm log tissue from egg to adult takes about 7 to 8 months after egg laying. In finer and more suitable media like cow dung and saw dust, the life cycle can be 5 to 6 months.

Apart from coconut and oil palm, the recorded host plants include the date palm and a variety of palms grown for ornamental purpose, including *Roystonea regia*, *Livistona chinensis*, *Corypha umbraculifera*, and *Raphia ruffia*; also recorded are pineapple, sugarcane, pandanus, and banana (ISSG).

Damage Symptoms

The beetles feed on the palm spears (unopened spindle leaf) with their chewing mouth parts. The burrowing into the spear cluster causes damage to the palms. Several symptoms of damage on the canopy of oil palm caused by rhinoceros beetles are recognized and summarized below:

a. Section of spear cut off with distinct chewing marks on remaining portion
b. Spear dieback (dead spears are easily pulled out) or snapping with chewing marks on basal end
c. Snapping of newly opened fronds
d. Fan-shaped cut fronds
e. Little leaf complex from palms recovering from severe attack
f. Oval-shaped bore holes on the rachis and/or petiole of fronds
g. Cluster of chewed materials close to frond bases
h. Complete absence of many upper fronds and rotting in the growing point (cabbage)

Severely damaged immature palms have many fan-shaped cut fronds (Fig. 6.14).

Fig. 6.14. Immature palm severely damaged by rhinoceros beetle with many fan-shaped cut fronds.

Management Approaches

The management of rhinoceros beetles in oil palm is a combination of ecological control, cultural control, physical control, biological control, chemical control, and semiochemical control. The current emphasis is preventive control, with emphasis on crop residue management and ground cover management at replanting. These steps are essential in areas with past history of serious infestation. The monitoring of pests and reduction of beetle populations are required in such areas by installation of vane traps baited with aggregation pheromone, oryctalure (ethyl 4-methyloctanoate). In replants with low pest population, mass trapping of beetles can remove large number of rhinoceros beetles and prevent serious damage to immature palms (usually no spraying of insecticide is needed). Even with high pest populations, causing serious risks to canopy damage or even palm death, immature palms are protected by targeted spraying with short residue synthetic pyrethroids to protect spear and new fronds. This combination has proven to be effective in solving many outbreaks. Commercial formulated biopesticide products containing spores of *Metarhizium anisopliae* are available for field application (FASSB, 2010). There has been no report of its commercial application.

Table 6-E. Effect of Damage at Immature Phase Caused by *O. rhinoceros* on FFB Yield in First Year of Harvesting.

Harvesting	Minimum damaged palms	Moderately damaged palms	Severely damaged palms
FFB yield in 12 months	100%	34.5%	8.3%
Crop loss	0%	65.5%	81.7%

Effect of Pest Damage on FFB Yield

In cases of continuous repeated attacks by the rhinoceros beetle, seedlings in polybags, young, and even mature palms can be killed. The loss caused by the death of seedlings and immature palms can be accounted for by basing total loss on the cost of producing the seedlings and the cost of supplying in immature fields. The death of a mature palm, especially at the young mature stage, equal to total loss as supply is no more feasible. High yield palms produce 185 kg FFB/palm, moderate yield palms produce 169 kg FFB/palm, and low yield palms yield 155 kg FFB/palm (Chiu, 2011). Using the low yield palm category, the potential crop loss is equal to 155 kg multiplied by 20 years of harvesting (i.e., 3100 kg).

In severely damaged, immature plantings with no control measure applied, 25% reduction in the crop from the first two years has been reported (Liau & Ahmad, 1993). In Indonesia, such damage also reduced the yield up to 79% in the first year of production (PPKS, 1996). Results of yield recording (Table 6-E) comparing 3 categories of palms with different severity of damage at immature stage have been reported (Chung et al., 1999). The results indicate that moderate to severe damage caused by rhinoceros beetles had significant adverse effect on early FFB yield, even though the affected palms had recovered after introduction of chemical control. The reduction in yield was mainly due to the delay in maturity of most damaged palms.

Red Palm Weevil

Red palm weevils are important pests in many growing areas of oil palm in the world. The species occurring in Africa is *Rhynchophorus phoenicis* (Coleoptera: Curculionodae). *R. palmarum* is the species found in the Americas. In Asia, the species is *R. ferrugineus*. In Malaysia, it is described by the synonym *R. schach,* or commonly known as red striped weevil. Incidence is low in Asia and Africa, but much higher incidences are reported in Central and South America.

The main host plants recorded are *Cocos nucifera, Elaeis guineensis, Euterpe edulis, Metroxylon sagu, Phoenix canariensis, Phoenix dactylifera*, and sugarcane *Saccharum officinarum* (EPPO Bulletin, 2007).

Brief Description of Pests

Red palm weevils (4–5 cm in length) are potentially lethal pests (Fig. 6.15). It has been reported that infested palms have been killed. The red palm weevils are vectors of a nematode [*Bursaphelenchus* (*Rhadinaphelenchus*) *cocophilus*)] that causes red ring disease in the Americas. The life cycle can last about 3 months. The adult female lays approximately 200–300 eggs. The eggs (2–3 mm long) are laid in damaged surfaces (after *Oryctes* attack/ after rat attacks/after other physical damage). Eggs hatch in 3 days into white legless larva. The grubs breed in felled palms, but newly cut stumps are preferred. The larva will feed on the soft fibers and terminal buds (Fig. 6.16), tunneling through the internal tissue of the tree for about a month. The larval stages last for 2 months. The larva will leave the tree and form a cocoon built of dry palm fibers in leaf litter at the base of the tree and the pupal stage takes about 25 days. Thus the life cycle is about 3 months.

Damage Symptoms

Almost all of the damage is caused by the larvae that feed on the inner tissues of host palm on upper parts of the trunk region near the crown. The larvae tunnel into the crown and trunk, causing surrounding tissues to decay (and may eventually kill the palm). The damage caused by the larvae is only visible long after infestation, and by the time the first symptoms of the attack appear, they are so serious that they generally result in the death of the tree.

Fig. 6.15. Red palm weevil (*Rhynchophorus schach*).

Fig. 6.16. Legless larva (right) and pupa (left) of red palm weevil (*Rhynchophorus schach*).

Management Approaches

The management of red palm weevils in oil palm is a combination of cultural control, chemical control, and semio-chemical control. Control of this weevil is not easy. Prevention is better than control. Wounding the palm must be avoided in order to prevent egg laying. For sanitation purposes, all dead and felled palms should be destroyed within 3 months (the life cycle of the red palm weevils). Effective control of rhinoceros beetles will give control of red palm weevil indirectly. In severe infested areas in the Americas, insecticide application to cut frond butts after harvesting becomes a routine measure. The most effective method is the mass trapping of weevils using bucket traps baited with pheromone and pieces of insecticide-treated sugarcane cubes (Chinchilla et al., 1995; Oehlschlager et al., 1993). The aggregation pheromone is insect specific, which attract both sexes, "rhyncholure" for *R. palmarum*; "ferrugineol"

Table 6-F. The Trap Density Recommended According to the Number of Weevil Trapped.

Number of weevil / trap / week	Trap density	In the center of grid
More than 8	1 trap / 2 ha	16 palms × 18 rows
8 or less	1 trap / 4 ha	22 palms × 26 rows
2 or less	1 trap / 10 ha	35 palms × 41 rows

& "ferrugineone" for *R. ferrugineus*. The trap density varies according to the number of weevil trapped per week. The recommended trap densities are given in Table 6-F.

Effect of Pest Damage on FFB Yield

Crop loss is a direct result of palms killed by the development of the grubs in the trunk, especially when the apical meristem tissues are destroyed. Serious losses of up to 5% have been reported in mature plantings caused by the transmission of a nematode resulting in red ring disease (Mexzon & Chinchilla, 1993).

There have also been reports of this nematode (causal agent of red ring disease) transmitted by red palm weevils causing losses up to 80%; however, the losses typically range from 10% to 15% on coconut palms and oil palms (Esser & Meredith, 1987).

Cockchafer Beetles

Cockchafer beetles are important pests of oil palm in Malaysia, usually causing serious damage in nursery, immature plantings, and some newly mature plantings. The smaller species, such as *Adoretus compressus* (Fig. 6.17), and *Adoretus borneensis*

Fig. 6.17. Brown color cockchafer beetle (*Adoretus compressus*), about 1–1.5 cm.

(Coleoptera: Rutelidae), have been found to cause severe damage to immature palms (Chung, 1998; Wood, 1968). These beetles are night flying and feed on leaves of a wide variety of plants, namely legume covers, *Melastoma malabathricum*, *Tetracera* spp., *Hedyotis* sp., *Stenochlaena palustris*, volunteer oil palm seedlings, and so on.

Brief Description of Pests

Adoretus spp. are small beetles of about 1 to 1.5 cm in length (Fig. 6.17). They hide in the soil during daytime and emerge flying at night. The eggs are laid in the soil and the grubs feed on plant roots or vegetable detritus. The fully developed grubs will become prepupae first and then become pupae. The complete life cycle from eggs to adults requires 3.5 to 4 months.

Damage Symptoms

These small cockchafer beetles feed by chewing irregular portions and usually from the edge of the pinnae. *Adoretus* beetles usually eat small holes in the leaves. When the fronds are badly damaged with many holes the remaining green tissues usually dry up slowly, which is especially serious during dry weather. During severe attacks by a large number of beetles young palms have been skeletonised, leaving the midribs and main veins. In most nurseries and immature plantings, some damage caused by *Adoretus* spp. can be detected on seedlings (Fig. 6.18) or immature palms, usually starting from the lower fronds. During serious outbreaks, a large number of *Adoretus* beetles can cause up to 80% defoliation of newly planted seedlings within 2 weeks (Fig. 6.19).

Fig. 6.18. Irregular shaped small holes caused by *Adoretus* feeding on seedling.

Fig. 6.19. Damage to fronds of newly planted seedling in the field caused by cockchafer beetle (*Adoretus compressus*).

It is not uncommon to find the lower fronds of immature palms badly damaged and skeletonized.

Management Approaches

The management of cockchafer beetles in oil palm is a combination of ecological control and chemical control. The current emphasis is on regular pest surveillance to detect damage symptoms on oil palm/legume cover crops. Night inspection using a torch light is carried out to confirm the identity of pest species and to gauge the pest population. At high pest population, immature palms are protected from damage by targeted spraying with short-lived synthetic pyrethroids. A few rounds of spraying need to include the spraying of legume cover crops, which are alternative host plants. Other alternative host plants, which include several species of common weeds and volunteer seedlings inside the immature fields or in the surrounding mature fields, have to be eradicated by spot spraying with a suitable herbicide in order to reduce cockchafer numbers in the plantations.

Effect Of Pest Damage on FFB Yield

The results of a field experiment carried out to monitor the FFB production of 3 categories of cockchafer beetle damage indicate that light damage (< 25% defoliation) to moderate damage (25% to < 50% defoliation) had no effect on FFB yield after successful control (Chung, 2007, 2008). However, severe damage (> 50% defoliation) caused by cockchafer beetles had a significant adverse effect on FFB yield, even though the affected palms had recovered after introduction of chemical control. The reduction in FFB yield by 27% in 27 months of harvesting was mainly due to the delay in maturity of most damaged palms. The corresponding reduction in the number of fruit bunches harvested is 18%.

Rats

Rats are important pests in oil palms (Chung, 2000; Wood, 1969). Three species of rats are common pests of oil palms in Malaysia, namely *Rattus tiomanicus, Rattus argentiventer* (Fig. 6.20), and *Rattus rattus diardii* (Rodentia: Muridae). Other species

Fig. 6.20. Close up view of *Rattus argentiventer* and loose fruits.

are occasionally found, usually in oil palm areas next to a secondary forest, jungle reserve, or similar habitats (e.g., *R. rajah, R. annandalei, R. whiteheadi*, and *R. exulans*). Rat species reported as pests of oil palms in Sabah (East Malaysia) are *Rattus muelleri* and *Rattus rattus diardii.*

Brief Description of Pests

The common species of rats can be identified by looking at the several features and measurements given in Table 6-G.

Rats occur in nurseries, new plantings, replants, immature, and mature plantings. Their presence may also be deduced by the appearance of the following:

a. burrows
b. clear runways
c. nests on canopy
d. nests under frond piles
e. feeding platforms of loose fruits and debris near the bunches
f. feeding of loose fruits under the frond piles

Food and shelter are two important factors that influence the rapid buildup of rat populations, besides the adaptability of local rat species to the mono cropping and perennial oil palm agro-ecosystem. In most oil palm plantations, food and shelter are abundant and the rat population can increase to a relatively high level causing serious damage. In the more stable environment of oil palm in the tropics, rat populations do not increase indefinitely, nor fluctuate markedly with season. Rat numbers change slowly within broad limits. This stability of an undisturbed population is illustrated in a long term study of *R. tiomanicus* in oil palms (Wood & Liau, 1984a, 1984b). The estimated population size fluctuated slowly over a period, between about 200 and 600 per hectare.

Table 6-G. Features and Measurements for Identification of Common Oil Palm Rats.

Rat species	*R. tiomanicus*	*R. r. diardii*	*R. argentiventer*
Head + Body length (mm)	100–180	110–200	110–220
Tail length (mm)	125–198	113–246	130–192
Weight (gm)	110	180	190
Belly color	white	brown	Silver grey
No of mammae—pectoral	2 pairs	2 pairs	3 pairs
No of mammae—inguinal	3 pairs	3 pairs	3 pairs

Damage Symptoms

Rats feed and damage germinated seeds and seedlings in nurseries and immature palms/ mature palms in field plantings. The following damage symptoms have been recorded:

a. pulling out germinated seeds and feed on the tender portion of plumule only
b. chewing into the bulb (basal portion of seedling) and feeding on the growing apical point (cabbage)
c. chewing on the lower fronds (Fig. 6.21) and feeding on the growing apical point (cabbage)
d. chewing and feeding on unripe fruits (Fig. 6.22; fresh damage in green color)
e. chewing and feeding on ripe fruits (Fig. 6.23; fresh damage in orange color)
f. chewing through the hard shell and feeding on the kernel (Fig. 6.22; white in color)
g. chewing on old male flower after shedding pollens and feeding on grubs of pollinating weevils (*Elaidobius kamerunicaus*)
h. chewing on unopened flowers (male and female) and developing fruit bunches at high population

Fig. 6.21. Badly damaged lower fronds caused by rat chewing.

Fig. 6.22. (A) Fresh damage, (B) kernel, and (C) old damage caused by rat on unripe fruits.

Fig. 6.23. Fresh damage caused by rat on ripe fruits.

Management Approaches

Management of rats in oil palm includes the combination of ecological, cultural/agronomic measures, biological, and chemical controls. In ecological control, the plantations have to implement good agricultural control in weed control and ground cover management and maintain all habitats for natural predators. Hunting of predators must be prohibited. To reduce the food supply to rats, several cultural/agronomic standards must be put in place, namely regular and complete harvesting, efficient loose fruit collection, good pruning standards (short frond butts), and good ground cover management. For long-term biological control, barn owl nest boxes can be set up at 1 unit per 10 ha and the owl population carefully monitored by regular assessment on nest box occupancy rates. These are preventive controls, and plantation personnel must be vigilant to assess their effectiveness. When these measures fail to keep rat populations low and prevent serious damage, rat baiting has to be carried out.

In areas of low rat populations and pockets of palms with more damage, rat damage censuses are carried out and baiting commences when the percentage of palm with fresh damage of 5% or more is noted. In areas of high rat population and palms with severe damage, baiting is carried out in all oil palm areas. Baits taken are replaced at intervals of 4 days until bait acceptance reaches 20% or less (indicating low rat population after successful baiting). Baiting campaigns are repeated at intervals of every six months to keep the rat population low. When low rat population (as indicated by low damage on oil palm fruits) is achieved, the plantation can revert back to a census and baiting technique.

Effect of Pest Damage on FFB Yield

Field-collected rats confirmed that the bulk of their diet is from oil palm fruit. Therefore, loss can be estimated from the known population size multiplied by the average amount eaten in captivity. This indirect method found that the amount of mesocarp eaten by the rats equal about 5% loss of the oil crop (Wood, 1976b; Wood & Liau, 1978). However, even if only ripe fruit is eaten, other parts of the fruits are also destroyed subsequently (affecting both oil and kernel production).

The losses from the removal of detached loose fruits by rats have been estimated, and this results in the loss of 7–10% of the oil crop (Liau, 1990).

Important Diseases

Many species of fungi, bacteria, viruses, phytoplasma, nematodes, and other agents are known to attack oil palm and cause many important deadly diseases in different areas of the world. Diseases of oil palm in Southeast Asia and the South Pacific are reported by Turner (1976). Renard (1976) reported diseases in Africa and South America. A more detailed report on oil palm diseases and disorders was given by

Turner (1981), and major diseases of oil palm were reported by Ariffin (2000). More reports on diseases of oil palm are given in books by Basri et al. (2003), Corley & Tinker (2003), and Turner & Gillbanks (2003). Important diseases of oil palm are listed accordingly in 3 regions (Tables 6-H to 6-J).

Table 6-H. Some Important Diseases of Oil Palm in Africa.

Disease name	Causal agent	Organism	Reference
Vascular wilt or Fusarium wilt	*Fusarium oxysporium* f. sp. *elaeidis*	Fungi imperfecti	1, 2, 4, 5,
Dry basal rot	*Ceratocystis paradoxa*	Ascomycete	1, 2,
Basal stem rot	*Ganoderma boninense*	Basidiomycete	3,
Cercospora leaf spot	*Cercospora elaiedis*	Fungi imperfecti	1, 3, 4,

Note. 1 Corley & Tinker, 2003; 2: Turner & Gillbanks, 2003; 3: Turner, 1981; 4: Renard, 1976; 5: de Franqueville & Diabate, 2005.

Table 6-I. Some Important Diseases of Oil Palm in Central and South Americas.

Disease name	Causal agent	Organism	Reference
Red ring disease	*Bursaphelenchus cocophilus*	Nematode	1, 6,
Spear rot / Bud rot	Various causes	Fungi, bacteria	1, 3, 4, 6,
Sudden wither / *Marchitez sorpresiva*	*Phytomonas staheli*	Protozoa	1, 2, 3, 4, 6,
Vascular wilt or Fusarium wilt	*Fusarium oxysporium* f. sp. *elaeidis*	Fungi imperfecti	1, 2, 3, 5,

Note. 1 Corley & Tinker, 2003; 2: Turner & Gillbanks, 2003; 3: Turner, 1981; 4: Renard, 1976; 5: de Franqueville & Diabate, 2005; 6: Gomez et al., 2005a.

Table 6-J. Some Important Diseases of Oil Palm in Asia and Pacific Regions.

Disease name	Causal agent	Organism	Reference
Basal stem rot	*Ganoderma boninense*	Basidiomycete	1, 2, 3, 4, 5, 6, 7, 8
Upper stem rot	*Ganoderma boninense, G. zonatum*	Basidiomycete	1, 2, 3, 4
Bunch rot	*Marasmius palmivorus*	Basidiomycete	1, 2, 3, 4,

Note. 1 Corley & Tinker, 2003; 2: Turner & Gillbanks, 2003; 3: Turner, 1981; 4: Renard, 1976; 5: Singh, 1991; 6: Ariffin et al., 2000; 7: Flood & Hasan, 2005; 8: Idris & Ariffin, 2005.

Basal Stem Rot (BSR)

Basal stem rot has certainly emerged as the most important disease of oil palm in Malaysia, as well as in Indonesia and PNG. Oil palms planted in the coastal clay soils in Peninsular Malaysia have higher incidence of BSR. Serious incidences of BSR in inland, lateritic and peat areas have been reported (Benjamin and Chee, 1995; Rao, 1990). BSR can be seen from young mature palms to tall old palms. Recently it was reported to be serious in palms planted in peat soils (Lim & Udin, 2010).

Brief Description of Disease

BSR is a root disease of oil palm, which includes the infection on the basal stem (Fig. 6.24). The roots and the affected basal stem are killed by the fungal pathogen. Water and nutrient transport to the upper parts of the palms, especially the foliage, are severely restricted. The causal pathogen of BSR is *Ganoderma boninense* (Fig. 6.25). All the *Ganoderma* isolates from diseased oil palms from various locations in Penin-

Fig. 6.24. Dry rot lesion at the basal stem.

Fig. 6.25. Fruiting bodies (sporophores) of *Ganoderma boninense.*

sular Malaysia were all the same species, *G. boninense* (Ho & Nawawi, 1985). Based on the morphology of the basidiomata, other species of *Ganoderma*, namely *G. boninense*, *G. miniatocinctum,* and *G. zonatum* have been reported to be associated with BSR (Idris & Ariffin, 2004; Khairudin, 1990).

Root contact with a source of inoculum is believed to be the method of infection and spread. Infection by root contact has also been described for BSR in Zaire. Several field experiments investigating the role of diseased crop residues on disease spread by planting indicator seedlings at close distances have proven beyond doubt that root contact is an important method of infection to immature seedlings. A report by Rees et al. (2009) confirms that root invasion is probably the main mode of infection by *G. boninense Ganoderma* spores, which can disperse by air movement and strong winds. The spore will germinate upon landing on a suitable substrate on the potential host plant and begins the infection process. It was suggested that basidiospores colonize new substrate and become new infection foci. The basidiospores of *G. boninense* are readily trapped in BSR infected fields. This mode of spreading is most apt to account for the occurrence of USR (upper stem rot) on the higher part of the stem. However, its overall role in *Ganoderma* diseases requires further research.

Damage Symptoms

The earliest visual symptoms occur on the canopy. Diseased palms thus exhibit visual symptoms of frond wilting and malnutrition. These symptoms are

1. several unopened new fronds (several spears)
2. wilting of green fronds hanging downward, like a skirt
3. yellowing fronds due to nutrient deficiency
4. small canopy due to production of smaller fronds

The visual symptoms are not specific to BSR. The detection of the disease lesion (dry rot) at the palm base (Fig. 6.24) and the appearance of one or more fruiting bodies (Fig. 6.25) are the two important diagnostic symptoms for confirmation of BSR in the field. Palms in advanced stages of BSR have very small canopies with numerous yellowing fronds and flowering/fruit production greatly impaired or none at all before dying (Fig. 6.26). In some cases, a palm with a healthy canopy and full of fruit

Fig. 6.26. Advanced symptoms of basal stem rot.

bunches may topple over due to the windblown effect on the fairly badly damaged stem tissues caused by lesion of BSR.

Management Approaches

Many measures have been tested and practiced for the integrated management of BSR. Several papers have reported the cost effectiveness of soil mounding on BSR (George et al., 2000; Ho & Khairudin, 1997; Lim et al., 1993). Soil from adjacent areas is collected for heaping at the palm trunk up to a height of 75 cm and 1 m radius wide at the ground. Soil mounding only helps to prolong the economic life of yielding palms affected by BSR. It is not effective in control of BSR. Sanitation is required if there are any fallen palms or dead palms.

The removal of diseased tissues by cutting using a hand held chisel (Turner, 1981) or mechanically using the blade of a back-hoe (Singh, 1991) is practiced in some plantations. In most cases of late detection, large disease lesion will occur and the lesion will extend to below ground, including the infected root masses. Moreover, it has been observed that a BSR palm with large lesions always toppled after the surgery.

The use of isolation trenching to contain the spread of BSR has been reported (Hasan & Turner, 1998; Lim & Udin, 2010). The area for trenching is 2 m × 2 m to adequately isolate the diseased palm or disease stump. The trench is 0.5 m wide and 1 m deep. Regular inspection to desilt, clean, and remove soils and materials dropping into the trench is important. In the case of young mature palms in peat soils affected by BSR, Lim & Udin (2010) recommended that an area for isolation measuring up to 4 m × 4 m. The trench is 30 cm wide and 75 cm deep. Isolation trenches need to be desilted every 6–12 months since there will be partial filling up due to erosion of the trench edges.

Sanitation at present is the most important disease-management measure, which requires the removal of all diseased materials from the field. Sanitation aims at preventing roots of newly planted palms from coming into contact with diseased stumps or root masses of BSR palms. This involves two operations. The first operation deals with advanced BSR palms, dead palms, and vacant points at least 6–12 months before felling of old stands. A large hole of 2 m × 2 m × 1 m deep is dug in this sanitation operation. The diseased materials are dug out, shredded into small pieces, and left in the inter-row areas for decay. The second operation deals with all BSR palms and diseased material during felling and chipping of standing palms. Similarly, large holes and treatment to remove diseased materials are required. Palm logs are cut and shredded into small pieces for placement in the inter-row areas (to speed up decay). In some plantations, the crop residues are pulverized using a tractor-mounted rotary cutter and converted into powder form to speed up decay (to minimize risks of *Ganoderma* colonization and also rhinoceros beetle breeding). The supervision by plantation personnel at the replanting field to ensure the proper execution of this sanitation is critical for the success of this measure. This is now widely adopted as the standard replanting practice in most large plantations.

Ploughing of land preparation and planting of legume cover crops are measures that enhance the good agricultural practice of sanitation. These measures lend further help in the reduction of inoculum potential and fast decay of crop residues which indirectly contribute to disease management. Chemical control using fungicide is not practiced due to its inability to cure the disease and troublesome logistical requirements. Biological control and breeding resistant planting materials are two active approaches that are being intensified.

Effect of Damage on FFB Yield

In coastal estates where the BSR has reached epidemic proportions, up to 85% of the original stands were killed by the time palms were 25 years old (Singh, 1991). In an estate in Perak, the BSR incidence had increased from 5.4% to 44.1% over a span of 14 years (Khairudin, 1990). Yield reduction of 26% to 46% has been recorded in fields showing 31% to 67% disease incidence (Singh, 1991). BSR palm (15 years old) on Jawa series soil produced only 43% (loss of 57%) and 19% (loss of 81%) FFB yield of healthy palms in the first and second year of detection, respectively (Khairudin, 1995).

Upper Stem Rot (USR)

USR is the stem disease of oil palm that causes the fracture at the upper portion of the stem/trunk or usually one to few meters below the palm crown (Fig. 6.27). This stem disease is more common on mature palms at older ages. USR are seen in both coastal and inland environment. Recently, it has been reported USR (or mid stem rot) is also increasingly spotted in prime age palms planted in peat areas in Indonesia (Lim & Udin, 2011).

Brief Description of Disease

Turner (1981) reported that fruiting bodies of two fungi (*Phellinus noxious* and *Ganoderma* sp.) were present in USR. Recently, *Ganoderma* fruiting bodies were readily seen on incidences of USR occurring on tall palms both in coastal and inland environments. Researchers from Malaysian Palm Oil Board (MPOB) had isolated two *Ganoderma* species (*G. boninense* and *G. zonatum*) from USR (Idris & Ariffin, 2004).

Damage Symptoms

No visual symptom is seen until the palms affected by USR are detected. The lesion at the breaking portion of the trunk is typically the dry rot lesions of *Ganoderma*. It is also quite common to find the fruiting bodies of *Ganoderma* around the affected trunk region (Fig. 6.28, on stem proper or frond butts).

Fig. 6.27. Upper stem rot in tall palm.

Fig. 6.28. Fallen stem section with fruiting bodies of *Ganoderma boninense*.

Management Approaches

Currently, no control measure is available. Even the early detection of USR has not been found for preventive treatment.

Effect of Damage on FFB Yield

It is only possible to conclude that the loss of FFB yield is the loss of the production potential of the palm for the many years it can live if not killed by USR. The loss for low yield palm is the FFB yield (155 kg/palm/year), multiplying the remaining years before replanting. Similarly, yield loss for moderate yield palm and high yield palm are 169 kg/palm/year or 185 kg/palm/year, multiplying the remaining productive years.

Fusarium Wilt or Vascular Wilt

Fusarium wilt or Vascular wilt is the most serious disease of oil palm in Africa, especially in replantings. It has also been recorded in Ecuador and Brazil. In its acute form, affected palms are killed rapidly. In the chronic form, affected palms produce zero to very low yield.

Brief Description of Disease

The causal agent of this disease is a fungal pathogen, namely *Fusarium oxysporium* f. sp. *elaeidis* (Foe in short from). The infection occurs from the soil-borne fungus (produces macro- and micro-conidia and long-lived chlamydospores) infecting the roots and growing towards the stem. Infection can also occur through wounds in the stem. The disease is detected in the field by looking at the external symptoms of dry fronds and hanging dry fronds. Sample of tissue showing vascular browning can help to diagnose infection on visually healthy palms.

Damage Symptoms

The external symptoms of the chronic form of the disease on mature palms are that the older fronds become desiccated and the rachis bends near the bases with the fronds hanging downward. New erect fronds of affected palms gradually become smaller in size and sometimes turn chlorotic. In the acute form of the disease, the affected frond is killed rapidly and dried out while still in an erect position. Remaining fronds also die quickly. The internal symptom of the stem is that the vascular bundles become brownish grey or black in color, caused by the fungus growing in the xylem vessels. The drying of the fronds is caused by both death of infected roots and the blocking of the xylem vessels.

In immature palm, the disease is diagnosed by a "lemon frond" symptom. A frond in the middle canopy develops a bright lemon yellow color before drying out completely. More leaves at the same level turn yellow and dry out quickly. Thus the canopy has many upper young leaves that dry out and lower older leaves that remain green.

Management Approaches

There are no easy methods for the control of *Fusarium* wilt or Vascular wilt. Replanting fields with high incidence of vascular wilt is discouraged. Sanitation before and during replanting is recommended for the control of this important disease. Palms showing chronic wilt symptoms (including palms not producing even 1 fruit bunch per year) must be destroyed at least 5 years before the mass felling of palms at replanting. All diseased palms and their neighbors must be destroyed by fire after felling during replanting. Breeding for plant resistance is considered the only practical solutions. Research by conducting screening of resistant planting materials in nursery trials is in progress. Exploitation of genetic diversity from the African center of diversity is essential, so movement of seeds and planting materials will continue under strict supervision. In areas free of vascular wilt (Malaysia, Indonesia, and the Far East), strict quarantine to prevent the importation of the disease pathogen to unaffected areas is clearly the most effective control measure.

Effect of Damage on FFB Yield

The decline in production of oil palm in African countries is caused by Fusarium wilt; a 25% loss of palms has been reported in Zaire (Turner, 1981). Dumortier et al. (1992) reported yield from palms with acute wilt in the year before death as 54% (loss of 46%) and from palms with chronic wilt as 30% (loss of 70%) that of healthy palms. Renard et al. (1993) reported yield losses as 15% in a susceptible cross and 6% in a tolerant cross. Renard and de Franqueville (1989) recorded 6–16% yield reduction in young palms, of which 5.5% showed external symptoms in field only 6 years post-planting.

Red Ring Disease

This is an important disease of oil palm found in Central and South America. The disease can increase very fast and can cause considerable damage. The disease is characterized by a red ring or red circular band discoloration (1–2 cm broad) inside the stem (more distinct towards the base). An affected palm may die within a few months.

Brief Description of Disease

The coconut nematode, *Bursaphelenchus (Rhadinaphelenchus) cocophilus* is the causal agent of the red ring disease. It is only able to reproduce in palm tissue. The red palm weevils, *Rhynchophorus palmarum,* is the vector. Other weevils reported by Gomez et al. (1996) as vectors of red ring disease in Colombia are *Dynamis borassi, Metamasius hemipterus,* and *Limnobaris calandriformis.*

Damage Symptoms

According to Chinchilla & Escobar (2007), the most striking symptom is a progressive yellowing that starts on the lower leaves and advances toward the youngest ones, which may be reduced in size. In the acute form of red ring disease, the symptom progression may be very fast and the affected palm may die within a few months after the first symptoms are noticed (progressive yellowing starting from lower leaves). On the other extreme is the chronic form of red ring disease with continuous symptoms—the youngest leaves emerge short and with several types of malformations. The new leaves emerging are much shorter than normal and the canopy appears like a flat top. The palms so affected usually do not die but may remain showing symptoms for several years.

Management Approaches

Disease management in an integrated approach considers the plantation itself and all surrounding areas. Monthly visits must be made to inspect all fields on any disease incidence and take phytosanitary action. Reduction of the vector's breeding sites and reduction in physical damage help to reduce infection. The sanitation removal of infected palms is important in order to reduce the sources of nematodes.

The mass trapping of weevils using pheromone traps helps to break disease transmission, which results in decline in disease incidence. The use of rhynchophorol (6-methyl-2-hept-2-en-4-ol), the aggregation pheromone, allows improvement captures by a factor of between 6 and 30 in traps with sugar cane as a food source. The red ring disease has been effectively controlled with the use of rhyncophorol in mass trapping of red palm weevils in an integrated approach in disease management (Chinchilla et al., 1995; Chinchill & Escobar, 2007; Oehlschlager et al., 1993).

Effect of Damage on FFB Yield

In palms less than 6 years old, incidence was normally very low, but 20% or more of the palms could be affected in plantations of intermediate age (around 12–15 years). An accumulation of 50% or more of affected palms was reached in some plantations (Chinchilla & Escobar, 2007). In one case, 22% death in palms was reported by Malaguti (1953).

The distributional range of red ring nematode mirrors that of its vector, the red palm Weevil, and the affected countries report palm crop loss of 10–15% or higher due to this nematode (Anon, 2010).

Sudden Wither

This deadly disease in Central and South America is also known as *Marchitez sorpresiva*, sudden wilt, fatal wilt, or "hartrot."

Brief Description of Disease

The sudden wilt disease is caused by the infection of protozoa, *Phytomonas staheli*. The protozoa has been found in association with sudden wilt in the phloem of roots, meristem zone, spear base, and stalks of inflorescence. Palms have been attacked from the age of one year. Several insects have been suggested to be the vectors of the protozoa, namely the root miner, *Sagalassa valida*, and the bugs *Myndus crudus* & *Lincus lethifer*. The alternate host plant of the bugs is a grass, *Panicum maximum*.

Damage Symptoms

The characteristic symptoms are (1) sudden rotting of all developing bunches, (2) reddish discoloration on top of petioles, and (3) rapid drying of fronds from lower older fronds upwards. The affected palms die in 2 to 3 weeks due to the advanced rotting in the root system.

Management Approaches

There is no known cure, and early eradication of infected plants is important. It is often difficult to recommend control due to the difficulty in providing the correct diagnosis of this disease. Measures to control the suspected vectors are used to break disease transmission. Application of insecticide around the base of palms is recommended to control the root miner, *Sagalassa valida*, which is considered a vector of the protozoa. In areas where *Lincus lethifer* is considered the vector, application of insecticide in combination with the spraying of herbicide to eradicate *P. maximum* is practiced. Chinchilla & Escobar (2007) indicated that if very basic control practices are used (such as an early eradication of infected plants, insecticide treatment of affected and neighbor palms, and, above all, a better agronomic management—weed control, drainage, and palm nutrition), sudden wither becomes a disease of secondary importance. It has been reported that some oil palms remained healthy and survived severe infestation of sudden wilt in the plantations. Selection of resistant planting material is therefore possible.

Effect of Damage on FFB Yield

The results of several papers reported sudden wither to be potentially very serious in oil palm (cited by Turner & Gillbanks, 2003). In young fields of 2 to 5 years old, 15–20% mortality has been reported. In the same estate, the sudden wilt destroyed two-thirds of the estate when the palms were 12 years old. Therefore, the effect of damage caused by sudden wither on FFB yield of the estate is very serious.

Bunch Rot

Bunch rot or *Marasmius* bunch rot is the most serious disease of oil palm fruit. The outbreaks of this disease in Malaysia and Indonesia have caused severe losses in the early days of oil palm expansion. It is potentially of economic importance in all countries, since the pathogen has been recorded in all oil palm growing countries (Turner, 1981).

Brief Description of Disease

The causal agent, *Marasmius palmivorus,* is widespread in the oil palm ecosystem, which can be seen on over-ripe fruit bunches and rotten bunches. In estates with good harvesting standards and up-to-date pruning of old fronds, *Marasmius* bunch rot is not a problem. However, when rotten bunches are available for colonization and coupled with wet seasons of high humidity it can attack healthy tissue like developing fruit bunches and frond bases and gradually spread to younger bunches and inflorescence. Serious outbreaks have been recorded in plantations with inadequate pollination, poor sanitation in wet seasons, damaged/aborted bunches due to pest attack and acid sulphate soils with high incidence of bunch abortion (Turner & Gillbanks, 2003).

Damage Symptoms

During the initial stage of infection, strands of mycelium are seen on top of the fruit bunches (Fig. 6.29). When conditions are very moist or humid, fruit bunch rapidly covered by many whitish mycelium and rhizomorphs. In the later stage, the mycelium grow over the fruits and penetrate the mesocarp to produce a wet rot in light brown color. The fruit bunches finally rot and abort due to *Marasmius* infection and secondary infection by other micro-organisms. After some time, *Marasmius* sporophores in the form of whitish mushroom—like fruiting bodies—can be seen (Fig. 6.30).

Management Approaches

In view of its widespread presence in the oil palm ecosystem, it is not possible to eradicate *Marasmius*. When an outbreak occurs, it is proper to look at the underlying cause which leads to palm infection. Some of these measures are listed below:

1. palm sanitation to remote rotten bunches
2. improve pollination
3. improve nutrition in acid sulphate soils

Many fungicides have been reported to be effective against *Marasmius palmivorus* in laboratory testing. Some of these fungicides may be useful if this disease becomes more serious in the field.

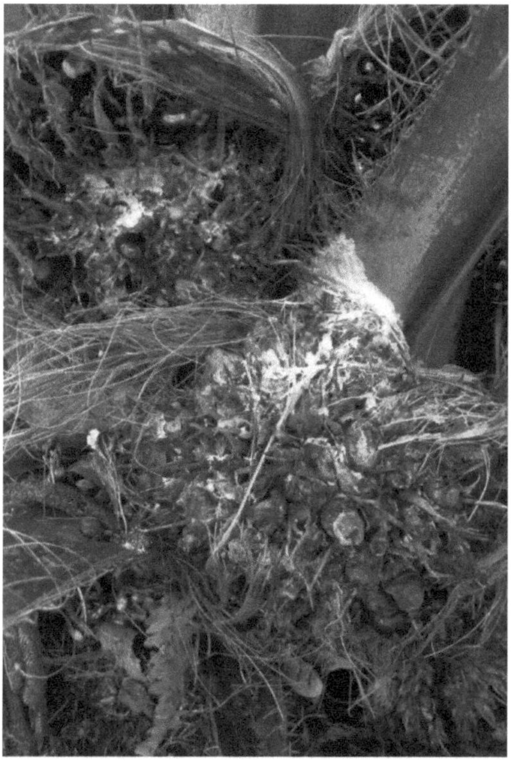

Fig. 6.29. White mycelium growing on rotten fruit bunches and frond base.

Fig. 6.30. White fruiting bodies of *Marasmius palmivorus* growing on rotten fruit bunches.

Effect of Damage on FFB Yield

Marasmius bunch rot tends to be serious in palms from 3 to 9 years of age. It can be serious in palms just reaching harvesting at maturity where sanitation has been poor or the level of pollination is inadequate. Recently, many plantations in peat soils were affected by attacks of *Tirathaba* bunch moth. *Marasmius* infection on frond bases has been seen on isolated palm, and this may initiate outbreaks of bunch rot during the wet seasons.

Conclusions

The effect of damage caused by pests and diseases are selected and summarized in Table 6-K for quick reference and to highlight their importance.

Important pests and diseases have an adverse effect on oil palm yield. Healthy growing palms are slowly or rapidly weakened by pests and diseases, and in some cases affected palms are killed. The loss of palms amounts to the loss of fruit produc-

Table 6-K. Summary of the Effects of Pests and Diseases of Oil Palm.

	Effect of damage	
Pests or Diseases	Palm mortality	Reduction in FFB yield
Leaf eating caterpillars	–	40–50% over 2 years
Leaf miner	–	40% over 2 years
Long horn grasshopper	–	50–100% over 2–3 years
Bunch moth	–	50% or more
Rhinoceros beetle	yes	91.7% for severe damage 65.5% for moderate damage
Red palm weevil	10–15%	–
Cockchafer beetle	–	27% for severe defoliation
Rats	Yes	7–10% oil crop
Basal stem rot	85% at 25 years of planting	57% in year 1 and 81% in year 2
Upper stem rot	yes	155kg/palm/year (low yield) 169kg/palm/year (medium yield) 185kg/palm/year (high yield)
Fusarium wilt	25%	46% (acute wilt) 70% (chronic wilt)
Red ring disease	22%	10–15% or higher
Sudden wither	15–20%	–
Bunch rot	–	serious

tion. The loss can be estimated by multiplying the yield potential/palm/year with the number of productive years. In cases where palms are not killed, FFB yield of palms affected by pests and diseases is reduced considerably, depending on the negative impacts of these agents and duration taken for amelioration to full recovery. Therefore, for maximization of productive palms and FFB production, efficient and timely control of pests and diseases using cost-effective control measures in line with integrated pest management are very important and critical for profitable plantation business. Monitoring and surveillance on pest and disease incidence are the basis of integrated management in oil palm plantations.

In some growing areas of the world, some key pests and diseases remain the limiting critical factor preventing the expansion of oil palm plantation industry, even in the 21st century.

References

Anon. Federal Import Quarantine Order: Palm Pests Rhynchophorus ferrugineus Olivier (Red Palm Weevil), *R. palmarum* Linnaeus (Giant Palm Weevil), and *Bursaphelenchus cocophilus* Cobb (Red Ring Nematode). January 25, 2010.

Ariffin, D. Major diseases of oil palm. Chapter 15. In Advances in Oil Palm Research, Volume 1.; Yusof, B., Jalani, B.S., Chan, K.W., Eds.; Malaysian Palm Oil Board: Malaysia, 2000; pp. 598–622.

Ariffin, D.; Idris, A.S.; and Singh, G. Status of Ganoderma in oil palm. In Ganoderma Diseases of Perennial Crops; Flood, J., Bridge, PR.D., Holdererness, M., Eds; CAB Publishing: United Kingdom, 2000; pp. 49–68.

Basri, M.W. Life history, ecology and economic impact of the bagworm, *Metisa plana* Walker (Lepidoptera: Psychidae) on the oil palm, *Elaeis guineensis* Jacquin (Palmae) in Malaysia. Ph.D. Thesis. University of Guelph, 1993.

Basri, M.W.; Norman, K. Insect pests, pollinators and barn owl. Chapter 13. In Advances in Oil Palm Research, Volume 1; Yusof, B., Jalani, B.S., Chan, K.W., Eds., Malaysian Palm Oil Board: Malaysia, 2000; pp. 466–541.

Basri, M.W.; Norman, K.; Idris, A.S.; Darus, A.; Shamala, S.; Ramle, M.; Siti Ramlah, A.A. Handbook of pests and diseases of oil palm. Malaysian Palm Oil Board: Malaysia, 2003p.

Benjamin, M.; Chee, K.H. Basal stem rot of oil palm—a serious problem on inland soils. MAPPS Newsletter 1995, 19(1), 3.

Berti, N.; Mariau, D. *Coelaenomenodera lameensis* n. sp., ravageur du palmier a huile (Coleoptera, Chrysomelidae). Nouv. Rev. Ent. 1999, 16, 253–267 (cited in Corley & Tinker, 2003).

Chinchilla, C.; Escobar, R. Red ring and other diseases of the oil palm in Central and South America. ASD Oil Palm Papers, 2007, 30, 1–27.

Chinchilla, C.M.; Oehlschlager, A.C.; Gonzales, L.M. Management of red ring disease in oil palm through pheromone-based trapping of *Rhynchophorus palmarum* (L.) In Proc. 1993 PORIM Int. Palm Oil Congr.—Agriculture; Jalani, B.S., et al. Eds.; Palm Oil Research Institute of Malaysia: Kuala Lumpur, 1995; pp. 428–441.

Chiu, S.B. Independent Consultant in Oil Palm Agronomy & Production. Penang, Malaysia., pers comm., 2011.

Chung, G.F. Spraying and trunk injection of oil palm for pest control. The Planter 1989, 65(764), 500–524.

Chung, G.F. Oil palm pest control. Paper presented in Symposium on Recent Advances in the Oil Palm Sector, 15–17 October, 1998, organized by Malaysian Oil Palm Scientists and Technologists Association, Lumut, Perak, 1998.

Chung, G.F. Vertebrate pests of oil palm. Chapter 14. In Advances in Oil Palm Research, Volume 1. Yusof, B., Jalani, B.S., Chan, K.W., Eds.; Malaysian Palm Oil Board: Malaysia, 2000; pp. 542–595.

Chung, G.F. Integrated management of insect pests of oil palm in Malaysia. Paper presented in MOSTA Seminar 23 October 2007, Prime City Hotel, Kluang, Johore, 2007.

Chung, G.F. Principles and practices of integrated pest management in oil palm. In Proceedings ACT 2008 Agronomic Principles & Practices of Oil Palm Cultivation; Goh Kah Joo, et al., Eds., 2008; pp. 211–243.

Chung, G.F.; Cheah, S.S; Balasubramaniam, R. Effects of pest damage during immature phase on the early yields of oil palm. In Proceedings 1999 PORIM Int. P.O. Cong. Emerging Technologies and Opportunities in the Next Millenium. Palm Oil Research Institute of Malaysia, 1999; pp. 454–476.

Chung, Y.C.A. Forest Research Centre, Forestry Department, Sandakan, Sabah, Malaysia. pers comm, 2010.

Corley, R.H.V.; Tinker, P.B. The oil palm. Fourth edition. Blackwell Science: United Kingdom, 2003.

De Chenon, D.; Sudharto, P.S.; Poeloengan, Z.H. Nettle caterpillars and grasshoppers of oil palm in Indonesia. In Proc. Inter. Conf. on Pests & Diseases of Importance to OP Industry; Mohd Bsari Wahid, Eds.; Malaysian Palm Oil Board: Kuala Lumpur, Malaysia, 2005; pp. 75–95.

De Franqueville, H.; Diabate, S. Status on oil palm vascular wilt. In Proc. Inter. Conf. on Pests & Diseases of Importance to OP Industry; Mohd Bsari Wahid Eds.; Malaysian Palm Oil Board: Kuala Lumpur, Malaysia, 2005; pp. 30–36.

Dumortier, F.; Van Amstel, H.; Corley, R.H.V. Oil palm breeding in Binga, Zaire, 1970–1990. Unilever Plantations: London, 1992. Cited in Corley & Tinker (2003).

Esser, R.P.; Meredith, J.A. Red ring nematode. Fla. Dept. Agric. & Consumer Serv., Div. Plant Ind., Nema., 1987, 141 (Cited in *Bursaphelenchus cocophilus:* http://en.wikipedia.org/wiki/Bursaphelenchus_cocophilus).

EPPO Bulletin. December 2007, 37(3), 571–579.

FASSB. ORY-X for effective control of rhinoceros beetle. Pesticide advertisement insert in The Planter Feb. 2010, 86, 1007.

Felda Agricultural Services Sdn Bhd, Kuala Lumpur. February 2010.

George, S.T.; Chung, G.F.; Zakaria, K. Benefits of soil mounding tall palms in a high ganoderma incidence areas in Lower Perak. In Proceedings of the International Planters Conference 2000;

Pushparajah, E., Ed.; The Incorporated Society of Planters: Kuala Lumpur, Malaysia, 2000; pp. 565–576.

Global Invasive Species Database. http://www.issg.org/database/species/ecology.asp?si=173&fr=1&sts.

Gomez, P.L.; Calvache, H.; Aldana, R.C. Stem and root borers: two important pests of oil palm in South America. In Proc. Inter. Conf. on Pests & Diseases of Importance to OP Industry; Mohd Basri Wahid, et al. Eds.; Malaysian Palm Oil Board: Malaysia, 2005; pp. 23–29.

Gomez, P.L.C.; Calvache, H.G.; Nieto Paez, L.E.; Garcia, A.A.; Alvarez, F.A.; Toquica, S.M. Pests and diseases of oil palm in South America: the American experience. In 1996 PORIM int. P.O. Cong. "Competitiveness for 21st Century"; Ariffin Darus, Eds.; Palm Oil Research Institute of Malaysia: Malaysia, 1996; pp. 303–316.

Gomez, P.L.; Munevar, F.; Tovar, J.P. Characteristics and management of the main diseases of oil palm in Colombia. In Proc. Inter. Conf. on Pests & Diseases of Importance to OP Industry; Mohd Basri Wahid, et al. Eds.; Malaysian Palm Oil Board: Malaysia, 2005a; pp. 53–66.

Hasan, Y.; Turner, P.D. The comparative importance of different oil palm tissues as infection sources for basal stem rot in replantings. The Planter 1998, 74(864), 119–135.

Ho, C.T. Ecological studies on *Pteroma pendula* Joannis and *Metisa plana* Walker (Lepidoptera: Psychidae) towards improved integrated management of infestation in oil palms. Abstract of PhD thesis, Universiti Putra Malaysia, 2002 (FP_2002_26_A.pdf).

Ho, C.T.; Khairudin, H. Usefulness of soil mounding treatments in prolonging productivity of prime-aged Ganoderma infected palms. The Planter 1997, 73, 239–244.

Ho, Y.W.; Nawawi, A. *Ganoderma boninense* Pat. from basal stem rot of oil palm (*Elaeis guineensis*) in peninsular Malaysia. Pertanika 1985, 8, 425–428.

Idris, A.S.; Ariffin, D. Basal stem rot—biology, detection and control. Paper presented in International Conference on Pests and Diseases of Importance to the Oil Palm Industry. 18–19 May 2004, Kuala Lumpur. Organized by Malaysian Palm Oil Board, 2004.

Idris, A.S.; Ariffin, D. Basal stem rot–biology, detection and control. In Proc. Inter. Conf. on Pests & Diseases of Importance to OP Industry; Mohd Basri Wahid, Eds.; Malaysian Palm Oil Board: Malaysia, 2005; pp. 134–165.

Invasive Species Specialist Group (ISSG). (http://www.issg.org/database/species/distribution.asp?si=173&fr=1&sts=&lang=EN).

Khairudin, H. Basal stem rot of oil palm: incidence, etiology and control. Master Agriculture Science Thesis. Universiti Pertanian Malaysia, Selangor, Malaysia, 1990.

Khairudin, H. Basal stem rot of oil palm by *Ganoderma boninense:* an update. In 1993 PROM Int. P. O. Cong. Uptake and Vision (Agri); Jalaini, S. et al. Eds.; Kuala Lumpur, 1995; pp. 739–749.

Liau, S.S. Rat population in oil palm replants and crop loss assessment. In Proceedings of the 3rd International Conference on Plant Protection in the Tropics. Volume IV. Malaysian Plant Protection Society, 1990; pp. 8–18.

Liau, S.S.; Ahmad, A. Defoliation and crop loss in young oil palm. In Proceedings of 1993 PORIM International Palm Oil Congress, Update and Vision (Agri); Jalani Sukaimi et al., Eds.; Palm Oil Research Institute of Malaysia, 1995; pp. 408–427.

Likhitekaraj, S.; Chayopas, T.; Athipunyakom, P.; Sienglew, P. Key pests and diseases of oil palm—biology, epidemiology and control methods. In Proc. Inter. Conf. on Pests & Diseases of Importance to OP Industry; Mohd Basri Wahid, et al. Eds.; Malaysian Palm Oil Board: Malaysia, 2005; pp. 67–74.

Lim, K.H. BMPs on peat soils. Paper presented in International Conference & Exhibition of Palm Oil, Jakarta, 11–13 May 2011, oral presentation slides.

Lim, K.H.; Chuah, J.H.; Ho, C.Y. Effects of soil heaping on Ganoderma infected oil palms. In Proceedings of the 1993 PORIM International Palm Oil Congress. Update and Vision. Agriculture. PORIM, Kuala Lumpur, 1993; pp. 735–738.

Lim, K.H.; Udin, W. Management of Ganoderma in peat soil in Indonesia. Paper presented in 2010 Second International Seminar Oil Palm Diseases: Advances in Ganoderma Research and Management. Yogyakarta, Indonesia, 31 May 2010.

Lim, K.H.; Udin, W. Ganoderma stem rot and its management on first generation oil palm on peat. The Planter, 2011, 87(1025), 585–597.

Malaguti, G. Agronomia Trop. 1953, 3, 13–31 (Cited in Corley & Tinker, 2003).

Mariau, D. Insect pests in West Africa. Chapter 26. In Oil Palm Research; Corley, R.H.V., Hardon, J.J., Wood, B.J., Eds.; Elsevier Scientific Publishing Company: Netherlands, 1976a; pp. 369–384.

Mariau, D. Insect pests in South America. Chapter 27. In Oil Palm Research; Corley, R.H.V., Hardon, J.J., Wood, B.J., Eds.; Elsevier Scientific Publishing Company: Netherlands, 1976a; pp. 385–394.

Mexzon, R.G.; Chinchilla, C.M. An inventory of the oil palm pests and their natural enemies in Central America. In 1991 PORIM International Palm Oil Conference–Agriculture (module 1); Yusof Basiron et al., Eds.; Palm Oil Research Institute of Malaysia: Kuala Lumpur, 1993; pp. 443–450.

Norman, K.; Mohd Basri, W.; Moslim, R.; Siti Ramlah, A.A.; Zulkifli, M. Bagworms, nettle caterpillars and rhinoceros beetle—biology, life cycle and control in oil palm in Malaysia. In Proc. Inter. Conf. on Pests & Diseases of Importance to OP Industry; Mohd Basri Wahid et al., Eds.; Malaysian Palm Oil Board: Malaysia, 2005; pp. 96–113.

Oehlschlager, A.C.; Chinchilla, C.M.; Gonzales, L.M.; Jiron, L.F.; Mexzon, R.; Morgan, B. Development of a pheromone-based trapping system for *Rhynchophorus palmarum* (Coleoptera, Curculionidae). J. Econ. Entomol. 1993, 86, 1381–1392.

Page, W.W. Integrated pest management strategies used by the oil palm industry of Papua New Guinea. http://www.scribd.com/doc/29346944/Integrated-Pest-Management-Strategies-Used-by-the-Oil-Palm-Industry-of-Papua-New-Guinea.

Philippe, R. Practical chemical control methods against *Coelaenomenodera minuta* Uhlmann (Coleoptera—Chrysomelidae—Hispinae) Oleagineux 1990a, 45, 165–173.

Philippe, R. Studies of Evisect S action on *Coelaenomenodera minuta* (Coleoptera—Chrysomelidae—Hispinae) Oleagineux 1990b, 45, 143–163.

Philippe, R.; Desmier De Chenon, R.; Lecoustre, R.; Mariau D. Contribution to the biological control of *Coelaenomenodera:* introduction to the Ivory Coast of parasites of *Hispinae* larvae. Oleagineux 1979, 35, 277–279. (Cited in Corley & Tinker, 2003).

Philippe, R.; Diarrassouba, S. Method of control of *Coelaenomenodera* by introduction of systemic insecticide into the oil palm trunk. Oleagineux 1979, 34; 229–233.

PPKS. New control of rhinoceros beetles with pheromone. Pusat Penelitian Kelapa Sawit (PPKS): Medan, Indonesia, 1996; pp. 1–7.

Ramesh, V. Addressing issues and challenges in the management of professional planters. In Proceedings of ISP NATSEM 2011–Enhancing value chain of the plantation industry "Seed to flying pan". 20–21 June 2011, Batu Ferringgi Penang, Malaysia, 2011; pp. 59–69.

Rao, A.K. Basal stem rot 'Ganoderma' problem in oil palm smallholders—the West Johore experience. In Proceedings of the Ganoderma Workshop, 11 September, 1990; Arifin, D., Jalani, S., Eds.; Palm Oil Research Institute of Malaysia: Malaysia, 1990; pp. 36–48.

Rees, R.W.; Flood, J.; Hasan, Y.; Potter, U.; Cooper, R.M. Basal stem rot of oil palm (*Elaeis guineensis*); mode of root infection and lower stem invasion by *Ganoderma boninense.* Plant Pathology 2009, 58, 982–989.

Renard, J.; de Franqueville, H. Oil palm vascular wilt. Oleagineux 1989, 44, 342–347.

Renard, J.J. Diseases in Africa and South America. Chapter 31. In Oil Palm Research; Corley, R.H.V., Hardon, J.J., Wood, B.J., Eds.; Elsevier Scientific Publishing: Netherlands, 1976; pp. 447–466.

Renard, J.L.; De Franqueville, H.; Diabate, S.; Ouvrier, M. Study on the impact of vascular wilt on FFB production in oil palm. Oleagineux 1993, 48, 495–504.

Singh, G. Ganoderma—the scourge of oil palm in the coastal areas. The Planter 1991, 67(786), 421–444.

Timti, I.N. Control of *Coelaenomenodera minuta* Uhlmann with *Crematogaster* species. Trop. Pest. Manage. 1991, 37, 403–408.

Turner, P.D. Oil palm diseases in South-East Asia and the South Pacific. Chapter 30. In Oil Palm Research; Corley, R.H.V., Hardon, J.J., Wood, B.J., Eds.; Elsevier Scientific Publishing: Netherlands, 1976; pp. 427–446.

Turner, P.D. Oil palm diseases and disorders. Incorporated Society of Planters: Kuala Lumpur, Malaysia, 2003.

Turner, P.D.; Gillbanks, R.A. Oil palm cultivation and management. Second edition. Incorporated Society of Planters: Kuala Lumpur, Malaysia, 2003.

Wood, B.J. Pests of oil palm in Malaysia and their control. Incorporated Society of Planters: Kuala Lumpur, 1968.

Wood, B.J. The extent of vertebrate attacks on the oil palm in Malaysia. In Progress in Oil Palm; Turner, P.D., Ed.; Incorporated Society of Planters: Kuala Lumpur, 1969; pp. 162–182.

Wood, B.J. Insect pests in South East Asia. Chapter 25. In Oil Palm Research; Corley, R.H.V., Hardon, J.J., Wood, B.J., Eds.; Elsevier Scientific Publishing: Netherlands, 1976a; pp. 347–368.

Wood, B.J. Vertebrate pests. Chapter 28. In Oil Palm Research; Corley, R.H.V., Hardon, J.J., Wood, B.J., Eds.; Elsevier Scientific Publishing: Netherlands, 1976a; pp. 395–418.

Wood, B.J.; Corley, R.H.V.; Goh, K.H. Studies on the effect of pest damage on oil palm yield. In Advance in Oil Palm Cultivation; Wastie, R.L., Earp, D.A., Eds.; The Incorporated Society of Planters: Kuala Lumpur, 1973, pp. 360–379.

Wood, B.J.; Liau, S.S. Rats as agricultural pests in Malaysia and the tropics. The Planter 1978, 46(537), 414–424.

Wood, B.J.; Liau, S.S. A long term study of *R. tiomanicus* population in an oil palm plantation in Johore, Malaysia, II—Recovery from control, and economic aspects. J. Appl. Ecol. 1984a, 21, 465–472.

Wood, B.J.; Liau, S.S. A long term study of *R. tiomanicus* population in an oil palm plantation in Johore, Malaysia, III—Bionomics and natural regulation. J. Appl. Ecol. 1984b, 21, 473–495.

7

The Economics of Malaysian Palm Oil Production, Marketing, and Utilization

Kelly Wong Kai Seng[1], Mad Nasir Shamsudin[1], and Amna Awad Abdel Hameed[2]
[1]Faculty of Agriculture, Universiti Putra Malaysia and [2]Institute of Agricultural and Food Policy Studies, Universiti Putra Malaysia

Introduction

The rapid growth of Malaysian palm oil industry has played an important role in the Malaysian economy. In 2010, the industry contributed 4.67% of the country's Gross Domestic Product (GDP) (Department of Statistics, 2010). During the last five years (2006–2010), the total area under oil palm has increased at the rate of 15.8% to reach 4.69 million hectares. In 2010, Malaysia's export earnings from oil palm products increased by 20.4% to RM 59.77 billion from RM * 49.66 billion in 2009.

Due to the oil palm economic potential and rapid expansion during the 1960s, the Malaysian government recognized oil palm as a good complementary crop to rubber in the poverty eradication program. In 1961, the Federal Land Development Authority (FELDA) first introduced 375 hectares of oil palm planted area in order to help in raising the income of the landless farmers. Owing to the decrease in rubber and tin prices, the old rubber estates land has been overtaken by the oil palm estates due to the high yields and profitability of palm oil. In 1996, Malaysia became the world's leading exporter of palm oil, surpassing Nigeria, the previous top world exporter of this oil.

In the late 1970s, the Malaysian government began to encourage a shift from crude palm oil (CPO) exports to that of its refined products through taxation and incentive policies. The "Malaysianization" of three major plantation companies previously run by the British, namely, Sime Darby, Guthrie and Harrison, and Cross field (later, Golden Hope Plantations) started in the 1980s. At that same time, the Kuala Lumpur Commodity Exchange (KLCE) was established in 1980 as an important key instrument on the price setting, hedging, and dissemination of market information to reduce market risk in the palm oil trade.

The First Industrial Master Plan (IMP1) was launched in 1986 to support and encourage the palm oil industry to sustain its upstream development. The rationalization of refining and fractionation was emphasized in the IMP1 in order to increase the palm oil industry efficiency and competitiveness in the global market. As a result,

*As of 18 Jan 2012, 1USD ≈ RM 3.13.

Malaysia turned into a hub of palm oil downstream processing, and it is now occupying a leading position in the world market of refined palm oil.

The Second Industrial Master Plan (IMP2) (1996–2005) led to the expansion of oil palm hectarage in Sabah and Sarawak. Furthermore, IMP2 encouraged the private sector to look for raw materials from abroad and stimulated participation in R&D to achieve productivity gains and more value-added product enhancement along the value chain. Meanwhile, the Malaysian Palm Oil Council (MPOC) was entrusted with the task of developing a comprehensive strategy to place Malaysia as an international leader in the oils and fats market through promotional activities. Since 2007, Indonesia has replaced Malaysia as the top producer of palm oil due to its vast land expansion and labor opportunities; however, the palm oil industry is still thriving in Malaysia, and it is still a leading palm oil exporter to the major consumers in China, EU, and India.

The Performance Management and Delivery Unit (PEMANDU) was formally established in 2009 under the Malaysian Prime Minister's Department. The main role of PEMANDU is to oversee the implementation and assess the progress of the Economic Transformation Programme (ETP) and the Government Transformation Programme (GTP). The primary target of ETP is to transform Malaysia into a high-income nation by the year 2020. The palm oil industry is among the targeted industries by PEMANDU, and its contribution to total GNI is predicted to increase by RM 52.7 billion to be RM 178 billion in 2020. In order to achieve this target increment, PEMANDU will implement eight core entry point projects (EPP) spanning the palm oil value chain to capture the fast-growing international demand for palm oil, which registered a growth rate of 10% between 2000 and 2009.

Policy Environment

Due to the economic potential of palm oil, the Malaysian government has implemented various policies to develop the palm oil industry through promoting the palm oil upstream and downstream activities in order to increase the national income. In the upstream activities, the government has implemented the replanting grant to promote the industry productivity. However, the export duty has been implemented to enhance the downstream activities and encourage the local companies to increase their export volume of both the crude and processed palm (CPO and PPO).

Replanting Grant

Until the 1970s oil palm had been planted commercially in Malaysia alongside rubber, which was then the dominant enterprise. However, major developments in the oil palm industry took place as a result of the policy of agricultural diversification that implied a shift of emphasis from rubber to palm oil. This shift was accelerated, apart from the relative profitability of palm oil compared to rubber, by the regressive nature of the tax policies on rubber and the relative low taxes on palm oil. The government also supported the production of palm oil by allowing the use of rubber replanting grants for planting oil palm. The grants amounted to RM 4,447 per hectare for holdings less than 4.05 hectares. For holdings larger than 4.05 hectares, the grant is RM 3,459.47 per hectare.

The Malaysian Government plans to boost the palm oil industry's contribution to the economy by raising its output to the gross domestic product (GDP) to RM 21.9 billion, and RM 69.3 billion of exports earnings, during the 10th Malaysia Plan period (2011–2015). To that end, the Oil Palm Smallholders Replanting and New-Planting Schemes were launched. Both schemes are among the Eight Entry Point Projects (EPP) under the Palm Oil National Key Economic Area (NKEA), introduced with the purpose of increasing the productivity and income of the smallholders.

Under the palm oil NKEA, the contribution of palm oil industry to the Gross National Income was expected to increase to RM 178 billion, compared to RM 52.7 billion in 2009. The palm oil NKEA comprises five EPPs for upstream activities, and three EPPs focused on downstream activities.

The replanting scheme is to help independent smallholders to replant old and unproductive trees, for planted area size below 40 hectares. Under this scheme, RM 7,000 per hectare will be provided for land clearing, supply of high quality seedlings, and agricultural input. Besides, smallholders with land size below 2.5 hectares will receive Farm Maintenance Assistance of RM 500 per month for two years. This is to assist smallholder while waiting for production of fresh fruit bunches.

In the case of new planting schemes, application is open to independent smallholders in Peninsular Malaysia with land area of 5 hectares and below, whereas in the case of Sabah and Sarawak, application is open to smallholders with land area of 7 hectares and below. However, assistance under this scheme is limited to 5 hectares.

Export Duty

In export, the major policy instrument used to simulate the downstream sector of the Malaysian palm oil industry has been the fiscal measure of imposing export duty on palm oil exports. Since the 1950s the exports of CPO and PPO have been taxed by the Malaysian government. In the 1950s, palm oil was taxed at a flat rate of 5% ad valorem that increased to 7.5% in 1960. In 1972, the tax was then converted to a graduated tax based on a rate of 2.5% for every RM 50 per ton increase in the Free On Board (FOB) price of CPO above the threshold price of RM 350, up to the price of RM 700, where the tax reached 30% ad valorem. In 1978, a surcharge was incorporated into the export tax, and the starting point in this tax was raised to 30% *ad valorem* for prices higher than the threshold price of RM 400. In 1980, the tax structure was changed again. The principle of cost-plus was used in determining the export tax payable from the threshold price of RM 500. With this tax structure, the portion of the price above RM 500 is taxed at a rate that increases with price starting at 30% for prices up to RM 549, and rising to 45% of the price until it reaches RM 1,000.

The next change in the tax structure was made in 1986. The starting point for the payment of the export tax was still RM 500, but the tax rate started at 10% on the next RM 50 above the threshold price. The tax rate increased at 15%, 20%, and 25% for every increment of RM 50. Thereafter, the tax rate was 30% on amount above RM 700. Currently, the tax structure, which came into effect from May 1, 1998 until today, the threshold price was increased into RM 650. For the price above RM 650, the tax structure is similar to that in 1986. The tax structure under the Customers Act 1967 Customer Duties (Amendment) (No.) 1998 is shown in Table 7-A.

The CPO export duty is based on the weekly gazette value and will be in effect every Monday. For instance, based on Table 7-A, if CPO gazette value is RM 1,000 per tonne, then the export duty is RM 80 per tonne (Table 7-B).

Table 7-A. Export Duty on Crude Palm Oil.

Gazatted FOB Price	Tax Rate (%)
On the First RM 650	Nil
Plus on the next RM 50.00 per tonne ad valorem	10
Plus on the next RM 50.00 per tonne ad valorem	15
Plus on the next RM 50.00 per tonne ad valorem	20
Plus on the next RM 50.00 per tonne ad valorem	25
Plus on the balance	30

Source: Statistics on Commodities 2010, Ministry of Plantation Industries and Commodities, Malaysia, 24th edition, 2010, p. 55.

Table 7-B. Calculation of Export Duty Based on CPO Gazette Value (RM 1,000/Tonne).

	RM
At RM 650 @ 0%	0.00
At RM 700.00: Plus RM 50.00 @ 10%	5.00
At RM 750.00: Plus RM 50.00 @ 15%	7.50
At RM 800.00: Plus RM 50.00 @ 20%	10.00
At RM 850.00: Plus RM 50.00 @ 25%	12.50
Sub Total	35.00
Plus (RM 1,000 – RM 850) RM 150 @ 30%	45.00
Total Duty Payable	80.00

Production of Palm Oil

Production System

The Malaysian oil palm plantations are mainly based on a private estate management system, government schemes, and independent smallholders. Government schemes involve Federal Land Development Authority (FELDA), Rubber Industry Smallholder Development Authority (RISDA), Federal Land Consolidation and Rehabilitation Authority (FELCRA), and state schemes (Table 7-C). This has proved to be successful as it provides better utilization of resources. According to Mohammad (1994), it has been essential in the application of advanced management, planting techniques, and high yielding material throughout most of the plantations and organized smallholders.

Table 7-D displays the distribution of oil palm planted area in Malaysia according to production category. Of the total planted area of 4,854 thousand hectares in 2010,

Table 7-C. Malaysian Government Organization Authorities.

Government Organization	Established	Authority
FELDA	1956	Develops settlement schemes on jungle land, provides a comprehensive package of services to the settlers on the schemes, deals with processing and trade in CPO and palm kernels, and is involved in the production and supply of inputs. These schemes are usually more than 2,000 hectares. The initial development activities of a scheme such as jungle clearing, establishment of the oil palms, building of houses and basic social amenities is done by contract labor. Fresh Fruit Bunches (FFB) are sold to a central oil mill run by a FELDA subsidiary. Provides intensive supervision at all stages.
FELCRA	1961	Deals with the rehabilitation of earlier developed low-cost State Schemes in addition to increasing the basic land resources of existing villages by opening up jungle or unused land. To provide a comprehensive package of advise, inputs, finance, and services to participating smallholders
RISDA	1972	Replant rubber as well as oil palm and is financed by rubber replanting grants (Block Plantings).

Source: Moll, 1987.

Table 7-D. Distribution of Oil Palm Planted Area in Malaysia by Category: 1980–2010 ('000 Hectares) and Percentage Shares (%).

	1980		1990		2000		2005		2009		2010	
Category	Area	%	Area	%	Area	%	Area	%	Area	%	Area	%
Private Estates	558	52	912	45	2,2024	60	2,413	60	2,807	60	2,935	60
Government Schemes												
FELDA	317	30	608	30	598	18	654	16	706	15	706	15
FELCRA	19	2	119	6	154	4	161	4	161	3	163	3
RISDA	21	2	33	2	37	1	80	2	79	2	79	2
State Schemes	86	8	175	9	242	7	318	8	330	7	321	7
Smallholders	70	6	184	9	321	10	425	10	609	13	651	13
Total	1,070	100	2,030	100	2,540	100	3,377	100	4,691	100	4,854	100

Source: Malaysian Oil Palm Statistics, various issues.

about 60% was under private estates, most of which were managed by the plantation companies such as IOI plantation and Sime Darby Plantation *Sdn. Berhad*. This shows that the private sector has been the main driver for the growth in the palm oil development and production in the last three decades. The palm oil companies' sizes vary considerably, depending on the size of their plantation estates which ranges from a few hundred hectares to more than 100 thousand hectares. Most of these companies are listed on the Main Board of the Bursa Malaysia. For example, Kuala Lumpur Kepong *Berhad* and Highlands and Lowlands *Berhad* are also listed on the London Stock Exchange, while United Plantations *Berhad* is listed on the Copenhagen Stock Exchange.

In 2010, the government schemes comprised about 20% of the total area under oil palm, and the rest was owned by the state schemes. The state schemes are generally smaller in area than FELDA schemes and the infrastructure provided is limited. The state schemes usually play a major role in Sabah and Sarawak (Amna & Fatimah, 2007; Moll, 1987). and independent smallholders. (The smallholders are private independent owners with smaller planted area, normally, in the range of 2–4 acres, and they are registered under the MPOB.) The government schemes were operated mainly by the Federal Land Development Authority (FELDA), Rubber Industry Smallholder Development Authority (RISDA), and Federal Land Consolidation and Rehabilitation Authority (FELCRA).

Over 30 years, the Malaysian state schemes' oil palm planted area increased from 86 thousand hectares in 1980 to be 321 thousand hectares in 2010. On the other hand, the planted area by smallholders made up 13% of the total area (about 0.6 million hectares) in 2010, up from around 10% (about 0.4 million hectares) in 2005 and it has been increased, as well. Due to the expansion in the area planted by Malaysia's independent smallholders, the FFB collection also increased from 228,313 tonnes in 1982 to 9.4 million tonnes in 2010.

Area Planted

The oil palm plantings in Malaysia during the past four decades have shown a dramatic growth in terms of the regional distribution of the area (Fig. 7.1). Before 1997, about 90% of the planted area approximately was in Peninsular Malaysia, and the rest was in East Malaysia. However, due to the limited land in Peninsular Malaysia, the expansion took place in Sabah and Sarawak (Ming & Chandramohan, 2002; Amna & Fatimah, 2007). In 2010, the regional distribution of planted area was 52.01% in Peninsular Malaysia, 29.04% in Sabah, and 18.94% in Sarawak.

Output and Productivity

The rapid expansion in oil palm cultivation resulted in a matching increase in palm oil production from less than 1.3 million tonnes in 1975 to about 17 million tonnes in 2010 (MPOB, 2010). During the five years ending in 2010 the total CPO production has increased at an average annual growth rates of 3%. With an equal growth rate achieved

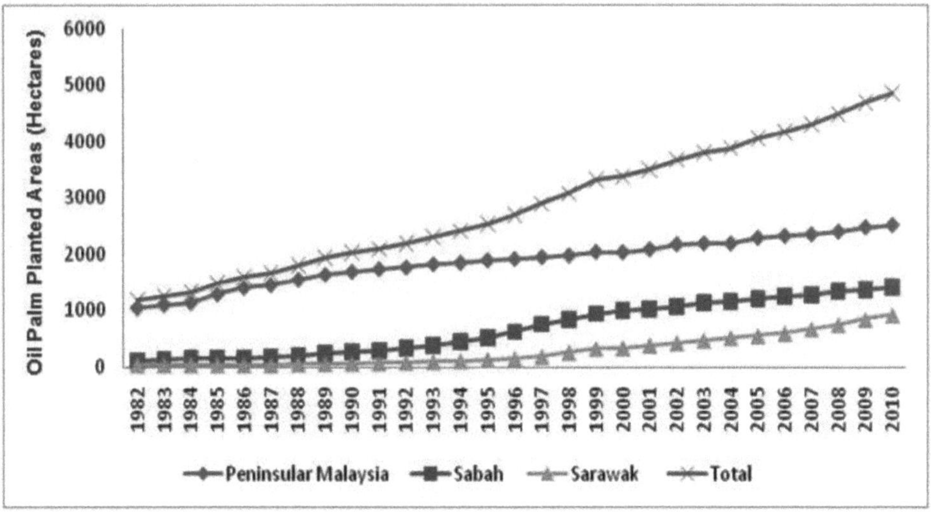

Fig. 7.1. Oil palm planted areas: 1982–2010 (hectares).

in Peninsular Malaysia, a lower rate of 13% was in Sabah and a higher one of 10% in Sarawak (Fig. 7.2).

Currently, the peninsular region is the largest palm oil producing region in Malaysia, which accounted for 58% of the average total production during the period of 2006–2010. Malaysian FFB yield has embarked on a steady increasing trend in Peninsular Malaysia and Sabah. During the years 2000–2010, Sarawak FFB Production

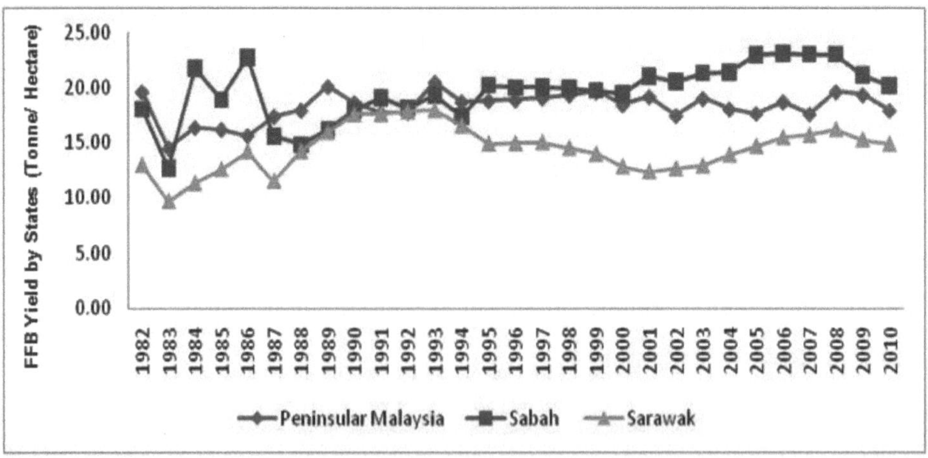

Fig. 7.2. Fresh fruit bunches (FFB) yield by states: 1982–2010 (tonne/hectare). Source: MPOB, 2010a.

has shown an increasing trend because of the slightly increasing yield, along with an expansion of the oil palm area planted there (Figs. 7.1 and 7.2).

This dramatic expansion in the palm oil production qualified Malaysia to displace Nigeria and occupy the top position list of major producers of this oil over a period of about four decades, extending from the 1970s until 2005, when it accounted for 44% of the global production, with a 41% share contributed by Indonesia (Table 7-E). However, the two latter countries exchanged places after 2005 (referring to 2010 figures.) Indonesia tops the list of the world major producers of palm oil with a 49% share compared with 37% share for Malaysia. Nigeria managed only a slow growth and it is now surpassed by Thailand, which achieved an amazingly high growth to be a distant third producer of palm in the world.

Fig. 7.3 shows the trend of the oil extraction rate (OER) on a regional basis. The OER in Peninsular Malaysia has increased from 19.18% in 1985 to 24.7% in 2009,

Table 7-E. World Major Producers of Palm Oil ('000 Tonnes).

Country	1965	1975	1985	1995	2000	2005	2010
Malaysia	150	1258	4,134	7,811	10,842	14,962	16,993
Indonesia	157	397	1,243	4,220	7,050	14,100	22,300
Nigeria	687	500	386	660	740	800	885
Thailand	1	1	89	354	525	700	1,500
Colombia	2.2	39.2	125	388	524	524	770
Others	579	934	1,060	1,769	2,186	2,890	3,140
Total	1576	3129	7,037	15,202	21,867	33,976	45,588

Sources: FAOSATAT Database and MPOB, various issues.

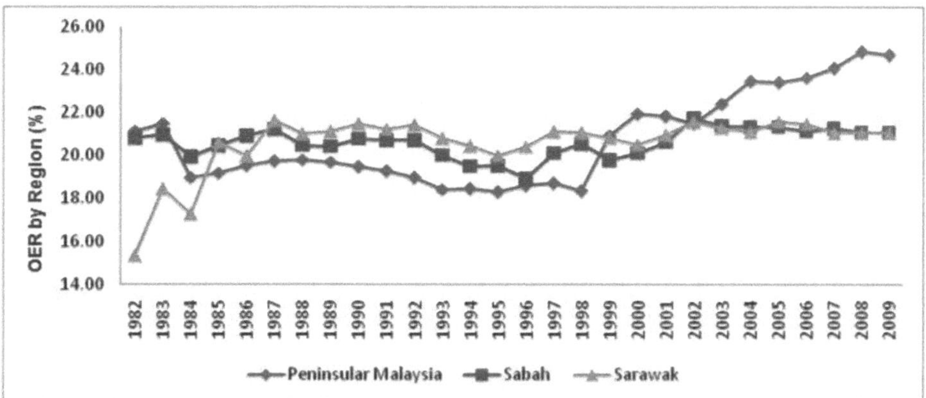

Fig. 7.3. Oil extraction rate (OER) by regions, 1982–2009 (%). Source: Malaysian Oil Palm Statistics, 2009.

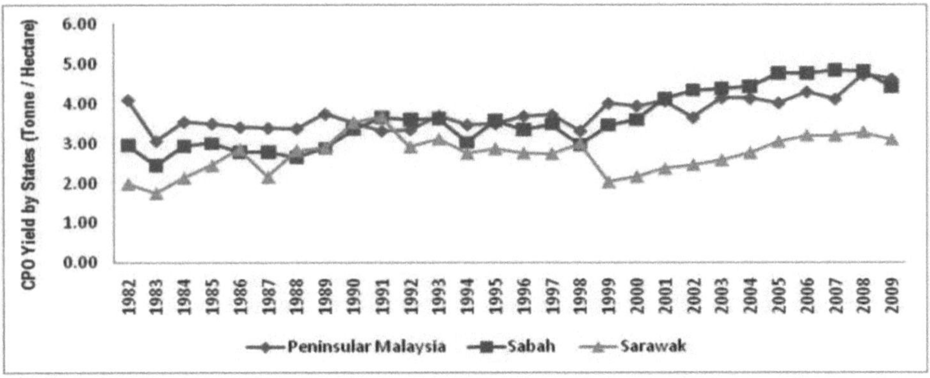

Fig. 7.4. Crude palm oil (CPO) yield by regions: 1982–2009 (tonnes/hectare).

partially due to the technology development. The OER in Sabah and Sarawak has been on the same rate, partly due to the age of the planted trees and also due to the, approximately, similar weather patterns of these two states (MPOB, 2010a).

As shown in Fig. 7.4, the productivity of Malaysian CPO has been on the increasing trend in Peninsular Malaysia and Sabah. However, in Sarawak CPO the yield dropped down from 3.0 tonnes per hectare per year in 1998 to 2.03 tonnes per hectare per year in 1999. During a decade ending in 2009, the Sarawak CPO yield showed an increasing trend to reach an average of around 3.08 tonnes per hectare per year in 2009. The increasing trend of the Sarawak CPO is attributed to the transformation of land area from Peninsular Malaysia to Sabah and Sarawak since 1999.

Marketing and Trade

Palm oil tends to have a lower producing cost and also command lower prices than the major alternative vegetable oils such as soybean, sunflower, rapeseed, and coconut oils (Carter et al., 2007). Compared to the other vegetable oils, palm oil has a strong comparative advantage in the global market. Since 1990, the global palm oil consumption has quadrupled. This increase in palm oil consumption is mainly due to food demand stimulated by the world population and wealth growth. On the other hand, the palm oil industry has also made significant inroads into the burgeoning trade in biofuels. According to Lim (2010), the rush to demand palm oil is primarily driven by two factors: (1) production cost, and (2) the multiple utilization or versatility of palm oil. Being the highest yielding vegetable oil, palm oil is the most cost-effective choice for consumers. In fact, palm oil can be used in numerous capacities such as oleochemicals and bio-fuels to satisfy consumers' evolving needs.

Malaysia plays an important role in the global palm oil trade as a leading exporter of this oil while Indonesia kept the second place (Table 7-F). The contribution of the Malaysian palm oil in the global export of the oil increased from 61% in 1980

Table 7-F. World Major Exporters of Palm Oil ('000 Tonnes).

Country	1980	1990	1995	2000	2005	2010
Malaysia	2,271	5,727	6,513	9,081	13,445	16,890
Indonesia	511	1,163	1,856	4,139	10,436	16,664
Papua New Guinea	33	143	220	336	295	465
Hong Kong*	0	51	275	158	39	42
Singapore*	679	679	399	240	205	187
Côte d'Ivoire	96	156	120	72	122	153
Others	129	283	811	993	1,960	16,786
Total	3,719	8,202	10,194	15,019	26,502	36,187

*re-exporting countries. Source: Malaysian Oil Palm Statistics, various issues.

to 70% in 1990. Although the export volume of palm oil by Malaysia sustained its upward growth, its percentage share in the world trade decreased to 47% in 2010. That was only due to the expansion of the Indonesian exports from 511 thousand tonnes (14% of the world exports) in 1980 to 16,664 thousand tonnes (46% of the world exports) in 2010, mainly attributed to the increase of the country's production of the oil. The government of Indonesia has implemented different policies concerning palm oil export. Since 1978, with the increased concern of the government about the allocation of palm oil for domestic purposes, it controlled the exports of the oil by imposing export restrictions in the form of quotas. The trade liberalization policy in 1991 resulted in an increase in both domestic price of cooking oil and the volume of palm oil export. Export tax policy has been imposed since 1994, and since that time the tax rates and calculation method changed to protect both the producers and consumers (Rifin, 2010).

Hong Kong and Singapore sustained their roles as re-exporters of processed palm oil to benefit from the value added of the oil products. Major exporters outside Asia (Papua New Guinea and Côte d'Ivoire) provide smaller shares in the world exports, but those from Papua New Guinea showed great expansion from 33 thousand tonnes in 1980 to 465 in 2010, while those of Côte d'Ivoire remained in the range of 72 thousand and 156 thousand tonnes in the same period.

The oil palm products in Malaysia are traded before and after processing. Nowadays, the private estate companies and FELDA have close links with refineries, crushing factories, and bulking installations through direct ownership or through participation in joint ventures. In July 1975, the FELDA Palm Industries *Sdn. Berhad* was established as an agency under FELDA to engage in purchasing and processing FFB coming from FELDA schemes, estates owned by FELDA Group and external suppliers. Table 7-G shows the performance status of the industry in 2009 and 2010. In 2010, the number of operating fruit mills plants in Malaysia has increased from 417 to 419 with the capacity of 96.99 million tonnes per year. However, the increase of kernel

Table 7-G. Industry Performance Status 2009–2010 (Thousand Tonnes/Year).

	2009		2010		Difference (%)	
Sector	Total	Capacity	Total	Capacity	Total	Capacity
Fruit Mills	417	95,538.6	419	96,988.4	0.00	0.02
Kernel Crusher	43	7,987.4	45	6,961.2	0.05	–0.13
Refineries	51	22,802.2	54	22,886.2	0.06	0.00
Oleo Chemical	17	2,598.9	17	2,598.9	0.00	0.00
Biodiesel	15	1,981.6	17	2,598.9	0.13	0.31
Bulking Installations	34	1,491.4	34	1,491.4	0.00	0.00
Total	577	132,400.2	586	133,525,185	0.24	0.20

Sources: Malaysian Oil Palm Statistics, 2009 & 2010.

crusher plants from 43 plants in 2009 to 45 plants in 2010 did not lead to capacity increase; capacity was even lower by 0.13%. In 2010, the biodiesel plants increased to 17 plants with total capacity of 2.59 million tonnes per year. Due to the industry performance status, the total exports of oil palm products, including palm and palm kernel oils, palm kernel cake, oleochemicals, biodiesel, and finished products, increased by 2.8% or 0.63 million tonnes to 23.06 million tonnes in 2010. In 2010, the higher export prices increased the Malaysian palm oil earnings by 20.4% or RM 10.13 billion to RM 59.77 billion. However, aside from finished products and biodiesel, most of the oil palm products registered growth in their export volumes (MPOB, 2010).

Road tankers with 10 to 20 tonnes of capacity (Moll, 1987) are used to transport CPO and refined products. Before being exported, palm oil products are stored in the bulking installations at ports. In 2010, Malaysia had 34 bulking installations with a total capacity of 1.49 million tonnes per year. However, the basic bulking installations are only suitable for crude and processed palm oils. For some of the refined products that are solid at ambient temperatures, an extra heating system must be installed (Moll, 1987).

Pricing

Due to the lower price of palm oil and the huge growth of palm oil consumption, the palm oil trade volume has surpassed the other fats and vegetable oils in the global market. The international palm oil prices tend to be lower than the soybean oil prices, sunflower oil prices, and rapeseed oil prices (Fig. 7.5). In 2009, the average annual CPO prices was selling at only US $683 per tonne in the global market, which was much cheaper than the soybean oil (US $849 per tonne), sunflower oil (US $855 per tonne), and rapeseed oil (US $859 per tonne). This is because the palm oil products have a relatively low production cost and are more cost-competitive than others

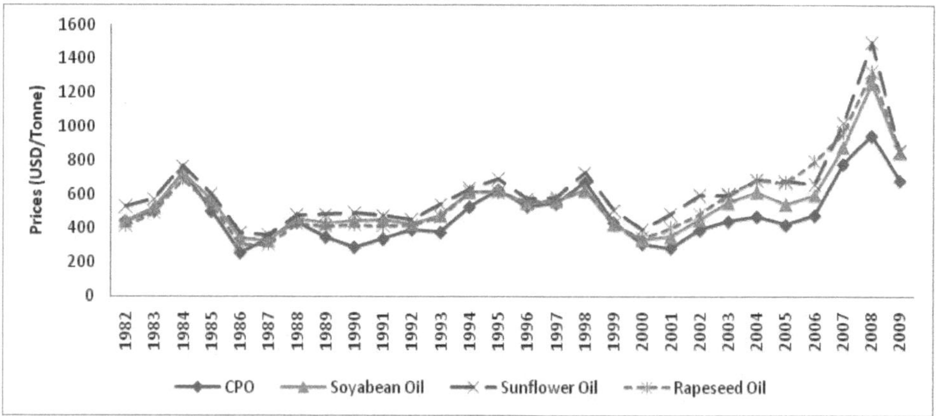

Fig. 7.5. Annual world average prices of selected vegetable oils: 1982–2009 (USD/tonne). Source: Malaysian Oil Palm Statistics, 2009.

(Carter, 2007). In the past 3 decades (1982–2009), the four selected competitive vegetable oil prices seemed to simultaneously adjust together in the global market, which implies that these vegetable oils are close substitutes for each other.

The supply and demand situations as well as other price-determining factors are usually considered in the process of setting palm oil price. Brokers bargain with their buyers and sellers until they arrive at agreed prices. As Fig. 7.6 indicates, the FOB price of CPO, as provided by the Department of Statistics, closely follows the world market prices. The

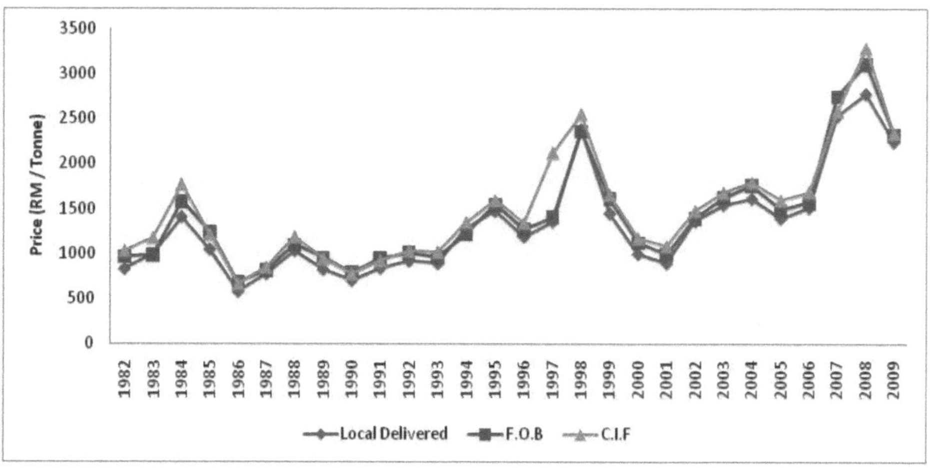

Fig. 7.6. Annual Malaysian crude palm oil prices: 1982–2009 (USD/tonne).

export duty is based on the "gazetted FOB prices," which are calculated by the Royal Custom and Excise Department. The gazetted price sometimes diverges slightly from the average FOB price assessed by the Department of Statistics (Moll, 1987).

Export Destinations and Performance

Most countries in the world have huge deficits in oilseed and fats products (Fig. 7.7). China, the European Union, and India are the world top three net importers. However, only three countries, namely Malaysia, Indonesia, and Argentina, are able to supply a large volume of oils in the global market at the current time. Palm oil, mainly exported by Malaysia and Indonesia, is the highest exported oil capable of meeting the world deficit in fats and oils. Therefore, the Malaysian palm oil industry is playing a dominant role in the international vegetable oils market.

In 2010, Malaysia became the world's second largest exporter of palm oil, which includes CPO and PPO. Compared to 2009, Malaysia exports of palm oil increased by about 4.9% from 15.88 million tonnes to 16.66 million tonnes in 2010. Fig. 7.8 shows that, during 1975 to 2000, the processed palm oil (PPO) accounted for the biggest stake of the total Malaysian palm oil exports (around 97%) and the rest is accounted for by the CPO. However, this ratio changed recently as Malaysia's export of CPO increase from 0.4 tonnes (or 4% of the total palm oil exported) in 2000 to 2.5 million tonnes (or 16% of the total palm oil exported) by 2009. In 2010, palm oil stood in the vicinity of around 17 million tonnes, which was 47% of the total world palm oil exports in that year, thereby, topping the list of the major world exporters of this oil. Indonesia followed closely with a share of 46%. Malaysia alone contributed

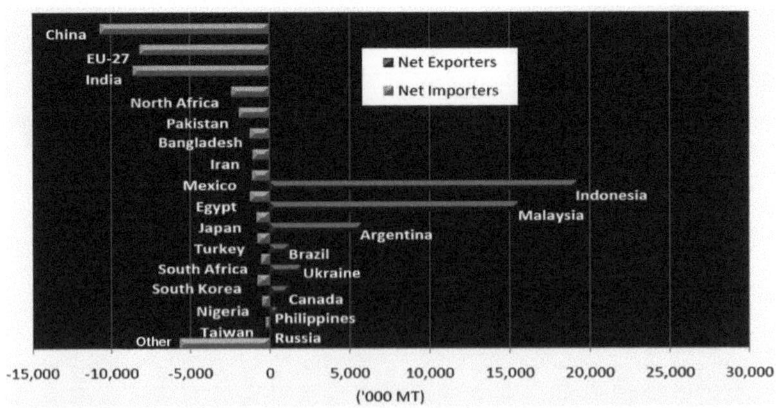

Fig. 7.7. Global oils and fats scenario: distinctive supply and demand balance in 2009. Source: Basiron (2010).

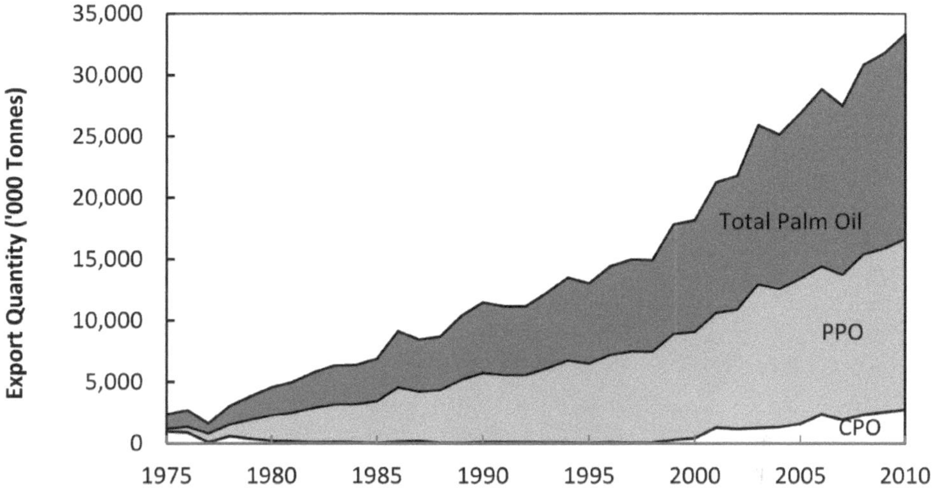

Fig. 7.8. Malaysian palm oil exports: 1975–2010 ('000 tonnes). Source: Malaysian Oil Palm Statistics, 2010a.

27% to the global oils and fats exports in 2010, qualifying it to be the largest oils and fats exporting country in the world.

In 2010, most of Malaysia's palm oil (59%) was exported to top five importing countries, namely, China, the European Union, Pakistan, the United States, and India. Around 20.9% of the total Malaysian palm oil exports went to China, and the rest was exported to Pakistan (12.8%), the European Union (12.4%), India (7%), and the United States (6.2%).

The changing trend of the palm oil exports to major importing countries is depicted in Fig. 7.9. China has been the major importer of Malaysia's palm oil, and it continues to maintain its position as the largest palm oil export market for nine consecutive years, with off-take totaling 3.48 million tonnes or about 21% of total palm oil exports (MPOB, 2010). Fig. 7.10 also shows that in 2010 the exports of palm oil to China decreased by 13.5% or 0.54 million tonnes from its previous year's level to 3.48 million tonnes. The major contributing factor for this significant decline was the higher imports of soybean for local crushing activities in China (MPOC, 2010). Indian vegetable oil use swings dramatically between palm and soybean oil since the import demand there is sensitive to the changes of palm oil prices and the country trade policy (Carter et al., 2007). In India, the import duty is applied to the palm oil products. This duty is based on the agreed rate with the World Trade Organization (WTO) members and it is changing due to the palm oil price changes. For example, if the price of palm oil decreases, the import duty "bound rate" will be raised up by the Indian government in order to control the palm oil import demand. Consequently, the prices of palm oil exported to India will be affected by the rate of import duty. In

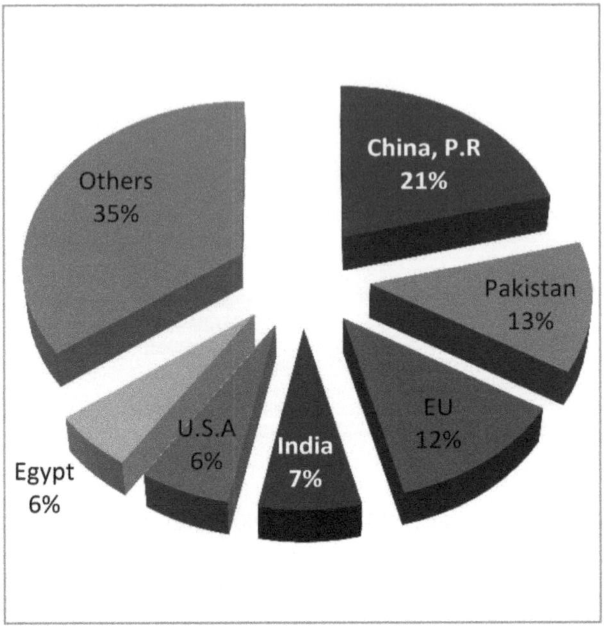

Fig. 7.9. Malaysian palm oil export to major importing countries in 2010. Sources: MPOB, 2010a.

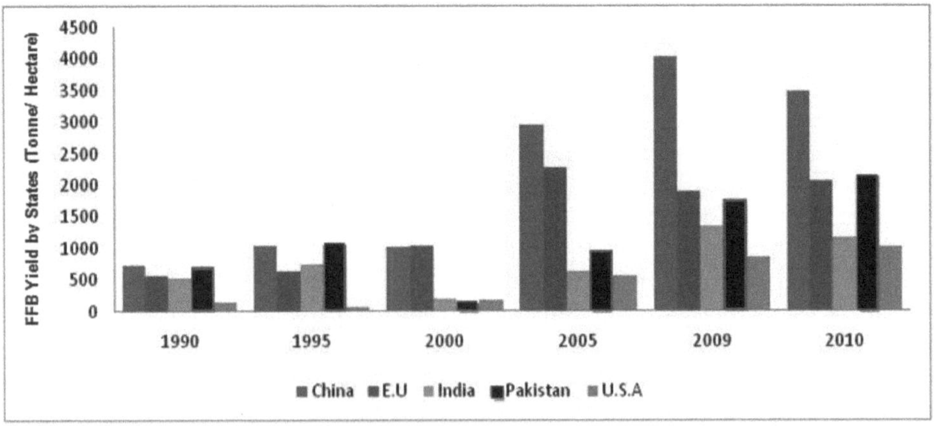

Fig. 7.10. Malaysia's palm oil exports to major importing countries: 1990–2010 ('000 tonnes). Sources: Malaysian Oil Palm Statistics, various issues.

Fig. 7.10 the volume of Malaysia's palm oil traded to India went down in 2010 mainly due to an increase in domestic rapeseed oil production by 15.1% to 2.12 million tonnes in 2010 (MPOB, 2010a).

Malaysian palm oil exports to Pakistan increased sharply from 1.07 million tonnes in 2000 to 2.13 million tonnes in 2010, mainly due to further reduction in import duty of Malaysian palm oil based on the Margin of Preference (MoP) by 10% in 2008 and 2009 and 15% for the year 2010 until 2014, under the Malaysia-Pakistan FTA Agreement (MPOB, 2010). Higher palm oil demand by both edible and non-edible sectors had contributed to the increase the Malaysian export of this oil to the European Union (EU) to reach an amount of 2.06 million tonnes in 2010. Since 2006, the trans-fatty acid labeling enforced in the United States has continued to push up the palm oil usage in the American food sector. This situation led to the expansion of the exports of palm oil to the United States from about 0.7 million tonnes in 2006 to 1.03 million tonnes in 2010 (MPOB, 2010).

Import Demand of CPO

Due the palm oil refining industry development, refined and processed palm oil accounts for almost 90% of the total palm oil exports. The rapidly increasing palm oil refining and fractionation capacity strengthened Malaysia's position not only as a leading producer but also as a major marketing player in the international trade of oils and fats. To sustain this position, more CPO is needed to meet the requirement of the processing sector, as its capacity exceeds the local production. To fill this gap, Malaysia imports CPO mainly from Indonesia. Fig. 7.11 displays the movements in the import quantities of this commodity. Over the period of 1985–2010 the imports showed high fluctuation with peak amount in 2010.

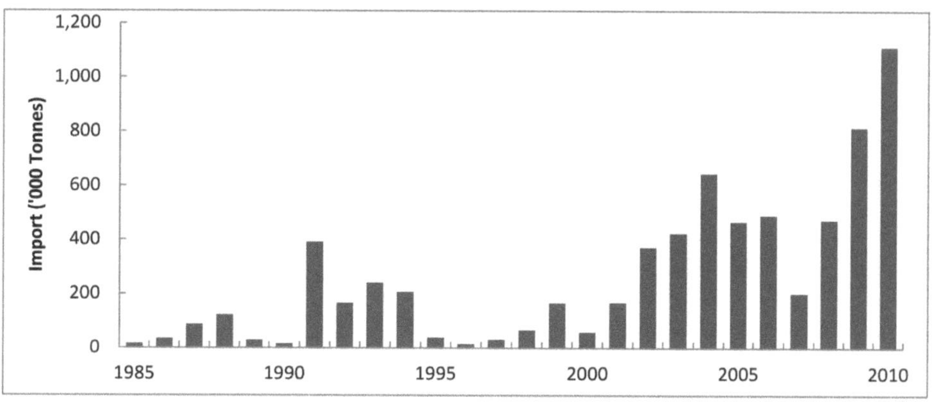

Fig. 7.11. Annual crude palm oil import demand by Malaysia, 1985–2010 ('000 tonnes).

Palm Oil Utilization

In order to satisfy the consumers' unlimited needs, research and development (R&D) must continuously develop the palm oil versatility. An oil palm tree is 10% oil and 90% biomass. Fig. 7.12 shows that palm oil needs to go through many processing steps such as milling, crushing or extraction, refining, pressing, bleached, and deodorization to obtain the final products. After the milling process, the fresh fruit bunches produce CPO, palm kernels oil, palm kernel cake, shell, and fiber. In Malaysia, the oil palm products comprise the following categories: (1) palm oil and palm kernel oil products, (2) oleochemicals, (3) bio-fuels, and (4) biomass.

Malaysia's CPO is processed into refined products including refined bleached deodorized (RBD) palm oil, RBD palm olein, and RBD palm stearin. The RBD palm oil is normally used to produce margarine, shortening, ice-cream, and so on. However, the RBD palm olein is used to make frying and cooking oils, shortenings, and margarines. Besides that, the RBD palm stearin can be used to manufacture fatty acids, soaps, fuel, emulsifiers, and so on. Palm kernels can be crushed to produce palm kernel oil and palm kernel meal. The fractionation and refining of the palm kernel oil will give palm kernel stearin and palm kernel olein.

Oleochemical plants registered considerable expansion both in terms of number of plants in operation and plant capacity. In Malaysia, the first oleochemical plant was set up in 1980; other ones followed, and by 1985 there were 5 plants and at the end of 2010 the number increased to 17 plants operating with a total capacity of 2.6 million tonnes per year. Oleochemical plants in Malaysia comprise local, foreign, and several joint-venture enterprises and all of them are located in peninsular Malaysia. The oleochemicals are mainly used in the manufacture of soaps, detergents and surfactants, toiletries, cosmetics, and so on.

Home-produced palm biofuel technologies, including winter grade biodiesel, have successfully been commercialized after years of research and development (R&D) in palm oil biodiesel that began in the early 1980s. During the period of 2000 to 2005, Malaysian energy consumption grew at a fast rate of 5.6% to reach 38.9 million tonnes in 2005. In response to that high energy consumption growth rate and sustainability issues, the government encouraged all the initiatives towards the development of green technology and renewable alternatives like hydro, biomass, and solar energy. In August 2005, Malaysia's National Biofuel policy was launched by the government, and it primarily aimed to reduce the country's fuel import dependence and further promote the demand for palm oil that is expected to be the primary commodity for biofuel production, as well as stabilizing its prices at a remunerative level, especially during periods of low export demand. Nowadays, Malaysia has succeeded to issue about 91 biofuel production licenses with installed production capacity of 10.2 million tonnes. Among these, about 18 companies are in operation with 1.86 million tonnes per annual production.

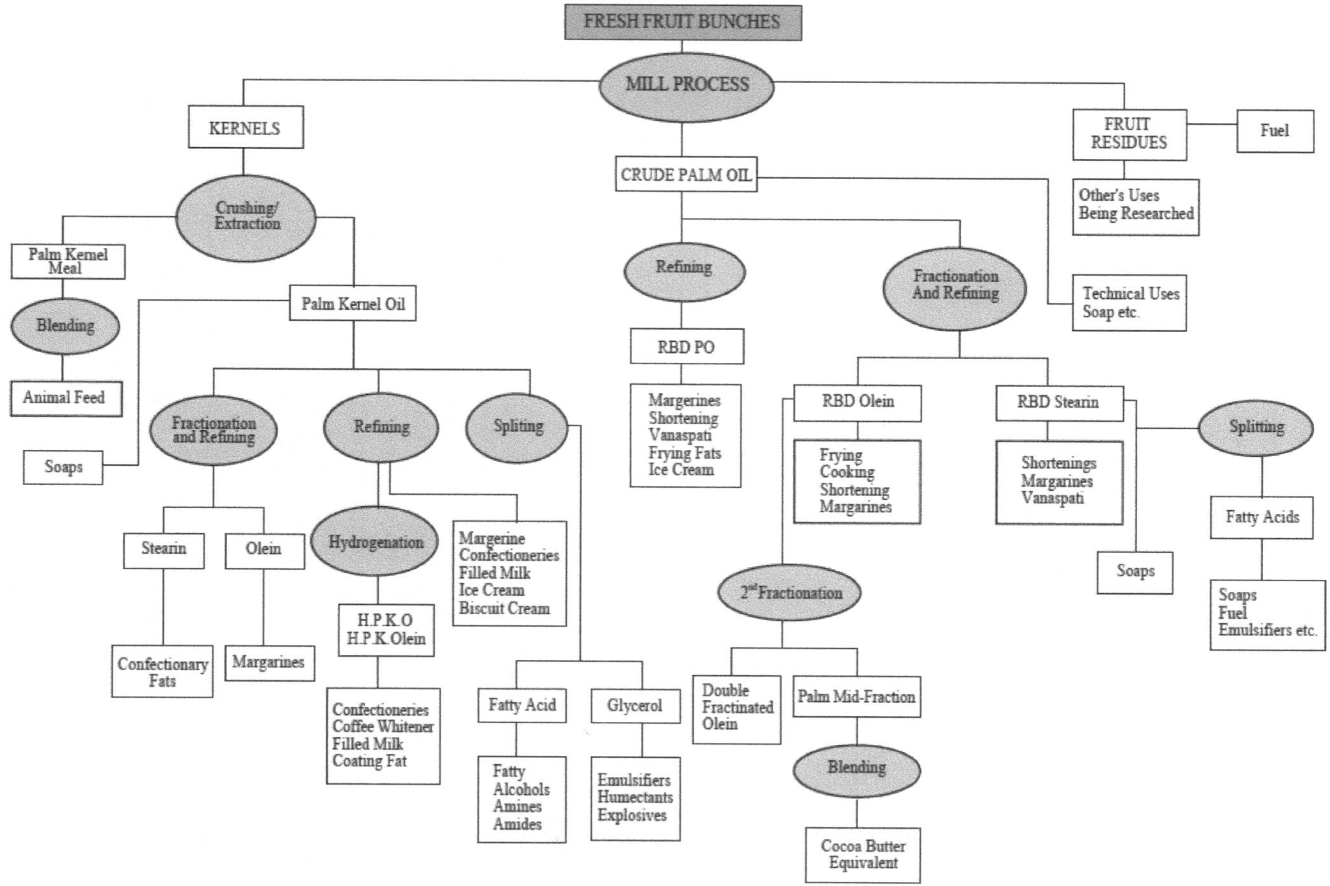

Fig. 7.12. Palm oil utilization. Source: Ministry of Plantation Industries and Commodities, Statistics on Commodities, 2010, Chap 8–1, p. 163.

Issues and Challenges

Despite the impressive achievements in Malaysia's palm oil industry, it is still facing a number of problems and challenges that will affect Malaysia's palm oil industry's market position as a major global palm oil producer. These hurdles include the following:

1. Low productivity due to expansion into marginal areas with unsuitable soils, inadequate use of the agronomic inputs required for maintaining yields, (especially during low palm oil prices), and lack of access to best practices. Another factor that affects the yield is the low replanting rate, as there is a total of 449 thousand hectares of non-productive and low yield plantation areas that require replanting. Out of this amount, 109 thousand hectares belongs to independent smallholders.
2. The problems of limited scope of mechanization and skilled labor shortage in the Malaysian palm oil industry have threatened the Malaysia palm oil productivity. This hurdle will probably make Malaysia's palm oil industry fall far behind the global competitors.
3. Protectionist measures and trade barriers adopted by the importing countries. In the global market, there are many tariff- and non-tariff barriers implemented by the importing countries to restrict market access to protect their local oils and fats industries. For instance, the Renewable Energy Directive adopted by the European Union, which imposes strict regulations on carbon emissions, is distorting the palm oil commodity price. India, is implementing an import tariff policy on all oils products based on the "bound rates" in agreement with the World Trade Organization (WTO). These bound rates set an agreed maximum tariff level for all WTO members, including Malaysia (Carter et al., 2007). Recently, the parliament of the commonwealth of Australia senate announced the Food Standards Amendment (Truth in Labelling-Palm Oil) Bill, which would compel the parties involved in the food industry to list palm oil on the content label of the food products containing it. This non-tariff barrier can give some negative impacts to the Malaysian palm oil industry such as increasing the costs of labeling as well as decrease of export demand from the United States and European market.
4. NGOs campaign against the expansion on palm oil area. The current global anti-palm oil campaign is an on-going effort and its roots can be traced to the early 1980s. In recent decades, the Western NGO's anti-palm oil campaigns highlighting its carbon footprint or raising health concerns have generated an increasing unwarranted negative perception of palm oil. Moreover, the American soybean lobby groups are claiming that palm oil is an unhealthy product as its saturated fat raises blood cholesterol levels, leading to heart

disease. There are also claims of extensive deforestation, greenhouse gas emissions, loss of biodiversity, and wildlife decimation, especially the orangutans, which are often not credibly supported (Basiron, 2011).

5. The Roundtable on Sustainable Palm Oil (RSPO), a certification body whose objective is to "promote the growth and use of sustainable palm oil through co-operation within the supply chain and open dialogue with its stakeholders" was established in 2004 in response to global market pressure to enhance the sustainability of palm oil (RSPO 2010). The process for starting certification was completed in August 2008. Malaysian palm oil industry is a founding member of RSPO and has been actively participating in RSPO discussions. The issue of the impact of RSPO raised many arguments. While some parties consider it as a "phenomenal success story because it is a market-driven instrument working to use resources responsibly" (Chandran, 2010), others suspect its credibility (World Rainforest Movement, 2010). More information on RSPO can be found in Chapter 25 of this book.
6. The infant biodiesel industry faces some challenges, including uncertainty in oil and feedstock prices, sustainability issues, in addition to second generation and non-food biofuels.

The Way Forward

Forward looking, to solve the low productivity problems, the Malaysian palm oil industry primary action is to accelerate the replanting of oil palm and improve the fresh fruit bunch yield. MPOB is playing a key role to ensure that replanting is accelerated. For example, the independent and organized smallholders will not be allowed to preserve palms older than 25 years with yield of less than 10 tonnes per hectare per year for the past 3 years. Additionally, the independent smallholders are required to take an initiative to learn the management skills and get involved into the MPOB, Malaysia Palm Oil Council (MPOC), as well as Malaysian Palm Oil Association (MPOA) activities such as the course of management and maintenance nursery palm organized by MPOB in 2011. Enhancing the mechanization level is also important to reduce the labor shortage problems and directly augment the FFB yield in the Malaysian palm oil industry.

Due to the world anti-palm oil campaign and the sustainable agriculture issues, the Malaysian palm oil industry urgently needs some right track solutions. According to Basiron (2011), the Malaysian palm oil industry needs to develop a strong set of certification standards and "fair trade" label in order to increase the palm oil products credibility in the international market. However, many of the currently existing certification schemes reflect the priorities of politically biased and unaccountable NGO groups. For instance, many of the available NGO's sustainability certification schemes lack transparency on the rules and standards of settings. They often set benchmarks that change randomly and without explanation. Accordingly, the Malaysian palm oil industry has established the own national best-in-class certification standard schemes to help in communicating and reinforcing best practices. The benchmark of standard certification system in Malaysia must be based on the best international practices, high transparency, accepted and recognized across the world. Additionally, Malaysia's palm oil industry needs to take the "fair trade" label that aims to guarantee good prices for smallholder producers' product, protecting them to some degree from ups and downs of global commodity prices. At the same time, the "fair trade" label also supports the local communities and encourages the oil palm industry to make some positive contributions on the environment and rebuilding of societies. All in all, only a fair and more complete certification standard will provide the Malaysian palm oil industry the credit it deserves. More information on the trade and traceability of palm oil can be obtained from Chapter 24 of this book.

References

Amna, A.A.; Fatimah, M.A. The Malaysian palm oil industry: development and challenges. In 50 years of Malaysian Agriculture: Transformational Issues Challenges & Direction; Arshad, F.M., Abdullah, N.M.R., Kaur, B., Abdullah, A.M., Eds.; Universiti Putra Malaysia Press: Serdang, 2007; pp. 371–400.

Basiron, Y. Pakistan-Malaysia Palm Oil Trade: Why Buy Malaysian Palm Oil. Presentation slide downloadable from http://cfile213.uf.daum.net/attach/20172A574DAFE5E4179D63, retrieved on 08 January 2011.

Basiron, Y. Current global anti-palm oil campaign—where and what needs to be done from palm oil's perspective. Paper presented at the Reach and Teach Friends of The Industry Seminar: Challenges and Opportunities in 2011. Organized by the Malaysia Palm Oil Council (MPOC), Sime Darby Plantation Sdn. Berhad., and Modipalm Engineering Sdn. Berhad, 25 January 2011.

Chandran, M.R. Eco-friendly credentials–moving from prattle to practice. Global Oils and Fats Business Magazine, 2010, 7(4), 12–15.

Carter, C.; Willa, F.; James, F.; David, J.; Lynn, W. Palm oil markets and future supply. Eur. J. Lipid Sci. Technol. 2007, 109, 307–314.

Claire, F.; Lan, C. China's policy dilemma: Balancing interests. Global Oils and Fats Business Magazine, 2010, 7(4), 16–20.

Lim, K.Y. A call to arms: to defend palm oil. Global Oils and Fats Business Magazine, 2010, 7(4), 9–11.

Ming, K.K.; Chandramohan, D. Malaysian palm oil industry at crossroads and its future direction. Oil Palm Industry Econ. J. 2002, 2(2), 10–15.

Mohammad, J.A. The Malaysian palm oil a dynamic industry. In Technical Committee of 1994 Palm Oil Familiarization Programme, Selected Readings on Palm Oil and its Uses. Palm Oil Research Institute Malaysia, 1994, 2.

Moll, H.A.J. Oil palm in Malaysia. In The Economics of Oil Palm: Economics of crops in developing countries, No 2; Pudoc Wageningen, 1987; pp. 135–166. MPOB Malaysian Oil Palm Statistics, 29th edition, 2009.

MPOB Malaysian Oil Palm Statistics, 30th edition, 2010a.

MPOB Overview of the Malaysian Oil Palm Industry 2010, 2010b. Downloadable from: http://econ.mpob.gov.my/economy/Overview_2010_final.pdf, retrieved 08 January 2011.

PEMANDU Economic Transformation Programme: A Roadmap for Malaysia, Putrajaya: Pusat Bahasa, 2009.

Rifin, A. The effect of export tax on Indonesia's crude palm oil (CPO) export competitiveness. ASEAN Economic Bulletin, August 2010, 27(2), 173–184.

Shamsudin, M.N.; Fatimah, M.A.; Fauziah, A.H. The effect of export duty liberalization on the Malaysian palm oil industry. Malaysian Oil Sci. Technol. 1997, 6(2), 79–82.

8

Palm and Palm Kernel Oil Production and Processing in Malaysia and Indonesia

Khairudin Hashim, Shawaluddin Tahiruddin, and Ahmad Jaril Asis
Sime Darby Berhad, Malaysia

Introduction

The oil palm industry has proven to be one of the most important industries in Malaysia and Indonesia. As of 2009, a total of 4.69 million hectares were planted with oil palm in Malaysia, with 2.4 million hectares in Peninsular Malaysia, 1.36 million hectares in Sabah, and 0.71 million hectares in Sarawak (MPOB, 2009). Due to the scarcity of new land for plantation development in Peninsular Malaysia, the focus on plantation development has now shifted to the East Malaysian states of Sabah and Sarawak, with a significant increase in planted areas of 28% and 101%, respectively, over the period of 2000 to 2009.

Correspondingly, the palm oil milling industry has seen a tremendous expansion over the past 10 years. The number of palm oil mills in Malaysia currently stands at 395, with 240 mills in Peninsular Malaysia and 115 and 40 mills in Sabah and Sarawak, respectively. Another 28 oil mills were in various stages of construction as of August 2010. Total milling capacity amounts to 89.8 million tonnes of fresh fruit bunches (FFB) annually (MPOB, 2009).

The total FFB and crude palm oil (CPO) production for January and December 2010 was 74.5 million tonnes and 15.4 million tonnes, respectively. A total of 3.9 million tonnes of palm kernel oil (PKO) was produced during the same corresponding period.

In the case of Indonesia, the industry has grown rapidly with the oil palm plantation expanding at an annual rate of more than 12% from 1990 to 2005. As of 2010, the total oil palm plantation in Indonesia stands at 7.82 million hectares (Direktorat Jenderal Perkebunan, 2010), with more than 60% of the oil palm plantations located in Sumatera Island, while the majority of the remaining are planted in Kalimantan and Sulawesi. The CPO production in 2009 peaked at 18.64 million tonnes produced at more than 400 palm oil mills (USDA, 2009).

Indonesia surpassed Malaysia as the biggest palm oil producer in 2006 and as the biggest exporter of palm oil in 2008 (Table 8-A, Fig. 8.1). With its vast land bank for

Table 8-A. Malaysian and Indonesian Oil Palm Hectarage and Palm Products Production, 1994–2009.

	Malaysia[1]			Indonesia[2]		
Year	Area, ha	CPO, tonnes	PKO, tonnes	Area, ha	CPO, tonnes	PKO, tonnes
1994	2,411,999	7,220,631	978,143	1,804,149	4,008,062	796,537
1995	2,540,087	7,810,546	1,036,538	2,024,986	4,479,670	942,063
1996	2,692,286	8,385,886	1,107,045	2,249,514	4,898,658	1,084,676
1997	2,893,089	9,068,729	1,164,697	2,922,296	5,448,508	1,095,273
1998	3,078,116	8,319,682	1,110,745	3,560,196	5,930,415	1,186,083
1999	3,313,393	10,553,918	1,338,905	3,901,802	6,455,590	1,291,118
2000	3,376,664	10,842,905	1,384,685	4,158,077	7,000,508	1,400,102
2001	3,499,012	11,803,788	1,531,917	4,713,435	8,396,472	1,675,676
2002	3,670,243	11,909,298	1,472,932	5,067,058	9,622,345	1,831,069
2003	3,802,040	13,354,769	1,644,126	5,283,557	10,440,834	2,104,722
2004	3,875,327	13,976,182	1,643,021	5,284,723	10,830,389	2,267,271
2005	4,051,364	14,961,654	1,842,628	5,453,817	11,861,615	2,474,532
2006	4,165,215	15,880,786	1,955,634	6,594,914	17,350,848	3,470,170
2007	4,304,923	15,823,745	1,907,613	6,766,836	17,664,725	3,532,945
2008	4,457,987	17,734,441	2,131,399	7,363,847	17,539,788	3,507,958
2009	4,691,150	17,564,951	2,097,061	7,508,023	18,640,881	3,728,176

[1]Malaysian Oil Palm Statistics, 2009; [2]Direktorat Jenderal Perkebunan, 2011.

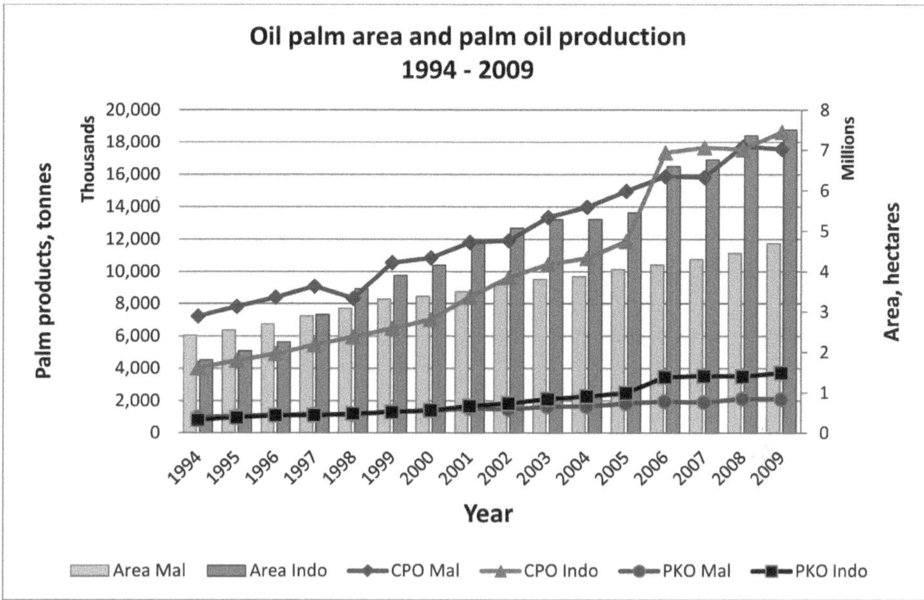

Fig. 8.1. Oil palm hectarage and crude palm oil, palm kernel oil production.

agricultural activities, the country has set a target to produce more than 40 million tonnes of palm oil by the year 2020. The major privately-owned producers besides the smallholders in Indonesia are PT Astra Agro Lestari TBK, PT Sinar Mas Agro Resources and Technology TBK, PT Perusahaan Perkebunan London Sumatra Indonesia TBK, and PT Bakrie Sumatera Plantation TBK.

Oil Palm Industry Value Chain

The whole oil palm value chain covers the growing and production of FFB in plantations, extracting CPO and palm kernel and PKO in palm oil mills and kernel crushing plants, and producing various types of refined palm and palm kernel oil products and blends for edible and non-edible applications. The versatility of palm oil also allows the extension of its uses into oleochemicals and biodiesel production.

Many plantation companies see vertical integration as an opportunity for expansion of their resource-based operations. Downstream operations are expected to provide higher value addition to the commodity products. In view of the variability in commodity prices, downstream manufacturing can provide a balance in revenue and returns. Increasingly, the major plantation companies have started to venture into downstream manufacturing activities both locally and internationally. These companies have operations in plantation, milling and crushing, and refining, as well as

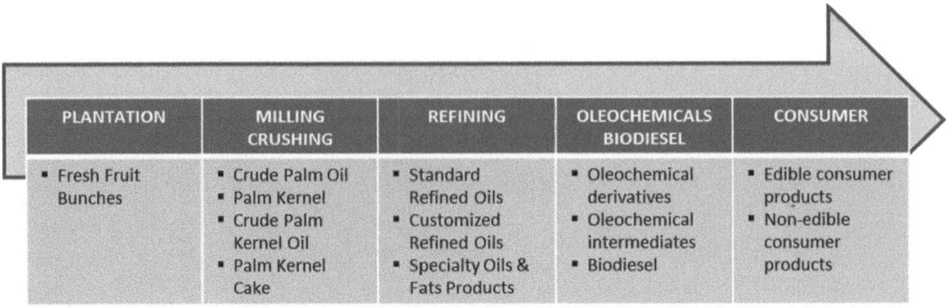

Fig. 8.2. Integration across value chain among major oil palm producers.

oleochemicals and biodiesel. A number of Malaysian companies operate edible oil refineries in places such as Thailand, Vietnam, China, Netherlands, Morocco, Bangladesh, and Africa, while oleochemical units operate in China, Europe, and the United States.

The generalized oil palm industry value chain and extent of integration among the major players are shown in Fig. 8.2.

Oil Palm Milling Process

Harvesting of oil palm fruits is usually conducted within a 7 to 14 day harvesting interval. The FFB are transported to palm oil mills that are usually located within the oil palm plantation for processing. The harvested FFB need to be processed immediately to minimize the rise of free fatty acids (FFA) that determine palm oil quality.

The basic oil palm milling process has remained relatively unchanged for many years. In simple terms, it involves the "sterilization" or steaming of FFB to arrest the rise of FFA and facilitate stripping of the fruits from the bunch stalk as well as conditioning the palm fruit mesocarp for oil extraction. The FFB received by the mill are loaded into sterilizers and steamed at pressures up to 3 bars for periods of sixty to ninety minutes (PORIM, 1985).

The sterilized bunches are then stripped of the fruits in a rotating thresher drum. The empty fruit bunches (EFB) are either transported to the plantation for mulching or composting as organic fertilizer.

The stripped fruits are conveyed into the digester where the fruits are heated using live steam and continuously stirred to loosen the mesocarp from the nuts as well as to break open the oil cells and release the crude oil before being pressed in continuous screw presses.

The liquor ex-screw press is diluted with hot water before being screened, using double deck vibrating screens to remove remaining fibers and coarse contaminants. The diluted oil is pumped to a clarifier tank where crude oil is separated from the water/solid fraction. The clarified oil is further cleaned in high speed purifiers to remove the remaining dirt and some moisture before being dried further in a vacuum drier. The clean and dry oil is ready for storage and despatch.

The sludge from the clarifier still contains residual oil which is recovered using bowl centrifuges or decanters. The recovered oil is recycled to the clarifier while the sludge and solids are discharged as palm oil mill effluent (POME) for waste water treatment.

The press cake, made up of pressed mesocarp fiber and nuts then undergoes a winnowing process to separate the nuts from the fiber. The fiber is sent to the boiler as fuel. The nuts are cleaned in the depericarper to remove remaining fiber before being cracked to release the kernels. The kernels and shells are separated in a multi-stage winnowing system, followed by claybath or hydrocylone. Palm kernels are dried prior to storage while shells are sent to the boiler as fuel.

The overall palm oil milling efficiency in terms of oil recovery is typically about 90–93%, depending on plant machinery and process control. The typical crude palm oil extraction rate (OER) during the milling process is in the region of 18–23% by weight to FFB, while the palm kernel extraction rate (KER) ranges between 4 and 6%. The simplified palm oil extraction process is shown in Fig. 8.3. The basic quality parameters for palm products are tabulated in Table 8-B.

Developments in Oil Palm Processing Technologies

The oil palm processing industry is currently facing major challenges. These include the sharply escalating production costs due to high costs of machinery and consumables, labor scarcity, stagnating productivity, and safety and environmental concerns. Developments in processing technologies need to effectively address these issues. Nevertheless, despite the increasing constraints faced by the palm oil milling industry, there are new opportunities to be exploited. The trends in environmental protection offer new business and revenue generating options. Bio-energy options such as biomass and biogas power generation not only reduce dependency on fossil fuel at the oil mills, they also offer an additional income stream through Clean Development Mechanism (CDM).

Table 8-B. Basic Quality Parameters for Crude Palm Oil and Palm Kernels.

Quality parameters	Crude palm oil	Palm kernels
Free fatty acid (FFA)	5% max	Not applicable
Moisture	0.2% max	6% max
Dirt and impurities	0.05% max	7% max
Deterioration of bleachability index (DOBI)	2.5 min	Not applicable

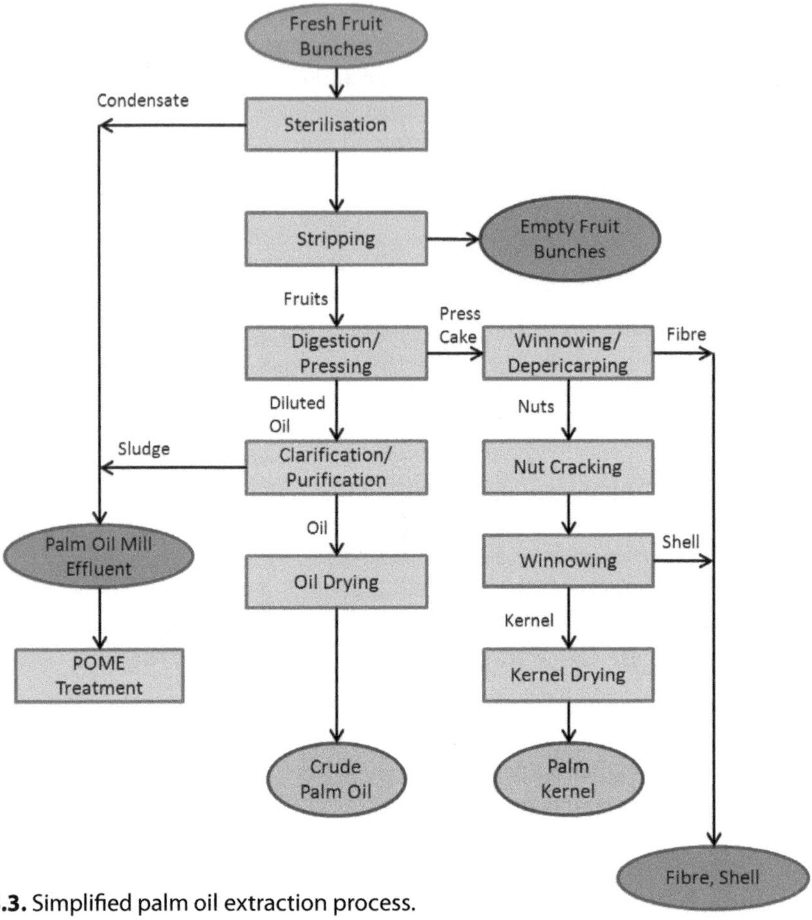

Fig. 8.3. Simplified palm oil extraction process.

There is also an increasing trend in the utilization of by-products such as palm oil mill effluent (POME), EFB, palm fiber, and shell for the production of compost bio-fertilizer and fiber materials for various applications, as well as activated carbon. The industry is well placed to diversify the current product range of CPO and palm kernel to include new value-added co-products.

Some of the process technologies that have been developed and introduced in palm oil milling industry are as follows:

1. FFB handling and transfer

   The introduction of the new technologies is to ease the FFB handling from the FFB ramp into the sterilizer and to increase the productivity; the operation previously was labor intensive. This includes the introduction of a new conveyor system to replace the capstan and bollard system, FFB weighing before entering into sterilizer cages or bin, indexer system for sterilizer cage feeding and discharge into/from the sterilizer, and the use of cage tippler to replace the use of overhead crane.

2. Sterilization

   The new technology focused on the FFB handling during the sterilization to eliminate the need for using sterilizer cages or bins as well as replacing the batch sterilization process with continuous process. The invention includes the introduction of continuous sterilization process at atmospheric steaming pressure (Sivasothy, 2009), vertical sterilizer, tilting sterilizer (Loh, 2009), inclined (oblique), and spherical sterilizers (Lim, 2007), as well as modification of the sterilization cages or bin design and steam distributor to improve the sterilization efficiency. Laboratory and pilot scale trials on microwave sterilization have also been attempted (Chow & Ma, 2007; Nazarulhisham & Kaida, 2009).

3. Oil Extraction

   The physical oil extraction is still focused on a screw press design that will enable maximized oil extraction while keeping nut breakage during the pressing to a minimum. The single screw press has now been replaced with a double screw press with adjustable release pressure, with a move towards bigger screw pressing capacity to reduce the number of screw presses in operation.

   The high price of palm oil and more efficient solvent recovery technology has enabled the introduction of oil recovery from the pressed palm fiber using solvent extraction. The pressed palm fiber is transported into the palm fiber solvent extraction plant for the oil recovery. The extracted fiber is still used as boiler fuel to generate the energy for the milling process. The extracted oil recovered from the palm fiber is rich in carotene and palm vitamin E and is used for the animal feed industry and not for direct human consumption (Goh, 2009).

4. Oil Recovery

   The introduction of two-phase decanter that eliminates the hot water dilution required during the clarification has significantly reduced the amount of palm oil mill effluent (POME) produced (Schuchardt et al., 2008; Wong & Sivasothy, 2007). The POME reduction is almost fifty percent lower than the conventional oil recovery using vertical clarifier. Membrane filtration technology is still at the developmental stage, and many membrane suppliers still have not been able to develop suitable materials and cleaning protocols to overcome the membrane fouling problems for efficient filtration.

5. Waste water treatment

   The need to eliminate biological treatment using a ponding system that requires very long holding or treatment period, usually longer than 100 days of retention and with more stringent discharge quality requirements imposed by the authority, is pushing the industry into developing systems with higher rates of biological treatment (Poh & Chong, 2009). This includes the use of anaerobic digester tanks, some combined with biogas production, mechanical assisted aerobic process, and the introduction of a tertiary treatment plant to meet the statutory requirements on discharge standards. Increasing numbers of palm oil mills have produced biogas for energy production as well as for methane capturing that will reduce the carbon footprint of the CPO production.

Extensive R&D activities in the palm oil industry in Malaysia and Indonesia have shown tremendous progress in developing and introducing new technologies to the whole plantation industry and the palm oil milling industry in particular. While there are several well-established engineering and fabrication companies with the capability to design, innovate, and manufacture oil palm milling machineries, new players are constantly emerging, bringing new ideas either developed internally or adapted from existing technologies in other industries. Most, if not all, of the new technologies are aimed at increasing productivity and enhancing product recovery rates.

Current and new technologies in the palm oil milling industry are summarized in Table 8-C. Second Generation technologies are those technologies that have been recently introduced in some of the palm oil mills and will be the industry norm for future new oil mills, while Third Generation technologies are technologies expected to be developed and commercialized on a large scale within the next 5 to 10 years.

By-product Utilization

Among the new developments in the oil palm milling industry are the rising interests in environmental sustainability. Waste materials generated from the milling processes are now being further examined for conversion into value-added co-products. With the ratification of the Kyoto Protocol under the United Nations Framework Convention for Climate Change (UNFCCC), which requires the reduction of emissions among the Annex I (industrialized) countries, projects on the reduction of Green House Gases (GHGs) in non-Annex I countries have become financially attractive as Certified Emission Reductions (CERs) that are now tradable. These projects which generate value-added co-products offer additional revenue streams to the palm oil mill, as well as enhance the image of the oil palm milling industry as a sustainable business. The potential applications of these waste materials are outlined in Table 8-D.

Table 8-C. Current and New Technologies in Palm Oil Milling Industry.

Process	Palm Oil Milling Technology		
	Conventional	2nd Generation	3rd Generation
FFB handling and transfer	• Cages of various sizes • Rope, capstan, and bollard • Prime movers • Overhead cranes	• Cages of various sizes • Indexer system	• Transfer conveyors
Sterilization	• Horizontal batch sterilizers	• Continuous sterilizers • Vertical/tilting/inclined/ spherical sterilizers	• Microwave sterilizer • Mobile sterilizer
Sterilized FFB transfer	• Cages of various sizes • Rope, capstan, and bollard • Prime movers • Overhead cranes	• Cages of various sizes • Indexer system • Tippler system	• Transfer conveyors
Oil extraction	• Screw presses	• Double pressing	• Solvent extraction • Enzymatic assisted extraction process
Oil recovery	• Vertical clarifiers • Centrifuges • 3-phase decanters	• Two-phase decanters	• Membrane filtration
Kernel recovery	• 2 stage winnower • Claybath/hydrocyclone	• Multi-stage winnower • Claybath/hydrocyclone	• Total dry separation
Boiler operation	• Conventional furnace	• Fuel feeding system • Moving grate furnace	• Fuel/air ratio control system • Pulverized boiler fuel system • High efficiency boiler
Waste water treatment	• Conventional biological anaerobic and aerobic process using ponding	• Faster rate anaerobic process in digester tanks • Tertiary treatment for BOD-20 ppm	• Rapid Anaerobic Biological reactor (RABR) • High rate anaerobic digestion process

Table 8-D. Value Added Applications for Palm Oil Mill Wastes.

Waste materials	Value added applications
Empty fruit bunches	• Fuel for additional energy and steam generation • Organic fertilizer through composting process • Dried fiber for various industrial applications • Biofuels (e.g., Bio-oil, bioethanol)
Palm shell	• Sold to external buyers as fuel or for conversion into other carbon products
Palm oil mill effluent	• Organic fertilizer, either through composting with EFB or on its own • Biogas generation for additional energy

Organic Fertilizer from EFB

With the escalating prices of fertilizers brought about by the steep increases in crude petroleum prices, alternative sources of nutrients for the plants need to be found. Biomass in the form of EFB can now be converted into organic fertilizer through composting. Several major plantation companies such as Sime Darby Plantation and Kulim in Malaysia have taken the lead in this field, as co-composting of EFB biomass with POME also assists in reducing the volume of liquid waste to be treated. Typical plant nutrient content in mature compost is shown in Table 8-E.

The EFB biomass composting process, when combined with POME, qualifies for registration as a CDM project. The system is maintained in aerobic conditions and therefore no methane gas is produced, in contrast to the conventional POME anaerobic treatment system. The potential revenue from CERs generated from the EFB biomass-POME composting process for a 90 tonne/hr palm oil mill is estimated at RM 1.0 million per year, assuming a CER price of USD 10 per CER (Khairudin et al., 2008).

Biogas from POME

An alternative renewable energy source is the methane-rich biogas produced during the anaerobic digestion of POME. While there had been attempts to harness this source of energy as early as the 1980s, interest has only gathered momentum lately,

Table 8-E. Compost Nutrient Content.

Compost Nutrient Content (dry basis)			
N	P	K	Mg
1.70–2.0%	0.25–0.42%	1.30–2.5%	0.70–1.0%

Source: Khairudin et al., 2008.

Table 8-F. Energy from Oil Palm Biomass.

Biomass	Moisture, %	Calorific Value, kJ/kg	Calorific Value, kJ/kg (dry)
EFB	65	6,578	18,795
Palm fiber	45	10,480	19,055
Shell	10	18,084	20,093
Palm kernel cake	7	17,562	18,884

Basri Wahid, 2008.

primarily due to the high fossil fuel costs and also partly due to the potential CDM revenue.

Based on data available, biogas production of between 20 to 28 m^3 per tonne POME is possible, equivalent to about 0.07 tonne methane per tonne FFB processed (Basri Wahid, 2008; Tan, 2008). In terms of power generation, this translates to about 33 to 37 kWh electrical energy that can be generated per tonne FFB processed.

Energy from EFB Biomass

EFB has been utilized as fuel for additional energy since the early 1980s when a few palm oil mills started burning EFB for steam and power generation for supplying related manufacturing facilities within the vicinity of the palm oil mill. Average values of energy content for the various biomass produced are shown in Table 8-F.

As of August 2010, the number CDM projects registered in Malaysia with the CDM was 55, with oil palm related projects making up 45 projects (CDM Energy Secretariat, 2010). Of these, 16 were biomass (EFB/shell) energy utilization projects, while 29 were POME biogas energy projects.

Palm Kernel Crushing Process

The typical palm kernel oil extraction uses mechanical extraction. The mechanical extraction method is widely practiced since the technology is much cheaper in capital investment and operating costs, even though it has limitations in oil recovery and produces poorer quality of palm kernel cake. Palm kernels delivered from palm oil mills are stored in kernel silos. These kernels are then fed directly into a series of screw presses that will extract out the crude palm kernel oil (CPKO) from the palm kernel. The first stage pressing will extract out the oil from 45–49% to less than 15% in the palm kernel expeller cake (PKC). The palm kernel expeller cake, which still contains approximately 15% of oil, is further fed into another series of screw presses to extrude the remaining oil until the oil in palm kernel expeller cake is about 6 to 7%.

The crude palm kernel oil, after the screw press, is passed through vibrating screens to remove coarse solids present in the oil before the filtration process, where either

Niagara filters or membrane filter presses are used to remove residual dirt and impurities. The filtered oil is stored in a crude kernel oil tank while the remaining solids from the filtration are recycled back into the screw press to recover the remaining oil.

Palm kernel solvent extraction initially was popular in the 1980s, but the majority of these solvent extraction plants have ceased operations due to high solvent loss and concerns about safety issues. To date, only one palm kernel solvent extraction plant is still operating in Malaysia. The solvent extraction process consists of a kernel flaking process to increase the solid surface area combined with the steaming for more efficient extraction before the hot flaked kernel is fed into the solvent carousels for the extraction process. The oil will dissolve in the solvent before being separated in the evaporator to separate the oil from the solvent. The simplified palm kernel oil mechanical extraction process is shown in Fig. 8.4. The basic quality parameters for palm kernel oil and palm kernel cake for export from Malaysia (MEOMA specifications) are shown in Table 8-G.

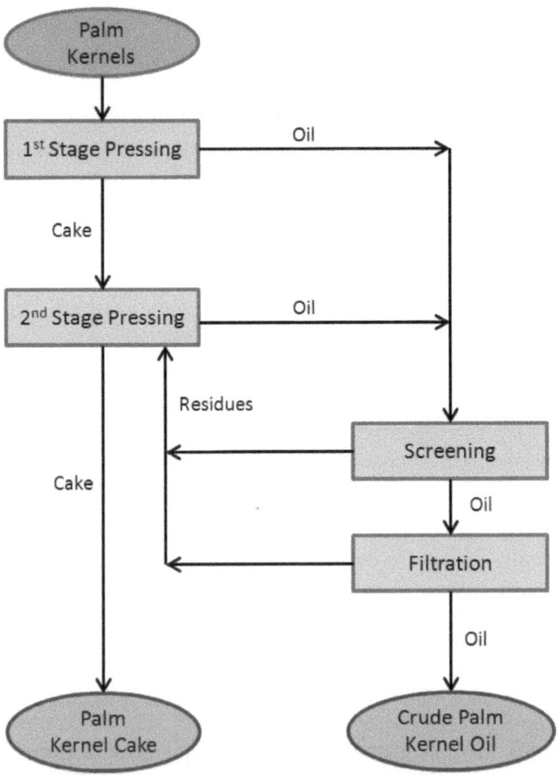

Fig. 8.4. Simplified palm kernel oil mechanical extraction process.

Table 8-G. Basic Quality Parameters for Crude Palm Kernel Oil and Palm Kernel Cake.

Quality parameters	Crude palm kernel oil
Free fatty acid (as lauric acid)	5% max
Moisture & impurities	0.5% max
Iodine value (Wijs)	19 max
Quality parameters	Palm kernel cake
Profat	21% min
Moisture	10% max
Protein	14% min
Oil	8% min
Shell & dirt	15% max
Raw fiber	18% max
Ash	6% max
Sand & silica	1.5% max

Developments in Processing Technologies

There is not much development in palm kernel processing technology apart from some development revolving around producing higher quality palm kernel cake suitable for the non-ruminant or poultry industry. Improvement in oil extraction is limited due to the specification of oil remaining in PKC, which shall be above 8% for the ruminant feed meal industry. Some of the improvement is achieved by the addition of a winnowing unit to remove some of the remaining shell and dirt in palm kernels before the extraction process and the use of a kernel chopper to improve the extraction.

By-product Utilization

The palm kernel cake (PKC) that is high in protein and fats content is mainly used as one of the main components for feedmeal formulation for the ruminant industry. The use of PKC as part of feedmeal formulation for the non-ruminant and poultry industry is limited due to its high fiber content (Wan Zahari & Alimon, 2004). It is recommended that the PKC component for poultry feedmeal formulation be capped at 15 to 25%. Additionally, the dirt and shell content needs to be reduced to below 7%. This could be done by removal of most of the shell and dirt present in palm kernel before the extraction process.

Oil Palm Sustainability Issues

The Roundtable on Sustainable Palm Oil (RSPO) was formed by a diverse group of stakeholders in the palm oil industry in 2004, based in Switzerland, to promote sustainable agriculture and to address the environmental impact of palm oil production.

The RSPO has described sustainable production of palm oil as "comprising of legal, economically viable, environmentally appropriate and socially beneficial management and operations" (RSPO, 2007).

The RSPO has developed a set of standards called the Principles & Criteria (P&C) that define the practices for sustainable palm oil production. These standards address the legal, economic, environmental, and social requirements of producing sustainable palm oil. RSPO's Principles and Criteria (P&C) for sustainable palm oil production are based on the following principles:

1. Commitment to transparency
2. Compliance with applicable laws and regulations
3. Commitment to long-term economic and financial viability
4. Use of appropriate best practices by growers and millers
5. Environmental responsibility and conservation of natural resources and biodiversity
6. Responsible consideration for employees and for individuals and communities affected by growers and mills
7. Responsible development of new plantings
8. Commitment to continuous improvement in key areas of activity

To-date, 123 palm oil mills and 28 oil palm growers have been certified to produce Certified Sustainable Palm Oil (CSPO). Total CSPO production capacity was 3.5 million tonnes, with sales of 1.3 million tonnes in 2010, as compared with 2009 production of CSPO at 1.5 million tonnes and sales of 0.34 million tonnes of CSPO (RSPO, 2011).

Indonesia is proposing to form its own Indonesia Sustainable Palm Oil (ISPO) to legislate and make it mandatory for all the oil palm plantations to be sustainable in their operation. Major plantation players such as Sime Darby and United Plantations continue to lead the CSPO production in Malaysia.

The Way Forward

The success of the oil palm industry in Malaysia and Indonesia owes greatly to the dedication of the planters and support and encouragement from both governments. Both governments have put in place initiatives to further spur growth and competitiveness in the industry.

In recognition of the industry's role in the economic development of the nation, the Malaysian government has set a target of a Gross National Income contribution (GNI) of RM 170 billion by 2020 from RM 52 billion currently (MPOB, 2011). Eight projects have been identified to improve both upstream and downstream productivity and sustainability, covering enhancements in yield and extraction rates, value addition through high value oleochemical derivatives, food and health based products, and bio-energy. The growth is expected to generate an additional 41,000 jobs.

The Indonesian government meanwhile is facilitating the oil palm industry to expand with progressive liberalization of land use permits. It has been estimated that potentially there are 32 million hectares (USDA, 2009) of land suitable for oil palm cultivation throughout Indonesia. Current areas of expansion are focused on Kalimantan, Sumatera, and Sulawesi. At the same time, the government is encouraging export and local downstream processing investment activities through various measures, including reviewing tariffs and duties on crude palm oil and refined products and derivatives exports. All of these governmental initiatives are expected to further spur the growth and development of the oil palm industry in the region.

References

Basri Wahid, B.; Biofertilizer and effluent management for the oil palm industry. National Seminar on Biofertiliser, Biogas and Effluent Treatment in the Palm Oil Industry. 2008.

CDM Energy Secretariat, Malaysia, 2010. http://cdm.greentechmalaysia.my/up_dir/CDM%20 Statistic%20as%20of%20August%20 2010.pdf, accessed August 8, 2011.

Chow, M.C.; Ma, A.N. Processing of fresh palm fruits using microwaves. J. Microwave Power Electromagnetic Energy 2007, 40(3), 165–173.

Direktorat Jenderal Perkebunan, 2010 http://ditjenbun.deptan.go.id/cigraph/index.php/viewstat/ komoditiutama/8-Kelapa%20Sawit, accessed November 28, 2011.

Goh Kee, S. Palm pressed fibre oil extraction (PFOE) technology and deoiled fibre applications. PIPOC International Palm Oil Congress, 2009.

Khairudin, H.; et al. Opportunity and potential of EFB composting. National Seminar on Biofertiliser, Biogas and Effluent Treatment in the Palm Oil Industry, August 2008.

Lim, C.L. The use of spherical sterilizer for sterilization of fresh fruit bunches. PIPOC International Palm Oil Congress, 2007.

Loh, T. The way forward in the palm oil milling process with the advent of TILTING STERILIZER. PIPOC International Palm Oil Congress, 2009.

Maycock, J.H., Ed. Palm oil factory process handbook part 1—general description of the palm oil milling process. Palm Oil Research Institute of Malaysia: Malaysia, 1985.

MBOP. Malaysian Oil Palm Statistics, 2009, http://econ.mpob.gov.my/economy/annual/stat2009/ EID_statistics09.htm, accessed July 7, 2011.

Nazarulhisyam, S.; Kaida, K. Effectiveness of sterilization of oil palm bunch using microwave technology. Ind. Crops Prod. 2009, 3(2), 179–183.

NKEA Palm oil: the way forward. Malaysian Palm Oil Board, 2011.

Poh, P.E.; Chong, M.F. Development of anaerobic digestion methods for palm oil mill effluent (POME) treatment. Bioresource Technology 100. 2009, 1–9.

RSPO Principles & Criteria for Sustainable Palm Oil Production—Including Indicators & Guidance, October 2007. RSPO, 2007.

Roundtable on Sustainable Palm Oil, 2011 www.rspo.org, accessed November 28, 2011.

Schuchardt, F.; et al. Effect of new palm oil mill processes on the EFB and POME Utilization. J. Oil Palm Res. Special Issue October 2008, 115–126.

Sivasothy, K. Recent developments in continuous sterilisation. PIPOC International Palm Oil Congress, 2009.

Tan, B.W. Electricity generation with biogas. National Seminar on Biofertiliser, Biogas and Effluent Treatment in the Palm Oil Industry, August 2008.

USDA Foreign Agricultural Service, 2009, http://www.pecad.fas.usda.gov/highlights/2009/03/Indonesia/, accessed November 28, 2011.

Wan Zahari, M.; Alimon, A.R. Use of palm kernel cake and oil palm by-products in compound feed. Palm Oil Develop. 2004, 40, 5–9.

Wong, P.W.; Sivasothy, K. Emerging trends in palm oil milling technology. PIPOC International Palm Oil Congress, 2007.

9

Palm and Palm Kernel Oil Production and Processing in Brazil

Ana Cristina Pinheiro do Prado and Jane Mara Block
Santa Catarina Federal University, Department of Food Science and Technology, Rodovia Admar Gonzaga l346, Itacorubi, Florianópolis, Santa Catarina, Brazil

Introduction

Oil palm trees belong to the family of plants known as *Palmae* or *Palmaceae* and to the genus *Elaeis,* which is composed of two species: *Elaeis guineensis* and *Elaeis oleifera,* the former being the most commonly found and cultivated in Brazil, comprising the major Brazilian oil palm tree (*dendezeiro*) crop (Bora et al., 2003; Sambanthamurthi et al., 2000). The tree is a monocotyledon and its structure consists of a stalk 40–45 cm thick and a crown with approximately 40 leaves (Pérez & Nápoles, 2000), reaching a height of up to 20 m (Bockisch & Germany, 1993).

The oil palm tree is characterized as a tropical tree and is one of the greatest oil producing plants in the world, and its commercial-scale production is on the increase, mainly in countries which lie in tropical zones with hot and humid climates (Ekpa, 1995; Poku, 2002; Tandom et al., 2001).

Since it is a large perennial plant, the adult oil palm tree offers a perfect covering of the soil and can be considered a cultivation system of acceptable ecological stability with low negative environmental impact. The plant begins producing fruits 3 years after initial planting, and fruit production is evenly distributed throughout the year for approximately 25 consecutive years. Under favorable ecological conditions, the oil palm tree produces an average of 5 tons of oil/hectare/year, approximately ten times more oil /hectare/year in comparison to soybean (Biodieselbr, 2006; Embrapa, 2011a).

Elaeis guineensis can be divided into three basic varieties: Dura, characterized by a fine mesocarp and a dense endocarp, generally with a larger number of seeds; *Tenera,* characterized by a dense mesocarp and a fine endocarp, with a moderate number of seeds; and *Pisifera,* with a dense mesocarp, the absence of an endocarp, a lower number of seeds, and lower quantity of oil (Umerie et al., 2004). The cross breeding of the varieties *Dura* and *Pisifera* (D × P), known as *Tenera,* represents the variety most cultivated in Brazilian crops (Biodieselbr, 2011).

Bunches of palm fruit can reach 10 to 40 kg, the individual fruits weigh around 6–20 mg. The fruits contain an external membrane (exocarp), pulp (mesocarp) from which the palm oil is extracted, and a central region (endocarp) which contains the seed (kernel). Oil can also be extracted from the kernel, although the composition differs significantly from that extracted from the mesocarp (Poku, 2002). Thus, there are two types of oil, which originate from the fruit of the oil palm: palm oil (also known as dende oil) and palm kernel oil, extracted from the mesocarp and the seed, respectively (Akpanabiatu et al., 2001).

The oil palm and its kernel are characterized by high oil contents (73% and 32.6%, respectively) and dry matter. The remaining nutritional composition of the palm fruit pulp produced in Brazil comprises approximately 13.3% carbohydrates, 4.3% total fibers, 4% moisture, 3.4% proteins, and 1.9% ash or total minerals. The palm kernel has a remaining nutritional composition of approximately 35.1% carbohydrates, 15.6% total fibers, 10.9% proteins, 4.4% moisture, and 1.7% ash or total minerals (Bora et al., 2003).

The physical and chemical properties of palm oil and palm kernel oil extracted from plants cultivated on Brazilian soil show a considerable variation in terms of the identification and quality parameters such as acid value (2.36 and 1.09% linoleic acid), peroxide value (2.96 and 1.37 meq/kg), iodine value (45.6 and 18.9 g/100 g), and saponification value (183.8 and 221.4 mg KOH/1.0 g). Although these indices indicate significantly different quality and stability characteristics, some physical properties have similar values, such as a refraction index of ~1.46 and a specific gravity of ~0.92 g/cm^3, respectively, for both palm oil and palm kernel oil (Bora et al., 2003).

In agreement with the indices for the physical and chemical properties cited above, the oil contained in the fruit pulp (known as palm oil) is composed of significant contents of monounsaturated and polyunsaturated fatty acids, with a high potential for use in cooking, whereas the Brazilian palm kernel oil contains approximately 92% saturated fatty acids. When the fatty acid profiles of Brazilian palm oil and palm kernel oil are compared with plants cultivated in other regions, there is a significant difference in the quantity and quality of fatty acids present (Bora et al., 2003).

The Brazilian fruit has approximately 24 different fatty acids making up the composition of the palm oil and 18 in the palm kernel oil. The main saturated fatty acids present are palmitic acid (36.9%) in Brazilian palm oil and lauric (53.3%) and myristic (19.3%) acids in Brazilian palm kernel oil, and oleic acid is the predominant monounsaturated fatty acid in the palm oil (45.29%) and palm kernel oil (5.5%) (Bora et al., 2003). The palm kernel oil obtained, given its significantly differentiated composition, has particular characteristics in terms of hardness at ambient temperature, yielding characteristics similar to those found only in coconut oil (Bora et al., 2003; Pantzaris & Ahmad, 2001). The typical triglyceride composition of crude oil obtained from palm pulp, known as palm oil or dende oil, can be observed in Table 9-A.

Table 9-A. Typical Composition of Crude Palm Oil.

Major constituents	> 99%	Principal components
Triglycerides	87–92	Palmitic-Oleic-Palmitic Palmitic-Oleic-Oleic Palmitic-Oleic-Linoleic
Diglycerides	3–8	
Monoglycerides	0–0.5	
Free fatty acids	1.5–4	Palmitic and Oleic
Minor constituents	< 1%	

Source: Grupo Agropalma (2002).

As can be observed in Table 9-A, besides the fatty acids of the triglycerides of the oil, minor constituents are also present, which represent less than 1% of the total crude oil. Notable among these constituents are the plant sterols, known as phytosterols including beta-sitosterol present in the range of 300–600 ppm in Brazilian palm oil, and in lesser amounts gramisterol (~360 ppm). Other minor components present in palm oil produced in Brazil include phospholipids (20–80 ppm), glycolipids (mono-glycosyl-diglyceride, 1000–3000 ppm), and triterpene alcohols (cycloartenol, ~320 ppm) (Grupo Agropalma, 2002).

The minor constituents of palm oil also include some vitamins such as vitamin E (comprising the natural antioxidant substances tocopherols and tocotrienols) (Arora et al., 2006; Berger, 2000). In fact, crude palm oil is considered to be a major source of vitamin E, comprised of 70% tocotrienols and 30% tocopherols (Al-Saqer et al., 2004; Sanagi et al., 2005). In oil palm tree cultivated in Brazil, average contents of 600–1000 ppm tocopherols have been reported, the predominant form being gamma-tocotrienol (Grupo Agropalma, 2002). Besides vitamin E, Brazilian palm oil contains significant concentrations (500–700 ppm) of vitamin A (carotenes), conferring a characteristic red color (Fernandez et al., 2000; Grupo Agropalma, 2002).

The combination of vitamins (carotenoids, tocopherols and tocotrienols) and the high content of saturated fatty acids confer a high oxidative stability to palm oil compared with other plant oils, as well as a better nutritional quality (Arora et al., 2006; Poku, 2002; Umerie et al., 2004).

Production and Commercialization of Palm Oil and Palm Kernel Oil in Brazil

The cultivation of oil palm tree (*E. guineensis*) originated in the west region of Guinea, and significant quantities of palm oil continue to be produced by traditional producers in West Africa, although the growth has been less pronounced compared with other regions such as Southeast Asia (Indonesia and Malaysia), where palm cultivation has expanded greatly (Basiron, 2005; Cottrell, 1991).

The expansion of palm cultivation also extended to Latin America soon after the commercialization of slaves began in the 16th century. Since that time, the region has demonstrated a high potential for production due to the climatic characteristics found in the Latin American tropics. Some South American countries have significantly expanded the cultivation of oil palm, with great potential for future expansion (Basiron, 2005; Cottrell, 1991).

The extensive development of the palm oil industry in many tropical Latin American countries is due mainly to its high productivity. Palm oil has the highest oil yield per area cultivated with palm crops growing well in tropical climates 5° north and south of the equator. The ideal conditions for growth include annual rainfall of around 2000 mm distributed throughout the year, with approximately 2000 sunshine hours and moderately high temperatures of around 25–33°C. Countries that do not offer ideal growth conditions for the oil palm reportedly have higher production costs and are not typically producers for the market since large-scale cultivation is unviable (Basiron, 2005; Poku, 2002).

In Brazil, the oil palm tree (*dendezeiro*; *E. guineensis*) was introduced around the 17th century. At that time, the slaves transported seeds on the ships and were responsible for the formation of the first sub-spontaneous plantations along the coast of the state of Bahia. The crop adapted well to the tropical and humid climate of coastal Bahia (Seagri, 1999). Despite having large geographical areas with conditions favorable for the cultivation of oil palm and the production of palm oil, Brazil still has not achieved its full potential in the market. Of the palm oil producing countries worldwide, Brazil was ranked 13th in oil production in 2008, and data published in 2010 indicated that Brazil ranked 11th (Oil World Annual, 2011; Portal do Meio Ambiente, 2011). In Latin America, Brazil was ranked 3rd in 2008 after Columbia and Ecuador (Oil World Annual, 2011; Parente, 2011). Although Columbia is notably the major representative of Latin America with 2% of the total global production in 2009, Indonesia and Malaysia are the largest producers responsible for 44% and 43% of the global palm oil production, respectively (Bazmi et al., 2011; Parente, 2011).

Of the species cultivated in Latin America, *E. oleifera* (also known as *Elaeis melanococca*), represents a native species and since the 1960s has shown a particular increase due to its oils being specifically suited for food production; high iodine value (70–80); and content of unsaturated fatty acids, mainly oleic and linoleic acids (Kiple & Omelas, 2000; Sambanthamurthi et al., 2000). Although the species *E. oleifera* is native to the region, its participation in the cultivated area has not been significant. This is mainly due to a long history of seeds from the species *E. guineensis* in Central America introduced by the United Fruit Company during the decades prior to the 1960s. The same company bought seeds from Sierra Leone for cultivation in Guatemala in 1920 and from Malaysia in order to establish plantations in Panama (1926) and Honduras (1927). After some years, other seeds of the species *E. guineensis* were introduced to start a commercial plantation of 250 hectares in Guatemala in 1940. In 1992, the total planted area of *E. guineensis* in Latin America reached 390,000 hectares (Kiple & Omelas, 2000), and in 1996 the main producers in the region were Colombia followed by Equador, Brazil, Costa Rica, and Honduras (BASA, 2011).

In South America, the agricultural policy of Columbia involves a significant expansion in the area of cultivation, particularly in regions like the eastern plains (Phosri et al., 2010). In 1998, Columbia had approximately 125.000 hectares planted, corresponding to 31% of the area cultivated on the American continents. The increase in the area of cultivation and production was intense, representing 12.57% and 8.42%, respectively, per year. This growth contributed to Columbia being ranked the 4th and 5th largest palm oil producer globally in 1998 and 2008, respectively (BASA, 2011; Oil World Annual, 2011). Fig. 9.1 shows the palm oil production (× 1.000 tons) on the American continent up to 2009.

Continuous investment in palm oil production in Southeast Asia, Africa, and Central/South America, particularly in the late 1980s and early 1990s, strongly contributed to the increased production (Basiron, 2005). During this period, palm oil became the second largest plant-produced oil in the world and was expected to overtake the production of soybean oil by the beginning of this century due to its high productivity at low cost, good quality, and widespread use (Embrapa, 2011a). It has been estimated that by 2012, the annual average production of palm oil will reach 30 million tons compared to 29.1 million tons of soy oil, 15.5 million tons of canola (rapeseed) oil, and 12.1 million tons of sunflower oil (Parente, 2011).

In Brazil in 1995, the harvested area of oil palm was 68,000 hectares and the production of palm oil was around 80,000 tons. The country consumed 280,000 tons of palm oil and its derivatives, imported approximately 180,000 tons, with a potential internal market of 400,000 tons per year of palm (dende) oil and its derivatives (Seagri, 1999). By 2004/2005, the Brazilian production of palm oil had reached

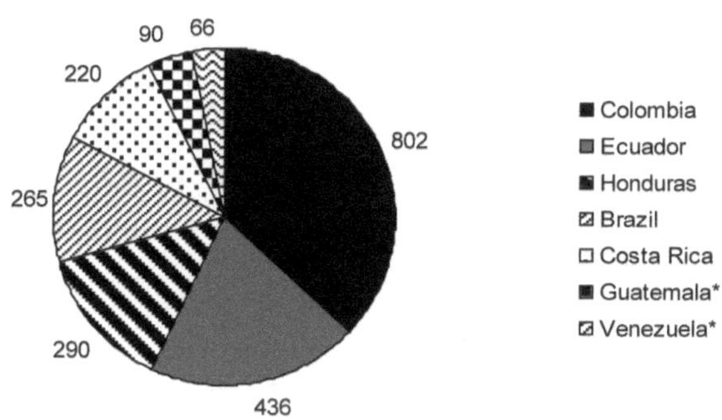

Fig. 9.1. Palm oil production (× 1.000 tons) on the American continent up to 2009 (except * data from 2005). Source: Oil World Annual (2011).

115–132 mil tons/year with a cultivated area of almost 60,000 hectares (0.1% of the potential) (FBDS, 2005; Parente, 2011).

The Brazilian states, which are major producers of palm oil include Pará, responsible for approximately 85% of national production (78,000 tons in 1999) and 0.6% of the global production, followed by Bahia and Amapá (Embrapa, 1995; Parente, 2011). However, there are areas with the potential for palm cultivation in Brazil yet to be exploited, and these lands ideal for cultivation were catalogued in 2000 and can be observed in Table 9-B.

In 2002, with less than 1% of the world production, Brazil had the largest agricultural area for the expansion of this crop, notably in states such as Amazonas. The production of palm oil is an excellent option as an anchor activity in programs related to spreading agriculture and maintaining populations in farming areas, with proven economic, social and ecological benefits, generating permanent employment (Embrapa, 2011a). Despite the significant potential for the development of oil palm cultivation, the Brazilian production is still not at consumption levels and in 2009, Brazil imported approximately 128,000 tons of palm oil/year to satisfy the internal market demand (Oil World Annual, 2011).

Production in the Amazon Region

In the Amazon region, the introduction of palm cultivation occurred at the beginning of the 1950s in the state of Pará, by the Agronomic Institute of the North (*Instituto Agronômico do Norte*—IAN), a precursor to the Brazilian Agricultural Research Com-

Table 9-B. Brazil—Potential Areas for Planting Oil Palm.

State	Area with potential for cultivation (ha)	2000 (hectares—ha)	
		Area planted	Productive Area
Acre	2,500,000	–	–
Amapá	1,500,000	2,000	–
Amazonas	50,000,000	1,200	1,200
Bahia	–	6,050	6,050
Maranhão	–	–	–
Mato Grosso	500,000	–	–
Pará	10,000,000	45,213	27,359
Rondônia	1,000,000	–	–
Roraima	4,000,000	–	–
Tocantins	500,000	–	–
Total	70,000,000	54,463	34,609

Source: Grupo Agropalma (2000).

pany (*Empresa Brasileira de Pesquisa Agropecuária—Embrapa Amazônia Oriental*), which imported some varieties from the African continent with the aim of assessing their potential for cultivation in the Amazon region (Pandolfo, 1981).

The Amazon region has around 70 million hectares considered areas appropriate or exploitable for oil palm cultivation. Of this potential area, only 39,000 hectares are used effectively for cultivation and almost 85% of this area lies in the state of Pará. Amazonas is the state that has the largest potential area for planting oil palm with around 50 million hectares, corresponding to 71.4% of the area appropriate for cultivation in the Amazon region (Barcelos, 1993; Barcelos et al., 1995; Parente, 2011).

Oil palm cultivation on a commercial scale in the Amazon region began in 1967, mediated by an agreement between the Bureau for the Economic Enhancement Plan for the Amazon Region (*Superintendência do Plano de Valorização Econômica da Amazônia*—SPVEA), later renamed the Bureau for the Development of the Amazon Region (*Superintendência de Desenvolvimento da Amazônia*—SUDAM), and the French Research Institute of Oils and Oil-producing Plants (*Institut Recherches Pour Les Huiles et Les oleagineux*—IRHO). The project at that time considered the planting of 3,000 hectares (BASA, 2011).

In 1974, a private initiative was incorporated into the economic exploitation of the cultivation of oil palm with the creation the company Dendê do Pará S.A (DENPASA), providing greater dynamics to the activity. In the 1980s, the production expressed a significant increase stimulated by new projects, which expanded the planted area mainly in the state of Pará. Furthermore, a large number of processing agro-industries were installed in this state, which up to 1998 were concentrated over 70% of the area cultivated for the production of palm oil in Brazil, reaching 90% of the national production in 2004 (BASA, 2011; FBDS, 2005). By 2005, Pará was the leader in terms of national palm oil production with 760,000 tons/year (Sagri, 2011).

The other states located in the west of the Amazon region, for instance, Acre, Amapá, Rondônia, and Roraima, had 9 million hectares in total corresponding to 12.9% of the total area potentially exploitable for oil palm cultivation (Parente, 2011). By 1998, the state of Amapá had allocated an ecologically favorable area of approximately 1,500,000 hectares available for potential cultivation; however, the actual area cultivated with oil palm was only 4,000 hectares. The agro-industry company COPALMA retained, up to that time, all of the area of palm oil production of the state with an installed capacity to process 12 tons of fresh fruit bunches per hour. In 1996, COPALMA of Amapá extracted approximately 9,000 tons of palm oil, representing the use of 22.5% of the extraction capacity of the company (BASA, 2011).

Currently, with good prospects being indicated by the international market of palm oil, the cultivation of oil palm tree in the state of Amapá is likely to gain a new impetus. BRASEX, a holding company from Sao Paulo State specializing in international commerce, carried out a project in collaboration with the Brazilian Development Bank BNDES (*Banco Nacional de Desenvolvimento*), forecasting the implantation of

10,000 hectares of oil palm in the municipality of Porto Grande, 80 km from Macapá, the capital of the state of Amapá (BASA, 2011).

As is the case for many areas in the Amazon, the cultivation of oil palm tree has a great potential for exploitation in the state of Roraima, mainly in the south region of the state (municipalities of São João da Baliza, São Luiz, and Caroebe), which has the most suitable climatic conditions for adaptation (annual rainfall close 2000 mm). In addition, the results available from scientific research carried out by Embrapa Amazônia Ocidental allow technological support for the cultivation in this region. Also, in Roraima, experiments are currently being conducted with oil palm hybrids in the Cerrado (tropical savanna region) and in an altered forest area in the south region of the state. The private initiative has also invested in the state of Roraima, installing businesses and initiating planting in rural areas (Cordeiro, 2009).

In order to enhance the expansion of planting, the Agro-climate Zonation of Oil Palm (*Zoneamento Agroclimático do Dendê*—ZonDendê) in the Amazon region was concluded in 2009/2010, constituting a landmark tool for the advancement of public policies for the cultivation of oil palm tree in Roraima. In the current setting, the agro-ecological zonation provides a basis for the implantation of oil palm tree cultivation on a sustainable basis, both for large companies and for small-scale farmers. This tool aids in the classification of the adaptability of land for oil palm tree cultivation, evaluating and comparing the eco-physiological requirements of the plant and environmental conditions of the area to be planted, seeking the most favorable cost-to-benefit ratio as well as providing information, which aids the implantation of the crop with less negative impact on the environment (Embrapa, 2010).

For the area known as Amazônia Legal, in particular, ZonDendê constitutes a fundamental tool in terms of technical-scientific knowledge seeking sustainability on a social, economic, and environmental basis through indicating which lands are most suitable for the agricultural production of the oil palm. The Agro-ecological Zonation of Oil Palm for Deforested Areas of Amazônia Legal (*Zoneamento Agroecológico do Dendê para as Áreas Desmatadas da Amazônia Legal*—Zon-Dendê) was commission by the Brazilian Government with financial support from FINEP. The area designated as Amazônia Legal, established in article 2 of the Brazilian Law nº 5.173, of October 1966, covers the states of Acre, Amapá, Amazonas, Mato Grosso, Pará, Rondônia, Roraima, Tocantins, part of Maranhão, and five municipalities of Goiás, comprising approximately 59% of the Brazilian territory. The project, led by Embrapa Soils (*Embrapa Solos*), was carried out in cooperation with Embrapa Units (*Unidades da Embrapa*) in the north region of Brazil and also with other public and private institutions which act in the region such as Protection Systems of the Amazon Region (*Sistema de Proteção da Amazônia*—SIPAM-Manaus), Ministry of the Environment (*Ministério do Meio Ambiente*)—through IBAMA, FUNAI and the Departments of Sustainable Development and of Biodiversity and Forests (*Secretarias do Desenvolvimento Sustentável e de Biodiversidade e Florestas*) –, CEPLAC/CEPEC, EMATER-RO, FEMACT-RR, SEPLAN-MT, the Diocese of Tabatinga-AM, the Federal University of

Pará (*Universidade Federal do Pará*—UFPA), and the companies Agropalma and Marborges. The data obtained from ZonDendê from 2010 indicated that 585,885 km^2 of land (which corresponds to 11.59% of the area known as Amazônia Legal) could be considered appropriate for oil palm tree cultivation, with the need to adopt different levels of soil management (Embrapa, 2010).

The incentive for oil palm tree cultivation in the Amazon region, besides the high potential for the development of the plants, is also due to the fact that this type of cultivation can be considered an activity consistent with environmental preservation since it is not associated with strong adverse effects on the native forest. The oil palm tree can be planted in degraded areas and/or to replace existing crops, enabling the complete recovering of these areas. In both the young and adult phases, the oil palm tree can be combined with soil-covering legumes, thus the crop will not become excessively aggressive, negatively affecting the forest biodiversity (Parente, 2011; Phosri, et al., 2010). Moreover, the oil palm tree can be cultivated on poor soils, as is the case of most of the land in the Amazon region, thereby remediating the land and imitating the tropical forest, thus protecting the soil against leaching and erosion. Besides restoring the microclimate and water balance, the oil palm tree cultivation contributes to fixing CO_2 and releasing O_2 to the atmosphere (FBDS, 2005).

In the current scenario, in which the integration of markets and the phenomenon of the globalization of economies predominate, the development and growth of oil palm tree cultivation assumes crucial importance for the socio-economic development of Brazil, especially for the Amazon region, since it could generate employment and income in this rural sector (BASA, 2011). Few economic options offer guaranteed permanent income for the Amazon population without environmental destruction. Oil palm tree cultivation therefore proves to be an excellent option, with a large capacity to generate jobs and income, since 10 planted hectares can generate one direct job (FBDS, 2005).

The capacity for the integration of agricultural production with industrial processing, enhancing the generation of jobs, is due to the fact that the fruit undergoes rapid acidification and needs to be processed within 24 hours of harvest. Thus, oil extraction processing plants need to be installed in the vicinity of the plantations (FBDS, 2005). With the agro-industrialization of the production in loco, it becomes possible to structure the palm oil production chain within the region itself, generating distributive effects within the economy of the region and therefore acting as a vector for the sustainable development of the Amazon region (BASA, 2011).

Furthermore, considering that the oil palm tree begins its commercial production 3 years after planting and continues for 25 consecutive years, requiring the use of intense labor, without the need for off-season periods of unemployment, it is possible to integrate oil palm tree cultivation within a highly profitable economic system. In this regard, research to genetically improve the oil palm tree has been carried out by Embrapa at the experimental field station in Rio Urubu, in the state of Amazonas. At this field station, there are 412 hectares of experimental crops where testing to improve the genetic profile, soil management, plant nutrition, and the collection of the germoplasm of the oil palm *Elaeis guineensis* (African origin) and *Elaeis oleifera* (American origin) are being conducted. The categories of seeds produced by Embrapa/Cirad, of the type Deli × La Mé, have slow growth, on average 45 cm/year, allowing a minimum 25 years of economic lifespan. When planted on marginal or optimal zones for the cultivation of the plant, they have a productivity of 15–40 tons of fruit bunches/hectare/year (Embrapa, 2011b).

Production in Northeast of Brazil

Brazil currently has the greatest potential globally for the production of palm oil, with approximately 75 million hectares of land appropriate for oil palm tree cultivation. The state of Bahia has approximately 900,000 hectares of this total, and is the only state in the northeast of Brazil with climatic conditions along the coastal strip suitable for the planting of oil palm tree (CEPLAC, 2011).

Bahia has an exceptional diversity of soils and climate for the cultivation of oil palm trees; the suitable land available in coastal regions extends from the Recôncavo to Tabuleiros do Sul. Most of the palm oil production (10,000 tons) originates from subspontaneous oil palm trees of low productivity which total approximately 19,650 hectares, and the area of cultivated oil palm trees is 11,500 hectares (extraction plants and independent producers), this being very low when compared with the total area available. These oil palm trees have low productivity particularly because they have surpassed the economically viable period of production (25 years) and also because they are in poor sanitary conditions (Seagri, 1997).

The state government of Bahia has dedicated itself to increasing the value of oil palm tree cultivation in the region. In this regard, they have promoted a project to implant and consolidate the commercial exploitation of oil palm tree cultivation in Bahia, on land in 29 municipalities grouped into the centers of Nazaré/Santo Amaro, Valença/Itacaré, Ilhéus/Canavieiras, and Belmonte/Prado (CEPLAC, 2010; Seagri, 1997).

The overall aim of the state government of Bahia is to reach a production of 48,000 tons of palm oil in 8 years, with 12,000 hectares planted in the first 3 years. In December 1999, the state government of Bahia finalized an agreement with the Bahian Agricultural Development Company (*Empresa Baiana de Desenvolvimento Agrícola*—EBDA); Executive Commission on the Cocoa Cultivation Plan (*Comissão Executiva do Plano da Lavoura Cacaueira*—CEPLAC); Bank of Northeastern Brazil (*Banco do Nordeste do Brasil*); and the companies Jaguaripe, Mutupiranga, OLDESA, and OPALMA aiming to produce 2,400,000 pre-germinated seeds and install nurseries to produce 1,724,000 hybrid (tenera) oil palm tree seedlings in areas covered by the program Development of Bahian Oil Palm Cultivation (*Desenvolvimento da Dendeicultura Baiana*) (CEPLAC, 2010; Seagri, 1997).

In 2004, Bahia ranked second in terms of palm oil production in Brazil, producing 9% of the total oil (FBDS, 2005).

Processing of Palm Oil and Palm Kernel Oil in Brazil

The oil palm tree is notable among oleaginous species due to its high oil production capacity per unit of land area. The average yield attains levels in the range of 3,000–5,000 kg of oil/hectare/year for palm oil, its main product, and a further 300–500 kg of palm kernel oil/hectare/year (BASA, 2011).

In Brazil, the palm oil and palm kernel oil extractors tend to be located strategically close to the plantations, in order to facilitate the transport of the fruit to the extraction plant to achieve optimal quality (Biodieselbr, 2011).

Several operational processes are used to obtain the final product. The first step of the processing produces the crude oil, extracted from the mesocarp (pulp) of the fruit. In the second step this can be refined or fractionated using a process of crystallization and simple separation where the solid (stearin) and liquid (olein) fractions are obtained (Biodieselbr, 2011).

The development of the Brazilian palm oil agro-industry has been significant in recent decades. In Brazil, the first palm oil extraction plants were established in Bahia in the 1950s. The organization of the agro-industry occurred later at the beginning of the 1970s in Pará, after which time the cultivation of oil palm was identified as an important source of vegetable oil and fat. Pará has a significant industrial complex and is notably the largest Brazilian producer, responsible for approximately 85% of the total palm oil processed and produced in Brazil. In decreasing order, other notable states are Amapá, Bahia, and Amazonas (Biodieselbr, 2011).

In 2003, in Brazil, there was a total of ten palm oil processing plants, with a total installed capacity reaching 241 tons of fresh fruit bunches per hour—a capacity of around 54,000 tons/month (241 tons/hour × 8 hours/day × 28 work days) or 647.8 tons/year (Parente, 2011). Table 9-C shows the agro-industrial capacity for the processing of palm oil and palm kernel oil installed in the north and northeast regions of Brazil.

The current palm oil production capacity in the Brazilian agro-industrial sector is modest. According to the Brazilian Foundation for Sustainable Development (*Fundação Brasileira para o Desenvolvimento Sustentável*—FBDS, 2005), only eight extraction plants in three Brazilian states remain active. It is also important to note that up to 2005, the production of palm oil had been mainly carried out by Grupo Agropalma, which was responsible for 72% of the national production (FBDS, 2005). In 2001, Grupo Agropalma do Pará alone produced approximately 105,000 tons of palm oil, equivalent to around 90% of the present market demand, exemplifying the concentration of the production capacity of this oil in Brazil (Parente, 2011). Currently, besides Grupo Agropalma leading the production of palm oil in Brazil, Yossam Ltda, in the state of Pará, has also invested in the processing of this oil. Other companies, such as OPALMA and Jaguaripe (both in the state of Bahia), and Marborges, Palmasa, Dentauá, and Condenpa (in the state of Pará) also process palm oil and palm kernel oil, although in substantially lower quantities.

Table 9-C. Palm Oil Production Capacity in Brazil.

Region / *Company*	Plant installed capacity	Municipality
Amapá	**12**	
Copalma	*12*	
Amazonas	**6**	
Caiaué	*6*	Presidente Figueiredo
Pará	**179**	
Agropalma	*128*	Tailândia and Acará
Codenpa	*15*	Santa Isabel do Pará
Dentauá	*12*	Santo Antônio do Tauá
Palmasa	*12*	Igarapé-Açú
Marborges	*12*	Mojú
North Region Total	**197**	
Region / *Company*	Installed capacity	Municipality
Bahia	**44**	
Oldesa	*20*	Nazaré
Jaguaripe	*12*	Munis Ferreira
Opalma	*12*	Uma
Northeast Region Total	**44**	
Brazil Total	**241**	

Source: Grupo Agropalma (2000).

The main problems inhibiting the growth in this sector associated with oil palm tree cultivation and the palm oil agro-industry in Brazil relate to two factors: the lack of experience in oil palm tree cultivation in the region and the insufficient infrastructure for energy and communications in the processing regions. The latter factor is directly related to the economic and social difficulties intrinsic to the producer regions, given that the processing is carried out in the vicinity of the cultivation locations since the product needs to be processed within a maximum period of 24 hours after harvest (Parente, 2011).

Use of Palm Oil and Palm Kernel Oil

Palm oil (dende oil) is the number one vegetable oil in terms of volume commercialized on the world market, and it serves as a support for two large industrial sectors in Brazil: the food industry and the oleochemical industry with countless uses and applications of palm oil and palm kernel oil, both in products involving human and animal diets and other non-alimentary uses (Embrapa, 2011a; Mamat et al., 2005; Parente et al., 2003).

In recent decades, 90% of the world production of palm oil was destined for alimentary use in different forms, such as margarines, fats for bread making, cooking oils, and powders for ice-cream or industrialized fried foods, among others. The other 10% is represented by diverse uses, which vary from the iron and steel industry, to soaps, cosmetics, and even as a source of vitamins A and E for the pharmaceutical industry (Embrapa, 2011a).

Since it has an average of around 50% saturated fatty acids, the oil obtained from the oil palm is suitable for use in commercialized fried foods, since the oils and fats produced for this application require high stability when subjected to high-temperature processes where oxidation, polymerization, and hydrolysis are a concern (Mamat et al., 2005). Millions of tons of palm oil are used worldwide for frying processes with major applications including the preparation of frozen pre-fried potatoes, potato chips, and the production of fast foods (Ismail, 2005).

The versatility of palm oil, which makes it appropriate for different applications is mainly due to its chemical composition (Mamat et al., 2005), high stability, and low price (Fernandez et al., 2000). Modern physico-chemical processes have also improved this versatility through the modification of the oil composition and, consequently, its properties (Mamat et al., 2005). Due to the high stability and flexibility of palm oil, the industrially-fractionated forms (palm olein and stearin) can be used for the production of a large number of products (Arora et al., 2006).

In animal feed, palm oil is used mainly for the preparation of nutritionally balanced feeds obtained from a mixture of byproducts resulting from oil extraction processes. For non-alimentary applications, it can be used as a component in the manufacturing of candles, cosmetics, detergents, diesel substitutes, fatty alcohols, glycerin, hair conditioner, sheet metal, paints, lubricants, plasticizers, polishes, resins, shampoos, oleochemicals, and steel plates (Parente et al., 2003).

Due to the high concentration of saturated fatty acids and high oleic-to-linoleic ratio, palm kernel oil differs significantly from palm oil in terms of thermal stability. Palm kernel oil, with a high composition of lauric and myristic acids, has similar applications as coconut oil applicable in the manufacturing of soaps, detergents, ointments, and mayonnaise, among other products. It can also be used in the production of chocolates, as a substitute for cocoa butter (BASA, 2011; Bora et al., 2003). In the palm kernel oil extraction process, a cake composed of 15% protein is obtained as a byproduct, which is widely employed as animal feed (Embrapa, 2011a).

Besides the use of palm oil for alimentary purposes and application in the oleochemical, iron and steel, cosmetics and pharmaceutical industries, the production of Brazilian palm oil has been receiving attention and investment from the government for the production of biodiesel.

Governmental Incentives for Oil Palm Tree Production in Brazil

The existence of immense areas in Brazil with the appropriate edaphoclimatic conditions and the promising market for its products, as highlighted previously, indicate the great potential for the expansion of oil palm tree cultivation, mainly in the Brazilian Amazon region (BASA, 2011). In an attempt to take advantage of this market, the Brazilian government has developed and given support to several programs for the expansion of oil palm tree cultivation in Brazil.

Governmental Project for the Promotion of Oil Palm Tree Production (1998–2002)

The Brazilian government has established projects for the promotion of oil palm tree cultivation in regions with appropriate climates through various incentives to producers, mainly in the north region of the country, in which the state of Pará is located, the major Brazilian producer. For this purpose, the Strategic Projects Study Group (*Grupo de Estudo de Projetos Estratégicos*—GEPE) was formed by the Department of Agriculture of Pará State (*Secretaria da Agricultura do Estado do Pará*—SAGRI). The aim of setting up GEPE was to evaluate the viability of the establishment of agricultural centers, in regions selected for the exploitation of cultivation, given the advantageous edaphoclimates and locations (BASA, 2011).

The government's proposal was to establish 3,000 hectares of oil palm tree by the year 2000, using the resources of the State Development Fund (*Fundo de Desenvolvimento do Estado*—FDE), administrated by the State Bank of Pará (*Banco do Estado do Pará*—BANPARÁ) and the Constitutional Fund for Financing of the North (*Fundo Constitucional de Financiamento do Norte*—FNO), administrated by the Amazon Region Bank (*Banco da Amazônia*—BASA). The objective was to install modules of 10, 20, 50, and 100 hectares in areas of selected producers, aiming, in the first instance, to fulfill the existing requirements of the agro-industrial companies, which had been operating with excess capacity. In the second phase, the proposal was to aid in the structuring of new agro-industrial companies to absorb overflow of excess product due to the increase in production in the following years (BASA, 2011).

In the period of 1989 to 1996, BASA financed, with the resources of the FNO, 4,519 hectares of oil palm crops in the state of Pará, representing approximately 13.7% of the area of the state cultivated by 1998. In the same period, BASA made several programs available for use with the aim of encouraging the oil palm cultivation, both in the form of investment or financing (BASA, 2011).

Notable within the FNO-Special (FNO-*Especial*) was the Agrarian Reform Support Program (*Programa de Apoio à Reforma Agrária*—PROCERA), The National Program to Strengthen Small-scale Farming (*Programa Nacional de Fortalecimento da Agricultura Familiar*—PRONAF), and the Support Program for Organized Rural

Small-scale Production (*Programa de Apoio à Pequena Produção Familiar Rural Organizada*—PRORURAL), directed toward small-scale rural farmers. These programs support the diversification of crops within the current production systems through the introduction of permanent crops such as oil palm tree (BASA, 2011).

In the modality FNO-Normal, the Support Program for Agricultural Development (*Programa de Apoio ao Desenvolvimento da Agricultura*—PRODAGRI) was developed, which aided medium and large-scale producers with the implantation of temporary and permanent crops, among them oil palm tree (BASA, 2011).

In 2002, a PRODAGRI project for oil palm was launched in Amazonas. The aim of this program was to support, through governmental policies and actions, the development of oil palm tree cultivation in the state of Amazonas as a means to achieve rural development, generate good quality jobs in the rural zone, diversify the economy and sources of income for the state, generate foreign currencies from the exportation of oil, and contribute to the development of the region, reducing the regional inequalities (Embrapa, 2011a).

The aim of the PRODAGRI project for oil palm in Amazonas was the expansion of oil palm tree cultivation to the state of Amazonas, initially of the order of 5,000 hectares of crops per year. Initially, this goal would be reached with the implantation of business projects concentrated in the areas of the Manaus Industrial District (*Pólo Industrial de Manaus*) (Manaus, Manacapuru, Presidente Figueiredo, Rio Preto da Eva, Itacoatiara, and, most notably, the farming district of SUFRAMA—DAS) (Embrapa, 2011a).

National Oil Palm Program: Oil Palm—Sustainable Production Program (2010)

In 2010, a bill was proposed in Brazil (*Projeto de Lei* nº 7.326) for the establishment of the commissions of Environment and Sustainable Development (*Meio Ambiente e Desenvolvimento Sustentável*), Agriculture, Livestock and Supply and Rural Development (*Agricultura, Pecuária e Abastecimento e Desenvolvimento Rural*), and Constitution and Justice and of Citizens (*Constituição e Justiça e de Cidadania*) of the Brazilian government. Article 1 of this bill establishes the creation of a Program for the Sustainable Production of Oil Palm in Brazil (*Programa de Produção Sustentável da Palma de Óleo no Brasil*) and establishes guidelines for national agro-ecological zoning for this crop (PL 7326, 2010).

The aim of this program is to promote the sustainable cultivation of oil palm tree observing the following guidelines: protection of the environment, conservation of biodiversity, and rational use of natural resources; respect for the social function of property; expansion of oil palm tree cultivation exclusively in areas already subjected to anthropogenic activity; stimulation of oil palm tree cultivation for the remediation of areas in different stages of degradation; social inclusion; and environmental regularization of rural properties (PL 7326, 2010).

In order to implement the program, the government relies on some instruments, for instance, Federal Government action in relation to land-use planning, aimed at agricultural land tenure regularization and the indication of areas destined for the sustainable production of oil palm, and governmental action aimed at social inclusion and increased productivity and competitiveness through scientific and technological development. In addition, the Program for the Sustainable Production of Oil Palm in Brazil (*Programa de Produção Sustentável da Palma de Óleo no Brasil*) operates in partnership with other programs instituted by the government, aimed at the environmental regularization of rural properties as well as modes of financing within the scope of the national system of rural credit. Besides these partnerships, the program relies on the agroecological zonation for oil palm cultivation and on the Agro-business Council (*Conselho do Agronegócio*—CONSAGRO) responsible for the promotion of dialog between the different segments of the production chain (PL 7326, 2010).

According to the proposed bill (*Projeto de Lei* nº 7.326), from the moment it is passed into law and enforced, the suppression of native vegetation for the expansion of oil palm tree planting is prohibited throughout the Brazilian national territory. The national agro-ecological zonation will involve the outlining of potential agricultural areas for oil palm tree cultivation without environmental restrictions in anthropogenic use, as well as land areas registered on maps as covered by native vegetation of the Brazilian biomes, and also indigenous land and conservation units, aimed at their exclusion. Thus, the objective of the zoning is to organize, in a rational and sustainable way, land occupation for the cultivation of oil palm tree (PL 7326, 2010).

The areas prioritized by the program are the degraded areas of Amazônia Legal and areas used for planting sugar cane in the northeast regions of Brazil, complying with the technical requirements for the adaptation of the species, as well as the legal aspects and environmental and land-use agendas (CMADS, 2011).

Credit incentives for the execution of the program include financing with PRONAF-Eco, for farmers participating in PRONAF, PROFLORA, for rural producers (individuals or companies), associations and cooperatives, and PRODUSA, for rural producers and cooperatives, including for passing on to its members (CMADS, 2011).

The investments in research and innovation involve approximately R $60 million for Embrapa, for the 2-year period 2010/2011. These resources are primarily destined towards the genetic improvement of palm seedlings and seeds and the increased modernization of the production of seedlings with a defined genetic profile. Furthermore, they should aid the articulation of international agreements and partnerships of excellence related to oil palm and encourage offers of technical assistance, in partnership with the MDA, Embrapa, and state governments. The investments are also intended to create the Oil Palm Sector Chamber (*Câmara Setorial de Palma de Óleo*), comprising representatives from the Federal Government, producers, and consumers, with the aim of promoting the identification of development opportunities along every stage in the production chain (CMADS, 2011).

The ambitious target of the national oil palm program is, by 2014, to reach 10 million hectares planted and elevate the country to the position of principal oil palm producer worldwide, overtaking the current leaders. To achieve this, the program predicts, initially, the participation of 900 small-scale farmers and 300 medium-to-large producers. Moreover, the program will rely on a strong presence and the incentives of Petrobras Biocombustíveis (PBio), since much of the oil, which will be produced in the north and northeast of the country, is destined for state power stations (Freitas, 2010).

Future Developments Related to Palm Oil In Brazil

Based on Brazilian government programs for the development and expansion of oil palm tree plantations, the potential of Brazil in terms of this crop are enormous. The country has the land, climatic conditions, labor, and market potential to surpass Indonesia and Malaysia as top producers. Companies linked to the government such as Petrobras and Vale have also entered this sector. The expectation of the Brazilian program is to cultivate an area of one million hectares (exclusively degraded areas or pasture lands) and to reach a level of palm oil production of 6 tons/hectare/year, with small-scale farming programs, achieving production of 6 million tons/year, which would rank the country third in terms of global production (Portal do Meio Ambiente, 2011).

This estimate is ambitious; however, considering that production will grow more substantially as access to technology, seeds, and technical assistance is better distributed among producers. The growth estimates are very promising considering the projections of both the government and private companies in this regard. In addition to Embrapa–Amazônia Ocidental, Brazil has strong partners in the private sector, such as Grupo Agropalma, the largest individual producer of palm oil in Latin America (Portal do Meio Ambiente, 2011). The projects to stimulate the production of dende oil, together with the constant search for alternative energy sources with less negative environmental impact and the preoccupation concerning degraded areas of the Amazon forest, encourage research into technology for the production of biodiesel from palm oil.

Production of Biodiesel

Currently, due to the increased environmental concerns related to the emission of atmospheric pollutants originating from fossil fuels, the search for alternative energy sources has been receiving considerable attention. In this context, the use of biodiesel produced from vegetable oils has been seen as an important alternative to fossil fuels, mainly in countries with a large area for oil production (Angarita et al., 2009).

Since 2005, programs for the production of biodiesel have already encouraged the structuring of the biodiesel production chain in Brazil, seeking the best solutions from the economic, social, and environmental viewpoints. To this aim, the Brazilian government created the *Selo Social* program for the production of biodiesel. The program is strongly focused on incentives for the production of palm oil for small-scale farmers, in particular offering the exemption of a significant amount of taxes. This policy is valid, especially in a country like Brazil with severe social inequalities (FBDS, 2005).

As part of the creation of the government program *Selo Social* for the production of biodiesel, the Brazilian government approved in January 2005 a new federal law (*Lei Federal* Nº 11.097). This law establishes guidelines for the commercial use of this new biofuel in the country, and as of January 1, 2006, the addition of 2% of biodiesel (B2) to conventional diesel was obligatory for companies which had opted for the *Selo Social* program. This represents a market potential of approximately 800 million liters of biodiesel (B100) per year. As of 2013, the obligatory percentage of biodiesel to be added to conventional diesel will be raised to 5% of biodiesel (B5), increasing to a market of 2.4 billion liters per year (FBDS, 2005).

Vegetable oil is responsible for approximately 80% of the total production cost of biodiesel, and it is thus of extreme importance to identify the best oleaginous species with the aim of increasing productivity and yields. This includes detailed studies on the species from which it is possible to obtain the best energy balance in the biodiesel production. The main producers worldwide of agricultural and industrial products have worked in Brazil, together with Embrapa, to investigate and develop technology suitable for the production of oleaginous plants adapted to the Brazilian conditions (FBDS, 2005).

Brazil has the production potential, consumer market, and agricultural capability, besides having extensive land areas available in the central-west, northeast, and north regions of the country, enabling vast expansion with high growth rates. The rate of arable land use in these regions indicates significant space for expanding the cultivation of oleaginous crops in Brazil (FBDS, 2005).

Of the oleaginous plants, soybean has played an important role during the initial years of the implantation of the Brazilian biodiesel program, since Brazil is notably one of the largest producers of this oleaginous crop. However, the relatively lower oil content in the seed is one of the main problems associated with soybean, leading to a low energy balance in the biodiesel production and the generation of limited employment. Moreover, ecologically, sensitive regions with high biodiversity can suffer adverse effects from the soybean production rate and the methods employed for the expansion of this crop. The Brazilian Cerrado (tropical savanna), which is the habitat of half of the wildlife species native to Brazil with a quarter endangered species, is among the areas worst affected by the increase in soybean cultivation. The expansion of agricultural production in these areas with complex ecosystems can result in irreversible damage and should be the object of considerable concern (FBDS, 2005).

In this context, the oil palm tree is notably a promising species for the production of biodiesel with a productivity of 4,500 liters of vegetable oil per hectare, and it also can be produced by small-scale farmers and companies in degraded areas of the Amazon region. Given this perspective, recently the Brazilian Federal Senate Commission on the Environment (*Comissão do Senado Federal sobre o Meio Ambiente*) approved a project permitting the replanting of degraded forest areas in the Amazon with oil palm species (FBDS, 2005).

The Amazon region is notable since it has the greatest potential for the planting of oil palm tree in the world, corresponding to an annual production of 350 million cubic meters of petroleum and offers the possibility for the occupation and socioeconomic development of 7 million families directly involved in the cultivation. Considering that the annual Brazilian consumption of diesel is of the order of 35 million cubic meters (ANP, 1999), allocating only 10% of the area appropriate for oil palm tree cultivation would surpass the needs for the domestic cargo transport fleet with ecologically correct fuel from a renewable source, employing 700,000 families in this activity. If 15% of palm oil were to be added to diesel (with no structural modification to motor vehicles) consumed in the north region or in the entire country, it would be possible to not only reduce undesirable gas emissions but also to generate jobs for thousands of families (Miranda & Moura, 2000).

Along with the appropriate edaphoclimatic characteristics for planting in Brazil, the energy balance of the biodiesel obtained from palm oil results in a much higher value, and thus oil palm tree offers the best potential for biodiesel production in Brazil (FBDS, 2005). When comparing the energy balance values for biodiesel obtained from palm oil with those of methyl esters obtained from other biodiesel sources in Europe, the United States, and even in Brazil, it can be observed that palm oil is even energetically more efficient than fossil fuels (Angarita et al., 2009).

Besides biodiesel representing an ecologically attractive alternative to fossil fuels, it offers an alternative means of generating energy and jobs for low-income communities in remote regions of Brazil. According to the Permanent Forum on Renewable Energies (*Foro Permanente de Energias Renováveis*—1994), in 1994, approximately 15% of the Brazilian population (25 million people) did not have access to a reliable electricity supply, and most were of very low income. These people live in rural communities in areas where access to conventional electrical power is very high (Almeida et al., 2002). One example is the Amazon region, where there is a particular demographic distribution which is characterized by the presence of small conglomerates of the population along the banks of the major rivers. For the generation of electricity in these remote regions, engines fed by diesel fuel coupled to alternate current generators are used, providing electricity for periods of 12 hours per day (from 6 p.m. to 6 a.m.) (Almeida et al., 2002). However, the long distances which the fossil fuel has to travel from the center of production to the place of consumption result in the diesel becoming more expensive, which together with the low demand, due to the low number of inhabitants in the community (around 100), makes the application of conventional diesel unfeasible (Miranda & Moura, 2000; Zylberstajn et al., 1996).

In an attempt to improve the access of these low-income communities to energy, some scientific studies have investigated the possibility of using pure, unmodified palm oil in diesel engines to provide energy to small villages in the Amazon Basin, in cooperation with different Brazilian government institutions (CEPEL/ELETROBRAS). The research studies have been directed towards evaluating the performance, emission characteristics and contamination of a lubricating oil for a diesel engine using 100% refined palm oil. These studies have obtained promising results, demonstrating the viability of adapting a diesel generator to run on palm oil (Almeida et al., 2002). The use of palm oil produced in the vicinity of the place of consumption is an alternative of great relevance to its application as a fuel for vehicle or stationary engines in these communities. However, this technology has not yet been fully developed since its efficiency needs to be improved and a better understanding of the socio-economic impact resulting from its use for electrical power generation is required (Miranda & Moura, 2000).

Conclusions

Due to its climate and vast land area, Brazil has been outstanding in the production of palm oil (*Elais guineensis*), especially among Latin American countries. The major producing areas in Brazil are found in the states of Pará, Amazonas, Amapá, and Bahia. Pará is the largest producer of palm oil in the country, having more than 80% of the planted area, mostly under the family farm system.

Strategically, the plants from which oil is extracted have been located near to the cultivation areas. After extraction, crude oil is subjected to various processes for the obtaining of oil for food products, feed, and other nonfood uses. The processing of crude oil is accomplished through continuous physical refining comprising acid pre-treatment, bleaching/filtration, and deodorization. The palm represents a yield 5 to 10 times higher than any other commercial crop, producing an annual oil yield of 3700 kg/hectare. The first palm oil extraction plants in Brazil were established in Bahia in the 1950s. The organization of the palm tree agro-industry began in the 1970s and is currently the largest industrial park dedicated to that product.

There are still many challenges for the production of palm oil in Brazil. Globally, productivity has been low and the culture has shown falling prices and rising costs, especially for family farms. Concurrently, in order to achieve sustainable production of palm oil (RSPO—Roundtable on Sustainable Palm Oil) in Brazil, it is needed to evaluate various social, environmental, political, and economic aspects directly or indirectly related to the cultivation of palm tree/palm tree oil.

The list of social aspects to be considered include the right to land (use and acquisition), respect for indigenous communities living in the Amazon rainforest, enforcement of labor laws to the benefit of workers in the field, illegal use of child labor, and safety issues.

Besides those social aspects, environmental aspects have been widely discussed as well including the following subjects: deforestation, biodiversity loss, climate change (occurring because of the changes undergone by the local micro-climates), use of pesticides, and fertilizers. The Brazilian government has also taken initiatives in order to try to solve problems related to regulations and enforcement of local laws.

The concern with issues related to the principles adopted by the RSPO has been reflected in the Brazilian cultivation and processing of palm. Up to April 2010, the RSPO aggregated/had 410 members from 38 countries, including four Brazilian partners (Agropalma, Amazon Refinery, IBD—Biodynamic Institute, and Natura), and 42 foreign associates doing business in Brazil. In November 2010, the RSPO Executive Board approved the set of Principles and Criteria for Sustainable Production in Brazil (RSPO, 2011). In addition, the Brazilian government (Commission for the Environment and Sustainable Development) has prepared a draft legislation for the creation of the production of sustainable palm oil in Brazil, establishing guidelines for ecological zoning for the cultivation of oil palm.

Acknowledgments

The authors acknowledge Bruce Manion of Cereal Laboratory, Department of Food Science, University of Guelph (Canada) for reviewing the manuscript in English and Homero Sousa (R & D Manager—AGROPALMA) for providing information about palm oil production in Brazil.

References

Akpanabiatu, M.I.; Ekpa, O.D.; Mauro, A.; Rizzo, R. Nutrient composition of Nigerian palm kernel from the dura and tenera varieties of the oil palm (*Elaeis guineensis*). Food Chem. 2001, 72, 173–177.

Almeida, S.C.A.; Belchior, C.R.; Nascimento, M.V.G.; Vieira, L.S.R.; Fleury, G. Performance of a diesel generator fuelled with palm oil. Fuel 2002, 81, 2097–2102.

Al-Saqer, J.M.; Sidhu, J.S.; Al-Hooti, S.N.; Al-Amiri, H.A.; Al-Othman, A.; Al-Haji, L.; Ahmed, N.; Mansour, I.B.; Minal, J. Developing functional foods using red palm olein. IV. Tocopherols and tocotrienols. Food Chem. 2004, 85, 579–583.

Angarita, E.E.Y.; Lora, E.E.S.; Costa, R.E.; Torres, E.A. The energy balance in the Palm Oil-Derived Methyl Ester (PME) life cycle for the cases in Brazil and Colombia. Renew. Energ. 2009, 34, 2905–2913.

ANP—Anuário Estatístico da Agência Nacional do Petróleo: 1990–1998; Rio de Janeiro, 1999.

Arora, S.; Manjula, S.; Gopala Krishna, A.G.; Subramanian, R. Membrane processing of crude palm oil. Desalination 2006, 191, 454–466.

Barcelos, E.; Chaillard, H.; Nunes, C.D.M.; Macêdo, J.L.V.; Rodrigues, M. do R.L.; Cunha, R.N.V.; Tavares, A.M.; Dantas, J.C.R.; Borges, R.S.; Santos, W.C. A cultura do dendê; Embrapa-CPAA; Embrapa-SPI; Coleção Plantar, 32, Brasília, 1995, 68 p.

Barcelos, E. Dendeicultura no Brasil–Diagnóstico. In × Conferência Internacional de Palma Eceiteira, Santa Marta: Colômbia, 1993.

BASA—Banco da Amazônia S. A. Consultoria Técnica–COTEC / Coordenadoria de estudos especiais–COESP. O comportamento do mercado do óleo de palma no Brasil e na Amazônia. Estudos Setoriais, 11, 1998. Downloaded from: http://www.bancoamazonia.com.br/bancoamazonia2/includes/institucional/arquivos/biblioteca/artigos/agronegocios/EstSetorial11.pdf (accessed January 2011).

Basiron, Y. Palm oil. In Bailey's Industrial Oil and Fat Products: Specialty Oils & Oil Products, 6 ed.; Shahidi, F., Ed.; Wiley–Interscience: EUA, 2005, v.6; pp. 333–429.

Bazmi, A.A.; Zahedi, G.; Hashim, H. Progress and challenges in utilization of palm oil biomass as fuel for decentralized electricity generation. Renew. Sust. Energ. Rev. 2011, 15, 574–583.

Berger, K. Minor components of palm oil. MOST 2000, 9, 1, 56–59.

Biodieselbr. Palma. Downloaded from: http://www.biodieselbr.com/plantas/palma/palma.htm (accessed January 2011).

Bockisch, M.; Germany, H. Fats and oils handbook. AOCS Press: Champaign, 1998; pp. 206–213.

Bora, P.S.; Rocha, R.V.M.; Narain, N.A; Moreira-Monteiro, A.C.; Moreira, R.A. Characterization of principal nutritional components of Brazilian oil palm (*Elaeis guineensis*) fruits. Bioresource Technol. 2003, 87, 1–5.

CEPLAC—Dendê–Potencial para produção de energia renovável. Downloaded from: http://www.ceplac.gov.br/radar/Artigos/artigo9.htm. (accessed May 2011).

CEPLAC—Ministério da Agricultura. Folder "Alternativas para Diversificação Agronômica da Região Sudeste da Bahia"—Bahia. Downloaded from: http://www.seagri.ba.gov.br (accessed December 2010).

CMADS. Comissão de Meio Ambiente e Desenvolvimento Sustentável. Relatório referente ao Projeto de Lei n.º 7.326, de 2010. Downloaded from: http://www.camara.gov.br/sileg/integras/816507.pdf (accessed May 2011).

Cordeiro, A.C.C. Informações Técnicas para a Implantação de Lavoura de Dendê em Roraima / Antônio Carlos Centeno Cordeiro, Admar Bezerra Alves, Francisco Clemilto da Silva Maciel. Boa Vista: Embrapa Roraima, 2009, 15p. (Embrapa Roraima. Documentos, 25).

Cottrell, R.C. Introduction: nutritional aspects of palm oil. Am. J. Clin. Nutr. 1991, 53, 989–1009.

Craveiro, A.A.; Craveiro, A.C.; Queiroz, D.C. Alimentos funcionais: a nova revolução. Fortaleza: PADETEC/UFC, 2003.

Ekpa, O.D. Bio-inorganic constituents and possible uses of the female inflorescence of the oil palm fruit (*Elaeis guineensis*). West Afr. J. Biol. App. Chem. 1995, 40, 13–18.

Ekpa, O.D.; Ekpe, U.J. Effects of coconut oil concentration on the melting point profile and free fatty acid formation of palm oil. Niger J. Chem. Res. 1996, 1, 1, 8–12.

Ekpa, O.D.; Peter, M.A.; Udo, A.E. Quality assessment for rurally processed palm oil and the effect of storage of palm fruits before processing on the quality of palm oil. West Afr. J. Biol. App. Chem. 1994, 39, 9–13.

Embrapa–Empresa Brasileira de Pesquisa Agropecuária. Série Vermelha Fruteiras—Dendê, 1995, Brasil–DF.

Embrapa–Zoneamento agroecológico do dendezeiro para as areas desmatadas da Amazônia Legal (Relatório-Síntese). Projeto realizado por encomenda do governo federal e financiado pelo MCT– FINEP (Convênio Código 01.05.0110.00), Rio de Janeiro, Abril, 2010. Downloaded from: http://www.cnps.embrapa.br/zoneamento_dende/ZonDende.pdf (accessed December 2010).

Embrapa–Amazônia Ocidental. Projeto de dendê / palma Amazonas. Programa de apoio ao desenvolvimento da agroindústria do dendê no Amazonas, 2002. Downloaded from: http://www.cpaa.embrapa.br/portfolio/sistemadeproducao/dende/projetodedendepalmaamazonas.pdf (accessed February 2011a).

Embrapa–Amazônia Ocidental. DENDÊ: Alternativa de Desenvolvimento Sustentável para Agricultura Familiar na Amazônia Brasileira, Folder, Manaus—AM, 2002. Downloaded from: http://www.cpaa.embrapa.br/servicos/livraria/arquivos_gratis/folderdendejackson.pdf (accessed May 2011b).

FBDS. Fundação brasileira para o desenvolvimento sustentável. Liquid biofuels for transportation in Brazil. Study commissioned by the German Technical Cooperation (GTZ); November 2005.

Fernandez, X.E.; Shier, N.W.; Watkins, B.A. Effect of alkali saponification, enzymatic hydrolysis and storage time on the total carotenoid concentration of Costa Rican crude palm oil. J. Food Compos. Anal. 2000, 13, 179–187.

Firestone, D. Physical and chemical characteristics of oils, fats, and waxes. AOCS Press: Champaign, IL, 1999.

Freitas, R.C. Palma de óleo. Biodieselbr. 2010, 3, 17, 22–23.

Foro Permanente de Energias Renováveis. Declaração de Belo Horizonte—Encontro para Definição de Diretrizes para o Desenvolvimento de Energias Solar e Eólica no Brasil, Belo Horizonte, Brasil, Abril 1994.

Grupo Agropalma. Óleo de palma: um produto natural. G.A. Editora: Belém, 2002.

Ismail, R. Palm oil and palm olein frying applications. Asia Pac. J. Clin. Nutr. 2005, 14, 4, 414–419.

Kiple, K.F.; Omelas, K.C., Eds. The Cambridge world history of food; Cambridge, MA: Cambridge University Press, 2000.

Mamat, H.; Aini, I.N.; Said, M.; Jamaludin, R. Physicochemical characteristics of palm oil and sunflower oil blends fractionated at different temperatures. Food Chem. 2005, 91, 731–736.

Miranda, R. de M., Moura, R.D. Óleo de dendê, alternativa ao óleo diesel como combustível para geradores de energia em comunidades da Amazônia.. In ENCONTRO DE ENERGIA NO MEIO RURAL, 3., 2000, Campinas. Anais eletronicos. Downloaded from: http://www.proceedings.scielo.br/scielo.php?script=sci_arttext&pid=MSC0000000022000000200042&lng=pt&nrm=abn (accessed January 2011).

Oil World Annual. Downloaded from: http://www.oilworld.biz/app.php?fid=1060&fpar=0&isSSL=0&aps=0&blub=65707639e9020b063148896592bf6f2e (accessed January 2011).

Pandolfo, C.A. cultura do dendê na Amazônia. SUDAM: Belém, 1981.

Pantzaris, T.P.; Ahmad, M.J. Palm kernel oil (2001). Palm Oil Research Institute of Malaysia. Downloaded from: http://www.fosfa.org/resources/res_seeds_palm.htm (accessed January 2011).

Parente, V.M.; Oliveira Jr., A.R.; Costa, A.M. Potencialidades Regionais–Estudo de viabilidade econômica Dendê–Sumário Executivo. (2003). v.5. Ministério do Desenvolvimento, Indústria e Comércio Exterior. Downloaded from: http://www.suframa.gov.br/publicacoes/proj_pot_regionais/sumario/dende.pdf (accessed January 2011).

Pérez, G.R.A.; Nápoles, A.L.D. Procesamiento de aceite rojo de palma africana *Elaeis guineensis* Jacq. para consumo humano em frituras. Escuela de Agricultura de La Región Tropical Húmeda. Trabalho de Conclusão de Curso. Costa Rica, dez., 2000.

Phosri, C.; Rodriguez, A.; Sanders, I.R., Jeffries, P. The role of mycorrhizas in more sustainable oil palm cultivation. Agr. Ecosyst. Environ. 2010, 135, 187–193.

PL 7326. Projeto de Lei n.º 7.326. Programa de Produção Sustentável da Palma de Óleo no Brasil. Câmara dos Deputados, Brasília, 03 May 2010.

Poku, K. Small-scale palm oil processing in Africa. FAO Agricultural Services Bulletin, Roma, 2002.

Portal do Meio Ambiente. Óleo de palma: maior commodity de óleo vegetal do mundo é desafio para o Brasil. Downloaded from: http://www.portaldomeioambiente.org.br/guia/ecoreleases/5148-oleo-de-palma-maior-commodity-de-oleo-vegetal-do-mundo-e-desafio-para-o-brasil.html (accessed January 2011).

Rodriguez-Amaya, D.B. Assessment of the provitamin A contents of foods—the Brazilian experience. J. Food Comp. Anal. 1996, 3, 196–229.

RSPO—ROUND TABLE ON SUSTAINABLE PALM OIL–IBD certificações. Downloaded from: http://www.rspo.org/sites/default/files/Brazilian%20Local%20Indicators%2011201 0_Portuguese.pdf (accessed September 2011).

Sagri—Secretaria Executiva de Agricultura do Estado do Pará-. Produtores Familiares de Dendê. Downloaded from: http:// www.sagri.pa.gov.br (accessed February 2011).

Sambanthamurthi, R.; Sundram, K.; Tan, Y.A. Chemistry and biochemistry of palm oil. Prog. Lipid Res. 2000, 39, 507–558.

Sanagi, M.M.; See, H.H.; Ibrahim, W.A.W.; Naim, A.A. Determination of carotene, tocopherols and tocotrienols in residue oil from palm pressed fiber using pressurized liquid extraction-normal phase liquid chromatography. Anal. Chim. Acta 2005, 538, 71–76.

Seagri / Aiba / Banco Nordeste/CREDICOOGRAP/IMIC—Revista Negócios Agrícolas 1999, 2, 9.

Seagri / Aiba / BANCO NORDESTE/CREDICOOGRAP/IMIC—Revista Negócios Agrícolas 1997, 1, 1.

Tandon, R.; Manohara, T.N.; Nijalingappa, B.H.M.; Shivanna, K.R. Pollination and pollen-pistil interaction in oil palm, *Elaeis guineensis*. Ann Bot-London 2001, 87, 831–838.

Umerie, S.C.; Ogbuagu, A.S.; Ogbuagu, J.O. Stabilization of palm oils by using *Ficus exasperata* leaves in local processing methods. Bioresource Technol. 2004, 94, 307–310.

Zylberstajn, D.; Coelho, S.T.; Ieno, G.O. Potencial de geração de eletricidade na Amazônia a partir de resíduos agrícolas; Instituto de Eletrotécnica e Energia; Universidade de São Paulo; 1996.

10

Palm and Palm Kernel Oil Production and Processing in Nigeria

Isona L. Gold, Celestine E. Ikuenobe, Omorefe Asemota, and Dere A. Okiy
Nigerian Institute for Oil Palm Research (NIFOR), Benin City, Edo State, Nigeria

Introduction

The African oil palm (*Elaeis guineensis* Jacq.) is native to West Africa as shown by fossil, archeological, historic, and linguistic records (Corley & Tinker, 2003; Hartely, 1977, 1988; Rees, 1965; Sowunmi, 1999; Zeven, 1964,). It was introduced to South America in the 16th century and in the 19th century to the Southeast Asia. Recent studies of oil palm germplasm collections employing molecular DNA techniques, such as restriction fragment length polymorphism (RFLPs), have shown that the oil palm germplasm collections from Nigeria exhibited the highest level of polymorphism with highest number of rare alleles, suggesting that Nigeria was likely to be the center of diversity of the oil palm (Maizura et al., 2001; Rafii et al., 2002).

In Africa, the oil palm grows in the lowland humid tropics and is found in natural groves or under cultivation between latitude 4° and 11°N and 10°N and 7°S of the equator from the coasts of the Gulf of Guinea through the southern latitudes of Sierra Leone, Liberia, the Côte d'Ivoire, Ghana, Togo, Benin Republic, Nigeria, Cameroon, and Zaire. The growth area is limited by rainfall distribution and altitude. It can occur from sea level to 300 m, but under sufficient rainfall it can be found up to 1000 m as on the Fouta Djallon in Guinea and the Cameroon and to 1300 m on the Cameroon Mountains. The oil palm is widely distributed in Nigeria, Ghana, Côte d'Ivoire, Cameroon, Democratic Republic of Congo (formerly Zaire), Republic of Benin, Togo, Liberia, Sierra Leone, Gambia, Senegal, Guinea Bissau, the Central African Republic, Gabon, Angola, Equatorial Guinea, Tanzania, Burundi, Uganda, and Madagascar.

The oil palm belt in Nigeria covers approximately 18 million hectares where it exists in natural and semi natural groves (Kilby, 1967). Nigeria has the largest natural/semi-natural groves worldwide, estimated at 2.1–2.3 million hectares. These groves supplied the world market with palm oil and palm kernels before the crop became cultivated in plantations in Southeast Asia and the Democratic Republic of Congo.

Trade in palm oil between Nigeria and the countries of Europe assumed great importance even as early as the 19th century such that up to 1965, foreign exchange earnings from palm produce was a major source of revenue for Nigeria's economic development. In 1961, for example, oil palm production from Africa, the bulk of which came from Nigeria, constituted over 78% of total world production. However, domestic palm oil production volume began to decline from 1965 (Table 10-A). In the late 1960s and 1970s domestic palm oil production increased at an average annual rate of 1.6%, which was less than the population growth rate of 2.5% per annum and consumption rate of 4% (Udom, 1989). Other factors that contributed to declining domestic production were low production from the wild groves due to declining productivity, uncoordinated and lack of execution of development programs, stagnation of organized planting, application of inefficient processing technologies, high taxes by the Marketing Boards, and the civil war in production areas especially in Eastern Nigeria. The major production area contributing more than 80% of total output became epicenters of civil conflicts and later the national fortunes from petroleum which led to total neglect of the agricultural sector of the Nigerian economy. Subsequently, Nigeria began importation of palm oil from Malaysia and Singapore. Furthermore, local prices per tonne of crude palm oil were higher than the international market prices by more than 150%, hence the impetus for importation which started with 5,000 tonnes per annum in 1976 and rose within a decade to 179,000 tonnes in 1986.

As part of efforts to improve the performance of the nation's agricultural sector, the Federal Government in 1986 banned the importation of palm oil and its fractionated products as a policy instrument to stimulate increase in domestic production of palm products. During the ban, there was high rate of smuggling to bridge the demand-supply gap of more than 400,000 tonnes of vegetable oil annually (Olayinka and Alegieuno, 1989). Furthermore, imported crude palm oil had better quality with free fatty acid content of between 3% and 5% compared to local palm oil with more than 7% free fatty acid. Also in the early and mid-1980s, refined palm oil was available as refined, bleached, and deodorized palm oil without fractionation. Unfractionated refined palm oil was not marketable to industrial end users in Nigeria hence the boom in the smuggling of refined bleached and deodorized palm olein from Malaysia and Singapore and the extraction of palm kernel oil for household frying. Indeed, the local market became too sophisticated for local product quality, so the ban was not effective. However, the ban was sustained till 2008 when the embargo on crude palm oil was lifted as a result of the poor quality products available in the local market which were not suitable for refining. Presently, crude palm oil is imported for refining in Nigeria, although some quantities of refined bleached and deodorized palm olein are being disguised as crude palm oil and acid oils. Sometimes RBD palm olein is tainted with about 0.03% of crude palm olein and imported into the country as crude palm oil. Despite the lifting of embargo on crude palm oil importation, local prices of crude and refined products are still very high because the landed cost of one tonne of imported crude palm oil as of the first half of 2011 was about US$1687.33

Table 10-A. Palm Oil and Palm Kernel Oil Production (1000 metric tons) between 1948 and 1987.

Palm oil production ('OOOT)	1948–50	1951–55	1956–60	1961–64	1969–71	1979–81	1985–87
Nigeria	1027.4	1961.7	2151.9	1595.9	528.5	667.2	751.7
Indonesia	301.8	762.5	751.5	596	218.5	667.2	1503.7
Malaysia	151.4	257	351.5	450.6	456.9	2529.4	4472
Côte d'Ivoire	105	318	526	676	45.7	164.3	178
Palm kernel production							
Nigeria	1113.8	2064.4	2205.1	1632.2	287.4	342.5	358.3
Indonesia	741	196.2	182.6	134.8	49.5	121.4	298.3
Malaysia	308	661	920	113.7	98.9	540.1	1343
Côte d'Ivoire	214	334	732	464	18.8	29	41

Sources: FAO (1966); Udom (1989).

(~NGN260,000) compared to local prices of US$1428.57 (NGN220,000). However, very few investors could import and store large quantities of crude palm oil due to limited bulk storage infrastructure at the nation's sea ports, especially in Lagos. Other Ports at Port Harcourt, Calabar and Warri are relatively small with maximum draught of between 8–11 meters. The minimum import capacity considering the bureaucracy of importation, docking, and clearing of such products should be about 10,000 tonnes per import because that would guarantee 3 months of raw materials supply and re-order planning for a 100 tonnes per day refinery. Also, efficiently operating a 100 tonnes per day refinery, which is the average vegetable oil refinery capacity in Nigeria, requires proximity to the ports where crude palm oil can be pumped directly to the tank farm in the factory premises rather than transportation with trucks over long distances with the attendant challenges of inadequate road network. These limitations have so far militated against the envisaged domestic crude palm oil availability for refining. Up until this date, there is a crude and refined palm oil supply deficit in the Nigerian market.

Palm and Palm Kernel Oil Production Structure in Nigeria

Palm oil and palm kernel oils are produced in Nigeria from three production systems: natural/semi natural groves, smallholders, and small/medium sized estates and large estates.

The groves are characterized by varied planting population from between 10–40 palms per hectare and 120–200 palms per hectare (Omoti, 2009). The fresh fruit bunch yield is between 1 and 5 tons per hectare per year and the extraction rate is 9% for palm oil and 5% for palm kernel oil. Palm oil output from the groves is estimated at 567,000 tonnes from an estimated area of 1,704,607 hectares. However, these groves have been depleting in size as a result of alternative land use such as urbanization (Omoti, 2009).

The small holder system comprises farm settlements, National Agricultural Land Development Agency plantings, Cooperative farms, out-grower schemes, and small/medium sized estates. They are usually 1–5 hectares in size and in most cases intercropped with other food crops such as pineapple, cassava, and maize. This is to enhance optimum land use and ensure provision of food resource to the grower. Yield of fresh fruit bunches is between 6 and 10 tons per hectare per year and extraction rate is about 20% due to repeated extraction. Estimated palm oil output is about 204,000 tonnes per annum from a total area of about 313,000 hectares.

The small holders together with the grove account for about 90% of national palm oil output (Omoti, 2009).

The estates which were initially owned by the Federal or State Governments have now been privatized and a small proportion is owned by private entrepreneurs. Fresh fruit bunch yield in estates is between 12 and 20 tonnes per hectare per year and extraction rates are 18–22%, based on the Nigerian Institute for Oil Palm Research (N.I.F.O.R.) *tenera* planting materials. Estimated oil palm area under the large industrial estate is about 118,264 hectares. Output of palm oil from these estates is estimated at about 256,000 tonnes of crude palm oil and about 57,000 tonnes of palm kernels.

Total national output of crude palm oil is 604,969.58 tonnes while palm kernel oil is 136,836.16 tonnes. Therefore, current national output of palm oil and palm kernel oil would be about 741,805.74 tonnes (Omoti, 2009).

In Nigeria, palm oil production is still very much dominated by the smallholder producers. Production from the natural groves account for about 53% and 65% of total palm oil and palm kernel output, respectively, while the smallholders account for 28% and 24% and the large estates 19% and 11%, respectively. Taken together, the smallholder producers who also exploit the natural groves account for 81% and 89% of the total palm oil and palm kernel output, respectively.

Palm and Palm Kernel Oil Market Structure in Nigeria

There are two major market segments for palm oil in Nigeria, namely: the domestic and industrial markets.

Domestic market represents local markets where traders sell palm oil products directly to consumers for household and commercial frying purposes. In this market, crude palm oil of free fatty acids content up to 20% is freely marketed and consumed. All the crude palm oil produced by cottage, small/medium scale producers, and some fractions of estate products, the total of which is more than 70% of domestic palm oil disappearance, are sold in this market without stringent product quality demands. Also more than 70% of refined palm oil and all the refined palm kernel oil are consumed in the local market. The fraction of refined palm oil sold to producers of extruded dough of wheat, rice, potato, etc. (noodles) which are discarded at free fatty acid content of less than 1% are recycled to the domestic market for consumption.

The Nigerian vegetable oil market uniquely features palm kernel oil packaged for household and commercial food frying. More than 60% of refineries in Nigeria are built on palm kernel oil as raw materials. This is because processing of palm kernel oil involves degumming, bleaching, deodorization, cooling, and packaging. This product is liquid at ambient conditions except during the rainy season when the product congeals due to low temperatures (<20°C). Processing of palm oil requires degumming, bleaching, and deodorization followed by fractionation. Fractionation of refined bleached and deodorized (RBD) palm oil is necessary because there is little or no domestic market for unfractionated RBD palm oil. Even after fractionation, the local market for palm stearin is not developed and therefore stearin is difficult to sell locally. Consequently, a producer of RBD and fractionated palm oil will need to acquire in addition to his refinery, a fractionation plant and sometimes a shortening plant whereas production of RBD palm kernel oil only requires the refinery. Also, the cost of refining palm kernel oil is about 50% of what is required to refine palm oil. As a result of the above, Nigerian vegetable oil refiners prefer palm kernel oil as raw material.

Demand and Supply Dynamics

Demand for vegetable oils in Nigeria is estimated at 1,968,500 tonnes, taking into account a per capita consumption of 12.7 kg and a population of 155 million in 2009 (WDI, 2011).

Production in 2009 was estimated at 1,137,580 (Oil World annual, 2008; Omoti, 2009). Palm oil and palm kernel contribute about 65% to the total vegetable oils and fats consumed in Nigeria. This implies that palm oil and palm kernel oil demand and supply is 1,279,525 tonnes and 739,427 tonnes, respectively, leaving a gap of 540,098 tonnes annually. Assuming an average oil extraction rate of 18% and fresh fruit bunch production rate of 12 tonnes per hectare, total land area required for cultivation to bridge the demand supply gap as at 2009 was about 250,000 hectares. The demand–supply deficit in the Nigerian market is met through importation from Malaysia, Indonesia, and neighboring countries in the West African region such as Ghana, Togo, and Côte d'Ivoire. Fats and oils imported into these West African countries ultimately end up in the Nigeria market either by outright smuggling or by leveraging on the ECOWAS Trade Liberalization Scheme (ETLS). In 2009 the Federal Government lifted the ban on bulk importation of vegetable oils and permitted the importation of crude palm oil at 35% import duty. This allowance was made so that the idle refineries would have raw materials to operate as the quality of most locally produced crude palm oil is low. Difficulties in locating tank farms at the local ports, high landing cost of crude palm oil at about 19% higher than local prices, lack of fractionation equipment to produce palm olein in most refineries, and low demand for stearin seem to have been disincentives to potential importers. Consequently, some companies import refined bleached and deodorized palm olein directly in quantities of up to 14,000 tonnes per import under Special Presidential Concession, while some other enterprises import palm olein disguised as acid oils for soap production. Therefore despite the lifting of ban on importation, the local prices of palm and palm kernel oil have remained stable and the supply deficit has not been bridged.

Production Cost of Palm and Palm Kernel Oil

Production cost analysis involves establishment costs and operating costs.

Establishment Cost

The establishment cost include cost of land validation studies, survey, field demarcation (blocking), road construction, drainage construction, land preparation, planting, seedlings, cover crops, fertilizers, herbicides, insecticides, and labor costs from the year of planting to the third year.

Pre-nursery and Nursery Seedlings

In Nigeria, planting materials are produced by a 2-stage nursery process in which sprouted seeds are nurtured in polybags for 3 months in a pre-nursery before transplanting to the main nursery where the seedlings are further kept for an additional 6–9 months. As of 2009, the production cost of pre-nursery and nursery seedlings for the first year were $1.74 and $0.73 per unit, respectively, but by the second year of seedling production, pre-nursery and nursery seedlings costs were $0.4 and $0.32 with a culling rate of about 20%. However, the market price of pre-nursery and nursery seedling in 2009 were $0.39 and $0.97, respectively. Tables 10-B and 10-C shows the labor rates for nursery and other field operations in Nigeria.

Field Development

Road requirement in a 1000 hectare area is about 13.5 km of main roads and 64.8 km of collection and perimeter roads. In 2009, land preparation, which involves block demarcation, under-brushing, chainsaw felling of trees, cross-cutting, preparation of

Table 10-B. Work Rates for Estate Labor.

Grade	Rate/day (US$)
Unskilled daily pay	4.00
Semi-skilled daily pay	4.20
Skilled daily pay	4.40
Harvester	5.70
Operator daily pay	5.20

Table 10-C. Work Rate for Nursery Labor.

Job	Rate (US$)	Per
Planting of nursery seeds	4.20	Task set by Manager
Transplanting of seedlings	4.20	Task set by Manager
Rodent control	4.10	Task set by Manager
Seedling culling	4.10	Task set by Manager
Irrigation day	4.20	Day shift
Irrigation night	6.80	Night shift
Pesticide spraying	4.20	Task set by Manager
Loading of nursery seedlings	0.03	Seedling
Loading and unloading of seedlings in field	0.06	Seedling

Table 10-D. Work Rates for New Development and Replanting.

Job	Rate (US$)	Per
Survey: Blocking out	4.10	1 hectare
1st tracing	4.10	1 hectare
2nd tracing	4.10	5 hectares
Peg preparation	4.10	1,000 pegs
Pegging	4.10	5 hectares
Measuring, Pillaring, leveling	4.20	Task set by Manager
Palm felling	2.00	tree
Opening of planting lines	4.10	Line
Holing	1.00	Line
Seedling distribution	1.00	Line
Planting	1.10	Line
Decollaring of wire nets (remove and bundle)	0.36	Line
Ablation	0.29	Line
Cover crop seed purchase: *Puereria*	1.77	kg
Centrosema	1.77	kg
Mucuna	0.48	kg
Cover crop winnowing and packing	4.03	Manday
Cover crop planting	0.52	Line
Cover crop broadcasting	0.29	Line

pegs, pegging of windrows and lines, windrowing, clearing of planting lines, and supervision was computed at $785 per hectare excluding depreciated costs of heavy duty equipment used in the land preparation. Planting cost was computed at $150 per hectare. Planting involves digging a hole, placing of the seedling, installation of protective wire mesh netting, and planting of cover crop. Some cost components of field development are shown in Table 10-D.

Seedlings are available in Nigeria through the Nigerian Institute for Oil Palm Research and other private nursery operators. Companies and individuals are also permitted by the government to import planting materials from any country of their choice following plant quarantine regulation. Wire mesh and fertilizers are available in the Nigerian local markets.

Cover crop seeds are not abundantly available in Nigeria as they are not widely used among smallholders. Currently, *Puereria* seeds cost $3.3 per kg (6 kg is required per hectare of oil palm).

Table 10-E. Work Rates for Upkeep Operations.

	US($)/Line	
Job	Manual	Chemical
Interrow slashing; Mature	1.50	
Interrow slashing; Yr 0, Yr 1 & Yr 2 replant area	1.60	
Interrow slashing; Yr 0, Yr 1 & Yr 2 new area	2.10	
Ring weeding; Mature and immature	1.40	0.20
Guinea grass pulling	Task by Manager	
Fertilizer application	0.33 per line	

Immature Crop Maintenance

The computed cost of maintenance of immature crops per hectare in 2009 was $335, $379, and $375 for years 1, 2, and 3, respectively. Maintenance during the immature years requires herbicides, fertilizers, herbicide sprayers and other accessories, and labor (Table 10-E).

The overall cost from field development through planting and maintenance of the immature crop constitutes the capital cost of the investment in oil palm production and this cost is usually amortized over the commercial life of the oil palm.

Operating Costs in Nigerian Oil Palm Estates

This includes costs of weeding around the palms, in-between the palms, and within palm lines by manual or chemical methods, cost of fertilizers and applications, cost of pruning, phytosanitary inspection, harvesting, and processing (Tables 10-F to 10-O).

Maintenance cost for the mature area is computed at $250 per hectare per year. Harvesting is mainly done by contract in most estates and the approximate cost is between $9.4 and $28.2 per tonne of fresh fruit bunches, depending on the age of the palm, area topography, and degree of field sanitation. In some cases, harvesting cost is based on number of harvested bunches. In young fields (i.e., 3–6 years), one tonne of bunches is equivalent to an average of 120 bunches. In fields of 7–9 years, an average of 80 bunches equal one tonne of fresh fruit bunches while 10 years and above produce an average of 60 bunches per tonne. The production cost of a tonne of fresh fruit bunches in Nigeria considering the value chain is about $29, excluding overhead costs. Fresh fruit bunch collection and loading in the field to the mill is about $1.74 per tonne. The cost of producing one tonne of palm oil including field cost, overheads, and milling cost is between $542 and $833.

Table 10-F. Work Rates for Harvesting, Carrying, Pruning and Empty Fruit Bunch Spreading.

Age of field	Total (US$)/Bunch High/Low Season	Age of field	Total (US$)/Bunch High/Low Season
34	0.19	13	0.12
31	0.19	12	0.10
30	0.19	11	0.10
29	0.19	10	0.08
23	0.19	9	0.08
22	0.16	8	0.06
21	0.16	7	0.06
20	0.14	6	0.06
16	0.14	5	0.05
15	0.12	4	0.05
14	0.12	3	0.04

Table 10-G. Loose Fruit Picking.

Job	Rate (US$)/kg
Loose fruit collection in bags	0.05

Table 10-H. Fresh Fruit Bunch Loading.

Job	Rate (US$)
FFB collection in the field	1.03 per tonne
FFB loading into trucks	0.71 per tonne

Table 10-I. Fresh Fruit Bunch Recorders.

Job	Rate (US$)
Fresh fruit bunch recorders	per manday

Table 10-J. Pruning Rates.

Age of field	Total (US$)/Palm	Age of field	Total (US$)/Palm
34	0.22	13	0.13
31	0.22	12	0.12
30	0.22	11	0.12
29	0.22	10	0.12
23	0.19	9	0.10
22	0.19	8	0.10
21	0.19	7	0.10
20	0.19	6	0.08
16	0.15	5	0.08
15	0.15	4	0.06
14	0.13	3	0.04

Table 10-K. Empty Fruit Bunch Mulching.

Job	Rate (US$)	Per
Mulching (task is 2 Mandays per heap)	4.10	½ heap

Table 10-L. Special Jobs.

Job			Rate (US$)	Per
Operator	Tractor	12 hour working day	5.16	Manday
Operator	Lorry	12 hour working day	5.16	Manday
Operator	HDM	12 hour working day	5.16	Manday
Mate	HDM	12 hour working day	5.16	Manday
Operator	Power tiller	12 hour working day	5.16	Manday
Operator	Lawn mower	12 hour working day	5.16	Manday
Operator	Motor Sprayer	12 hour working day	5.16	Manday
Operator	Chainsaw	12 hour working day	5.16	Manday
Rodent control		12 hour working day	4.19	Manday
Hunting grass cutters		12 hour working day	4.19	Manday
Sanitary observers/recorders		12 hour working day	4.35	Manday

*HDM=Heavy duty machinery

Table 10-M. Headmen.

Job	Rate (USD)	Bonus
Contract headman; Upkeep (no overtime)	Nil	100% of average gang daily earnings × 1.1
Contract headman; Harvesting (no overtime)	Nil	50% of average gang daily earnings × 1.1
Staff headman; Upkeep (Overtime on Sat & Sundays)	Salary	10% of average gang daily earnings
Staff headman; Harvesting (Overtime on Sat & Sundays)	Salary	5% of average gang daily earnings

Table 10-N. Transport Rates.

Job	Rate (₦)	Per
Staff tractor operators and truck drivers (rates in lieu of overtime)	0.11	Tonne FFB
Contract transport FFB field to mill	4.19	Tonne FFB
Contract transport FFB less than 100 km	5.00	Tonne FFB
Contract transport FFB 100–200 km	9.03	Tonne FFB

Table 10-O. Unloading Fertilizer.

Job	Rate (₦)	Per
Unloading fertilizers, Kaolin, etc.	1.26	Tonne

Processing of Palm and Palm Kernel Oil

The palm oil and palm kernel oil processing in Nigeria is carried out by small/medium scale (Table 10-P) and large scale processes employing rudimentary techniques and sophisticated systems.

The small/medium scale (1.5 ton/hr) processing technology comprises the batch equipment fabricated by local fabricators and others adapted from the small scale processing equipment developed by the Nigerian Institute for Oil Palm Research. Some of these equipment sterilize whole bunches while others sterilize loose fruits. The NIFOR equipment units include a fruit screen, sterilizer, digester screw press, clarifier, and storage tanks heated with steam pipes.

In small-scale milling of fresh fruit bunches (FFB), fruit bunches are received in the mill and stored for about 3 days to allow natural detachment of the palm fruit from the stalk. The bunches are then divided into four parts using an axe (quartering) or a guillotine. Quartered bunches are knocked with hard objects to release the

Table 10-P. Characteristics of Palm Oil Produced from Small/Medium Scale Technologies.

Parameter	Value
Free fatty acids (%)	2.5–5
Moisture (%)	<2
Dirt (%)	<1
Oil loss in dry matter (%)	7.5–8.5
Moisture in kernels (%)	7.5
Efficiency in kernel station (%)	>90

fruits. Usually the calices are attached to the fruits after knocking. These calices were believed to be responsible for poor quality palm oil, such as emulsified palm oil, especially while processing during the rainy season. Consequently, the calices were sieved off the fruits with a fruit screen prior to sterilization. However, studies at the Nigerian Institute for Oil Palm Research have shown that calices are not responsible for emulsified palm oil production. Although oil extracted from calyx is high in phosphatidyl choline and phosphatidyl ethanolamine, these substances are not in the chemical form to effect emulsification of palm oil and water during processing. Experiments have shown that palm fruit calyx oil does not emulsify vegetable oil and water. Moreover, the digester or screw press used in the FFB milling process are unable to extract calyx oil.

The emulsification observed in the clarifier during small scale processing of palm oil is probably caused by partial glycerides (mono- and diglycerides), resulting from fresh fruit bunch degradation in the field prior to evacuation to the mill. It has been recognized for a long time that fungal lipases can bring about almost total hydrolysis of palm oil (Fickendey, 1910). Finkendey (1910) observed free fatty acid levels of 67.2% after 48 hours of bruising a fruit. In small farms in Nigeria, harvested bunches are left in the field for more than 72 hours due to lack of evacuation equipment.

A simple test of the absence of phospholipids in emulsified palm oil is by heating a little quantity in a glass beaker on a hot plate. The color of the oil changes towards black if there are phospholipids but will remain unchanged if there are no phospholipids.

Following the sieving of palm fruit calyx, the fruits are sterilized in an atmospheric sterilizer and transferred manually to a digester screw press where the sterilized fruits are pressed to release oil and presscake. The presscake is a mixture of palm fruit fiber and palm nut. The oil is transferred with buckets to the clarifier which contains water heated to 90°C. The oil is water washed in the clarifier to release the clean palm oil called special palm oil. The water layer containing dirt, carbohydrates, nitrogenous materials, and other components of the sludge is discharged through the waste water channel into a waste pit.

In the medium scale continuous process, whole bunches are sterilized in atmospheric sterilizers and conveyed to the stripper where the fruits are removed from the bunches. Usually the stripping process is poor because of atmospheric sterilization. Therefore, in some cases, pressurized sterilizers with manual discharge are installed. Pressurized sterilization in fruit cages is achievable within one hour at a pressure of 2.5 to 3 bars and a temperature of between 125°C and 140°C. Stripped fruits are conveyed with augers to the digester and screw press. Prior to this time, digested fruits were processed for palm oil in a hydraulic press, which was prone to breakdown due to high seal wear requiring large quantities of lubricating oil to operate. Palm oil released from the screw press is transferred to the clarifier where it is steam washed and transferred into storage. The presscake is transferred through the cake breaker to the cyclone where the nuts are separated from the fiber. The nuts are further processed into kernels and the fiber used to fire the boiler. These medium scale technologies are upgradable to 3 tons fresh fruit bunches per hour.

The NIFOR and other small/medium scale fabricated equipment are employed by processors who account for more than 60% of palm oil produced in Nigeria.

Large Scale Mill

Process Description

The following process description is for a 20 tonne per hour fresh fruit bunch mill.

Fresh Fruit Bunch Reception

The bunches are collected from the fields and delivered to the mill in tipping trucks and tractors with trailers. These trucks are weighed on arrival (loaded) and departure (empty) on a weighbridge of 40 tonnes capacity. This enables the exact bunch quantity to be recorded upon entry into the mill.

The collection trucks tip the bunches into the loading hoppers of 45 tonnes capacity. These hoppers, fitted with hydraulically-operated doors, are used to fill the 2.5 tonnes capacity sterilization cages (60 units) made of perforated iron plates.

Sterilization

Sterilization is carried out in two 20 tonnes sterilizers where the fruit stays for a certain period in steam at 2.3 kg/cm^2. Each sterilizer takes one train of 8 cages and are fitted with fast closing, self-locking doors equipped with security devices (valves and vacuum breakers) and measuring instruments. They are thermally insulated to avoid excessive heat loss.

The sterilization cages are placed on bogies running on tracks of 600 mm gauge. The concrete pavement for the rail-tracks for cage transportation between the loading ramp and the sterilizer line measures 210 meters long and 9 meters wide.

The bogies are formed into trains of eight cages, or a tonnage of $8 \times 2.5 = 20$ tonnes of bunches. The trains are pushed into the sterilizers and removed from the sterilizers by a capstan.

Supply to the Stripper

The sterilization cages are taken to the stripper by an overhead crane (hoist) of 2.5 tonnes capacity. The bunches are dumped into the hopper of the stripper which is fitted with an adjustable speed device for regulating the delivery of bunches to the processing line.

Stripper

The bunches are tipped into the 7.5 tonnes static capacity stripper, which is a horizontal squirrel cage revolving at 20 rpm. The fruits pass through the bars of the cage, fall into a recovery hopper, and are taken up by a screw conveyor. The stalks remain in the cage and pass into another hopper.

Treatment of Stalks

The stalks are conveyed through an empty bunch conveyor to an incinerator and burnt to ashes.

Fruit Wash

The screw conveyor carries the fruit to a vibrating screen where it is washed in hot water, which removes the dirt. The water is then drawn off and recycled.

Digesting

On leaving the vibrating screen, the fruits are taken up by a bucket elevator, then by a horizontal feed screw. Two 10-tonnes digesters are fed by a single elevator. The digesters are vertical cylindrical containers with a central shaft onto which a series of horizontal arms are attached. The container is fixed; the shaft and digesting arms rotate. The digesters are fed continuously from the top and the fruit leaves continuously from the bottom.

The digester is fitted with steam injectors and thermometers for recording the temperature of the mixture. It is thermally insulated to prevent heat loss and fitted with a warm water supply and discharge measurer. The maintenance of the digester involves replacing the arms and liners of the digester wall.

Pressing

The presses are of continuous screw type. The mixture obtained in the digesters passes into a perforated horizontal cage where it is pressed. The pressure inside the cage is regulated so as to obtain the best yield (minimum losses) under optimum working conditions (minimum damage and breakdowns). Usually, oil loss to fiber is about 5 to 7% or 2.5% oil in proportion to the total theoretical oil content of the fruit. At optimum conditions, life span of a set of screws is 1000 hours while that of the cage is 2000 to 3000 hours.

On leaving the presses, the crude juices are diluted by high injection of hot water (2 to 3 m^3/h) which facilitates clarification.

Clarification

Clarification is carried out to separate the oil from the crude juice and reduce moisture content to about 0.1% and dirt to about 0.01%. The clarification station consists of the following equipment.

Vibrating Screen

This receives the crude juices leaving the presses and separates the juices from any fibers which have passed through the perforated cage of the press. These fibers are returned to the input of the digesters. The filtered crude oil is collected in a tank where it is reheated before being evacuated to the primary settling tank.

Primary Settling Tank

From the crude oil tank, the oil is pumped by a crude oil pump into a large tank where it undergoes natural settling. Before flowing into the settling tank, the oil passes through a boiling water column with live steam injection. This raises the temperature of the crude oil to 95°C and breaks up the fibrous matter still containing oil. The tank has three outlets: one for oil, one for light sludge containing oil, and one for heavy sludge and water containing traces of oil. The first two outlets operate continuously while the third operate intermittently.

Oil Outlet

The oil leaving the primary settling tank still contains few impurities. It is therefore fed through purifiers and then vacuum dryers before passing onward to the storage tanks prior to loading the oil in tankers.

Sludge Tank

The sludge tank contains the sludge from the primary settling tank with some trapped oil. The sludge is supposed to be centrifuged to separate the oil, water, and sludge.

The Sludge

The sludge is drained through a gutter measuring 310 meters long with average width of 0.055 meters and average depth of 0.65 meters into the sludge pit from where it is discharged into the plantation as a source of water and nutrient for the palms.

Fiber Separator

This serves to separate the oil-free fibers from the nuts contained in the oil cake leaving the presses. It consists of the cake-breaker conveyor and the fiber separator.

Cake-breaker Conveyor

This is a pallet-screw conveyor with heated side walls which collects the cake from the presses, breaks it up, and conveys the mixture of fibers + nuts to the fiber separator.

Fiber Separator

The mixture of fibers + nuts passes into a very light air-stream. The heavier nuts fall to the bottom, and the fibers are blown upwards and carried to a fiber/air separating cyclone.

Treatment of Fibers

The fiber leaving the cyclone separator is taken to a boiler to be consumed as fuel. Energy resulting from burning fibers is sufficient to provide the necessary steam for processing and also via a turbine and alternator, the electricity required for operating the equipment, both inside and connected to the mill, as well as the villages for mill and plantation workers.

Kernel Station

This function to separate kernels from the shells and produce kernels with the following characteristics: Moisture = between 5.5% and 7.5%, Impurities = less than 2.75%, Broken kernels = less than 15%.

Polishing Drum

The nuts fall by gravity from the fiber separator directly into the polishing drum, whose function is to detach the fibers still clinging to the nuts.

Nut Silo-Dryers

The nuts are carried by an air-stream or mechanically from the polishing drum to the nut silo-dryers. The function of this silo is to reduce the moisture content of the nuts to an acceptable level for the next polishing stage. Air-drying is carried out by blowing hot and dry air.

Crackers

The system operates by grading the nuts in a drum with horizontal shaft and then feeding them into the centrifugal crackers; once in the center of the crackers, the nuts are thrown at high velocity against the sides and are cracked.

Separation of Cracked Nuts

The mixture leaving the cracker consists of uncracked nuts, whole kernels, broken kernels and fairly large proportion of shell fragments. These three components are separated by a pneumatic dust extractor functioning as a fiber separator and two successive hydrocyclones and disposed of as follows

a. *Kernels:* These are dampened superficially in the hydrocyclones and sterilized in live steam as they pass through a vibrating screen. They are then blow-dried in hot air in a silo-dryer. Finally, they are stored loose in silos ready to be loaded into special truck containers.
b. *Uncracked Nuts:* These are recycled into the cake breaker conveyor.
c. *Shells:* The shell fragments are taken to a shell silo. Subsequently, they are used as start-up furnace fuel for the boiler. They can also be used during processing if for some reason the supply of fiber is interrupted.

Mill Service Installations

Clean Oil Storage Station

This consists of bulk oil pumps, 100 and 1000 tonnes capacity storage tanks. Each tank is equipped with a steam heating coil and a side entry agitator, as near to the bottom as possible, to fluidify the oil prior to pumping. The bottom is sloped towards a sump for thorough drainage. The tanks are also fitted with oil level gauges.

Water Supply

Water is supplied to the mill through bore-hole. The untreated water is stored in an overhead storage tank and then conducted to the mill through pipes. Untreated water can damage the boiler. Ideally, the water for the boilers should be treated by filtration and addition of defoaming agents where necessary and chemicals to inhibit corrosion and increase the pH value.

Boiler Room

Steam requirement for fresh fruit bunch processing is about 600 kg of steam per hour. The boiler room is usually equipped with appropriate working pressure boilers with complementary accessories and piping.

Power House

Electricity for the mill and outbuildings can be supplied by power generating sets. By design, two 625 KVA steam turbines (one to be in operation and the other on stand-by) could be installed to provide electricity.

Nigerian Fresh Fruit Bunch Processing Capacity

According to Omoti et al., (2009), the estimated fresh fruit bunch milling capacity per hour of industrial mills, integrated small/medium scale mills, and NIFOR small scale processing equipment are 281, 51, and 92 tonnes fresh fruit bunches per hour, respectively. This gives a total capacity of about 423 tonnes of fresh fruit bunches per hour. Currently, the milling capacity utilization is low due to poor palm fruit production volume. Mills of 6 tonnes per hour and below can be locally fabricated, but capacities of 10 tonnes fresh fruit bunches per hour and above are imported. Mills with capacities of 1–3 tonnes fresh fruit bunches per hour are mainly the technologies of NIFOR, Stork, Amsterdam, Pioneer, VDK, Rek, and De-Wecker. Mills with capacities of 5–40 tonnes fresh fruit bunches per hour are technologies of De-Wecker, UDW, Stork, Speichim, Pioneer, and Salcon. Total national fresh fruit bunch capacity of industrial scale mills used by medium and large scale farms in Nigeria is about 244 tonnes of fresh fruit bunch per hour.

Palm Kernel Oil Production in Nigeria

Small/Medium Scale Production

When fresh fruit bunches are processed, press-cakes are generated as by products. The press-cake is a combination of fruit fiber and palm nuts. The mixture is allowed to air-dry and processed in a nut fiber separator to remove the nut. In integrated small/medium scale fresh fruit bunch mills, the nut and fiber are separated in a fiber cyclone. Thereafter, the nut is further air dried for cracking in a ripple mill to generate kernel and shell. This mixture is separated in a kernel shell separator to recover the kernel and shell. The kernels are sold to crushing factories and the shells are generally used for road maintenance and solid fuel in the boiler house. Prior to crushing, kernels are dried in the factories to a moisture content of about 10% and then fed through the hopper of the mechanical expeller to extract the oil and cake simultaneously. The operating temperature of the extruder further dries the kernel during operation and the cake is released dried. Residual oil in cake is about 10%. The equipment for small/medium scale processing as described above are locally fabricated in Nigeria. They are often single stage expellers. The major drawback of the kernel shell separation system is that up to 25% of shell is unrecoverable from the mixture (i.e., separation efficiency is about 75%), hence the wear rate of the extruder worm is very high resulting in frequent worm maintenance. In some factories, mechanically separated kernels are further processed in a clay bath system consisting of a rectangular vat with a slurry of kaoline and water. The cracked mixture of shell and kernels are poured into the slurry and mixed together at a ratio of 600 kg cracked nut to 50 kg of kaoline. The shells which are less dense float and are scooped leaving the kernels in the slurry. The kernels are subsequently water washed free of the kaoline and air dried before bagging. During the rainy season, air drying is difficult due to high relative humidity. A silo dryer is therefore required, which is often beyond the financial capacity of the operators. As a result, the kernel quality is reduced by microbial degradation and this affects the quality of extracted oil. Expelled kernel oil is stored in drums where the dirt settles to the bottom and the upper palm kernel oil layer drained into clean oil storage containers.

Palm kernel oil extrusion mills of 1–15 tonnes per day (24 hrs) of kernels are fabricated locally in Nigeria. Additional equipment to complement the extruder include a kernel drying silo integrated with heated oil radiators as source of heat supply. Steam heating is expensive because of the initial capital outlay for pressurized steam boiler.

Large Scale Palm Kernel Oil Production

Palm kernel oil production at large scale mills is carried out by estates such as Okomu oil Palm Plc and PRESCO Plc. These mills have fresh fruit bunch processing capacities greater than 40 tonnes per hour. The kernel handling section has been described earlier. Dried kernels are fed into a crusher of about 15 tonnes of kernel per day to crush the kernels and then into the expeller with capacity of 17 tonnes of crushed kernel per day to expel the oil and also generate palm kernel cake. The expelled oil is transferred through plate and frame or pressure leaf filters into storage. Extraction rates of these imported machines are usually about 86% compared to locally fabricated variants with extraction rates of about 80%. The total capacity of the major large industrial crushers is estimated at about 3,718 tonnes per day or 1,115 million tonnes per annum (Omoti, 2009).

Total national output of palm kernel as of 2009 was 234,133 tonnes, and production by large estates accounted for about 11% (25,982 tonnes per annum). Palm kernel production from the wild groves and small/medium scale farmers was about 208,151 tonnes per annum (89% of total annual production). This accounts for the reason palm kernel oil production is dominated by small scale operators managing factories with throughput of 1.5 tonnes kernel per day. Palm kernel production factories are as dispersed as oil palm trees in the wild grove of the oil palm production belt of Nigeria.

Palm and Palm Kernel Oil Refining

Palm oil is refined in Nigeria by the physical refining process while palm kernel oil is refined by physical and chemical refining processes. From a survey of refineries in Nigeria, about 22% of the local refineries use the chemical refining technology while 88% use the physical refining technology. The refining technologies are those of De Smet, Belgium, Kumar Metal Works, Mumbai, India, and Jenazi, Italy. Seventy-seven percent of the local refineries were built to use palm kernel oil as raw materials. The reason for this trend is that palm oil requires fractionation and there is little local market for palm stearin. Second, the cost of fractionation equipment is equivalent to the cost of refining equipment. Therefore, investors are content refining palm kernel oil since there is local demand for it rather than refining and fractionating palm oil when they can only sell refined bleached and deodorized palm olein. Palm kernel oil deodorization and fatty acid distillation are carried out under pressure of about 3 torr and temperature of about 245°C while palm oil refining conditions are pressure of 1–2 torr and temperature of 260–270°C.

The other factor for the preponderance of palm kernel oil refining equipment in Nigeria is that more than 70% crude palm oil produced locally are not refine-able because they are from the wild groves where processors use milling equipment without considerations for industrial refining quality. Since 1986, bulk importation of crude palm oil into Nigeria was banned. This ban was lifted in 2010. Even at this time, the landing cost of a tonne of crude palm oil at Nigerian Ports was about $1,687.33, whereas local price (ex-works) was about $1,419 per tonne. Although the good quality crude palm oil is not sufficient locally, most of the refineries are located inland sometimes 100 to 400 km from the Ports. The poor state of roads and the attendant high cost of transportation to feed a 100 tonnes per day refinery makes the business economically unviable. Due to the above factors, refineries are not about to leverage on the permit to import crude palm oil.

Blending of Palm and Palm Kernel Oil

A unique feature of the Nigerian vegetable oil market is a household product in which palm oil and palm kernel oil are blended at ratios 20:80 to 40:60. This product evolution arose from the need to reduce production cost because palm oil is always cheaper than palm kernel oil in the Nigerian market. At palm kernel oil: palm oil blending ratio of 80:20, 70:30, and 60:40, savings in raw material cost is 2.2%, 3.3%, and 4.4%, respectively, and this amounts to $960,000; $7,596,774; and $7,636,708, respectively, for a 100-tonnes per day refinery operated at 60% capacity utilization.

Studies by Gold (2006) showed that mixtures of palm kernel oil and palm olein produced eutectic compositions (mixtures with lower melting point than the individual parent). These mixtures were also found to have lower solid fat content at the same temperature than the pure components. This finding corroborates those of other authors (Bigalli, 1988; Timms, 1984).

These blended products are directed to the open market where the only quality assessment is the fluidity of the oil.

Fresh Fruit Bunch Yield Fluctuations in Nigeria

Several authors (Boekmans, 1957; Bredas & Scuvie, 1960; Devuyst, 1948; Hemptinne & Ferwerda, 1961; I.R.H.O. 1957; Michaux, 1961) have reported that annual yield variations in oil palm originate from variations in climatic conditions which influence both the sex-differentiation in the inflorescence primordial a few years before the opening of the flower bud and the development of the fruit bunches at a later stage.

Devuyst (1948) observed that useful rainfall 33 months earlier cause variation in fresh fruit bunch yield. I.R.H.O. (1957) showed that rainfall from 24–30 months before the beginning of year affects fresh fruit bunch yield. According to Broekmans (1957), rainfall from November to April, 24–30 months earlier influence fresh fruit bunch yield.

There is, however, an agreement that carbohydrate status of the oil palm at the time of sex differentiation, determined by the cumulative amount of effective sunshine, is the most important factor in annual yield variations (Sparnaaj et al., 1963).

Uses of Palm and Palm Kernel Oil in Nigeria

More than 70% of crude palm oil produced in Nigeria is consumed without refining. Crude palm oil is used for various food recipes including frying. Refined bleached and deodorized palm olein is mainly used for food frying in the household and commercial food frying outlets.

Crude palm kernel oil is used for soap making. Refined bleached and deodorized palm kernel oil is used for food frying. Some refineries blend palm kernel oil and palm oil at ratios of 8:2 to 6:4 for food frying. Indeed, locally produced frying oils are mainly blends of palm kernel oil and palm oil. Palm olein found in the Nigerian market is smuggled products from Ghana, Cameroon, and Republic of Benin.

Banga Sauce

Banga sauce is used for the preparation of *Banga* soup, a local food delicacy. Banga sauce is prepared by squeezing sterilized and digested palm fruit in a hydraulic press. The crude oil that exudes is a mixture of oil and carbohydrates with some protein. This mixture is allowed to settle in a drum in order to recover the solid bottom layer. This bottom layer is an international product referred to as *Banga.* The product is cherished by West Africans. Gold et al. (2003, 2007) showed the commercial viability and characteristics of the product.

Conclusions

In conclusion, the oil palm originated from West Africa. Since the beginning of its exploitation, the wild grove has been the major source of palm produce and currently supplying more than 60% of locally produced palm oil in the Nigerian market. Efforts to increase production through plantation programs have not been very successful. Government-led development programs have not been successful but little successes have been recorded by private investors. Expansion by private investors has been limited by the high cost of credit and lack of policies to protect the industry. The current supply deficit, especially in Nigeria, is due mainly to the poor quality materials in the grove, the old age of the grove, inefficient cultural practices and processing methods, employment of land that were once cultivated to oil palm for other agricultural and industrial purposes, high population density and complex land tenure systems in the oil palm growing areas, rural-urban drift of farming population, and systemic erosion of oil palm cultivation and processing skills from the growing areas, among other factors. Nigeria has a land bank of more than 300,000 hectares in the oil palm belt and a market of over 150 million consumers with a per capita consumption of 12.8 kg per annum. Compared to industrialized nations with per capita consumption of more than 23 kg per annum, the fats and oils growth potential in Nigeria is enormous. However, long term credit with interest rate of less than 5% and mechanized farm operations including harvesting will be required to produce enough palm oil for local consumption. The current methods of oil palm development which is largely manual is not attractive to modern generations and, if allowed to continue and not modernized, would require Nigeria to import more than 60% of its fats and oils requirement in the future.

References

Aladewolu, C. A small scale three tons fresh fruit bunches per hour mill. Proceedings of the International Conference on Palms and Palm Products by the Nigerian Institute for Oil Palm Research, November 21–25, 1989, 2, 772–773.

Bigalli, G.L. Practical aspects of the eutectic effect on confectionery fats and their mixtures, Manuf. Confect. 1988, 53, 65–80.

Bredas, J.; Scuvie, L. Aperçu des influences climatiques sur les cycles de production du palmier à huile. Oléagineux 1960, 15, 4, 211–222.

Broekmans, A. Growth, flowering and yield of the oil palm in Nigeria. J. W. Afric. Inst. Oil Palm Res. 1957, 2. 7, 431–437.

Corley, R.H.V. & Tinker P.B. World agricultural series: the oil palm, 4th ed. Blackwell Publishing: UK, 2003.

Devuyst, A. Influence des pluiessur le rendements du palmier à huile en registrés à la station de La Mé de 1938 a 1946. Oléagineux 1948, 3, 3, 137–144.

Fickendey, E. Uber die verwendungsfahigkeit des Pamolsalsspeisefett, Tropenflanzer, 1910, 14, 566–567, cited in Enzymes in Food Processing; Tucker, G.A., Woods, L.F.J., Eds.; Blackie Academic and Professional, an imprint of Chapman and Hall: Glasgow, 1995; pp. 278.

Food and Agricultural Organization World Crop Statistics: Area, Production and Yield 1948–1964; 1966.

Gold, I.L.; Okpefa, E.; Enonuya, D.O.M.; Ojomo, E.E.; Bafor, M.E.; Okiy, D.A. Techno-economic analysis of banga sauce production. Riv. Ital. Sostanze Grasse LXXX, Sept./October, 2003; pp. 329–332.

Gold, I.L.; Okpefa, E.; Ojomo, E.E.; Iyasele, J.U.; Bafor, M.E. Production and characteristics of banga sauce, a concentrated oil palm fruit mesocarp juice. Proceedings of Product Development and Nutrition Conference, PIPOC 2007, Kuala Lumpur Convention Center: Malaysia, 26–30 August 2007.

Hartley, C.W.S. The oil palm, 2nd edition; Longman: London, 1977.

Hartley, C.W.S. The oil palm, 3rd edition; Longman: London, 1988.

Hemptinne, J.; Ferwerda, J.D. Influence de precipitations sur les productions du palmier à huile. Oléagineux 1961, 16, 7, 431–437.

I.R.H.O. Rapport annuel 1957 de l'Institute de Recherches pour les Huiles et Oléagineux, 1957.

Kilby, P. The Nigerian oil palm industry. Food Research Institute Studies, 1967.

Maizura, I.; Cheah, S.C.; Rajanaidu, N. Genetic Diversity of oil palm germplasm collection using RFLPs. Proceedings of the 2001 PIPOC International Palm Oil Congress (Agriculture), 2001; pp. 526–534.

Michaux, P. Les composants climatiques du cycle annuel de productivité du palmier à huile. Oléagineux, 1961, 16, 8–9, 523–538.

Olayinka, O. Alegieuno, J. Oil palm projects: Financing and marketing. International Conference on Palms and Palm Products, 1989, 558.

Omoti, U. Oil palm sector analysis. United Nations Industrial Development Organization, 2009.

Rafii, M.Y.; Rajanaidu, N.; Jalani, B.S.; Kushiri, A. Performance and heritability estimations on the oil palm progenies tested in different environments. J. Oil Palm Res. 2002, 14(1), 15–24.

Rees, A.R. Evidence for the African origin of the oil palm. Principes 1965a, 9, 30–36.

Sowunmi, M.A. The significance of the oil palm (*Elaeis guineensis* Jacq.) in the late Holocene environment of west and west central Africa: A further consideration. Veg. Hist. Archeobotany 1999, 8, 199–210.

Sparnaaj, I.D.; Rees, A.R. Annual yield variations in the oil palm. J. West African Inst. Oil Palm Res. 1963, 4, 14, 111–125.

Timms, R.E. Phase behaviour of fats and their mixtures. Prog. Lipid Res. 1984, 23, 1–38.

Udom, D.S. A review of recent oil palm production policies in Nigeria. International Conference on palm and palm products 1989, 2, 534.

World Bank Development Indicators. Data-worldbank.org/data-catalog/world-development-indicators.

Zeven, A.C. On the origin of the oil palm. Grana Palynol. 1964, 5, 121–123.

Zeven, A.C. Oil palm groves in Southern Nigeria. Part I. Types of groves in existence. J. Nigerian Inst. Oil Palm Res. 1965, 4, 226–249.

11

Oil Recovery from Palm Fruits and Palm Kernel

Jude Uche Obibuzor, Emmanuel Arabame Okogbenin, and Roland Doye Abigor
Biochemistry Division, Nigerian Institute For Oil Palm Research (NIFOR), Km 7 Benin-Akure Road, P.M.B. 1030 Benin City, Edo State, 300001, Nigeria

Introduction

Despite the wide diversity of today's sources of vegetable oils, all recovery processes, whether for soybeans, palm fruits, or other oilseeds, are designed to obtain triglycerides in high yield and purity and to produce co-products of maximum value (Johnson, 2008). In the very early period of palm oil processing the oil recovery efficiency of the chosen process was not considered a major production factor. This may be due to the fact that the oil palm was not a plantation crop and the primitive processing equipment used. Secondly, the palm oil was not an article of international trade. However, with the elevation of oil palm to a plantation crop and the advent of mechanized processing methods, which are driven by profit and quality, oil recovery efficiency became a major factor. At present, environmental factors and considerations as well as sustainable exploitation of the palm tree have been added to the list of criteria to be met. These criteria require that in the course of exploiting natural resource by any means the foot prints of that process on the nature's pathway must be efficiently and sustainably covered.

This chapter therefore sets to highlight the economic importance of efficient palm oil recovery and source of oil loss along the processing chain/route both in modern and traditional methods. The major challenges and problems of oil recovery to the major stakeholders as well as the quality and use of the recovered oil will also be discussed.

Since the creation of life, fats and oil have been playing a major role in the activities of human kind especially in the areas of food, cosmetics, lubrication, pharmaceutical (nutraceuticals), and folk medicine (Edem, 2002; Rozman et al., 2001). Fats and oils are usually recovered from oil bearing seeds and fruits and fatty animal tissues (Johnson, 2008). Oil palm is a very important crop to the economies of tropical countries in Africa, Asia, and Latin America. Although the oil palm is native to West Africa, at present Malaysia and Indonesia account for about 90% of palm oil world export trade and contributes 25.5% of the global fats and oils production of 37 million metric tons in 2006 (Rupilius & Ahmad, 2007). The African oil palm (*Elaeis guineensis*) is used in a wide variety of ways by the local population such as a source of foodstuffs and medicines, its sap is used to make palm wine, and its fibers are also used for various purposes. However, the large-scale monoculture oil palm plantations being promoted today are geared towards two main objectives. Their primary purpose until now has been the extraction of palm oil (from the flesh of the oil palm fruit) and palm kernel oil (from its kernel or seed) for the production of edible and industrial oils. More recently, a second major objective has emerged: the production of biodiesel from crude palm oil (Carrere, 2006).

The essence of heavy investments in oil palm plantation establishment by the investors is ultimately to economically and efficiently recover, in high quality state, oil and the necessary minor components from the oil-bearing cells. In other words, all recovery methods and processes aim to recover 100% of oil residing in the oil-bearing cells by minimizing the sources of oil loss. One of the ways to attempt the complete recovery of the oil is to first trace the entire production and processing path. This tracing strategy identifies all the points where oil is lost and in what form the oil is lost with the view to developing appropriate recovery methods and procedure or modifying that part or section of the processing equipment where the loss is highest.

The Palm Fruit

The primary feedstock in palm oil mill is the palm fruit. Palm fruit is a drupe, oval in shape, and contains a kernel. The kernel, surrounded by the fruit wall (pericarp), is the true seed. The pericarp is made up of the hard shell (endocarp), oil bearing tissues (mesocarp), and the skin (exocarp). In the early stages of fruit formation, the oil content of the fruit is very low, but as fruit approaches maturity the formation of oil increases rapidly to about 50% of mesocarp weight. Two distinct types of oils, palm oil and palm kernel oil, respectively, are extracted from the mesocarp and the seed. Palm oil processing methods and procedure at village, small/medium or industrial scale whether in Africa, Asia, or America are fundamentally similar. The differences in the recovery efficiency and quality of the oil in these countries are generally due to feedstock (*dura, pisifera,* or *tenera*), efficiency in the pretreatment of fresh fruit bunch (FFB), and the level of sophistication in the type of equipment and their designs as well as waste management and disposal. Generally, developments in the milling technology over the years have been fairly minor (Menon, 2001).

Fig. 11.1 shows a generalized palm oil and palm kernel processing flow chart, whereas Fig. 11.2 shows the material balance of palm oil processing expressed in percentages.

There are several methods of palm oil extraction, but generally it is classified into two: the wet and dry processes. In the wet process a liquid, usually water, is used to extract the oil from the digested palm fruits, while in the dry method, which could be either batch or continuous, a hydraulic press or a screw press is employed. The screw press is generally more applicable to a continuous extraction system while the hydraulic press is usual to a batch or semi-batch extraction system. After pressing out

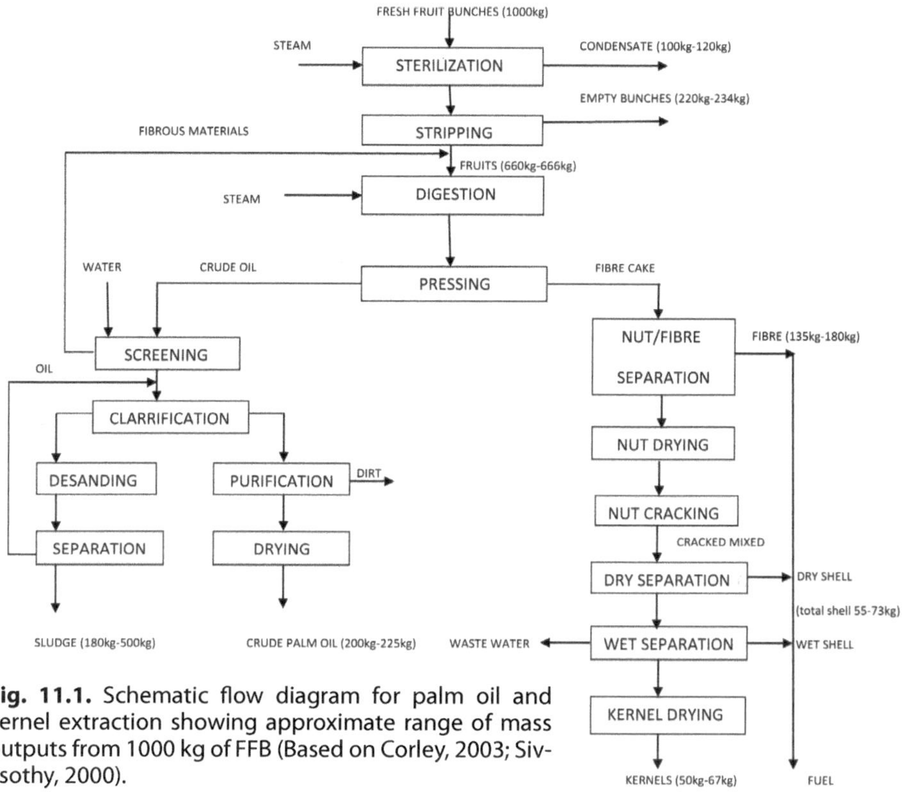

Fig. 11.1. Schematic flow diagram for palm oil and kernel extraction showing approximate range of mass outputs from 1000 kg of FFB (Based on Corley, 2003; Sivasothy, 2000).

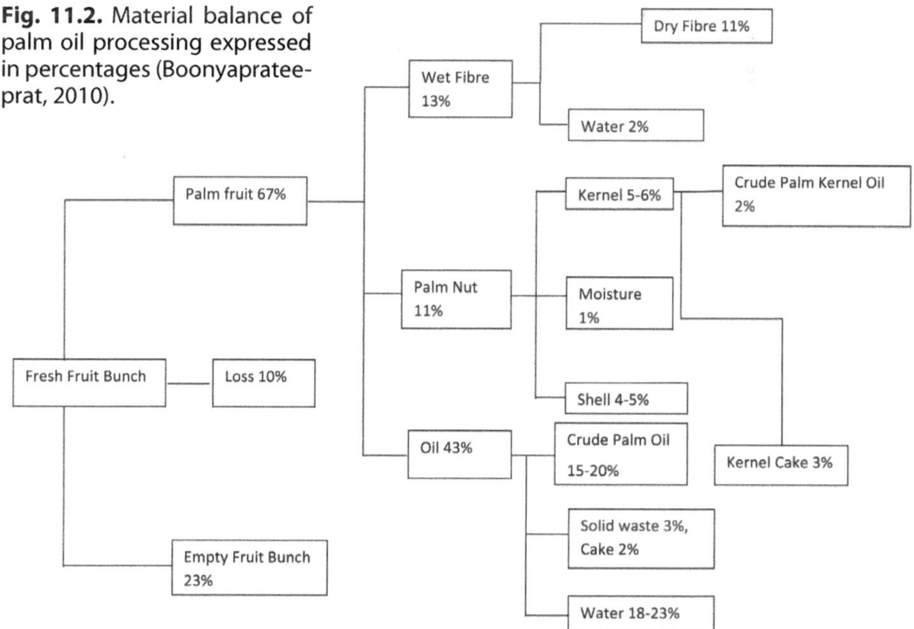

Fig. 11.2. Material balance of palm oil processing expressed in percentages (Boonyaprateeprat, 2010).

the crude palm oil from the digested mass, the mesocarp fiber still retains about 5–6% of oil. But to avoid cracking of palm kernels, the pressure is reduced and oil retention increase to 10–12%.

The oil extraction ratio (OER) and kernel extraction ratio (KER) are calculated from the tonnage of FFB coming into the mill and the weight of oil kernel produced (Corley & Tinker, 2003). In order to press out more oil from the fiber, the pressure must be increased. This increase, however, will lead to some complication in the process. In the first place, the increased pressure will crack the kernel and damage the perforated cage. Secondly, the kernel oil may be expressed from the broken kernels into palm oil. This will lead to contamination and deterioration of the crude palm oil quality.

The factory efficiency is calculated as:

Factory efficiency = Oil produced ÷ (oil produced + known losses). Equation 11.1. (Corley & Tinker, 2003)

This means that as the value of known losses approaches zero, the factory efficiency approaches one (that is, 100%). Therefore, it is important to minimize known oil losses in order to improve the factory efficiency.

Economic Importance of Efficient Oil Recovery from the Palm Fruit and Kernel to the Various Stakeholders

The oil palm is grown for its mesocarp oil and kernel oil, which, even though obtained from the palm fruit, are quite different in chemical compositions. The goals and objectives of an oil palm farmer and major stakeholders in the oil palm industry is basically sustainable profitability. This sustainable profitability is closely linked to the respective and entire efficient production process along the entire value chain. This is because if any of the production processes are identified as an environmental pollutant the entire systems may be considered not to be environmentally sustainable. Therefore, in order to maximize profit and ensure good practice, the oil palm farmer must optimize fruit yield and employ efficient and standard harvesting systems. These factors have direct bearing on the process (factory) of oil extraction efficiency and extraction ratio (low oil retention loss). These require huge investment in plant breeding and other agronomic and crop protection programs.

From the traditional methods to the highly automated power operated techniques, the process of producing the two types of oils (palm oil and palm kernel oil) from palm fruit in the mills essentially involves steaming and squeezing. These processes can be broken into several stages (Fig. 11.1), with each stage playing an important role in producing the processed oil. Despite improved technologies to increase extraction efficiencies, losses are often encountered at the different stages of processing. Also, some quantities of the oil are lost to the large amount of by-products and wastes generated during processing. Several techniques have or are being developed to recover and put these lost oils to use.

In spite of the fact that most times the quality and quantity of the recovered oil from the retaining components or by-product is low, the recovery process is still usually embarked upon basically because of some stringent environmental requirements or the desirability of some components in the oils. For example, palm pressed fibers contain 5–6% residual oil and contain 4,000–6,000 parts per million (ppm) of carotenes, six times the concentration in crude palm oil. The carotenoids content of the residual oil in the pressed fiber from the hybrid oil palms is even higher, 6,000–7,000 ppm (Neoh et al., 2011).

Fruit Miller and Kernel Crusher

Palm oil resides in the oil cells of the mesocarp of the fruit while the palm kernel is located in the palm kernel nut. Releasing the oil from the native cell requires mechanical rupturing and series of other treatments (the detailed description of processing have been given in another chapter of this book). The palm fruit miller is responsible for extracting palm oil from the FFB by using any of the various methods of processing available to him. The most widely used processing methods adopted by palm fruit millers often lead to the generation of massive amount of waste. However, with the advancement in science and technology of waste management some of this "waste" may be raw materials for another industry, hence could be called a co-product. The ultimate aim of a miller or a processor is to extract all the oil nature has placed in the oil bearing cells of fruit mesocarp. However, because there are no man-made systems or piece of machinery that is 100% efficient, the miller will only attempt to maximize the oil extraction or processing systems. The miller is required to meet both the local and international standards for palm oil and palm kernel oil and waste management and disposal. These are challenges that require long-term planning, funds, and human capital. The miller is also faced with the twin problems of meeting the set standards of the primary products, palm oil and palm kernel oil, and the stringent environmental conditions both local and international with respect to waste management and the use of non-polluting machineries.

Environment

Palm oil mill effluent (POME) generated from palm oil milling processes to produce crude palm oil (CPO) is the most polluting agro-industrial effluent. It is estimated that 0.5–0.75 tonne of raw POME were produced for every tonne of fresh fruit bunches (FFB) processed (Yacob et al., 2006). POME is a highly polluting wastewater with typical properties: biochemical oxygen demand (BOD) of 25,000 milligram per liter, chemical oxygen demand (COD) of 50,000 milligram per liter, oil

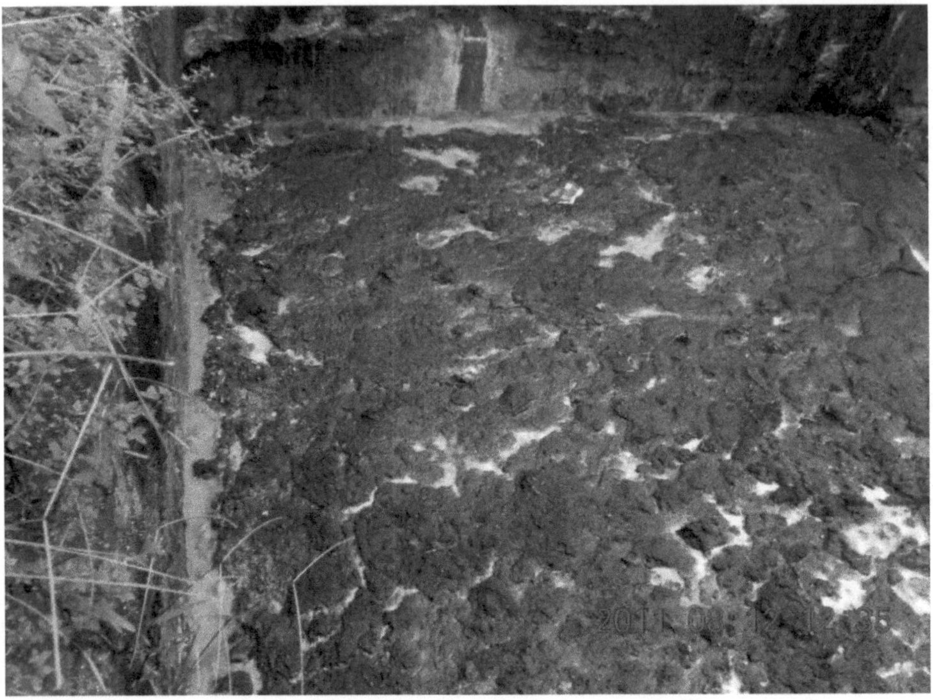

Fig. 11.3. Palm oil mill effluent with high oil content collected in a pond in NIFOR.

and grease of 8,000 milligram per liter, suspended solids of 20,000 milligram per liter and total solids of 40,000 milligram per liter (Nik et al., 2001). Therefore, POME also contains significant amount of oil (Fig. 11.3) and must be pre-treated before discharging it into the environment. The oil must be removed to avoid problems in water treatment units and biological treatment stages. The removal of the oil from POME before anaerobic treatment is desirable, as the oil can be recovered for sale as low-grade oil and the POME then becomes more suitable for subsequent anaerobic treatment. Investors in the nutraceuticals industries are now turning to these low grade oils as a potential source for the valorisation of important bioactive compounds.

Sources of Oil Loss Along the Palm Fruit and Palm Kernel Processing Chain

The aim of commercial cultivation of the oil palm and ensuring proper husbandry and management is to convey 100% of the fruits produced by the palms to the extraction plant and process these fruits into premium palm oil and the co-product, palm kernel. The palm kernel can be further processed into palm kernel oil by cracking, crushing, pressing, and sedimentation. However, it is absolutely impossible to harvest and convey all the fruits borne by the palms to the processing plant without any loss. The stages involved in the processing of palm fruits into palm oil are lengthy and complicated. Therefore, given the fact that no machine has 100% efficiency and the long processing chain of palm fruits into palm oil, the aggregate oil loss along the processing chain is considerable. This loss can be due to a number of factors, namely size of the fruits, state of ripeness, method of conveyance to the plant, age of the palm, and so on. The sources of oil loss are traced from the plantation right through the mills to the wastes.

Oil losses in mills vary from mill to mill, and much attention is given to the control of oil loss. The main oil loss are from sterilizer condensate, empty bunches, fruit loss in unstrapped bunches, press cake fiber, nuts, and sludge. Over-ripe bunches will lose more oil during sterilization. To minimize this, shorter sterilizer cycles are used, or better control of bunch ripeness and quality will help ensure less wastage. A typical oil loss in sterilizer is estimated to be 0.1% to FFB. Oils recovered from sterilizer condensate should be used as technical oils, as they often contain higher iron content and would reduce oil stability if mixed with the crude oil (Wai-Lin, 2011).

Field/Plantation and During Conveyance to the Mill

The palm oil mills or factory can be compared to the computer jargon "garbage in garbage out." In the same vein, the quality of the palm oil obtained from the mill is very much dependent on the quality and state of the palm fruits (bruised versus whole fruits and ripe versus unripe fruits) fed into the mill. Processing of high amounts of bruised fruits translates to oil with high free fatty acids, FFA, while processing of high amounts of unripe fruits will translate to highly oxidizable oil due to increased amounts of chlorophyll associated with unprocessed fruits. Hence, proper scrutiny of palm fruits fed into mills is an integral aspect in the production of premium quality palm oil chain. In other words, excellent quality palm oil is made in the seed production (planting material) and oil palm plantation, not during milling processes in the mills/factory. This means that good seed production and cultivation practice is an important aspect of the palm oil production chain. The loss of oil begins in the fields and during the period of conveyance to the mill and this loss is associated with poor cultivation practice—poorly maintained field, use of inappropriate conveyance methods, or overloading of the conveyor.

Conveyance of Palm Bunches

In Africa most bunches are conveyed to the processing site in baskets carried on the head. To dismount the load, the tendency is to dump contents of the basket onto the ground. This results in more bruises. Sometimes trucks and push carts, unable to set

Table 11-A. Output-Balance of Standard Wet Process Showing Total Loss.

Medium	Type of material	Individual mass (kg/t FFB)	Individual oil content kg/t FFB	Oil loss relative to total loss %
Solids	EFB	230	4.5	26
	Fiber	145	5.0	30
	Shell	60	0	0
	Kernel (incl. PK oil)	60	0	0
	Centrifuge cake (oil separation and purification.)	0	0	0
Subtotal		495	9.5	56
Liquids	Raw oil PO	163	0	0
	Washing/cooling water. (except indirect cooling water)	Depending on local conditions 150	0.5	3
	Sterilizer effluent underflow of settling tank after centrifugation (separator) with a suspended solid load of > 30kg/t FFB	742	7	41
Subtotal		1055	7.5	44
Gas/vapor	Water vapor	250	0	0
Subtotal		250	0	0
Leakage	Process instability, breaks downs inefficient equipment	Depending on local conditions	–	–
Subtotal			–	–
Grand Total		1800	17	100
Summary of Total oil loss		–	17	100

bunch down gently, convey the cargo from the villages to the processing site. Again, tumbling the fruit bunches from the carriers is rough, resulting in bruising of the soft mesocarp (http://oil-palm-fruit-singapore.blogspot.com/2007_08_01_archive.html).

Palm fruit bunches on the tree is said to be ripe and ready for harvesting when about 6–8 loose fruits are found at the base of a particular palm (Corley & Tinker,

2003). When the ripe bunch is not harvested, the fruits will continue to fall until the bunch is empty. However, when the plantation is not properly maintained by ring weeding the palms, pruning of the palm, slashing of the weeds, or the plantation is infested with rodents, a considerable amount of fruit (which is processed into oil) is lost. Ring weeding enables the farmer or harvester to identify the ripe fruit by counting the loose fruits at base of the palm in addition to completely picking up all the loose fruits from the bunch after harvesting the bunches. When this ring weeding is not done, a lot of fruits are lost to the bush in the field and the rodents that feed on palm fruits, especially after harvesting. This loss may seem insignificant at face value. It will be better appreciated when a loss of 100 g of palm fruits per palm in a 100 hectare oil palm plantation containing 150 palms per hectare is considered. This will result in the loss of 1.5 metric tonnes of loose palm fruits. When a palm is not pruned the loose fruits may not be able to fall to the ground at all, hence the ripe bunch may not be detected for harvesting until it is rotten. This means that the oil due or expected of that palm fruit bunch is totally lost.

Furthermore, harvested fresh fruit bunch (FFB) are usually conveyed to the mills by various means of transportation—human head, wheel barrow, bucket mounted on a tractor, lorries and pick up vans and, conveyor belts. During the transportation of harvested FFB from the plantation to the mill, a lot of loose fruits are lost by falling off the vehicle (at times whole bunches are lost in a trip to the mill), retention at the crevices of the vehicle or belt. Again, sometimes fresh fruit bunches are harvested prematurely by some farm managers in order to avoid loss of the fruits when the bunch falls. This action will lead to the rejection of the bunch at the mill because some of the physiological processes of ripening have not been completed, besides the possibility of affecting the quality of processed oil. The rejection of these immature bunches if not properly controlled and managed amount to heavy loss of oil per hectare of plantation. Therefore, the harvesting efficiency and proper harvesting standards are important in maximizing product value and oil recovery efficiency.

Sterilizer

The types of equipment used in sterilizing the fresh fruit bunch (FFB) vary with the size of the mill and configuration. It sometimes varies also from location to location and country to country, the configurations and operation conditions of a sterilizer have been adequately described by Corley & Tinker (2003). Figs. 11.4 and 11.5 are horizontal cylindrical sterilizer and a loaded trolley, respectively, en route the sterilizer. The sterilizer in the Fig. 11.4 is the type currently used in NIFOR Oil Mill Division to process FFB. A cursory look at the foot of the trolley will reveal possible loss of palm fruits or even a whole bunch which may not be recovered completely. Sterilization is the first step in the pre-treatment process of FFB. The FFB is sterilized in order to enhance oil extraction from the mesocarp. It involves autoclaving of the FFB with live steam [in small or local mills cooking for about 1 hour or less, depending

Fig. 11.4. Conventional horizontal cylindrical sterilizer units at NIFOR.

Fig. 11.5. Loaded trolleys filled with FFB en route the sterilizers.

Fig. 11.6. Small scale sterilizer units at NIFOR.

of course on the efficiency of the heating source (Ilechie, 2011)]. Fig. 11.6 shows a sterilization process in a small-scale palm oil processing unit of NIFOR, the process involved cooking and not steaming. That method and the sterilization equipment are adequate and very popular among the small-scale processors in Nigeria. The essence of sterilization process is to stop the activities of the native lipolytic enzymes in the fruit to avoid FFA build up in the oil when processed. Secondly, sterilization loosens the fruit from the stalk, softens the mesocarp, and loosens the nut from the kernel to facilitate stripping, digestion, and cracking, respectively. During sterilization the moisture introduced by the steam acts chemically to break down gums, resins, and starches which cause the oil to foam during frying because of their solubility in water. These broken down materials are removed during oil clarification. The expansion of the nut during cooking or sterilization and contraction during cooling, respectively, cause the nut to detach from the shell wall (loosening). The benefit of this action comes during cracking of the nuts. However, poor sterilization due to inefficient evacuation of may cause secondary oxidation leading to discoloration of palm kernels, poor deterioration of bleachability index (DOBI) of the extracted oil, and reduction of the protein value of the press cake.

The major concern here is what enhances oil loss during sterilization. During harvesting, transportation, and quartering of FFB, some oil bearing cells of the fruits are heavily bruised or macerated. Therefore, at the sterilization conditions temperature 130°C, pressure 2.1 bars, and for a duration of 1 h, these bruised or partially digested fruits easily released the oil to steam, and when the steam eventually condenses the oil will float. Ultimately the amount of oil loss depend largely on the state of the FFB feed to the sterilizer, less bruises will translate into low oil loss. In the case of the small scale unit where the quartered or loose fruit are cooked, the oil loss will be visible in the cooking water after the fruits have been transferred to the digester, as against the condensate in the case of steaming. Though the loss due to boiling has not been estimated, it should be appreciable, considering the fact that quartering inflicts more bruises or outright cutting of the fruits.

Bunch Stripper

The function of bunch stripper (a rotated drum) is to separate the sterilized fruits from bunches. The process of stripping is also called threshing. The major oil loss in this segment of FFB processing is a result of incomplete stripping of fruit from the sterilized bunch and some mesocarp mass normally stocked on the stripper and the crevices. The unstripped bunches (hard bunches) which usually are detected by visual inspection are collected and recycled where possible and is cost effective. In the case of small scale mills, bunches are quartered, allowed to stand for between 4 and 5 days (Ilechie, 2011) for the fruit to loosen, and then threshed manually by beating with a slightly heavy sticks. The threshed fruits together with the calyx are transferred to a calyx/fruit separator where the calyx and other extraneous dirt are separated from the fruit. In some cases the quartered bunches are boiled prior to threshing and separation but the economy of scale favors the cooking of loose fruits because more fruits will be cooked per batch and the cooking process will be more efficient. Fig. 11.7 shows a heap of stripped bunches in which some unstripped fruits could be easily seen. Fig. 11.8 shows a calyx separator, smaller palm fruits are equally lost at this stage of operation, while Fig. 11.6 shows the front view of a small scale palm oil processing shed in NIFOR, and Fig. 11.9 shows threshed and cleaned loose fruits ready to be cooked. It is obvious from Fig. 11.9 that the quality of the oil expected from those loose fruits will be highly compromised due to the visible microbial growth on the fruits.

Digester/Extractor

Digestion process is considered as the heart of palm oil production process and very critical to the efficiency of oil recovery. The digester is simply a vessel in which rotating shafts with attached stirring arms and digestion is achieved when the sterilized stripped fruits are rotated about, causing the loosening of the pericarps from the nuts. In other words, it involves mechanically detaching or pounding (in the case of the traditional methods) mesocarp of the sterilized and stripped fruits to a homogenous oily pulp. The digestion process is more efficient when the digestion vessel is filled properly. This is because the useful action of the process is based on mechanical friction among the fruits, the wall of the vessel, and the stirring arms of the shaft. The digested pulp is reheated either by live steam or addition of boiling water to aid

Fig. 11.7. Stripped bunches with retained fruits.

Fig. 11.8. NIFOR small scale bunch stripper with threshed calyx.

Fig. 11.9. Threshed fruits ready for cooking.

homogenization. In some cases the pulp are retained in the crevices of the digester or not completely transferred to the pressing cage and these result in possible loss of oil.

The oil recovery process is continued by transferring the digested mash for pressing where the crude palm oil is extracted. This segment is the key to all the other oil recovery processes. The processing could be done by the use of screw, hydraulic, or centrifugation in decreasing order of efficiency. The types, advantages, and efficiencies of screw press, which is generally more acceptable than other pressing methods, have been described by Corley & Tinker (2003). Corley & Tinker (2003) also stated the major drawbacks of other methods of oil recovery from the digested mash. The major source of oil loss in the press is oil retention in the pressed fiber due to absorption or improper/incomplete digestion of the fruit inherited from the digestion process. The type of palm fruit mix may increase the oil loss too. For example, when a batch of palm fruit with high proportion of *Pisifera* to *Tenera* or *dura* ratio is digested, the tendency here will be high oil retention in the pressed fiber and sludge. This is because the *Pisifera* has high mesocarp to kernel ratio and generates a lot of sludge, which is difficult to press, hence results in the inefficient oil recovery. Fig. 11.10 shows the NIFOR small scale combined digester screw press, which is very efficient and popular among the small scale palm oil processors in Nigeria.

Fig. 11.10. NIFOR small scale combined digester screw press.

Clarifier and Dryer

The pressed out liquid from the press is a mixture of water, oil, dirt, and fruit debris in varying concentrations depending on the kinds of treatments and feedstock digested. This mixture is fed into the clarifier/decanter with the objective of separating the crude palm oil from the other listed constituents of the mixture. The separation of the oil from the debris is targeted to yield a crude palm oil composition of 90% oil and 10% water. This process, besides being critical to the quality of the oil, is a major source of oil loss and environmental pollution. This separation which can be done either in batch or continuous mode is usually achieved by the use of clarifiers or decanters. In the NIFOR Small Scale Processing Equipment (SSPE), clarifiers are used and the composition of the liquid received from the press ranged from 55 to 62% oil, 20 to 30% water, and 8 to 16% debris which are within the range of other clarification equipment in that size (Ilechie, 2011).

Another widely used method of oil clarification is the separation tank system. The oil separation tank method is a conventional procedure of separation of oil from water and suspended solids. In this procedure, the principle of separation of the oil (main product) from the other constituents from the mother liquor is principally based on gravity setting. And the process generally involves pumping (transferring), dilution with hot water (80–95°C) to reduce the viscosity, and settling (1.5–5 h) of the mother liquor in a tank. The removal of the debris from the mother liquor during settling process is usually facilitated by a vibrating screen. At the end of the settling period, there will be a separation of the mixture into two major fractions with oil on top and water at the bottom. Then the oil fraction (called crude oil at this stage) is skimmed off and subjected to further purification via centrifugation and drying. The drying is performed under vacuum to avoid further degradation in the quality parameters of the oil especially the free fatty acid content. The dried oil is then cooled and pumped to storage tanks or other suitable containers.

However, depending on the surface loading and retention time (90–120 h) this procedure exhibits separation efficiency as low as 50%. This means that the top fraction (the oil) still contains high concentration of debris or the bottom debris contains high percentage of oil. When the separation is about 50% then the lower layer from the clarification tank is sent to the centrifugal separator where the remaining oil is recovered. The oil from this second separation is also dried in vacuum dryers, cooled, and sent to storage tanks.

More efficient equipment for separating the debris from the mother liquor involves the use of decanters. The decanter, available in various designs, requires frequent maintenance from wear and tear, which ultimately results in high overall maintenance cost compared to the settling tank method. However, in some mills it is still being used in spite of this cost-related drawback.

Othman & Ng (2003) had estimated that the oil content in the sterilizer condensate and sludge/decanter is about 0.16% and 0.46% of the fresh fruit bunch (FFB), respectively. These two effluents are normally mixed together prior to discharge into a pond or other collection systems, depending on the disposal method available to the processor. This discharge effluent is generally known as palm oil mill effluent (POME). According to Ho et al. (1992), the residual oil, 10% (w/v) which is strongly associated with the POME (sludge), is discharged alongside the sludge. In view of the enormous volume of sludge (3.8 cubic metre per ton of crude palm oil produced) in the milling process, this represents a considerable oil loss. In order to minimize the oil loss during the milling process, recovery of oil from POME must be properly studied and implemented.

Palm Kernel Crushing Processes

The crushing of palm kernel begins with the kernel production from the fiber/nut pressed fiber separation. The press fiber from the digester is fed to a vertical column (depericarper) where air is channelled to lift the fiber, thus separating the fiber from the nuts. The processing chain of palm kernel oil (PKO) is shorter than the palm oil processing chain. Fig. 11.11 shows a schematic diagram of the palm kernel processing methods. Like palm oil, palm kernel oil loss begins in the plantation. This is because

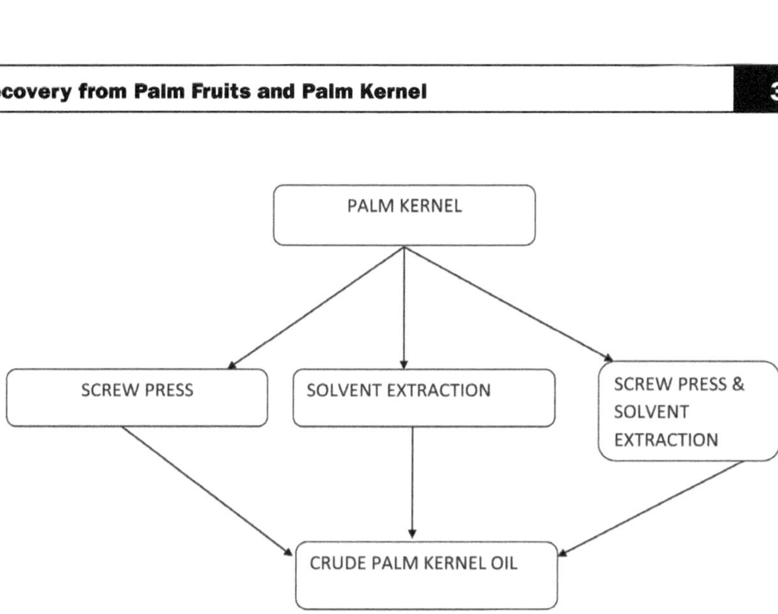

Fig. 11.11. Flow diagram methods of palm kernel oil extraction.

any fruit not picked by the harvesters during harvesting amounts to the loss of PKO bearing nut to the bush. Again, whole palm fruits are also lost during transportation to the mills, sterilization, and stripping and fiber nut separation. These losses have not been studied and estimated with respect to PKO. When the size of the plantations is considered, the loss may be appreciable.

Nutcracker

The separated nuts are passed to a polishing drum at the bottom of the depericarper, where pieces of stalks are removed. Fresh kernels have a moisture content of about 20% and cannot be stored without deterioration (moisture content of above 7% can encourage hydrolysis of the oil through autocatalytic and by lipolytic microorganism) (Coley & Tinker, 2003). A nutcracker cracks the nuts after the conditioning and drying process. A ripple mill is also used instead of nut cracker. Then the mixture of cracked nuts and shells are separated via a winnowing system, followed by a hydrocyclone or a clay bath. The clay bath separation technique is in the realm of traditional method while modern mills almost exclusively use a hydrocyclones for separation even though the clay bath system is cheaper (Corley & Tinker, 2003). A hydrocyclone uses centrifugal force to separate the kernel from the shell using water. The clay bath principle works on the specific gravity of kernel of 1.07 g per milliliter and the shell of 1.17. The kernels will float while the shells sink in a clay bath mixture of specific gravity 1.12. The kernels are then dried in hot air silos to moisture content of less than 7%. About 0.4 mt of kernels are produced with every metric tonne of crude palm oil (CPO) (Wai-Lin, 2011). The sources of oil loss in this process are due to incomplete cracked, un-cracked, and non-recovery of cracked nut.

Nut Crusher and Extractor

Palm kernel, the by-product of the oil palm industry, has a great potential for a source of oil and dietary protein. The extraction procedure of oil from the palm kernel is distinct from the palm oil extraction. The palm kernel is the true seed of the palm fruit, hence it shares similar properties and processing or milling equipment with oil seeds such as shea nuts, groundnuts, rapeseed, cottonseed, or coconut copra.

The oil extraction from the kernel is achieved by different methods (Fig. 11.11) under different settings, with emphasis on the cost and safety aspects. The principal methods include screw pressing, solvent extraction, and a combination of the former and the latter (screw pressing followed by solvent extraction). The stages involved in the palm kernel extraction is made up of grinding the kernels into smaller particles, heating (cooking), and extracting the oil using an oilseed expeller or petroleum-derived solvent (the use of volatile solvents is under threat as was highlighted in "Modern and Traditional Methods of Oil Recovery from the Retaining Components, Solvent Extraction" section). The palm kernel extraction is a fully fledged industry comprising many stakeholders, both at national and international levels globally along the value chain. The required equipment for the extraction of palm kernel oil in various sizes and capacities (from 10 kg to several tonnes per hour) are readily available both in the local and international markets.

Solvent extraction is very expensive to venture into and must be large capacity. According to Johnson (2008), processing plant using direct solvent extraction (soybean) needs to be capable of processing at least 2700 metric tons per day to be commercially viable in the United States, representing an investment of over 75 million USD.

Another method of extracting PKO from the cracked and separated nuts involves the use of oil presses. There are many types of oil presses in the engineering market today, but all work on a similar principle. The oil press consists of a heavy perforated or slotted metal cage and a plunger while a screw press/oil expeller consist of a horizontally-rotating screw contained in a barrel-shaped outer casing with perforated walls. A screw press is a continuous screw auger designed to subject the oil-rich material to increasing pressure as it is conveyed through the barrel. The barrel forms a cage of bars surrounding the screw in a parallel fashion. The bars of the cage are separated by spacers, which allow the oil to drain from the cage while the solids are conveyed down the barrel toward higher pressure. The discharge opening for the reduced-oil solids and, thus, the back pressure are controlled by a choking device. A plug of compressed oil-lean solids, termed *cake*, forms at the discharge. The pressure increases down the length of the barrel by increasing the root diameter of the screw, decreasing the pitch of the screw flights and controlling the cake discharge opening by means of a choking device (Johnson, 2008).

In the oil expeller, kernel is continuously fed to the expeller which grinds, crushes, and presses out the oil as it passes through the machine. The oil flows through the perforated casing and is collected underneath while the residue (oilcake) is pushed out of the end of the unit (Fellows & Hampton, 2000). The extracted oil is classified as crude PKO because it still contains some extraneous materials such as seed debris and minor chemical constituents like gum, hence requiring clarification in a filter press or by sedimentation.

Settling and Clarification Tank

The crude PKO is pumped or flows by gravity to a sedimentation basin and is either allowed to stand undisturbed for a few days and the upper layer is removed by the use of clarifier (Fellows & Hampton, 2000). The settled foots is recycled to the press to recover the entrained oil.

Traditional Milling Process

The traditional milling processes of FFB in Africa have been extensively and comprehensively researched by the FAO (2007). The changes in the traditional milling process since 1995 is not significant apart from the fact that more processors are including or introducing some levels of mechanisation along the processing chain.

Sterilizer

There are different ways of sterilizing palm fruits in the rural village setting in Africa. Generally, the bunch stripping process comes before the sterilization in the village milling process. This is because the bunches are first quartered and allowed to stay for a few days (see the "Sources of Oil Loss Along the Palm Fruit and Palm Kernel Processing Chain, Bunch Stripper" section). The sterilization process where it is practiced is performed by boiling the loosened fruits in pots or drums, whichever is applicable. In some places the loose fruits are packed in translucent polyethylene bags and placed under the sun for 3 to 5 days. The aims of these treatments are very similar to those in the industrial processing discussed earlier in this chapter. A colander, basket, or a vessel with fine perforated holes in the bottom is used to filter out fiber and nuts. The wet mixture is then put on the fire and brought to a vigorous boil. After about 1 or 2 hours, depending on the volume of material being boiled, the firewood is taken out and the boiled mixture allowed to cool. Herbs may be added to the mixture at this point just before reducing the heat.

Bunch Stripper

This is achieved by manually quartering the palm fruit bunches individually with the aid of a cutlass. The quartered bunches are packed in a heap in a corner of the compound and covered with banana leaves or palm fronds or polythene material. The covering material depends on availability and not any other considerations or objectives.

Digester/Extractor

The loose fruits are pounded in a mortar with a long heavy pestle either in the compound or in a central location in a particular village. The kernels are then separated from the fibrous mash. This is then scooped into a cylindrical hollow press. The wrench is then turned slowly and gradually, as this is being done, the extracted oil flows through the perforated holes in the press and guided through a duct at the bottom of the press into a large bowl, trough or container. This process is carried out several times until oil is drained from the marshy mixture. Where there is no press, the kernels are manually separated from the digested mass. The digested mass is then transferred to a large frying pan heated and stirred occasionally with a hand paddle for about one hour depending on the intensity of the heat. The oil is squeezed out between palms from a handful of the hot mass in batches into a clean pot.

Clarifier and Dryer

There is no definite clarifier or dryer used in the rural village. The extracted oil is heated to reduce the moisture content allowed to stand overnight for the debris to settle at the bottom of the container into which the oil is poured. Thereafter the clear oil is decanted into a storage container from where the oil will either be sold or used for domestic cooking. The sludge at the bottom of the pot still contains some amount of oil which has not been estimated.

Traditional Palm Kernel Crushing Processes

In Africa, the traditional palm kernel extraction is not as wide spread as the palm oil processing. In Nigeria, often the palm kernel processors may not necessarily be palm oil processor. This is because palm kernel is generated in small batches usually at family levels which may not economically feasible to process into PKO. In this case they may either use the kernel as fire wood, cooking or selling to prospective buyers. The stages involved in the traditional processing of palm kernel included cracking, separating, and extraction. The cracking of the kernel is done manually between two stones. The current practice now is that a prospective kernel buyer takes a mechanical cracker from family to family or from one palm oil processing site to another to crack the kernel at a fee or negotiated terms. The cracked nut and shell are moved to a convenient place in the compound where the nut is separated from the shell by hand picking, and depending on the quantity or the intention of the owner, the nut may either sold or processed into PKO.

Crusher

Prior to extraction by any methods, the kernel must be cleaned to ensure that the kernels are free from shell and stones, both of which will result in high wear and possible breakage of machine parts. They should be simply broken, aiming to keep the production of fines to a minimum. The kernels are preheated in an improvised sand bath in a container to about 60°C. In the situation where the improvisation is not possible, the kernels are warmed in the sun before processing. The extracted crude oil is allowed to settle overnight, keeping warm to avoid solidification and decanted. The yield on the basis of 50 kg nut is low 37% crude oil, 33% settled oil, and 60% cake. Where several batches are ran, it will then be more economical to combine sediments from different runs and recover as much oil as possible by subsequent standing and decantation.

Settling/Clarification Tank

The expelled oil invariably contains a certain quantity of fines and foots that need to be removed. The oil from the presses is drained to a reservoir. It is then either pumped to a decanter or revolving coarse screen to remove a large part of the solid impurities. The oil is then pumped to a filter press to remove the remaining solids and fines in order to produce clear oil prior to storage. The cakes discharged from the presses are conveyed for bagging or bulk storage.

Overall, the major sources of the loss in the palm kernel production chain are during nut cracking and separation and nut crushing/solvent extraction. Solvent extraction of crushed palm kernel is used where crushing is done centrally and the PKO targeted. This method is used in Malaysia because of its large export market for PKO. The oil loss to the cake is 2–3%. In smaller markets like Nigeria where the PKO is for local consumption only, crushing and pressing is done on site (in the mill or plantation) using screw press. The screw press method, although cheaper and simpler to maintain and operate, is not as efficient as the solvent methods because it retains 8–13% of residual oil (Corley & Tinker, 2003).

Modern and Traditional Methods of Oil Recovery from the Retaining Components

From the discussion in "Sources of Oil Loss Along the Palm Fruit and Palm Kernel Processing Chain" section, it was demonstrated that oil is loss to some of the fruit components and processing equipment along the production chain. This means that the total oil biosynthesized and accumulated in the mesocarp and seed by the oil palm per hectare of land are not totally recovered during processing into oil. The non-recovery of the entire oil from the fruits or retention of the oil in some of the process co-products besides decreased income to the farmer may be a source of environmental pollution, as the case with POME. This section looks at how the retained oil is recovered from the retaining components by the millers/processors.

Pressing

With a small scale, which does not imply compromise in extraction efficiency, after digestion of the fruits and pressing with the screw press or any other pressing technique, the pressed fiber/nut mass is sent to the nut–fiber separator where the nuts are separated from the fiber. The fiber, which still contains high percentage of oil, is then collected and sent back to the screw press for further pressing. In the second pressing of the separated fiber, very high pressure is usually applied because there are no dangers of cracking the nut and contaminating the palm oil with the extraneous palm kernel oil any longer. This method is widely practiced in rural Nigeria where small processors abound, although some amount of oil is still retained even after the second pressing. In the large scale mills, the nut and fiber is pressed once and the fiber is separated from the nut. The resulting pressed fiber is used to fire the boilers.

Solvent Extraction

Another method of oil recovery from the retaining component of the fruit is via solvent extraction. For example the separated pressed fiber described in the "Modern and Traditional Methods of Oil Recovery from the Retaining Components, Pressing" section of this chapter can be extracted with a solvent rather than pressing a second time.

Again, oil can be recovered from POME by the addition of solvent such as food grade hexane or petroleum spirit with the aid of soxhlet extraction (Ahmad et al., 2008; Bassim et al., 2003). And because of the high solubility of oil in these highly volatile and inflammable non-polar solvents, it will extract more oil than pressing. However, the POME has to be pretreated by filtration and drying in order to remove water, and this will definitely add to the cost palm oil production. The health and safety considerations as well as huge investment in infrastructure, therefore, may be a huge disincentive to this method of oil recovery (Corley & Tinker, 2003), especially the rural processors.

However, other safer but high-cost and promising alternatives, such as the use of supercritical carbon dioxide, are currently being investigated for possible industrial oil extraction applications. In practice, carbon dioxide is compressed to >350 bar (>5,100 psi) to increase oil solvency. When the pressure is reduced in two stages (<50 bar), oil solvency declines, and the free oil can be trapped and removed, while the oil-lean carbon dioxide can then be recompressed and oil solvency restored for recycling back to the extraction vessel. This technology has been commercially used to decaffeinate coffee and extract hops for beer production and other flavorings. Unfortunately, no one has yet devised a practical means of getting large masses of feed solids into and out of the high-pressure vessel in a continuous manner, and consequently all supercritical fluid extraction, SFE processes thus far have been batch systems that are too expensive to apply to an inexpensive commodity like soybeans (Johnson, 2008). However in the case of oil recovery from the retaining components wastes such as POME, there may be a trade-off between the benefit to the environment and the cost of the techniques involved.

The major interest in the renewed investigations of supercritical fluids as a clean technology was due to the possibility of separation of multi-component mixtures such as palm oil and palm kernel oil. The use of supercritical fluids also helps to circumvent or meet the ever increasing stricter regulations and provisions on solvent residues in pharmaceuticals, medical products, and nutraceuticals. As a result of the preliminary success in the use of supercritical fluids (SCF), they are beginning to replace the organic solvents (hexane and methylene chloride) that are used in industrial purification and recrystallization operations due to regulatory and environmental pressures on hydrocarbon and ozone-depleting emissions (Marentis 2001; Mohamed et al., 2001).The application of supercritical fluid technology for the extraction and purification of oil has a number of processing advantages over conventional processing methods. These include low-temperature operation, inert solvent, selective separation, and the extraction of high-value product or new product with improved functional or nutritional characteristics.

The supercritical carbon dioxide extraction method used by Norhuda & Omar (2009) for the extraction of palm kernel oil from palm kernel is illustrated in the flowchart (Fig. 11.12). The various components of supercritical carbon dioxide extraction equipment consist of a carbon dioxide cylinder (MOX, Penang) with 99.99% purity of carbon dioxide, a chiller (Yih Der Bl – 730) to liquefy the carbon dioxide gas, high pressure 100 DX syringe pump with maximum operating pressure of 689.5 bar (Isco, Inc., Lincoln, NE), an SFX 220 extractor with size 22.7 cm by 21.2 cm by 24.4 cm (Isco, Inc., Lincoln, NE), equipped with a 2.5 mL extraction vessels, a heated capillary restrictor 50 μm with maximum operating temperature of 150°C use to minimize analyte deposition, and a 30 mL vial to collect the analyte. The SC-carbon dioxide technique of extraction involves less unit operations and is a fast process as opposed to the current conventional method. According to Norhuda & Omar (2009a), using SC-carbon dioxide method, extraction, fractionation, and separation of PKO in palm kernel could be done simultaneously. Norhuda & Omar (2009b) used the extraction apparatus to study the mass transfer of PKO under supercritical conditions and

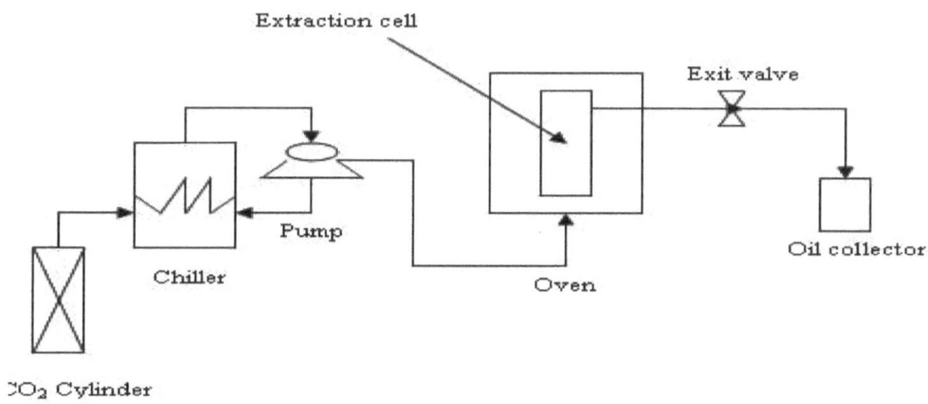

Fig. 11.12. Flow diagram of supercritical fluid extraction system. Source: Norhuda & Omar (2009a).

reported that within the system studied, the overall mass transfer coefficient of PKO obtained at supercritical conditions were found to increase with pressure. The overall mass transfer coefficient was found to be in the range of (1.21×10^{-4} to 1.72×10^{-4}) m per min at constant temperature of 50°C, in the range of (2.02×10^{-4} to 2.43×10^{-4}) m per min at constant temperature of 60°C and in the range of (1.77×10^{-4} to 3.64×10^{-4}) m per min at constant temperature of 70°C within pressure range from 27.6 MPa to 48.3 MPa.

Use of Nanotechnology

Sterilization process causes emulsification of oil and water and emulsified oil cannot be separated by mechanical means. Sovereign Innovation Sdn Bhd (S.I.), a Malaysian Bio-Tech Company in collaboration with eOx international b.v. of the Netherlands, has developed through research and development a Bio-Based Nano-Surfactant using palm-based oils and the successfully tested Bio-Surfactant solutions are being commercially used in the agricultural, fisheries, industrial, and pharmaceutical sectors locally and internationally.

The Bio-Surfactant was made from palm oil fatty acids and converted to nano particles (4–7 nm). The introduction of this bio-surfactant helps in breaking up emulsion layer and allows more oil to flow to the top. It also helps in reducing oil loss, which in turn helps in reducing biological oxygen demand, BOD, and chemical oxygen demand, COD, in waste. The use of the biosurfactant increased palm oil production by 0.5–0.75% (2–4 tonnes/day) in a typical 40 to 60 ton mill running for 15 hours a day, enhancing deterioration of bleachability index (DOBI), and reduces free fatty acid (FFA) values. It improves POME cleansing, reduced total solids (TS), BOD & COD levels, and retention time needed in anaerobic pond and improve biogas quality and economics.

The company had carried out tests in various mills to increase oil extraction at clarifier and at the same time cleanse the POME. These tests were conducted at the Jugra Palm Oil Mill Sdn. Bhd., Golden Hope West oil mill at Carey Island, KLK Tuan Mee Mill, Ijok, Yee Lee Mill, Bidor, Pt United Kingdom Indonesia Plantations, Medan and TelokSengat, Kota Tinggi. The Mills have a capacity of 50, 25, and 60 tons per hour, respectively. Trial results from Golden Hope, West Mills, and other mills consistently showed increased OER even when fruits with low oil content and less loose fruits, lower oil losses, higher oil recovery, and ultimately increased revenue. Generally, small-scale using machinery is as follows. The whole fruits bunches are sterilized by steaming or boiling in water. This operation destroys enzymes responsible for the formation of free fatty acids. The fruits are then separated from the bunch, either mechanically or by hand, and digested by stirring at high temperature. Digestion renders the mass homogenous, detaches the pulp from the nuts and makes the kernel removal easier. A mixture of oil and water is extracted from the mass using a curb press, a hydraulic press, or continuous low-pressure screw expellers. The mixture is allowed to stand and the oil separated from water by decantation.

Trapping and Skimming

Pressing is more applicable to oil recovery from the pressed fiber than trapping and skimming. Trapping and skimming involves the pooling together of all the liquid effluents from the processing stages in a suitable container such as a pond and allowed to stand for days, and then the oil component due to difference in density compared to water will float on the surface and skimmed off. Conventionally, POME is collected in pond (Fig. 11.3), and the oil is recovered by skimming the surface. This floating oil is hence skimmed off into another container and subjected for further purification. This method, however, recovers potentially free oil but not those retained intact or partially digested fruits.

Fermentation and Enzymatic Digestion

With the increased focus on the use of more environmentally friendly techniques in industries globally in order to minimize pollution and energy consumption, biotechnological methods are receiving greater attention. The use of biological methods to recover oil from POME is considered to be friendly to the environment.

Generally, the unit operations involved in the enzymatic oil recovery from POME consists of mixing/agitation, solid-liquid separation, centrifugation, oil purification, and the application of suitable temperature and pH. Chow et al. (1987) in their earlier work had shown that treatment of sludge with enzyme is able to degrade the remaining intact or partially macerated oil cell, hence freeing the entrapped oil. The freed oil is then dislodge from the matrix of the cell material by the use of surface active agent such as sodium dodecyl sulphate (SDS) (Ho et al. 1992). Fundamentally, in enzymatic treatment about 0.3 to 0.5% weight/volume of the suitable enzymes (mainly cellulase) are added to the sludge during the mixing stage at a suitable pH, temperature, and mixer speed. The optimal conditions for oil recovery from sludge according to Ho et al. (1992), at laboratory scale are digestion at a pH 4.6, temperature 30°C, time 5 h, and the enzyme concentration of 0.5% weight per volume followed by washing with 0.03 mol per dm cubed SDS solution. The economic viability of this technique at the commercial scale is not well established but has very high potential of application in other related areas other than mere oil recovery.

The commercial viability of the process is affected by the lengthy processing chain and additional equipment that is needed. For example, the sludge will be cooled from 90°C to ambient temperature, thereby increasing the retention time of sludge in the holding tanks for a day or two. This is one of the drawbacks of this recovery technique at the moment.

Tertiary systems have been developed as new systems to provide effluent treatment in a more sustainable manner, as well as achieving the standards required by regulatory bodies. These include biological sequencing batch reactors, bio-filtration systems, systems with high aeration rates, activated sludge plants with aerobic reactors, bio-flow polishing plants, as well as membrane bioreactors. More palm oil plantations are investing into these technologies to harvest the biogas for fuel, and re-use other biomass materials for fertilizers, bio-composite materials, etc.

In Situ Reactions

The quality of extracted oil is paramount and this directs the use to which the oil will be put to. However, there are instances when the process of oil recovery is not economically viable when compared with the revenue or utility derivable from it. *In situ* exploitation of the oil then becomes a veritable way of improving the economics of the oil recovery. By *in situ* exploitation, the authors refer to a situation where the resident (retained) oil is directly converted to other products. For example, if the recovered oil is intended to be converted to alkyl esters, using transesterification reaction technique, there will be no need to pre-extract the oil. This is because the retained oil can be transesterified with suitable alcohols and catalysts with the alcohol being both an extracting and a transesterifying agent. In this way the economics of oil recovery will be greatly improved because the process of pre-extraction is long, expensive, and labor intensive. Obibuzor et al. (2002, 2003a, 2003b) had demonstrated that *in situ* transesterification could be applied in oil recovery from palm fruit processing wastes at the laboratory scale. Transesterification involves the reaction between an alcohol and a triacylglycerol molecule in the presence of a catalyst acid, base, or enzymes. In order to circumvent the long process of oil extraction and purification prior to transesterification, *in situ* transesterification where the alcohol perform dual purpose will be more economical (compare solvent extraction with hexane).

Feeding Stuff for Animal

In the NIFOR, herd of cattle are currently maintained. The aim of keeping these animals is to help in the maintenance of the plantations by feeding on the grasses.

As discussed earlier in this chapter, about 20–30% of the palm fruit bunches that enters some the mills are rejected basically on the degree of ripeness, size, etc. in order to enhance extraction efficiency and quality of oil. These rejected fruits could be fed to the cattle, which will convert the oil into protein in the form of meat and the droppings as manure for the palms. The cattle can equally be useful in picking up some or all the fruit omitted by pickers in the plantation and treat them as the rejected fruits. In this way the oil is recovered not as oil but as meat.

Direct Combustion for Energy

Some of the oil retaining part could be converted to energy by combustion in suitable and environmentally friendly appliances. For example, the POME can be moulded into briquettes and used to cook with a suitable stove.

Major Challenges and Problems of Oil Recovery to the Oil Processors

In general, the essence of oil palm plantation establishment in countries where the climatic factor favor its growth is to generate palm fruits that will be milled into palm oil. The profitability and economy of this enterprise is dependent on the efficient extraction of the oil from the fruit. In other words, the oil extraction ratio (OER) of the processing/milling set up must be very high. Therefore one of the major challenges of oil recovery facing by the processor is in the area of availability and selection of processing equipment and methods that will extract approximately 100% of the oil in the feedstock at a competitive price at both the small and large scales.

Secondly, the processor, especially those using modern techniques, must consider the planting materials to be used in the plantation because the oil extraction ratio is closely correlated to the quality of the feedstock mix fed to the mill. For instance, the *Tenera* fruit (a cross between *Dura* and *Pisifera*) form yields more oil and are easily processed than the other forms *Dura* and *Pisifera*. Therefore, the maximization of palm oil yield by planting high-yielding oil palm varieties such as *Tenera* developed by breeding programs and the production of high quality palm oil are challenges to palm oil processors in various countries. Under ideal climatic conditions and good management, they are capable of producing in excess of 20 tonnes of bunch per hectare per year, with palm oil in bunch content of 25%. This is equivalent to a yield of 5 tonnes oil per hectare per year (excluding the palm kernel oil), which far outstrips any other source of edible oil.

Other challenges faced by the palm oil processor include alleged threats to the environment by poor management of generated wastes, high energy consumption, and increasingly high cost of energy in the global market, affordability of capital equipment and access to credit facilities, lack of proper policies especially in Africa, and minimization of oil loss along the entire process chain.

Developing strategies on how to reduce working capital and operating costs without compromising product quality especially in aspect of conveyance of the FFB to the mills is also a major challenge. The location of mills, especially the small mills, is important for increased accessibility in order to reduce fruit spoilage and consequent post-harvest losses.

Quality and Uses of the Recovered Oils

The quality of extracted palm oil determines the premium payable and the grades. Palm oil that has low free fatty acid (FFA), low moisture and dirt contents, and good bleachability index is considered in general terms to be of good quality. The palm fruit from which the oil is extracted is considered to be fragile because it is prone to microbial and enzyme attacks. These microbial and enzymatic activities on the fruits, contributes greatly to the lowering of the quality of the extracted oil, especially when it is prolonged.

Again, process equipment, conditions, methods, and number of process steps also affect the quality of oil. For example, palm oil recovered from POME or the pressed fiber will not be the same with one from the first press in the screw press given the same feedstock.

The quality of palm oil dictates and directs its use and the quality of palm oil recovered subsequent to clarification is lower. Broadly speaking, the high quality palm will find use in the edible oil industry while the lower quality oil will be used in the non-edible industry—candles, soap, biofuels, cosmetics etc.

Conclusions

The already competitive global oil palm industry is currently facing new and emerging environmental challenges, such as climate change that may affect its sustainability. The oil loss along all the palm oil and palm kernel recovery systems especially in the plantations is generally significant and has direct bearing on the overall production economy. The current relatively low milling efficiencies are affected by use of inefficient milling methods, particularly by the small/medium scale processors. The statistics of the oil loss during production and processing of FFB is very weak. There are no deliberate and concerted efforts to calculate the amount of oil loss in the entire production system, especially in the plantations, on an annual basis. The data available on oil loss is generated from the installed milling efficiencies, which decrease with age of the respective mills. The current extraction ratio (overall recovery efficiency) can still be further improved through nanotechnology and low power consuming enzyme strategies to enhance the competitiveness of the oil palm industry. Although the loss of fruits in the plantation is not estimated, the highest oil loss usually occurs at the clarification stage of the milling process, hence improvement in its efficiency by redesigning or modification will ultimately minimize the oil loss and raise the profitability of oil processing. The avenues to add value to the retained oil (*in situ* or pre-extracted) should be explored to meet some of the environmental protocols, policies, and provision.

References

Afshin, K.; Johari, B.E.; Haniff, H.; Desa, A.; Farah, S. Palm oil quality monitoring in the ripening process of fresh fruit bunches. IJAEST 2011, 4(1), 26–52.

Ahmad, A.L.; Chan, C.Y.; AbdShukor, S.R.; Mashitah, M.D.; Sunarti, A.R. Isolation of carotenes from palm oil mill effluent and its use as a source of carotenes. Desalination Water Treatment 2009, 7, 251–256.

Bassim, H.H.; Abdul, L.A.; Ng, A.H. Removal of residual oil from palm oil mill effluent using solvent extraction method. J. Teknologi 2003, 38(F), 33–42.

Boonyaprateeprat, W. Thai oil palm situation in globalization. www.gtz.de/de/.../enboonyaprateeprat-thai-oil-palm-2010.pdf. Accessed 23 April 2011.

Carrere, R. Oil palm: the expansion of another destructive monoculture: oil palm from cosmetics to biodiesel colonization. Hersilia Fonseca Edition. World Rainforest Movement: Uruguay, 2006; pp. 6–12.

Chow, M.C. Nature of oil in sludge discharge. Eng. News 1996, 41 (April–June), 4–5.

Corley, R.; Tinker, P. The products of the oil palm and their extraction. The Oil Palm, fourth edition. World Agricultural Series. Blackwell Science: UK, 2003; pp. 451–460.

Edem, D.O. Palm oil. Biochemical, Physiological, Nutritional, Hematological and Toxicological Aspects. A Review. Plant Foods Hum. Nutr. 2002, 57, 319–341.

Fellows, P.; Hampton, A. Vegetable oil. In Small-Scale Food Processing: A Guide to Appropriate Equipment. Chapter 8, ITDG Publishing: UK, 1992; pp. 40–45.

Ho, C.C.; Chow, M.C.; Ong, S.H. Recovery of residual oil from the centrifuge of a palm mill: Effect of enzyme digestion and surfactant treatment. J. Am Oil Chem. Soc. 1992, 69(3), 276–282.

Ilechie, C.O. Head Agric. Engr. Res. Div. NIFOR. Personal communication, 2011.

Johnson, L.A. Oil recovery from soybeans. In Soybeans: Chemistry, Production and Utilization; Johnson, L.A., White, P.J., Galloway, R., Eds.; AOCS monograph series on oilseeds, 2008, 2; pp. 331–374.

Menon, N.R. Advances in milling technology: evolutionary changes. In Strategic Direction for the Sustainability of the Oil Palm Industry; Pushparajah, Ed.; Corp Soc. Planters: Kuala Lumpur, 2001; pp. 1–19.

Moahammad, R.; Ssrmidi, M.R.; Hesham, A.; El Enshasy, H.; Mariana, A.H. Oil palm: The rich mine for pharma, food, feed and fuel industries. American-Eurasian J. Environ. Sci. 2009, 5(6), 767–776.

Mohamed, R.S.; Mansoori, G.A. The use of supercritical fluid extraction technology in food processing. Business Briefing Food Tech. 2001.

Neoh, B.K.; Thang, Y.M.; Zain, M.Z.M.; Junaidi, A. Palm pressed fiber oil: A new opportunity for premium hardstock? Int. Food Res. J. 2011, 18, 769–773.

NikNorulaini, N.A.; Ahmad Zuhairi, A.; Muhamad, H.; Mohd Omar, A.K. Chemical coagulation of setileable solid-free palm oil mill effluent (pome) for organic load reduction. J. Ind. Technol. 2001, 10(1), 55–72.

Norhuda, I.; Jusoff, K. Supercritical carbon dioxide as a clean technology for palm kernel oil extraction. J. Biochem. Tech. 2009a, 1(3), 75–78.

Norhuda, I.; Mohd Omar, A.K. Mass transfer of palm kernel oil under supercritical conditions. World Academy of Science, Engineering and Technology. 2009b, 49, 169–172.

Obibuzor, J.U.; Abigor, R.D.; Okiy, D.A. In situ transesterification of *Raphia hookeri* mesocarp oil with lower aliphatic alcohols. La Rivista Italiana Delle Sostanze Grasse 2003, 80, 393–396.

Obibuzor, J.U.; Abigor, R.D.; Okiy, D.A. Recovery of oil via acid catalysed transesterification. J. Am. Oil Chem. Soc. 2003, 80, 77–80.

Okonkwo, E.U. Hazard analysis and critical control points in palm oil processing in Anambra State, Nigeria. African J. Agric. Res. 2011, 6(2), 244–247.

Othman, W.; Ng, S.B. OER and total oil losses. Palm Oil Eng. Bull. 2003, 66, 7–11.

Roger, M. Supercritical fluid extraction of nutraceuticals product. Paper presented at the 4th Brazilian Meeting on Supercritical Fluids EBFS, 2001.

Rozman, H.D.; Tay, G.S.; Abubakar, A.; Kuwar, R.N. Tensile properties of oil palm empty fruit bunch-polyurethane composites. Eur. Polymer J. 2001, 37(9), 1759–1765.

Rupilius, W.; Ahmad, S. Palm oil and palm kernel oil as raw material for basic oleochemicals and biodiesel. Eur. J. Lipid Sci. Technol. 2007, 109, 422–439.

Sovereign Innovations Sdn Bhd (S.I.) Malaysia and eOX international b.v., Netherland. SPEC; A bio-surfactant that enhances oil extraction & cleanses pome. A paper presented at a forum of large and small scale millers. June 2011.

Wai-lin, S. Edible oil processing—production palm oil milling process http://lipidlibrary.aocs.org/processing/palmoil/index.htm. (Accessed on 7 March 2011).

Yacob, S.; Shirai, Y.; Hassan, M.A.; Wakisaka, M.; Subash, S. Startup operation of semi-commercial closed anaerobic digester for palm oil mill effluent treatment. Process Biochem. 2006, 41, 962–964. http://www.fao.org/DOCREP/005/Y4355E/y4355e08.htm#TopOfPage. (Accessed 23 April 2011).

12

Palm Oil and Palm Kernel Oil Refining and Fractionation Technology

Véronique Gibon
Desmet Ballestra Group, Desmet Ballestra R&D, Minervastraat, 1, B-1930 Zaventem, Belgium

Introduction

Crude oils contain non-triacylglycerol components (free fatty acids, partial acylglycerols, phosphatides, metallic compounds, pigments, oxidation products, pesticides, glycolipids, hydrocarbons, sterols, tocopherols, waxes, etc.) that must be partially or totally removed in order to become acceptable for human consumption. Crude oils are therefore submitted to several treatments, the objective being to remove the objectionable minor components with the least possible damage to the oil fraction (*trans* fatty acids, polymeric and oxidized triacylglycerols, etc.) and minimal losses of desirable constituents (tocopherols, tocotrienols, sterols, etc.). There are two major processing methods, known as chemical and physical refining (Fig. 12.1). In chemical refining, the free fatty acids and most of the phosphatides are removed during alkali neutralization; during physical refining, the free fatty acids are distilled during deodorization (De Greyt & Kellens, 2000; Kellens & De Greyt, 2000). Advantages of physical refining lie in higher overall yield, use of less chemicals, and production of less effluent. The quality of the crude oil is a crucial parameter for selecting the most appropriate refining route; indeed, a wider range of undesirable products are better removed by an alkali treatment than by advanced degumming. The acidity of the crude oil, the ability to get rid of the soapstock, and local environmental legislation are other important parameters. In terms of direct refining costs, physical refining becomes interesting only when the acidity of the crude oil is sufficiently high. For crude palm and palm kernel oils with high initial free fatty acid (FFA) content and relatively low phosphatides, physical refining is preferred, and processing parameters can possibly be adapted for the best retention of desirable minor components like tocopherols and tocotrienols and the minimal production of unwanted *trans* fatty acids; however, chemical refining is still used at a limited capacity (Gibon et al., 2007). Some oils have a restricted applicability in their original form mainly due to their triacylglycerol composition; the growing demand for food products with specific physical properties has led to the development of chemical or physical modification technologies. The

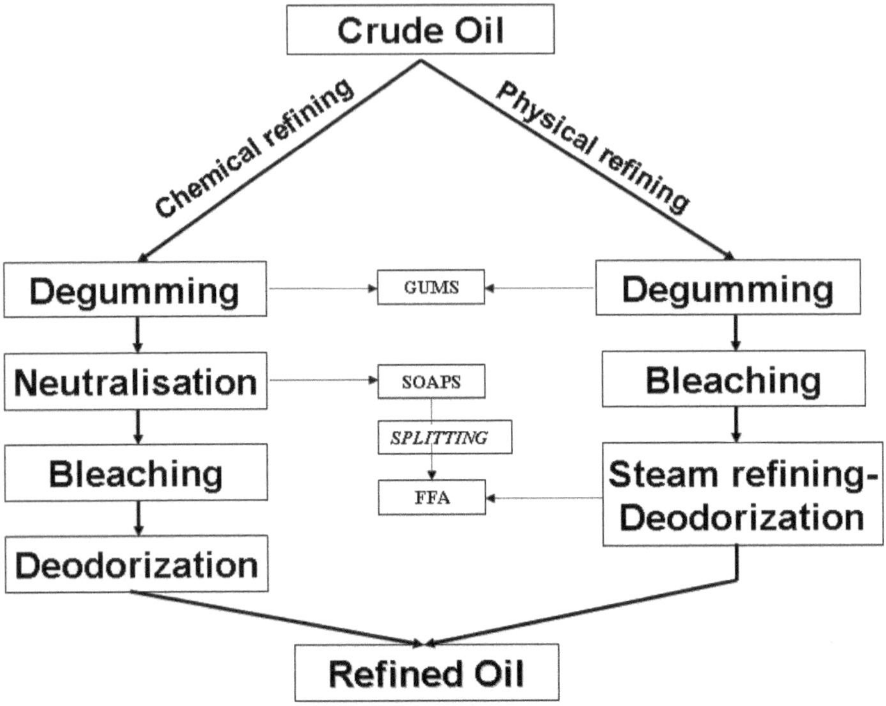

Fig. 12.1. General overview of chemical and physical refining lines. FFA: free fatty acids (Kellens & De Greyt, 2000).

best known strategies in the edible oil industry are hydrogenation, fractionation, and interesterification; however, the formation during partial hydrogenation of *trans* fatty acids and unnatural *cis* isomers, both suspected of having adverse effects on health (Lin Khor & Mohd Esa, 2007), can explain why there is today an increasing interest for fractionation and interesterification compared to hydrogenation (mainly partial hydrogenation). A large variety of food products can be formulated based on fractionated and/or interesterified fats. Due to several reasons like environmental issues and operating costs, dry fractionation and enzymatic interesterification (De Greyt & Dijkstra, 2007; Gibon, 2006; Xu et al., 2006) are now preferred to solvent/detergent fractionation and chemical interesterification.

Palm oil is by far the most important fractionated oil in the world (Kellens et al., 2007); crude, semi- and fully refined palm oil can be fractionated in multi-stage, giving access to several products for specific applications [margarines, shortenings, frying fats, cooking oils, salad oils, cocoa butter equivalents (CBE)]. Identically, palm kernel oil is a popular feedstock for the formulation of cocoa butter substitutes (CBS) after fractionation and full hydrogenation (deMan & deMan, 1994).

Compositional Properties of Palm and Palm Kernel Oils

The fruit of the palm tree *Elaeis guineensis* is the source of two distinctively different oil types. Palm oil is obtained from the mesocarp of the palm fruit; the outer pulp contains kernels that are the source of palm kernel oil. Even though they are derived from the same tree, palm and palm kernel oils considerably differ in their characteristics and properties. Both are known as tropical oils but palm kernel oil (like coconut oil) belongs to the class of lauric fats (Rognon & Wuidart, 1996; Wuidart, 1996). Palm oil is rich in C16:0 and C18:1 fatty acids while palm kernel oil has a high content in C12:0 fatty acid (Table 12-A); this acid is accompanied by significant proportions of other short and medium chain acids (C8:0, C10:0, and C14:0) but modest levels of C16:0 and C18 (C18:0, C18:1, C18:2) fatty acids. Palm kernel oil is indeed highly saturated compared to palm oil; the saturated fatty acid content is generally above 80% in palm kernel oil while it is only ~50% in palm oil. Mono-unsaturated fatty acids count for ~40% and ~15% and poly-unsaturated for ~10% and ~3%, respectively, in palm and palm kernel oils. This typical composition of palm kernel oil is reflected in its iodine value (IV: 16–19), which is significantly less compared to palm oil (IV: 51–53). In consequence of high acidity in the crude oils, both palm and palm kernel oils have relatively high diacylglycerol levels. While this level can easily be

Table 12-A. Fatty Acid Composition (FAC) of Palm and Palm Kernel Oils.

FAC (% by gas chromatography)	Palm oil	Palm kernel oil
C8:0	N.D.	3.6
C10:0	N.D.	3.5
C12:0	0.2	47.8
C14:0	0.9	16.3
C16:0	44.9	8.5
C16:1	0.1	N.D.
C18:0	4.1	2.6
C18:1	40.6	15.3
C18:2	9.0	2.4
C18:3	0.2	N.D.
SFA	50.1	82.3
MUFA	40.7	15.3
PUFA	9.2	2.4

SFA: saturated fatty acids; MUFA: monounsaturated fatty acids; PUFA: polyunsaturated fatty acids; N.D. not detected.
Source: Desmet Ballestra Group (2011).

determined in palm oil, accurate results are difficult to obtain for short chain fats like palm kernel oil (and coconut oil), due to partial peak overlap when the official gas-liquid chromatography method is used (Petrauskaite et al., 2000). The diacylglycerol content of palm oil is ranging from 4–5% in some premium oils to 10–11% in the worst cases; typical values in crude palm oils processed in Europe are 7–8%.

In consequence of completely different fatty acid profiles, the triacylglycerol composition of palm and palm kernel oils also greatly differs. Expressed as carbon number (Fig. 12.2), the distribution is narrow in palm oil (Pantzaris, 1997), with a majority of C50 and C52 triacylglycerols (>80% in total); on the contrary, there is a larger distribution in palm kernel oil (deMan & deMan, 1994) with C36 and C38 triacylglycerols contributing ~40% in total. Table 12-B reports the typical triacylglycerol composition of both oils (Chen et al., 2007; Desmet Ballestra Group, 2011). The trisaturated triacylglycerols (StStSt) are 10 times more abundant in palm kernel oil and mainly made of lauric fatty acids (LaLaLa), while it is made of palmitic fatty acids (PPP) in palm oil. Palm oil is an important source of POP (dipalmito-olein), which is the most abundant triacylglycerol (~30%) in total; the most abundant triacylglycerol in palm kernel oil is LaLaLa (~22%) directly followed by LaLaM (17%).

In consequence of typical composition, the physical properties of palm and palm kernel oils are completely different; the solid fat content profiles and the melting

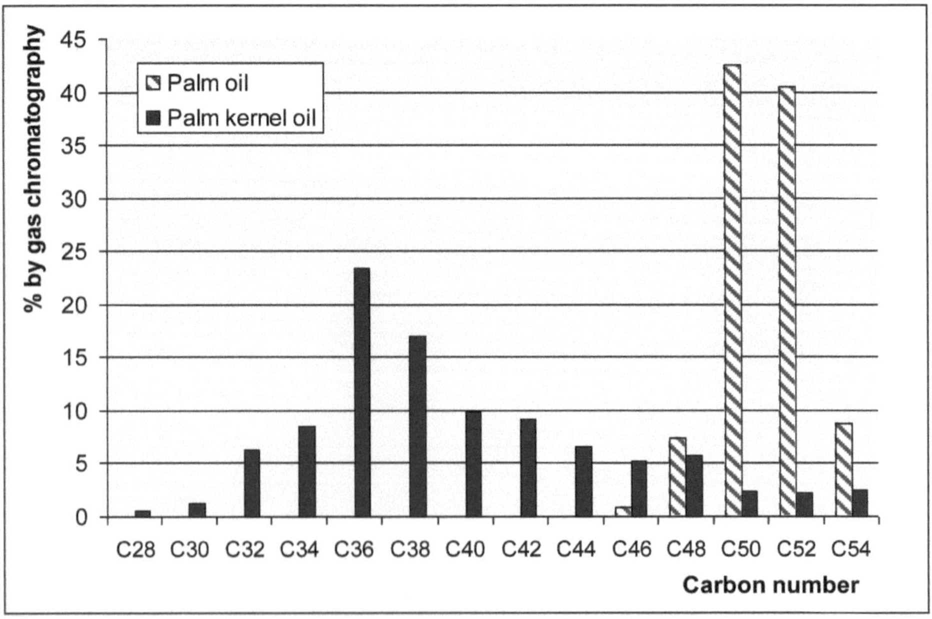

Fig. 12.2. Triacylglycerol composition (carbon number by gas chromatography) of palm and palm kernel oils (deMan & deMan, 1994; Pantzaris, 1997).

Table 12-B. Triacylglycerol (TAG) Composition of Palm and Palm Kernel Oils.

TAG composition (% by HPLC)	Palm oil	Palm kernel oil
CCLa	N.D.	6.1
CLaLa	N.D.	9.0
LaLaLa	N.D.	21.8
LaLaM	N.D.	17.0
LaLaO	N.D.	5.8
LaMM	N.D.	8.9
LaLaP	N.D.	0.8
LaMO	N.D.	5.0
LaPM	N.D.	4.6
LLO	0.4	N.D.
LaOO	N.D.	3.8
PLL	2.5	N.D.
MLP	0.8	N.D.
OOL	1.9	N.D.
LaOP	N.D.	4.6
MMO	N.D.	N.D.
LaPP	N.D.	2.4
MPP	0.4	N.D.
MOO	N.D.	1.9
POL	10.7	N.D.
PLP	10.1	N.D.
MOP	N.D.	2.1
OOO	4.1	2.0
POO	23.5	2.0
POP	29.6	1.1
PPP	6.0	0.3
SOO	2.6	0.5
POS	5.4	0.3
PPS	1.2	N.D.
SOS	0.8	N.D.
StStSt	7.6	70.9
StUSt	46.7	18.9
StUU	39.3	8.2
UUU	6.4	2.0

C: capric acid; La: lauric acid; M: myristic acid; P: palmitic acid;
S: stearic acid; O: oleic acid; L: linoleic acid; St: saturated fatty acid;
U: unsaturated fatty acid; N.D. not detected.
Source: Chen et al. (2007), Desmet Ballestra Group (2011).

Table 12-C. Solid Fat Content (SFC) Profile by p-NMR of Palm and Palm Kernel Oils.

SFC (%@°C)[1]	Palm oil	Palm kernel oil
0	58.1	73.0
5	60.9	68.0
10	52.9	64.3
15	42.2	54.2
20	28.9	37.3
25	17.6	13.3
30	10.2	0.2
35	5.4	0.0
40	1.4	
45	0.0	
Slip melting point (°C)	36.0	27.5

[1]According to IUPAC 2.150 serial method without special pre-treatment.
Source: Desmet Ballestra Group (2011).

points are presented in Table 12-C (Desmet Ballestra Group, 2011). Palm kernel oil is significantly harder at low temperatures (below 20°C) but quickly melts above; its melting point is therefore 10°C below the one of palm oil.

Minor Components in Crude Palm and Palm Kernel Oils

The minor components in crude palm oil have extensively been reviewed by Berger (2000). Beside partial acylclycerols and free fatty acids (FFA), the remaining components consist in a multitude of chemical entities with actual or potential commercial value, known as phytonutrients. Crude palm oil contains about 1–2% of valuable molecules that are mostly beneficial to health (Nesaretnam, 2009). These phytonutrients include tocopherols, tocotrienols, carotenoids, sterols, squalene, co-enzyme Q, polyphenols, and phospholipids (Table 12-D). In over-ripe fruits or during harvesting, a very active lipase, most probably originating from yeast cells, will be responsible for increased production of FFA and partial acylglycerols. The acidity of crude palm oil is on average about 3.5%. There is no strict correlation between the free fatty acid content and the amount of partial acylglycerols; diacylglycerols found in crude palm oil are also residual by-products of the biosynthesis of triacylglycerols (Jacobsberg & Oh, 1976). While the content of monoacylglycerols is generally low (< 1%), the amount of diacylglycerols is typically 4–8%, mainly palmitoyloleyl (PO), dipalmitoyl (PP), and dioleylglycerols (OO); 1,3-diacylglycerols predominate on 1,2-diacylglycerols in a proportion

Table 12-D. Major Phytonutrients in Crude Palm Oil.

Phytonutrients	Concentration (ppm)
Tocopherols/tocotrienols	600–1000
Carotenoids (α-carotene, β-carotene, lycopene, phytoene)	500–1000
Phytosterols (sitosterol, stigmasterol, campesterol)	300–620
Squalene	250–800
Phospholipids	20–100
Co-enzyme Q10/ubiquinones	10–80
Polyphenols (phenolic acids, flavonoids)	40–70

Source: Nesaretnam (2009).

of about 2:1 (Goh & Timms, 1985; Siew & Ng, 1995). In commercial crude palm oil, total phosphorus is reported present at ~20 ppm (Goh et al., 1982). However, detailed studies showed that most of the phosphorus determined in the crude oil is in fact inorganic phosphate rather than from phospholipids. This inorganic phosphate is arising presumably from the degradation of phosphatides by phospholipases and further chemical transformation (Goh et al., 1984). Major components of phosphatides are phosphatidylcholine, phosphatidylethanolamine, phosphatidylinositol, and phosphatidylglycerol. Typical fatty acid composition of phosphatides is palmitic (~38%), oleic (~37%), and linoleic (~32%). Glycolipids, although present at larger amounts than phospholipids, do not appear to contribute too much to the quality of the crude palm oil. Monogalactosyl diglycerides are reported as the most abundant while other glycolipids like steryl glycosides, esterified sterol glycosides, diglycosyldiglycerides, monoglycerosyldiglycerides, and cerebrosides are also mentioned (Berger, 2000). The total sterol content in crude palm oil is 300–600 ppm. Phytosterols are present, sitosterol, campesterol, and stigmasterol being the major constituents while a minor amount of cholesterol is observed (Goh et al., 1985). Sterols are observed as free, as esters, as free sterol glycosides, and as acetylated sterol glycosides. Methylsterols and mainly triterpene alcohols are present at concentrations of 800 ppm. Isoprenoid alcohol is ~80 ppm. The content in squalene, the major hydrocarbon in crude palm oil, is 200–500 ppm, and the content in non-terpene hydrocarbon is 30–50 ppm. Crude palm oil also contains 10–80 ppm of ubiquinone-10 and 5 ppm of ubiquinone-9, which are related to vitamin K and have some anti-oxidant properties (Berger, 2000). Tocopherols and tocotrienols are detected at 600-1000 ppm levels (Goh et al., 1985). The main constituents are ~44% γ-tocotrienols, ~22% α-tocopherols and ~12% δ-tocotrienols, the rest being α- and β-tocotrienols and β-, γ-, and δ-tocopherols. Tocopherols are important antioxidants which also function as vitamin E. Vitamin E activity is generally expressed as α-tocopherol + 0.25 β/γ-tocopherol + 0.01 δ-tocotocopherol, meaning that α-tocopherol has the higher vitamin E activity. Tocotrienols are powerful anticarcinogenic substances and help against thrombosis. Tocopherols and tocotrienols can reduce cell damage caused by toxic substances and environmental pollution, which are known to speed up the development of certain diseases (Sundram et al., 2003). As its deep red color testifies, crude palm oil is a rich source of carotenoids components (500–2000 ppm) (Goh et al., 1985). β- and α-carotenes are the major constituents with about 90% of the total (Ooi et al., 1994); the rest are γ–carotene, lycopene, and xanthophylls. The highest pro-vitamin A activity is due to β-carotene, which is strongly implicated in the prevention of some cancers. Unfortunately, most of the carotenoids are destroyed during the refining process and for that reason various methods for carotenoids recovery from crude palm oil are described.

Phytonutrients are definitely less abundant in crude palm kernel oil. The typical free fatty acid content is ranging between 2.0 and 5.0%, with on average values around 3.0%. Partial acylglycerols are present but quantity is difficult to estimate. The phosphatide content is relatively low, generally below 1–2 ppm as phosphorus. A normal content of few ppm only of carotene can be expected; however, there is a chance of carotene penetrating the thinner shells of certain kernels or by breakage of kernels so that this normal content can sometimes be exceeded (Pritchard, 1983). Identically, the tocopherol and tocotrienol content is usually very low (30–100 ppm) in crude palm kernel oil with a majority of α-, β-, or γ-tocopherols and possibly some tocotrienols. The sterol content is higher (600–1200 ppm) and like in crude palm oil, β-sitosterol is the most abundant (~60%) (Gordon & Griffith, 1992).

Quality Parameters of Crude Palm and Palm Kernel Oils

Important aspects of the quality of crude oils have to be considered in that they affect the refining process (effect of ripeness, storage, transportation, etc.) (Gibon et al., 2007). The chemical and physical properties of crude palm and palm kernel oils have been thoroughly determined from several surveys and the data incorporated in standards (Tables 12-E and 12-F). Freshly expelled crude palm oil, promptly dried and cooled, will show low hydrolysis and oxidation. In some cases, crude palm oil can be produced at less than 2.5% FFA and 0.15% moisture; at this level, both hydrolysis and oxidation are slowed (Berger, 1983). Overall, a FFA content below 3% gives some guarantees that fresh and unbruised fruits have been used and that the crude oil was stored and transported under good conditions. Too high acidity levels together with high iron and copper (picked up during mechanical wear and corrosion at the oil mill, storage, or transportation) will lead to too high oxidation parameters (Willems & Padley, 1985). Values as low as peroxide value (PV) of 4 meq O_2/kg, anisidine value (AnV) of 4, iron of 3 ppm, and copper of 0.1 ppm are desirable in crude palm oil. Copper is particularly unwanted as very few quantities are capable of encouraging oxidative

Table 12-E. Quality Standards for Crude Palm Oil.

	Special quality grade	Standard quality I	Standard quality II	Special grade	Lotox	Standard	Contractual
FFA (% palmitic max.)	2.5	3.5	5.0	2.5	2.5	3.5	5.0
Moisture and impurities (% max)	0.25	0.25	0.25	0.22	0.22	0.22	Basis pure
Peroxide value (meq O_2/kg max.)	2.0	–	–	3.0	3.0	5.0	Basis pure
Anisidine value (max)	4.0	–	–	4.0	3.5	–	–
DOBI (min.)	2.8	2.5	2.2				–
Carotene (ppm max)				–	600–700	–	–
Fe (ppm max)				4	4	5	–
Cu (ppm max)				0.02	0.2	0.2	–

DOBI: deterioration of bleachability index.

Table 12-F. Quality Standards for Crude Palm Kernel Oil.

	FOSFA 54	British standard	PORIM	PORIM data
	GMQ	Typical	Typical	–
FFA (% lauric max.)	5.0	6.0	–	–
Moisture and impurities (% max)	0.5	–	–	–
Lovibond color	–	1.5R/20Y (1″)	–	5.5R/50Y (5[1/4″])
Iodine value	–	13–23	16.2–19.2	17–20
Carotene content (ppm)	–	–	–	7–10

GMQ: good merchantable quality.

attack; it is considered 10 times more active as pro-oxidant compared to iron (Patterson, 1992; Swoboda, 1985). Refiners using good quality palm oil can easily fulfill the requirements of good quality refined oil based on FFA, iron, copper, and color. However, the TOTOX value (2* peroxide + anisidine value) seems to be a critical parameter; because transport conditions are governed more by commercial requirements than by quality requirements, the TOTOX can greatly increase during transportation obliging the European refiners to re-treat the already processed oil. Good quality crude palm oils of different origins show intense absorption peaks in the region of 450 nm (UV spectroscopy) due to high levels in carotenoids. Modest deviations from a precise absorption pattern are explicable by some variation in the carotenoids present in different species. If crude palm oil has suffered from damage, the height of the absorption peak is diminished and a displacement towards 450 nm can be observed (Patterson, 1992). Low grade crude palm oils may suffer auto-oxidation of the carotenoids giving rise to off-flavors; nitrogen blanketing of storage tanks during transportation retards oxidation. Antioxidants can be added to the crude oil (combination of TBHQ and citric acid), assuring a better color of the oil after refining and allowing the reduction of bleaching earth consumption because the amount of oxidized crude oil has been made substantially less. Deterioration of bleachability index (DOBI) is a good test to assess the quality of a crude palm oil (Lal & Gasper, 1991; Siew, 2001). DOBI is the ratio between the content of carotenes (absorbance measured at 446 nm) and secondary oxidation products (absorbance measured at 269 nm).

$$\text{DOBI} = \frac{\text{Absorbance at 446 nm}}{\text{Absorbance at 269 nm}}$$

An arbitrary scale was suggested (Patterson, 1992). DOBI = 3.25 and above (very satisfactory quality); DOBI = +/- 2.75 (average quality); DOBI < 2.0 (very poor oil).

However, other sources propose less severe classifications like DOBI above 2.3 (more easily refined), DOBI between 2.0 and 2.3 (unpredictable), and DOBI below 2.0 (difficult to refine). Because of its low degree of un-saturation, palm kernel oil has a good oxidative stability but on hydrolysis gives a soapy off-flavor. The acidity of crude palm kernel oil can vary, depending on quality, from 2.0 to 6.0%. Poor crude palm kernel oil may show 1.0R/10Y (Lovibond 1″) while good crude palm kernel oil may be no more than 0.4R/4Y (Lovibond 1″); this depends largely upon the way the kernels have been pressed and extracted, shipped, and stored (Patterson, 1992). From commercial practices, peroxide values of 7 meq O_2/kg and anisidine values of 3 maximum have been put forward (Pritchard, 1983); iron and copper contents should not exceed 2 ppm and 0.2 ppm, respectively.

Refining Practices for Crude Palm and Palm Kernel Oils

Crude oils are refined in order to remove all impurities such as undesirable odor, flavor, and color but at the same time retaining the beneficial components such as vitamins, pro-vitamins, and anti-oxidants. Refining can be operated according to two main routes: the chemical refining or the physical refining. The principal difference between these two routes is how the free fatty acids are taken off. In a physical operation, most of the free fatty acids are carried through the deodorizing unit; as distillation requires a higher temperature treatment, the crude oils must carefully be degummed and bleached. The physical refining process can offer important advantages such as better oil yield, reduction of use of chemicals (phosphoric acid, citric acid, caustic soda, and sulfuric acid), reduction of water and effluent, and considerable reduction of the environmental impact; unfortunately, bleaching earth consumption is usually higher. At the opposite, in a chemical operation, the oil is degummed and de-acidified by alkali pre-treatment and the deodorization is usually performed at lower temperature (Kellens & De Greyt, 2000). For crude palm and palm kernel oils, with high initial free fatty acids and relatively low phosphatides content, physical refining is preferred today, but chemical refining is still used mainly for some specific applications.

The Chemical Route

When the crude oil is refined according to the chemical route, free fatty acids and phosphatides are removed by alkali neutralization (Gibon et al., 2007). The non-hydratable phosphatides are first activated with acid and further washed out together with the free fatty acids during alkali neutralization with caustic soda. This acidic pre-treatment is necessary for crude palm oil which contains higher quantities of non hydratable phospholipids; most of the time, acidic pre-treatment can be skipped for crude palm kernel oil due its very low phosphatide content. It is assumed that caustic soda converts the non-hydratable phosphatides into sodium salts, which are less oil soluble. The free fatty acids are converted into sodium soaps which improve, in combination with intense water washing, the efficiency of phosphatides removal. However, the drawback of this caustic pre-treatment is the production of high quantities of soapstock which needs to be treated accordingly. Practically, the non-hydratable phosphatides are activated by adding 0.03% to 0.1% (w/w) of concentrated phosphoric acid to the crude oil. In order to avoid emulsion, the quantity of caustic solution is calculated based on the initial free fatty acid content with some excess (10–30%) and a 20–22° Bé caustic solution is thoroughly mixed to the oil at 90–95°C. The contact time is generally short (~30 seconds) to prevent too high saponification losses. The mixture is delivered to centrifuges where it is separated into a light phase (the neutralized oil still containing traces of soaps, phosphatides, free caustic, and other soluble impurities) and a heavy phase (the soapstock). The total fatty matter of diluted soapstock is ~10–15%. The light phase is further washed (~90°C) and separated with a second centrifuge. The range of residual free fatty acids in the neutralized oil is usually between 0.08 and 0.25% and the soaps are below 50 ppm. The neutralized oil is consequently bleached before deodorization. The soapstock splitting for acid oil production is usually closely connected to the caustic treatment. Concentrated sulfuric acid is mixed to the soapstock; the reaction takes place in a decantation tank to break down the emulsion (due to the presence of phosphatides, proteins, and other mucilaginous substances). The acid water is removed leaving the acid oil to pass through a separator for the removal of the last traces of water and mucilaginous material. Extra washing is done in order to assure no residual sulfuric acid in the acid oil. The free fatty acid content of palm acid oil is usually 40–70%.

The Physical Route

When the crude oil is refined according to the physical route, the phosphatides are removed by a specific degumming process and the free fatty acids are distilled during the steam refining/deodorization (Gibon et al., 2007). The first step of crude palm oil and palm kernel oil physical refining is the degumming, which consists of the removal of the phosphatides. When this content is low, as is the case for these oils, the dry degumming is the common way. Classically, the hot crude oil is first mixed with 0.05–0.1% (w/w) concentrated phosphoric acid; after a short retention time, acid activated bleaching earth is contacted under vacuum at 80–120°C. After suitable contact time, the bleaching earth is removed by filtration. The phosphoric acid disrupts non-hydratable phosphatides by decomposing magnesium and calcium complexes; it coagulates the phosphatides and sequestrates iron and copper, which are all further adsorbed on bleaching earth. Citric acid, or, alternatively, a mix of phosphoric and citric acids, is sometimes preferred. The amount of acid and bleaching earth added to the oil is perfectly dosed in order to assure total removal of the phosphatides and to avoid residual traces of acid. Indeed, the excess of remaining phosphatides can break down at the deodorization temperature, leading to darkening and off-flavor problems

(Patterson, 1992); identically, the excess of phosphoric acid can promote darkening during steam refining/deodorization, probably due to phosphorylation with the triacylglycerols (Thiagarajan & Tang, 1991). Adsorptive cleansing by bleaching earth should reduce the phosphorus level to an acceptable limit (preferably < 2 ppm), lower the trace metal content (Fe < 0.1 ppm, Cu < 0.05 ppm), and minimize the oxidation products. Part of the carotenes are also adsorbed on the bleaching earth, but for palm oil the color after dry degumming alone is not so critical as the majority of the carotenoids are thermally bleached under the conditions of steam refining/deodorization. The wet acid degumming is sometimes preferred to dry degumming, especially when the phosphatides and iron contents in the crude oil are higher than expected. The hot crude oil is thoroughly mixed with phosphoric or citric acids; after a retention time of ~5–20 minutes, 2–5% of water is added before separation on a centrifuge. One property of phosphatides is that they form hydrates in contact with water and these hydrates are less soluble in oil. Phosphatidylcholine (PC) and phosphatidylinositol (PI) hydrate very fast while phosphatidic acid (PA) and phosphatidylethanolamine (PE) react considerably slower with water. This means that when the phosphatides in the crude oil contain about 55% PC, they all can be removed by fast hydration at elevated temperature. If the phosphatides contain only about 30–35% PC, fast hydration will give only partial removal of the slowly hydrating phospholipids. An acceleration of the rates of hydration of the calcium and magnesium salts of PA and PE can be obtained by addition of strong acid prior to hydration. It explains the varied performance of wet acid degumming according to the phosphatides composition in the crude oil. Soft degumming is an alternative to acid wet degumming to remove high phosphatides and iron from the crude palm oil (Gibon & Tirtiaux, 1998). The hot crude oil is thoroughly mixed with water (2–5%) containing a limited quantity of ethylene diamine tetracetic acid (EDTA). EDTA is forming strong chelates with magnesium, calcium, and iron, making hydratable phospholipids easier removal. After ~20 minutes contact time, the oil is separated by centrifugation. The wet acid or soft degummed oils are further bleached before steam refining/deodorization.

Bleaching Principles

The removal of minor components through adsorptive bleaching is based on several adsorption mechanisms (Gibon et al., 2007). Part of the coloring pigments (carotenoids in the case of palm oil) are physically adsorbed on the bleaching clays, involving "Van der Waals" surface attraction forces; other components are chemically bound to the bleaching clay surface via covalent or ionic bonds or removed through molecular entrapment in the porous structure of the clays. Some minor components are also chemically altered due to the catalytic activity of the clays. A typical example is the decomposition of the hydroperoxides into unsaturated conjugated products. Factors affecting the bleaching are temperature and humidity, structure (particle size distribution), and type (neutral or acid activated) of bleaching earth. The bleaching process is the most expensive step in refining when considering the utility costs: relatively high cost of the bleaching earths, oil in earth losses (30–45%), and spent bleaching earth disposal costs. Bleaching is usually performed in several stages. Prior to bleaching, the oil is properly de-aerated (to avoid oil oxidation and deterioration) and dried (to avoid hydrolysis) under reduced pressure (< 100 mbar) down to a moisture content of 0.1%. De-aerated bleaching earth is added to the oil and the whole is further heated under reduced pressure (~30 mbar) and high temperature (80–120°C) and intensively mixed (sometimes by injection of steam) to ensure intimate contact. The reaction takes place for a specific residence time (~20–30 minutes) after which it is transferred to a filter, usually a hermetic leaf filter (horizontal or vertical) with stainless steel mesh elements. The first quantity of oil that passes through the leaf filter is normally recycled back to the bleacher vessel, as it may still contain some bleaching earth. After polishing filtration, the bleached oil is diverted to the bleached oil storage tanks and eventually kept under nitrogen. The spent bleaching earth cake is flushed by steam or nitrogen to recover as much of the entrained oil from the cake as possible. Bleaching is usually performed in one single step, but in order to reduce the consumption, alternative multi-step bleaching processes have been developed; most known are the two-stage co-current and counter-current bleaching processes. The co-current process is the least complicated. Bleaching earth is added in two consecutive stages in the oil, with a filtration after every stage. In the counter-current bleaching process, the spent bleached earth from the second bleaching stage is re-used in the first stage. The efficiency of the counter-current process is largely affected by the way the filtration with spent bleaching clay is performed. The easiest way is to use a third filter operating in counter-current at the inlet of the bleacher. The pre-filtration of the oil through the spent bleaching clay cakes in the filter, built up during the second bleaching stage, already results in a saving of 10–15% bleaching earth as compared to single stage bleaching. Even more efficient than counter-current pre-filtration process is the Oehmi process which guarantees savings up to 40% (Transfeld & Scheinder, 1996). This process requires two bleachers and two sets of filters; the spent bleaching clay of the second filtration is mixed back into the degummed oil in a specific manner and subsequently filtered. For palm oil, bleaching earth consumption has been quantified with conventional single stage and counter-current processes. A bleaching earth consumption of 1% can be reduced to 0.8% with a conventional pre-filtration and up to 0.6% with the full counter-current system. Continuous bleaching units are often operating by means of the wet bleaching concept instead of conventional dry bleaching. With wet bleaching, the cleansing effect is more effective and the consumption of bleaching earths can be reduced. The presence of water increases the cation exchange capacity of the bleaching earth, which improves hydration of the non-hydratable phosphatides. The best performance is observed for amount of water about 50–100% of amount of bleaching earth, which is not always economically acceptable. In wet bleaching, the oil and bleaching earth are contacted under atmospheric pressure for 20 minutes at about 70–90°C, which is followed by the usual bleaching process under vacuum for 15–30 min. During this step, the excess water evaporates allowing the bleaching earth to be filtered without problems.

Steam Refining/Deodorization Principles

Deodorization is basically a vacuum steam stripping at elevated temperature during which free fatty acids and volatile odoriferous components are removed to obtain bland and odorless oil. Although the process is commonly named "deodorization", it is a combination of three different operations: distillation (stripping of volatile components), actual deodorization (removal of odoriferous components), and heat bleaching (thermal destruction of pigments) (De Greyt & Kellens, 2000; Gibon et al., 2007; Kellens & De Greyt, 2000). Optimal deodorization parameters (temperature, vacuum, and amount of stripping gas) are determined by the type of oil and the selected refining process (chemical or physical refining) but also by the deodorizer design. The stripping medium requirements are described by a mathematical equation derived from Dalton and Raoult laws.

$$\frac{S}{O} = \frac{P_t}{\mathrm{E}.P_i^0}.\mathrm{h}\frac{V_\alpha}{V_0} + \left(\frac{P_t}{\mathrm{E}.P_i^0} - 1\right)\left(V_\alpha - V_O\right)$$

S = Total moles of steam;

O = Total moles of oil;

P_t = Total pressure;

P_i^0 = Vapor pressure of a given fatty acid i;

E = Vaporization Efficiency;

V_α = Initial molar concentration of the volatile component in the oil;

V_0 = Final molar concentration of the volatile component in the oil.

Physical refining requires more severe conditions than chemical refining, as the initial free fatty acid content is considerably higher. It is important to make a distinction between odoriferous components already present in the crude oil and odoriferous components formed by thermal degradation of flavor precursors. Removal of the first group of compounds is quite similar to free fatty acid stripping and can be considered not as nearly time dependent. However, perfect deodorization is more complex and minimum time is demanded to avoid some flavor precursors remaining in the deodorized oil resulting in the development of off flavors during storage. Besides a bland odor and taste, a light color is also a quality parameter for most refined oils. In the refining process, the color components can be removed during the bleaching process by adsorption on bleaching earth or alternatively by thermal degradation during deodorization. This latter phenomenon, also known as heat bleaching, is purely time-temperature dependent and not affected by the deodorizing pressure. Literature data on heat bleaching of palm oil indicate furthermore that the rate of carotene breakdown doubles per 20°C increase in temperature. For illustration, heat degradation of carotene takes a few hours at 210°C but only a few minutes at 270°C (Maclellan, 1983). However, if the oil is not properly pre-treated (too high residual phosphatides, iron, and other impurities) or if traces of bleaching earth are still present, the oil can become darker during deodorization. This phenomenon is also known as color fixation since it is nearly impossible to remove these color pigments.

Latest Developments in Palm and Palm Kernel Oils Refining

The quality of fully refined palm and palm kernel oils is completely inseparable from the quality of the crude oils. As said before, the high acidity of crude palm and palm kernel oils places physical refining as the first option. The latest developments in physical refining technology have been driven by the increased attention to nutritional parameters that can be improved through optimized process conditions for a maximum retention of natural components. Due to the market demand for improving tocopherols and tocotrienols retention and minimizing *trans* fatty acids content, a dual temperature deodorization concept was developed (Gibon et al., 2007). This concept is fully integrated in the Qualistock FS design (Fig. 12.3) (Desmet Ballestra Group, 2011), which is a compact system where de-aeration, heat exchanges, gas stripping, deodorization, cooling, and scrubbing all take place in one single vessel. The trays are modular and tailor-made designs can be built up out of standard modules in order to suit the customer's requirements. In this system, the next possibilities are available: a low pressure drop structured packing (LPDSP), a dual temperature deodorization with the options low-high and high-low temperatures, and variable deodorization tray oil bed (shallow–deep) allowing variable residence times. The key processors of palm oil are interested in a more economical solution for stripping the high free fatty acids of palm and palm kernel oil, and the purpose of having a structured packing prior to deodorization trays is to allow this initial fast stripping; the low pressure drop LPDSP concept allows much less pressure drop compared to other strippers available on the market (Desmet Ballestra Group, 2011). Its surface is subdivided into a multitude of small lamellas and the overall surface is therefore subdivided into a large number of surface elements. In a dual temperature low-high system, the incoming oil is heated in the first trays to moderate temperature, after which it passes through the next deodorization trays. In these trays, further de-acidification and deodorization take place under mild conditions (moderate temperatures and moderate stripping). After the first trays, the oil passes a second heating tray in which it is heated to higher temperatures. In the last trays, final stripping and heat bleaching occur. Process parameters can be adjusted in order to arrive at a desired tocopherol and tocotrienol level in the refined oil. The dual temperature deodorizer shows a number of benefits which include low *trans* fatty acid formation (because time at high temperature is restricted), adjustable tocopherol and tocotrienol retention, and optimized heat bleaching effect. In this deodorizer, the sparge steam is introduced into the oil through special distributors that can be sparge coils with very fine holes or sintered

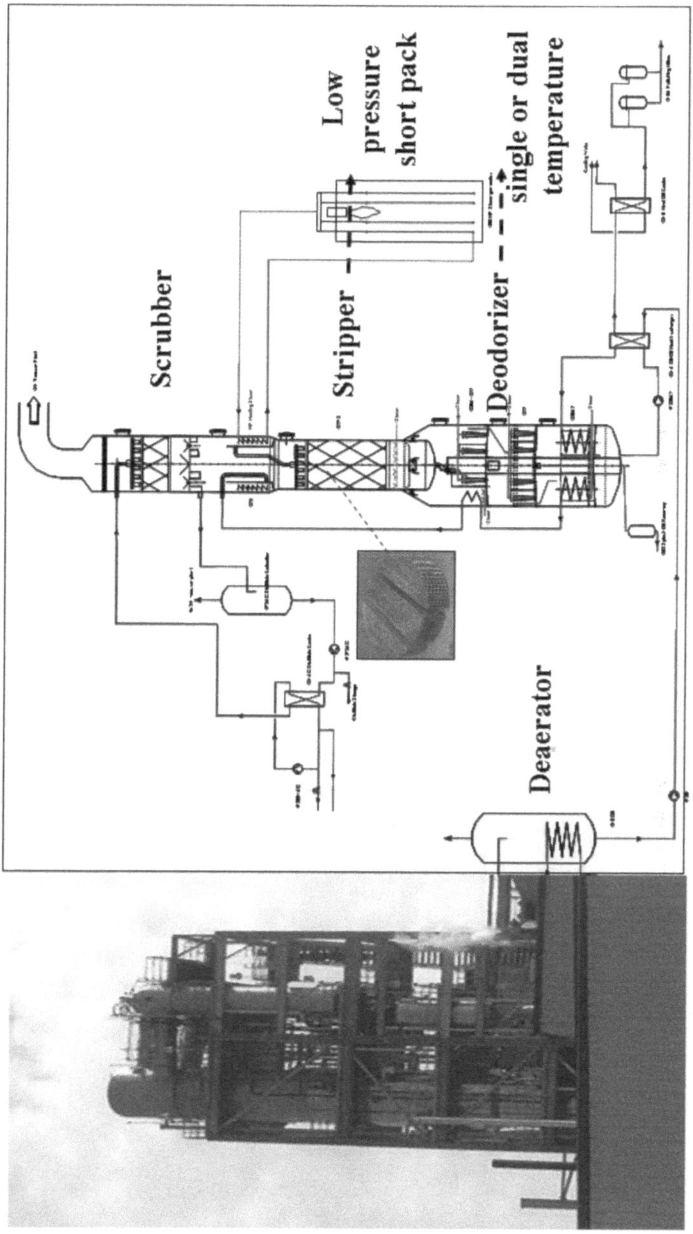

Fig. 12.3. General view of the Qualistock FS deodorizer (Desmet Ballestra Group, 2011).

metal pipes. Steam lift pumps have been introduced (Fig. 12.4) in order to improve agitation and overall deodorization efficiency by continuously refreshing the oil in the top layer (Desmet Ballestra Group, 2011). Steam enters the gas lift pump through a central tube and rises up the tube drawing oil with it. Oil is deflected downwards by a conical hat and steam carries away all impurities. Steam containing fatty matter comes down to a central chimney; fatty matter is condensed in a separate scrubber by spray of cooled fatty matter (distillate). The spray nozzles break up the flow of cooled fatty acid distillate into fine droplets that contact the vapor and condense the fatty acids. Vapors pass through a demister made from knitted wire pads, and droplets are collected on the wires. In some cases, an extra structured packing is installed in the scrubber to improve condensation and to reduce fatty matter carry-over to the barometric water condenser system. The dry condensing system is becoming more and more the standard in the refining plants. The Sublimax is a new generation vertical ice condensing system optimized for deodorizing duties and drastically reducing the energy consumption and pollution (Desmet Ballestra Group, 2001). It allows to easily reach pressures of 2 mbar or less instead of 3–4 mbar with the conventional ejectors. It is made of two shell and tube heat exchangers or sublimators that operate alternately. A film of liquid ammonia falls down the inside of the tubes and process vapors freeze on the outside. Two vessels are used alternately so that one is being cleaned while the other is icing (Fig. 12.5). The overall energy consumption of the ice condensing system is less than half of the chilled water system and less than 20% of the conventional steam ejector system. In addition to energy savings, ice condensing gives low deodorizing pressure, low volume of effluent, and less odor and emissions to the atmosphere. Counter flow is more and more used for heat recovery; oil flows up each set of coils, over the baffle into the next compartment, and then down to the bottom of the next set of coils (Desmet Ballestra Group, 2011). Each coil contains

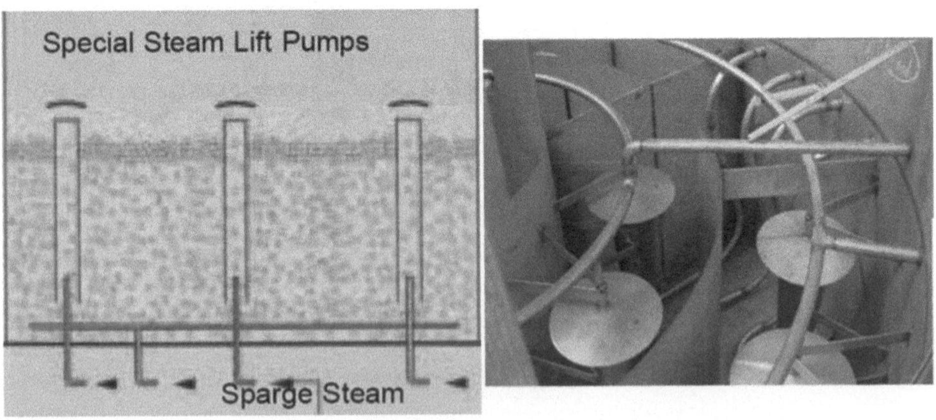

Fig. 12.4. Lift pumps for sparge steam distribution (Desmet Ballestra Group, 2011).

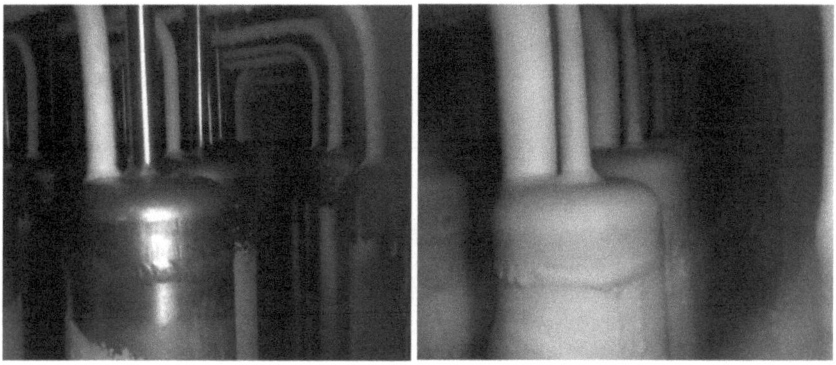

Fig. 12.5. Ice condenser system (Sublimax) with vertical ice condensing system: a) cleaned tubes b) frozen tubes (Desmet Ballestra Group, 2011).

either 2 or 4 parallel pipes depending on the capacity of the deodorizer (Fig. 12.6). About 80% of the heat required to raise the oil to the deodorizing temperature is provided by the heat recovery system. The rest has to be supplied by a high temperature heater. In the past low pressure organic fluids were used for high temperature heat transfer. Today, most oil processors have changed to high pressure steam because they are concerned about leaks contaminating the oil. The free fatty acid content of

Fig. 12.6. Heat recovery trays (oil/oil countercurrent) (Desmet Ballestra Group, 2011).

a palm deodorizer distillate is usually 80–85% when the oil is refined according to the physical process. If two condensing units are introduced in the scrubber section, each of them set at an optimized temperature, purity of the distillate can be improved. With this double condensing system, the content of acylglycerols and other minor components in the condensed fatty acids is reduced and the free fatty acid purity is significantly improved (Fig. 12.7).

Quality Parameters of Refined Palm Oil and Palm Kernel Oil

Free Fatty Acids

The efficiency of deodorization is usually quantified by the stripping of the FFA. Experience has shown that flavor and odor removal correlates well, to a certain extent, with the FFA reduction. Nevertheless, some differences among the different oils exist. In general, chemically refined oils are easier to deodorize than physically refined oils, not only because of the lower acidity before deodorization but also because a whole range of polar minor components, oxidation products, and pigments are already removed during alkali neutralization. For most refined oils, like palm oil and palm kernel oil, the targeted residual FFA content is 0.03–0.05%.

Color of the Deodorized Oils

Color specifications for refined palm and palm kernel oils are not identical. A final palm kernel oil color of 1.0R $5^{1/4''}$ may be acceptable for many purposes; however, less than 0.4R $5^{1/4''}$ is quite feasible from good crude feedstock. For refined palm oil, acceptable color is generally 2.5–3.0R $5^{1/4''}$(Patterson, 1992). However, color reversion can occur that is normally associated with poor quality of the crude oil or with improper degumming and bleaching processes. The reversion could also be due to colored pigments present in the crude palm oil or as a result of the oxidation of carotenoids during deodorization; also, oils derived from badly damaged palm fruits contain brown pigments from decomposed protein and carbohydrates that are resistant to bleaching. Moreover, oxidized crude palm oils may contain colored compounds of quinoid nature developed from oxidation of colored materials. It seems that oxidation not only develops new pigments but also stabilizes the pigments against their removal by adsorption. A bleachability test was developed at laboratory scale by which the final color of refined palm oil could be predicted (Gibon et al., 2007). Six crude palm oils from different origins with an initial FFA content between 1.9 and 7.2% were prepared. Afterwards, the samples were immediately deodorized at 260°C for 50 min at a pressure of 3 mbar, with the addition of only small amounts of steam (to ensure good mixing and heat transfer). The low DOBI values combined with elevated levels of fatty acids and peroxides in some samples

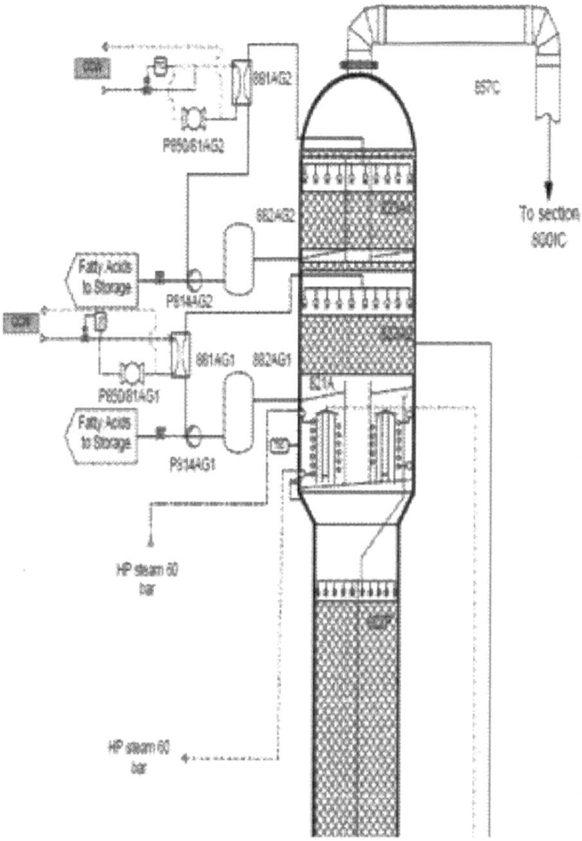

PFAD (Palm Fatty Acid Distillate)	Chemical Refining	Physical Refining		
FFA (%)	33-50	80-85	88.0	98.1
Neutral oil (%) MAG (%) DAG (%) TAG (%)	25-33	5-10	 2.0 0.7 8.0	 0.7 0.05 0.1
Unsap. (%) (tocos, sterols ...)	25-33	2-8	2.0	<1

Fig. 12.7. Composition of the palm fatty acid distillate (chemical refining and physical refining with and without double condensing system). MAGs: monoacylglycerols; DAGs: diacylglycerols; TAGs: triacylglycerols (Desmet Ballestra Group, 2011).

Table 12-G. Final Color Achieved After a Bleachability Test (260°C, 3 mbar, 50 min.) Starting from Different Crude Palm Oil Grades.

Oil grade	Bad	Poor	Poor	Poor	Fair	Good
FFA (% as C16:0)	7.21	4.32	3.54	4.91	2.79	1.90
DOBI	1.34	1.76	2.02	2.23	2.67	3.13
Peroxide value (meq O_2/kg)	2.33	3.79	3.15	3.54	0.36	1.03
Final color Lovibond $5^{1/4''}$	3.1R/33Y	3.0R/35Y	2.2R/20Y	2R/20Y	1.3R/15Y	1.4R/15Y

Source: Gibon et al. (2007).

(Table 12-G) gave an indication that these oils were difficult to bleach (final color $5^{1/4''}$ above 3R). Moreover, a direct relationship between the DOBI and the final color of the heat bleached palm oil was observed. A heat-bleached palm oil with color below 2R could only be produced when the crude palm oil had a DOBI above 2.5.

Trans Fatty Acids

Trans fatty acids are mono- or polyunsaturated fatty acids with one or more double bonds in the *trans* configuration. *Trans* fatty acids have a structure comparable to that of saturated fatty acids and have a higher melting point than the corresponding *cis* isomers. The activation energy for the thermal *cis-trans* isomerization is rather low, meaning that *trans* fatty acids are relatively easily formed at elevated temperatures of deodorization. Therefore, there is more risk to form *trans* fatty acids if the oil is refined according to the physical route. The isomerization rate for linolenic acid is significantly higher compared to linoleic acid. Highly polyunsaturated oils are more sensitive to *cis-trans* isomerization during deodorization compared to palm and palm kernel oils which contain relatively low amounts of poly-unsaturated fatty acids; values of 0.5% *trans* fatty acids or below are typically achievable if palm and palm kernel oils are physically refined especially with the dual temperature concept. While in commercial refined palm oil products, a mean value of 0.6% of total *trans* fatty acids is reported for processing conditions of 260–275°C and short residence times, it was demonstrated that the formation of *trans* fatty acid significantly accelerated at higher temperatures and longer residence times; values as high as 2.1% and 1.5% were detected in palm olein and palm mid fraction, respectively, processed at 280°C with a residence time of 4 hours (Siew & Mohammad, 1989). Identically, intra- or interesterification may occur during drastic deodorizing conditions, resulting in a significant increase in the saturated fatty acids at sn 2-position of the triacylglycerols. The negative effects of time and temperature on the POP/OPP ratio of palm oil was extensively discussed (Willems & Padley, 1985).

Polymeric and Oxidized Triacylglycerols

Polymeric and oxidized triacylglycerols can be formed during deodorization. The increasing evidence that these compounds may be detrimental to health means that their presence needs to be restricted to the lowest possible levels. Polymerization and more specifically dimerization may be of thermal or oxidative origin. Contrary to *cis-trans* isomerization, polymerization already takes place during crushing and pretreatment; oxidized triacylglycerols act hereby as catalysts. The level of oxidized triacylglycerols should therefore be kept to a minimum by restricting oil-oxygen contact as much as possible in the different refining stages.

Tocopherols and Tocotrienols

Tocopherols and tocotrienols are important compounds of the unsaponifiable fraction; they are the most important natural antioxidants of phenolic nature present in vegetable oils. As said before, crude palm oil contains 800–1000 ppm tocopherols and tocotrienols while crude palm kernel oil has a lot less (30–100 ppm). In consequence, refined palm kernel oil is particularly depleted in these minor components, whatever the process conditions applied during refining. During deodorization, part of the tocopherols and tocotrienols is lost as a result of distillative removal and also due to thermal degradation. The major factors affecting distillative losses are deodorization temperature, pressure, and quantity of stripping gas. Tocopherols and tocotrienols are more stripped with increasing temperature and quantity of stripping gas and with decreasing pressure. Pure thermal degradation depends on time and temperature and only becomes significant at elevated temperatures (>260°C).

Sterols

Contrary to tocopherols and tocotrienols, sterols can be partially present in the esterified form. Deodorization results in the reduction of the total sterol content which is attributed to the free sterol fraction since esterified sterols are not so volatile. In case of physical refining, an increase of the steryl ester content can sometimes be observed, probably due to a heat-promoted esterification reaction between free sterols and fatty acids. This phenomenon will not strictly take place during deodorization of chemically refined oil since the initial FFA concentration is much lower in that case.

Contaminants, Pesticides, and Polycyclic Aromatic Hydrocarbons

Most all crude oils may contain some contaminants. The most hazardous contaminants are the so-called "Persistent Organic Pollutants." Contaminants need to be removed during the refining process in order to achieve low amounts of residual levels. In practice, this can be achieved by selective adsorption on suitable adsorbents or during the deodorization process. Many studies have been published regarding the removal of pesticides; all concluded that organo-chlorine and organo-phosphorous pesticides are fully removed during deodorization, provided that the process temperature is high enough (>230°C) and the applied pressure sufficiently low (<4 mbar). Due to their relatively high volatility, most pesticides are removed by distillation. Some pesticides like captan and dichlorvos are thermally decomposed into more volatile products that are subsequently stripped from the oil. When applying appropriate deodorization conditions, the residual pesticide content in refined oils is generally below the limit of detection (10–50 ppb). Crude palm oil contains ~2–42 ppb of light polycyclic aromatic hydrocarbons (PAH) and only ~0.7–10 ppb of heavy PAH; contents in palm kernel oil are varying between ~6–81 ppb for the light fraction and 0–10 ppb for the heavy fraction. These levels are a lot less compared to some other crude oils (especially crude coconut oil) and therefore no special treatment is necessary (Patterson, 1992). Note that only the light polycyclic aromatic hydrocarbons are volatile enough to be removed during deodorization; the heavy ones should be removed with a carbon active pre-treatment.

3-MCPD and Glycidol Esters

3-MCPD esters represent a group of contaminants that are mainly formed during refining of oils and fats. 3-MCPDs (3-monochloro, 1,2-propanediols) were identified in free form and as mono- and diesters. The alkyl group in 3-MCPD esters typically consists in C18 fatty acids (stearic, oleic, or linoleic). Apart from the free 3-MCPD and esters, glycidol esters (characterized by an epoxy ring in the structure) can also be present in refined oils and fats. Free 3-MCPD content in food oils is generally very low (< 10 ppb) while 3-MCPD esters and glycidol esters are found in significantly higher quantities. The 3-MCPD ester content was determined (Van Duijn, 2009) in 37 samples of refined palm oil; the average content was 4.5 ppm, with a maximal level of 13 ppm. The capability of oils to form 3-MCPD esters and glycidol esters by heating 2 hours at 240°C was studied (Matthäus et al. 2011); analyzed levels for 70 different palm oil samples varied largely between 1.3 ppm and 14 ppm in total. No straight correlation could be made between any specific parameters (extractable chloride, mono-, and diacylglycerols) and the 3-MCPD/glycidol ester content but it was shown that origin of palm oil was a factor of importance. Highest levels were found in palm oil from Malaysia and Indonesia while much lower in palm oil from Ghana and Columbia (different soil and climate conditions, use of different fertilizers, and different handling procedures after harvesting). Correlation between the diacylglycerol content in oils and the formation of glycidol esters was clearly identified (Haines et al., 2011) and it was shown that the formation of 3-MCPD esters was mainly limited to palm oil while glycidol esters could be found in various edible oils in correlation with the diacylglycerol content. For palm oil with a diacylglycerol content of ~4%, there is already a high chance for a 3-MCPD and glycidol content above 5 ppm. Higher levels (up to 16 ppm) were observed for diacylglycerol contents varying between 4 and 10% (Matthäus et al., 2011). In general, three different strategies can be followed for the reduction of low 3-MCPD/glycidol esters in refined oils: the removal of critical precursors from the crude oil prior to the refining process, the modification of the refining process conditions, or the removal from the refined oils. Today, there is still limited understanding of the effect of refining conditions and only a few studies have been published (Hrncirik & Van Duijn, 2011; Matthäus et al., 2011; Pudel et al., 2011; Ramli et al., 2011). It was confirmed (Pudel et al., 2011) that chemically refined palm oil had significant lower glycidol ester content mainly due to lower deodorization temperature. Initially, bleaching was also considered as crucial in the formation of 3-MCPD/glycidol esters; however, it was shown (Schurz, 2010) that the content in refined palm oil increased from 3.2 to 5.2 ppm by either reducing the bleaching earth dosage or increasing the bleaching temperature; it was surprisingly reported that non activated bleaching earth gave higher content in palm oil. Today, it is generally accepted that 3-MCPD/glycidol esters are mainly formed during deodorization, being time and temperature dependent (Hrncirik & Van Duijn, 2011). For example, the formation of glycidol esters is very limited at 230°C while a steep increase is observed at higher temperatures (240–260°C), which is the typical refining temperature of palm oil. A possibility to remove 3-MCPD/glycidol esters from the refined palm oil was investigated (Strijowski et al., 2011) by adsorbing them on a suitable material; it was shown that the glycidol ester content could be reduced with approximately 40% while the 3-MCPD esters were not absorbed.

Specially Refined Palm Based Products

Red Palm Oils with High Vitamin Content

When starting with very good quality crude palm oil (low FFA and high DOBI), chemical refining is able to purify the oil so effectively that the bleaching step can be skipped or performed with very small amounts of non-activated bleaching earths or with silica. Low temperature and low pressure deodorization will limit the destruction of carotenoids and keep high amounts of tocopherols and tocotrienols in the refined oil. Dedicated products issued from special refining procedures and labeled as Red Cooking Oils are available on the Asian market. Nutrolein Golden Palm Oil and Carotino Cooking Oil are two representative products. Nutrolein Golden Palm Oil, for example (which is produced by Unitata Berhad in Malaysia), is a high IV palm superolein issued from the dry fractionation of chemically refined crude palm oil. Carotino Cooking Oil is a blend of physically refined palm and canola oils. Main characteristics of both products are presented in Table 12-H (Desmet Ballestra Group, 2011). Similarly, Siaoma oil (produced by Danec S.A. in Ecuador), which is a more unsaturated variety of palm oil (a palm hybrid), is available on the Latin American market; processing conditions are similar to the ones of Nutrolein Golden Palm Oil.

Table 12-H. General Quality Parameters of Nutrolein Golden Palm Oil and Carotino Cooking Oil.

	Nutrolein Golden Palm Oil	Carotino Cooking Oil
FAC (% gas chromatography)		
C 12:0	0.1	0.1
C 14:0	0.5	0.3
C 16:0	29.4	11.2
C 16:1	0.3	0.2
C 17:0	0.1	0.1
C 17:1	0.1	0.1
C 18:0	2.8	2.4
C 18:1	52.9	57.8
C 18:2t	0.1	0.6
C 18:2c	12.9	18.2
C 18:3t	N.D.	2.6
C 18:3c	0.4	4.5
C 20:0	0.3	0.5
C 20:1	0.1	1.1
SFA	33.2	14.6
MUFA	53.5	58.9
PUFA	13.3	25.8
IV (Wijs)	69.9	101.3
β-carotene (ppm)	913	197
Tocopherols (ppm)		
α-tocopherol	202	177
β-tocopherol	N.D.	50
γ-tocopherol	25	20
Total tocopherols	227	435
Tocotrienols (ppm)		
α-tocotrienol	188	55
γ-tocotrienol	407	74
δ-tocotrienol	61	17
Total tocotrienols	656	146

SFA: saturated fatty acids; MUFA: monounsaturated fatty acids; PUFA: polyunsaturated fatty acids; N.D. not detected.
Source: Desmet Ballestra Group (2011).

Physically Refined Palm Oil with Maximum Retention of Tocopherols and Tocotrienols

As said before, palm oil can physically be refined in such a way that retention of tocopherols and tocotrienols can be maximized. Dual temperature (low/high) deodorization combined with low operating pressure (easily achieved with the ice condensing system) is able to produce a high-grade physically refined palm oil with higher tocopherols and tocotrienols content and a bit more natural color (β-carotene) (Table 12-I) (Gibon et al., 2007); almost 92% of tocopherols and tocotrienols were retained in the physically refined palm oil at the deodorizing conditions of 200/220°C (dual temperature low/high) and at a pressure of 1 mbar. In such conditions, some residual β-carotene was still observed and the acidity was decreased down to 0.07%.

Production of White Soaps from Palm and Palm Kernel Oils

Production of white soaps from palm and palm kernel oils is possible when starting with fully deodorized oils. Saponification color should be 2.0 R max. (Lovibond $5^{1/4''}$), which is difficult to achieve with refined palm oil, usually more colored than other oils (like palm kernel and coconut oils). Starting with a refined palm oil having a 2.5 R color (Lovibond $5^{1/4''}$), the saponification color is usually too dark (> 5.0 R). Good results (saponification color < 3.0 R) can only be expected when starting with a high DOBI crude palm oil and when using adapted physical refining conditions. It was shown that very good saponification color (2.5R Lovibond $5^{1/4''}$) could be be

Table 12-I. Physically Refined Palm Oil (Dual Temperature) with Maximum Retention of Tocopherols and Tocotrienols and More Natural Color.

	Crude palm oil	RBD palm oil dual A	RBD palm oil dual B
FFA (% as C16:0)	4.4	0.07	0.07
Lovibond (R)	29 (1″)	4.1 ($5^{1/4''}$)	6.2 ($5^{1/4''}$)
β-carotene (ppm)	520	3	9
α-tocopherol	219	205	200
α-tocotrienol	178	167	167
γ-tocotrienol	245	239	229
Total tocos (ppm)	730	699	671

A: Dual temperature: 220–240°C, 3 mbar, 1.5 % steam, 60 min.
B: Dual temperature: 200–220°C, 1 mbar, 1.5 % steam, 60 min.
Source: Gibon et al. (2007).

Table 12-J. Production of White Soaps Through Physical Refining of Palm and Palm Kernel Oils.

Physical refining	RBD oil color Lovibond 5¼″	Saponification color Lovibond 5¼″
Crude palm kernel oil (Lovibond 5¼″: 9.9R)		
Bleached with 1.0% BE and refined at 230°C	1.0 R	2.0 R
Crude palm oil (DOBI 3.3; Lovibond 1″: 25R)		
Bleached with 1.5% BE and refined at 280°C	0.9 R	3.6 R
Bleaching with 1.5% BE/0.4% AC and refined at 280°C	0.7 R	2.8 R
Bleached with 2.5% BE/0.5% AC and refined at 280°C	0.6 R	2.5 R

BE: Tonsil optimum 210 FF
AC: active carbon
Source: Gibon et al. (2007).

obtained from a very light RBD palm oil (0.6R) that was produced from crude palm oil having a DOBI above 3 (Table 12-J) (Gibon et al., 2007).

Fractionation Principle

Several modification processes exist in order to modify the compositional properties of native oils. The main purpose of these processes is to change their physicochemical properties either by reducing the degree of unsaturation of the acyl groups (hydrogenation), by redistributing the fatty acids chains on the glycerol backbone (interesterification), or by selectively separating the triacylglycerol components with crystallization and filtration (fractionation) (Kellens, 2000). Fractionation is a thermo-mechanical separation process in which a multi-component mixture is physically separated into two or more fractions with distinct physicochemical properties (Gibon, 2006; Gibon et al., 2007; Krisnamurthy & Kellens, 1996). The separation can be based on a difference in solubility or volatility in between the different compounds. Fractional crystallization is based on differences of solubility of the solid fatty molecules (triacylglycerols, fatty acids, fatty acid methyl esters, etc.) in the liquid phase depending on their molecular weight and degree of unsaturation. Beside fractional crystallization, lipids can also be separated according to other physical principles, fractional distillation, short path distillation, supercritical extraction, liquid-liquid extraction, adsorption, urea complexation, and membrane separation being some of the main practiced techniques.

Solvent Fractionation

In solvent fractionation, fractional crystallization is performed in dilute conditions, thereby reducing viscosity; mainly acetone or hexane is used as solvent. The process is characterized by a short crystallization time and an easy filterability. The main advantage of solvent fractionation is the high separation efficiency and hence improved yield and high purity of the finished products. The major reason for this can be explained by the fact that with any type of separation technique, it is not possible to remove all liquid from the solid portion; part of the liquid always remains entrapped within the solid phase. In a dilute solution, as is the case in solvent fractionation, this liquid portion consists of a considerable quantity of solvent that hinders oil occlusion. After separation (generally on vacuum filters), the solvent is evaporated leaving behind a much lower quantity of liquid oil in the solid phase. Due to the high production costs and capital investments, as well as the possible fire hazards and environmental issues, solvent fractionation is becoming less interesting. Today, most of the solvent fractionation plants still in operation produce some specialty fats as, for example, cocoa butter replacement fats; only a few solvent processes are still operating industrially.

Detergent Fractionation

Detergent fractionation was developed to improve the separation of the crystallized phase from the remaining liquid by adding an aqueous detergent solution to the crystallized oil. The wetting agent, usually sodium lauryl sulfate, in combination with an electrolyte, usually magnesium sulfate, allows the crystals to be easily suspended in the aqueous phase. Separation of the water phase from the remaining liquid oil is performed by means of a centrifuge. The water phase is subsequently heated and the melted stearin is recovered in a second centrifugation step. After separation, the olein and stearin fractions are washed and dried in order to remove traces of detergent. Today, the technique has lost interest due to the high costs and contamination of the end products. However, some detergent fractionation plants are still in operation mainly for fatty acid processing.

Panning and Pressing Technology

The panning and pressing technology has been extensively used for the fractionation of lauric oils and more specifically of palm kernel oil, the main objective being to produce cocoa butter substitutes (CBS). Due to its particular triacylglycerol composition, palm kernel oil cannot be easily processed with a classical dry fractionation technology, and panning and pressing was for a long time the only alternative to solvent or detergent fractionation. In this process, the palm kernel oil is poured in pans and stored in cold rooms to crystallize. After hard setting, the cakes are wrapped in filter cloths (usually cotton), stacked between perforated plates in hydraulic presses, and pressed up to 200 bar. By this method, good quality final products can be obtained at reasonable yield compared to solvent fractionation; the capital cost is much lower but operation is very labor intensive. In order to overcome these disadvantages, the process has undergone several improvements with years. The use of continuous belt instead of pans in the cold rooms and the introduction of automated high-pressure membrane press filters instead of the hydraulic manual presses had led to considerable reduction in manual operation. Remaining disadvantages of this process can now be by-passed thanks to the development of an innovative technology particularly suited to the fractionation of palm kernel oil, the Statolizer.

Dry Fractionation

Initially, dry fractionation was used to improve the cold stability of liquid oils by removing small amounts of solid impurities (waxes or saturated triacylglycerols); this operation is better known as winterization. Dry fractionation is often identified as winterization, but although both processes are based on the same basic principles (crystallization by cooling and separation), they have different purposes. In winterization, the oil is kept at low temperatures for some time, after which it is filtered in order to remove some solids. From a technological point of view, the process is less sophisticated than dry fractionation; the main differences between both processes is the distinct change in physicochemical properties of the liquid and solid fractions produced by dry fractionation compared to winterization. As much larger amounts of solids are crystallized during dry fractionation, control of crystallization is more critical and therefore filtration is more elaborated. The dry fractionation process is an environmentally friendly and cost-effective process (no chemicals, no solvent, no effluent, and no losses) which consists in a controlled crystallization from the melt by slow cooling. Crystallization operates in the bulk and the degree of crystallization is therefore limited. Multi-step operations are most of the time necessary to achieve final specifications which has also the advantage to produce a large range of distinct fractions usable for specific applications. This technology largely improved with an increased knowledge in fat crystallization, with the design of new crystallizers and with the development of highly efficient equipment to separate the fractions (high pressure membrane press filters). In dry fractionation, crystallization is achieved from the melted state and is carefully conducted to approach selectivity and fraction quality similar to solvent fractionation. Practically, the first step consists in fully melting the oil in order to erase its thermal memory; the oil is further cooled down and kept under super-cooling in order to initiate the crystallization; crystal growth is followed, and during this period the heat of crystallization must be perfectly released and viscosity of the crystal slurry perfectly controlled. Different cooling strategies can be used and the crystallizer design principally determines the conditions under which the crystallization is controlled. The cooling rate largely depends on the technical features of the equipment, more specifically the heat exchange properties of the crystallizer. One strategy is to cool down the oil according to a differential temperature profile between the oil and the cooling water circulating in coils and in double jackets; this strategy is cumulating delta oil/water temperature control, oil temperature control and water temperature control. Another strategy is simply to control the cooling water temperature. Both strategies have proven efficiency and result in crystallization executed in relatively short or alternatively longer times. Most of the crystallizers are processed batch wise, which has the advantage of an exact control of the cooling sequence and of the crystallization; it is also more adapted to frequent feedstock changes. The crystallization is conducted up to a certain degree, which is generally expressed as the solid fat content of the crystallizing slurry; depending on the processed oil, this solid fat content will range between 5 and 30%. After partial crystallization, the oil is pumped to high pressure membrane press filters for the separation of the solid (stearin) from the liquid (olein), each fraction having its proper physicochemical characteristics for particular applications. Today, the majority of the dry fractionation industrial plants operate with automatic membrane press filters, fitted with airtight membranes; the main advantage is the higher percentage of liquid it yields compared to previous vacuum technologies (belt filters or rotary drums) or centrifugal separators. The liquid physically trapped on the crystals is easily removed by applying a pressure to the cake during each filtration cycle.

Equipment for Dry Fractionation of Palm Oil and Palm Kernel Oil

A dry fractionation plant consists of a series of crystallizers operating independently and interconnected with one or more high pressure membrane press filters. Crystallization is performed with successive steps of melting, nucleation, and crystal growth. Crystal shape and size are principally determined by the way the oil is cooled and agitated. Oils being poor heat conductors, the heat transfer is largely determined by mass transfer to and from the cooling surface and the crystallizer design is therefore important to guarantee a perfect heat transfer and accurate control of the temperature.

Crystallizer Design

To obtain necessary performance, different types of crystallizers with appropriate and specific design (Fig. 12.8) (Desmet Ballestra Group, 2011) have been developed; some of them operate at relatively high agitation speed (high shear), some at lower agitation speed (low shear), and some in fully static conditions. One of these is a concentric crystallizer (high shear) in which the oil is contacted with the cold medium through concentric cooling walls properly designed to assure relatively short crystallization time. The Tirtiaux type is a crystallizer where the cooling medium is circulating through vertical fins or bundles; in this system, the crystallization time is generally longer. One of the latest developments is a crystallizer which is operating with low agitation speed and known as Mobulizer. This crystallizer reflects the integration of cooling and agitation in one single system and particularity is its rectangular design. The cooling medium is continuously circulating within the horizontally moving bundles;

a)

b)

Fig. 12.8. Crystallizer designs: a) concentric, b) Tirtiaux, c) Mobulizer, d) Statolizer, e) tubular, and f) contube (Desmet Ballestra Group, 2011).

c)

d)

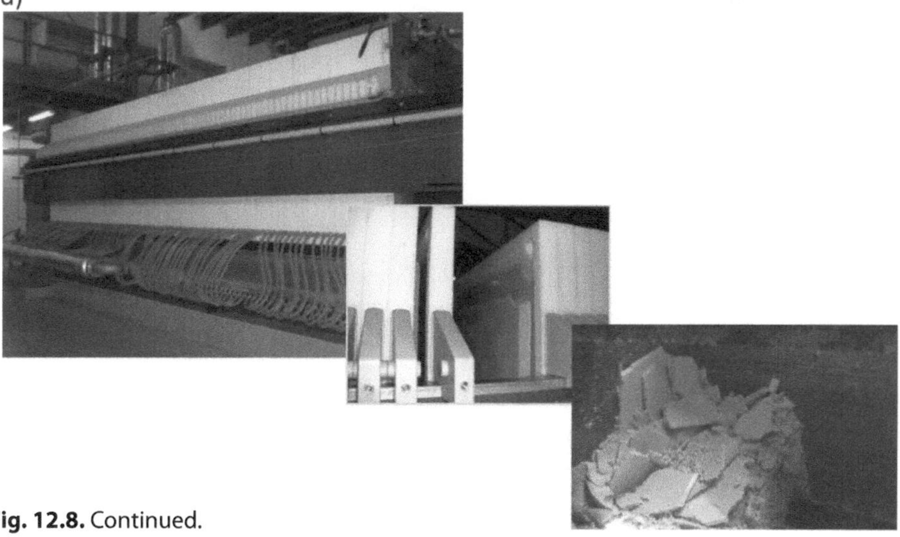

Fig. 12.8. Continued.

e)

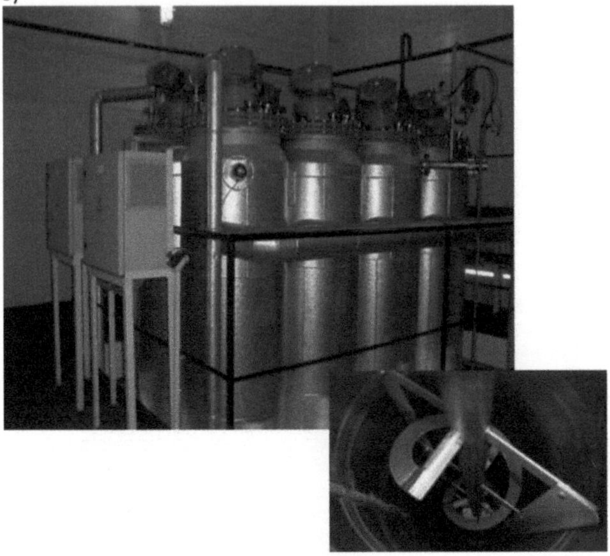

f)

Fig. 12.8. Continued.

this integrated agitation/cooling system makes it possible to move slowly through the oil maintaining a low and uniform speed through the oil volume which is translated into excellent heat and mass transfer and a well-controlled process. The Mobulizer was designed in such a way that it requires minimal construction and material costs, which offer a considerable saving on both investment and energy consumption. Lauric fats, like palm kernel oil, cannot easily be processed under shear. Up to a few years ago, the panning and pressing was the preferred technological alternative to solvent or detergent fractionations. More recently, a fully automatic static crystallization system was developed, the Statolizer; it was initially especially foreseen for this application but it is also extendable to other oils like coconut oil or palm mid fractions. After a short pre-crystallization stage, the crystallizing slurry is sent in between cooling plates for static crystallization. After a time, soft cakes are formed and consequently discharged and sent via a mono-screw pump to a high pressure membrane press filter operating at high squeezing pressure. Compared to the conventional processes, it is more secure, less labor intensive, and very flexible regarding the production of different qualities of stearin. Crystallizers of the first generation (tubular and contube) are still under operation in a limited number of plants. Particularity of these crystallizers is that they can process very highly viscous materials; today, they are still proposed for the dry fractionation of the fatty acids.

Membrane Press Filter

A membrane press filter consists of series of chamber and membrane plates kept together by means of a hydraulic cylinder (Fig. 12.9) (Desmet Ballestra Group, 2011). After being concentrated upon filling the filter chambers, the solid part is additionally squeezed together by means of the inflatable membrane, which results in better removal of the entrained liquid phase. The standard membrane press filters normally operate at a maximum squeezing pressure of 16 bars, which is, in most of the cases, more than sufficient. Higher squeezing pressures (up to 30 bars) become indispensable for the separation of high viscosity crystallized fats like for the production of special cuts (cocoa butter equivalents, cocoa butter substitutes, etc.). Practically, the filtration is made through a sequence of mechanical operations, each carefully controlled and monitored via a computer. During the filling, the important point is to make sure all chambers are evenly filled with the crystalline slurry. The squeezing follows by applying behind each membrane a pressure by means of a gas or a liquid. During filling and squeezing, the olein flows through flexible holes and is collected in an intermediate tank before being sent to storage tanks. The next operation takes care of draining the feed core and the olein channels. The filter is then opened and the stearin cakes are orderly discharged by gravitation into a melting hopper where the stearin is melted and further pumped to the storage tanks. The filter is regularly washed by a flow of hot olein in order to melt away any stearin traces smeared on the filter clothes.

a) b)

Fig. 12.9. General view of a side bar membrane press filter a) squeezing side b) olein holes side (Desmet Ballestra Group, 2011).

Multi-step Dry Fractionation of Palm Oil

Palm oil is by far the most fractionated oil in the world; dry fractionation can be conducted according to a multi-step sequence, as shown in Fig. 12.10.

Commodity Fats

Palm olein has become a cheap primary commodity because it is formed in high proportions in one single step. Palm oil fractionation is carried out to produce an olein IV 56–59 having the characteristics to be still liquid below 25°C. The objective here is simply to remove the majority of the tri-saturated triacylglycerols (mainly PPP), aiming at an improvement of the cold stability of the liquid fractions.

Special Cuts

Today, attention has been focused on the multiple-step dry fractionation process to manufacture higher added-value products known as super stearin, hard palm mid fraction (HPMF), and high IV super- and topoleins. The so-called solid route is aim-

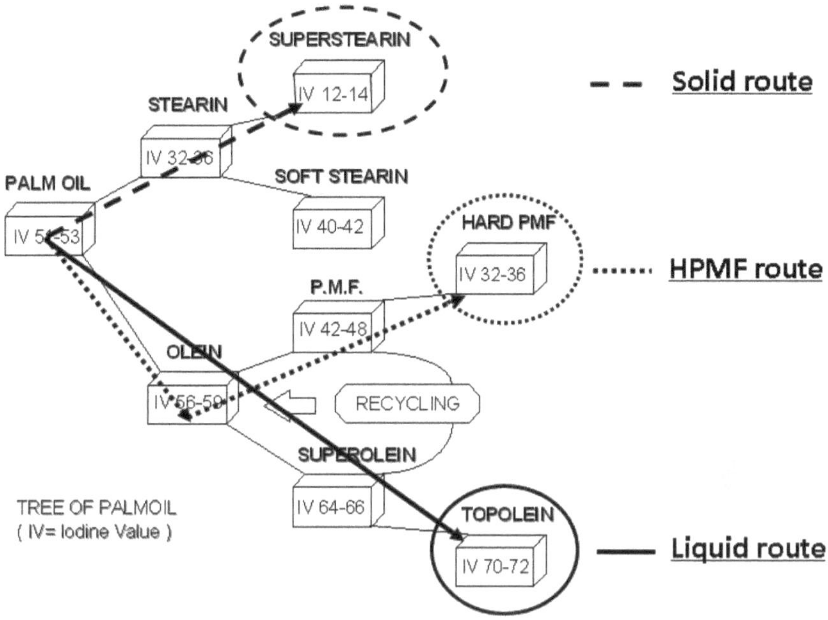

Fig. 12.10. Multi-step dry fractionation of palm oil: the solid route, the Hard Palm Mid Fraction (HPMF) route, and the liquid route (Desmet Ballestra Group, 2011).

ing, in two steps (palm oil > palm stearin > palm super stearin), to the production of very low iodine value stearins (IV 12–14). This super stearin contains ~90% of saturated fatty acids, predominantly palmitic; PPP particularly concentrates in this fraction (> 65%) which is very hard below 40–45°C and progressively melts at 65–70°C (Fig. 12.11) (Desmet Ballestra Group, 2011). Super stearin is an excellent hard stock to substitute hydrogenated products and is also usable in the formulation of zero or low *trans* margarines and shortenings. The corresponding soft stearin (IV 41–43) can find applications also as ingredient for margarines and shortening fats or as dairy replacer fat. The hard palm mid fraction (HPMF) route generally consists of four fractionation steps (Fig. 12.12) (Desmet Ballestra Group, 2011). Palm oil is first fractionated into an olein IV 56–57; this olein is further re-fractionated into a super olein IV 64–66, and a soft palm mid fraction IV 44–45 which may contain some PPP in excess. This excess is simply removed by a quick topping-off of the soft palm mid fraction. The topped soft palm mid fraction IV 46–47, which is an appropriate starting material for the production of good quality hard palm mid fraction, should contain ~48% POP which is concentrated in the hard palm mid fraction up to 65% min. (Danthine & Gibon, 2007). This hard palm mid fraction is an excellent ingredient for

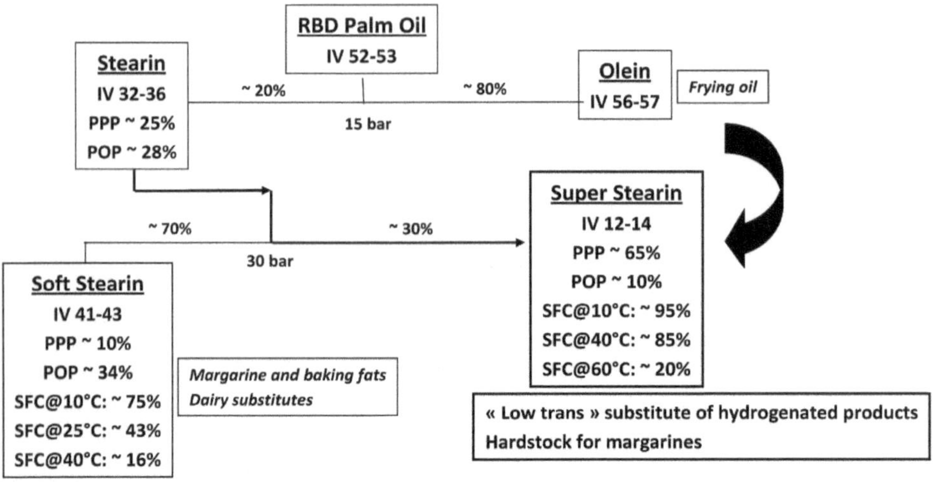

Fig. 12.11. Multi-step dry fractionation of palm oil—the solid route (Desmet Ballestra Group, 2011).

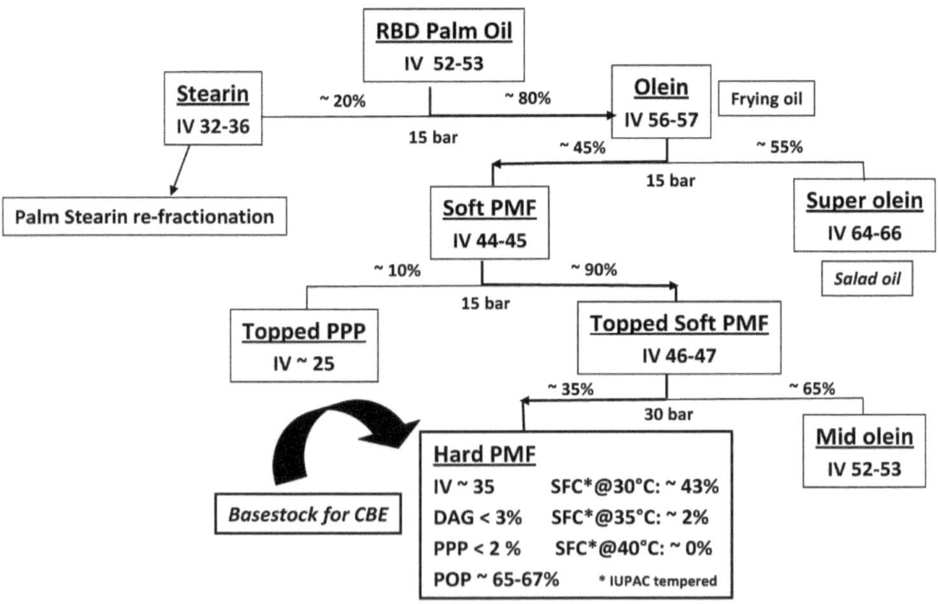

Fig. 12.12. Multi-step dry fractionation of palm oil—the Hard Palm Mid Fraction (HPMF) route (Desmet Ballestra Group, 2011).

manufacturing cocoa butter equivalents (CBE), and the quality is fully comparable to the one of solvent fractionation. Liquid palm fractions of increased cold stability can easily be obtained through consecutive fractionation steps (Calliauw et al., 2007; Calliauw et al., 2010) (Fig. 12.13). In some cases, a superolein IV 64–66 can even be produced in one single fractionation step from palm oil, but most of the time two steps are necessary (palm oil → palm olein → palm superolein). Higher IV 67–68 can possibly be obtained also in two steps, while topoleins IV 70–72 usually require a three step processing (palm oil → palm olein → palm superolein → palm topolein). The cold resistance of the liquid fractions can be evaluated by means of the cold test method, which is the time at which the oil remains clear at a certain temperature or the cloud point which is the temperature at which the oil begins to solidify when cooled at a certain rate. Fig. 12.14 is illustrating how the cloud points of the palm liquid fractions are decreasing when the IV is increased (Danthine et al., 2010). A major reduction of the cloud points is observed from palm oil to palm olein, which is due to the progressive removal of the trisaturated triacylglycerols (mainly PPP). When PPP is fully depleted (IV > 58), the next (less pronounced) decrease of cloud point is due to the removal of disaturated triacylglycerols (mainly POP) from the liquid fraction. At IV > 68, crystallization of POS (and PLP) allows an extra serious drop of the cloud point. Cloud points of the liquid fractions are also partially influenced by the

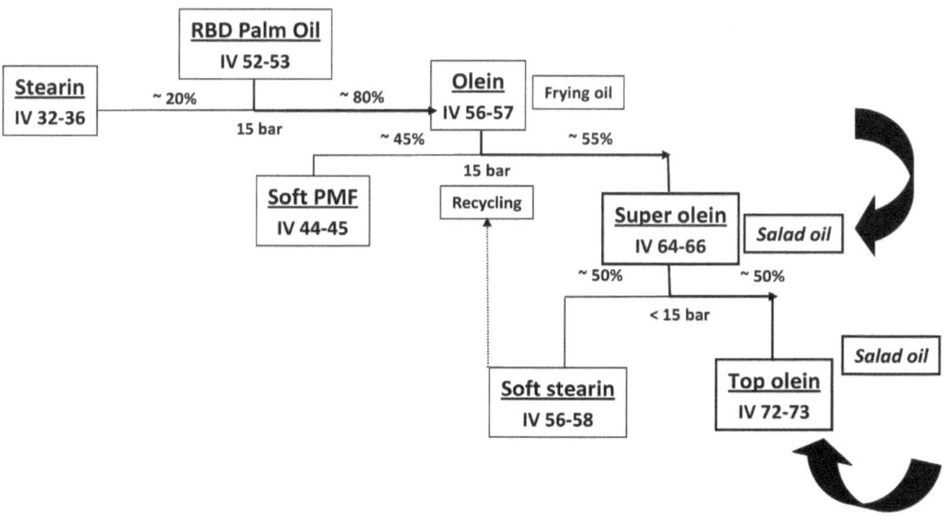

Fig. 12.13. Multi-step dry fractionation of palm oil—the liquid route (Desmet Ballestra Group, 2011).

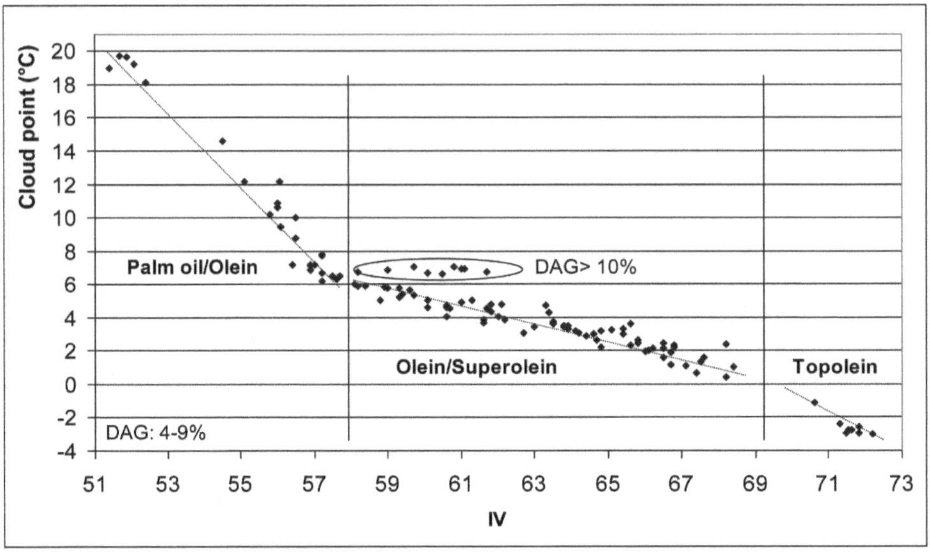

Fig. 12.14. AOCS cloud point (°C) as function of IV for palm oil, olein, superolein, and topolein (Danthine et al., 2010).

diacyglycerol content (DAG) (Cailliauw, 2008); in Fig. 12.14, the DAG content was fluctuating between 4 and 9% for majority of the points; however, some cloud points slightly separate from the others (circle at IV 58–62), these liquid fractions being indeed characterized by particularly higher DAG content (10–11%).

Red Fractions

Fractionation is classically operated on the fully refined oil, but the high vitamin content of crude palm oil makes the dry fractionation process also attractive for especially refined oils; it's a matter of fact that β-carotenes, tocopherols, and tocotrienols concentrate markedly in the olein, superolein, and topolein during the multi-step fractionation process (Gibon & Tirtiaux, 2002). The concentration of β-carotene in a topolein IV 72 is nearly twice the content in the corresponding palm oil IV 52 (Table 12-K).

Dry Fractionation of Palm Kernel Oil

Owing to its high content in lauric and myristic fatty acids, palm kernel oil is well known as a suitable feedstock for the production of confectionery fats. The highest quality and steepest-melting fats resulting after modification are called CBS (cocoa butter substitutes); high quality CBS possess a good flavor release and show sufficient bloom resistance. The modification from palm kernel oil to CBS is traditionally carried

Table 12-K. β-carotene Content of Chemically Refined and Fractionated Red Palm Oil.

	AOCS cloud point (°C)	β-carotene (ppm)
Palm oil IV 52	21.5	382
Olein IV 57	8.0	409
Superolein IV 63	3.3	670
Topolein IV 71	–2.4	854
Stearin IV 36	–	280
Soft palm mid fraction IV 48	–	235
Hard palm mid fraction IV 35	–	80

Source: Gibon & Tirtiaux (2002).

Table 12-L. General Characteristics of Palm Kernel Fractions Processed with Traditional Solvent, Detergent or Panning and Pressing Technologies (PKO: Palm Kernel Oil, PKS: Palm Kernel Stearin, HPKS: Hydrogenated Palm Kernel Stearin).

Fatty acid composition (% by gas chromatography)

	IV	C6:0–C10:0	C12:0	C14:0	C16:0	C18:0	C18:1	C18:2
PKO	17.5	7	48	16	9	2	15	2
PKS	9	6	53	21	9	2	8	1
	7	4	55	22	9	2	7	1
HPKS	4	4	55	22	9	6	5	0
	0.4	4	55	22	9	9	0.5	0

Solid fat content by p-NMR (%@°C)

	IV	20	25	30	35	40
PKO	17.5	44	20	0	–	–
PKS	9	70	48	8	0	–
	7	82	68	29	0	–
HPKS	4	87	70	24	1	–
	0.4	95	90	50	5	1

Source: Rossel (1985).

out via a combination of fractionation and post-hydrogenation. Table 12-L reports some characteristics of palm kernel stearins and hydrogenated palm kernel stearins traditionally processed via solvent, detergent, or panning or pressing technologies (Rossel, 1985). As said before, the Statolizer is an innovative static dry fractionation technology that was more recently developed for reliable and consistent production (high quality and high yield) of palm kernel stearin IV 7 and below, which, after full hydrogenation,

Table 12-M. General Characteristics of Some Crude, Degummed Bleached (DB), and Refined (RBD) Palm Kernel Stearin (PKS) Processed with the Statolizer Technology.

	Crude PKS					DB PKS			RBD PKS		
IV	7.5	7.0	5.1	4.7	4.0	7.3	6.5	5.3	7.0	6.7	5.9
Fatty acid composition (% by gas chromatography)											
C8:0	1.8	1.7	1.5	1.4	1.2	1.9	1.9	1.5	1.3	1.5	1.4
C10:0	2.7	2.6	2.5	2.5	2.4	2.7	2.7	2.52	2.2	2.5	2.6
C12:0	55.2	55.1	56.0	56.5	56.4	56.2	56.3	57.3	54.5	53.5	57.5
C14:0	22.2	22.4	23.9	24.2	25.3	22.2	22.4	24.2	23.2	23.3	22.7
C16:0	9.1	9.4	8.9	8.9	9.1	8.4	8.3	8.5	9.8	9.9	8.6
C18:0	2.0	1.9	1.9	1.8	1.8	1.6	1.5	1.4	2.2	2.1	1.7
C18:1	6.1	6.0	4.4	4.1	3.4	6.0	5.8	4.1	6.8	6.3	4.7
C18:2	0.9	0.9	0.6	0.5	0.4	0.8	0.81	0.5	0.1	0.8	0.6
SFA	93.0	93.1	94.8	95.4	96.2	93.1	93.3	95.3	93.1	92.8	94.5
MUFA	6.1	6.0	4.4	4.1	3.4	6.0	5.8	4.1	6.8	6.2	4.7
PUFA	0.9	0.9	0.6	0.5	0.4	0.8	0.8	0.5	0.1	0.8	0.6
Solid fat content by p-NMR (%@°C)[1]											
10	91.3	92.0	93.8	94.5	95.1	91.7	91.3	94.6	92.3	93.1	94.3
20	82.8	85.5	90.4	91.1	92.7	83.2	83.2	90.3	86.7	88.0	89.1
25	65.6	68.5	79.9	80.0	84.9	68.7	68.6	79.8	74.7	72.6	76.9
30	34.4	38.2	44.9	52.7	62.1	33.8	36.0	51.7	34.1	41.8	42.7
35	0.0	0.0	0.0	0.0	0.0	0.0	0.0	0.0	0.0	0.0	0.0

[1] IUPAC serial non tempered method.
Source: Desmet Ballestra Group (2011).

is an excellent cocoa butter substitute. Crude, degummed bleached, or fully refined palm kernel oils can be processed with this technology (Calliauw et al., 2005) (Table 12-M). The saturated fatty acid content is increased from ~80% in the palm kernel oil to 93–96% in the palm kernel stearins; the lauric and myristic fatty acid contents go up to 80%. More particularly, a palm kernel stearin IV 4 can even be obtained in one single fractionation step; the solid fat content profile of this stearin is particularly steep between 25 and 30°C, and, for some CBS applications, it can be used as such without post-hydrogenation. Most of the time, the dry fractionation of palm kernel oil is conducted in one single step with the Statolizer, the objective being a palm kernel stearin IV ~7; a double step process (Fig. 12.15) (Desmet Ballestra Group, 2011) permits in first stage the production of unhardened yet high quality CBS (IV ~4) and

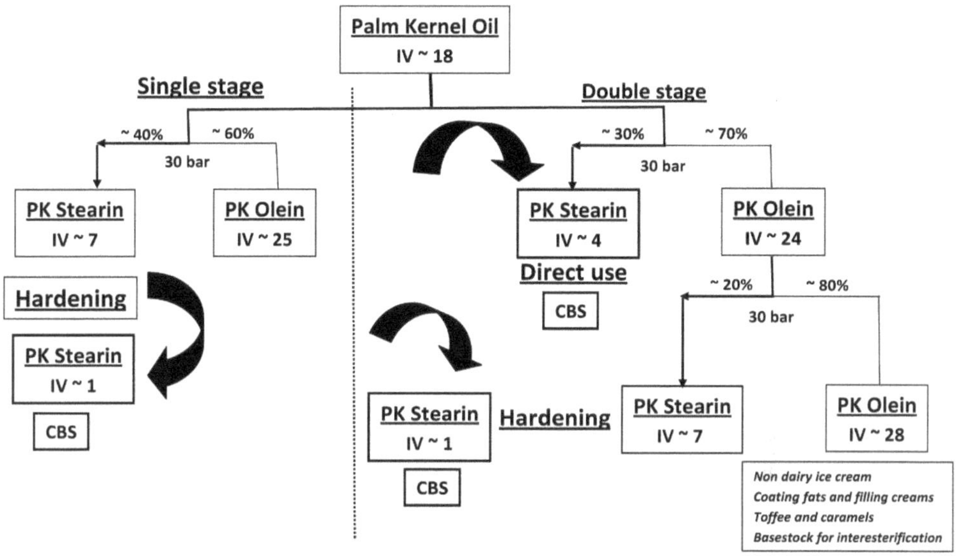

Fig. 12.15. Single and double step dry fractionation process of palm kernel oil (Desmet Ballestra Group, 2011).

successful re-fractionation of the corresponding palm kernel olein (second stage) allows significant increase of the overall yield. This second stage produces extra palm kernel stearin IV 7, which is post-hydrogenated to achieve the characteristic of a good CBS. Consequently, the two steps Statolizer fractionation of palm kernel oil results in the following three products: a palm kernel stearin IV ~4 (for direct use as CBS), a palm kernel stearin IV ~7 (for use as CBS after full hydrogenation), and a palm kernel olein IV ~27. The unhardened palm kernel stearin IV 4 having outstanding melting and crystallization properties comparable to the ones of traditional hydrogenated stearin fraction, the reduced hydrogenation capacity is most probably the most important benefit of this two-step approach.

Conclusions

Crude palm oil is rich in minor components (phytonutrients) that impart unique nutritional properties; the most relevant are tocopherols, tocotrienols, and carotenoids. Due to high acidity and relatively low phosphatide content, palm and palm kernel oil are generally physically refined, which is preferred to the chemical process. The quality of the crude oils has to be considered as it can greatly affect the efficiency of the refining process. In the case of palm oil, the DOBI is a good indicator of its capability to be successfully refined. Beside commodities, especially refined oils open a market for new high quality products like red palm oil with high vitamin content or physically refined palm oil with optimized tocopherol and tocotrienols contents. White soaps can also be produced from palm oil and palm kernel oil, within specific conditions especially for palm oil. Lastly, developments in refining technology are driven by the increased attention for the nutritional quality. The development of vacuum systems (ice condensing) capable to reach low operating pressure is very important as it allows a reduction of the deodorization temperature without affecting the vaporisation efficiency. Working at different temperatures, according to the dual temperature concept, is the best compromise between required residence time for deodorization (performed at low temperature) and heat bleaching/final stripping at higher temperature for a shorter period. A design which includes structured packing for oil stripping which is able to reduce the steam consumption perfectly fits with the requirement of high acidity palm and palm kernel oils. The introduction of a double condensing unit can further improve the fatty acid purity in the distillate. Thanks to continuous developments in the design of new crystallizers and to the technological improvement of the membrane press filters, a whole variety of products normally obtained by solvent or detergent fractionation are now processed with a high degree of selectivity using dry fractionation. Single and multi-step operations are used to tailor a wide range of palm oil fractions suitable for different applications. Beside commodities, new demands for special cuts drifted the industry towards a more sophisticated approach: production of super stearins, hard palm mid fractions (ingredient for CBE), high cold stability topoleins, and red fractions with increased carotene, tocopherol, and tocotrienol contents. The main feature of palm kernel oil is its difficulty for being dry fractionated under classical conditions. Up to short time, solvent, detergent and/or panning and pressing technologies were the only possibilities to achieve high yield and high quality CBS. New developments in static dry fractionation and the introduction of the Statolizer technology allow today to easily fractionate this oil; beside single stage production of PKS IV ~7, a double stage scheme ending to PKS IV ~4 (for direct use as CBS), palm kernel olein IV ~27, and PKS IV ~7 (used as CBS after full hydrogenation) is implemented on industrial scale. Latest developments have shown the applicability of the Statolizer technology to the production of high quality and high yield hard palm mid fraction; this route is also implemented on industrial scale. Last but not least, recent developments in continuous dry fractionation have shown serious advantages (product performance, capacity, and energy savings) when applied on palm oil.

Acknowledgments

The author thanks Kellens Marc, Hendrix Marc, Calliauw Gijs, and De Greyt Wim for their contribution in this chapter.

References

Berger, K.G. Production of palm oil from fruit. J. Am. Oil Chem. Soc. 1983, 60(2), 206–210.

Berger, K.G. Minor components of palm oil. Malaysian Oil Sci. Technol. 2000, 9(1), 56–59.

Calliauw, G. Molecular interactions affecting the phase composition during dry fractionation of palm olein. PhD Thesis, University of Ghent, 2008.

Calliauw, G.; Foubert, I.; De Greyt, W.; Dijckmans, P.; Kellens, M.; Dewettinck, K. Production of cocoa butter substitutes via two-stage static fractionation of palm kernel oil. J. Am. Oil Chem. Soc. 2005, 82(11), 83–89.

Calliauw, G.; Fredrick, E.; Gibon, V.; De Greyt, W.; Wouters, J.; Foubert, I.; Dewettinck, K. On the fractional crystallization of palm olein: solid solutions and eutectic solidification. Food Res. Intern. 2010, 43(4), 972–981.

Calliauw, G.; Gibon, V.; De Greyt, W.; Plees L.; Foubert, I.; Dewettinck, K. Phase composition during palm olein fractionation and its effect on soft PMF and superolein quality. J. Am. Oil Chem. Soc. 2007, 84(9), 885–891.

Chen, C.W; Chong, C.L.; Ghazali, H.M.; Lai, O.M. Interpretation of triacylglycerol profiles of palm oil, palm kernel oil and their binary blends. Food Chem. 2007, 100, 178–191.

Danthine, S.; Gibon, V. Comparative analysis of triacylglycerol composition, melting properties and polymorphic behavior of palm oil and fractions. Eur. J. Lipid Sci. Technol. 2007, 109, 359–372.

Danthine, S.; Lefébure, E.; Dijckmans, P.; Blecker, C.; Gibon, V. Compositional factors affecting the cold stability of palm liquid fractions: a statistical approach. Euro Fed Lipid Congress, Munich Germany, 2010.

De Greyt, W.; Dijkstra, A.J. Fractionation and interesterification. In Trans Fatty Acids; Dijkstra, A.J., Hamilton, R.J., Hamm, W., Eds.; Blackwell Publishing: Oxford, UK, 2007; pp. 181–202.

De Greyt, W.; Kellens, M. Refining practice. In Edible Oil Processing, First Edition; Hamm, W., Hamilton, R.J., Eds.; Sheffield Academic Press: Sheffield, England, 2000; pp. 79–127.

deMan, J.; deMan, L. Specialty fats based on palm oil and palm kernel oil, Malaysian Palm Oil Promotion Council, Kuala Lumpur Malaysia, 1994.

Desmet Ballestra Group, Internal data, Zaventem Belgium, 2011.

Gibon, V. Fractionation of lipids for use in food. In Modifying Lipids for Use in Food; Gunstone, F.D., Ed.; Woodhead Publishing Limited: Cambridge, England, 2006; pp. 201–229.

Gibon, V.; De Greyt, W.; Kellens, M. Palm oil refining. Eur. J. Lipid Sci. Technol. 2007, 109, 315–335.

Gibon V., Tirtiaux A., Soft degumming: the simplest route to physical refining of soft oils. Malaysian Oil Science and Technology 1998, 7(2), 48-54.

Gibon, V., Tirtiaux A. Latest trends in dry fractionation. Lipid Technol. 2002, 33–36.

Goh, E.M.; Timms, R.H. Determination of mono- and diglycerides in palm oil, olein and stearin. J. Am. Oil Chem. Soc. 1985, 62(4), 730–734.

Goh, S.H.; Choo, Y.M.; Ong S.H. Minor constituents of palm oil. J. Am. Oil Chem. Soc. 1985, 62(2), 237–240.

Goh, S.H.; Khor, H.T.; Gee, P.T. Phospholipids of palm oil (*Elaeis guineensis*). J. Am. Oil Chem. Soc. 1982, 59(7), 296–299.

Goh, S.H.; Tong S.L.; Gee, P.T. Total phospholipids in crude palm oil: quantitative analysis and correlations with oil quality parameters. J. Am. Oil Chem. Soc. 1984, 61(10), 1597–1600.

Gordon, M.H.; Griffith, R.E. A comparison of steryl esters of coconut and palm kernel oils. Fat Sci. Technol. 1992, 94(6), 218–221.

Haines, T.D.; Adlaf, K.J.; Pierceal, R.M.; Lee, I.; Venkitasubramanian, P.; Collison, M. Direct determination of MCPD fatty acid esters and glycidyl fatty acid esters in vegetable oils by LC-TOFMS. J. Am. Oil Chem. Soc. 2011, 88, 1–14.

Hrncirik, K.; Van Duijn, G. An initial study on the formation of 3-MCPD esters during oil refining. Eur. J. Lipid Sci. Technol. 2011, 113, 323–329.

Jacobsberg, B.; Oh, C.H. Studies in palm oil crystallization. J. Am. Oil Chem. Soc. 1976, 53(10), 609–615.

Kellens, M. Oil modification processes. In Edible Oil Processing, First Edition; Hamm, W., Hamilton, R.J., Eds.; Sheffield Academic Press: Sheffield, England, 2000; pp. 129–173.

Kellens, M.; De Greyt, W. Deodorization. In Introduction to Fats and Oils Technology, Second Edition; O'Brien, R.D., Farr, W.E., Wan P.J., Eds.; AOCS Press: Champaign, Illinois, 2000; pp. 235–268.

Kellens, M.; Gibon, V.; Hendrix, M.; De Greyt, W. Palm oil refining. Eur. J. Lipid Sci. Technol. 2007, 109, 336–349.

Krisnamurthy, R.; Kellens, M. Fractionation and winterization. In Bailey's Industrial Oil and Fat Products, Vol. 4 Edible Oil & Fat Products: Processing technology, Fourth Edition; Hui, Y.H., Ed.; John Wiley and Sons: New York, 1996; pp. 301–337.

Lal, V.K.; Gasper A. Crude palm oil, quality requirements for good refined products. Proceedings of PORIM International Palm Oil Conference, Kuala Lumpur Malaysia, 1991, 243–254.

Lin Khor, G. ; Mohd Esa, N. Trans fatty acids intake: epidemiology and health implication. In Trans Fatty Acids; Dijkstra, A.J., Hamilton, R.J., Hamm, W., Eds.; Blackwell Publishing: Oxford, UK, 2007; pp. 25–53.

Maclellan, M. Palm oil. J. Am. Oil Chem. Soc. 1983, 60(2), 368–373.

Matthäus, B; Pudel, F.; Fehling, P.; Vosmann, K.; Freudenstein, A. Strategies for the reduction of 3-MCPD esters and related compounds in vegetable oils. Eur. J. Lipid Sci. Technol. 2011, 113, 380–386.

Nesaretnam, K. Minor components in palm oil and their health benefits, Palm Oil—The Sustainable 21st Century Oil, SCI Conference, London UK, March 2009.

Ooi, C.K.; Choo, Y.M.; Yap, S.C.; Basiron, Y.; Ong, A.S.H. Recovery of carotenoids from palm oil. J. Am. Oil Chem. Soc. 1994, 71(4), 423–426.

Pantzaris, T.P. Pocketbook of palm oil uses; Palm Oil Research Institute of Malaysia: Kuala Lumpur, Malaysia, Fourth Edition, 1994; pp. 12–15.

Patterson, H.B.W. Bleaching and purifying fats and oils, theory and practice; Patterson, H.B.W., Ed.; AOCS Press: Champaign, Illinois, 1992.

Petrauskaite, V.; De Greyt, W.; Kellens, M. Physical refining of coconut oil: effect of crude oil quality and deodorization conditions on neutral oil loss. J. Am. Oil Chem. Soc. 2000, 77(6), 581–586.

Pritchard, J.R. Oilseed quality requirements for processing. J. Am. Oil Chem. Soc. 1983, 60(2), 322–332.

Pudel, F; Benecke, P.; Fehling, P.; Freudenstein, A.; Matthäus, B.; Schwaf, A. Eur. J. Lipid Sci. Technol. 2011, 113, 368–373.

Ramli, M.R.; Siew, W.L.; Ibrahim, N.A.; Hussein, R.; Kuntom, A.; Razak, R.A.A.; Nesaretnam, K. Effects of degumming and bleaching on 3-MCPD esters formation during physical refining. J. Am. Oil Chem. Soc. 2011, accepted.

Rognon, F.; Wuidart, W. Copra-palm kernel. In Oils and Fats Manual, a Comprehensive Treatise; Karleskind, A., Wolff, J.P., Eds.; Lavoisier Technique et documentation: Paris, 1996; Volume 1, pp. 196–206.

Rossel, J.B. Fractionation of lauric oils. J. Am. Oil Chem. Soc. 1985, 62(2), 385–390.

Schurz, K. Development of the bleaching process under the perspective of classic and new applications, 3rd DGF symposium, Leipzig Germany, March 2010.

Siew, W.L. Deterioration of bleachability index (DOBI). INFORM 2001, 12, 1183–1187.

Siew, W.L.; Mohammad, Y. Effects of refining on chemical and physical properties of palm oil products. J. Am. Oil Chem. Soc. 1989, 66(7), 1116–1119.

Siew, W.L.; Ng, W.L. Diglyceride content and composition as indicators of palm oil quality. J. Sci. Food Agric. 1995, 69(1), 73–79.

Strijowski, U; Heinz, V.; Franke, K. Removal of 3-MCPD esters and related substances after refining by adsorbent material. Eur. J. Lipid Sci. Technol. 2011, 113, 387–392.

Sundram, K.; Sambanthamurthi, R.; Tan, Y. Palm fruit chemistry and nutrition. Asia Pacific J. Clin. Nutr. 2003, 12(3), 355–362.

Swoboda, P.A.T. Chemistry of refining. J. Am. Oil Chem. Soc. 1985, 62(2), 287–292.

Thiagarajan, T.; Tang, T.S. Refinery practices and oil quality. Proceedings of PORIM International Palm Oil Conference, Kuala Lumpur Malaysia, 1991, 254–267.

Transfeld, P.; Scheinder, M. Counter-current system cuts bleaching costs. INFORM 1996, 7, 756–767.

Van Duijn, G. Potential ways of reduction of 3-MCPD esters in vegetable oils. ILSI Conference, Brussels Belgium, February 2009.

Willems, M.G.A.; Padley, F.B. Palm oil: quality requirements from a customer's point of view. J. Am. Oil Chem. Soc. 1985, 62(2), 454–459.

Wuidart, W. Palm oil and its fractions. In Oils and Fats Manual, a Comprehensive Treatise; Karleskind, A., Wolff, J.P., Eds.; Lavoisier Technique et documentation: Paris, 1996; Volume 1, pp. 233–243.

Xu, X.; Guo, Z.; Zhang, H.; Vikbjerg, A.F.; Damstrup, M.L. Chemical and enzymatic interesterification of lipids for use in food. In Modifying Lipids for Use in Food; Gunstone, F.D., Ed.; Woodhead Publishing Limited: Cambridge, England, 2006; pp. 234–272.

13

The Physicochemical Properties of Palm Oil and Its Components

Chin-Ping Tan[1] and Imededdine Arbi Nehdi[2]
[1]Department of Food Technology, Faculty of Food Science and Technology, Universiti Putra Malaysia, 43400 UPM Serdang, Selangor, Malaysia and [2]Department of Chemistry, College of Science, King Saud University, Riyadh 1145, Saudi Arabia

Introduction

Palm oil, the principal source of palmitic acid, is derived from the fruit of the oil palm tree, *Elaeis guineensis*. Palm oil is a liquid extracted from the fleshy orange-red mesocarp of the fruits of the oil palm tree, which typically contain 45 to 55% oil. Palm oil has surpassed soybean oil to rank first in the worldwide production of edible oils. Palm oil is commercially fractionated on a large scale, resulting in low-melting and high-melting fractions known in the trade as palm olein and palm stearin, respectively. Palm olein is used as a liquid cooking or frying oil, and palm stearin is used as a solid fat in margarine and shortening blends (Law & Thiagarajan, 1990). Palm kernels are a by-product of palm oil milling (Basiron, 1996). Palm kernel oil is obtained as a minor product during the processing of oil palm fruit and represents approximately 2–4% of the harvested palm fruit bunch by weight (Gascon et al., 1989). Palm kernel oil is extracted by mechanical-pressure screw processing, solvent extraction, or preprocessing followed by solvent extraction (Tang & Teoh, 1985). In the field of oils and fats, understanding the physicochemical properties of these materials is essential for the industrial manufacturing of edible oils and fats (Cebula & Smith, 1991). This chapter describes the important physicochemical properties of palm oil and its components.

The Physical Properties of Palm Oil and Its Components

An in-depth understanding of the physical properties of palm oil and its components is important for the application of these oil products in various food and non-food products. Research on phase transitions and polymorphism in palm oil has been widely reported (Che Man et al., 1999; Haryati et al., 1997; Kawamura, 1980; Rosselle, 1975; Tan & Che Man, 2002; Yap et al., 1989). Other important physical properties of palm oil include solid fat content, viscosity, density, specific gravity, and refractive index.

Phase Transitions of Palm Oil and Its Components

Phase transitions govern changes in the physical state of all materials (Roos, 1992). Phase transitions are important in determining the physical states of oils and fats during processing, storage, and consumption. Melting and crystallization, two phase-transition events commonly used to characterize the thermal behavior of oil samples, require the intake or release of thermal enthalpy. In the palm oil industry, melting and crystallization analyses are frequently and routinely performed for both quality control and product development. In addition, the melting and crystallization behaviors of palm oil and its components are important properties affecting their functionality in many prepared food products. The most important phase-transition characteristic of palm oil and its products is the melting point. Che Man et al. (1999) described the thermal profiles of palm oil and its products. The authors outlined the importance of the thermal behavior of various palm oil products and concluded that their thermal profiles can be used as guidelines for the fractionation of crude palm oil or refined, bleached, and deodorized palm oil.

Typical crystallization curves for palm oil and palm olein are illustrated in Fig. 13.1, and the melting curves for these two oils are displayed in Fig. 13.2. The higher-temperature region defines the crystallization of the stearin fraction, whereas the lower-temperature region indicates the crystallization of the olein fraction (Fig. 13.1, top). Fig. 13.2 illustrates that the higher-temperature region is absent in the crystallization curve of palm olein. The melting curve of palm oil shows two major endothermic regions (Fig. 13.2), corresponding to endothermic transitions of the olein (lower-temperature peak) and stearin (higher-temperature peak) fractions. The endothermic region at higher temperatures consists of a plateau with a pair of shoulder peaks, whereas the endothermic region at lower temperatures contains four overlapping peaks. However, palm olein shows only one major endotherm in the lower-temperature region (corresponding to the olein fraction).

Tan & Che Man (2000) studied the crystallization and melting curves of palm oil, palm olein, palm stearin, and palm kernel oil. For palm oil, the crystallization curve displays two major exothermic regions. The crystallization curve of palm stearin exhibits both of these regions, but the higher-temperature region in palm stearin is apparently the major feature. The crystallization curve of palm kernel oil exhibits two overlapping exothermic peaks. In addition, the melting curve of palm stearin exhibits both endothermic regions (similar to palm oil); the higher-temperature region is distinguished by a tall peak (and two completely overlapping peaks) preceding the low-temperature region (consisting of four small, overlapping peaks) (Tan & Che Man, 2000). The small, low-temperature peak exhibited in the melting curve of palm stearin indicates that a small amount of the olein fraction is trapped in this oil sample after fractionation. These results corroborate these authors' observations of the crystallization curves. In palm kernel oil, there is a major endotherm peak with a shoulder peak and two small fusion peaks.

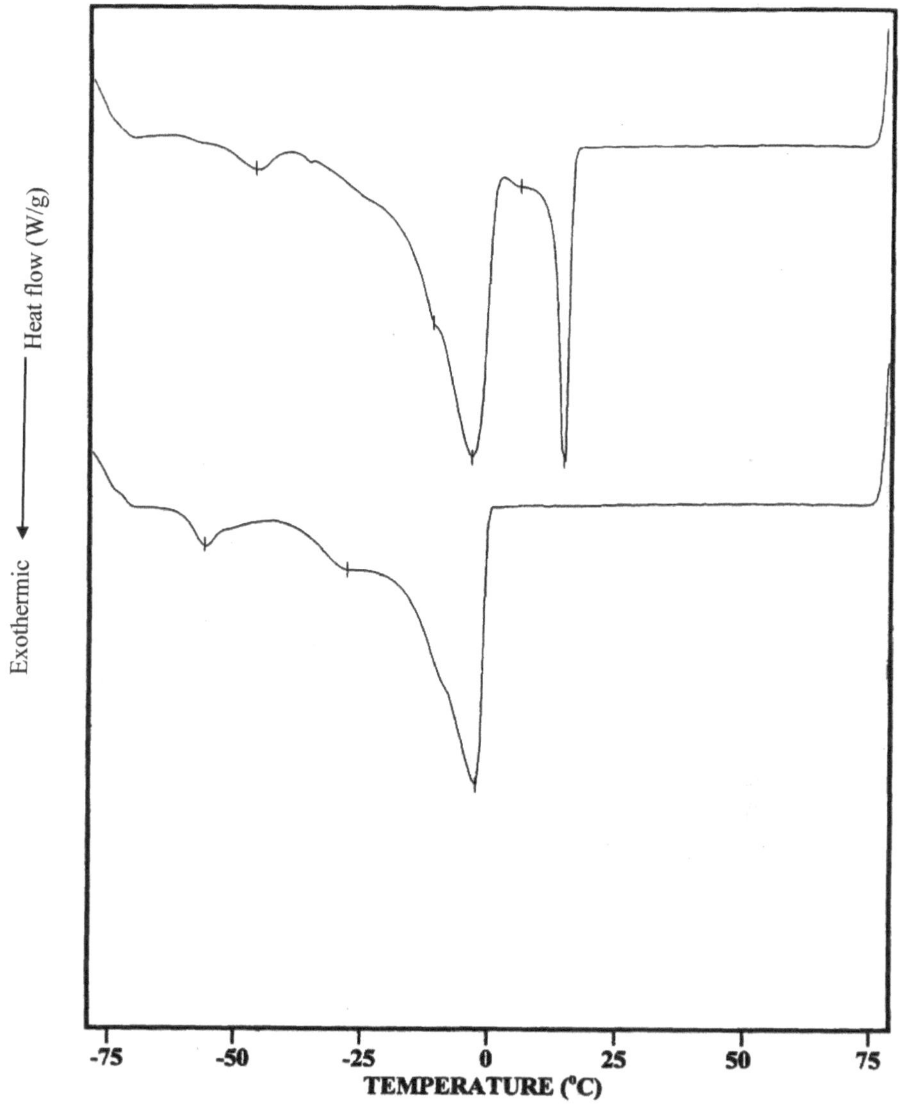

Fig. 13.1. Typical differential scanning calorimetry cooling curves of palm oil (top) and palm olein (bottom).

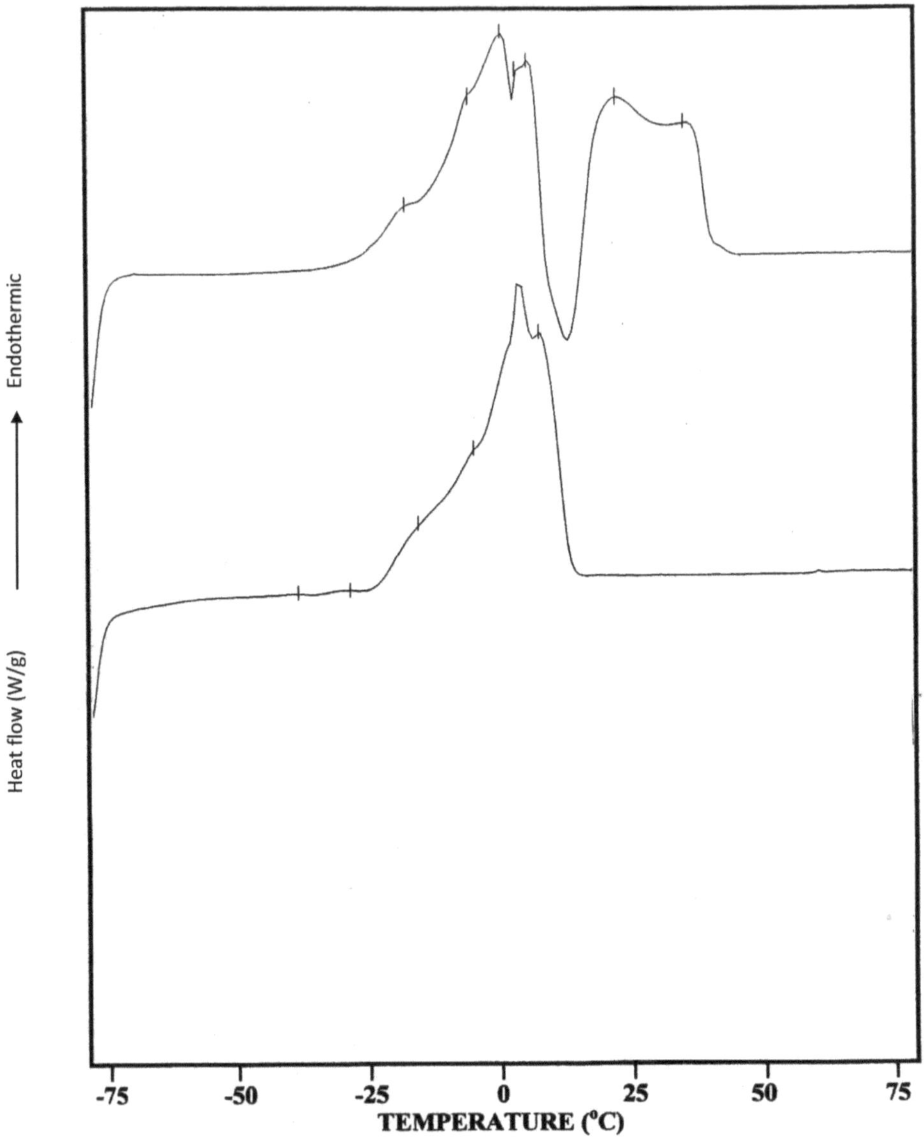

Fig. 13.2. Typical differential scanning calorimetry melting curves of palm oil (top), palm olein (bottom).

Differential scanning calorimetric techniques have been widely used to determine the phase transitions of various fat and oil products (Tan & Che Man, 2002). The phase transitions of palm oil and its components, as determined from differential scanning calorimetry melting and crystallization curves, are characterized by various transition temperatures. Tan & Che Man (2000) also reported the onset temperature (T_o), offset temperature (T_f) and temperature range (i.e., the difference between T_o and T_f) for both the melting and crystallization curves of palm oil, palm olein, palm stearin, and palm kernel oil.

Polymorphism and Crystal Structure

The systematic investigation of polymorphism is an important stage in the development of new oil- and fat-based products. Polymorphism describes phase changes and structural modifications in the solid-fat phase (Herrera & Marquez Rocha, 1996). There is abundant evidence in scientific literature indicating the importance of polymorphism in palm oil fats and products (D'Souza, 1990; Jacobsberg & Ho, 1976; Yap et al., 1989).

It has been extensively reported that oils solidify in more than one polymorphic form or crystalline type (Sato, 1999). Triacylglycerols (TAGs) exhibit three main crystal types: α, β′, and β, listed in order of increasing melting point and stability (Talbot, 1995). Depending on the conditions under which they are crystallized, palm oil crystals can also be found in these three polymorphic forms. The molecular conformations and packing in the crystals of each polymorph have been reported elsewhere. In the α form, the fatty acid (FA) chain axes of the TAGs are randomly oriented, and the α form has the greatest freedom of molecular motion and the most loosely packed hexagonal subcell structure (Lawler & Dimick, 1998). The β′ and β forms feature extended-chain conformations with orthorhombic and triclinic subcell structures, respectively. In the β′ form, alternating FA-chain axes are oriented in opposite directions, whereas in the β form, all FA-chain axes are oriented in one direction (Talbot, 1995).

In general, palm oil tends to display the β′ form, whereas palm kernel oil is β form (Mag, 1995). It also has been found that the addition of palm oil to fats destined for shortening and margarine production has beneficial effects on polymorphic stability of these fats (Berger, 1986). The desirable texture of fat-containing foods is dependent on the presence of the correct polymorphic form. For example, certain foods may require the fat crystals to either melt very quickly or very slowly to provide a desirable sensation upon being eaten (i.e., a good "mouth feel") (deMan, 1994). In addition, fat-crystal morphology determines the appearance of certain food products (e.g., margarine) which must be in the β′ form to have a smooth and glossy texture. Thus, the crystallization behavior of many readily available oils is of great concern to manufacturers of fats and oils.

Binary and ternary mixtures of palm-oil-based products contain fat crystals with different polymorphic forms than those of individual palm-based component oils.

This topic has been extensively studied by many researchers (Byung et al., 2008; Timms, 1984; Zhou et al., 2010). Zhou et al. (2010) studied the binary and ternary phase behaviors of mixtures of palm kernel oil, palm stearin, and palm olein using isosolid diagrams. In these ternary systems, these authors observed that the addition of palm kernel oil to the blends stabilized the β′ crystal form, whereas increasing the content of palm stearin caused the blends to undergo a β′-to-β polymorphic state transformation. The polymorphic forms of palm oil products can also be modified by the interesterification process (Rousseau et al., 1998).

In addition, the influence of minor components on polymorphic changes in palm oil products has been investigated by many researchers studying fats and oils (Fredrick et al., 2008; Miskandar et al., 2004; Okiy, 1978; Siew & Ng, 1995; Siew & Ng, 1996; Smith, 2000; Smith et al., 2011; Timms, 1985). Fredrick et al. (2008) observed that the nucleation of a particular polymorph in palm oil is not promoted by the presence of an unsaturated monoacylglycerol from sunflower oil but by a monoacylglycerol from palm oil itself. In addition, both types of monoacylglycerols accelerated the transition from the α to the β′ polymorph. Miskandar et al. (2006) elucidated the effects of emulsifiers on the fat crystals in palm oil and palm olein blends.

Solid Fat Content

Solid fat content is another important physical parameter for palm oil and its components, especially for the production of shortening, margarine, and other fat products for which temperature has a strong relationship to performance (Ong et al., 1982). Typical solid fat contents of palm oil at 10, 20, 30, and 40°C are 47–56%, 20–27%, 6–11%, and 1–6%, respectively (O'Brien, 1998). In general, no solid fat is found in palm oil at 50°C. However, for palm stearin, the solid fat content measured at 50°C ranges from 0 to 40% depending on the fractionation process. The solid fat contents of other palm oil products, such as palm olein, palm stearin, and palm kernel oil, have also been reported elsewhere (Ong et al., 1995).

Slip Melting Point

The slip melting point, also known as the softening point, is the melting point determined using an open capillary tube method. The slip melting point is defined as the point at which a fat changes from solid to a liquid. However, it is difficult to define the actual melting point of fats, as almost all fats are complex mixtures of TAGs with different melting profiles. In addition, this method of measuring the slip melting point is heavily influenced by the pretreatment of fats (e.g., tempering time and temperature). The typical slip melting points for palm oil and palm kernel oil range from 33 to 40°C and 24–26°C, respectively (Firestone, 2006). Timms (1985) reported values for the slip melting points of palm oil, palm olein, palm stearin, and palm kernel oil as 36.3, 22.8, 54.0 and 27.6°C, respectively.

Density

In the fats and oils industry, density data are essential for trading purposes because all fat and oil products are traded on a weight basis but are measured on a volume basis (Gunstone, 2008). The relative densities of palm oil, palm olein, palm stearin, and palm kernel oil are in the ranges of 0.891–0.899 (at 50°C), 0.899–0.920 (at 40°C), 0.881–0.891 (at 60°C), and 0.899–0.914 (at 40°C), respectively (Codex Alimentarius, 1999).

Refractive Index

The refractive index of an oil measures the angle at which a ray of light bends when it passes from the oil into air. The value of the refractive index is affected by temperature and is normally measured at 25°C. Because of the solid nature of palm oil and its components, the refractive indexes of these oils are normally measured at temperatures above 40°C. This value is also affected by internal factors (e.g., molecular weight, chain length, and the degree of unsaturation of the fatty acids in an oil sample). In general, the refractive index of palm oil measured at 50°C is in the range of 1.449–1.455 (Firestone, 2006).

The Chemical Properties of Palm Oil and Its Components

Palm oil and its components are increasingly used in foods such as cooking oils, margarines, shortenings, and confectionery products. The versatility and adaptability of palm oil to different food applications is the result of its chemical composition (Yap et al., 1989).

Fatty Acid Composition

Palm oil differs from many common vegetable oils in its high level of palmitic acid, which is approximately 44%. In general, palm oil contains approximately 50% saturated fatty acids, 40% monounsaturated fatty acids, and 10% polyunsaturated fatty acids. The FA compositions of palm oil and its components have been widely reported by many researchers (Che Man et al., 1999; O'Brien, 1998; Ong et al., 1995; Tan & Che Man, 2000; Tan et al., 1981). Table 13-A shows the typical FA compositions of palm oil and its components as tabulated by Tan & Che Man (2000) in comparison with Firestone's report (2006). The proportions of saturated fatty acid (SFA), monounsaturated fatty acid (MUFA), and polyunsaturated fatty acid (PUFA) data are listed in Table 13-B. Palm oil, palm olein, and palm stearin are the oils obtained from palm pulp. These oils are distinguished from other oils by their high levels (41.6–68.3%) of palmitic (C16:0) acid. The predominance of this FA is the main reason for the low iodine value (IV) of palm oil and its components. The fractionation of palm oil into olein and stearin fractions has a significant influence on FA composition. As shown in Table 13-A, the C16:0 component tends to migrate into the palm stearin fraction.

Table 13-A. Fatty Acid Composition of Typical Palm Oil and Its Components.

	PO		PO_O		PO_S		PKO	
Fatty acid (%)	Results[1]	Range[2]	Results[1]	Range[2]	Results[1]	Range[2]	Results[1]	Range[2]
C6:0	–	–	–	–	–	–	0.5	0–0.8
C8:0	–	–	–	–	–	–	6.4	1.9–6.2
C10:0	–	–	–	–	–	–	5.2	2.6–5.0
C12:0	0.5	0–0.4	0.7	0.1–0.5	0.4	0.1–0.4	55.8	40–55
C14:0	1.7	0.5–2.0	1.5	0.9–1.4	2.1	1.1–1.8	14.7	14–18
C16:0	48.7	40–48	41.6	38.2–42.9	68.3	48.7–73.8	5.8	6.5–10.3
C16:1	–	0–0.6	–	0.1–0.3	–	0.05–0.2	–	–
C18:0	3.9	3.5–6.5	3.8	3.7–4.8	4.0	3.9–5.6	1.3	1.3–3
C18:1	37.1	36–44	42.0	39.8–43.9	20.6	15.6–36.0	8.9	12–21
C18:2	8.1	6.5–12.0	10.4	10.4–13.4	4.6	3.2–9.8	1.5	1–3.5
C18:3	–	0–0.5	–	0.1–0.6	–	0.1–0.6	–	0–0.7
C20:0	–	0–1	–	0.2–0.6	–	0.3–0.6	–	0–0.3
C20:1	–	0–0.2	–	–	–	–	–	0–0.5
C22:0	–	0–0.1	–	–	–	–	–	–
C24:0	–	0–0.2	–	–	–	–	–	–

[1]Authors' own findings.
[2]Source: Firestone, 2006.
Abbreviations: RBDPO, palm oil; PO_O, palm olein; PO_S, palm stearin; PKO, palm kernel oil.

However, the FA composition of palm olein is relatively similar to palm oil despite the fractionation process. In palm oil, palm olein, and palm stearin, SFAs account for more than 47% of the total FA, whereas MUFAs, mainly oleic (C18:1) acid, are approximately 20–42% of the total FA. The PUFA content ranges from 4 to 11%, corresponding to the amount of linoleic (C18:2) acid.

The FA composition of palm kernel oil is similar to that of coconut oil (Young, 1983). Palm kernel oil is a coproduct of palm oil milling. Although palm kernel oil and palm oil are derived from the same plant, these oils differ considerably in their characteristics and properties. Palm kernel oil is similar to coconut oil in that both are high in lauric (C12:0) and myristic (C14:0) acids. However, palm kernel oil has a lower content of short-chain FA (C6:0–C10:0) and a slightly higher oleic (C18:1) acid content than coconut oil. In palm kernel oil, the SFA content accounts for almost 90% of the total FA, and the remaining 10% of the FAs are MUFAs and PUFAs (Table 13-B).

Acylglycerol Composition

Acyl esters of glycerol constitute the major storage lipids in palm oil and its components. Like other vegetable oils, palm oil and its components are complex mixtures of TAGs. Approximately 94–98% of palm oil and its components consist of mixtures of TAGs, each esterified with three FAs (O'Brien, 1998). The balance consists of mono- and diacylglycerols plus various other minor components (Nawar, 1996; Zambiazi & Przybylski, 1998). Consequently, the functional properties of palm oil and its components as ingredients in prepared foods are directly related to the type of TAGs in these oils and fats (Williams & Hron, 1996). TAG type is determined by the FA composition of the TAG and the distribution of the FAs in the individual TAG molecule, and the quantities of each TAG type depend on the proportions of the individual FAs, the fat or oil source, and the processing history of a product (Reske et al., 1997).

Table 13-B. Distribution of Saturated, Monounsaturated, and Polyunsaturated Fatty Acids in Typical Palm Oil and Its Components.[1]

	Fatty acid distribution (%)		
Oil	SFA	MUFA	PUFA
PO	54.7	37.1	8.1
PO_O	47.6	42.0	10.4
PO_S	74.8	20.6	4.6
PKO	89.7	8.9	1.5

[1]Authors' own findings.
Abbreviations: SFA, saturated fatty acid; MUFA, monounsaturated fatty acid; PUFA, polyunsaturated fatty acid. For other abbreviations see Table 13-A.

Table 13-C. Triacylglycerol Composition of Typical Palm Oil and Its Components[1]

Triacylglycerol (%)	PO	PO_O	PO_S	PKO
CCLa	–	–	–	6.8
CLaLa	–	–	–	9.9
LaLaLa	–	–	–	21.2
LaLaM	–	–	–	17.0
LaLaO	–	–	–	5.3
LaMM	–	–	–	8.8
MMM	0.4	0.7	0.2	–
LaLaP	–	–	–	1.2
LaMO	–	–	–	4.6
MPL	2.4	3.2	1.0	–
LaMP	–	–	–	4.6
LaOO	–	–	–	3.8
LaPO	–	–	–	4.3
LaPP+MMO	–	–	–	1.9
OOL	0.7	0.7	0.1	–
MMP	1.8	2.3	0.8	0.7
MOO	–	–	–	2.0
MPO+POL	–	–	–	2.1
POL	10.1	12.8	5.3	–
PPL	9.8	10.7	7.8	0.6
MPP	0.6	–	2.3	–
OOO	4.1	4.9	1.8	1.4
POO	24.2	29.1	12.0	1.9
PPO	31.1	27.2	29.8	1.1
PPP	5.9	–	29.2	0.1
SOO	2.3	3.1	0.8	0.4
PSO	5.1	5.0	3.8	0.4
PPS	0.9	–	5.2	–
SSO	0.5	0.4	–	–

[1]Source: Tan & Che Man (1999).
Abbreviations: TAG, triacylglycerol; C, capric; La, lauric; M, myristic; P, palmitic; S, stearic; O, oleic; L, linoleic. For other abbreviations see Table 13-A.

The TAG compositions of palm oil and its components have been reported in many studies (Che Man et al., 1999; Ghazali et al., 1995; Okiy, 1978; Tan & Che Man, 2000). Table 13-C shows the TAG compositions of typical palm oil, palm olein, palm stearin, and palm kernel oil reported by Tan & Che Man (2000). TAGs can be further divided into four component categories: trisaturated (SSS) TAGs, disaturated-monounsaturated (SSU) TAGs, monosaturated-diunsaturated (SUU) TAGs, and triunsaturated (UUU) TAGs. Tan & Che Man (2000) also reported these four TAG groups in palm oil and its components (Table 13-D). Every fat and oil possesses a natural FA distribution among the glycerides, which affects a product's consistency as either a solid or a liquid (Sonntag, 1979c). SSS TAGs provide structure, SSU TAGs provide both structure and lubricity, and the lower-melting unsaturated TAGs provide lubricity only.

In palm oil, palm olein, and palm stearin, Tan & Che Man (2000) reported that POO and PPO (P, palmitic; O, oleic) account for up to 56% of the TAGs (Table 13-C). In addition to POO and PPO, palm stearin also predominantly contains PPP (29.2%), a completely saturated TAG. Thus, in palm stearin, SSS and SSU constitute more than 75% of the total TAG (Table 13-D). In contrast, Tan & Che Man (2000) found that palm oil and palm olein consist mainly of SSU and SUU, with small quantities of SSS and UUU (Table 13-D). Palm kernel oil is characterized as a hard oil. The saturated TAGs (SSS) are the major TAG of palm kernel oil and account for 86% of the total TAG. Among these, LaLaLa and LaLaM (La, lauric acid; M, myristic acid) are the major SSS TAGs (Table 13-C).

In addition to TAGs, palm oil and its components have been reported to contain minor amounts of monoacylglycerols (MAGs) and diacylglycerols (DAGs). Che Man et al. (1999) reported that palm oil, palm olein, and palm stearin contain 5.20, 5.55,

Table 13-D. Distribution of Trisaturated, Monounsaturated, Diunsaturated, and Triunsaturated Triacylglycerols (TAG) in Typical Palm Oil and Its Components[1]

	TAG distribution (%)			
Oil	SSS	SSU	SUU	UUU
PO	9.8	48.8	36.5	4.8
PO_O	2.9	46.5	44.9	5.6
PO_S	37.7	42.3	18.1	1.9
PKO	70.3	16.3	6.0	1.4

[1]Source: Tan & Che Man (1999).
Abbreviations: S represents saturated fatty acids; U represents unsaturated fatty acids; SSS, trisaturated triacylglycerol; SSU, monounsaturated triacylglycerol; SUU, diunsaturated; UUU, triunsaturated triacylglycerol. The sequence does not necessarily reveal that on the glycerine moiety. For other abbreviations see Tables 13-A.

and 5.15% DAGs, respectively. However, in another study almost 15 years earlier than Che Man et al. (1999), Goh & Timms (1985) reported higher DAG contents of 6.44, 6.83, and 5.76% for palm oil, palm olein, and palm stearin, respectively. Goh & Timms (1985) also reported the MAG contents of palm oil, palm olein, and palm stearin as less than 0.078%.

Minor Components

Crude palm oil has a deep orange-red color resulting from its high carotene content of 0.03 to 0.15%, of which 90% consists of β- and β-carotene. In addition, palm oil products are rich in many phytonutrients such as tocopherols, tocotrienols, and phytosterols. The oxidative stability of palm oil is affected by the presence of high levels of carotene, which acts as a pro-oxidant even in the presence of high tocopherol and tocotrienol concentrations (Basiron, 2005). Therefore, palm oil is often bleached (which eliminates the carotene), refined, and deodorized. An in-depth discussion of the minor components of palm oil is presented in Chapter 16 of this book.

Other Chemical Properties

Other important chemical properties of palm oil and its components include the iodine value (IV) and the saponification value (SV). The IV measures the degree of unsaturation and has been used as an important quality parameter in the palm oil industry. The SV can be used to predict the types of acylglycerols in the oil sample by measuring the alkali-reactive groups in oil samples. Typical IV ranges for palm oil, palm olein, palm stearin, and palm kernel oil are 50.0–55.0, >56, <48, and 14.1–21.0 g of I_2/100 g oil, respectively (Codex Alimentarius, 1999). Typical ranges of SV for palm oil, palm olein, palm stearin, and palm kernel oil are 190–209, 194–202, 193–205, and 230–254 mg KOH/ g oil, respectively (Codex Alimentarius, 1999).

Conclusions

The physicochemical properties of palm oil and its components were extensively studied during the 1980s and 1990s. Most of the current knowledge in this area is based on studies conducted during that period. Two classic reports on the physicochemical properties of palm oil and its components were published by Tan & Oh (1981a, 1981b). However, continuous improvements in our understanding of the physicochemical properties of this major edible oil will play an important role in the sustainable development of the palm oil industry. The utilization of palm oil products in various complex food and nonfood systems requires the constant improvement of our knowledge of the physicochemical properties of palm oil and its components. In addition, product diversification in the palm oil industry also requires substantial understanding of the physicochemical properties of this globally important edible oil. Because the major portion of the world's palm oil production is for food consumption, changes in the physicochemical properties of palm oil during various food processing steps must also be closely monitored. Safety and health issues related to palm oil products are also closely related to the basic chemical components of this edible oil. Recent safety issues, such as the presence of 3-mono-chloropropane-1,2-diol (3-MCPD) esters in refined edible oils, should also be addressed through in-depth studies of the interaction of basic components in the unrefined oil with external precursors. Many novel physicochemical properties of palm oil may also be determined through the development of new and sophisticated analytical instruments.

References

Basiron, Y. Palm oil. In Bailey's Industrial Oil and Fat Products, Vol 2, 6th ed.; Shadidi, F., Ed.; John Wiley & Sons: New York, 2005; pp. 333–429.

Berger, K., Palm oil products—why and how to use them. Food Technol. 1986, 40(3): 72–79.

Byung, H.K.; Lumor, S.E.; Akoh, C.C. Trans-free margarines prepared with canola oil/palm stearin/palm kernel oil-based structured lipids. J. Agric. Food Chem. 2008, 56, 8195–8205.

Cebula, D.J.; Kevin, W.S. Differential scanning calorimetry of confectionery fats. Pure Triglycerides: Effects of Cooling and Heating Rate Variation. J. Am. Oil Chem. Soc. 1991, 68, 591–595.

Che Man, Y.B.; Haryati, T.; Ghazali, H.M.; Asbi, B.A. Composition and thermal profile of crude palm oil and its products. J. Am. Oil Chem. Soc. 1999, 76, 237–242.

Codex Alimentarius, Codex Standard For Named Vegetable Oils (CODEX STAN 210-1999), 1999, 1–16.

de Man, J.M. X-ray diffraction spectroscopy in the study of fat polymorphism. Food Res. Int. 1994, 25, 471–476.

D'Souza, V.; deMan, J.M.; deMan, L. Short spacings and polymorphic forms of natural and commercial solid fats: A review. J. Am. Oil Chem. Soc. 1990, 67, 835–843.

Firestone, D., Ed. In Physical and Chemical Characteristics of Oils, Fats, and Waxes, 2nd edition; AOCS Press: Champaign, IL, 2006; pp. 110–115.

Fredrick, E.; Foubert, I.; Van De Sype, J.; Dewettinck, K. Influence of monoglycerides on the crystallization behavior of palm oil. Cryst. Growth Des. 2008, 8, 1833–1839.

Gascon, J.P.; Noiret, J.M.; Meunier, J. Oil palm. In Oil Crops of the World, Robbelen, G., Downey, R.K., Ashri, A., Eds.; McGraw-Hill: New York, 1989; pp. 475–495.

Ghazali, H.M.; Maisarah, A.; Yusof, S.; Yusoff, M.S.A.M. Triglyceride profiles and melting properties of lipase-catalyzed transesterified palm stearin and coconut oil. Asia Pacific J. Mol. Biol. Biotech. 1995, 3, 280–289.

Goh, E.M.; Timms, R.E. Determination of mono- and diglycerides in palm oil, olein and stearin. J. Am. Oil Chem. Soc. 1985, 62, 730–734.

Gunstone, F.D. Physical properties. In Oils and Fats in the Food Industry; Wiley-Blackwell: New York, 2008; pp. 59–70.

Haryati, T.; Che Man, Y.B.; Swe, P.Z. Effect of repeated heating on thermal behavior of crude palm oil. J. Am. Oil Chem. Soc. 1997, 74, 393–396.

Jacobsberg, B.; Ho, O.C. Studies in palm oil crystallization. J. Am. Oil Chem. Soc.1976, 53, 609–617.

Kawamura, K. The DSC thermal analysis of crystallization behavior in palm oil. II, J. Am. Oil Chem. Soc. 1980, 57, 48–52.

Law, K.; Thiagarajan, T. Palm oil–edible oil of tomorrow. In Edible Fats and Oils Processing: Basic Principles and Modern Practices: World Conference Proceedings, Erickson, D.R., Ed.; AOCS Press: Champaign, IL, 1990, pp. 260–269.

Lawler, P.J.; Dimick, P.S. Crystallization and polymorphism of fats. In Food Lipids: Chemistry, Nutritional, and Biotechnology; Akoh, C.C, Min, D.B., Eds.; Marcel Dekker: New York, 1998, 229–250.

Mag, T.K. Food industry requirements for fats and oils: functional properties. In Development and Processing of Vegetable Oils for Human Nutrition; Przybylski, R., McDonald, B.E., Eds.; AOCS Press: Champaign, IL, 1995, 36–46.

Miskandar, M.S.; Che Man, Y.; Rahman, R.A.; Aini, I.N.; Yusoff, M.S.A. Effects of emulsifiers on crystallization properties of low-melting blends of palm oil and olein. J. Food Lipids 2006, 13, 57–72.

O'Brien, R.O. Raw materials, in fats and oils: formulating and processing for applications; Technomic Publishing Company: Lancaster, PA, 1998; pp. 1–46.

Okiy, D.A. Interaction of triglycerides and diglycerides of palm oil, Oleagineux 1978, 33, 625–628.

Ong, A.S.H.; Boey, P.L.; Ng, C.M. A spectrophotometric method for determination of solid fat content of palm oil. J. Am. Oil Chem. Soc. 1982, 59, 223–226.

Ong, A.S.H.; Choo, Y.M.; Ooi, C.K. Developments in palm oil. In Developments in Oils and Fats; Hamilton, R.J., Ed.; Blackie Academic and Professional: London, 1995; pp. 153–191.

Reske, J.; Siebrecht, J.; Hazebroek, J. Triacylglycerol composition and structure in genetically modified sunflower and soybean oils. J. Am. Oil Chem. Soc. 1997, 74, 989–998.

Roos, Y.H. Phase transitions and transformations in food systems. In Handbook of Food Engineering. Heldman, R., Lund, D.B., Eds.; Marcel Dekker: New York, 1992; pp. 145–197.

Rossell, J.B. Differential scanning calorimetry of palm kernel oil products. J. Am. Oil Chem. Soc. 1975, 52, 505–511.

Rousseau, D.; Marangoni, A.G.; Jeffrey, K.R. The influence of chemical interesterification on the physicochemical properties of complex fat systems. 2. morphology and polymorphism. J. Am. Oil Chem. Soc. 1998, 75, 1833–1839.

Sato, K. Solidification and phase transformation behavior of food fats—a review. Fett/Lipid 1999, 12, 467–474.

Siew, W.L.; Ng, W. Partition coefficient of diglycerides in crystallization of palm oil. J. Am. Oil Chem. Soc. 1995, 72, 591–595.

Siew, W.L.; Ng, W.L. Characterisation of crystals in palm olein. J. Sci. Food Agric. 1996, 70, 212–216.

Smith, K.W.; Bhaggan, K.; Talbot, G.; Van Malssen, K.F. Crystallization of fats: Influence of minor components and additives. J. Am. Oil Chem. Soc. 2011, 88, 1085–1101.

Smith, P.R. The effects of phospholipids on crystallisation and crystal habit in triglycerides. Eur. J. Lipid Sci. Technol. 2000, 102, 122–127.

Sonntag, N.O.V. Structure and composition of fats and oils. In Bailey's Industrial Oil and Fat Products, Vol I, 4th ed.; Swern, D., Ed.; John Wiley & Sons: New York, 1979c; pp. 1–98.

Talbot, G. Fat eutectics and crystallization. In Physicochemical Aspects of Food Processing; Beckett, S.T., Ed.; Chapman and Hall: London, 1995; pp. 142–166.

Tan, B.K.; Oh, F.C.H. Malaysian palm oil chemical and physical characteristics. PORIM Technol. 1981a, 3, 1–5.

Tan, B.K.; Oh, F.C.H. Oleins and stearins from Malaysian palm oil chemical and physical characteristics. PORIM Technol. 1981b, 4, 1–6.

Tan, B.K.; Oh, F.C.H.; Siew, W.L.; Berger, K.G. Characteristics of processed palm oils. In Palm Oil Product Technology in the Eighties, PORIM/Planters' Seminar; Incorporated Society of Planters: Kuala Lumpur, Malaysia, 1981; pp. 127–143.

Tan, C.P.; Che Man, Y.B. Differential scanning calorimetric analysis of edible oils and fats: comparison of thermal properties and chemical composition. J. Am. Oil Chem. Soc. 2000, 77, 143–155.

Tan, C.P.; Che Man, Y.B. Comparative differential scanning calorimetric analysis of vegetable oils: I. effects of heating rate variation. Phytochem. Anal. 2002, 13, 129–141.

Tan, C.P.; Che Man, Y.B. Differential scanning calorimetric analysis of palm oil based products and coconut oil: effects of scanning rate variation, Food Chem. 2002, 76, 89–102.

Tang, T.S.; Teoh, P.K. Palm kernel oil extraction—the Malaysian experience. J. Am. Oil Chem. Soc. 1985, 62, 254–258.

Timms, R.E. Phase behaviour of fats and their mixtures. Progress Lipid Res. 1984, 23, 1–38.

Timms, R.E. Physical properties of oils and mixtures of oils. J. Am. Oil Chem. Soc. 1985, 62, 241–249.

Williams, M.A.; Hron Sr., R.J. Obtaining oils and fats from source materials. In Bailey's Industrial Oil and Fat Products, 5th ed., Vol. 4: Edible Oil and Fat Products: Processing Technology; Hui, Y.H., Ed.; John Wiley & Sons: New York, 1996; pp. 61–156.

Yap, P.H.; deMan, J.M.; deMan, L. Polymorphism of palm oil and palm oil products. J. Am. Oil Chem. Soc. 1989, 66, 693–697.

Young, F.V.K. Palm kernel and coconut oils: analytical characteristics, process technology and uses. J. Am. Oil Chem. Soc. 1983, 60, 374–379.

Zambiazi, R.C.; Przybylski, R. Effects of endogenous minor components on the oxidative stability of vegetable oils. Lipid Technol. 1998, 12(5), 58–62.

Zhou, S.-L.; Zhang, F.-Q.; Jin, Q.-Z.; Liu, Y.-F.; Shan, L.; Zhang, T.; Zou, X.-Q.; Wang, X.-G. Characterization of palm kernel oil, palm stearin, and palm olein blends in isosolid diagrams. Eur. J. Lipid Sci. Technol. 2010, 1041–1047.

14

Polymorphic Properties of Palm Oil and Its Major Component Triacyglycerols

Kiyotaka Sato
Hiroshima University, Higashi-Hiroshima, Japan

Introduction

Palm oil is a fat and oil resource for many industrial uses, particularly for food applications as cooking oil, margarine, fat spread, shortening (Aini and Miskandar, 2007; Miskandar et al., 2002a, 2002b), confectionery fats (Timms, 2003), and others. Palm oil has several advantageous properties such as high productivity, high thermal and oxidative stability, and plasticity at ambient temperature. However, the crystallization properties of palm oil are disadvantageous because of a low rate of nucleation and crystal growth of granular crystals. The granular crystals are easily formed during long storage, causing sandy taste and an inhomogeneous fat crystal network in the end products (Ishikawa et al., 1980; Tanaka et al., 2007; Tanaka et al., 2009; Watanabe et al., 1992).

Although many triacylglycerols (TAGs), diacylglycerols (DAGs), free fatty acids, and so on are involved in palm oil (Braipson-Danthine & Gibon, 2007; Che Man et al., 1999; Goh & Timms, 1985), it exhibits polymorphic forms. α and β' forms occur under normal cooling rate, and β form also occurs at very slow crystallization rate. The presence of DAGs and the addition of food emulsifiers into palm oil are important external factors to modify the polymorphic crystallization of palm oil in bulk and emulsion systems (Arima et al., 2007; Basso et al., 2010; Frederick et al., 2008; Garbolino et al., 2005; Kuriyama et al., 2001; Miskandar et al., 2006; Siew & Ng, 1996; Smith et al., 2011).

The crystallization behavior of palm oil has been reviewed by many authors, who highlighted specific crystallization properties of palm oil in relation to industrial applications (Chong et al., 2007; Kellens et al., 2007; Smith, 2001). However, the polymorphic properties of palm oil are still under discussion, mostly because palm oil contains different types of TAGs containing saturated and unsaturated moieties. The main TAGs which can crystallize at ambient temperature (>10°C) are PPP (tripalmitoyl-glycerol), POP (1,3-dipalmitoyl-2-oleoyl-sn-glycerol), PPO (1,2-dipalmitoyl-3-oleoyl-rac-glycerol), and POO (1,2-dioleoyl-3-palmitoyl-rac-glycerol). The palm oil produced in Malaysia contains PPP (7.2%), POP (23.7%), POO (21.5%). In addition, PPS (S, stearic acid, 1%) and POS (3.1%) are included. Other TAGs such a trioleoyl-glycerol (OOO, 5.1%), PLiO (Li, linoleic acid, 6.0%), and POLi (3.1%) are also major TAGs. However, they do not crystallize at ambient temperature and act as a role of solvent when the crystallization of high-melting TAGs occurs during dry fractionation, and, therefore, we do not consider the polymorphic behavior of these low-melting TAGs in this chapter.

Fig. 14.1 shows DSC cooling and heating thermograms of palm oil taken at a rate of 2°C/min. On cooling, two large exothermic peaks appear around 19°C and 0°C, and quite complicated peaks appear on heating. The exothermic peaks noted by A and B in Fig. 14.1 corresponds to the crystallization of α form of high melting TAG fractions of PPP, POP, and PPO, and low-melting TAG fractions of POO, PLiO, POLi, and OOO, respectively. The exothermic peaks observed in the cooling experiments vary their occurrence temperatures, occurrence patterns, and polymorphic forms as the cooling rate decreased because of differences in the crystallization behavior of each TAG fraction. Similarly, the DSC heating thermograms also vary with decreasing heating rate because of the differences in the rates of polymorphic transformations from metastable forms to more stable forms of each TAG fraction.

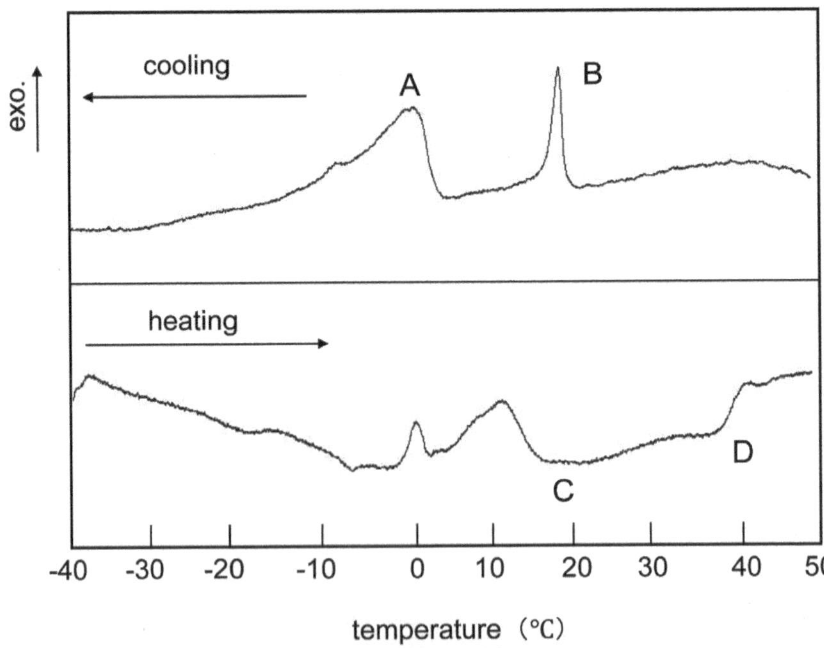

Fig. 14.1. DSC cooling and heating thermograms of palm oil.

Therefore, a simple understanding of "polymorphism of palm oil" may not have a unique physical meaning. This is because, in general, the behavior of crystallization and polymorphic transformation of major TAG fractions and their mixtures are largely influenced by the rate of cooling and heating and the crystallization temperatures (Larsson et al., 2006; Sato & Ueno, 2005; Sato & Ueno, 2011). These properties are also observed in palm oil and, as a result, quite complicated crystallization phenomena are observed in palm oil and palm oil-based hard stocks.

The main causes of such a complexity are (1) polymorphism of the major TAGs in palm oil is different from each other and (2) the mixing behavior of the major TAGs also depends on the combination TAG fractions. For example, the polymorphic structures of PPP and POP are largely different, and the binary mixture of PPP and POP exhibit immiscible eutectic property. In other examples of POP and PPO, their polymorphic structures are largely different as well, and the binary mixing phases of POP and PPO exhibit molecular compound forming property. A similar situation also appears in the cases of PPP-POO and POP-OPO, as we discuss these properties in this chapter in more detail.

Thus, it is evident that the polymorphic properties of palm oil should be understood in two folds: bulk properties of palm oil itself and analytical properties of every major TAG fraction and the mixtures of major TAG fractions. In this chapter we discuss the analytical properties of major TAG fractions and their mixtures of major TAG fractions, mostly based on our recent studies.

To begin with, we briefly summarize the basic concepts of polymorphism and mixing phase behavior of TAGs.

Polymorphism and Mixing Phase Behavior of TAGs

Molecular Properties of Polymorphism

Polymorphism is defined as the ability of a chemical compound to form different crystalline structures. α, β′ and β are typical polymorphic forms of TAGs (Sato & Ueno, 2005; Sato & Ueno, 2011). The melting and crystallization behavior will differ from one polymorph to another. Polymorph α is the least stable form, and it easily transforms either to the β′ form or to the β form, depending on the thermal treatment. Polymorph β′ is usually a meta-stable form but it can be the most stable form in some cases. The β′ crystals are used in margarine, shortening, confectionery fats, and so on because of its optimal crystal morphology and fat crystal networks, which give rise to optimal rheological and texture properties. Such properties of the β′ form are revealed by their tiny, plate-like crystals which enable the formation of a fine crystal network. The most stable β form tends to form large and plate-like crystal shapes (Acevedo & Marangoni, 2010a, 2010b; Acevedo et al., 2011), resulting in poor macroscopic properties in shortening and margarine. However, the β form is employed as cocoa butter equivalent (CBE) fats (Timms, 2003).

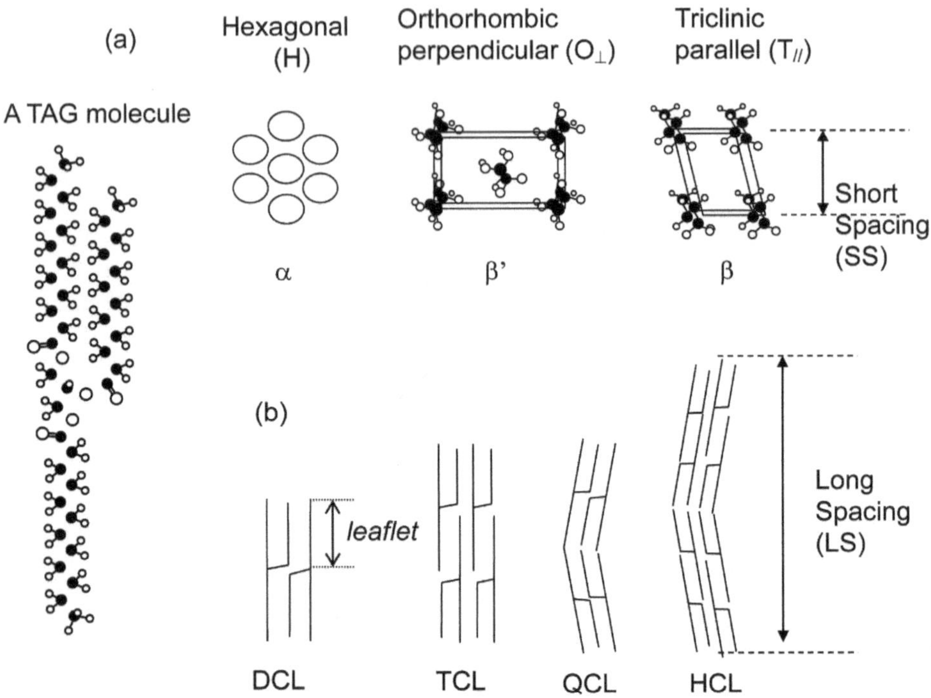

Fig. 14.2. Molecular aspects of polymorphism of triacylglycerols.

The α, β′ and β forms of TAGs are defined in accordance with subcell structure: α has a hexagonal subcell (H), β′ has an orthorhombic–perpendicular subcell ($O_{\perp}$), and β has a triclinic–parallel subcell ($T_{//}$) (Fig. 14.2). The subcell structures can be determined most clearly by measuring X-ray diffraction (XRD) short spacing values using poly-crystalline samples.

The structures of TAGs are also characterized by chain length structure, which illustrates the repetitive sequence of the acyl chains involved in a unit cell lamella along the long-chain axis. The chain length structure varies from double (DCL) to triple (TCL), quarto (QCL), or hexa (HCL) structures, as the fatty acid moieties of the fats become more complicated.

In addition to the subcell and chain length structures, glycerol group conformations such as tuning fork and chair conformations are also important (Fig. 14.3). Two conformations of chair-1 and chair-2 differ stereochemically between optical isomers, when the three fatty acid moieties are all different or the fatty acid moieties of sn-1 and sn-2 are different from that of sn-3, or the fatty acid moieties of sn-2 and sn-3 are

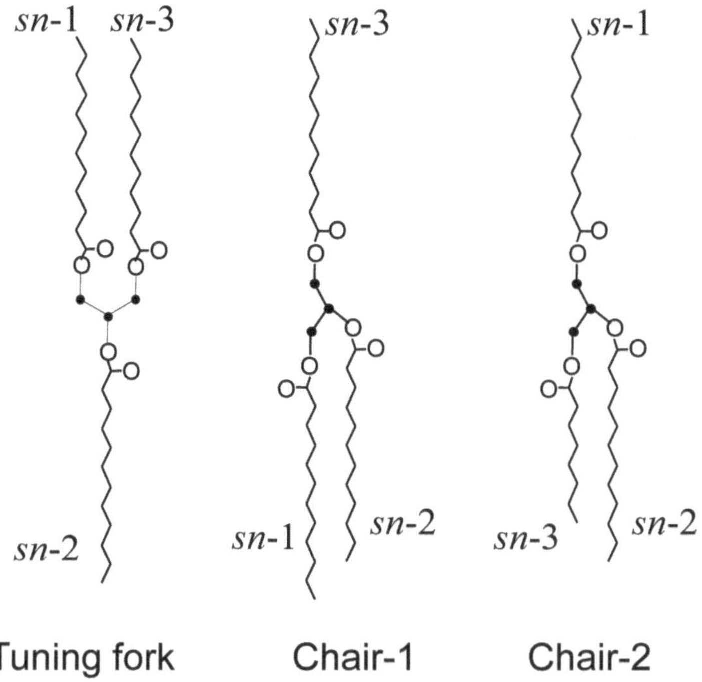

Fig. 14.3. Glycerol group conformations. Three carbon atoms of a glycerol group are noted by closed circles, and oxygen atoms are noted by open circles. Hydrocarbon chains are noted by zigzag lines, in which sn means stereospecific number.

different from that of sn-1. One may assume that the glycerol conformations must affect mixing behavior of fats having different glycerol conformations and fatty acid moieties, as discussed below.

To combine the polymorphic structures and the chain length structures, one can refer every polymorphic form in such ways to α-2 or β-3, which mean α form of DCL and β form of TCL, respectively.

In food fats, transformation from β′ to β often causes deterioration of the end product, mostly because of changes in crystal morphology and network (Marangoni, 2005). As bulky large crystals of β form deconstruct fine crystal network formed by β′ crystals, rheological and textural properties of the end products are damaged. For example, lowered spreadability, loss of oil binding property, and sandy taste may occur.

The β-type polymorph is found in confectionery fats made of cocoa butter (Timms, 2003). There are two β forms: a meta-stable β_2 form is more useful than the more stable β_1 form due to preferred crystal morphology and crystallization behavior for β_2.

It is a prerequisite for those who are engaged in the material production of oils and fats to know how the fatty acid composition influences the fat polymorphism. Much work has recently been done on the polymorphic behavior of mixed-acid TAGs including saturated and unsaturated fatty acid moieties (Arishima & Sato, 1989; Arishima et al., 1991; Boubekri et al., 1999; Goto et al., 1992; Miura & Konishi, 2001; Sato et al., 1989, 2001, 2009; Takeuchi et al., 2000; van Langevelde et al., 2000; Yano et al., 1993, 1999). It is important to understand such behavior because natural oils and fats always contain these mixed-acid TAGs. Palm oil is the most typical fat of this type.

The fatty acid compositions of TAGs are closely related to β′-tending properties. The TAGs containing different types of fatty acids are more stable in the β′ form, as exemplified in pure TAGs. Recent work on single-crystal structure analyses of the β′ form of 1,3-dilauroyl-*sn*-2-caproyl-glycerol and 1,2-dipalmitoyl-*sn*-3-myristoryl-glycerol has indicated that chain–chain interactions through methyl end stacking combined with glycerol group conformations (chair conformation) may stabilize the β′ structures (Sato et al., 2001; van Langevelde et al., 2000). In natural fats, milk fats having long-chain and short-chain saturated and unsaturated fatty acids are β′-tending; palm oil is also categorized as a β′-tending fat because of the presence of asymmetric mixed-acid TAGs such as POO and PPO.

Binary Mixing Behavior of Triacylglycerols

When the two TAGs are mixed in all proportions in a liquid phase, three types of mixing phases can occur: solid solution phase, eutectic phase, and molecular compound formation phase. The temperature-concentration (T-c) phase diagrams for the mixtures of A and B components are illustrated in Fig. 14.4.

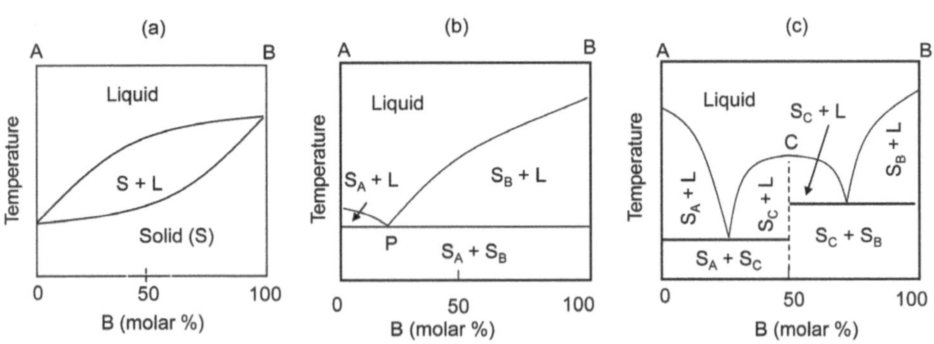

Fig. 14.4. Typical phase behavior of binary mixtures of triacylglycerols. (a) Miscible solid solution phase, (b) immiscible eutectic phase, and (c) molecular compound forming phase.

Solid Solution Phase

A solid solution phase occurs when two components are miscible at all concentration ratios, in liquid and solid phases. Only two phases can exist in such a system: a homogeneous liquid-solution and a solid-solution. As shown in Fig. 14.4(a), in this system, when a binary mixture liquid is cooled, neither component crystallizes without containing some of the other component. That is, both components are crystallized simultaneously. All mixtures of the two components have melting points intermediate between the melting points of the pure components.

This phase property is formed when the TAGs are very similar in melting point, molecular volume, and polymorphism, for example, mixtures of SOS/POS (Rousset et al., 1998), LLL/MMM (Takeuchi et al., 2003) and PPP/SSS (α and β' forms) (Kellens et al., 1991), and SOS/SLiS in α and γ forms (Takeuchi et al., 2002a), in which S, L, and M represent stearoyl, lauroyl, and myristoyl moieties.

Eutectic Phase

A eutectic phase exhibits when two components are miscible in liquid but not miscible in solid, as shown in Fig. 14.4(b). The straight line, named as a solidus line, represents the temperature at which solid mixtures of A and B begin to melt, or temperature below which mixtures of these two components are completely solid. The two curves, named as liquidus lines, represent the temperature at which homogeneous liquid-solutions begin to crystallize, or the temperature above which mixtures of these two components are completely liquid. The eutectic point, P, is the lowest melting or solidification temperature. When the eutectic point shifts along the solidus to endpoint, this phase is called monotectic.

The eutectic phases including monotectic phases are most commonly found in the binary mixtures of TAGs, such as PPP/SSS (β form) (Kellens et al., 1991), PPP/LLL (Takeuchi et al., 2003), PPP/POP (Minato et al., 1996), POP/POO (Zhang et al., 2007), SOS/SOO (Zhang et al., 2009), etc.

Molecular Compound Formation Phase

The phase diagram for a molecular compound (noted by C) formation demonstrates two eutectic phases. As shown in Fig. 14.4(c), it can be split into two phase diagrams of its sub-systems: A/C and C/B, and each of these sub-systems are eutectic. If a molecular compound can coexist in stable equilibrium with a liquid of the same composition, the compound is said to have a "congruent melting point," which is represented by point C.

A molecular compound formation is based on a possibility to achieve specific close-packing arrangements of the molecules in the solid state. The occurrence of compound formation in particular combinations of TAGs has been observed in the mixtures of SOS/OSO (Koyano et al., 1991), SOS/SSO (Engstrom, 1992; Takeuchi et al., 2002b), POP/OPO (Bayés-García et al., 2011b; Ikeda et al., 2010; Minato et al., 1997a, 1997b), and POP/PPO (Minato et al., 1997c). The presence of oleic acids moiety is commonly observed in the TAGs that are forming the molecular compound phases.

Polymorphism of Major Component Triacylglycerols in Palm Oil

Table 14-A summarizes the polymorphic forms and the molecular structure models of the most stable form of five major TAGs, which can crystallize above about 10°C.

PPP

PPP possesses three forms of α, β′ and β. As PPP is trisaturated-acid TAG, the chain length structures of the three forms are all DCL [(Fig. 14.5(a)], and their melting points are shown in Table 14-A. The glycerol conformation is of a tuning fork type. The long-spacing (LS) values are 4.60 nm (α), 4.23 nm (β′), and 4.09 nm (β). Some literature refers to the occurrence of two β′ forms (β'_1 and β'_2), but the presence of these two forms seems to be still under debate.

The metastable α and β′ forms transform to β′ and β in a consequential manner, or directly to β, through melt-mediation or solid state. The melt-mediated transformation occurs by raising the temperature of the crystal above the melting point of a particular polymorph. Recrystallization of the more stable forms soon after the melting of less stable forms causes the melt-mediate transformation. The solid-state transformation occurs below its melting point through rearrangement of the subcell structures and chain stacking sequences in the solid phase.

Table 14-A. Polymorphic Forms of Major Triacylglycerols in Palm Oil and Their Melting Points (°C).

PPP	POP	PPO	POO
α-2 (44.7)	α -2 (15.2)	α -3 (18.5)	α -3 (−13.0)
β′-2 (56.6)	γ-3 (27.0)	β′-3 (35.2)	β′-3 (20.0)
β -2 (66.4)	β'_1-3 (29.2)		
	β'_2-2 (30.3)		
	β'_3-2 (33.5)		
	β_2-3 (35.1)		
	β_1-3 (36.7)		

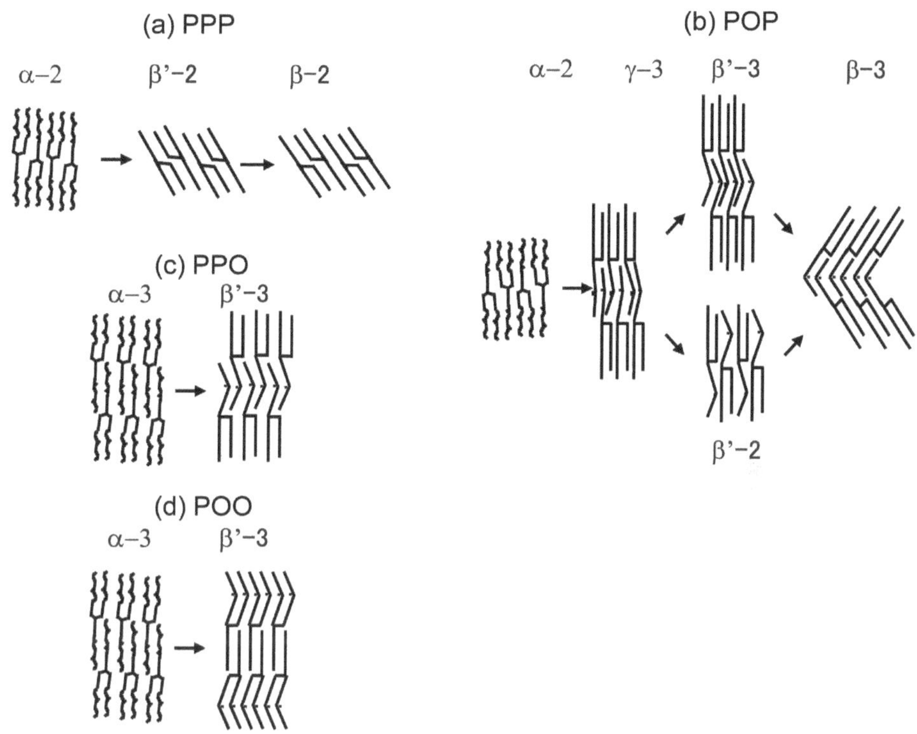

Fig. 14.5. Polymorphic transformation processes of four major triacylglycerols in palm oil.

The transformation pathways in PPP are different between the melt-mediated and solid state transformations. In the case of the transformation from the α form, the melt-mediated transformation directly forms the β form, whereas β′ and β forms sequentially occur in the solid state transformation.

POP

The polymorphic structures of POP are more complicated than that of PPP, as shown in Fig. 14.5(b) and Fig. 14.6. This may be caused by the chain-chain aliphatic interactions between palmitic acid moiety at the sn-1 and sn-3 positions and oleic acid moiety at the sn-2 position. We isolated seven forms which occur at ambient temperature, and their melting points are shown in Table 14-A, in which the forms of β'_1-3, β'_2-2, and β'_3-2 in Table 14-A were called as δ, pseudo- $\beta'_{1,}$ pseudoa- β'_2 in our first paper (Sato et al., 1989), respectively. The LS values are 4.65 nm (α), 6.54 nm (γ), 6.25 nm (β'_1-3), 4.24 nm (β'_2-2), 4.24 nm (β'_3-2), 6.10 nm (β_2), and 6.10 (β_1). These

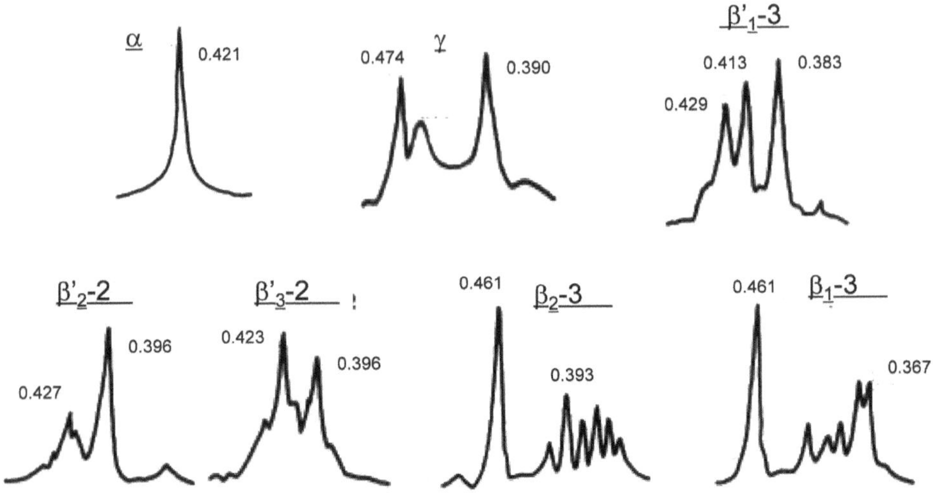

Fig. 14.6. Wide-angle X-ray diffraction patterns of seven polymorphs of POP (unit, nm).

polymorphic forms were classified according to their unique melting points and wide-angle X-ray diffraction (WAXD) patterns. In addition to the six forms, sub-α form is formed when the α form was cooled.

The α form is packed according to the H subcell, as revealed in its single sharp WAXD peak of 0.412 nm. The chain length structure is DCL, in which the palmitic acid and oleic acid moieties are packed and glycerol groups are also arranged neighbor-to-neighbor within the same lamellae. We may assume that such a mixed packing of palmitic and oleic acid chains is enabled, as the hydrocarbon chains in the α form are in randomly disordered conformation.

The α form converts to γ form by the solid-state and melt-mediated transformations. The palmitic acid and oleic acid moieties in γ are separated to form the TCL structure. The α–γ transformation may be caused by instability of the chain-chain packing of the palmitic acid and oleic acid moieties in the DCL structure of α, as the randomly disordered palmitic acid chains turn to be more ordered conformation and steric hindrance between the palmitic acid and oleic acid moieties does not allow them to maintain the DCL structure. The γ form having the molecular structure model shown in Fig. 14.5(b) are widely observed in Sat-Unsat-Sat TAGS, in which Sat and Unsat refers to saturated and unsaturated fatty acid moieties. Those TAGs are SOS (1,3-distearoyl, 2-oleoyl-sn-glycerol), SLiS (1,3-distearoyl, 2-linoeoyl-sn-glycerol), SRS (1,3-distearoyl, 2-ricinoleoyl-sn-glycerol), S-ALA-S (1,3-distearoyl, 2-α-linolenoly-sn-glycerol), S-EPA-S (1,3-distearoyl, 2-eicosapentanoyl-sn-glycerol),

S-DHA-S (1,3-distearoyl, 2-docosahexanloyl-sn-glycerol), AOA (1,3-arachinoyl, 2-oleoyl-sn-glycerol), and BOB (1,3-behenoyl, 2-oleoyl-sn-glycerol).

Quite interestingly, the γ form is the most stable form without β' and β forms in SLiS (Takeuchi et al., 2002a), S-EPA-S and S-DHA-S (Sato et al., 2009). Vibrational spectroscopic study using protonated and deuterated SOS indicated that the subcell structure of the stearic acid lamellae is parallel in arrangement, and disordered H-type subcell is revealed in the oleic acid lamellae in the γ form (Yano et al., 1999).

Further stabilization in the polymorphism caused the transformation from the γ form either to two β'-2 forms (β'_2-2 and β'_3-2) or to β'_1-3 form, whose WAXD patterns are quite similar to that of β' of PPP. According to the melting points, we named the three β' forms as β'_1-3, β'_2-2 and β'_3-2, having the subscript numerals with increasing melting temperature. We may reasonably understand the transformation mechanisms from γ-3 to β'_1-3, as this transformation involves the rearrangement of the subcell structure, keeping the same chain length structure. However, the transformation from γ-3 to β'_2-2 and β'_3-2 looks quite curious, since it involves simultaneous rearrangements in the chain length structures and subcell structures, and the chain length structure converts again to TCL when they transform to the more stable form of β_2-3. The mechanistic interpretation of the transformation pathway from γ-3 to β_2-3 through β'-2 forms is still open to question. We must note that the transformation pathways from DCL to TCL structures are also observed in the transformation from Form IV to Form V of cocoa butter, which includes major TAGs of POP, SOS, and POS (1-palmitoyl, 2-oleoyl, 3-stearoyl-rac-glycerol) (Sasaki et al., in press). Form IV is β'-type polymorph of DCL and Form V is β-type polymorph of TCL.

There are two stable β forms, both of which show the TCL structures. The two β forms are also observed in SOS, AOA, and BOB, whereas one β form having the WAXD pattern similar to β_1-3 is observed in POS.

PPO and POO

The polymorphism of asymmetric saturated-unsaturated mixed-acid TAG may differ among optical isomers, racemic mixtures, or racemic compounds. For example, in optical isomers of PPO, two palmitic acid moieties are placed either at the sn-1 and sn-2, or sn-2 and sn-3 positions, and the rest is oleic acid moiety, whereas two combinations are mixed in racemic PPO. As there is no precise work to compare the polymorphic properties of the racemic mixture and optical isomer, except for a recent work asymmetric saturated mixed-acid TAGs containing long and short chains (Craven & Lencki, 2011), this chapter reviews recent work on racemic PPO (Minato et al., 1997a, 1997c) and POO (Zhang et al., 2007).

PPO has two polymorphs, in which α is metastable and β' is the most stable form (Table 14-A). The two forms are stacked in the TCL structure. Their LS values are

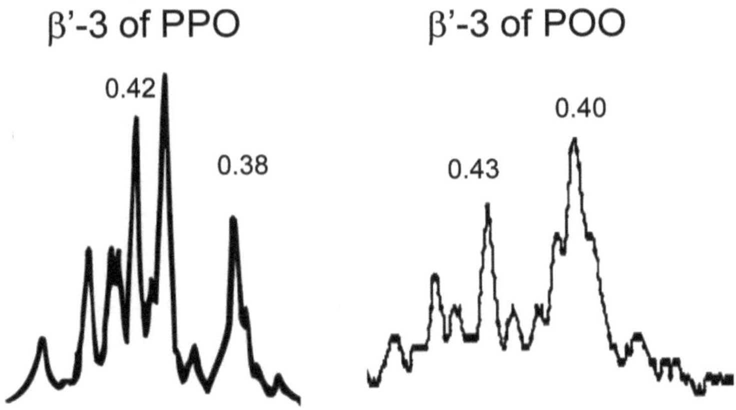

Fig. 14.7. Wide-angle X-ray diffraction patterns of stable polymorphs of racemic PPO and POO (unit, nm).

7.8 nm (α–3) and 6.60 nm (β′-3). No β form was observed in PPO. Quite similarly, POO has the metastable α–3 with the LS value of 5.7 nm and the most stable form of β′-3 with the LS value of 6.7 nm. No β form occurred in POO. The WAXD patterns of β′-3 forms of PPO and POO are shown in Fig. 14.7.

No clear reason has been given to explain two peculiar polymorphic properties of PPO and POO: (1) the occurrence of α with TCL structure, which is different from α-2 of PPP and POP, and (2) the absence of β form, which occurred in PPP and POP. In the case of polymorphism of OPO (1,3-dioleoyl-2-palmitoyl-glycerol), four forms of α-2, β′-2 and β-3 occur, and its polymorphic behavior is quite similar to that of POP as far as the chain length structures of α and β forms are concerned.

We may assume that glycerol group conformation of the asymmetric TAGs containing palmitic and oleic acid moieties may not allow the formation of the α form with H subcell packing in the DCL structure and the β form with $T_{//}$ subcell packing in the TCL structure. Further fundamental research may be necessary to answer this question.

Although not as a major TAG component in palm oil, OPO exhibits unique polymorphic structures and crystallization behavior (Baoyes-Garica et al., 2011; Minato et al., 1997b; Miura & Konishi, 2001).

Binary Mixing Behavior of Main Component Triacylglycerols in Palm Oil

In this section, we discuss polymorph-dependent mixing phase behavior between the main TAGs in palm oil. In the end, practical implication of the understandings of the mixing behavior will be discussed in relation to the applications of palm oil.

PPP/POP Mixture

We studied the mixing behavior of the PPP-POP binary mixture system in the metastable and stable conditions (Minato et al., 1996). Usually, the phase diagram is studied on the most stable forms by measuring the melting points of the stable forms of the component TAG crystals which were formed by incubating (ageing) the metastable forms to reach the thermal equilibrium. The reason for this is due to difficulty in determining the melting points of the metastable forms which easily transform to the most stable forms during the experimental runs with DSC or other methods.

However, metastable forms often crystallize in the actual crystallization processes, and mixing phase behavior of the TAGs for the metastable forms should be known prior to the crystallization experiments. For this purpose, we applied synchrotron radiation X-ray diffraction (SR-XRD) techniques to reveal the mixing properties of the metastable α and β′ forms of PPP-POP mixtures, in addition to the most stable β forms. By using the SR-XRD, we could observe the occurrence of metastable forms and their transformation pathways into the more stable forms and to the liquid phase (melting) by cooling and heating the mixed systems at different concentrations of the component TAGs. The time-resolved SR-XRD was measured during rapid cooling and heating processes, and the results were compared with those of DSC cooling and heating experiments to construct the phase diagrams of the metastable and most stable polymorphs.

Fig. 14.8(a) shows a bird's-eye view of the wide-angle SR-XRD patterns of the PPP/POP= 90/10 mixture taken during the heating process from 9 to 67°C, after

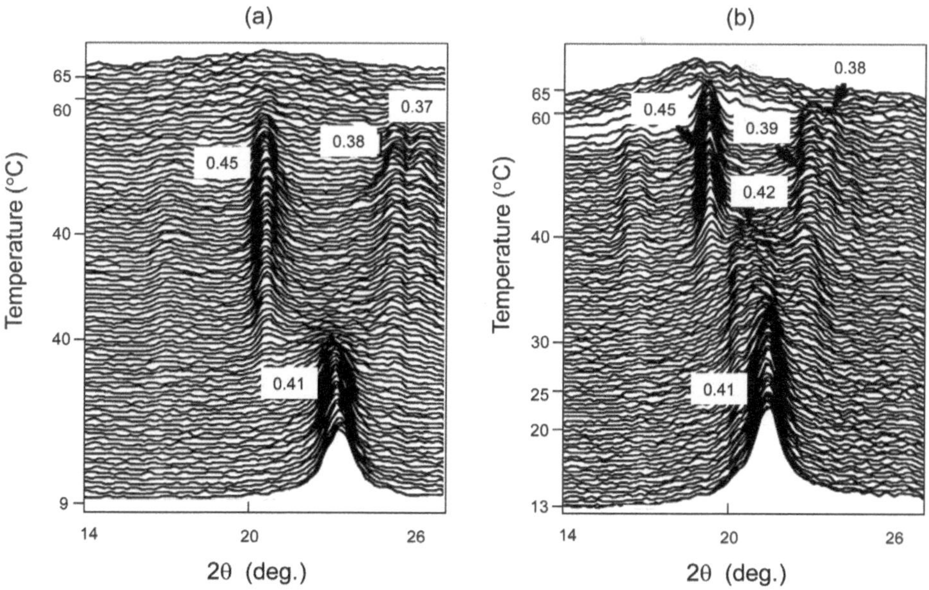

Fig. 14.8. Synchrotron radiation X-ray diffraction patterns taken during heating processes of binary mixtures of PPP and POP. (a) PPP/POP = 90/10 and (b) PPP/POP = 50/50. Unit, nm.

rapid crystallization at 9°C. When the mixture is cooled rapidly, the short-spacing pattern of 0.41 nm, which is specific to the hexagonal (H) subcell of PPP α and POP α, appeared. Then, upon heating, the short spacing pattern changed to that characteristic of the β-form. This change occurred at 40°C. It is clear that no β′ form occurred during the transformation from α to β.

By contrast, Fig 14.8(b) shows the SR-XRD pattern of the PPP/POP = 50/50 mixture on heating from 13°C The α form gradually transformed to β′ around 25°C in the first, and finally converted to β above 25°C. The α form was of PPP and POP. As for the β′ form, the first occurring fraction was POP, which converted to β above 30°C whose melting point is higher than that of POP β′. The β′ form, which was still present above 30°C, was of PPP. The finally occurring β form was of the PPP fraction.

Having made the SR-XRD experiments of the PPP/POP mixtures at different concentration ratios and corresponding DSC experiments, we determined the phase behavior of the metastable and stable forms as shown in Fig. 14.9.

As to the most stable form of β, the mixing behavior is of an immiscible monotectic nature, as β forms of PPP and POP occurred below the temperatures noted by a solid

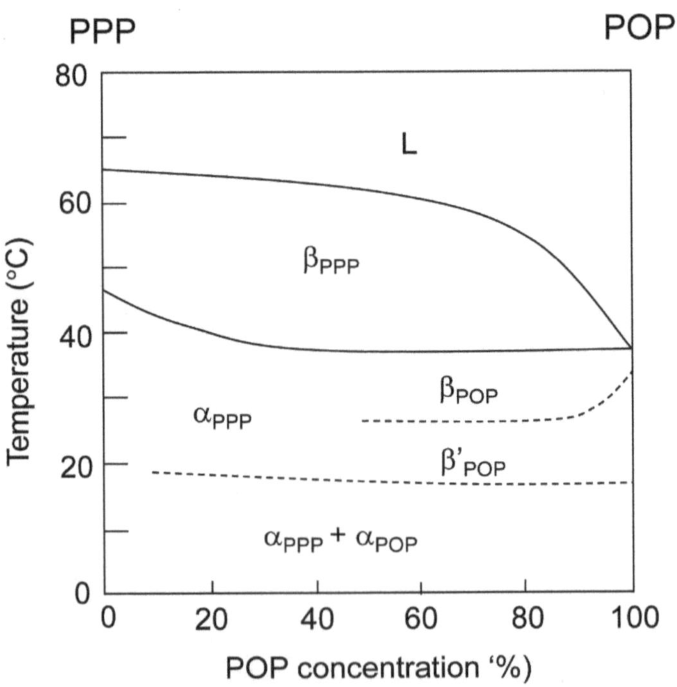

Fig. 14.9. Mixing phase behavior of binary mixture of PPP and POP (L: liquid). Solid and dotted lines refer to the most stable β forms, and metastable α and β′ forms, respectively.

line connecting between 48°C (pure PPP) and 38°C (pure POP). Above this temperature line, the β form of POP fraction melts, and the β form of PPP fraction melts at the temperature line connecting between 66°C (pure PPP) and 38°C (pure POP).

The kinetic phase diagram, which corresponds to the α and β′ forms, is divided into two regions at concentration ratios of PPP/POP between 60:40 and 50:50. In the PPP-rich region, the α forms of both PPP and POP crystallized below 20°C. Upon heating, both fractions transformed to the β form. In the POP-rich region, however, the POP fraction revealed a successive α-β′-β transformation. The melting points of α and β′-forms of PPP were not available above the POP 50% concentration.

To summarize, the PPP-POP binary system showed immiscible monotectic properties in the metastable as well as stable forms. The polymorphic behavior of PPP-POP mixtures resulted in the following: (1) in the POP ratio below 40%, the α-β solid-state transformation of PPP was observed; (2) in the POP ratio above 50%, the polymorphic transition proceeded from α to β through the intermediate β′ form, both for POP and PPP.

PPO/POP Mixture

We studied the phase behavior of PPO and POP binary mixture using DSC, conventional and synchrotron radiation XRD (Minato et al., 1997a, 1997c). In this mixture system, unlikely to the PPP-POP mixture, molecular compound crystals were formed for the three polymorphs of α, β′ and β, as summarized in Fig. 14.10.

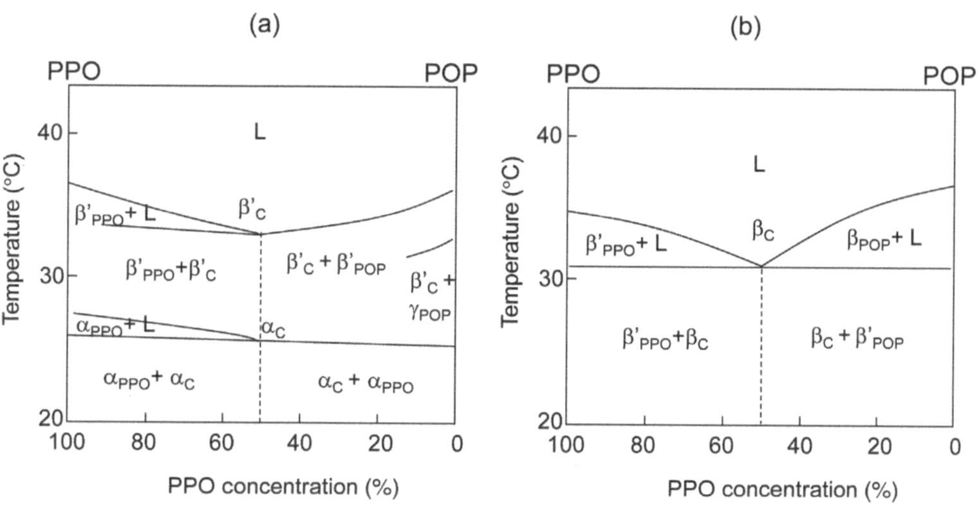

Fig. 14.10. Mixing phase behavior for (a) metastable forms and (b) most stable forms of binary mixture of PPO and POP (L: liquid).

Fig. 14.10(a) shows the kinetic phase behavior, which was constructed by using DSC and SR-XRD measurements in the same manner as those obtained for PPP-POP mixtures described in a previous section. Molecular compound crystals, α_C and β'_C, were formed at the 1:1 concentration ratio of PPO and POP, giving rise to two monotectic phases of PPO/compound and compound/POP in a juxtapositional way. Fig. 14.10(b) shows the phase diagram of the most stable form constructed by β'_{PPO}, β_{POP} and β_C.

The α_C form with the melting point of 15.5°C has a long spacing value of 4.6 nm of DCL structure. This makes a contrast to α_{PPO} having the TCL structure. In the PPO/compound region, the DSC melting peak increased with increasing PPO concentration, whereas the DSC melting peak increased with increasing POP concentration in the compound/POP region. Similarly, β'_C with the melting point of 28.0°C and β_C with the melting point of 31.2°C revealed two eutectic mixture phases with β'_{PPO} and β'_{POP} and with β'_{PPO} and β_{POP}, respectively.

Data of DSC and time-resolved SR-XRD study showed the formation of α_C and β'_C. Fig. 14.11 shows DSC heating thermogram of the PPO/POP = 90/10 mixture, which was quenched from 50°C, incubated for 15 min at 0°C and then subjected to heating DSC analysis at a rate of 2°C/min. Because of the complexity of the thermogram, we refer here to the peak-top temperatures of endothermic and exothermic peaks. A large endothermic peak appeared at 18.3°C, which was followed by an exothermic peak around 20°C. Further heating exhibited a large endothermic peak at 29.3°C. The time-resolved SR-XRD data have shown that the polymorphic occurrences and transformations of the molecular compound and PPO are simultaneously

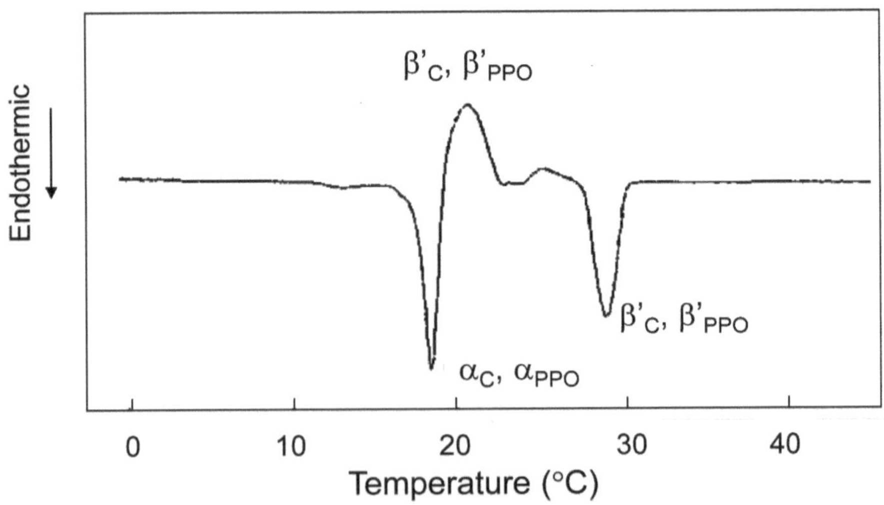

Fig. 14.11. DSC heating thermogram of the PPO/POP = 90/10 mixture.

involved in this DSC thermogram, in such a way that the endothermic peak around at 18.3°C means the melting of α_C and α_{PPO}, the exothermic peak around 20°C means the crystallization of β'_C and β'_{PPO}, and the large endothermic peak at 29.3°C corresponds to the melting of β'_C and β'_{PPO}.

Fig. 14.12 shows small and wide angle SR-XRD patterns of PPO/POP = 90/10 mixture, taken under the temperature variation profile shown in an inserted figure. Soon after 10°C was reached, a small peak appeared in small-angle region, having LS of 4.9 nm. Upon incubation at 10°C, the peak of 4.9 nm continuously shifted

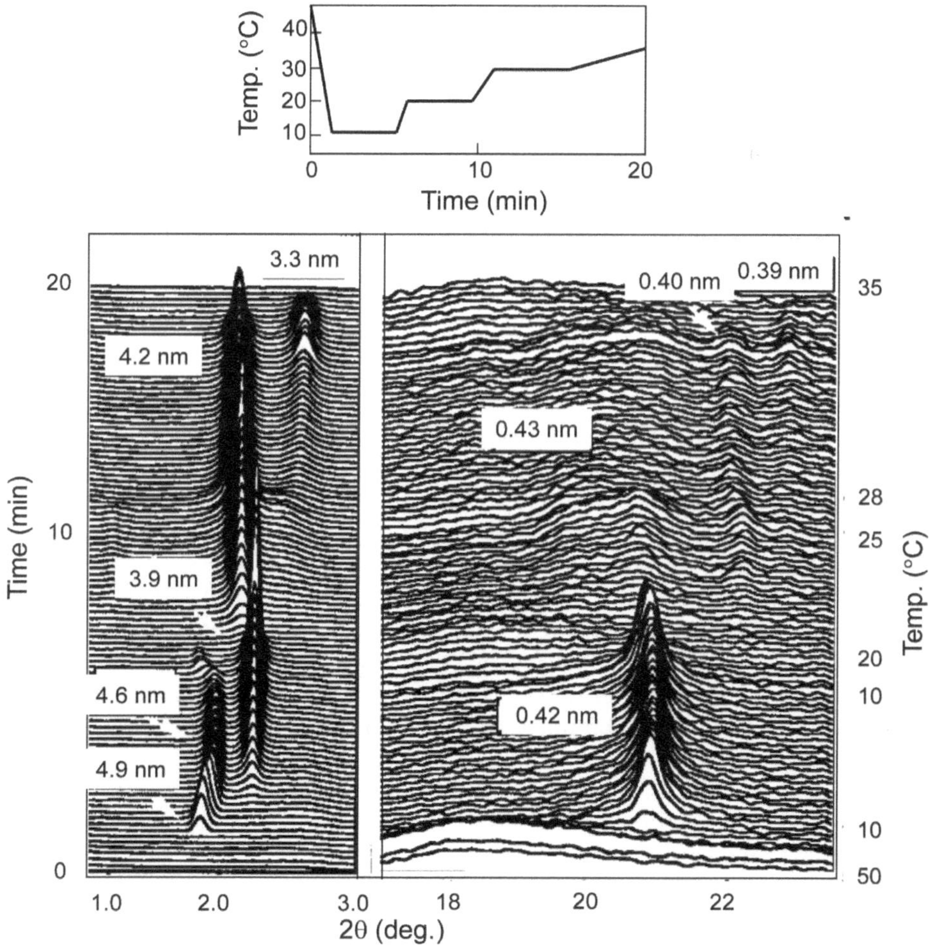

Fig. 14.12. Temperature variation profile (inserted) and small and wide angle SR-XRD patterns of PPO/POP = 90/10 mixture.

to 4.6 nm, and a new peak of 3.9 nm appeared 50 s after the occurrence of the 4.9 nm peak. Correspondingly, a short spacing peak of 0.42 nm appeared at 10°C in the wide angle region. Since the α_{PPO} has a long spacing value of 7.8 nm with the TCL structure, the peak of 3.9 nm corresponds to the (002) reflection of LS value of α_{PPO}. Thus, the peaks of 4.9 nm and 4.6 nm are due to α_C, which crystallized in the first, and α_{PPO} started to crystallize after then at 10°C. The temperature jump from 10 to 20°C induced the melting of the two α forms, as shown in the small- and wide-angle diffractions of Fig. 14.12. Hence, the DSC endothermic peak at 18.3°C in Fig. 14.11 corresponds to the melting of two α forms. Soon after the melting of α, new forms having a sharp long spacing peak of 4.2 nm and a broad peak around 3.5 nm appeared, both corresponding to the exothermic peak around 20°C in Fig. 14.11.

The temperature jump from 20 to 28°C induced the conversion from broad to sharp peaks at 3.3 nm, yet the peak at 4.2 nm was almost unchanged. Correspondingly, three short spacing spectra of 0.42, 0.40, and 0.39 nm gradually appeared and increased in their intensity in the wide-angle region, although their intensity was low. These peaks are characteristic of β′. Since β'_{PPO} has a LS value of 6.6 nm with TCL structure, the peak of 3.3 nm is the (002) reflection of β'_{PPO}. Thus, the peak of 4.2 nm is due to β'_C. A separation of β'_{PPO} and β'_C indicates the monotectic nature of the mixing system. The SR-XRD patterns of two β′ forms disappeared at 28°C upon further heating to 28°C, corresponding to the DSC endothermic peak at 28°C shown in Fig. 14.11.

Based on the data of time-resolved SR-XRD experiments of the PPO/POP mixtures at different concentration ratios, we constructed the kinetic phase diagram of α and β′ forms of PPO/POP mixture systems shown in Fig. 14.10(a). As to the most stable forms, it is not necessary to apply the SR-XRD, and conventional XRD and DSC methods can provide the data to construct the phase diagram.

The conventional XRD patterns are shown in Fig. 14.13. The small-angle peaks of 3.3 nm (PPO) and 3.1 nm (POP) correspond to the (002) reflection of LS values of 6.6 nm and 6.2 nm in the TCL structure. The mixtures of PPO/POP of 70/30 and 30/70 exhibited additional peaks at 4.1 nm. The 50/50 mixture only showed the 4.1 nm peak as shown in the small-angle XRD peak. This peak is the (001) reflection of β_C containing a 1:1 ratio of PPO and POP with the DCL structure. The separation of the two peaks of 4.1 and 3.3 nm at the 70/30 mixture and 4.1 and 3.1 nm at the 30/70 mixture is proof that the molecular compound is immiscible with POP and PPO. As evidence of this, it is notable that the intensity of 4.1 nm was higher than those of other peaks in the two mixtures. This can be explained by considering the concentration ratios of the compound and pure materials. For example, in PPO/POP = 70/30, the compound consists of 60% (compound: PPO 30% + POP 30%) and 40% of PPO. A similar interpretation can be applied to the POP 70% mixture. As for the wide-angle reflection of Fig. 14.13, pure PPO and POP exhibited the typical short spacing patterns of β′ and β_1, respectively. The short spacing patterns of the molecular compound (50%) resemble those of the molecular compounds of SOS/

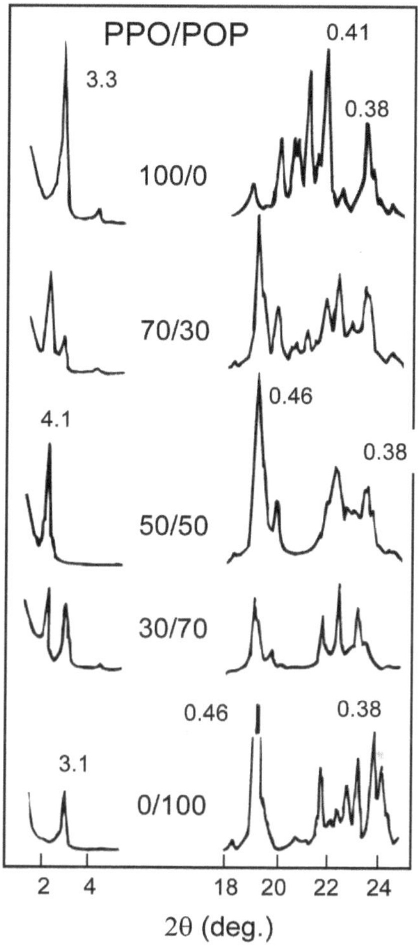

Fig. 14.13. X-ray diffraction patterns of the most stable forms of PPO/POP mixtures at five concentration ratios.

OSO and SOS/SSO, categorized as β having triclinic subcell structure. The short spacing patterns of the 70/30 and 30/70 mixtures are the superposition of those of PPO β′ and β_C, and POP β_2 and β_C, respectively.

The results shown in Fig. 14.10 through Fig. 14.13 have shown that the molecular compound crystals were formed at the 1:1 concentration ratio of PPO and POP, exhibiting the α_C, β'_C, and β_C forms. The phase behavior shown in Fig. 14.10 may be rather curious in such a sense that there is no increase in the melting point of α_C, β'_C, and β_C without showing eutectic points between the compound crystals and each pure component. This behavior is different from those of the mixtures exhibiting the molecular compound crystal formation, as illustrated in Fig. 14.4. We infer that the difference in the typical mixing behavior shown in Fig. 14.4 and the phase diagrams observed in PPO/POP mixture may be due to the racemic nature of PPO. One may expect that the phase typical diagram shown in Fig. 14.4(c) is formed when optical isomer sample of PPO is mixed with POP.

Although not presented here, the formation of α_C, β'_C, and β_C was observed in the mixtures of POP/OPO, SOS/SSO, and SOS/SSO, and the structure models of the molecular compound of POP/OPO and PPO/POP will be discussed below.

POP/POO Mixture

POP and POO are two major TAGs that account for roughly half of the total TAGs in palm oil. The binary phase behavior of a POP/POO mixture plays an important role in the crystallization of palm oil. We conducted thermodynamic and kinetic studies of POP/POO mixtures using DSC and XRD with a conventional generator and synchrotron radiation (Zhang et al., 2007). The POP/POO mixtures exhibited immiscible eutectic natures in both their metastable and their most stable states, in contrast to PPO-POP and POP-OPO, in which molecular compounds of the DCL structure were formed. Fig. 14.14 shows the phase behavior of metastable forms and the most stable forms of POP/POO(*rac*) mixtures, both of which show that POP and POO are immiscible, revealing neither solid solution nor molecular compound formation.

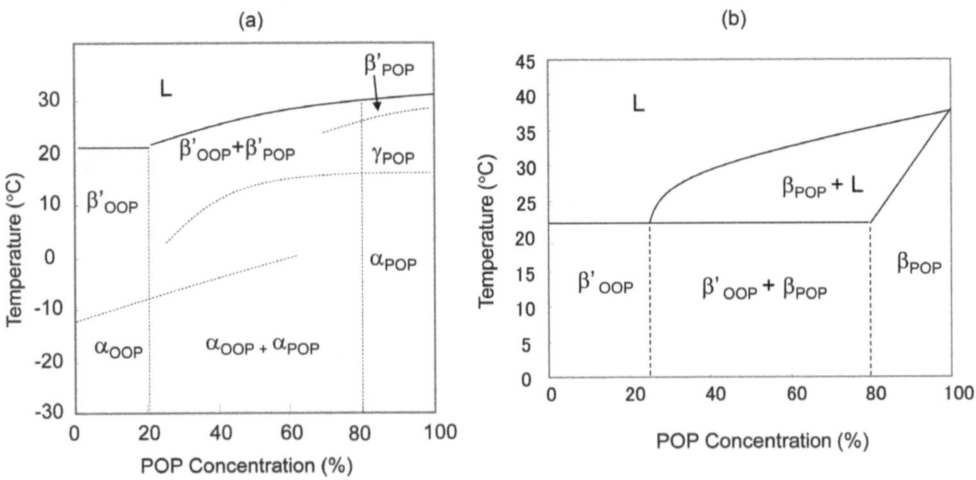

Fig. 14.14. Phase behavior of (a) metastable forms and (b) the most stable forms of POP/POO(*rac*) mixtures.

As to the metastable forms, Fig. 14.15 presents the SR-XRD patterns of the mixture of POP/POO = 30/70 obtained by cooling and heating between 30°C and –30°C when isothermal treatment was performed for 5 minutes before heating. The small angle XRD patterns clearly indicated separate crystallization and transformation of the POP and POO fractions, whereas the wide-angle XRD patterns did not

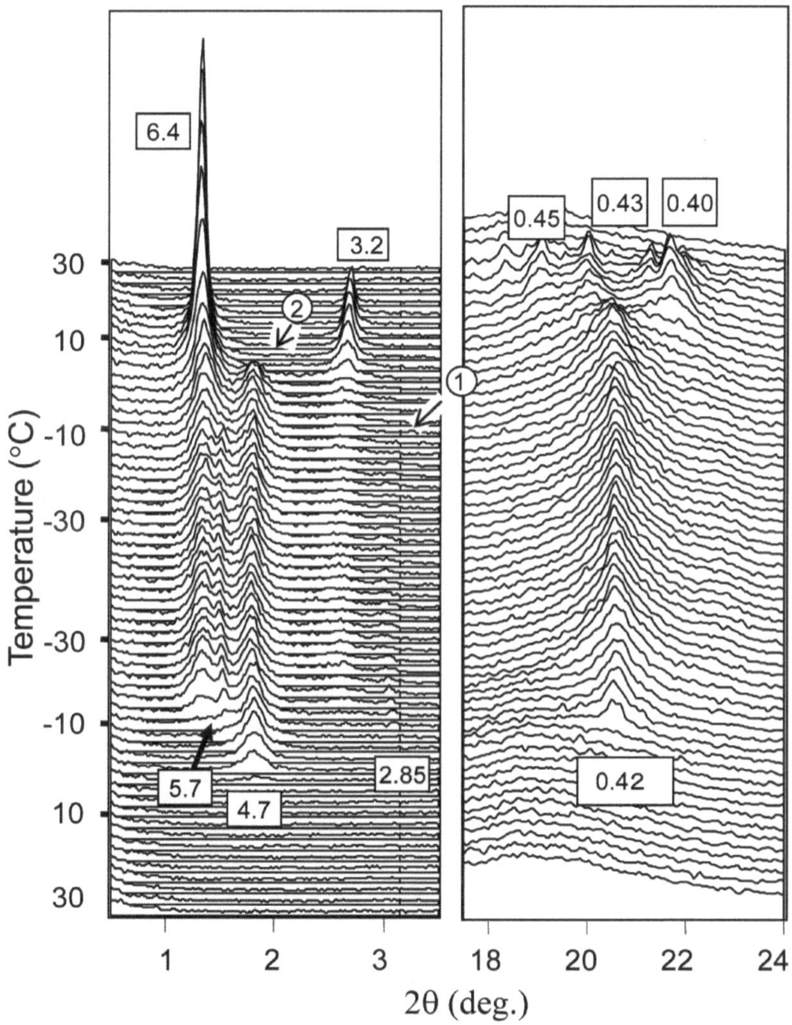

Fig. 14.15. Synchrotron radiation XRD patterns of crystallization, transformation, and melting of metastable forms of the mixtures of POP/OOP = 30/70 obtained during cooling and heating processes at a rate of 5°C/min. Unit, nm.

differ between POP and POO. The α form of POP with a LS value of 4.7 nm and short spacing of 0.42 nm crystallized around 5°C during the cooling process in the first. Subsequently, the α form of POO with LS value of 5.7 nm (001 reflection) and 2.85 nm (002 reflection) crystallized at –10°C. Further cooling caused crystallization of the β′ form of POO with LS value of 6.4 nm (001 reflection) and 3.2 nm (002 reflection) at –14°C. This pattern continued to occur throughout the cooling and heating processes.

The α form of POO transformed to β′ at –10°C during the heating process, as revealed by the disappearance of the XRD patterns of 5.7 nm and 2.85 nm (see arrow 1 in Fig. 14.15). The α form of the POP fraction transformed to β′ around 8°C, as indicated by arrow 2. The small angle XRD peak of 4.7 nm and a wide angle XRD peak of 0.42 nm disappeared at this temperature and the small angle XRD peaks of 6.4.nm and 3.2 nm increased their intensity, indicating that the β′ form of POP thus transformed from the α form was a TCL structure. At the same time, wide angle XRD patterns of 0.45 nm, 0.43 nm, and 0.40 nm appeared at the expense of the peak of 0.42 nm. Therefore, the β′ forms of the POO and POP fractions were present above 8°C, both melting around 24 to 27°C.

As to the most stable forms in the mixture of POP/POO, Fig. 14.16 presents XRD patterns of the most stable forms of POP/POO binary mixtures obtained by conventional X-ray beam generator. The small angle XRD patterns of 3.04 nm (POP) and 3.27 nm (POO) correspond to the (002) reflection of the long spacing values of 6.08 nm and 6.54 nm of a TCL structure. Pure POO exhibited distinct diffraction peaks in the wide angle XRD patterns of 2θ = 22.5° and 20.5°, revealing the presence of lattice spacing of 0.40 nm and 0.43 nm. These patterns are characteristic of the $O_{\perp}$ subcell structure, and thus the most stable polymorphic form of POO was the β′ form. Pure POP exhibited the wide angle XRD patterns of the β form.

Mixing the two samples produced no new long-spacing peak in any of the ratios of POP/POO examined. This makes a clear contrast with the XRD patterns obtained for mixtures of PPO/POP shown in Fig. 14.13, in which new LS patterns corresponding to the formation of the molecular compounds appeared and increased their intensity compared with those of the component TAGs when the concentration ratios of PPO/POP approached 1/1. The short-spacing pattern exhibited a strong peak of 0.46 nm in the mixtures with POP concentrations over 50%, including 100% POP, which corresponds to a triclinic parallel ($T_{//}$) subcell, indicating the occurrence of the β form of POP.

We constructed the phase diagram in Fig. 14.14 to illustrate the coexistence of the β′ form of POO and the β forms and POP, based on the DSC and XRD experiments.

The fact that the POP-POO mixture was immiscible, not forming a molecular compound like the POP/OPO and POP/PPO mixtures, can be explained by taking into account the molecular interactions between POP and POO in the mixture phases, as discussed below.

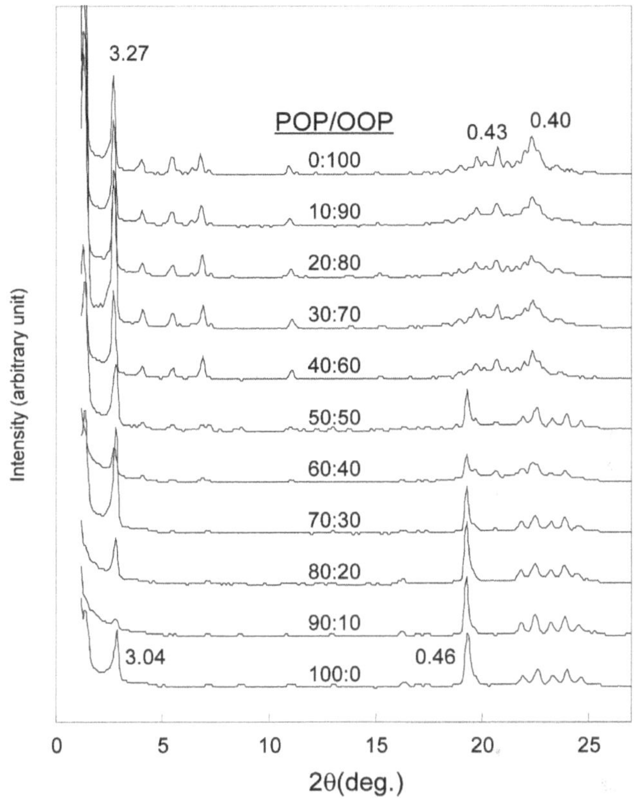

Fig. 14.16. X-ray diffraction patterns of the most stable states of the POP/OOP mixtures.

Structure Models of Molecular Compound Crystals

In general, the molecular interactions influencing the stabilization of the crystal structures of TAGs containing saturated and unsaturated fatty acid chains are aliphatic chain packing, glycerol conformation, and methyl end stacking (Sato & Ueno, 2011). The aliphatic chain packing results form the molecular interactions between saturated and unsaturated fatty acid chains, determining the subcell structures, olefinic conformations, and chain-length structures. The glycerol conformation among the glycerol groups determines the overall configuration of the TAG molecules; methyl-end stacking may play an important role in organizing the chain inclination and chain-length structure. The formation of the DCL structure in the molecular compounds of POP/OPO and POP/PPO may be due to combined effects of the above interactions. Furthermore, the fact that the POP/POO mixture does not form a molecular compound can be explained by the same consideration. Therefore, we start from the glycerol conformations observed in the stable polymorphs of representative TAG crystals (Zhang et al., 2007, 2009).

Fig. 14.17(a) illustrates the different types of glycerol conformations of two asymmetric units of TAG molecules present in a unit cell of the most stable forms of the TAGs, whose atomic-level crystal structures were determined by X-ray methods. We symbolized the glycerol conformation using the direction of the glycerol group indicated by an arrow, which is defined as the direction between the middle point of two glycerol carbons at the first and third positions and a glycerol carbon at the second position.

The directions of the glycerol groups of neighboring molecules in the β form of tricaproyl-glycerol (CCC) lay parallel to the chain axis with the opposite turn. The glycerol conformation of CCC is of tuning fork type. The same arrangements were observed in the asymmetric units of the β'_2 form of 1,2-dipalmitoyl-3-myristoyl-sn-glycerol (PPM), although there were two combinations in the directions of the glyc-

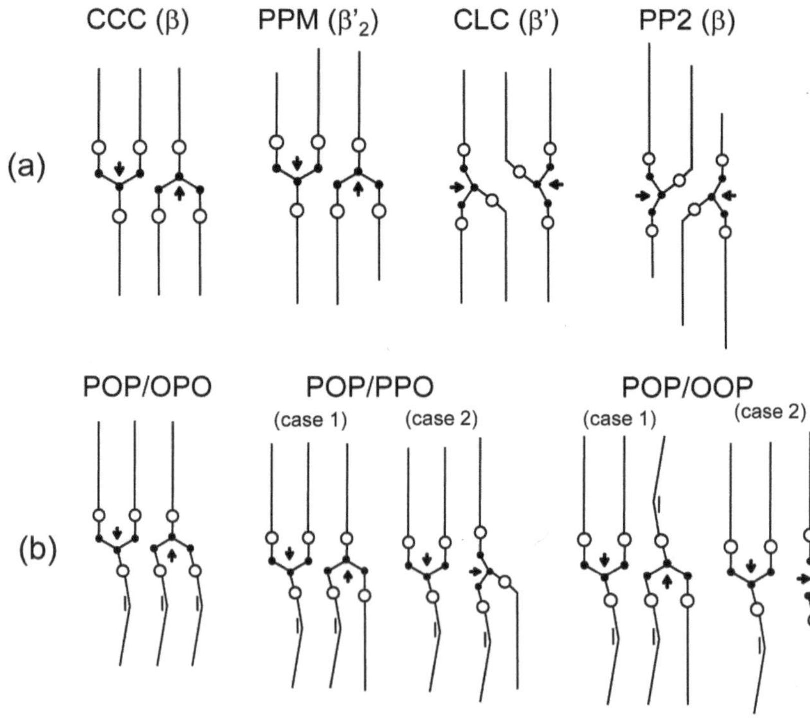

Fig. 14.17. Structure models of (a) principal triacylglycerols and (b) molecular compound crystals of POP/OPO, PPO/POP, and POP/POO.

erol groups having the chair coformation with an opposite turn and parallel turn (Sato et al., 2001). The directions of the glycerol groups of two TAGs in the β′ form of 1,3-dicaproyl-sn-2-lauroylglycerol (CLC), which has a chair type conformation, were parallel with an opposite turn, both making a right angle with respect to the chain axis (van Langevelde et al., 2000). The same arrangements were observed in the β form of 1,2-dipalmitoyl-3-acetyl-sn-glycerol (PP2) having the chair type conformation of the glycerol group (Goto et al., 1992).

The most important property, which was commonly observed in the four TAG crystals revealing the tuning fork and chair conformation of the glycerol groups, was the lack of arrangement of unparallel directions of adjacent glycerol groups making right angles to each other. This indicates that such an arrangement must cause instability of packing of the glycerol groups.

Fig. 14.17(b) illustrates possible glycerol conformations in the molecular compound crystals of mixtures of POP/OPO, POP/PPO, and POP/POO with the DCL structures. The neighboring glycerol groups in a parallel arrangement were directed along the chain axis with an opposite turn in the POP/OPO mixture. This glycerol conformation enables the oleoyl and palmitoyl chains to make separate leaflets with stabilized aliphatic interactions, which may be destabilized when the oleoyl and palmitoyl chains are placed in the same leaflet due to steric hindrance. The structure may be revealed in molecular compound crystals in the metastable and stable states of POP/OPO in which the glycerol conformations are both of tuning fork type.

In POP/PPO mixtures, the palmitoyl and oleoyl chains should be located in one leaflet in addition to being present in another leaflet composed of all the palmitoyl chains in the DCL structure. Two situations may occur; in case 1, the adjacent glycerol groups would be directed along the chain axis with an opposite turn, similar to those in the POP/OPO mixture. The glycerol conformations are both of tuning fork type. In case 2, the directions of adjacent glycerol groups would make right angles to each other. The glycerol conformations are of tuning fork type for POP and of chair type for PPO. Destabilization in the aliphatic chain packing may be caused in both cases by coexistence of the oleoyl and palmitoyl chains in the two leaflets. However, destabilization may be more enhanced in the structures in case 2, in which the glycerol conformation may be destabilized due to the unparallel directions of the glycerol groups, as the glycerol conformations are different.

Furthermore, the methyl end stacking may also be destabilized due to unequal chain lengths of the neighboring molecules of POP and PPO, which would again be caused by the unparallel directions of the glycerol groups. Therefore, the molecular compound structure of the POP/PPO mixture observed in the metastable and most stable states may be that of the structural model described in case 1.

We have considered two examples of the molecular compound crystals of POP and POO mixtures. The model of case 1 assumes a parallel direction for the glycerol groups having the tuning fork type conformation, which may be more stabilized than the unparallel direction in case 2. However, the structural model of case 1 causes destabilization of the acyl chain packing because of the coexistence of oleoyl and palmitoyl chains in the two leaflets. This situation significantly differs from the structural model of case 1 of the POP/PPO mixture, in which the coexistence of oleoyl and palmitoyl chains are limited in either leaflet.

An alternative model for POP/POO is described for case 2, in which the oeloyl and palmitoyl chains are separately placed in their own leaflets. This structure was postulated by Moran (Moran, 1963). However, this model must cause additional destabilization of the glycerol groups due to co-existence of chair-type and tuning fork-type conformations. Furthermore, unequal chain lengths of the neighboring molecules of POP and POO may occur, as discussed for case 2 of the POP/PPO mixture. Consequently, neither of the two possible models of the molecular compound structure for a POP/POO mixture may alleviate destabilization of the acyl chain packing, glycerol conformation, and methyl end stacking. Therefore, no molecular compound has been observed in the metastable and stable states in the POP/POO mixture, as actually proved in the experiments shown above.

Practical Implication of Mixing Phase Behavior of Principal TAGs in Palm Oil

The thermodynamic and structural properties of the mixtures of principal TAGs in palm oil (PPP, POP, PPO and POO) may have the following practical implications.

In the first, the fact that the TAG mixtures reveal the eutectic property for α, β′ and β forms, may lead to separate crystallization of the component TAGs, when they are cooled from the mixture liquid. This may indicate that the separated crystallization of the TAGs whose mixtures show the eutectic property may occur under highly supercooled state, as the metastable forms can crystallized under these conditions. This is applied to the cases of PPP, POP, and POO, making a contrast to the mixtures of PPP/SSS and PPP/MMM (M: myristic acid moiety), where α and β′ forms are miscible and β form is immiscible eutectic.

Although binary mixtures of PPP/POO and PPO/POO and ternary mixture phases of PPP, POP, and POO have never been studied in detail, the results of the binary mixtures of PPP/POP, PPO/POP, and POP/POO indicate that every TAGs can be crystallized without forming the solid-solution and molecular compound crystals. This property is applied to the concentration ranges except for the areas illustrated in Fig. 14.4(b), where the concentrations of B close to 0% or 100%. In these areas, the TAG molecules of B (or A) can be incorporated into the B (or A) crystals as impure molecules.

The TAGs, however, showing the mixtures of the molecular compound crystals cannot be separated by crystallization methods easily. This is applied to POP and PPO, which make the molecular compound crystals in α, β′ and β forms.

The second implication of the mixing behavior is that the separated crystallization also may occur during ageing processes of fat products which are stored over a long period. A typical example is the formation of granular crystals in fat spread or margarine, whose high-melting fats are based on palm oil or its fractions containing PPP, POP, PPO, and POO. We will discuss it in the next section.

Granular Crystal Formation in Palm-based Fat Spread

Margarine and fat spread are water-in-oil emulsions consisting of semi-solid fat, water, emulsifiers, salt, antioxidants, and other ingredients. The semi-solid fats are made of high-melting fats forming crystal network structures of β′ form to yield better physical properties (e.g., good texture in the mouth and high spreadability). In addition, the fat crystal network prevents destabilization of water droplets from coalescence with crystallized emulsifiers that are adsorbed at the oil-water interface.

Hydrogenated fats are high-melting fat materials including *trans*-fatty acids that have been reported to decrease HDL cholesterol and increase LDL cholesterol, resulting in a higher risk of arterial sclerosis. Therefore, trans-fat alternatives have been studied, and palm oil is one of the most promising of these alternatives. However, when palm oil is used in margarine and fat spread, granular crystals often crystallize, causing the margarine to deteriorate. Many studies have been conducted to clarify the formation mechanism of granular crystals in palm-oil-based margarine (Ishikawa et al., 1980; Tanaka et al., 2007; Watanabe et al., 1992).

Recently, we carried out DSC, optical microscopy, and XRD experiments on granular crystals in fat spread, which were purposely grown under thermal thawing over several months (Tanaka et al., 2009). Two types of XRD studies were examined. A conventional laboratory-scale technique was applied to examine thermal properties of high-melting fat fractions in the granular crystals. In addition, microstructures of the granular crystal were precisely observed with a synchrotron radiation small-angle XRD technique by using a microbeam having a cross section of 5 x 5μm^2 (SR-μ-SAXD). The SR-μ-SAXD analysis has recently been applied to fat crystallization in bulk and emulsion systems by our group (Arima et al., 2009; Bayes-Garcia, et al., 2011; Shinohara et al., 2008; Ueno, 2011; Ueno, et al., 2008).

We obtained the following results: (1) DSC indicated that the granular crystals are composed of high-melting fractions having a melting temperature of 2°C. (2) Conventional XRD of the granular crystals indicated that β-fat crystals of a TCL structure melted below the melting of β′-fat of a DCL structure. (3) The SR-μ-SAXD studies indicated that the fat crystals in normal margarine were the β′ form of the DCL structure. However, the fat crystals of the DCL and TCL structures were simultaneously observed at the center region of a granular crystal, whereas only the fat crystals of the TCL structure were observed at the outer region of a granular crystal. Considering the above results and the major TAGs of palm-blended margarine, we analyzed the microstructures and formation processes of granular crystals in relation to the fractional crystallization of the β form of POP promoted by crystallization and transformation of PPP and SSS fractions.

To form model fat spread, fully hydrogenated rapeseed oil (2wt%), soybean oil (48wt%), and palm oil (20wt%) were employed as oil and fat materials. Fully hydrogenated rapeseed oil mostly containing SSS was added to increase the hardness of the margarine. The other components were three emulsifiers (monoglyceride, sorbitan fatty acid ester, and lecithin), salt, skim milk, and β-carotene. β-carotene was added in the oil phase, and salt and skim milk were added in the water phase.

All the mixtures were blended, melted, and stirred at 70°C and then placed in a jacket. Next, the sample was quenched by inserting the jacket into ice water with strong agitation by a homogenizing mixer (8000 rpm). The prepared samples were kept at 5°C for 3 days. Thermal thawing between 7°C and 15°C every 12 hours was then applied to form granular crystals.

Fig. 14.18 shows the POM images of the palm-oil-blended fat spread samples. Before thermal thawing [Fig. 14.18(a)], no granular crystals appeared. After thermal

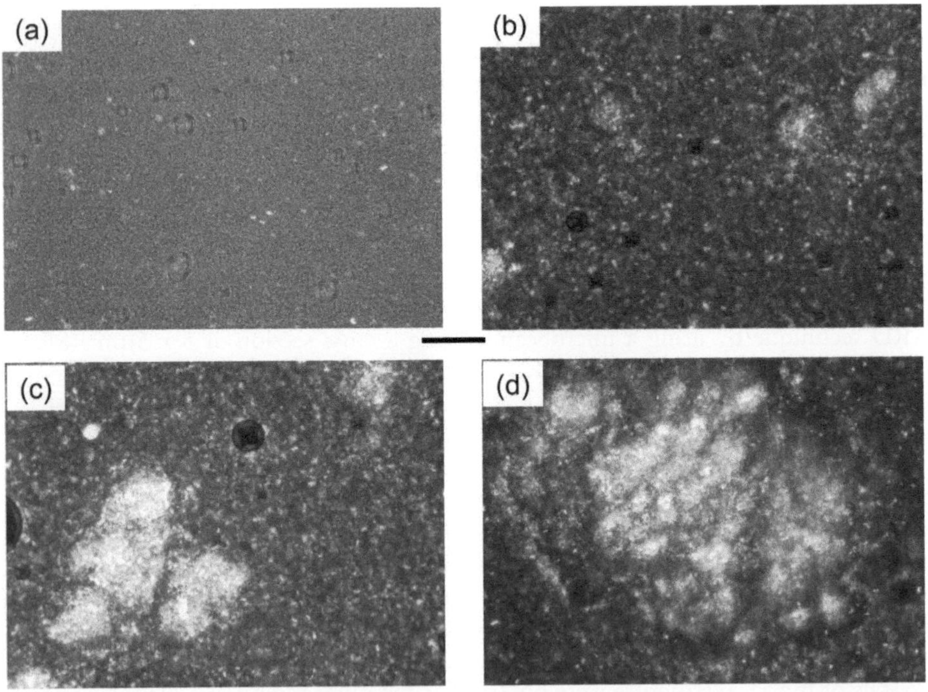

Fig. 14.18. Polarized optical microscope images of palm-oil-blended fat spread. (a) 1 day after keeping at 5°C. (b) 4 days after thermal thawing. (c) 5 days after thermal thawing. (d) 7 days after thermal thawing. Scale bar, 100 μm.

thawing, however, granular crystals (denoted as white spots) appeared one week after the sample preparation (b) and increased in size with increasing time over two weeks (c) and a month (d) of thermal thawing.

Fig. 14.19 shows the DSC heating thermopeaks of fat spread samples taken soon after the production (a) and two weeks after the thermal thawing (b). In Fig.14.18(a), a relatively wide and broad melting peak was observed below 10°C and small endothermic peaks appeared between 20°C and 40°C. The former peak corresponds to the melting of low-melting fat fractions, and the latter peaks include the melting of high-melting fat fractions and transformation of emulsifiers. X-ray diffraction analysis indicated that the endothermic peak noted by an arrow in Fig. 14.19(a) corresponds to the melting of β′ crystals. In Fig.14.19(b) for the granular crystals, a sharp endothermic peak appeared around 23°C (arrow 1) and a small peak also appeared around 30°C (arrow 2). This result is in good agreement with the optical micrographs of Fig. 14.18 showing the formation of granular crystals.

Fig. 14.20(a) shows the scanning SR-μ-SAXD patterns of the sliced granular crystal. Scanning step distance was 100 μm. Three regions were separated in accordance with the SR-μ-SAXD patterns. (1) The exterior region which is out of the solid line (region A). (2) The outside region which is in-between the broken line and solid line (region B). (3) The inner region inside of the broken line (region C). The region A is the area of normal sample around the granular crystal.

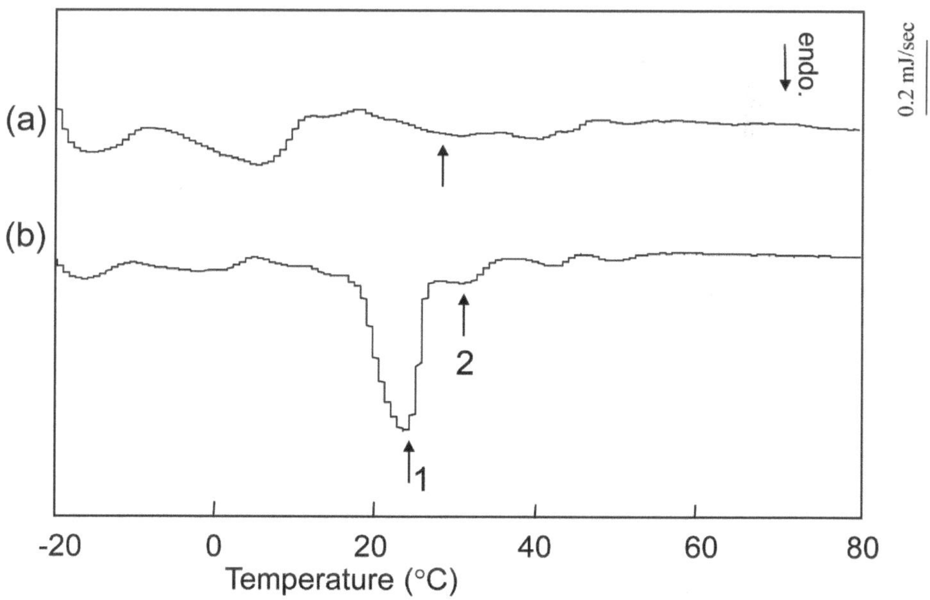

Fig. 14.19. DSC heating thermopeaks of (a) normal sample, and (b) granular crystals.

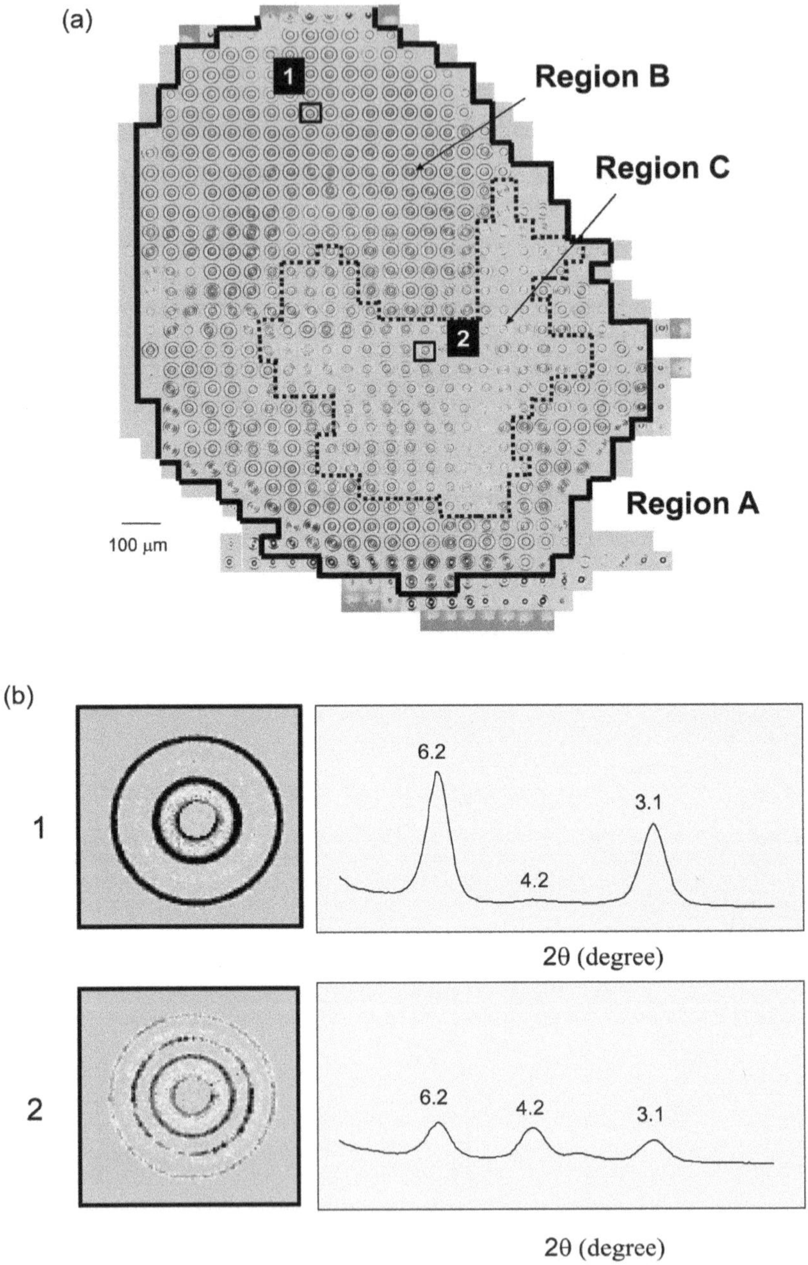

Fig. 14.20. SR-μ-SAXD patterns of a granular crystal; (a) overall patterns and (b) 2θ extension of the positions noted by arrows in (a).

The typical SR-μ-SAXD patterns, however, in the regions B and C are quite different from those in the region A, as revealed in position 1 and position 2 of regions B and C, respectively, as shown in Fig. 14.20(b). The diffraction intensity in the 2θ extension for positions 1 and 2 were calculated by integrating the diffraction intensity along the all azimuthal angles of the diffraction circle at the same 2θ angle in the two-dimensional μ-SAXD patterns. Position 1 in region B showed the strong diffraction peaks of 6.2 nm (001 reflection) and 3.1 nm (002 reflection) of the TCL structure. The peak of 4.2 nm of the DCL structure is quite weak. By contrast, the μ-SAXD peaks of 4.2 nm, 6.2 mm, and 3.1 nm appeared in position 2 of the region C. Compared with position 1, the intensity of the peak of 4.2 nm is high. Although some positions in region B showed the μ-SAXD peaks both of the DCL and TCL structures, the intensity of the peaks corresponding to the TCL structure was always higher than those of the DCL structure, as revealed in position 1 in Fig. 14.20(b).

Based on the results described above, we discuss the following points. (1) What are major TAGs and their polymorphic structures in the fat crystals comprising the normal fat spread and granular crystals? (2) How the granular crystals were formed during the thermal thawing?

Table 14-B shows concentration of the fatty acids in the fats and oil employed for the model sample and major TAGs in palm oil. Our main concerns are the fat

Table 14-B. Fatty Acid Compositions (%) of the Total Fats and Oil Phase in a Model Fat Spread and Component Fats and Oil.

Fatty acid	Fats and oil phase (70%)	Soy bean oil (48%)	Palm oil (20%)	Fully hydrogenated rapeseed oil (2%)	Major triacylglycerols in palm oil[a] (wt.%)	
C12:0	0.1	0.0	0.2	0.0		
C14:0	0.3	0.1	1.0	0.1		
C16:0	19.9	10.4	44.9	4.7		
C18:0	6.6	3.8	4.3	93.0	StStSt	8.8
C18:1 *cis*	27.5	24.6	39.5	0.1	StOSt	30.9
C18:2 *cis*	38.5	52.0	9.1	0.0	StStO	7.1
C18:2 *trans*	0.6	1.1	0.3	0.0	StLSt	8.8
C18:3 *cis*	4.3	5.4	0.1	0.0	StOO	22.2
C18:3 *trans*	1.1	1.7	0.0	0.0	OStO	2.1
C20:0	0.4	0.4	0.4	1.8	StLO	10.0
C20:1 *cis*	0.2	0.2	0.1	0.0	OOO	3.5
C22:0	0.3	0.4	0.1	0.3	> 3db	4.5

[a]St = saturated; O = oleic; L = linoleic; db = double bond.

components which melt above 20°C at the presence of liquid oil. Such fats are SSS which is included in fully-hydrogenated rapeseed oil, PPP, POP, and PPO which are included in palm oil. Having the concentration data displayed in Table 14-B, we may reasonably assume the concentrations (wt.%) of major high-melting fats as SSS (~2%), PPP (~1%), POP (~6%), and PPO (~2%) with respect to the total margarine. Table 14-C shows melting points and long spacing values of the major TAGS with different polymorphic forms.

The mixing behavior among the major TAGs can be summarized that SSS and PPP are miscible in the metastable β′-2 polymorph, but immiscible in the most stable β-2 polymorph, and that POP and PPP crystallize separately in β′-2 and β-3 polymorphs. POP and PPO form molecular compound crystals in β′-2 and β-2 polymorphs. Because of eutectic effects with the liquid oil, the melting points of every TAG must be much lower than the melting point data shown in Table 14-C.

From the data of fatty acid and TAG compositions and the experimental results, we may conclude that the major fats and their polymorphic forms in the normal samples are PPP, SSS, POP, and PPO, all of which are in the metastable β′ form of the DCL structure. The long spacing values of the β′ form of PPP, POP, and PPO are almost similar and therefore are not separated in the two XRD measurements of this study.

As for the granular crystals, comparing the results of SR-μ-SAXD, we can separate the granular crystal into two regions; the inner region simultaneously containing the fat crystals of DCL and TCL structures (region C in Fig. 14.20(a)) and the outside area predominantly containing the fat crystals of the TCL structure (region B). According to the TAG compositions shown in Table 14-B and the above discussion, the region B may consist of β-3 of POP as the major TAG, whereas the region C may consist of β-3 of POP, β′-2 of PPP, SSS, and POP.

Regarding the formation mechanism of the granular crystals, we think that heterogeneous structures as revealed in regions B and C may be the key and the forma-

Table 14-C. Physical Properties of Major High Melting TAG Contents of Palm Oil Blended Margarine.

	TAG (polymorph)	Melting point (°C)	Long spacing (nm)
Palm oil	POP β-3	36.7	6.10
	POP β′-2	33.5	4.24
	PPP β-2	66.4	4.09
	PPP β′-2	56.6	4.23
	POP+PPO β′-2	29.0	4.20
	POP+PPO β-2	31.2	4.10
Fully-hydrogenated rapeseed oil	SSS β-2	73.5	4.51
	SSS β′-2	63.2	4.68

tion mechanism of granular crystal can be thought as follows. When the temperature decreased, β′-2 crystals of PPP, SSS, POP, and PPO were formed in the first, and they maintained their structures until the granular crystal started to form. The thermal thawing provides the driving force for the transformation from β′-2 to β-3 of POP, which may be accelerated by the transformation from β′-2 to β-2 of the PPP and SSS fractions. The granular crystals might occur at specific portions in fat spread samples where agglomerated precipitation of PPP and SSS occurred. Once the transformation from β′-2 to β-3 occurred in the POP fraction, the β-3 form may grow at the expense of β′-2 and extend to grow through volume diffusion of POP molecules from surrounding areas (normal areas). Therefore, the granular crystals consists of two areas, POP β-3 in the outside area, and POP β-3, PPP β′-2 and β-2, and SSS β′-2 and β-2 in the inside area, as observed by the SR-μ-SAXD measurements.

Conclusions

The polymorphic properties of palm oil and mixing properties of TAGs comprising palm oil have been reviewed. We tried to show that the macroscopic crystallization properties of palm oil can partly be understood by looking at the polymorphism and mixing behavior its major component TAGs of PPP, POP, PPO, and POO.

However, further research is necessary to understand the effects of minor components of palm oil, the most important of which is the effects of DAGs (Siew & Ng, 1996). Smith et al. (2011) has recently given a comprehensive review on this problem. In addition, the crystallization behavior of palm oil and its fractions such as palm olein and palm superolein at chilled temperature must be affected by low-melting fractions such as POP, PPO, POO, OPO, and PLiO. In addition, OOO must be crystallized below 5°C. Further research is again necessary to assess the crystallization properties occurring at low temperature, which is crucial for the production of palm-based edible oils without clouding at chilled temperature, from the viewpoints of polymorphism and mixing phase behavior.

Acknowledgment

The author is highly indebted to Professor Satoru Ueno, Hiroshima University, for his cooperative work over many years.

References

Acevedo, N.C.; Marangoni, A.G. Characterization of the nanoscale in triacylglycerol crystal networks. Cryst. Growth Des. 2010a, 10, 3327–3333.

Acevedo, N.C.; Marangoni, A.G. Toward nanoscale engineering of triacylglycerol crystal networks. Cryst. Growth Des. 2010b, 10, 3334–3339.

Acevedo, N.C.; Peronel, F.; Marangoni, A.G. Nanoscale structure intercrystalline interactions in fat crystal networks. Curr. Opin. Coll. Interf. Sci. 2011, 16, 374–383.

Aini, I.N.; Miskandar, M.S. Utilization of palm oil and palm products in shortenings and margarines. Eur. J. Lipid Sci. Technol. 2007, 109, 422–432.

Arima, S.; Ueji, T.; Ueno, S.; Ogawa, A.; Sato, K. Retardation of crystallization-induced destabilization of PMF-in-water emulsion with emulsifier additives. Coll. Sur. B, 2007; 55, 98–106.

Arima, S.; Ueno, S.; Ogawa, A.; Sato, K. Scanning microbeam small-angle X-ray diffraction study of interfacial heterogeneous crystallization of fat crystals in oil-in-water emulsion droplets. Langmuir, 2009, 25, 9777–9784.

Arishima, T.; Sagi, N.; Mori, H.; Sato, K. Polymorphism of POS. I. Occurrence and polymorphic transformation. J. Am. Oil Chem. Soc. 1991, 68, 710–715.

Arishima, T.; Sato, K. Polymorphism of POP and SOS. III. Solvent crystallization of β_2 and β_1 polymorphs. J. Am. Oil Chem. Soc. 1989, 66, 1614–1617.

Basso, R.C.; Ribeiro, A.P.B.; Masuchi, M.H.; Gioielli, L.A.; Goncalves, L.A.G.; dos Santos, A.O.L.P.; Grimaldi, R. Tripalmitin and monoacylglycerols as modifiers in the crystallisation of palm oil. Food Chem. 2010, 122, 1185–1192.

Bayés-García, L.; Calvet, T.; Cuevas-Diarte, M.A.; Ueno, S.; Sato, K. Heterogeneous microstructures of spherulites of lipid mixtures characterized with synchrotron radiation microbeam X-ray diffraction. Cryst. Eng. Comm., 2011a, 13, 6694–6705.

Bayés-García, L.; Calvet, T.; Cuevas-Diarte, M.A.; Ueno, S.; Sato, K. In situ synchrotron radiation X-ray diffraction study of crystallization kinetics of polymorphs of 1,3-dioleoyl-2-palmitoyl glycerol (OPO). Cryst. Eng. Comm., 2011b, 13, 3592–3599.

Boubekri, K.; Yano, J.; Ueno, S.; Sato, K. Polymorphic transformations in sn-1,3-distearoyl-2-ricinoleyl-glycerol. J. Am. Oil Chem. Soc. 1999, 76, 949–955.

Braipson-Danthine, S.; Gibon, V. Comparative analysis of triacylglycerol composition, melting properties and polymorphic behavior of palm oil and fractions. Eur. J. Lipid Sci. Technol. 2007, 109, 359–372.

Che Man, Y.B.; Haryati, T.; Ghazali, H.M.B.; Asbi, A. Composition and thermal profile of crude palm oil and its products. J. Am. Oil Chem. Soc. 1999, 76, 237–242.

Chong, C.L.; Kamarudin, Z.; Lusieur, P.; Marangoni, A.; Bourgoaux, C.; Ollivon, M. Thermal and structural behaviour of crude palm oil: Crystallisation at very slow cooling rate. Eur. J. Lipid Sci. Technol. 2007, 109, 410–421.

Craven J.; Lencki, R.W. Crystallization, polymorphism, and binary phase behavior of model enantiotropure and racemic triacylglycerol. Cryst. Growth Des. 2011, 11, 1723–1732.

Engstrom, L. Triglycerides systems forming molecular compounds. J. Fat Sci.Technol. 1992, 94, 173–181.

Frederick, E.; Foubert, I.; Van De Sype, J.; Dewettinck, K. Influence of monoglycerides on the crystallization behavior of palm oil. Cryst. Growth Des. 2008, 8, 1833–1839.

Garbolino, C.; Bartoccini, M.; Floeter, E. The influence of emulsifiers on the crystallization behavior of a palm oil-based blend. Eur. J. Lipid Sci. Technol. 2005, 107, 616–626.

Goh, E.M.; Timms, R.E. Determination of mono- and diglycerides in palm oil, olein and stearin. J. Am. Oil Chem. Soc. 1985, 62, 730–734.

Goto, M.; Kodali, D.R.; Small, D.M.; Honda, K.; Kozawa, K.; Uchida, T. Single crystal structure of a mixed-chain triacylglycarol: 1, 2-dipalmitoy l-3-acetyl- sn-glycerol. Proc. Natl Acad. Sci. USA 1992, 89, 8083–8086.

Ikeda, E.; Ueno, S.; Miyamoto, R.; Sato, K. Phase behavior of a binary mixture of 1,3-dipalmitoyl-2-oleoyl-sn-glycerol and 1,3-dioleoyl-2-palmitoyl-sn-glycerol in n-dodecane solution. J. Phys. Chem. B 2010, 114, 10961–10969.

Ishikawa, H.; Mizuguchi, T.; Kondo, S. Studies on granular crystals growing in palm oil (in Japanese). J. Jpn. Oil Chem. Soc. 1980, 29, 235–242.

Kellens, M.; Gibon, V.; Hendrix, M.; de Greyt, W. Palm oil fractionation. Eur. J. Lipid Sci. Technol. 2007, 109, 336–349.

Kellens, M.; Meeussen, W.; Hammersley, A.; Reynaers, H. Synchrotron radiation investigation of the polymorphic transitions. Part 2: Polymorphic study of 50:50 mixture of tripalmitin and tristearin during crystallization and melting. Chem. Phys. Lipids 1991, 58, 145–158.

Koyano, T.; Hachiya, I.; Sato, K. Phase behavior of mixed systems of SOS and OSO. J. Phys. Chem. 1991, 96, 716–718.

Kuriyama, J.; Miyamoto, Y.; Sakamoto, M. Effect of polyglycerol fatty acid esters on resistance to crystallization of palm olein. J. Oleo Sci. 2001, 50, 831–838.

Larsson, K.; Quinn, P.; Sato, K.; Tiberg, F. Lipids: structure, physical properties and functionality. The Oily Press: Bridgwater, 2006.

Marangoni, A.G. Fat crystal networks. Marcel Dekker: New York, 2005.

Minato, A.; Ueno, S.; Smith, K.; Amemiya, Y.; Sato, K. Thermodynamic and kinetic study on phase behavior of binary mixtures of POP and PPO forming molecular compound systems. J. Phys. Chem. 1997c, B 101, 3498–3505.

Minato, A.; Ueno, S.; Yano, J.; Smith, K.; Seto, H.; Amemiya, Y.; Sato, K. Thermodynamic and structural properties of sn-1,3-dipalmitoyl-2-oleoylglycerol and sn-1,3-dioleoyl-2-palmitoylglycerol binary mixtures examined with synchrotron radiation X-ray diffraction. J. Am. Oil Chem. Soc. 1997b, 74, 1213–1220.

Minato, A.; Yano, J.; Ueno, S.; Smith, K., Sato, K. FT–IR study on microscopic structures and conformations of POP–PPO and POP–OPO molecular compounds, Chem. Phys. Lipids 1997a, 88, 63–71.

Minato, A; Ueno, S.; Yano, J.; Wang, Z.H.; Seto, H.; Amemiya, Y.; Sato, K. Synchrotron radiation X-ray diffraction study on phase behavior of PPP–POP binary mixtures. J. Am. Oil Chem. Soc. 1996, 73, 1567–1572.

Miskandar, M.S.; Che Man, Y.B.; Yusoff, M.S.A.; Rahman, R.A. Effect of scraped-surface tube cooler temperatures on the physical properties of palm oil margarine. J. Am. Oil Chem. Soc. 2002b, 79, 931–936.

Miskandar, M.S.; Man, Y.C.; Rahman, R.A.; Aini, I.N.; Yusoff, M.S.A. Effects of emulsifiers on crystallization properties of low-melting blends of palm oil and olein. J. Food Lipids 2006, 13, 57–72.

Miskandar, M.S.; Man, Y.C.; Yusoff, M.S.A.; Rahman, R.A. Effect of emulsion temperature on physical properties of palm oil-based margarine. J. Am. Oil Chem. Soc. 2002a, 79, 1163–1168.

Miura, S.; Konishi, H. Crystallization behavior of 1,3-dipalmitoyl-2-oleoil- glycerol and 1-palmitoyl-2,3-dioleoil-glycerol. Eur. J. Lipid Sci. Technol. 2001, 103, 804–809.

Moran, D.P.J. Phase behavior of some palmito-oleo triglyceride systems. J. Appl. Chem. 1963, 13, 91–100.

Rousset, P.; Rappaz, M.; Minner, E. Polymorphism and solidification kinetics of the binary system POS-SOS. J. Am. Oil Chem. Soc. 1998, 75, 857–864.

Sasaki, M., Ueno, S.; Sato, K. Polymorphism and mixing phase behavior of major triacylglycerols of cocoa butter. In Cocoa Butter and Related Compounds; Garti, N., Wildrak, N., Eds.; AOCS Press: Champaign, IL, 2012.

Sato, K.; Arishima, T.; Wang, Z.H.; Ojima, K.; Sagi, N.; Mori, H. Polymorphism of POP and SOS. I. occurrence and polymorphic transformation. J. Am. Oil Chem. Soc. 1989, 66, 664–674.

Sato, K.; Goto, M.; Yano, J.; Honda, K.; Kodali, D.R.; Small D.M. Atomic resolution structure analysis of β′ polymorph crystal of a triacylglycerol: 1, 2-dipalmitoyl-3-myristoyl-sn-glycerol. J. Lipid Res. 2001, 42, 338–345.

Sato, K.; Ueno, S. Pomorphism of triacylglycerols and natural fats, In Bailey's Industrial Oil & Fat Products, 6th Ed.; Vol.1; Shahidi, F., Ed.; J ohn Wiley & Sons: Hoboken, NJ, 2005; pp. 77–120.

Sato, K.; Ueno, S. Crystallization, transformation and microstructures of polymorphic fats in colloidal dispersion states. Curr. Opin. Coll. Interf. Sci. 2011, 16, 384–390.

Sato, K.; Kigawa, T.; Ueno, S.; Gotoh, N.; Wada, S. Polymorphic behavior of structured fats including stearic acid and ω-3 polyunsaturated fatty acids. J. Am. Oil Chem. Soc. 2009, 86, 297–300.

Siew, W.L.; Ng, W.L. Partition coefficient of diglycerides in crystallization of palm oil. J. Am. Oil Chem. Soc. 1995, 72, 591–595.

Siew, W.L.; Ng, W.L. Effect of diglycerides on the crystallization of palm oleins. J. Sci. Food Agric. 1996, 71, 496–500.

Shinohara, Y.; Takamizawa, T.; Ueno, S.; Sato, K.; Kobayashi, I.; Nakajima, M.; Amemiya, Y. Cryst. Growth Des. 2008, 8, 3123–3127.

Smith, K.W. Crystallization of palm oil and its fractions. In Crystallization Processes in Fats and Lipid Systems, Garti, N., Sato, K., Eds.; Marcel Dekker: New York, 2001; pp. 357–380.

Smith, K.; Bhaggan, K.; Talbot, G; van Malssen, K.F. Crystallization of fats; Influence of minor components and additives. J. Am. Oil Chem. Soc. 2011, 88, 1085–1101.

Takeuchi, M.; Ueno, S.; Floeter, E.; Sato, K. Binary phase behavior of 1,3-distearoyl-sn-2-oleoylglycerol (SOS) and 1,3-distearoyl-sn-2- linoleoylglycerol (SLS). J. Am Oil Chem. Soc. 2002a, 79, 627–632.

Takeuchi, M.; Ueno, S.; Sato, K. Crystallization kinetics of polymorphic forms of a molecular compound constructed by SOS (1,3-distearoyl-2-oleoyl- sn-glycerol) and SSO (1,2-distearoyl-3-oleoyl-rac-glycerol). Food Res. Intern. 2002b, 35, 919–926.

Takeuchi, M.; Ueno, S.; Sato, K. Synchrotron radiation SAXS/WAXS study of polymorph-dependent phase behavior of binary mixtures of saturated monoacid triacylglycerols. Cryst. Growth Des. 2003, 3, 360–374.

Takeuchi, M.; Ueno, S.; Yano, J.; Floeter, E.; Sato, K. Polymorphic transformation of 1,3-distearoyl-sn-2-linoleoyl-glycerol. J. Am. Oil Chem. Soc. 2000, 77, 1243–1249.

Tanaka, L.; Miura, S.; Yoshioka, T. Formation of granular crystals in margarine with excess amount of palm oil. J. Am. Oil Chem. Soc. 2007, 84, 421–426.

Tanaka, L.; Tanaka, K.; Yamato, S.; Ueno, S.; Sato, K. Microbeam X-ray diffraction study of granular crystals formed in water-in-oil emulsion, Food Biophys. 2009, 4, 331–339.

Timms, R.E. Confectionary fats handbook. The Oily Press: Bridgwater, 2003.

Ueno, S. A new method—synchrotron radiation microbeam X-ray diffraction for lipid crystallization. In Cocoa Butter and Related Compounds; Garti, N., Wildrak, N., Eds.; AOCS Press: Champaign, IL, 2012.

Ueno, S.; Nishida, T.; Sato, K. Synchrotron radiation microbeam X-ray analysis of microstructures and the polymorphic transformation of spherulite crystals of trilaurin. Cryst. Growth Des. 2008, 8, 751–754.

van Langevelde, A.; van Malssen, K.; Driessen, R.; Goubitz, K.; Hollander, F.; Peschar, R.; Zwart, P.; Schenk, H. Structure of CnCn+2Cn-type (n-even) β′-triacylglycerols. Acta Cryst B, 2000, 56, 1103–1111.

Watanabe, A.; Tashima, I.; Matsuzaki, N.; Kurashige, J.; Sato, K. On the formation of granular crystals in fat blends containing palm oil. J. Am. Oil Chem. Soc. 1992, 69, 1077–1080.

Yano, J.; Ueno, S.; Sato, K.; Arishima, T.; Sagi, N.; Kaneko, F.; Kobayashi, M. FT-IR study of polymorphic transformations in SOS, POP and POS. J. Phys. Chem. 1993, 97, 12967–12973.

Yano, J.; Sato, K.; Kaneko, F.; Small, D.M.; Kodali, D.R. Structural analyses of polymorphic transitions of sn-1,3-distearoyl-2-oleoylglycerol (SOS) and sn-1,3-dioleoyl-2-stearoylglycerol (OSO): assessment on steric hindrance of unsaturated and saturated acyl chain interactions. J. Lipid Res. 1999, 40, 140–151.

Zhang, L.; Ueno, S.; Miura, S.; Sato, K. Binary phase behavior of 1, 3-dipalmitoyl-2-oleoyl-glycerol and 1, 2-dioleoyly-3-palmitoyl-rac-glycerols. J. Am. Oil Chem. Soc. 2007, 84, 219–227.

Zhang, L.; Ueno, S.; Sato, K.; Adlof, R.O.; List, G.R. Thermal and structural properties of binary mixtures of 1, 3-distearoyl-2-oleoyl-glycerol (SOS) and 1, 2-dioleoyly-3-stearoyl-sn-glycerols (sn-OOS). J. Therm. Anal. Calorim. 2009, 98, 105–111.

15

Measurement and Maintenance of Palm Oil Quality

Chiew Let Chong
Product Development and Advisory Services Division, Malaysian Palm Oil Board, Bangi, Malaysia

Introduction

The oil palm tree was first introduced into Malaysia (then Malaya) in the early 1900s by European planters. After the independence of Malays from the British, the Malaysian Government diversified the country's agricultural practices and encouraged a shift from the cultivation of rubber to oil palm in plantations in the 1960s and 1970s in order to break the dependence on a single rubber crop, especially after the rubber price slump in the 1970s. Today, the oil palm is the biggest agricultural industry in Malaysia. In 2010, Malaysia produced 16.99 million tons of crude palm oil (CPO), bringing in a revenue of RM44 billion. In fact, the Malaysian oil palm industry has helped the country to weather the economic slump in the 1990s.

Unlike non-perennial crops like soybean, rapeseed, sunflower, and other vegetable oil crop, the oil palm is a tree crop which starts to bear fruit 3 years after planting. The oil palm tree bears fruit all year round and can be harvested for a period of 25–30 years before being replanted due to falling oil yields in the fruit bunches as the tree matures.

Palm oil is obtained from the mesocarp of the palm fruit while palm kernel oil is obtained from the kernel inside the fruit. Palm kernel oil is meant for propagation of the species. The actual purpose of the mesocarp palm oil is unknown. Since palm oil is a very nutritive oil, it is probably meant as a food enticement so that animals will eat the mesocarp and help to disperse the seeds.

The Malaysian palm oil industry can be broadly categorized into five major sectors, as follow:

1. The Plantation Sector

   This sector covers oil palm plantation management from the nursery or planting of the seedlings of oil palm trees through to the harvesting of the fresh fruit bunches.

2. The Milling Sector

   This sector is involved with the processing of the oil palm fruits (FFB) to extract crude palm oil (CPO) from the fruit mesocarp and the production of palm kernels (PK) for extraction of palm kernel oil from the kernel.

3. The Kernel Crushing Sector

   This sector is involved with the crushing of palm kernels by mechanical means and/or by solvent to extract palm kernel oil (PKO) with the resulting production of palm kernel cakes (PKC) or palm kernel expeller (PKE).

4. The Refining Sector

   Crude palm oil and crude palm kernel oil are refined and/or further processed to various refined products for export as a component for mainly downstream food and non-food applications, respectively, although some kernel oils are further processed to obtain certain fractions for edible applications.

5. The Oleochemical/Down Stream Sector

   The palm oil and palm kernel oil can be further processed using modification processes to produce value added products for industrial and oleochemical applications.

Malaysia is the second largest producer of palm oil after Indonesia but is the largest exporter of palm oil due to the fact that a large part of the Indonesian production is consumed domestically by its relatively larger population. As the largest exporter of palm oil products, safeguarding the image of palm oil is important to the Malaysian oil palm industry, especially the quality image.

Quality

Quality is a subjective attribute which is closely related to a function or object. The American Society for Quality (American Society for Quality, 2008) defines quality as the characteristics of a product or service that bear on its ability to satisfy stated or implied needs or a product or service that is free of deficiencies.

Joseph M. Juran (American Society for Quality, 2008) defines quality as fitness for use, with fitness being defined by the customer. Noriaki Kana, quality consultant and originator of the Kano model of quality, went one step further than the fitness for use concept by extending the meaning to products and services that meet or exceed customers' expectations (Kano, 1984). ISO 9001 defines quality as the degree to which a set of inherent characteristics fulfills requirements. The standard defines requirement as a need or expectation. From these definitions, it can be seen that there is no one universal definition of quality. Depending on the user or product, what is quality in one case may not be quality in another. For instant, the oxidative state of an oil or fat is important when the oil or fat is to be used for margarine production while oil color is of more concern if the same oil is to be used for the production of very pure white soap.

Quality in the Various Sectors of the Oil Palm Industry

Fruit Harvesting Quality

In the Malaysian oil palm industry, there is an adage that says that oil quality is made in the field and not in the mill. This is a reflection of the harvesting standards used to gauge fresh fruit bunch ripeness, handling of the bunches in the field, and the fruit reception area in the mill and the speed with which the fruit bunches are sent to the mill. Harvesting under-ripe bunches would increase the free fatty acid (FFA) of the extracted oil beside reducing the amount of extractable oil from the bunch [lower oil extraction rate (OER)]. This is due to the incomplete biosynthesis of triacylglycerols. Traditionally, the number of loose fruits detached from the bunch has been used as the criterion to determine the ripeness of the bunch for harvesting. In the early days of the industry, five to ten loose fruits around the base of the oil palm tree is the criterion for bunch harvest. In the late 1980s, Palm Oil Research Institute of Malaysia (later known as Malaysian Palm Oil Board, MPOB) research appears to suggest that one loose fruit is sufficient to determine the ripeness of the bunch. This was widely practiced by the industry for a while, but the trend has now slowly drifted back to the five to ten loose fruit criterion.

Handling of the bunch during and after harvesting is important in order to minimize fruit bruising. In the ripe fruit, the oils are contained within the oil vacuoles. When the fruit is bruised, the oils are released from the vacuoles and comes into contact with hydrolyzing enzymes, which will break the triacylglycerols into free fatty acid and partial triacylglycerols. The larger the extent of fruit bruising, the higher will be the FFA level. Work by Bek Nielsen (1975) has shown that the FFA level of crude palm oil extracted from heavily bruised fruits can rise to as high as 9% within 40 minutes and to about 10% two hours after bruising. The rate of acidification of unbruised fruitlets is from 0.1% to 0.3% over a 4-day period. Work by Ravigadevi et al. (1991) has shown that when bruised fruits are further subjected to cold treatment, the level of FFA could increase to more than 60% at freezer temperature as a function of exposure temperature and time. This is the reason why it is important to ensure minimal fruit bruising or damage during harvesting and collection in the field and to transport the fruit bunches to the mills as soon as possible.

Crude Palm Oil Quality at the Mill

At the mill, the ripeness of the fruit bunch is graded according to the guidelines set by the Malaysian Palm Oil Board (MPOB). Zero under ripe bunch is the target of the grading standard strongly enforced by MPOB. Work by Tan et al. (1995) has shown that oil extracted from under-ripe or unripe fruits contain high levels of chlorophylls. At high levels, chlorophylls can act as pro-oxidants instead of being an antioxidant and, hence, can affect the quality of crude palm oil. In the same report, the workers also showed that laser-induced fluorescence (LIF) analysis can be used as an alternative to analyzing all the common oxidative and hydrolytic parameters. LIF at 673 nm was indicative of the oxidative quality of crude palm oil. There was also positive correlation between fluorescence intensity and the carotene and the deterioration of bleach ability index of crude palm oil, and the Rancimat induction time of the final refined product.

Palm fresh fruit bunches should be processed as soon as possible to shorten the time of exposure of the exposed oil in bruised fruits to the enzymes and for the reversal of oil synthesis. If fruits have to be left overnight, it would be more preferable to sterilize the fruits before leaving them overnight in order to inactivate the enzymes. Research has shown that under the standard milling conditions, oil quality deterioration during milling is minimal if proper care and precautions are taken. In general, exposure of the oil during the milling process to high temperatures of water for long periods of time are kept to a minimum. Similar principles apply to the oil during storage in the tanks while awaiting dispatch. These tanks should be regularly cleaned to avoid the build-up of foots and moisture at the bottom of the tank. In addition, it would be ideal if the storage tanks could be epoxy coated internally to prevent rusting and to reduce or eliminate the contact of hot oil with bare mild steel. The use of side entry stirrers and sections of vertical heating coils would also help to reduce or eliminate localized overheating of the oil, especially if the oil has been stored over an extended period of time without any heating. Stainless steel heating coils and automatic temperature controllers could also contribute to quality preservation at the mill.

One of the earliest comprehensive research works carried out on palm oil milling and palm oil quality was in a trial at Mongana in Africa. The Mongana Report (1955) deals with palm oil milling and palm oil quality in great detail in a mill. In the early days during the establishment of the Malaysian oil palm industry, palm oil was mainly exported in the crude form. In those days crude palm oil quality was mainly defined in terms of percentage of FFA and total percentage of moisture and impurities (%M&I). Free fatty acid and moisture and impurities levels were then set by the Malaysian Oil Palm Growers Council (MOPGC) after talks with buyers of crude palm oil. Domestic sales were then based on the MOPGC crude palm oil contract where the levels of FFA and M&I were set at 5% maximum and 0.5% maximum, respectively. Any crude palm oil with FFA less than 5% would be entitled to an agreed premium per percentage reduction in FFA. On the other hand, every percentage FFA over the maximum contractual value would incur an agreed to penalty. With improvements in the industry over the years, the trading specifications for crude palm oil under the domestic sales contract was revised in the 1990s to 5%max for FFA and 0.25%max for M&I. For this, mills tend to aim for an FFA level of 3.5% ex-mill gate. If the initial FFA and moisture levels in crude palm oil are low, then the oil quality in terms of these two parameters are quite stable as can be seen in Table 15-A.

In the last few decades, it was realized that the production of a stable refined palm oil of good color is also dependent on characteristics other than FFA content, especially on the degrees of oxidation and contamination with catalytic traces of metals. Surveys carried out on Malaysian palm oil showed that the peroxide value (PV) of standard palm oil exported from Malaysian fell from 3.9 milligrams per kilogram oxygen to 2.3 milligrams per kilogram oxygen between 1974 and 1981 (Jacobsberg, 1974; Wong, 1981).

In Africa, after the Second World War, a more complex quality grading system was set up to deal with the various grades of oils produced by small village industries, as compared to the plantation industries established in Malaysia. The grading system used then was a five grade system with grade 1 having FFA of less than 1% and grade V having FFA over 36%. Grades II, III and IV covers the FFA ranges of 9–18%, 18–27%, and 27–36%, respectively (Berger & Martin, 2000).

In Malaysia, the system practiced was the carrot and stick system where the FFA level was capped at a maximum of 5% for crude palm oil, with a premium or penalty for each percentage FFA below or above 5%, respectively. The system established for the palm oil trade in Malaysia could have been based on the African experience, for just like the carrot and stick system set up in Malaysia, premium prices were paid for better quality oils. This system encouraged the village producers to improve their oil quality leading to the specification of a Special Grade palm oil in 1950 with a 4.5 % maximum FFA level at time of purchase. This special Grade palm oil commanded a significant premium. As a result, the amount of Special Grade palm oil jumped from 0.25% of production oil in 1950 to more than 50% by 1954. This led to a tightening of the standard to 3.5% in 1955, and by 1965 more than 80% of the export oil from Africa was of this Special Grade quality (Iwuchukwu, 1965).

As part of the continual development process, palm oil producers in Africa and Malaysia strove to develop premium products with exceptionally low FFA through strict harvesting practices and fast processing after harvesting. Special Prime Bleach (SPB) grade was developed in the Belgian Congo (Loncin & Jacobsberg, 1965), while

Table 15-A. Changes in FFA and Moisture of Crude Palm Oil over a Period of 12 Weeks.

	Moisture (%)			FFA (%)		
	Initial	Final	Decrease	Initial	Final	Increase
Purified sample	0.19	0.10	0.09	2.28	2.56	0.28

Source: Bek-Nielsen (1977).

in Malaysia two special grades of oil were commercially available, namely SQ and Lotox. SPB have a maximum FFA of 2% and reduced levels of iron and copper. The SQ and Lotox oils specify limits for oxidation values and trace metals, which satisfied the stringent requirements of the European market at that time (Johansson & Pehlergard, 1977).

Unfortunately, the oil palm industry in Africa declined rapidly, especially after the discovery of petroleum oil in Nigeria, for example. Today, the palm oil production from Africa is hardly significant compared to the total world palm oil production.

One of the more comprehensive works on PV development during milling was the work reported by Bek-Nielsen in 1972, together with test results from heating palm oil with varying surface areas exposed to air or inert atmosphere. Table 15-B shows the PVs of production oil at the various stages of oil extraction at the mill and storage thereafter.

From the results, it can be seen that the oil coming out from the milling process is of very good quality, with free fatty acids below 2% and moisture of less than 0.15%. Peroxide value of the fresh oil is zero and the value after about 2 months of storage is less than two. This could be due to the protective influence of the two natural antioxidants present in crude palm oil, namely the carotenoids and the tocopherols and tocotrienols.

At the mill, oxidation of palm oil can be reduced by avoiding the following (Bek-Nielsen, 1977):

- over-heating during extraction and during transfer to the purification plant
- aeration during purification and drying by using sealed purifiers and vacuum driers
- rapid oxidation during storage by cooling the oil to 50°C before pumping the oil to storage tank and to use small diameter despatch tanks to reduce oil surface contact with air
- over-heating in storage tanks by installing automatic steam cut off valves activated at 55°C
- prolonged heating of oil in a storage tank before despatch and to apply inert gas blanketing or sparging if necessary
- contamination with iron by using stainless steel tanks and equipment, where practical

Quality in the Refining Sector

With the setting up of the refining and down-streaming sectors of the oil palm industries in Malaysia in the late 1960s and 1970s, the Palm Oil Refiners Association of Malaysia (PORAM) and Malayan Edible Oils Manufacturers Association (MEOMA)

Table 15-B. Peroxide Values of Oil at Various Stages of Production and After Storage.

	Day 0		Peroxide value (mmole/kg oil)								
Source of sample	%FFA	% H2O	Day 0	Day 7	Day 14	Day 21	Day 28	Day 35	Day 42	Day 49	Day 56
Sterilizer under vacuum	1.37	0.064	0.00	0.00	0.00	0.00	0.18	0.74	0.99	1.11	1.37
Sterilizer with air release	1.42	0.067	0.00	0.00	0.00	0.00	0.21	0.72	1.02	1.17	1.42
Fruit elevator prior to discharge into digesters	1.26	0.064	0.00	0.00	0.00	0.00	0.65	0.76	0.99	1.25	1.59
Digesters (std press)	1.75	0.034	0.00	0.00	0.00	0.00	0.59	0.89	1.08	1.31	1.48
Crude oil from std presses	1.78	0.090	0.00	0.00	0.00	0.00	0.35	0.42	0.55	0.78	1.24
Crude oil from screw presses	1.87	0.057	0.00	0.00	0.00	0.00	0.21	0.32	0.53	0.81	1.15
Continuous tank	1.75	0.046	0.00	0.00	0.00	0.00	0.53	0.74	1.02	1.37	1.79
Sludge separator	2.07	0.125	0.00	0.00	0.63	0.79	0.92	1.00	1.10	1.42	1.86
Purified once	1.90	0.041	0.00	0.00	0.78	0.86	1.16	1.35	1.45	1.65	1.98
Purified twice	1.98	0.097	0.00	0.46	1.06	1.14	1.31	1.52	1.67	1.87	2.35

Source: Bek-Nielsen (1972A).

was formed to look after the interest of these sectors of the industry. PORAM established the quality specifications for refined palm oil products, and these formed the normal specifications for refined oil trade based on the Federation of Oils, Seeds and Feeds Association (FOSFA) Free On Board (FOB) contract for bulk oils. These normal specifications are generally referred to as the PORAM specifications, which are accepted for international trading of refined palm oil products in bulk. Appendices A and B give the respective specifications for crude and refine palm oil and their products (PORAM and MEOMA specifications). The PORAM Handbook is a good source of reference for those in the palm oil business. This reference book contains information on all contracts issued and jointly-issued by PORAM; Arbitration Rules; and other useful information related to palm oil specifications, analytical test methods, material safety datasheet, and other useful trade information pertaining to palm oil products.

Palm Oil Quality during Refining

Crude palm oil is refined to clean up the oil for edible purposes as well as for aesthetic acceptance by consumers who require a bland and light colored oil. Palm oil can be refined either by the chemical or physical refining process. The difference between the two processes is that the acid neutralization step is omitted in the physical refining process as compared to the chemical refining process. The free fatty acids are removed by the deodorization step in the physical refining process. Chemical refining is carried out at a lower temperature than the physical refining process and the residence time is also shorter. This would imply that there is a larger destruction of tocopherols and tocotrienols in the latter process. It also has a bigger chance of oxidative damage should there be an air leakage during the process. The higher processing temperature would lower the storage stability of the oil (Gapor & Berger, 1982; Loncin & Jacobsberg, 1963). In physical refining, it is important to eliminate phosphorus as much as possible during degumming and to prevent phosphoric acid slip-through, as this would affect the efficiency of heat bleaching later on in the process. Time-temperature effect is stronger in physical refining, as evidenced by the higher temperature used in the process and the longer residence time.

To obtain good refined palm oil, it is important to start from good crude palm oil. Besides increasing refining costs due to more severe refining conditions, poor quality crude palm oil will also reduce the long term keeping quality of the refined oil. For example, the high diglycerides content of crude palm oil with high free fatty acids will not only lead to more oil loss in terms of palm fatty acid distillates, the high diglycerides content cannot be reduced by refining. This will cause difficulty later on if the refined oil is to be used for fractionation. Drastic refining may appear to be able to refine oxidised crude oil to acceptable levels, but, again, the keeping quality is low due to the presence of precursors in the refined oils.

It is difficult to try to establish rules to forecast the quality of refined oils from the conditions of crude palm oils given the nature of the trade. Work by Unilever in the 1970s appeared to depend on free fatty acids, iron, and bleachability itself, in that order.

Quality during Storage and Handling

Common Quality Deteriorations during Handling and Storage of Palm Oil Products

Palm oil, like all other vegetable oils, is a natural product. Hence, it is subjected to natural deterioration reactions. In addition to these natural deterioration reactions, wrong handling and storage practices can also cause the oil quality to deteriorate. In general the most common quality deteriorations that can occur in oils/fats are

a. Oxidation (leading to oxidative rancidity)
b. Hydrolysis (leading to hydrolytic rancidity)
c. Cross-contamination between different grades of products and
d. Contamination with foreign substances

Oxidative Rancidity

This is the most common quality deterioration to occur in oils and fats. This is caused by the reaction of oxygen with the double bonds present in the fatty acid chains in the oil molecule. This is a two-stage reaction. The first stage is called primary oxidation. The main product of this first reaction stage is the formation of hydroperoxides in the oil. These hydroperoxides are unstable compounds and will breakdown into a complex mixture of aldehydes, ketones, fatty acids, etc., which will result in off-flavours and odors. This breakdown of the hydroperoxides into a mixture of aldehydes, ketones, and fatty acids is called secondary oxidation. It is the secondary oxidation products (a mixture of aldehydes, ketones, fatty acids, etc.) which gives rise to the rancid taste and smell of oxidized oil. Since the reaction is by oxygen on the double-bonds in the fatty acid chains, the more double-bonds that are present in the fatty acid chains of the oil molecules, the less stable is the oil or fat. A number of factors can promote and accelerate the oxidation reaction:

1. high temperature and excessive heating,
2. the presence of oxygen,
3. the presence of pro-oxidant metals such as copper and iron, and
4. light of ultraviolet wavelength.

The use of high temperatures in some palm oil products such as palm oil, palm stearins, and palm fatty acid distillates, is unavoidable due to the semi-solid or solid nature of these products at ambient temperatures. High temperature will increase the rate of the oxidation reaction. Under these circumstances, it is important to ensure that the oils are heated to within the recommended temperatures for handling and storage and at the recommended heating rate, in order to minimize or maintain the quality of the oil. Prolonged heating of the oil should also be minimized since the longer the oil is under heating the more chance there is that the oil quality will be damaged by the oxidation reaction.

Oxygen is soluble in oil to a certain extent and the higher the temperature of the oil, the larger will be the volume of oxygen dissolved in it. Since oxidation is the attack of the double bonds in the fatty acid chains of the oil molecules by oxygen, the greater the volume of dissolved oxygen in the oil the greater will be the degree of oxidation.

The presence of some metals such as iron and copper are known to increase the rate of the oxidation reaction, that is, they can act as a catalyst. For this reason it is important to remember that equipment used in contact with oil should never be made of copper or alloys of copper. The presence of trace amounts of iron in the oil is inevitable since a lot of the handling and storage equipment are made of iron and machinery wear and tear will contribute to some trace level of iron in the oil. However, the use of rusty equipment or tanks that can result in a high level of iron being present in the oil should be absolutely avoided.

The presence of ultraviolet light will result in photo-oxidation via a specialized reaction pathway for the oxygen to attack the double bonds of the oil molecule. Hence, it is important to protect the oil from light if this reaction is to be avoided. This is the basis of using amber containers for sample storage.

Basically, measures taken during handling, storage, and transportation to minimize quality changes due to oxidation are those of avoiding or minimizing the above factors.

Hydrolytic Rancidity

Under the right conditions, water can be solubilized into the oil, resulting in the hydrolysis of the oil into free fatty acid and partial glycerides. This is called chemical hydrolysis of the oil. Excessive hydrolysis will lead to the formation of more fatty acids and glycerol. The higher the temperature of the oil, the faster is the hydrolysis rate. The more water solubilized in the oil, the larger the extent of hydrolysis. In addition, the hydrolysis reaction is also catalyzed by the presence of acid, certain emulsifiers, and compounds that can form soaps. Hence, if the free fatty content of the oil is high, the free fatty acids present in the oil can help to catalyze the reaction, leading to more free fatty acids being formed. High levels of free fatty acids in the oil can also result in an increase of phosphorus content of the oil (Goh et al., 1982; Jacobsberg, 1983).

Hydrolysis of the oil can also occur due to the action of enzymes by microorganisms on the oil molecules, which will also result in the formation of free fatty acids. This is called enzymatic hydrolysis of the oil. For this to occur, a level of more than 0.25% of moisture and a level of dirt of more than 0.1% must both be present for the survival of the micro-organism in the oil. For refined oils these conditions will never be present, hence the chances for enzymatic hydrolysis is very low. For crude palm oil the maximum level specifications for these two parameters are set at levels too low for enzymatic hydrolysis to occur.

Contamination with Foreign Matter

Oils and fats may be contaminated by undesirable foreign substances during the storage, handling, and transportation process. Such contamination may occur due to negligence and bad handling practices. Contamination with water will lead to hydrolysis while contamination with pro-oxidant metals will lead to oxidation. However, the contamination with harmful substances such as leaded mineral oils and toxic chemicals forms the most serious contamination of oils and fats. These harmful chemicals could be the residues from previous cargoes that have been carried in the tanks or have been handled by the pumping equipment. To prevent this from happening, surveyors will have to be alert, vigilant, well trained, and experienced. The use of banned or previously allowed cargo lists is designed to prevent this contamination from occurring.

Cross-Contamination

Cross-contamination is the contamination of one oil grade or type by another grade or type (e.g., between olein and stearin, between palm oil and palm kernel oil, etc.). This cross-contamination between different grades and types of oils and fats will affect the quality specifications, such as iodine value and melting point. This will in turn affect the intended use of these oils and fats since they cannot be used in the products that they are intended for. For refined palm oil products, contamination with crude products or with lauric oils is considered bad. This cross-contamination of the oil is also due to negligence, bad handling, or faulty equipment and tanks.

In addition to cross-contamination between different grades or types of oils, the practice of not cleaning storage tanks regularly or between uses may cause contamination of the new consignment of oil by rancid residues left in the tank. This can affect the quality of the new oil in the tank. Commingle of oil consignments in ship tanks can also be another source of cross-contamination.

Evidences of Quality Deterioration during Handling and Transportation

Data collected to date have indicated that quality deteriorations are quite common occurrences given that palm oil products, like other vegetable oils, are natural products subjected to natural deterioration. However such deteriorations are quite slow if proper practices and equipment are adopted or used respectively. Quality deteriorations of the oils is said to have occurred if the arrival quality is significantly different from the loading or contractual quality specifications.

The PORAM standard specifications for RBD palm olein are FFA 0.1% maximum and color 3R maximum. In most cases, quality deterioration during the shipping period was observed to be mild. However, it was also observed that for 1985–1986 shipments, the arrival data in Europe showed that the FFA could range from 0.06% to 0.6%, color from 0.27 R to nearly 5 R. Peroxide value, which is not a PORAM specification, was between 1 to 10 mequiv/kg, usually on the higher side of the range. Iron, which is normally specified by the buyer for shipments at 0.3 ppm maximum, was seen to arrive at the destination on the higher side, sometimes even up to 10 times higher than the buyer's specification. Table 15-C shows the quality deteriorations that can occur during storage, handling, and transportation during the export process.

Table 15-C. Unusual Oil Quality Changes During Palm Oil Transfer.

Parameters	No.	Product	1*	2*	3*	4*	5*
Iron (ppm)	1	A	0.2	0.2	0.35	0.5	0.5
	2	A	0.3	0.3	0.4	0.5	0.8
	3	B	0.3	0.4	0.55	0.7	0.7
PV (meq/kg)	4	A	0.58	0.58	1.53	2.44	8.7
	2	A	2.29	1.83	3.7	4.0	8.3
	5	A	3.62	2.15	3.6	7.07	8.21
	6	B	1.45	1.55	4.61	5.95	8.47
	3	B	0.81	0.93	7.54	15.09	15.11
	7	C	3.47	4.3	1.82	13.06	12.9
	8	C	–	7.83	13.73	17.29	16.3
FFA (% P.A.)	1	A	0.08	0.07	0.09	0.12	0.11
	4	A	0.09	0.095	0.1	0.14	0.11
	3	B	0.07	0.6	0.11	0.15	0.15
	9	B	–	0.04	0.06	0.06	0.085
	8	C	–	0.12	0.14	0.21	0.21

A: RBD Palm Olein; B: RBD Palm Oil; C: RBD Palm Stearin; 1*: Malaysian Shore Tank; 2*: Ship Tank After Loading; 3*: Ship Tank At Receiving Port; 4*: Barge Tank; 5*: Destination Shore Tank.

Remark: a) In No. 2, there was 71 M/T of previous oil in the shore tank before loading.
b) In No. 3, the shore tank contained 203 MT of old oil before loading.
c) In No. 8, the shore tank contained 200 MT of old oil before loading.

Source: Leong Wan Leong (1991).

The same observation was made by Bek-Nielsen (1972b) where the PV of a palm oil shipment was seen to rise fourfold between the bulking installation tanks at the loading port in Malaysia and the discharge port oversea after a 6-week journey.

It is for this reason that the Malaysian trade in palm oil products is mainly on a fob basis. It will be risky for Malaysian suppliers to trade on a cost, insurance, and freight (cif)/cost and freight (cnf) basis if quality changes are outside their control during shipping, handling, and storage.

Some Causes of Oil Quality Deterioration of Palm Oil Products during Storage and Handling

Aeration (Dissolved Oxygen Leading to Increase in Peroxide and Anisidine Values)

Exposure to atmospheric oxygen or aeration will lead to a high level of dissolved oxygen in the oil, which will result in an increase in the oxidation of the oil, leading to increases in the peroxide and anisidine values of the oil. This can happen during the storage and handling processing if equipment used in the process are not properly enclosed or sealed and rendered air-tight due to negligence.

It was observed in a study that, in most cases, care was not taken to reduce aeration during loading in most refinery tank farms. It was observed that over-top loading into storage tanks could increase the dissolved oxygen content (DOC) in the oil up to as much as 35 ppm. In a survey of some of the tank farms in Malaysia, it was found that the DOC was generally high, up to about 46.0 ppm in RBD palm oil.

Over-top loading of a ship's tank, carried out to speed up loading and to avoid demurrage, can cause high increases in the level of DOC. Measurements of DOC after loading and discharge to and from ships' tanks can increase the DOC of the oil to more than 20%. This high increase in DOC during loading is subsequently observed to decrease with time, indicating that the dissolved oxygen was used up in the oxidation reaction during the voyage (Elias & Jaffar, 1990). This oxygen consumption during the voyage was found to be well correlated with PV increases, especially for RBD palm stearin.

Another cause of oxidation is the practice of sweeping the tank floors (especially in ship tanks). This may have to be carried out for palm oil during discharge for complete emptying of the tank. Sweeping is normally done by 5–7 people who will have to go down into the tank to physically push the oil towards the sump or discharge vent. The act of scraping the tank wall and floor will cause loose rust and tank coating to peel off, thereby increasing the iron contamination of the oil. Very often the sweeping is also carried out in conjunction with jet spraying or re-circulate hot oil. This will further cause aeration of the oil.

Aeration of the oil could also occur as a result of faulty or poorly maintained pumps with faulty glen seals or as a result of other leakages.

Trace Metal Contamination

Trace metal contamination could occur as a result of hot oil coming into contact with the following:

a. rusty tanks
b. tanks with flaking coatings
c. copper or copper alloy equipment, such as brass or yocalbro in gauging and sampling equipment, valves, and ship tank heating coils
d. faulty or poorly maintained equipment with a high wear and tear rate

Causes of Increase in Hydrolysis Products (FFA)

The use of dirty or unwashed tanks could cause the contamination of fresh oil with oxidized and hydrolysis oil. This will increase both the oxidation and hydrolysis contents of the oil.

The risk of sea-water contamination is high in sea voyage. It is common to find that for nearly 50–60% of the voyage duration, the deck of the ship is covered with water, especially if the ship is fully-loaded and in winter. The hatch of the ship's tanks and other enclosures for butterworth holes etc. should be properly sealed to prevent the seepage of water into the tank.

Unnecessary overheating and prolonged heating of the oil will increase the hydrolysis of the oil leading to increase in the level of free fatty acids. These free fatty acids in turn will act as a catalyst to the hydrolysis reactions, thereby promoting the reaction further, especially if the unnecessary heating is continued.

Causes of Contamination with Undesirable Substances

This could happen as a result of negligence, ignorance, and/or insufficient information; weaknesses in the surveying, storage, and handling procedures and practices; inadequate tank cleaning procedure; abnormal trade practices; etc. The chances of contamination with undesirable foreign substances is high because chemical tankers that are used to carry palm oil are also used to carry other chemicals that are undesirable and foreign to palm oil. This is definitely an important area in which vigilance is necessary.

Other Causes of Quality Deterioration and/or Contamination

Apart from the above causes of quality deteriorations, there are other causes which may be inherent in the present handling and storage practices which could lead to quality deteriorations or contamination:

a. difficulty in tank and pipeline inspection
b. difficulty in ensuring tank and pipeline cleanliness
c. inadequacy of tank cleaning procedure
d. possible migration of remnants of previous cargo from tank coating or blisters
e. non-availability of actual chemical names of previous cargo and their information on toxicity
f. the necessity to cope with working at odd hours and adverse weather conditions and limited time to avoid demurrage.

Measurement of Palm Oil Quality

As can be seen from the above discussion, quality is an important attribute of palm oil products and it is a very important attribute from the trade point of view. Non-compliance to quality specifications could involve huge discounts or rejection of the consignment if it cannot be used for the intended purpose. In order to measure the quality attributes of palm oil products with an amount of certainty, accurate and reproducible methods are important. The following section will outline the analytical methods used for measuring quality attributes of palm oil products, as specified in the trade contracts. Complete details of the analytical methods will not be given but the methods will be referenced.

Standardization of Analytical Techniques

As mentioned before, the global palm oil trade is based on quality specifications. With the growing innovations of product developments and improvements, there is an increasing demand for quality improvement from consumers and end-users of palm oil products. Hence, it can be expected that with time, trading specifications will become wider and more stringent albeit on a willing buyer/willing seller basis for the more discerning end-user. This will require reliable and commonly agreed methods of analysis to ensure that these contractual specifications are met. It will also enable analysts to have recourse to validated methods to meet trade dispute requirements. To meet this growing need for method standardization, there are many organizations around the world which are involved with development and publication of standard methods of analysis for oils and fats, some of which are listed in Table 15-D (Tan, 1990).

The MPOB Methods is a compilation of test methods where the special requirements of palm oil and palm kernel oils had been considered. Where there are any shortcomings in the standard methods for oils and fats in general, collaborative studies or cross checks were carried out to resolve these shortcomings. In general, the MPOB test methods are based more on the International Standards Organization

Table 15-D. List of Standard Organizations Involved in Development and Publication of Standard Methods of Analysis for Oils and Fats.

1. British Standards Institute (BSI)
2. Association Francaise de Normalisation (AFNOR)
3. Nederlands Normalisatie-Institut (NNI)
4. Deutsche Gessellschaft fur Fettwissenschaft (DGF)
5. American Oil Chemists Society (AOCS)
6. Federation of Oils, Seeds and Fats Association (FOSFA)
7. International Association of Seed Crushers (IASC)
8. International Standards Organization (ISO)
9. Oils, Fats and Derivatives Commission VI 3 of the International Union of Pure and Applied Chemistry (IUPAC)
10. Association of Official Analytical Chemists (AOAC)
11. Codex Alimentarius Commission for Oils and Fats
12. European Economic Commission (EEC)
13. International Dairy Federation (IDF)

methods for oils and fats analysis rather than the American Oil Chemists Society (AOCS) methods.

General Palm Oil Quality Characteristics

According to the Codex Alimentarius Committee's Named Vegetable Oil Standard (Codex Alimentarius, 2005), palm oil products must meet certain quality characteristics. In general, palm oil products must be characteristic of the designated product (i.e., palm olein, palm stearin, etc.) and be free of foreign and rancid odor and taste. Table 15-E shows the limits for some of the Codex Alimentarius quality requirements for edible oils and fats, including palm oil products.

Oxidation Analysis

Peroxide Value (PV) and UV233nm$_{corrected}$ Measurements for Primary Oxidation

Rancidity of an oil or fat refers to the oxidative state of the oil which may impact the nutritional value of the oil or its intended application. Oxidation of palm oil products can be measured using PV analysis and ultraviolet absorption analysis at a wavelength of 233nm (UV233nm$_{corrected}$) for primary oxidation products and p-anisidine values (p-AnV) analysis and ultraviolet absorption analysis at a wavelength of 269nm

Table 15-E. Codex Alimentarius Quality Requirements for Edible Oils and Fats (Including Palm Oil and Products).

Quality characteristics	Level	Analytical Method
Volatile matter at 105°C (% m/m, max)	0.2	ISO 662:1998
Insoluble impurities (% m/m, max)	0.05	ISO 663:2000
Soap content (% m/m, max)	0.005	BS 684 Section 2.5 AOCS Cc 17-95 (97).
Iron (mg/kg, max)	1.5 (refined oil) 5.0 (virgin oil)	ISO 8294: 1994; AOAC 990.05 AOCS Ca 18b-91 (03)
Copper (mg/kg, max)	0.1 (refined oil) 0.2–0.4 (virgin oil)	ISO 8294: 1994; AOAC 990.05; AOCS Ca 18b-91 (03)
Peroxide value (meq/kg, max)	10 (refined oil) 20 (cold pressed/ virgin oil)	AOCS Cd 8b-90 (03); ISO 3960: 2001
Acid value (mg/KOH/g, max)	0.6 (refined oil) 4 (cold pressed/ virgin oil)	ISO 660: 1996, amended 2003; AOCS Cd 3d-63 (03)

Source: CODEX STAN 210-1999.

($UV269nm_{corrected}$) for secondary oxidation products. Peroxide value is a good indicator of the quality of the oil. Fresh palm oil extracted from the fruit has a PV of less than 1. The method measures the amount of oxygen chemically bounded to the oil as hydroperoxides and, hence, the degree of oxidation of the oil. The PV of an oil is expressed as the number of milliequivalents of oxygen per kilogram of oil and is determined by the reaction of peroxides (hydroperoxides) with potassium iodide in a solvent mixture. The liberated iodine is then determined titrimetrically. The two main sources of error for this analysis are the absorption of iodine by the unsaturated bonds in the fatty acid molecules and the liberation of iodine from potassium iodide by oxygen in the solution leading to lower and higher titre values, respectively.

The $UV233nm_{corrected}$ measurements determine the amount of peroxides present via the unsaturated bonds double conjugated system of the hydroperoxides. PV and $Uv233nm_{corrected}$ results are observed to be strongly correlated.

P-Anisidine and $UV269nm_{corrected}$ Measurements for Secondary Oxidation

The peroxides or hydroperoxides formed in the primary oxidation of the oil are high unstable intermediates and are readily decomposed into various carbonyls and other compounds. The PV of an oil tends to go up and then down as the secondary oxidation step occurs before it builds up again together with secondary oxidation products. This breakdown is accelerated by increase in temperature. The secondary oxidation carbonyl compounds, principally aldehydes and ketones, give the oil the characteristic off-flavor smell of rancid oil. Therefore, an assessment of the aldehydes present in the oil will give a complete picture of the extent of oxidation of the oil together with PV. The p-AnV measurement determines the amount of secondary oxidation products present in the oil as a result of the decomposition of the hydroperoxides into aldehydes and keto-dienes, mainly 2-alkenals and 2,4-dienals. In the measurement, the oxidized oil is treated with p-AnV in an iso-octane solution. The anisidine reacts with the non-volatile portion of the fatty acids when the hydroperoxides break down and the level of reaction products is spectroscopically determined at a wavelength of 350nm. The UV269nm$_{corrected}$ method is the physical method for the determination of the same secondary oxidation compounds through the triple conjugated system of the aldehydes and keto-dienes formed from the decomposition of the hydroperoxides. Table 15-F shows the official methods that can be used for quality analysis in conjunction with the joint FOSFA-PORAM contract (FOSFA 81).

Peroxide value is the most widely used indicator of fat oxidation, as it measures the lipid peroxides and hydroperoxides formed during the initial primary stages of oxidation. It should be mentioned that the titration end point could be difficult to ascertain for low PV levels in the oil. Although the PV method is highly empirical, it is still nevertheless a good guide to the quality of the oil.

The p-AnV test is a more sensitive test than the TBA test. Even if the aldehydes are no longer detectable with the TBA test, the p-AnV test may still indicate that the fat has been oxidized because the residues detected are not volatile. The volatile aldehydes and ketones will also lead to some analysis errors. While this method may provide some indication of the oxidative history of an oil or fat, it is not suitable for the detection of fat oxidation in total. Average oxidation values for crude palm oil are shown in Table 15-G, which is the results for a Malaysian CPO survey over the production year 1997–1998.

Totox values (Bek-Nielsen, 1975) and discriminant function (Siew, 1988) have been used in the past in an attempt to obtain a total picture of the oxidation status

Table 15-F. Official Methods for Oxidation Quality Analysis.

Parameter	AOCS Method number	MPOB Test Method number
PV (meq/kg)	Cd 8b–90 (09): Peroxide value using isooctane	MPOB p2.3: Palm Oil p series: Determination of peroxide value
UV233nm$_{corrected}$ UV269nm$_{corrected}$	—	MPOB p2.14: Determination of ultraviolet absorption expressed as specific UV extinction in light
p-Anisidine Value(AnV)	Cd 18–90 (09): Anisidine value	MPOB p2.4: Determination of Anisidine Value (AnV)

Table 15-G. Quality of Malaysian CPO (1997–1998).

Test parameters	97–98 CPO Survey Average Values (Std dev)
Free fatty acid. as palmitic (%)	3.17 (0.747)
Moisture and volatile matter (%)	0.18 (0.073)
Insoluble impurities (%)	0.03 (0.013)
Peroxide value(meq kg^{-1})	0.65 (0.692)
Anisidine value	4.65 (1.767)
Totox Value (2PV+AnV)	6.03 (2.39)
UV E233	1.31 (0.136)
UV E269	0.30 (0.050)
Iodine value (Wijs)	52.0 (0.74)
Carotene (ppm)	582 (52.0)
DOBI value	2.7 (0.272)
Discriminant function (Yi-Yj)	15.7 (4.30)
Phosphorus (ppm)	13.30 (5.859)
Iron Fe (ppm)	4.89 (2.52)
Total chlorophyll (ppb)	1 879 (483)

Source: PORIM Technology No 22 (April 2000).

of the oil. However, these practices have been phased out probably due to the large uncertainties inherent in anisidine value analysis.

Thiobarbituric acid (TBA) test is used to measure the aldehydes formed during the oxidation process and the breakdown of the hydroperoxides (i.e., after the secondary oxidation process). Hence, it is an indicator of the oxidative stability or rancidity of fats, especially of the secondary oxidation of the oil. Some of the aldehydes formed are volatile, and these may be lost when the oils or fats are heated for extended periods of time leading to some errors in the analysis. In general, TBA tests are seldom carried out for palm oil products.

Ultraviolet light measurements are considered to be more accurate than the wet chemical PV and p-AnV methods, as they measure the non-volatile parts of the breakdown products. This has also been observed to be the case in practice.

Determination of the oxidation stability of palm oil products can be effected through methods for the measurement of induction times. The measured induction time characterizes the resistance of the oils to oxidation. In addition to the term induction time, the expression Oil Stability Index (OSI) is also commonly used. The most widely used instrument for carrying out this test is the Rancimat (Metrohm) equipment using AOCS Method Cd12b-92 (AOCS, 1998). The Rancimat method is also known as the automated Swift test or accelerated oxidation test. Today, the

Rancimat test is generally used as an automated version of the complicated Active Oxygen Method (AOM). Temperatures used in the Rancimat test ranges from 100°C to 140°C for vegetable oils in general. The most common temperature used for palm oil products is 110°C.

During the measurement a stream of purified air is passed through the oil sample contained in a sealed and heated reaction vessel resulting in oxidation of the oil. Hydroperoxides are formed as the primary oxidation products. As the oxidation reaction progresses, the unstable hydroperoxides will breakdown into secondary oxidation products, including low-molecular organic acids in addition to other volatile organic compounds. These are carried by the air stream to a second vessel containing distilled water. The conductivity of the water in this vessel is recorded continuously. The dissolution of the low molecular weight in the water will lead to an increase in conductivity of the water. The time that elapses until the appearance of these secondary oxidation products is the induction time, induction period, or Oil Stability Index (OSI). The Rancimat software is also equipped with a unique temperature extrapolation routine which allows for a rough estimation of the storage stability of a product, the so-called temperature extrapolation. This is an aid for estimating the shelf life of oils and fats. The extrapolation makes use of the relationship between the measured induction time and the temperature given by van't Hoff's law. Several measurements are made at different temperatures and then extrapolated to the storage temperature. The value obtained allows a rough estimation of the storage stability of the oil or fat.

Hydrolysis Analysis

The hydrolysis of oils and fats is an autocatalytic reaction whereby the FFA produced will further catalyze the reaction. Hence, free fatty acid content is the most important CPO characteristic used to measure quality of an oil consignment. The free fatty acid content of an oil is therefore an indication of the hydrolytic deterioration of the oil. Free fatty acid content is also an indication of process control during CPO production and the care taken during storage and handling post refining. Although contractual specification for FFA is set at 5% maximum for CPO and less than 0.1% for refined oils, values in the region of 3% are normally observed for CPO. The mean FFA value of CPO produced in Malaysia in 1997–1998 was observed to be 3.17% (Tan et al., 2000).The longer the oil has been kept, the higher the FFA content. Generally, in the palm oil trade free fatty acid content analysis is normally carried out as compared to acid value (AV) analysis for other vegetable oils. The free fatty acid content is determined by a simple acid-base titration where the oil in isopropanol is titrated with an alcoholic or aqueous solution of potassium or sodium hydroxide using phenolphthalein as an indicator. The results are expressed as a percentage of the molecular weight of an acid. For palm oil and its products, the method is based on the molecular weight of palmitic acid while for palm kernel oil and products, the method is based on lauric acid. The molecular weights of palmitic and lauric acids are 256 and 200, respectively.

Table 15-H. Methods for Determination of Free Fatty Acid Content of Palm and Palm Oil Products.

Parameter	AOCS Method number	MPOB Test Method number
Free fatty acid content (%)	Ca 5a–40 (09): Free Fatty Acids in Crude and Refined Fats and Oils	MPOB p2.5: Determination of acidity

For crude palm oil free fatty acids analysis using phenolphthalein as an indicator, the persistence of the permanent faint pink color change in crude palm oil may be difficult to ascertain due to the orange-red color of CPO. For this, it is recommended that the color change should be observed in the upper alcoholic layer after the separation in the mixture. Other factors that could affect the end-point detection in the titration include the temperature of the solution, the vigority of shaking, and the diffusion of carbon dioxide into the system. The MPOB test method recommends gentle heating to 40°C together with moderate shaking during titration.

For blended palm oils, uncertainty may arise as to the molecule of the fatty acid to be used for the calculation of free fatty acid. In this case, acid value can be used to define acidity where the acid value is defined as the number of milligrams of potassium hydroxide which is required to neutralize the free fatty acids in one gram of the blended sample.

Table 15-H list the AOCS and MPOB test methods for determining the free fatty acid content of palm oil products.

Contamination with Foreign Substances Analysis

Contamination with foreign substances can be detected depending on the contaminant. Common methods used are Gas Chromatography (GC), High Pressure Liquid Chromatography (HPLC), Gas Chromatography-Mass Spectroscopy (GC-MS), High Pressure Liquid Chromatography-Mass Spectroscopy (HPLC-MS), Atomic Absorption Spectroscopy (AAS) for trace and heavy metals, and Inductively Coupled Plasma Mass Spectroscopy (ICPMS), just to name a few.

In the early days of the palm oil trade, contamination of palm oil shipment with industrial chemicals are often the reason for trade disputes on quality of goods. This is due to the fact that chemical tankers carrying chemicals from Europe and the USA to Asia are using the same tanks after washing to carry palm oil and products back to Europe and the USA on their return journey. However, use of ban and acceptable last immediate and past three cargoes lists have alleviated these problems to a minimum. Presently, contamination from previous cargoes rarely occurs.

In terms of mineral oils, traces of mineral oil may be found in vegetable oils. A small part of these mineral oils are inherent to the oil while some are due to contact with process equipment and vessels where mineral oils are present. During transportation, the mineral oils can also be contaminants from an improperly cleaned ship tank which had been used for carriage of mineral oil previously. High levels of mineral oil in palm oil can be determined by measurement of the unsaponifiable matter content of the oil. One of the acceptable methods for determination of mineral oil contamination is the IUPAC semi-quantitative thin layer chromatography method, IUPAC 2.611(1987), with a limit of detection of 200 mg/kg. Modification of this method by FOSFA and MPOB has lowered the limit to 10 mg/kg. Besides being a contaminant to the oil, the presence of mineral oil, even in trace quantities, is undesirable as it contributes to the reduction of smoke, fire, and flash points of palm oil products.

Contamination by Pesticides and Weedicides

Pesticides and weedicides are used in oil palm plantations for the control of pests and weeds, respectively. Some of the pesticides and weedicides used in oil palm plantations are chlorpyrifos, monochrotophos, methamidophos, carbofuran, diuron, and cypermetrin, to name a few. These agrochemicals are mainly water soluble compounds and are generally not soluble in palm oil. Cypermetrin, which is oil soluble, is used mainly in the nursery and young palms before fruit bearing only. Work by Halimah et al. (Halimah, 2003–2011) have shown that these agrochemicals persist in the soil, leaves, and run-off water for only less than a week after their applications.

Extensive field trial studies by Halimah et al. (Halimah, 2003–2011) and Yeoh et al. (2010) on these agrochemicals have shown that these water soluble agrochemicals are not detected in palm oil, even for trunk injected agrochemicals. Using spiked palm oil samples, Yeoh et al. (2009) further showed that if accidental contamination of palm oil with these agrochemicals is to occur, these contaminations are effectively removed by the physical refining process, in line with observations by other researchers on other vegetable oils (Chaudry et al., 1978; Hashemy-Tonkabony &. Soleimani-Amiri, 1976; Muhamad et al., 2004; Pages-Xatart-Pares et al., 2004). A review on this has been published by Holland et al. (1994). Hence, it could be concluded that the usage of agrochemicals in oil palm plantations for the control of pests and weeds do not pose any contamination problem to the production of crude palm oil due to the water soluble nature of these agrochemicals. This is further enhanced if the manufacturers' recommendations for proper chemical application procedures are followed and the fruits are harvested after a safe time window from the time of application of the agrochemical.

Cross-Contamination Analysis

Cross contamination can be determine via fatty acid composition (FAC) analysis using GC or via native component analysis for component(s) that is/are present in one oil but not in the other.

Other Quality/Contractual Analysis

Moisture and Volatile Matter and Insoluble Impurities Analysis

Water is one of the most common impurities in palm oil. The removal of moisture by drying is the most convenient way to determine moisture content. However, the removal of moisture by heating will also remove other volatile matters in the oil. Hence, moisture analysis methods based on drying tends to quantify the loss of both moisture and volatile matters as a percentage weight of the palm oil samples used. These methods involve heating the oil sample to a temperature above 100°C followed by gravimetric measurement due to the moisture loss. This is repeated until a constant sample weight is obtained. For palm oil products, these methods are more acceptable compared to the case of more unsaturated oils where the constant heating above 100°C may lead to an increase in weight of the sample despite the water loss due to oxidation of the oil instead. These drying methods are also not suitable for palm kernel oil and its products because their low chain free fatty acids can volatilize slowly under the test conditions, making it difficult to dry the sample to constant weight.

Moisture in palm oil can also be quantified by the Dean and Stark entrainment method (AOCS Official Method Ca 2a-45 Reapproved 2009), where the water is distilled out of the sample as a constant boiling mixture with xylene or toluene. This distilling method is specific for water and does not include the volatile substances present in the oil. This method is applicable to samples containing larger quantities of moisture of over 0.5%. This method is nowadays seldom used due to the fact that it involves the usage of xylene or toluene.

For the determination of very small amounts of water in oil (<0.1%), the Karl Fisher titration method is preferred, as it is more accurate. The method involves dissolution of the oil in methanol containing sulphur dioxide and pyridine followed by potentiometric titration with iodine solution. The sulphur dioxide and iodine will react in the presence of water to form sulphur trioxide and hydroiodic acid. The water content is calculated from the iodine consumption expressed in percentage. The pyridine-based solution used previously has now been replaced by pyridine-free solutions such as Hydranal.

Controlling the water content in palm oil product is important, as it will affect the hydrolytic stability of the oil if the presence of water is high in addition to the economic losses upon refining. A study by Chong & Gapor (1983) has shown that a moisture level within the range of 0.15% to 0.25% is optimal for CPO in terms of oxidative and hydrolytic stability and depending on the amount of trace metals present. This is due to the chelation of trace metals by the water molecules. This will effectively reduce the catalytic activity of the trace metals present in CPO as well as reduce the moisture available for the hydrolytic reaction. This result is in line with the observation by Chan (1979) that the optimal moisture level for CPO is 0.18%.

For palm oil products, moisture and volatile content is usually measured by either the Karl Fisher method or by oven drying at 103°C to constant weight. The oven method consists of heating the oil sample in an oven at 103°C ± 2°C until moisture and volatile substances are completely eliminated, and the loss in mass is determined.

The MPOB impurities test specifies a method for the determination of insoluble impurities content of palm oil and palm oil products. These impurities include mechanical impurities, mineral substances, carbohydrates, nitrogen, various resins, calcium soaps, oxidised fatty acids, fatty acid lactones, and (in part) alkali soaps, hydroxy-fatty acids, and their glycerides. Impurities are defined as substances which remain insoluble and can be filtered off after the oil is dissolved in a specific solvent such as petroleum ether or diethyl ether. The solvent used will influence the solubility of the substances separating from the oil, hence, it is important to specify the method and the solvent or solvent system used in the test report. If it is not desired to include soaps (particularly calcium soaps) or oxidised fatty acids in the insoluble impurities content, a different solvent and procedure can be used to achieve it. In this case, the method should be the subject of agreement between the parties concerned. The insoluble impurities are expressed as a percentage by weight. The test procedure involves dissolution of the oil in an excess of solvent, filtration of the solution, drying the residue on the filter paper at 103°C ± 2°C, followed by repeated drying and weighing to constant weight.

The methods of analysis for moisture and volatile matter and the amount of impurities in oils and fats are given in Table 15-I.

Table 15-I. Methods of Analysis for Impurities and Moisture and Volatile Matters in Palm Oil Products.

Parameters	AOCS/ISO Method	MPOB Method
Moisture and volatile matters	Ca 2f–93 (09) Moisture and Volatile Matter, by Vacuum Oven, in Fats and Oils (except coconut) Ca 2b–38 (09) Moisture and Volatile Matter, by Hot Plate Method ,for butter, Fats, Margarines, Oils. Ca 2a–45 (09) Moisture, by Distillation Method in Fats and Oils Ca 2e–84 (09) Moisture, by Karl Fischer Method in Oils and Fats	MPOB p2.1 Part 1—Determination of moisture and volatile matter—Method using a drying oven MPOB p2.1 Part 2—Determination of moisture and volatile matter—Method using a sand bath or hot plate.
Impurities	ISO 663 Animal and vegetable fats and oils—Determination of insoluble impurities	MPOB p2.2 Determination of impurities

Iodine Value (IV)

Iodine value measures the unsaturations content or double bonds which can react with iodine. It is defined as the percentage of iodine absorbed by the oil under the test conditions. The most common methods used for iodine value determination are the Wijs and Hanus methods.

The Wijs method involves the addition of a solution of iodine monochloride in a mixture of acetic acid and cyclohexane to a test sample. After iodine addition to the double bonds has occurred, the excess iodine monochloride is reduced to free iodine by addition of potassium iodide solution and water. The liberated iodine is titrated with a standardized solution of sodium thiosulphate with starch as indicator. Results are expressed as the number of grams of iodine absorbed by 100 g of the oil sample. Precautions to be taken when carrying out Wijs analysis are to re-standardize sodium thiosulphate solutions with potassium dichromate solution at regular intervals and to carry out blank tests on the Wijs reagent. This is due to the long term instability of both solutions.

Melting Point—Slip Melting Point (SMP)

Palm oil and palm kernel oil are complex mixtures of triacylglycerols. Hence, unlike pure organic compounds, they do not possess sharp melting points but rather a melting range. There are various ways of defining and measuring the melting characteristics of an oil or fat such as the slip melting point, Wiley melting point (AOCS. Method Cc 2-38,1988) and the Mettler dropping point. For the palm oil trade, slip melting point using open capillary tubes are usually specified. In this method, the oil or fat column is solidified in open capillary tubes tempered at a fixed temperature for a specified time period and then heated in a water bath until the fat rises in the tube under hydrostatic pressure. The temperature at which the fat column begins to rise or slip is taken as the slip melting point. A study on the factors affecting slip melting point was reported by Berger et al. (1982), while a comparison of nine methods for determining melting points of oils and fats was discussed by De Man et al. (1983).

Cloud Point (CP)

The cloud point of an oil is an indication of its resistance to crystallization and is generally used as a specification for palm olein. The olein sample is cooled down under the test conditions until it begins to cloud due to crystallization. The temperature at which the temperature bulb is no longer visible to the eye due to clouding is termed the cloud point. Siew et al. (1985) investigated the factors affecting cloud point analysis. They found that the factors that could affect the measurement of cloud point are the cooling vessel, bath temperature, stirring rate, light source, and filtration of test sample.

Oil Color

Color is an important specification for refined palm oil products, as it could affect the end-use of the oil. Refined palm oil color is measured by the Lovibond Tintometer. The method involves the matching of the color of light transmitted through a specified depth of oil with the color transmitted from the same source through a set of colored glass slides. Oil color test is a subjective measurement, depending on the analyst's judgement and type of tintometer used. Due to this subjectivity, there is usually a large interlaboratory variation. The main sources of errors are (Tan, 1990)

1. presence of scratches, dirt, or oil films on cell windows or glass slides
2. fatigue of electric lamp light source
3. differences in color vision of analysts
4. potential variation in colors of the calibrated glass slides or of different Tintometer models
5. presence of suspended or opalescent matter in the oil which must be in a fully liquid and transparent condition
6. influence of light reflected within the instrument or within the glass sample cell.

Unsaponifiable Matter

Contamination by foreign matters such as mineral oils and oxidative damage to the oil can be checked by determining the unsaponifiable matter of the oil. This is due to the fact that the polymerized fatty acid of an oxidized oil can be extracted together with the unsaponifiable matters. The unsaponifiable matter is defined as the sum of those components of an oil or fat which, after saponification, can be extracted from the aqueous alkaline solution with petroleum ether and remain as non-volatile residues on drying at 103 ± 2°C. They include natural components such as sterols, hydrocarbons, tocopherols, alcohols, carotenoids or pigments, and the inorganic impurities such as mineral oils. The test involves saponification by boiling with ethanolic potassium hydroxide solution and extraction of the unsaponifiable matters with petroleum ether from the soap solution after dilution with water. The final residue is determined by evaporation of the solvent and weighing the residue after drying at 103 ± 2°C. The result is expressed as a percentage weight of the oil sample.

DOBI Analysis

Deterioration of Bleachability Index (DOBI) was added as a quality parameter at the turn of the millenium. The DOBI index was developed by Swoboda in MPOB (then PORIM) in the 1980s. The DOBI value was used as a guide as to how easily a sample of crude palm oil can be bleached. The DOBI value is a numerical ratio of the UV absorbance of the sample at 446 nm to the absorbance at 269 nm. Hence, it is an interplay between the amount of carotene present in the crude palm oil to the amounts of secondary oxidation products present in the oil. This is because the carotenes present in crude palm oil will be the first to be sacrificed in protecting the oil from oxidation, due to its high number of double bonds. The higher the DOBI value, the easier it is to bleach the oil through heat and absorptive cleansing bleaching. Crude palm oil over 3 is considered to be good while average quality palm oils have DOBI values above 2.3.

DOBI method is described in ISO 17932:2005 as "Animal and vegetable fats and oils—Determination of the Deterioration of Bleachability Index (DOBI)", while the corresponding method under the MPOB Test Method is MPOB p2.9: Determination of the Deterioration of Bleachability Index (DOBI).

MPOB Analytical Methods

Table 15-J below summarizes the MPOB test methods for chemical and physical characteristics and quality analyses of palm oil products.

Possible Solutions to Overcoming Quality Deteriorations

Overcoming Quality Deterioration Per Se

Based on present infrastructures and facilities, the reduction of quality deterioration could be achieved to a certain extent by proper management of handling, storage, and transportation practices. Such efforts will reduce the frequency of bad quality deterioration. To completely eliminate the possibilities of quality deteriorations, large investments may have to be incurred which will lead to drastic increases in transportation costs. Therefore, a balanced approach based on a combination of the following is possible.

Improving the Quality of Crude Palm Oil for Refining

Improving the quality of crude oil feed into refineries is important, as a bad crude will result in unstable refined products which may suffer rapid oxidation, hydrolysis and color reversion, and fixation during subsequent storage and transportation. Color fixation can be a deterrent towards the wider use of palm oil in white products such as creams, shortenings, etc. This is a severe problem because color fixation can cause difficulty even after re-refining whereby the best color one may be able to achieve is 1.5 to 2.0 R. This is not competitive when compared to other oils such as soybean oil, rapeseed oil, fish oil, lard, etc., the color of which can be reduced to less than 1.0 R.

Table 15-J. MPOB Test Methods for Chemical and Physical Characteristics and Quality Analyses of Palm Oil Products: Palm Oil P-Series.

Test method Number	Title of test
MPOB p1.1 part 1	Sampling from bulk consignments
MPOB p1.1 part 2	Sampling from packages
MPOB p1.2	Preparation of test samples
MPOB p2.1 Part 1	Determination of moisture and volatile matter—Part 1: Method using a drying oven
MPOB p2.1 Part 2	Determination of moisture and volatile matter—Part 1: Method using a sand bath or hot plate
MPOB p2.2	Determination of impurities
MPOB p2.3	Determination of peroxide value
MPOB p2.4	Determination of anisidine value (AnV)
MPOB p2.5	Determination of acidity
MPOB p2.6	Determination of carotene content
MPOB p2.7 Part 2	Determination of unsaponifiable matter: Part 2—Rapid method using hexane extraction
MPOB p2.8 Part 1 (a)	Determination of phosphorus—Part 1 (a): Colorimetric method Molybdenum blue method
MPOB p2.8 Part 1 (b)	Determination of phosphorus—Part 1 (b): Colorimetric method (Phosphovanadomolybdic method)
MPOB p2.8 Part 2	Determination of phosphorus—Part 2: Method using graphite furnace atomic absorption spectrometry
MPOB p2.9	Determination of the deterioration of bleachability index (DOBI) of crude palm oil
MPOB p2.10	Determination of chromium, copper, iron and nickel—Method by atomic absorption spectrophotometry
MPOB p2.11	Determination of chromium, copper, iron, nickel and manganese—Method by atomic absorption spectrophotometry using a graphite furnace
MPOB p2.12	Determination of bleachability
MPOB p2.13	Determination of soap content (Alkalinity)

Table 15-J. continued.

Test method Number	Title of test
MPOB p2.14	Determination of ultra-violet absorption expressed as specific UV extinction in light
MPOB p3.1	Determination of saponification value
MPOB p3.2	Determination of iodine value (Wijs)
MPOB p3.3	Determination of triacylglycerol by gas chromatography: Section
MPOB p3.4—Part 1	Preparation of methyl esters of fatty acid—Part 1: Boron trifluoride method
MPOB p3.4—Part 2	Preparation of methyl esters of fatty acids—Part 2: Alkaline methanolysis method
MPOB p3.4—Part 3	Preparation of methyl esters of fatty acids—Part 3: Acidic methanolysis method
MPOB p3.4—Part 4	Preparation of methyl esters of fatty acids—Part 4: Rapid method
MPOB p3.5	Determination of fatty acid composition (FAC) as their methyl esters
MPOB p3.6	Determination of ash
MPOB p3.7	Determination of total fatty matter (TFM)
MPOB p3.8	Determination of total saponifiable (TSM)
MPOB p4.1	Determination of lovibond colour
MPOB p4.2	Determination of slip melting point
MPOB p4.3	Determination of cloud point
MPOB p4.4	Determination of refractive index
MPOB p4.5	Determination of conventional mass per volume ("Litre weight in air)
MPOB p4.6	Determination of titre
MPOB p4.7	Determination of cold test
MPOB p4.8	Determination of solid fat content by pulsed nuclear magnetic resonance (PNMR) Section 1: Direct method
MPOB p4.9	Determination of solid fat content by pulsed nuclear magnetic resonance (PNMR) Section 2: Indirect method

Reducing Aeration

This can easily be done without incurring high costs. Important areas where aeration can possibly be reduced are

a. During loading into storage or ship's tank extend the drop-line to the bottom of the tank or below the oil surface if overtop leading is required. Overtop loading resulting in cascading of oil through air should be discouraged. Loading into a ship's tank should be through the manifold at all times, where possible.
b. For shore tanks without a drop-line, the oil should be allowed to flow down the wall. This will help to minimize aeration as compared to spraying the oil into the tank.
c. For loading into lorry-tankers at the dispatch bay, a telescopic pipeline, which can be extended to below the oil surface, should be used. This retractable telescopic pipeline can be slowly retracted upwards as the tanker is filled.
d. Pipeline blowing for pigging and clearing should be carried out with nitrogen instead of with air. This is to avoid the air used for blowing the pig from aerating the oil as the oil in the pipeline is blown into the tank.
e. During storage in shore or the ship's tank, exposure to air can be minimized by the use of inert gas blanketing such as nitrogen which is cheap (Leong et al., 1982). Reducing the concentration of oxygen with nitrogen in the tank headspace will reduce the solubility of oxygen in the oil by causing the oxygen to migrate out of the oil into the headspace. Blanketing therefore reduces exposure to oxygen as well as providing a means of oxygen adsorption from the oil. Floating lids on oil surface will also help to reduce exposure to atmospheric oxygen. However this practice has been discontinued.
f. De-aeration and the removal of dissolved oxygen in the oil can also be carried out effectively by pipeline sparging of the oil with nitrogen during pumping, such as during loading into the ship's tanks or during discharge. Oxygen desorption by sparging has the advantage of breaking up the nitrogen atmosphere into tiny bubbles within the oil itself, thus increasing the surface area of the oil in contact with nitrogen, thereby bringing the nitrogen gas into closer contact with the oil. The slow migration of these tiny bubbles upward into the tank headspace to form a blanket of nitrogen over the oil surface will reduce the contact of oil with atmospheric oxygen.

Protection from Oxidation

While de-aeration and removal of dissolved oxygen directly will contribute to reducing the risks of oxidation, a more direct protection from oxidation can be affected by the use of chemical antioxidants. Some of these antioxidants also have antimicrobial activity, which is an added advantage. Some of the points to note when using chemical antioxidants are as follows.

a. The use of antioxidants is restricted by the importing country's food act or regulations. A concentration range of 100–200 ppm is well within the legal maximum of 200 ppm for most countries.
b. Common antioxidants used are butyl hydroxyl anisole (BHA), butyl hydroxyl toluene (BHT), and tertiary butyl hydroquinone (TBHQ). The use of citric acid synergist is a common practice, although it is commonly used to chelate or inactivate pro-oxidant metal. Citric acid is universally permitted in edible oils and is used as a good practice in refineries all over the world. It is advantageous to use antioxidants in conjunction with citric acid; where it is legally permitted, it is recommended to use TBHQ or TBHQ/Citric acid (110 ppm/360 ppm) (Chong, 1988), which has proven to be the most effective combination for the oxidative protection of the oil.
c. It is recommended that, where possible, antioxidant should be added to the bulk oil immediately after deodorization in order to obtain maximum protection.
d. It is important to ensure complete dispersion and dissolution of antioxidants in the oil. This can be achieved by dissolving the antioxidant either in hot oil (ca 60°C) or in a solvent miscible with oil before addition.
e. Where possible, a dosing pump should be used. Complete mixing of the antioxidant can be ensured by the incorporation of a static mixer in the loading line.
f. Consideration should also be given to the provision of an antioxidant dosing facility at bulking installations.

Conclusions

In conclusion, it can be said that some of the quality deteriorations highlighted can be avoided by proper management of handling, storage, and transportation practices by all parties concerned. Further reduction in risks of excessive increases in PV and FFA can be achieved (e.g., by nitrogen sparging) without incurring high investment costs. This is especially important in the future, in view of the fact that more and more countries are becoming more and more concerned about chemical additives in food products.

It is obvious from the foregoing discussion that quality deterioration does affect the application and marketability of palm oil products. Efforts to minimize quality deteriorations and to preserve quality must be exerted during processing of palm oil from the start till the storage, transportation, and handling during the export of the oil at the tail end of the chain in order to make palm oil products more competitive in the international markets.

Analytical methods and techniques are important to the palm oil trade, as they are used to determine adherence to contractual specifications. It is critical that the methods used are acceptable to both buyers and sellers in case of any trade disputes. It is of most importance that the methods used are thoroughly evaluated by collaborative trials and ring testing. It would be ideal for the palm oil trade if there is a set of detailed and agreed upon analytical methods to fall back on should trade disputes arise.

References

AOCS Official Methods and Recommended Practices for Commercial Oils and Fats, 2009. American Oil Chemists' Society: Champaign, IL, 2009.

Bek-Nielsen, B. Factors responsible for the development of peroxides during production and handling of palm oil. Oleagineux 27, 7, July, 1972a.

Bek-Nielsen, B. Factors responsible for the development of peroxides during production and handling of palm oil. Oleagineux 27, 9, September, 1972b.

Bek-Nielsen, B. Technical aspects of marketing quality palm oil; Seminar Proceedings on Market Development of Palm Oil Products. 27 March–1 April, 1975, Kuala Lumpur.

Bek-Nielsen, B. Quality preservation and testing of Malaysian palm oil from fresh fruit bunches to the oil refinery. Oleagineux 32, 10, October 1977, 437–441.

Berger, K.G.; Martin, S.M. The Cambridge World History of Food; Kiple, K.F., Ornelas, K.C., Eds.; Cambridge University Press, 2000.

Berger, K.G.; Siew, W.L.; Flingoh, C.H.O. Factors affecting slip melting point of palm oil products. J. Am. Oil Chem. Soc. 1982, 59, 5, 244–249.

Chan, K.S. IKM Conference on Chemical Research, Kuala Lumpur, 1979.

Chaudry, M.M.; Nelson, A.L.; Perkins, E.G. Distribution of chlorinated pesticides in soybean, soybean oil, and its by product during processing. J. Am. Oil Chem. Soc. 1978, 55, 851–853.

Chong, C.L.; Gapor Mohd, A. Top. Effects of moisture and trace metals on oil quality. Proceedings of the workshop on quality in the palm oil industry, PORIM, August, 1983.

Chong, C.L.; Ong, A.S.H. Effect of antioxidant (TBHQ/citric acid) on the oxidative stability of palm oil. Poster presented at the Asean Food Conference, October 1988, Bangkok, Thailand.

Chong, C.L.; Sambanthamurthi, R. Effects of mesocarp bruising on the rate of free fatty acid release in oil palm fruits. Int. Biodeterior. Biodegrad. 1993, 31, 1, 65–70.

Codex, A. Codex standard for named vegetable oils CODEX STAN 210-1999; amended 2005, 2001.

Deman, J.M.; Deman, L.; Blackman, B. Melting-point determination of fat products. J. Am. Oil Chem. Soc. 1983, 60, 1, 91–94.

Elias, B.A.; Mohd. Jaaffar, A. PORIM Report G(141)90 Restricted, Transhipment studies at the Port of Rotterdam, 1990.

Gapor Md Top, A.B.; Berger, K.G. Effects of processing on the content and composition of tocopherols and tocotrienols in palm oil. Proceedings of Palm Oil in the Eighties, Kuala Lumpur, 1982.

Goh, S.H.; Khor, H.T.; Gee, P.T. Phospholipids of palm oil. JAOCS 1982, 59(7), 296–299.

Halimah, M. The leaching of thiram and benomyl in oil palm nursery ecosystem during wet season. 16th Malaysian Chemical Congress 2010, 12-14 October, 2010 PWTC, Kuala Lumpur.

Halimah, M.; Badrul Hisyam, Z.; Nor Kartini, A.B. Determination of λ-cyhalothrin in crude palm oil and crude palm kernel oil by solid phase extraction and gas chromatography electron capture detection. MPOB June 2010, 82.

Halimah, M.; Ismail, B.S.; Mehdi, S.; Nashriyah, M. Adsorption study of ^{14}C-paraquat in two Malaysian agricultural soils. Environ. Monit. Assess. 2011, 176, 43–50.

Halimah, M.; Maznah, Z.; Ismail, B.S.; Idris A.S. Determination of hexaconazole in crude palm kernel oil by solid phase extraction and gas chromatography electron capture detection. MPOB, June 2010, 78.

Halimah, M.; Maznah, Z.; Ismail, B.S.; Idris, A.S. Determination of hexaconazole in field samples using gas chromatography-electron capture detection (GC-ECD). MPOB June 2010, 79, 80, 81.

Halimah, M.; Maznah, Z.; Ismail, S.; Tan, Y.A. Determination of hexaconazole in field samples of an oil palm plantation. The Saskatoon International Workshop Validation and Regulation Analysis, 19–22 June 2011, Saskatoon, Saskatchewan, Canada.

Halimah, M.; Mohd Izwarie, R.; Zuriati, Z.; Ismail, B.S. Determination of diuron in crude palm oil and crude palm kernel oil by solid phase extraction and high performance liquid chromatography using ultra violet detection. MPOB TS No. 83. June 2010.

Halimah, M.; Nashriyah, M.; Tan, Y.A.; Ismail, B.S. Adsorption and desorption study of 14C-chlorpyrifos in two Malaysian agricultural soils. J. Nucl. Relate. Technol. 2004, 1(1), 31–40.

Halimah, M.; Nashriyah, M.; Tan, Y.A.; Ismail, B.S. The adsorption coefficient (k_{oc}) of chlorpyrifos in clay soil. J. Nucl. Relate. Technol. 2005, 2(2), 23–30.

Halimah, M.; Osman, H.; Ainie, K.; Pauzi, A. Determination of chlorpyrifos in refined palm olein by GC-ECD and GC-FPD. J. Oil Palm Oil Res. 1999, 11(2), 89–97.

Halimah, M.; Osman, H.; Ainie, K.; Tan, Y.A.; Pauzi, A. Optimization of extraction and clean-up procedures for chlorpyrifos residue determination in oil matrix. Malaysian J. Chem. 2002, 4(1).

Halimah, M.; Pauzi, A.; Chian, S.S. Optimization of sweep co-distillation clean-up method for the determination of organochlorine pesticide residues in palm oil. J. Palm Oil Res. 2004, 16(2), 30–36.

Halimah, M.; Tan, Y.A.; Ismail, B.S. Method development for determination of fluroxypyr in soil. J. Environ. Sci. Health Part B-Pest. Food Contam. Agric. Wastes 2004, B39(5), 765–777.

Halimah, M.; Tan, Y.A.; Ismail, B.S. Fate of fluroxypyr in soil in an oil palm agroecosystem. Weed Biol. Manag. 2005, 5(4), 184–189.

Halimah, M.; Tan, Y.A.; Ismail, B.S. Determination of herbicide fluroxypyr in oil matrices. J. Environ. Sci. Health Part B-Pesticides, Food Contam. Agric. Wastes 2008, B34(2), 184–189.

Halimah, M.; Tan, Y.A.; Ismail, B.S.; Nashriyah, M. Downward movement of chlorpyrifos in the soil of an oil palm plantation in Sepang, Selangor, Malaysia. J. Oil Palm Res. April 2010, 22, 721–728.

Halimah, M.; Tan, Y.A.; Kuntom, A.; Ismail, B.S. Method development for determination of fluroxypyr in water. J. Environ. Sci. Health 2003, B(38), 429–440.

Halimah, M.; Zulkifli, M.; Tan, Y.A.; Hasnol, O.; Ismail B.S. Leaching of chlorpyrifos in peat soil of an oil palm plantation in Malaysia. American-Eurasian J. Sustain. Agric. 2011, 5(2) 209–215.

Hashemy-Tonkabony, S.E.; Soleimani-Amiri, M.J. Detection and determination of chlorinated pesticide residues in raw and various stages of processed vegetable oil. J. Am. Oil Chem. Soc. 1976, 53, 752–753.

Holland, P.T.; Hamilton, D.; Ohlin, B.; Skidmore, M.W.; Effects of storage and processing on pesticide residues in plant products. Pure Appl. Chem. 1994, 66, 335–356.

Iwuchukwu, A.N. Influence of marketing boards system and preliminary purification on Nigerian palm oil exports. In Proceedings of Oil Palm Conference at Tropical Products Institute, London, 1965; pp. 73–77.

Jacobsberg, B. Survey of the palm oil production in 1973. In Proceedings of the First Malaysian Agricultural Research and Development Institute Workshop on Palm Oil Technology; Kuala Lumpur, 1974; pp. 72–119

Jacobsberg, B. Quality of palm oil. PORIM Technology, October 1983, 10.

Johansson, G.; Pehlergard, P.O. Aspects on quality of palm oil. In International Developments in Palm Oil; Earp, D.A., Newall, W., Eds.; MARDI, Kuala Lumpur, 1977; pp. 203–223.

Juran, M.J. American Society for Quality, Glossary Entry: Quality http://www.asq.org/glossary/q.html, accessed 07/20/2008.

Kano, N. Attractive quality and must-be quality. J. Japanese Soc. Qual. Contr. 1984, April, 39–48.

Leong, W.L. Quality deterioration during transport: their effect on marketability and methods of overcoming them. Lecture notes 11, Fifth palm oil ship-shore surveyors course, 1991.

Leong, W.L.; Berger, K.G. Storage, handling and transportation of palm oil products. PORIM Technol. August 1982, 7.

Loncin, M.; Jacobsberg, B. Studies on Congo palm oil. JAOCS 1963, 40(6), 18–19, 40–45.

Loncin, M.; Jacobsberg, B. Research on palm oil in Belgium and the Congo. In Proceedings of Oil Palm Conference at Tropical Products Institute, 85–95, London, 1965.

MEOMA Handbook. Malayan Edible Oil Manufacturers' Association (Meoma), No. 134, 2008–2010, Kuala Lumpur.

Metrohm, A.G. Information Brochure 2.743.0014 743 Rancimat for Oils and Fats.

Mongana Report. Research on Production and Storage of Palm Oil (Book 1 and 2). Institute for the Advancement of Scientific Research in Industry and Agriculture, 1955.

Oil Palm Grading Manual. Malaysian Palm Oil Board (MPOB), 2003.

MPOB Methods of test for palm and palm oil products. Malaysian Palm Oil Board: Selangor, 2004.

Muhamad, H.; Hassan, O.; Abdullah, M.P.; Tan, Y.A.; Sahid, I. Investigation of the effectiveness of pesticide removal by laboratory scale refining. Proceedings SKAM-17, August 24–26, Kuantan, Malaysia) 2004.

Pages-Xatart-Pares, J.; Laur, C.; Birot, J.; Arnaud, F. Lactose: Influence of the refining process on pesticide residues removal. Poster presentation during 95th AOCS Annual Meeting & Expo, May 9–12, Cincinnati, OH, 2004.

PORIM. Recommended Practices for Storage and Transport of Edible Oils and Fats; Berger, K.G., Ed.; Palm Oil Research Institute of Malaysia (PORIM), 1985.

PORAM Handbook. Palm Oil Refiners Association of Malaysia, Selangor, Malaysia, 2011.

Sambanthamurthi, R.; Let, C.C.; Cheang, O.K.; Huat, Y.K.; Rajan, P. Chilling-induced Lipid hydrolysis in the oil palm (*Elaeis guineensis*) mesocarp. J. Exp. Bot. 1991, 42(9), 1199–1205.

Siew, W.L. Quality index of crude palm oil using discrimination functions. Malaysian Chemical Conference 1988, August 9–10, 1988.

Tan, Y.A. Analytical techniques in palm oil and palm kernel oil specifications. Lecture notes, PORIM Palm Oil Familiarization Programme (POFP), 1990.

Tan, Y.A.; Chong, C.L.; Low, K.S. Relationship between laser-induced fluorescence intensity and crude palm oil quality. J. Sci. Food Agric. March 1995, 67, 3, 375–379.

Tan, Y.A.; Ainie, K.; Siew, W.L.; Mohtar, Y.; Chong, C.L. PORIM crude palm oil survey 97/98 quality and identity characteristics. PORIM Technol. April 2000, 22.

Wong, C.Y. Malaysian crude palm oil quality delivered to refiners. In Palm Oil Product Technology in the Eighties; Pushparajah, E., Rajadurai, M., Eds.; Institute of Planters, Kuala Lumpur, 1981; pp. 87–87.

Yeoh, C.B. Personal communication, 2010.

Yeoh, C.B.; Chong, C.L. Acephate, methamidophos and monocrotophos residues in a laboratory-scale oil refining process. Eur. J. Lipid Sci. Technol. 2009, 111, 593–598.

Appendix A

Table A-1. Identity Characteristics for Crude Palm Oil.

Identity Characteristics	Observed min. to max.
Apparent density, g/ml at 50°C	0.8896 to 0.8910
Refractive index, nD 50°C	1.4544 to 1.4550
Saponification value, mg KOH/g oil	190 to 202
Unsaponifiable matter, %	0.15 to 0.99
Fatty acid composition, (wt % as methyl esters)	
C12:0	0.1 to 0.4
C14:0	1.0 to 1.4
C16:0	40.9 to 47.5
C16:1	0 to 0.6
C18:0	3.8 to 4.8
C18:1	36.4 to 41.2
C18:2	9.2 to 11.6
C18:3	0 to 0.5
C:20:0	0 to 0.8
Iodine value (Wijs)	50.1 to 54.9
Slip Melting Point, °C	33.0 to 39.0
Total carotenoids (as β-carotene), mg/kg	500 to 1000

(From Malaysian standard MS 814: palm oil)

Appendix B

Palm Oil Products Specifications

Table B-1. Crude Palm Oil (Malaysian Palm Oil Association [MPOA] Domestic Contract).

FFA (as palmitic)	5.00% max
M&I	0.25% max
DOBI	2.3 min

Table B-2. PORAM Standard Specifications for Processed Palm Oil.

Neutralized Palm Oil	
FFA (as Palmitic)	0.25% max
M&I	0.1% max
Iodine value (Wijs)	50–55
M.Pt °C (AOCS Cc 3–25)	33–39

Table B-2. continued.

Neutralized and Bleached Palm Oil	
FFA (as Palmitic)	0.25% max
M&I	0.1% max
Iodine value (Wijs)	50–55
M.Pt °C (AOCS Cc 3–25)	33–39
Color (5 ¼ in. Lovibond cell)	20 Red max
Refined, Bleached, and Deodorized (RBD) / Neutralized, Bleached, and Deodorized (NBD) Palm Oil	
FFA (as Palmitic)	0.1% max
M&I	0.1% max
Iodine value (Wijs)	50–55
M.Pt °C (AOCS Cc 3–25)	33–39
Color (5 ¼ in. Lovibond cell)	3 Red max
Crude Palm Olein	
FFA (as Palmitic)	5.0% max
M&I	0.25% max
Iodine value (Wijs)	56 min
M.Pt °C (AOCS Cc 3–25)	24 max
Neutralized Palm Olein	
FFA (as Palmitic)	0.25% max
M&I	0.1% max
Iodine value (Wijs)	56 min
Neutralized and Bleached Palm Olein	
FFA (as Palmitic)	0.25% max
M&I	0.1% max
Iodine value (Wijs)	56 min
M.Pt °C (AOCS Cc 3–25)	24 max
Color (5 ¼ in. Lovibond cell)	20 Red max
Refined, Bleached, and Deodorized (RBD) / Neutralized, Bleached, and Deodorized (NBD) Palm Olein	
FFA (as Palmitic)	0.1% max
M&I	0.1% max
Iodine value (Wijs)	56 min
M.Pt °C (AOCS Cc 3–25)	24 max
Color (5 ¼ in. Lovibond cell)	3 Red max

Table B-2. continued.

Crude Palm Stearin	
FFA (as Palmitic)	5.0% max
M&I	0.25% max
Iodine value (Wijs)	48 max
M.Pt °C (AOCS Cc 3–25)	44 min
Neutralized Palm Stearin	
FFA (as Palmitic)	0.25% max
M&I	0.15% max
Iodine value (Wijs)	48 max
M.Pt °C (AOCS Cc 3–25)	44 min
Neutralized and Bleached Palm Stearin	
FFA (as Palmitic)	0.25% max
M&I	0.15% max
Iodine value (Wijs)	48 max
M.Pt °C (AOCS Cc 3–25)	44 min
Color (5 ¼ in. Lovibond cell)	20 Red max
Refined, Bleached, and Deodorized (RBD) / Neutralized, Bleached, and Deodorized (NBD) Palm Stearin	
FFA (as Palmitic)	0.2% max
M&I	0.15% max
Iodine value (Wijs)	48 max
M.Pt °C (AOCS Cc 3–25)	44 min
Color (5 ¼ in. Lovibond cell)	3 Red max
Palm Acid Oil	
Total Fatty Matter	95% min (basis 97%)
M&I	3% max
Palm Fatty Acid Distillate	
Saponifiable Matter	95% min (basis 97%)
M&I	1.0% max
FFA (as Palmitic)	70% min

Note. For slip melting point (SMP): Slip Point, Softening Point or Rising Point; For free fatty acid FFA: Molecular Weight of Palmitic Acid is taken as 256; Color: Color measurement based on Tintometer Model 'E' AF 900 and Model 'D' AF 702; Specifications for palm stearin depend on the process used; Specifications are shipped quality final.
Source: PORAM Handbook.

Table B-3. MEOMA Lauric Oil Specifications.

Crude Palm Kernel Oil	
FFA (As Lauric)	5.0% max
M&I	0.5% max
Iodine Value (Wijs)	19 max at time of shipment
Crude Palm Kernel Olein	
FFA (As Lauric)	5.0% max
M&I	0.5% max
Iodine Value (Wijs)	21 min
Crude Palm Kernel Stearin	
FFA (As Lauric)	5.0% max
M&I	0.5% max
Iodine Value (Wijs)	8 max
Refined Bleached and Deodorized (RBD) Palm Kernel Oil	
FFA (as Lauric)	0.1% max
M&I	0.1% max
Iodine Value (Wijs)	19 max at time of shipment
Color 5 ¼ in. (Lovi Cell)	Red 1.5 max
Refined Bleached and Deodorized (RBD) Palm Kernel Olein	
FFA (as Lauric)	0.1% max
M&I	0.1% max
Iodine Value (Wijs)	21 min
Color 5 ¼ in. (Lovi Cell)	Red 1.5 max
Refined Bleached and Deodorized (RBD) Palm Kernel Stearin	
FFA (as Lauric)	0.1% max
M&I	0.1% max
Iodine Value (Wijs)	8 max
Color 5 ¼ in. (Lovi Cell)	Red 1.5 max
Palm Kernel Fatty Acid Distillate	
FFA (as Lauric)	50% min
M&I	1% max
TFM	95% min
Crude Coconut Oil	
FFA (as Lauric)	4% max, basis 3%
Moisture and impurities combined	1% max
Color (AOCS Method Cc 13b-45)	100 Yellow, 15 Red max
Iodine value	12 max

Premium or discount to be allowed for FFA variances at the rate of 1% of contract price for each 1% FFA, fractions in proportion.
Source: MEOMA Handbook

Table B-4. Recommended Temperature Range During Transportation and at Time of Loading/Discharge.

Product	Temperature During Voyage		Temperature at Time of Loading / Discharge	
	Min (°C)	Max (°C)	Min (°C)	Max (°C)
Palm Oil (Crude/Processed)	32	40	50	55
Palm Olein (Crude/Processed)	25	30	30	35
Palm Stearin (Crude/Processed)	40	45	55(a)	70(a)
Palm Kernel Oil/Coconut Oil (Crude/ Processed)	27	32	40(b)	45(b)
Palm Kernel Olein (Crude/Processed)	25	30	30	35
Palm Kernel Stearin (Crude/ Processed)	32	38	40	45
Palm Fatty Acid Distillate/Palm Acid Oil	45	50	55	70
Palm Kernel Fatty Acid Distillate	27	32	35	45
Coconut Oil (Crude/Processed)	27	32	40(b)	45(b)
Hydrogenated Oils	Refer to note (c) below			

Note: a. Different grades of palm stearin may have wide variations in their slip melting points and the temperature quoted may need to be adjusted to suit specific circumstances.
b. For warmer climates, the loading and discharge temperatures for palm kernel oil and coconut oil are Min 30°C, Max 39°C, or ambient temperature.
c. Hydrogenated oils can vary considerably in their slip melting point, which should always be declared. It is recommended that during the voyage, the temperature should be maintained at around the declared melting point and that this should be increased prior to discharge to give a temperature of between 10°C and 15°C above that point to effect a clear discharge.
From PORIM (1985).

16

Minor Constituents of Palm Oil: Characterization, Processing, and Application

Yanping Zou[1], Yuanrong Jiang[1], Tiankui Yang[1], Peng Hu[1], and Xuebing Xu[1,2]

[1]Wilmar Global R&D Center, Shanghai 200137, China; [2]Department of Engineering, Aarhus University, Aarhus C 8000, Denmark

Introduction

Palm oil is a lipid extract from the fresh mesocarp of the fruits from the oil palm tree (*Elaeis guineensis*), which belong to the Palmae family. The world production of palm oil had a huge increase in the last decades due to the vast applications in industry. Palm oil has overtaken soybean oil as the most important vegetable oil in the world, the world production of palm oil is 46.0 million tons in 2009–2010, which accounted for 33.1% of the total world vegetable oil production of 139 million tons (Gunstone, 2010). The largest supplier of palm oil is from Southeast Asia countries such as Malaysia and Indonesia. Palm oil is the raw material used for the production of cosmetics, waxes, lubricants, toothpaste, and biodiesel. Throughout the world, 90% of palm oil is used for edible purpose, while the remaining 10% is used for soap and oleochemical manufacturing (Edem, 2002).

The major constituents of palm oil is triacylglycerols (TAG), while the crude palm oil contains approximately 1% of minor constituents including carotenoids, tocopherols and tocotrienols, phytosterols, phospholipids, glycolipids, terpenic, and aliphatic hydrocarbons (Table 16-A) (Goh et al., 1985). Some minor components, especially the carotenoids and tocopherols and tocotrienols, not only maintain the stability and quality of palm oil but also possess significant biological properties including anti-oxidation (Stahl & Sies, 2002; Wu & Ng, 2007), anti-cancer (Guthrie et al., 1997), anti-inflammatory (Wu et al., 2008), anthrosclerosis-controlling, and cholesterol-lowering effects (Das et al., 2007). The growing interest in the bioactivities of those minor components has led to the development of functional foods or nutraceuticals incorporated with phytosterols, tocopherols, and tocotrienols.

In this chapter, general information on the characterization; source; processing effect on the content; analytic methods; biomedical properties; and applications of minor components, in particularly, carotenoids, tocopherols and tocotrienols, and phytosterols, are reviewed.

Table 16-A. Minor Components of Crude Palm Oil.

Minor components	Content (ppm)
carotenoids	500–700
tocopherols and tocotrienols	600–1000
phytosterols	326–527
phospholipids	5–130
tritenpen alcohols	40–80
methyl sterols	40–80
squalene	200–500
alphatic alcohols	100–200
ubiquinones	10–80
alphatic hydrocarbons	50

Source: Goh, et al. (1985).

Carotenoids

Carotenoids are 40-carbon isoprenoids which consist of eight isoprene units joined in such a manner that the arrangement of isoprenoid units is reversed at the center of the molecule so that the two central methyl groups are in a 1,6-position relationship and the remaining nonterminal methyl groups are in a 1,5-position relationship. Chemically, carotenoids can be divided into two groups, carotenes and xanthophylls. Carotenes are hydrocarbon compounds which include β-carotene, α-carotene, and lycopene,while xanthophylls are oxygen-containing carotenoids including β-cryptoxanthin, lutein and zeaxanthin, etc. (Fig. 16.1). Carotenoids are a family of compounds of more than 600 fat-soluble plant pigments that provide much of the color we see in nature. They produce the beautiful colors of many birds, insects, and marine animals as well as the colors of many flowers and fruits (Wald, 1968; Carotenature). α-Carotene, β-carotene, and cryptoxanthin are also demonstrated for their provitamin A activity, with β-carotene being the most potential provitamin carotenoids (Hendler & Rorvik, 2008). However, only a few carotenoids, primarily in fruits and vegetables, are ingested in sufficient quantity to be detected in human plasma (Khachik et al., 1992). The estimation of carotenoids intakes has been made possible through the publications of the qualitative and quantitative carotenoids content of commonly consumed foods (Mangels et al., 1993).

Source of Carotenoids

Since plants are able to synthesize carotenoids de novo, the carotenoids composition of foods and plant origin is variable (Delia et al., 2008), so it is not strange for it to occur in various vegetable oil, for example, yellow maize oil, groundnut oil, soybean oil, rapeseed oil, linseed oil, olive oil, barley oil, sunflower seed oil, cottonseed oil, and

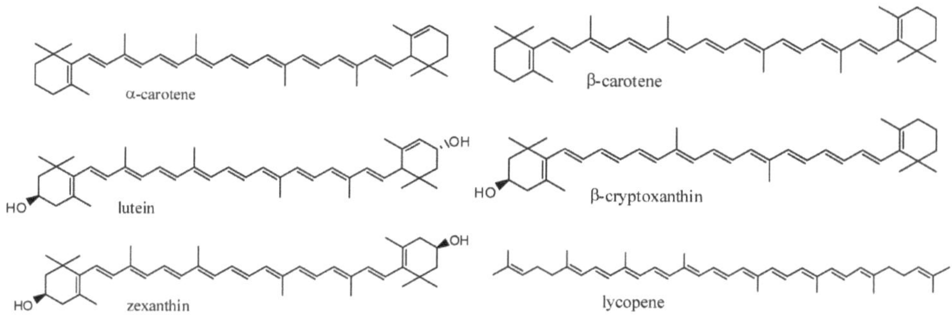

Fig. 16.1. The structures of the predominant carotenoids found in human plasma.

so on. However, the concentration of carotenoids in these vegetable oils are generally very low, which is possibly less than 100 ppm with the sole exception of palm oil (Harold & Purcell, 1967; Ong & Tee, 1992). Palm oil is the richest source of carotenoids. Table 16-B shows the retinol equivalents comparison between red palm oil and other fruits (Zachman, 1988).

Carotenoids content differs a lot due to plant species, varieties and the locations where they grow, etc. As known to all, palm forests are mainly set out in East Africa, South America, Malaysia, Java, Indonesia, etc. (Hartley, 1977). The literature showed its concentration of crude palm oils (CPO) from Malaysia and Zaire ranges between 500 and 700 ppm (Jacobsberg, 1974); larger amounts (800–1600 ppm) from Dura species have been reported from Nigerian sources (Ames et al., 1960), while the yield is low. Malaysian palm kernel oil generally contains lower than 12 ppm (Siew, 1981). An extremely high carotenoid-containing oil palm species called as *Elaeis oleifera,* which contained about 4600 ppm (Yap et al., 1991).

A typical analysis of the composition of the carotenoids in crude palm oil shows that α- and β-carotenes are the major components and the rest are γ-carotene and

Table 16-B. Retinol Equivalents Comparison.

Sources	Carotene derived (ppm)
red palm oil	30,000
carrot	2,000
leafy vegetables	685
apricots	250
tomatoes	100
bananas	30
orange juice	8

Source: Zachman (1988).

Table 16-C. Typical Carotenoids Composition of Crude Palm Oil (CPO) and Pressed Fiber Oil (PFO).

	Compositions (%)	
Carotene	CPO	PFO
phytoene	1.27	11.87
phytofluene	0.06	0.40
cis-β-carotene	0.68	–
β-carotene	56.02	30.95
α-carotene	35.16	19.45
cis-α-carotene	2.49	1.17
ζ-carotene	0.69	7.56
γ-carotene	0.33	2.70
δ-carotene	0.83	6.94
neurosporene	0.29	3.38
β-zeacarotene	0.23	0.37
lycopene	1.30	14.13
total (ppm)	500–700	4520–5600

Source: Ooi et al. (1994), Choo et al. (1996).

lycopene, etc. (Table 16-C) (Ooi et al., 1994). As said above, crude palm oil contains the highest concentration of natural carotenoids, which proves it to be the most promising plants for carotenes sources in natural (Tan, 1989).

Another potential natural resource might be found in the recovered fiber from pressed palm fruits, which is generally the energy source for palm mill while it contains about 5–6% residual oil. It was noted that various solvent extracted fiber oils contain 4000–6000 ppm carotenoids and the oil from hybrid oil palm is even as high as 5000–7000 ppm (Table 16-D). Furthermore, it was also reported that although the major carotenoids are still β-carotene and α-carotene, they only occupied about 50% total, which results in other compositions being much higher than its content in commercial crude palm oil (Table 16-C). For example, higher amounts of phytoene (11.9%), lycopene (14.1%), τ-carotene (7.6%), and δ-carotene (6.9%) were found (Choo et al., 1989, 1996). Another study also showed about 1900 ppm carotenoids existed in oil palm leaflets (Gapor et al., 1987).

Changes of Carotenoids During Processing and Storage

In order to satisfy consumption, CPO need to be treated in several steps, which normally include degumming, neutralization, bleaching, and deodorization. However, besides removal of undesired components, some other minor constituents in

Table 16-D. Carotenoid Content in Different Solvents Extracted Palm Fiber Oil (PFO).

Sample name	Carotenoid content (ppm)
Chloroform extracts	3800–5300
Hexane extracts	4000–5500
Liquid carbon dioxide extracts	4100–6000
Palm pressed fiber oil from hybrid palms	5000–7000

Source: Choo et al. (1989).

the oil are also damaged or taken away during the refining process, one of which is carotenoids.

It is noted that carotenoids are a thermal pigment so that it will be easily decomposed during the heating process. In the oil refining process, bleaching and deodorization or de-acidification steps are mainly responsible for the loss of carotenoids. Generally speaking, there are two factors affecting its content during the bleaching process. First, the bleaching earth will partly absorb the carotenoids physically due to "Van der Waals" surface attraction force (Gibon et al., 2007). Moreover, different concentrations and kinds of bleaching earths have been studied and confirmed to have different carotenoids removal ability (Table 16-E) (Rossi et al., 2001). Acid-activated bleaching clays were more powerful on removing carotenoids and carotenoids amounts decreased clearly with the rise of its dosage.

Table 16-E. Influence of Different Concentration and Kinds of Bleaching Earth on Carotenoids (Rossi et al., 2001).

Treatment	Carotenoids (ppm)
Crude palm oil	455.8
Acid degummed palm oil	415.0
Bleached with acid-activated clay	
0.5% bleaching clay 1	297.0
0.75% bleaching clay 1	245.8
1.0% bleaching clay 1	223.4
Bleached with acid-activated clay 2	
0.5% bleaching clay 2	327.9
0.75% bleaching clay 2	313.9
1.0% bleaching clay 2	258.6
Bleached with neutral clay	
0.5% neutral clay	324.4
0.75% neutral clay	329.9
1.0% neutral clay	304.0

Studies on thermal degradation of carotenoids and chemical changes induced by heating CPO have been published. When temperature is below 100°C, it is generally accepted that carotenoids degrade by oxidation. While at higher temperatures, the effect of heat becomes more important. The loss of β-carotene in palm oil when heated at 160°C, 180°C, and 200°C for 20 min, were above 35%, 50%, and 85%, respectively. The destruction rate of β-carotene by heat doubles every 20°C and when palm oil is used as a frying medium, only a very few of the carotene is retained in the food. MacLellan (1983) showed that palm oil was heated after 20 minutes at 180°C, half of the carotene had disappeared, at 200°C about 70% and at 240°C more than 98% was destroyed. Therefore, very limited amounts of carotenoids would exist in refined commercial palm oil.

Carotenoids are well-known antioxidants that could trap free radicals or quench singlet oxygen in lipids (Subagio & Morita, 2001). As exposure to light and pro-oxidants, there are also prone to be loss on carotenoids.

Yi et al.(2011) evaluated the storage stability of carotenoids in red palm olein, its mixture with fish oil and different concentrations of ascorbyl palmitate (Fig. 16.2). As time passed, there occurred certain amounts of degradation of carotenoids in these

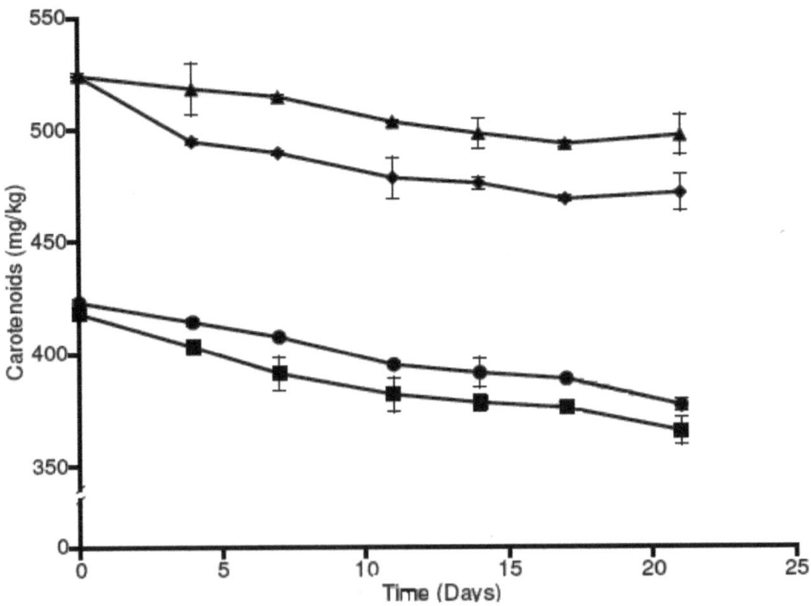

Fig. 16.2. Stability of carotenoids during storage at 30°C in the dark. (▲) Red palm olein, (●) Red palm olein/fish oil (4:1), (◆) Red palm olein with ascorbyl palmitate (200mg/kg), and (■) Red palm olein/fish oil (4:1) with ascorbyl palmitate (200mg/kg). Source: Yi et al., 2011.

four combinations. The results also demonstrated the degradation of carotenoids seems to be unaffected by the presence of ascorbyl palmitate. Oppositely, as Quackenbush & Smallidge (1986) reported, the application of antioxidants may accelerate autooxidation of β-carotene. Moreover, carotenoids might be regenerated by tocopherols or tocotrienols as reported in liposomes (Mortensen & Skibsted, 1997).

In addition, Scita (1992) studied the stability of β-carotene in different storage conditions. β-Carotene was found to be fairly stable in 48 hours in the dark; ultraviolet and visible light were also found to destroy β-carotene easily in the presence of atmospheric oxygen and the rate of which came to be slower with a lower air-oxygen exchange.

In order to meet the demands of market, fractionation is commonly adopted into its refining process and the products are known as olein (lower melting fraction) and stearin (higher melting fraction). One study confirmed that fractionation would give a distribution and carotenoids in palm olein were raised up to about 20% (Table 16-F) (Basiron et al., 2000).

Analytic Methods of Carotenoids

Due to the high importance of these minor components in edible oils, such as antioxidants and health benefits, it had attracted great interest to study the rapid and reliable analytical methods of carotenoid types.

From the late 1970s, high-performance liquid chromatography (HPLC) was widely adopted into various studies on the carotenoids and retinoids because of its complex composition. It is quite obvious that the technique is replacing other methods, including the once widely used column chromatography procedure (Tee & Lim, 1991). Detection limits of HPLC method for all-trans β-carotene and lycopene were reported to be 10 and 50 fmol, respectively, which was 10 to 100 times lower than the conventional UV-Vis detection (Ferruzzi et al., 1998).

According to AOCS Ce 9-01 method, analysis of minor components in edible oil thereof such as carotenoids by HPLC should enrich them into the unsaponifiable fraction while HPLC should be combined with various modes of detection by setting ultraviolet-visible spectroscopy at 450 nm typically (AOCS, 2009). Among that,

Table 16-F. Carotenoids Contents of Palm Oil and Its Fraction.

Name	Carotenoid contents (ppm)
Crude palm oil	630–700
Crude palm olein	680–760
Crude palm stearin	380–540

Note. Total carotenoids estimated at 446 nm.
Source: Basiron et al. (2000).

saponification is often used to eliminate lipids and chlorophylls and to release esterified xanthophylls. However, since the high degree of unsaturation property, carotenoids are very sensitive to heat, light, acid, and oxygen. Great loss of carotenoids will occur in that step, especially the epoxycarotenoids and xanthophylls as literature shows (Khachik et al., 1986; Lietz & Henry, 1997). Moreover, it will also result in isomerization of carotenoids. Hence, plenty of efforts had been done in the following years.

With the development of high-performance liquid chromatography (HPLC) and numerous methods, both normal- and reversed-phase (RP) have been described to separate xanthophylls and carotenes. In HPLC analysis of carotenoids, the most widely employed detector is ultraviolet-visible (UV-Vis), and more recently the photodiode array detector (DAD) has been used which allows a continuous collection of spectrophotometric data during the analysis (Huck et al., 2000). Findings also show liquid chromatography-mass spectrometry (LC-MS) is a more sensitive and powerful tool for carotenoids analysis, which mainly includes atmospheric pressure ionization interfaces (APCI) or electrospray ionization interfaces (ESI) (Rodríguez-Bernaldo de Quirós & Costa, 2006).

Recovery of Carotenoids

Nowadays, carotenoids are becoming increasingly popular due to not only their antioxidant function but also many other health-promoting activities including lowering blood pressure and preventing heart disease (Baker & Gunther, 2004). With the change of the concept of consumers, it becomes extremely important to recover carotenoids from some sources to meet the market demands.

Developed recovery methods may include saponification (Blaizot, 1953), adsorption (Baharin et al., 1998), selective solvent extraction (Mamuro et al., 1986), molecular distillation (Ooi et al., 1986), and transesterification followed by distillation (Hama et al., 1986; Hara et al., 1988). Although these processes can produce concentrated palm carotenes, most of them are hard to perform because they are inefficient or costly. Only transesterification and distillation processes had been commercialized, but they require large amounts of energy for evaporation because of the high latent heat of the esters (Darnoko & Cheryan, 2006). The process first needs to convert the palm oil triglycerides to esters, leaving the valuable minor components intact and allowing for recovery of carotenoids.

The study from Ooi et al. (1994) demonstrated that 50–90% carotenes could be recovered by one-stage distillation (18,000 ppm and above) from palm oil alkyl esters (mainly methyl and ethyl esters) with 2.5% refined bleached deodorized (RBD) palm olein, RBD palm oil and neutralized bleached deodorized (NBD) palm oil. Moreover, when two-stage distillation was carried out, the carotenoids concentration was increased to 75,000 ppm and the yield was about 75%. A similar study was carried out to recover carotenoids by molecular distillation (Batistella & Maciel, 1998). The technically viable process was to recover carotenoids from palm oil through transesterification and molecular distillation using falling film and centrifugal molecular distillators, and more than 30,000 ppm carotenoids-contained distillate could be concentrated. Unfortunately, no edible oil is obtained in these processes due to the chemical conversion of TAG to methyl esters.

In order to recover the oil in its natural form, an adsorption chromatographic process using synthetic polymers has been developed (You et al., 2001). However, the recovery rate is relatively lower. Another method was carried out which involved an additional pretreatment: hydrolysis to produce free fatty acid rich oils. More than 95% carotene recovery rate could be obtained with the increasing of polarity of the palm oil (You et al., 2002).

One method involves obtaining red palm methyl esters (RPME) from red palm oil by transesterification; after that RPME is nanofiltered to recover carotenoids with DS-7 membrane. The carotene yield could reach 85.6% while the concentration could be achieved at 11,900 ppm (1.19g/L) (Darnoko & Cheryan, 2006). More recently, Chiu et al. (2009) also studied the enrichment of carotenoids from the red palm ethyl esters (RPEE) which were obtained from the crude palm oil by using NF-10 membrane (a polymeric membrane) at 40°C and 2.5 MPa (Fig. 16.3). The results were positive and showed many advantages when compared with conventional processes: low temperatures, minor waste of energy, retention of nutritionally desirable compounds, and no effluent production. However, this only presented a potential for industrial application because higher fluxes are needed to be achieved (Chiu et al., 2009).

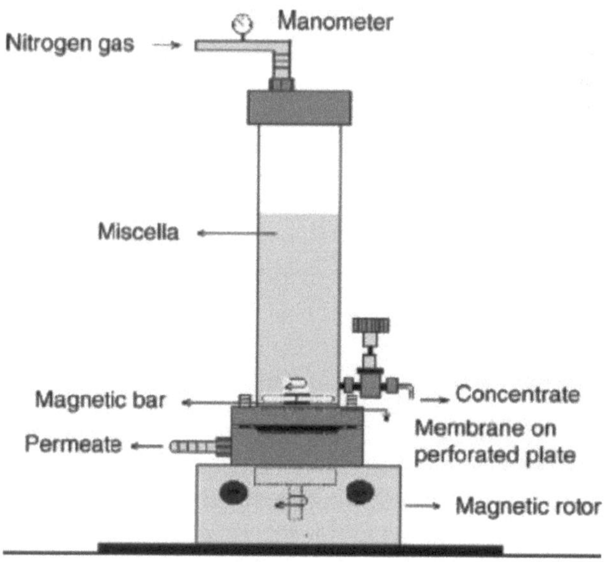

Fig. 16.3. Membrane separation device. Source: Chiu et al., 2009.

Bioactive Properties and Application of Carotenoids

Carotenoids have long been known for their pro-vitamin A activity, as they can be transformed into vitamin A in vivo. Vitamin A is necessary for vision, growth, cellular differentiation, morphogenesis, and several other cellular and physiologic functions (Olson, 1996). Their activity as antioxidants and their potential roles in the pathogenesis of several different diseases have been examined and continue to intrigue researchers. However, only those carotenoids that have at least one unsubstituted β-ionone ring attached to an intact conjugated polyene structure from C-7 to C-15 can be metabolized to retinal, which is then reduced to retinol (Goodwin, 1986). Because of these structural requirements, β-carotene is the most important pro-vitamin A carotenoids, although others (e.g., α-carotene, β-cryptoxanthin) also meet these requirements (Rock, 1997).

The antioxidant properties of carotenoids have been suggested as being the main mechanism by which they afford their beneficial effects. In view of their antioxidant properties, carotenoids have received considerable interest by researchers, health professionals and regulatory agencies in recent decades (Krinsky, 1998, 2003; Paiva & Russell, 1999; Stahl & Sies, 2002). Several kinds of carotenoids are reported as efficient quenchers of singlet oxygen and can neutralize thiyl radicals and combine with and stabilize peroxyl radicals, like α-, β-carotene, and lycopene. The rate constants for singlet oxygen scavenging by carotenoids are reported in the order of 10^{10} $M^{-1}S^{-1}$, and are even greater than tocopherol (10^8 M^{-1} s^{-1}) (Ojima et al., 1993).

However, some skepticism has appeared as to whether carotenoids have any antioxidant action in vivo (Briviba et al., 2004; Rice-Evans et al., 1997). The reason why ambiguous results came out was concluded as using different methodologies: (1) in dissolving the carotenoids to be evaluated, (2) in initiating oxidant stress, (3) in the presence of other antioxidants, (4) in the type of animal used for in vivo studies, and (5) in the evaluation technique used to determine the efficacy of the various carotenoids (Krinsky & Johnson, 2005).

In fact, various animal species have been used for many years in attempt to evaluate the in vivo antioxidant effect of carotenoids. However, these studies are marred by the fact that most experimental animals are very poor absorbers of carotenoids, and only large, pharmacological doses permit absorption of carotenoids into these animals. Some animals that can absorb dietary carotenoids, such as the ferret, gerbils and pre-ruminant calves, have been used to study carotenoids absorption, but virtually nothing has been done with respect to antioxidant effectiveness in these species (Lee et al., 1999). In mice treated for 15 days with canthaxanthin (4,40-diketo-β-carotene), a direct antioxidant effect was not observed, inasmuch as the endogenous levels of MDA were not changed, but indirectly, a significant over-expression of the MnSOD gene was observed, strongly suggesting that this carotenoid was capable of altering the antioxidant protection in this species (Palozza et al., 2000).

The definitive studies therefore must deal with humans, who have an almost unlimited capacity to absorb dietary carotenoids (Parker et al., 1999) but have a somewhat limited ability to indicate their oxidative stress status. It has been suggested that the evidence for an antioxidant role for carotenoids in vivo is not very strong (Rice-Evans et al., 1997).

There also have been human studies supporting the pro-oxidant properties of β-carotene. Supplementation of β-carotene at pharmacological levels increased lung cancer incidences in smokers in the Alpha-Tocopherol-Beta-Carotene (ATBC) trial and increased mortality from cerebrovascular disease (CVD) in a group of smokers, former smokers and asbestos exposed individuals in the β-carotene and retinol efficiency trial (CARET) (Omenn et al., 1996).

Overall, these observations suggest a possible biphasic response of β-carotene that promotes health when taken at dietary levels but may have adverse effects when taken in higher amounts.

Carotenoids are becoming popular by providing great amounts of health benefits, while the increased awareness and use of these ingredients in dietary supplements warrants comprehensive studies on their safety. Four groups of rats (12 rats per group) were fed a diet supplement with 0.2% palm oil-based carotenoids concentrate (20,000 ppm), methyl ester, ethyl ester for 16 weeks. Heart, lungs, adrenals, kidneys, liver, and spleen were all reported to be normal for all groups, suggesting that palm oil carotenoids concentrate is not harmful (Choo et al., 1992). In 1992, the Food and Agriculture Organization (FAO) has accepted and included palm oil carotenoids as a permissible food colorant (JECFA).

Tocopherols and Tocotrienols

Tocopherols and tocotrienols (known as tocochromanols) are fat soluble 6-hydroxychroman compounds, which exhibit the biological activity of α-tocopherol. The tocopherols and tocotrienols are usually called vitamin E. The structures of the tocochromanols, as illustrated in Fig. 16.4, have a chromanol head, formed by phenolic and heterocyclic rings, and a phytyl tail. The number and position of methyl substitutions on the chromanol nucleus give rise to the subfamily of α-, β-, γ-, and δ-tocopherols/tocotrienols. The difference in the structure of tocopherols and tocotrienols is only in the phytyl tail, the tocopherols have a saturated tail, while the tocotrienols have an unsaturated chain with three isolated double bonds (Hood, 1998; Kamal-Eldin & Appelqvist, 1996).

The tocopherols have three chiral centers of asymmetry at position 2 of the chromanol head and at positions 4′ and 8′ of the phytyl tail, which make eight steroisomeric forms possible. In natural forms, all tocopherols have the 2R, 4′R, 8′R (d- or RRR) configuration. On the other hand, the tocotrienols have one chiral center at position 2, only making two steroisomeric forms. A total of eight isomers could be present for each of the tocotrienols since four cis/trans geometrical isomers are possible due to the existence of two double bonds at positions 3′ and 7′. However, only the 2R, 3-trans and 7-trans isomers exist in nature. Synthesis may lead to stereoisomerism of tocochromanols. Synthetic α-tocopherol (referred as *all-rac*-α-tocopherol) is an equal mixture of all eight stereoisomeric forms (Drotleff & Ternes, 1999).

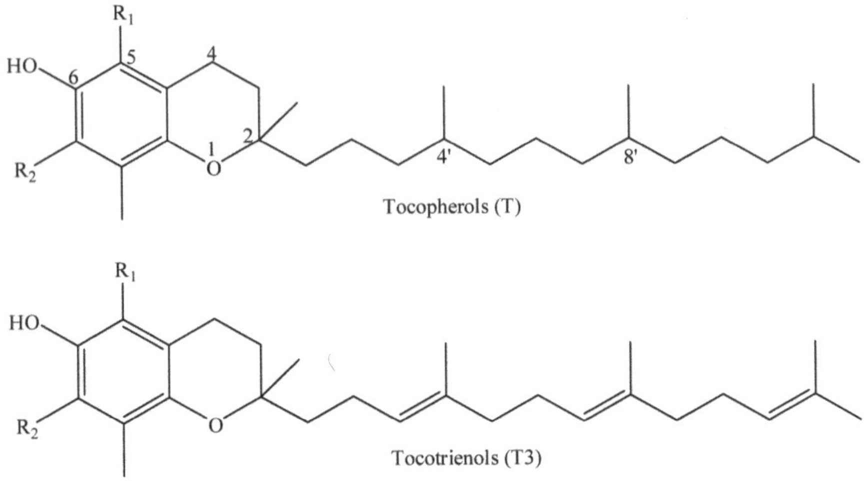

Tocopherol/tocotrienols	R_1	R_2
α	CH_3	CH_3
β	CH_3	H
γ	H	CH_3
δ	H	H

Fig. 16.4. Structures of tocopherols and tocotrienols.

Source of Tocopherols and Tocotrienols

Tocopherols have been found in all photosynthetic organisms, but the tocotrienols are found only in certain plant families. The major source of tocopherols and tocotrienols is vegetable oils. The contents of tocopherols and tocotrienols in some vegetable oils are shown in Table 16-G. The tocopherols and tocotrienols present at different concentrations depending on the oil origin, and the order of total tocopherols and tocotrienols in some commercial oils is as follows: wheat germ (> 2000 ppm) > soybean, corn, palm (800–1000 ppm) > cottonseed, sunflower, rapeseed (550–800 ppm) > peanut, olive (150–350 ppm) > coconut, palm kernel (<50 ppm) (Tan, 1989). Most oils contain only tocopherols, while palm oil and rice bran oil contain significant amounts of tocotrienols.

Palm oil is one of the richest sources of vitamin E. The content of vitamin E in palm oil is unique since it is composed of tocotrienols rather than tocopherols. Nor-

Table 16-G. The Contents of Tocopherol (T) and Tocotrienol (T3) in Some Vegetable Oils.

oil	α-T	α-T3	β-T	β-T3	γ-T	γ-T3	δ-T	δ-T3	total
soybean	95		13		689		239		1036
	120		10		610		360		1000
	179		28	4	604		371		1186
corn	257	15	9		752	20	32		1085
	120		10		400		20		550
	272	54		11	566	62	25		990
cottonseed	403			9	383		4	–	799
sunflower	622		23		27		–	–	672
	610		10		30		10	–	660
	564		24		4		–	–	592
safflower	449		12		26		6	–	493
rapeseed	189				486		12	–	687
	200				430		10	–	640
linseed	5				573		7	–	585
wheat germ	1507	36	312		527	18			2400
	1330	26		181					1537
olive oil	119				13				132
	90			4	5				99
rice bran oil	324	236				349			909
CPO	420	260				360		80	1160
	190	131				339		113	773
RBD palm oil									
palm kernel oil	13	21							34
Coconut oil	5	5		1		19	6		36

Source: Lampi et al. (2002); Ng et al. (2004); Ong & Choo (1997); Sheppard et al. (1993).

mally, CPO containing 600–1000 ppm vitamin E, in which tocopherols accounts for 18–22% and tocotrienols accounts for 78–82%. Amongst the tocotrienols, the major ones are γ-tocotrienol (46%), α-tocotrienol (22%), and δ-tocotrienol (12%), while amongst tocopherols, α-tocopherol is the most abundant.

Oil extracted from palm-pressed fiber (PFO) has been found to be another rich source of vitamin E. Palm-pressed fiber, which is normally burned as fuel to provide energy for the palm oil mill, contains 5–6% residual oil (based on dry weight). PFO extracted with different solvents contained higher concentration of vitamin E than normal commercial CPO (Table 16-H). The content of vitamin E ranged from 2000 ppm to 3200 ppm compared with those of CPO. However, the profiles of vitamin E were much different from CPO. In PFO, the major component was α-tocopherol (57–68%), and the rest is α-tocotrienol (15–19%), γ-tocotrienol (15–21%), and δ-tocotrienol (1–6%) (Choo et al., 1996).

Palm phytonutrient concentrate (PPC), the residue from the vacuum distillation of CPO methyl esters in the palm biodiesel industry, could be a good source of vitamin E. During the biodiesel-producing process, crude palm oil is transesterified in the alcoholic medium with an alkyl catalyst to produce a colored ester, which could be used as fuel. Moreover, the colored ester could be distillated by molecular distillation to give distillate, a less colored liquid for the production of biodiesel, and residue, known also as PPC, which is a thick red concoction and is rich in minor compounds (Table 16-I) (Chandrasekaram et al., 2009). Ng et al. (2004) also reported the concentration and composition of vitamin E in the CPO, PFO, and PPC. The major constituent of vitamin E in CPO and PPC is γ-tocotrienols, whereas in PFO the major constituent is α-tocopherol (Table 16-J).

Tocotrienols are extracted commercially from palm oil and rice bran oil. Unfortunately, the common commercial source of natural vitamin E (soy, corn, cottonseed, sunflower, and canola oil) contained little or no tocotrienols. The practical method to

Table 16-H. Composition of Tocopherol (T) and Tocotrienols (T3) in Different Solvent Extracted PFO.

PFO	α-T (%)	α-T3 (%)	γ-T3 (%)	δ-T3 (%)	Total(ppm)
$CHCl_3$ extract	67.7	18.6	15.7	trace	2300–3200
hexane extract	61.1	15.4	18	5.5	2020–2640
ligquid CO_2 extract	57.0	17.4	20.5	5.1	2870–3220
CPO	22.0	20.0	46.0	12.0	600–1000

Source: Choo et al. (1996).

Table 16-I. Concentration (ppm) of Minor Compounds in Crude Palm Oil (CPO) and Palm Phytonutrient Concentrate (PPC).

Compounds	CPO	PPC
carotenoids	530	23,710
vitamin E	1020	14,640
sterols	910	15,780
squalene	440	3,020

Source: Chandrasekaram et al. (2009).

Table 16-J. Concentration (ppm) and Compositions (%) of Vitamin E in Crude Palm Oil (CPO), Palm Phytonutrient Concentrate (PPC), and Palm Fiber Oil (PFO).

Sample	α-T	α-T3	γ-T3	δ-T3	α-T1*	Total (ppm)
CPO	420 (36)**	260 (22)	360 (31)	80 (7)	40 (4)	1,160
PPC	1,400 (10)	2,430 (18)	8,320 (60)	1,280 (9)	350 (3)	13,780
PFO	1,810 (44)	760 (19)	930 (23)	110 (3)	430 (11)	4,040

*α-T1: α-Tocomonoenol; ** numbers in the parentheses represents the percentage compositions of vitamin E.
Source: Ng et al. (2004).

extract vitamin E is from palm fatty acid distillate (PFAD), in which the total tocopherols and tocotrienols are concentrated to 0.4–0.5% (w/w). The typical compositions of PFAD are shown in Table 16-K (Posada et al., 2007). The production of CPO in Malaysia was 20.9 million metric tons (MT) for 2009, and almost 700,000 MT of PFAD were generated (Cheah et al., 2010). Theoretically, 2800–3500 MT of palm vitamin E will be produced using PFAD as raw material.

α-Tocomonoenol, an tocopherol analogue with a double bond at carbon-11′ of the phytyl chain, was first isolated and structurally elucidated by mass spectroscopy and proton nuclear magnetic resonance spectroscopy (Matsumoto et al., 1995). The levels of α-tocomonoenol in palm oil were reported to be 12.8 and 40 mg/kg, respectively (Matsumoto et al., 1995; Ng et al., 2004).

Other sources of tocotrienols are cereal grains, legumes, and fruits, including barley, rye, nuts, strawberry, etc., but the content are much lower than that in palm and rice bran oil (Sookwong et al., 2010).

Changes of Tocopherols and Tocotrienols during Processing and Storage

The effect of the refining process on the loss of vitamin E has been reported. For example, only 69, 72, and 76% of the original level of vitamin E in the CPO were retained in RBD palm oil, palm olein, and palm stearin, respectively (Sundram & Gapor, 1996). Rossi et al. (2001) reported the effect of bleaching and physical refining on the levels of vitamin E in palm oil. Five concentrations (0.5–1.0% of the oil mass) of one natural and two activated clays (pH = 6–7, 2.5, 3.2, respectively) mixed with 0.125% (of the oil mass) synthetic silica were used for bleaching. The results (Table 16-L) showed the total concentrations of vitamin E increased during the bleaching in correlation with the amount of clay used. The authors assumed that the acid clay caused the release of tocopherols from the linked forms such as dimeric or

Table 16-K. Composition of PFAD.

Component	Weight (%)
squalene	1.03
vitamin E	0.5
sterols	0.24
free fatty acids	90.03
glycerides	7.33
unknown*	0.88
Vitamin E composition (wt%)	
a-tocopherol	23.79
a-tocotrienol	23.76
g-tocotrienol	38.76
d-tocotrienol	13.69
Sterol composition (wt%)	
campesterol	24.24
stigmasterol	18.07
sitosterol	57.68
Free fatty acid composition (wt%)	
myristic acid	1.31
palmitic acid	50
stearic acid	4.39
oleic acid	36.48
linoleic acid	7.82
Glycerides composition (wt%)	
triglycerides	2.78
mono- and diglycerides	97.22

*Unknown may include other hydrocarbons such as phytoene and aliphatichydrocarbons, glycerol, and low-volatility compounds such as ketones andaldehydes.
Source: Posada et al. (2007).

Table 16-L. Tocopherol Concentrations (ppm) in Crude Oil, in Samples Bleached with Different Levels of Clay and 0.125% of Synthetic Silica, and in RBD Palm Oil Samples.

Tocopherols	Crude	Bleached					RBD				
		0.50%	0.625%	0.75%	0.825%	1.00%	0.50%	0.625%	0.75%	0.825%	1.00%
α-T	190 (25)[a]	220 (28)	247 (31)	233 (29)	276 (31)	223 (27)	147 (27)	200 (26)	125 (25)	176 (26)	150 (25)
α-T3	131 (17)	155 (20)	172 (21)	162 (20)	177 (20)	156 (19)	140 (25)	178 (23)	112 (22)	154 (22)	147 (24)
γ-T3	339 (44)	291 (37)	283 (35)	286 (35)	311 (35)	250 (30)	213 (39)	316 (41)	201 (40)	287 (42)	251 (41)
δ-T3	113 (15)	127 (16)	101 (13)	135 (17)	139 (15)	206 (25)	51 (9)	72 (9)	69 (14)	72 (10)	62 (10)
total	773	794	803	815	901	834	551	766	506	688	610

[a]Relative concentrations (g/100g) are reported in parentheses.
Source: Rossi et al. (2001).

other esterified compounds. After physical refining, the RBD oil only retained 80% of vitamin E in crude oil. The loss mainly occurred in the steam-refining treatment. In addition, the relative compositions of vitamin E were altered. Bleaching increased the concentrations of α-tocopherol and α-tocotrienol while lowering γ-tocopherol level. The RBD oil contained higher concentrations of α-tocotrienol and lower concentration of δ-tocotrienol. However, in another study, Puah et al. (2007) took all the oil samples from the processing line of a local refinery and analyzed the vitamin E content. The total content of vitamin E in crude, degummed, bleached, and deodorized were 1273, 1134, 1094, and 1029 ppm, respectively. It seemed the loss mainly occurred in the degumming and bleaching process. The compositions of all vitamins were similar throughout the physical refining process. The discrepancy might lie in the different design and parameters for each step of processing.

An investigation demonstrated that the hydrogenation process of γ-tocotrienol-rich palm oil does not produced any detectable γ-tocotrienol side chain isomers, although hydrogenation may destroy tocotrienols, and thus cause loss of γ-tocotrienol (Drotleff &Ternes, 1999).

Interesterification, either by chemically or enzymatically catalyzed, could alter the ratio of tocopherol. In chemical interesterification of olive/palm blend oil (50/50, w/w), a minor reduction in the tocopherols and tocotrienols was found. The total loss of tocopherols was 0.5 and 3% at 90 and 120°C, respectively, with δ-tocotrienol during the interesterification at 120°C being the highest (35%), as compared with their content in the starting oil blend (Azadmard-Damirchi & Dutta, 2008). During enzymatic interesterification of palm stearin and rice bran oil, a 4.0–8.0% decrease in total tocopherols and tocotrienols was observed (Reshma et al., 2008).

Fractionation of palm oil into liquid (olein) and solid (stearin) fractions is normally effected either on the CPO or RBD. Vitamin E tends to partition preferentially into the olein fraction during fractionation of RBD palm oil (Sundram et al., 2002). A red palm oil with 2130 ppm vitamin E was dry fractionated into a stearin fraction (31%) and an olein fraction (69%), in which the vitamin E was 1199 and 1889 ppm, respectively. After further refining, the vitamin E reduced to 659 and 826 ppm, respectively (Khatoon & Krishna, 2007).

Fat frying is an important method of food preparation. In an experiment of repeat, discontinuous deep frying with French fries and chicken nuggets, the total vitamin E level of RBD palm olein decreased from 840.06 in fresh oil to 69.15 and 158.81 mg/kg in PO-FF (palm olein with French fries) and PO-CH (palm olein with chicken nuggets), respectively, after 45 cycles of frying, while the vitamin E was destroyed totally in the control frying oil without any food (Bansal et al., 2010).

In a storage test of palm olein/fish oil (4:1, w/w) mixtures, the concentration of α-tocopherol and α-tocotrienol decreased after storage at 30°C in the dark for 21 days, whereas δ- and γ-tocotrienol remained stable. Adding ascorbyl palmitate into the mixture reduced the losses of α-tocopherol and α-tocotrienol (Yi et al., 2011).

Analytic Methods of Tocopherols and Tocotrienols

Large varieties of techniques have been published for determination of vitamin E; they can be analyzed either by gas chromatography (GC), both with flame-ionization and mass spectrometric detections, or by high-performance liquid chromatography (HPLC) coupled with fluorescence or DAD-UV detectors. The quantification of tocopherols by GC technique has been reported (Slover et al., 1983), but it has several drawbacks such as (1) derivatization of sample prior to injection, (2) destruction of vitamin E which is heat-sensitive, and (3) a longer analysis time. Therefore, at present the prevalent methods for vitamin E analysis are HPLC, with sensitivity and specificity being much higher than those obtained with GC methods. Moreover, the sample preparation is simpler. The analytical methods used for the characterization and determination of tocopherols and tocotrienols are summarized in Table 16-M.

HPLC analysis can be performed by normal-phase (NP) or reverse-phase (RP) liquid chromatography. The separation is based on the number and position of methyl substituent on the chromanol ring. Mobile phase include hexane, isopropanol, ethyl acetate, dioxane, and so on, with hexane being the major component. The flow rate is 1 to 1.5 ml/min with isocratic elution. The elution order of eight isomers is usually as follows: α-tocopherol, α-tocotrienol, β-tocopherol, β-tocotrienol, γ-tocopherol, γ-tocotrienol, δ-tocopherol, and δ-tocotrienol. In RP-HPLC, the separation is based on the structure of the side chain and the number of methyl substituents. The complete separation of β- and γ-tocopherol and β- and γ-tocotrienol by RP-HPLC is difficult because the side-chain structure and methyl substituent on the chromanol ring are the same. Methanol and acetonitrile are the main components in the mobile phase, and water, methylene chloride, and acetic acid are minor components. The flow rate is varied between 0.2 and 1 ml/min with isocratic or gradient elution. Usually only six peaks appear in the HPLC chromatogram since the complete separation of β- and γ-tocopherol and β- and γ-tocotrienol by RP-HPLC is difficult to achieve. The elution order is as follows: δ-tocotrienol, β- and γ-tocotrienol, α-tocotrienol, δ-tocopherol, β- and γ-tocopherol, and α-tocopherol.

Silica columns are frequently used in NP-HPLC (Kamal-Eldin et al., 2000), while C18-ODS columns are most commonly used in RP-HPLC (Bruii, 2007). The advantages of NP- over RP-columns in the separation of vitamin E include: (1) the ability to operate with organic solvents, allowing a high solubility for lipids to be possible; (2) the ability to tolerate high loads of lipids, which are easy to wash-out by non-polar solvents; and (3) the ability to provide a wide range of selectivity through the inclusion of different polar modifiers in the mobile phase (Kamal-Eldin et al., 2000). However, the packing material silica in NP column is very reactive to strong polar chemicals. Any high-polarity compounds in sample and mobile phase will diminish the column performance and shorten the column lifetime (Xu, 2002). The RP-columns have better stability and longer durability than NP-column, but they are not able to separate

Table 16-M. HPLC Analysis of Tocopherols and Tocotrienols in Palm Oils and Its Products.

Sample	Sample preparation	Parameters of HPLC					Reference
		Standards	Detector	Column	Mobile phase	FR (ml/min)	
palm oil	diluted with EtOAc, dried under nitrogen and diluted with MeOH	δ-T	EC	RP-ODS 2 (250 × 2)*	MeOH:water (80:20, v/v) isocratic	0.2	Monde et al., 2009
palm olein soybean oil corn oil	oil samples were dissolved with appropriate dilution	ES	FL Ex: 290, Em: 330 nm	NP-Lichrosorb Si60 (250 × 4)	0.9% 2-propanol in hexane, isocratic	1	Simonne & Eitenmiller, 1998
CPO		–	UV: 295 nm CIS-MS NMR	RP-polymeric C30 bonded phase (250 × 4.6)	MeOH	1	Strohschein et al., 1999
red palm oil	Saponified with KOH, extracted with diethyl ether, dried under nitrogen and diluted with mobile phase	ES	DAD-UV 295 nm	(1) NP-Prodigy 100A silica (150 × 4.6) (2) RP-C30 YMC Carotenoid (250 × 4.6)	(1) iso-octane/THF (97.5/2.5 v/v) isocratic (2) A: MeOH:methyl tertiarybutyl ether (MTBE): H_2O (81:15:4), B: MTBE: MeOH (91:9). gradient: 0–50% B, 0-45 min; 50–100% B, 45–55 min; 100–0% B, 55–60 min	(1) 1.5 (2) 1	Darnoko et al., 2000
Mixture of oat extract, palm oil, and vitamin E			FL Ex: 294, Em: 326 nm	(1) 6 silica columns (2) 3 amino columns (3) 1 diol columns	11 different mobile phase, isocratic	1	Kamal-Eldin et al., 2000

Table 16-M. continued.

Sample	Sample preparation	Parameters of HPLC					Reference
		Standards	Detector	Column	Mobile phase	FR (ml/min)	
plant seed oil	Saponified with KOH, extracted with hexane:EtOAc (85:15), dried and diluted with MeOH	IS 5,7-Dimethyltocol	FL Ex:298, Em:345	RP-pentafluoro-phenylsilica (PFPS) (250 × 4.6)	MeOH:water (90:10, v/v) isocratic	0.5	Abidi, 2003
palm oil and its products	dissolved in dichloromethane	ES: α-T	FL Ex:290, Em:330	NP- APEX DIOL II (250 × 4.6)	A: heptane, B: 2-propanol; gradient: 2-2% B, 0-2 min; 2-15% B, 2-10 min; 15-2% B, 10-11 min; 2-2% B, 11-18 min.	1	Moh et al., 2001
CPO, PFO, PPC	dissolved in EtOH, centrifuged, evaporated, and redissolved in EtOH	ES		RP-Develosil Aqueous	100% MeOH, isocratic	1	Ng et al., 2004
CPO and its products		ES	UV: 297 nm	NP-Shim-pack [LC-NH_2(M)] (250 × 4.6)	hexane: isopropanol (96:4)	1	Mayamol et al., 2007
CPO and its fractions	CPO was saponified with 60% KOH, extracted with diethyl ether; other products were diluted with hexane	ES	FL Ex:290, Em:330	NP-LiChrosorb Si 60	Hexane: dioxane (97:3)	1.5	Szulczewska-Remi et al., 2005
Palm stearin and rice bran oil blends	dissolved in hexane	ES	UV: 297 nm	NP-Phenomenex NH_2	hexane: isopropanol (92:8)		Reshma et al., 2008

IS, internal standard; FR, flow rate; EC, electrochemical; FL, fluorescence; ES, external standard β- and γ-isomers of vitamin E. Thus, RP-columns are recommended for samples with little or no β- and γ-isomers.

For refined vegetable oils, the pretreatment of samples is simple. The oil sample is dissolved and diluted at an appreciate concentration with mobile phase and injected directly to the HPLC system. However, for crude vegetable oil, a procedure including saponification (usually saponified with 60 to 80% KOH at 70°C for 30 min under the protection of an antioxidant or under nitrogen), extraction with organic solvent, dried under nitrogen, and re-dissolved in mobile phase for injection is needed to eliminate the interference of impurities. A detailed procedure for pretreatment of crude oil is well-documented by Xu (2002).

For quantitative analysis of vitamin E, both external standard and internal standard methods are adopted. Pure standard-grade vitamin E isomers can be available commercially from Sigma, Merck, etc. Internal standards include tocol and tocopheryl acetate (Bruii, 2007; Monde et al., 2009).

Detection of vitamin E isomers can be performed with a wide variety of detectors, including fluorescence detector, ultraviolet absorbance (UV) detector, electrochemical detector, evaporative light-scattering detector (Chase et al., 1994), MS, and NMR (Strohschein et al., 1999). Fluorescence and UV detectors are extensively used (Table 16-M). The fluorescence detector is ten times more sensitive and has less background noise than the UV detector, which make the fluorescence detector the first choice for the detection of vitamin E isomers.

Other analytic methods such as capillary electrochromatography (Fanali, 2007), and Supercritical Fluid Chromatography (Choo et al., 2005) are also reported, but their uses in the analysis of vitamin E are less frequent than HPLC.

Recovery of Tocopherols and Tocotrienols

Many attempts on recovery of tocopherols and tocotrienols have been made. The concentration isolated varies depending on the source and recovery techniques applied. CPO and PFAD are unique in that the content of tocotrienols are extremely high compared with other vegetable oils and FAD, which make them, especially the PFAD, a good source for extraction of tocotrienols commercially. The recovery processes include saponification, esterification, liquid-liquid extraction, crystallization, distillation, ion-exchange, and adsorption chromatography. Normally those methods are used in combinations to increase the purity of tocopherols and tocotrienols.

PFAD is an important source for production of tocopherols and tocotrienols. PFAD is a complex mixture of FFA, TAG, DAG, MAG, and minor components (Table 16-K). Removal of fatty components (FFA, partially acylglycerols, and triacylglycerols) from the raw materials is the first step in concentration of tocopherols and tocotrienols. For removal of FFA, the following ways can be considered. One is to distillate FFA directly using flash evaporator or molecular distillation (Jacobs, 2002). The other is to convert FFA into fatty acid methyl ester (FAME) catalyzed by acid or lipase (Top et al., 1993; Ghosh & Bhattacharyya, 1996), then the formed FAME is distillated, thus a concentrate rich in minor components and glycerides is obtained. The third is to react FFA with the intrinsic sterols in the FAD to form a high boiling-point steryl ester (Barnicki et al., 1996).

A process including conversion of free fatty acid and acylglycerols into methylesters by esterification, followed by removal of these esters by distillation was patented (Gapor et al., 1989). The overall yield is about 65% and the purity of the vitamin E is increased to 80–95%. However, this process is complex, and requires additional steps such as solvent extraction and solvent purification. Another process involves enzymatic hydrolysis of TAG in the PFAD, then the liberated FFA together with the existing FFA were neutralized using alkali, and the water-soluble salt formed was washed with water to give a vitamin E-concentrate fraction. When hydrolysis was performed using 2.5% (w/w) of immobilized Novazyme 435 and 45.2–47.3% (v/w) of water for 5.5–5.7 h, maximum vitamin E content increased from an initial amount of 0.37 to 4.32% (Chu et al., 2003a). After further subjected to a silica adsorption chromatography, and eluted with hexane and isopropanol successively, vitamin E concentration was 15.2% (Chu et al., 2003b).

Since the high cost of lipase, and large amount of waste water produced during saponification, the process involving hydrolysis with lipase and saponification with alkali is of little use for recovery tocotrienols and tocotrienols in the industry. One commercialized process has been developed by the Malaysian Palm Oil Board (MPOB) in association with Japanese researchers (Sundrum et al., 2002; Top et al., 1993). The process comprises the following steps: (1) conversion of FFA by esterification and acylglycerols by transesterification into methyl esters; (2) Distilling the product under reduced pressure to remove a major part of the alkyl esters and leaving behind a vitamin E concentrate that is about 5–10% pure; (3) Cooling the residue to bring about crystallization of higher melting point substances and other impurities such as sterols and waxes, filtering off the crystalline material to leave the tocopherols and tocotrienols in the filtrate; (4) Treating the filtrate from step 3 using an ion-exchange procedure with a high selectivity anionic resin to produce a concentrated (60–70% pure) tocopherol and tocotrienol fraction (an acidic solution is used for desorbing the tocopherols and tocotrienols from the resin); (5) Removing the solvent from the tocopherol and tocotrienol fraction by evaporation; (6) Washing and drying the product; (7) Subjecting the dried product to molecular distillation to produce a further concentrated tocopherol and tocotrienol product; (8) Deodorizing the tocopherol and tocotrienol product to remove any remaining odoriferous materials. The overall yield in this process is about 65% and final purity of the vitamin E up to 80–95% is achievable. The final production was dissolved in a precise quantity of palm super olein to make "Palm Vitee" capsule, which include 60 mg of vitamin E comprised of 32.1% α-tocopherol, 21.3% α-tocotrienol, 0.3% γ-tocopherol, 2% β-tocotrienol, 24.7% γ-tocotrienol, 15.7% δ-tocotrienol, and 3.7% other tocotrienols.

Other processes include the formation of complexes with the fatty acids present in the PFAD. It was discovered that urea forms an inclusion complex with C6 to C22 fatty acids and glycerides, in which hydrogen-bonded urea molecules are oriented in a helical lattice into which the fatty acid or the glyceride molecule fits. This complex does not entrap the sterols, tocotrienols, tocopherols, and other bulky molecules present in the deodorizer sludge or distillate. Using this fact, a process for recovering tocopherols from deodorizer sludge was developed (Sampathkumar, 1986).

Concentration of vitamin E by fractionation without chemical transformation of palm oil has been reported by Szulczewska-Remi et al. (2005). This process can maintain the oil quality and does not require converting CPO to methyl ester. The process included three steps of crystallization of acylglycerols from red palm oil at –6°C (neat), –80°C (in acetone), and –20°C (in methanol). The final product after the three-step crystallization process was characterized by having 10 times more tocopherols and almost 30 times higher amounts of tocotrienols.

Supercritical fluid extraction is an environmental-friendly process and seems a promising technology to concentrate vitamin E from palm oil. Puah et al. (2007) reported the solubility of tocopherols and tocotrienols from palm oil in supercritical carbon dioxide (SC-CO_2). The solubility of total vitamin E ranged from 0.94×10^{-3} to 0.018 mg/g CO_2 under the temperature of 40–80°C and pressure of 14–30 MPa. The selectivity of vitamin E followed the order of δ-tocotrienol > γ-tocotrienol > α-tocotrienol > α-tocopherol. This study could be helpful in design and optimization of SFE extraction of vitamin E. After the FFA of CPO was esterified by acid-catalyst followed by transesterification of TAG into FAME, the mixture was subjected to SFE extraction, a concentrate with 10.66% vitamin E was obtained, but the yield was low with only 16.2% recovery (Puah et al., 2008).

Bioactive Properties and Application of Tocopherols and Tocotrienols

Over the past 40–50 years, many studies specific for the tocopherols have been published, whereas the tocotrienols are poorly understand until the 1980s and 1990s when their cholesterol-lowering potential and anti-cancer effect were described (Qureshi et al., 1986; Sundram et al., 1989). Moreover, the tocotrienols exhibits powerful neuroprotective, anti-cancer, and cholesterol-lowering activities that are not possessed by tocopherols. Therefore, the interest in the tocotrienols is the main topic in this chapter.

Tocotrienols are able to scavenge free-radicals and prevent lipid peroxidation in biological membranes, and most of functions of tocotrienols are linked to its antioxidant activities. The mechanism of antioxidant of tocotrienols is similar to that of tocopherols. The antioxidant function per se is located in the chromanol head, where the phenolic hydroxyl group donates an H-atom to quench lipid radicals. The phytyl tail has no effect on the chemical reactivity but is responsible for the very strong lipophilic properties of these compounds,which determines their proper positioning in the biomembranes. However, whether the unsaturation of the phytyl tail makes any difference on the activity of the tocotrienols relative to the corresponding tocopherols is not known (Burton & Ingold, 1985).

Although α-tocotrienol is known to possess only 30% vitamin E activity of α-tocopherol, researches suggested that α-tocotrienol is a better antioxidant than α-tocopherol. α-Tocotrienol possessed 40–60 times higher antioxidant activity in vitro in rat liver microsomes against (Fe^{2+}+ascorbate)- and (Fe^{2+}+NADPH)-induced lipid peroxidation and 6.5 times better protection of the intrinsic membrane protein (Cytochrome P-450) against oxidative damage than α-tocopherol. The higher antioxidant potency of α-tocotrienol over α-tocopherol was hypothesized to be due to the combined effects of three properties: (1) α-tocotrienol was mentioned to have a higher recycling efficiency from its chromanoxyl radicals than α-tocopherol, (2) α-tocotrienol is significantly less associated in clusters and is more uniformly distributed in membrane bilayers than α-tocopherol, and (3) α-tocotrienol has a strong disordering effect on membrane lipids which makes interaction of the chromanols with lipid radicals more efficient (Serbinova et al., 1991).

The protective effect of tocotrienol rich-fraction from palm oil (TRF) on the copper-induced oxidation of plasma low density lipoprotein (LDL) and indices of lipid peroxidation in human umbilical vein endothelial cells (HUVEC) was compared with α-tocopherol. The antioxidant efficacy of tocotrienols, as manifested by delaying the formation of thiobarbituric acid reactive substances (TBARS) and reducing the formation of conjugated diene, was more potential than that of tocopherols (Mutalib et al., 2003). Another study showed that TRF can inhibit oxidative damage induced by ascorbate-Fe^{2+} and light photosensitization in rat liver microsomes. The protective ability of TRF, against both protein oxidation and lipid peroxidation, was significantly higher than that of α-tocopherol (Kamat et al., 1997). Further evidence of the greater antioxidant activity of tocotrienols compared with tocopherols include the improvement of the physiological condition and reduction of the exercise-induced oxidative stress in forced swimming rats (Lee et al., 2009), significant reduction of TBARS levels and increase of glutathione peroxidase activity in the rat femur (Maniam et al., 2008), and greater peroxyl radical scavenging potency than a-tocopherol in liposomal membranes (Suzuki et al., 1993).

The anti-cancer activity of palm tocotrienols was first discovered by Sylvester et al. (1986) in investigating the roles of high dietary fat intake on carcinogen-induced mammary cancer development and growth in rats. Since then, the antiproliferative effects of tocotrienols on wide varieties of tumor cells such as breast (Guthrie et al., 1997), colon (Agarwal et al., 2004), stomach (Liu et al., 2010), liver (Sakai et al., 2004), pancreas (Hussein & Mo, 2009), and prostate (Srivastava & Gupta, 2006) have been extensively studied in vitro and in vivo.

Accumulated evidence has demonstrated that individual tocopherol and tocotrienol isomers display different anti-cancer potencies. Mclntyre and colleagues reported the antiproliferative effects of tocopherols and tocotrienols in neoplastic mammary epithelial cells, and the antiproliferative effects of individual isomers can be characterized as δ-tocotrienol > γ-tocotrienol > α-tocotrienol > δ-tocopherol > γ-tocopherol > α-tocopherol (McIntyre et al., 2000). When tocotrienols were administered intraperitoneally to mice with transplanted tumors, it was found that tocotrienols were more effective than α-tocopherol, and γ-tocotrienol was more effective than α-tocotrienol (Komiyama et al., 1989). Furthermore, it is tocotrienols but not α-tocopherol that suppressed the growth of human breast and colorectal cancer cells (Eitsuka et al., 2006; Nesaretnam et al., 1995). Additional studies have demonstrated that combined treatment of tocotrienols with other traditional chemotherapeutic agents results in a synergistic anti-cancer response. Studies showed the synergistic effect of tocotrienols and atorvastatin, celecoxib, and so on in the antiproliferation of cancer cells (Shirode & Sylvester, 2010; Yang et al., 2010). In one in vitro study, tocotrienols were reported to enhance the antitumor activity of tamoxifen in human breast cancer cells (Guthrie et al., 1997), while in a double-blinded, placebo-controlled pilot trial to test the effectiveness of adjuvant tocotrienol therapy in combination with tamoxifen for 5 years in women with early breast cancer, no association seems to exist between adjuvant tocotrienol therapy and breast cancer-specific survival (Nesaretnam et al., 2010).

A number of mechanisms of anti-cancer effects of tocotrienols have been proposed. Tocotrienols can induce apoptosis of cancer cells via activation of both extrinsic and intrinsic pathways (Aggarwal et al., 2010). The extrinsic pathways involve induction of death receptors and activation of caspase-8, which leads to caspase-3 activation. The activation of intrinsic pathways by tocotrienols involves mitochondrial depolarization and is mediated through the upregulation of Bax, cleavage of Bid, release of cytochrome C, and activation of caspase-9, which in turn leads to activation of caspase-3. This unsaturated form of vitamin E also mediates apoptosis through DNA fragmentation and upregulation of p53 in certain cells. Another mechanism involved is suppression of angiogenesis through inhibition of VEGF expression and VEGF receptor signaling (Shibata et al., 2009; Weng-Yew et al., 2009). Suppression of the matrix metalloproteinase (MMP)-9 GENE could contribute to the angiogenesis-suppressive activity (Liu et al., 2010).

Tocotrienols have been suggested to have protective effects on the progression coronary heart diseases (CHD). The cardioprotective effects of tocotrienols are mostly mediated through inhibition of the HMG-CoA reductase enzyme involved in cholesterol biosynthesis. It was found that tocotrienols and not tocopherols suppress the HMG-CoA reductase, which directly inhibits the biosynthesis of cholesterol (Parker et al., 1993). The significant hypocholesterolemic effect of tocotrienols was compared with tocopherols in humans, chicken, hypercholesterolemic rat, swine, and hamster's plasma (Das et al., 2007). When large doses of TRF and its constituents supplemented diets were fed to female chickens for 4 weeks, no difference in weight gain and feed consumption were found, while supplemental TRF led to a dose-dependent lowering of serum total and LDL cholesterol levels, with γ and δ isomers being the most active forms (Yu et al., 2006). Other mechanisms are due to their antioxidant activities and their anti-inflammation effects (Watkins et al., 1993), inhibition of expression of adhesion molecules and monocyte-endothelial cell adhesion (Theriault et al., 2002), and inhibition of phospho-c-Src activation and stabilization of proteasome (Das et al., 2005). Dietary tocotrienols from TRF significantly reduced the concentration of plasma cholesterol and apolipoprotein B, platelet factor 4, and thromboxane B2, indicating its ability to protect against platelet aggregation and endothelial dysfunction (Qureshi et al., 1991). Palm tocotrienols could improve postischemic ventricular function and reduce myocardial infarct size, reduce ischemia-reperfusion-induced activation of c-Src activities, and thus stabilization of proteasomes by preventing the ischemiareperfusion-mediated reduction of 26S and 20S proteasomes (Das et al., 2005). Furthermore, the γ-isomer of tocotrienol was the most cardioprotective of all the isomers followed by the α- and δ-isoforms (Das et al., 2008).

Tocotrienols have been found to function as a neuroprotection agent against stroke, and they exert this neuroprotective effect through suppression of proinflammatory pathways. Sen and his group have investigated extensively the prevention of glutamate-induced neurodegeneration by tocotrienolsat nanomolar concentration. They found that tocotrienols fed orally does reach the brain, and the modulation of c-Src, 12-lipoxygenase, and PLA2 is involved in the neuroprotective effects of tocotrienols (Khanna et al., 2010; Sen et al., 2004).

Other benefits of dietary tocotrienols in animal models include protection of the skin against damage induced by UVB and sesamin enhances this effects (Yamada et al., 2008); lowering blood glucose and prevention of diabetes-associated cognitive deficits (Kuhad & Chopra, 2009; Wan Nazaimoon & Khalid, 2002); reduction of bone loss, stimulation of bone formation, and improve bone structure (Hermizi et al., 2009); immunoregulatory effect (Gu et al., 1999), radioprotective effect (Ghosh et al., 2009), hepatoprotective effect (Lee et al., 2010), as well as gastroprotective effects (Nafeeza et al., 2002).

Phytosterols

Phytosterols are important structural components of plant membranes. They belong to the family of triterpenes, which include the four-ring steroid nucleus and the 3β-hydroxyl group. Structurally, phytosterol are similar to cholesterol, however, cholesterol has a side-chain composed of 8 carbon atoms, whereas most of phytosterols contain 28 or 29 carbon atoms, with a side-chain composed of 9 or 10 carbon atoms and one or two carbon-carbon double bonds. Typically, a 5,6-double bond appears at the steroid nucleus, and sometimes the alkyl side chain may also contain a double bond. While more than 250 phytosterols from different plant class have been identified to date, β-sitosterol, campesterol, and stigmasterol are the most abundant (Poli

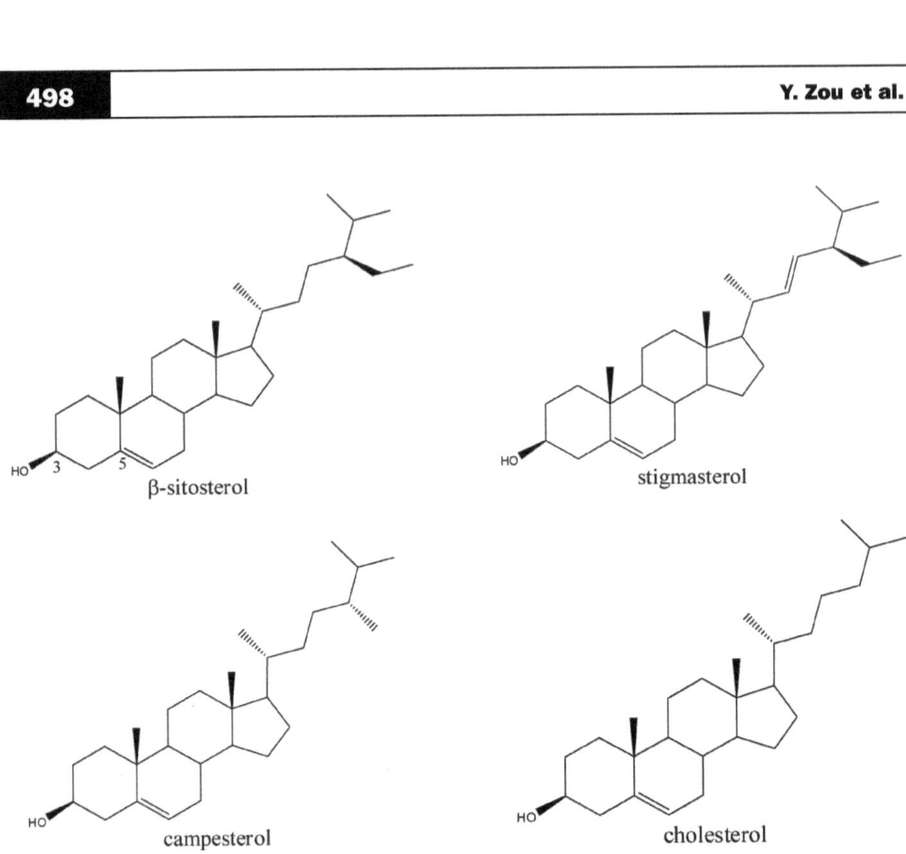

Fig. 16.5. Chemical structures of representative phytosterols and cholesterol.

& Marangoli, 2010). The structures of β-sitosterol, campesterol, and stigmasterol are illustrated in Fig. 16.5; the structure of cholesterol are also included for comparison.

Phytosterols have been the subject of many nutritionists because of their bioactivities such as cholesterol-lowering actions (Klingberg et al., 2008) and anti-cancer effects (Weyengo et al., 2009). There is growing interest in developing functional foods, including vegetable oils, spreadable fats, butter, yogurt, bakery products, juice, and milk, in which different types of phytosterols are intrinsic or just added as novel food ingredients to fortify the health-promoting effect of those products. At present, some products containing phytosterols are available in the market, especially in several European countries (Laakso, 2005).

Source of Phytosterols

Phytosterols cannot be synthesized by human beings, so they must be obtained from the diets. Phytosterols exist in the form of free sterols, fatty acid esters, steryl glycosides, and acetyl steryl glycosides in which the 3-β hydroxyl group is esterified (Phillips et al., 2002). When the double bond at the C-5 position of the ring is saturated, it is referred to as stanols, which are produced by the hydrogenation of sterols and are not abundant in nature (Law, 2000).

The most important natural resource of phytosterols comes from vegetable oils, although they are also found in a wide range of nuts, seeds, legumes, and vegetables (Normén et al., 2007; Ryan et al., 2007). Phytosterols contained in vegetable oils exist as natural occurring constituents, which make up the largest proportion of unsaponifiable part of oils. Phytosterols in fatty foods consumed in Sweden and the Netherlands have been reported. The phytosterols content varied from 39 mg/100 g in palm oil to 978 mg/100 g in corn oil (Normén et al., 2007). Philips and colleagues reported the free and esterified sterols content of 31 edible oils, with free sterols ranging from 32 to 94% of total sterols. The total sterols ranged from 58 mg/100 g in hydrogenated palm oil to 1098 mg/100 g in crude evening primrose (Philips et al., 2002). The content and compositions of phytosterols in common consumed oils are shown in Table 16-N. Unlike tocotrienols and carotenes, the content of phytosterols in palm oil is relatively lower than that in other oils. Sitosterol, campesterol, and stigmasterol are the most predominant phytosterols in all oils; cholesterol in vegetable oils are minimal and can be negligible.

The main source of phytosterols in large scale production is the by-products from the process of refining oils. Those by-products are known as deodorizer distillate (DD) or fatty acid distillate (FAD), which is a complex mixture of different compounds such as fatty acids, sterols, tocopherols, sterol esters, hydrocarbons, breakdown products of fatty acids, aldehydes, ketones, and acylglycerol species. Refining can be done by chemical refining or physical refining. For palm oil, it is generally processed by the physical refining since the non-hydratable phospholipids is low and the acidity of crude palm oil is high, which can lead to excessive loss of neutral oil in the soapstock after alkali neutralization. The deodorization step of refining process not only strip FFA and volatile odiferous components, but also remove minor components such as tocopherols, tocotrienols, and phytosterols into the distillate. Therefore, phytosterols are concentrated in the FAD, and this is the common industrial method to obtain phytosterols from vegetable oils since the concentration of phytosterols intrinsic in vegetable oils is low. The percentage and compositions of phytosterols in PFAD are illustrated in Table 16-K.

Another source of phytosterols is from the process of palm biodiesel. In this process, crude palm oil is transesterified in the alcoholic medium with an alkyl catalyst to produce a colored ester, which could be used as fuel. Moreover, the colored ester could be distillated by molecular distillation to give distillate, a less colored liquid for the production of biodiesel and residue, known also as palm phytonutrient concentrate (PPC), which is a thick red concoction and is rich in minor compounds. The content of phytosterols in PPC is 15780 ppm, 17 times higher than that in the crude palm oil (Chandrasekaram et al., 2009).

Phytosterols also can be obtained from kraft pulping tall oil, a by-product of the wood pulp industry from soft woods of the *Pinus* genus (Quilez et al., 2003). The

Table 16-N. Content (mg/100g) and Compositions of Phytosterols in Commonly Consumed Oils.

Oils	Origin	Sito	Campe	Stigma	Brassica	5-AS	Sitostanol	Campestanol	Chole	Total
Palm	S	24	9	6	0	0	0	0	–	39
	U	39.5	14.8	8.7	nd	1.6	< 1	nd	1.5	66
	B	36.5	13.7	7.1	2.6					59.9
Hydrogenated palm	U	29.7	9.6	6.5	nd	< 1	nd	3.8	1.5	48
Coconut	S	61	17	15	6.3	21.6	3.6	0	–	125
	U	42.2	5.7	11.1	nd	10.9	< 1	1.5	< 1	70
	B	48.6	7.8	12.5		–				68.9
Olive	S	141	9	4	0	22.9	0	0	–	177
	U	129.2	4.3	2.3	nd	17.2	2.8	0.7	< 1	156
	B	127.7	5.3	–	60.1					193.1
Grape seed	S	160	24	24	nd	nd	6.5	0	–	215
Peanut	D	203	52	28	6	22.8	3.3	0	–	315
	S	166	37	21	0	22.2	5.8	0	–	251
	U	115	23.7	12	1.2	12.9	2.5	1.4	< 1	167
Soybean	S	189	68	56	2.8	11.1	18	5.0	–	349
	D	172	67	59	0	7.7	5	2.2	–	313
	U	144.7	52.3	50.1	nd	5.8	3.6	2.8	1.1	257
	B	160.2	69.5	63.9	–	13.8				307.4

Table 16-N. continued.

Oils	Origin	Sito	Campe	Stigma	Brassica	5-AS	Sitostanol	Campestanol	Chole	Total
Sunflower	D	309	52	47	0	24.5	3.0	0	–	436
	U	194.1	27.1	17.7	2.1	18.7	2.9	1.2	< 1	263
	B	237.7	43.8	30.8	55.5					367.8
Cottonseed	U	256.1	20.2	5	nd	7.5	2.9	0.9	0.7	292
	B	357.2	32.1	3.9	15					408.2
Linseed	S	226	120	38	13	70.7	0	3.1	–	471
Sesame	S	335	100	37	nd	nd	0	0	–	472
	U	335.4	76.3	36.7	nd	53.4	2	1.1	nd	504
Rapeseed	S	377	279	6	81	24.7	2.2	3.1	–	773
	F	395	202.6	3.5	91.5	17.1	2.1	0.6	3.1	715
	B	390	299.7	77.4						767.1
Corn	S	586	215	66	6.7	62.1	28	14	–	978
	U	509.4	138.9	58.5	nd	41.3	15.9	10.6	1.9	766
	B	541	168.9	58.3	11.4					779.6
Wheat germ	S	613	239	37	1	46.9	20	11	–	967
Evening primrose	D	814.4	70.4	3.6	nd	122.7	14.4	9.6	nd	1026

Abbreviations: S, Sweden; D, The Netherlands; F, Finland; U, USA; B, Belgium; Sito, Sitosterol; Campe, campesterol; Stigma, stigmasterol; Brassica, brassicasterol; 5-AS, Δ5-avenasterol; Chole, cholesterol.
Source: Normén et al. (2007); Philips et al. (2002); Verleyen et al. (2002a).

compositions of crude tall oil is typically composed of 35–50 fatty acids, 15–55% diterpenic resin acids, and 5–35% unsaponifiable matters such as waxes, long chain alcohols, fatty acid esters, sterols, and steryl esters. Sterols from tall oil contains up to 80% β-sitosterol (Wang et al., 2002).

Changes of Phytosterols during Processing and Storage

The effect of processing, especially the refining process, on the phytosterols in oils have been extensively studied. However, only a few studies, especially focusing on the phytosterols of palm oil, have been published.

Generally, loss of phytosterols during the refining of palm oil is unavoidable. The different steps of refining, either chemically or physically, gradually reduced the content of phytosterols. Siew (1990) analyzed the sterol composition of several samples including crude palm oil, their processed products, and fractions. The results are illustrated in Tables 16-O and 16-P, respectively. Crude palm oil contained the highest content of sterols (200–600 mg/Kg), while processing, such as degumming, bleaching, and deodorization, decreased the unsaponifiable matter and the content of sterols in both oil and their olein products and resulted in the removal of some of the sterols into the FAD. Fractionation of palm oil resulted in the migration of the unsaponifiable matter to the palm olein. Verleyen et al. (2002b) studied the influence of physical refining on esterified and free phytosterols in palm oil. The results showed that acid degumming and bleaching hardly affected the ratio of esterified and free phytosterols, but a significant decrease in the content of free phytosterols and increase in the ratio of esterified and free phytosterols was observed after deodorization, partly due to the heat-promoted esterification reaction between FFA and free phytosterols in the process of deodorization.

To widen the commercial use of vegetable oils, they are often modified physically, by fractionation or blending, or chemically, by hydrogenation or interesterification.

Table 16-O. Unsaponifiable Matter and Sterol Content in Palm Oil Products.

Samples	Unsaponifiable matter content (wt. %)	Sterol in unsaponifiable matter (wt. %)	Sterol content in oil (wt. %)
crude palm oil	0.38–0.47	4.7–13.6	0.02–0.06
degummed and bleached palm oil	0.34–0.41	9.3–13.7	0.02–0.05
RBD palm oil	0.28–0.32	2.5–10.8	0.01–0.03
crude palm olein	0.48–0.66	4.1–9.5	0.03–0.05
degummed and bleached palm olein	0.35–0.43	3.7–5.7	0.02
RBD palm olein	0.22–0.26	4.3–6.7	0.01
RBD palm stearin	0.18	10.7	0.02

Source: Siew (1990).

Table 16-P. Sterol Compositions of Crude, Refined Palm Oil, Palm Olein, and Palm Stearin (mg/Kg).

Samples	Cholesterol	Campesterol	Sigmasterol	Sitosterol	Unknown	Total
crude palm oil	2.7–13.0	46.4–150	26.3–65.7	120–369.5	2–21	210–620
degummed and bleached palm oil	4.8–9.9	49.4–116	22–51.2	113.8–285.7	0–7.5	190–470
refined palm oil	1.2–5.5	15.3–65.4	8.5–36.9	45–198	0–10.5	70–316
crude palm olein	6.2–7.5	56.7–103.8	29.9–51.0	149–253	24.6–28.1	270–440
degummed and bleached palm olein	3.2–3.8	36.2–43.2	21.4–25.2	99–123	0–5.2	160–200
refined palm olein	2.1–2.4	25.6–30.4	12.4–23.3	67.7–114	0–1.2	109–170
crude palm stearin	7.4	90.3	42	210	–	350

Source: Siew (1990).

In a study of chemical interesterification of a blend of refined olive oil and palm stearin, the content of sterols in the oil blend reduced slightly by 3.2% and 5.5% at 90°C and 120°C, respectively. However, the distribution of sterols in the esterified sterols were unchanged, but interesterification caused an increase in esterified sterols content of the oil blend of 4.0% and 6.6% at 90°C and 120°C, respectively, compared with the starting blend. The dominant phytosterols oxidation products (POPs) were 6-hydroxysitostanol and 6-hydroxycampestanol, and the total POPs content were generally unchanged (Azadmard-Damirchi & Dutta, 2008). Enzymatic interesterification can also cause a reduction in phytosterols content. During enzymatic interesterification of palm stearin and rice bran oil blends using a 1,3-specific immobilized lipase, a 3.0–12.0% decrease in phytosterols levels was observed (Reshma et al., 2008).

The content of phytosterols in palm oil could be affected by the extraction solvent. Supercritical carbon dioxide (SC-CO_2) and hexane extracted relatively more phytosterols compared with conventional screw-press extraction. Lau and colleagues reported that the phytosterol content in a commercial crude palm oil and palm oil extracted with hexane and SC-CO_2 was 674, 768, and 826 ppm, respectively. A significant amount of phytosterols is expected to be bound deeply inside the solid fibrous materials that can only be extracted by hexane or SC-CO_2. The smaller molecules of phytosterols permit phytosterols to solubilize easily in SC-CO_2 as compared to TG with molecular weight of more than 800 g/mol (Lau et al., 2006).

Few studies are available on quantitative changes in phytosterol contents during storage of palm oil. Minor components in palm oil are naturally bioactive compounds, which are sensitive to heat and light. Storage temperature definitively affects the stability of those bioactive compounds. The normal room temperature itself is enough to cause those compounds to disintegrate. Chandrasekaram et al. (2009) studied the effect of storage temperature (room temperature: 30±2°C vs refrigerated temperature: −16±2°C) on the stability of those compounds in palm concentrate. It was found that 24% of carotenes, 37% of vitamin E, 36% of sterols, and 73% of squalene were deteriorated after 3 months of storage at room temperature. On the other hand, only 5% of carotenes, 6% of vitamin E, 8% of sterols, and 1% of squalene was lost at refrigerated temperature. Therefore, the nutritionally valuable phytochemicals are recommended to be refrigerated and stored in a cold and dry place to prevent their loss.

Analytic Methods of Phytosterols

The analyses of phytosterols in foodstuff or functional food products are comprehensively reviewed (Abidi, 2001; Lagarda et al., 2006; Moreau et al., 2002). Regarding the analysis of sterol fractions of fats and oils, some international reference methods such as the ISO 6799 (1991), IUPAC methods 2.401 (1987) and 2.403 (1987), ISO 12228 (1999) and AOCS Ch. 6–91 (1997) could be available.

When extracting of lipids using non-polar solvents, phytosterols are simultaneously extracted. In vegetable oils, phytosterols exist mainly as free phytosterols (FS) and phytosteryl fatty-acid esters (SE), while the phytosteryl glycosides are removed during the refining process. Typically, the analysis of phytosterols in vegetable oils, which is the main focus in this chapter, involves (1) alkaline hydrolysis or saponification, (2) extraction of the non-saponifiable matter, (3) possible cleanup of the extract, (4) derivatization of the sterols, and (5) separation and quantification of the sterol derivatives by capillary gas chromatography (Piironen et al., 2000).

Although saponification is laborious and time consuming, it is still in use for the analysis of total phytosterols in crude or refined vegetable oil samples. In the process of saponification, phytosterol esters are alkaline hydrolyzed to liberate the free sterols, therefore, the determination are expressed as the total content of phytosterols. Usually, an aliquot of the oil sample is stirred overnight at room temperature with 1 M ethanolic potassium hydroxide. The mixture is diluted with water and extracted with organic solvents such as diethyl ether, ether, light petroleum, or heptane. The combined ether extract is saponified again with ethanolic potassium hydroxide, washed with several batches of distilled water until neutral to pH paper and then dried sequentially with short columns of anhydrous sodium sulfate, deactivated alumina and anhydrous sodium sulfate. Removal of solvent yields an unsaponifiable residue suitable for chromatographic quantification of sterols (Abidi, 2001; Abidi et al., 1999; Nourooz-Zadeh & Appleqvist, 1992). Saponifications under heating (hot saponification) with ethanolic potassium hydroxide in order to reduce the time of saponification are also described.

The unsaponifiable matter contains not only phytosterols but also a number of other compounds including tocopherols, carotenoids, and other hydrocarbons. It is necessary to clean up the unsaponifiable matter. Thin layer chromatography (TLC) and column chromatography (CC) are two common techniques to clean up. The choice depends on the amounts of samples to be processed; samples more than 200 mg can be conventionally evaluated by CC, whereas relatively simple and/or small-quantity samples of less than 200 mg can be assessed by TLC (Piironen et al., 2000). Since TLC and CC procedures are very time-consuming and tedious, they have been gradually replaced by the solid-phase extraction (SPE) technique, which provides faster fractionation and small volume of solvent is needed. Reverse-phase mode, normal-phase mode, neutral alumina cartridge, and silica SPE cartridge are employed for clean-up procedures (Breinhölder et al., 2002; Phillips et al., 2002; Toivo et al., 1998).

Phytosterols are often derivatised before GC analysis in order to make them more stable and to improve analytical sensitivity. The most common derivatisation procedure used for phytosterols is to derivatise them to their trimethylsilyl (TMS) ethers, which offers higher thermostability, lower polarity, and improved peak shape. An example of the derivatization procedure is as follows: the dried unsaponifiable matter was dissolved in dry pyridine (0.5 ml), and 1 ml *N,O*-bis (trimethylsilyl)-trifluoroacetamide containg 1% trimethylchlorosilane solution. The mixture was placed in a test tube, and put in an oven at 70°C for 20 min for completion of the silylation (Verleyen et al., 2002a).

After derivatization, the derivatives of sterols is separated and analyzed by capillary gas chromatography, which is currently recognized as the most suitable method for sterol analysis. The capillary column (0.1–0.3 mm I.D.) in some respects is advantageous over pack column (3–8 mm I.D.) because it offers shorter analysis times and lesser peak interference, improvement in component resolution, and high thermal stability. Glass and fused-silica capillary column coated with non-polar stationary phases (100% polysiloxane phases) or slight polar stationary phases (5% diphenyl–95% dimethyl polysiloxane) are employed. Since FID has the best GC features in terms of detection sensitivity, response linearity, and response generality, GC is frequently interfaced with flame ionization detector (FID) to monitor the effluent. Sometimes, mass spectrometry (MS) is connected to GC for the structural identification and peak purities evaluation of the eluting phytosterols (Abidi, 2001; Dutta and Normén, 1998; Piironen et al., 2000).

Analysis of phytosterols by high-performance liquid chromatography (HPLC) has been developed with the advent of HPLC technology. HPLC is performed under milder column temperatures and under non-destructive detection conditions. Therefore, it is suitable to analysis of thermally unstable phytosterols. Both normal phase (NP) and reverse-phase (RP) HPLC have been used for the analysis. RP-HPLC uses less volatile polar organic solvents in water and offers ready equilibration in a bonded silica stationary phase with the mobile phase solvents, compared with NP-HPLC; in general, RP-HPLC has been more frequently and more widely used than NP-HPLC for the separation of individual sterols. The detectors include UV (200–210 nm), photodiode array detector (DAD), refractive index (RI) detector, evaporative light scattering (ELSD) detector, and mass spectrometry (MS). One shortcoming with HPLC is in that it cannot exhibit significant selectivity for the separation of sterols and its corresponding stanols (Abidi, 2001).

Supercritical fluid chromatography (SFC) has been reported since the 1980s. The unique properties of supercritical fluid as a mobile phase in SFC overcome the difficulties of solute thermal instability and volatility encountered in GC and also shorten the relatively longer analysis times of HPLC separations. Using SFC coupled with UV detector, Choo et al. (2005) analyzed the minor components from palm oil. After saponification, the unsaponifiable matter was dissolved in dichloromethane and injected directly into a SFC. Content of phytosterols in crude palm oil was 230 ppm, which was comparable with those determined by conventional GC method.

A rapid technique for the quantitative determination of a mixture of FFA, acylglycerols, sterols, and squalene in a single GC-FID injection of vegetable oils, with special reference to palm oil, has been exploited recently (Lau et al., 2005, 2006). The oil sample (0.02 g) was mixed with 1.2 mL of dichloromethane and 0.3 mL of the silylating reagent BSTFA, and heated at 60°C for 2 h. Compared to the conventional GC, higher temperature with detector (370°C) and oven (final temperature: 360°C for 16 min) was employed. This procedure eliminated the laborious and time-consuming sample preparation of sterols in conventional GC method, and it gave a good separation of major free sterols. The individual sterols determined by the rapid technique was consistent with those determined by conventional method.

Recovery of Phytosterols

Recovery of phytosterols is basically based on the difference either in the solubility or in the volatility between phytosterols and other components. In vegetable oils, some phytosterols partially exist as phytosterol esters; therefore, the phytosterol esters must be hydrolyzed via hydrolyzing with water under pressure (1.5–50 MPa) and high temperature (200–260°C) or alkali hydrolysis with sodium or potassium hydroxide to liberate free sterols (Rohr et al., 2005).

Phytosterols, FFA, TAG, and other minor components are concomitant in vegetable oils and its refining by-products. Recovery of phytosterols is a combination of processes such as (trans)esterification, saponification, short-path distillation, fractional extraction, and crystallization. Many studies and patents on how to recover phytosterols have been published. A typical process for the recovery of phytosterols from FAD is illustrated in Fig. 16.6 (Fernandes & Cabral, 2007).

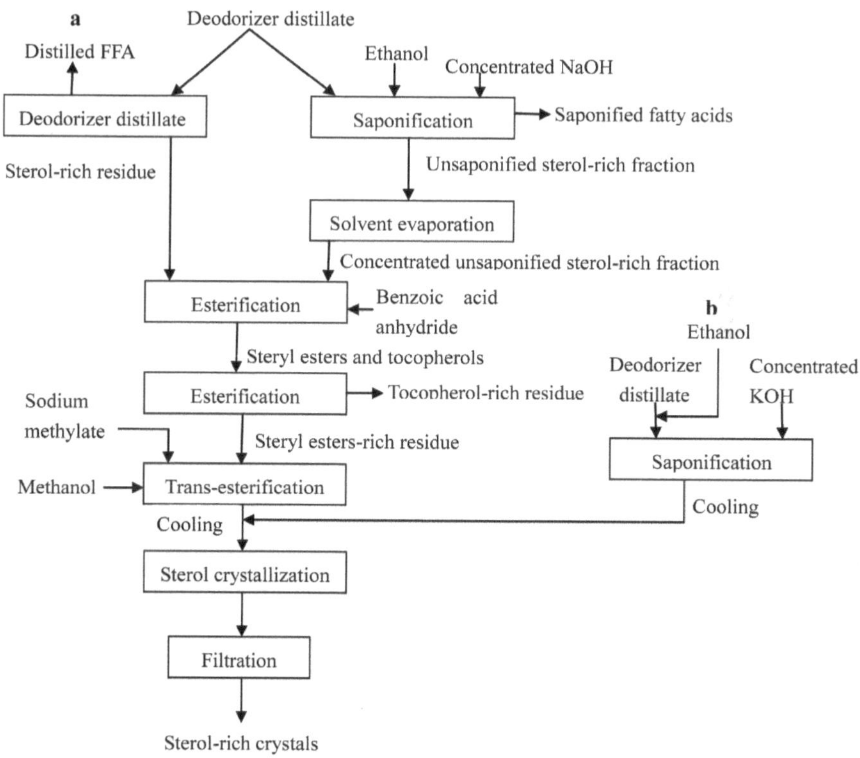

Fig. 16.6. Process for the recovery of phytosterols from deodorizer distillates (Fernandes & Cabral, 2007). (a) low sterol content and high acid value, and (b) high sterol content and low acid value.

An integrated process for the simultaneous extraction of phytosterols, vitamin E, and squalene from CPO has been patented by Choo et al. (2009). The process comprised the following steps: (1) conversion of the CPO into palm oil methyl esters by acid/alkaline catalyzed esterification/transesterification; (2) performing three stages of SPD on the CPO methyl esters to yield a phytonutrient concentrate; (3) saponification of the phytonutrient concentrate to give a saponified product, and then extraction of unsaponifiable materials by hydrocarbon solvents; (4) crystallization of phytosterols using water/alcohol/hydrocarbon system by heating and cooling; and (5) solvent partitioning of vitamin E and squalene from the remaining saponified product following crystallization. The purity of phytosterol crystal was claimed to be as high as 99%, but the process yield was relatively low.

An alternative method to recover minor constituents from CPO is to use SFE. Chuang & Brunner (2006) reported the enrichment of minor constituents with supercritical CO_2 in a pilot-scale mixer-settler apparatus. FFA in the CPO was first removed by neutralization with KOH, and the remaining oil was then transesterified into FAME since FAME has solubility magnitudes higher than TAG in CO_2. The FAME was extract using mixer-settler apparatus at 60°C and 140 bar. After three steps of extraction, a raffinate with 6% of tocochromanols, 10.5% of β-caroten, and 3% of sterols could be obtained. The concentration of those minor constituents increased to 100-fold, 200-fold, and 100-fold, respectively. Later, a unique combination of counter current separation with selective adsorption and desorption using supercritical fluids was developed by Brunner (2009). The first enrichment comprises one or more separation steps by application of a near critical or a supercritical fluid in a single or multistage separation process, and a second step comprises selective adsorption and desorption by directly adsorbing the fraction from the countercurrent separation, and subsequently eluting the mixture from the adsorbent by the same or another supercritical or near critical fluid.

Bioactive Properties and Application of Phytosterols

Phytosterols play a major role in pharmaceuticals, nutrition, and cosmetics. They are known to have bioactive properties with possible implication for human health for many years.

Increased plasma concentrations of cholesterol and higher low-density lipoprotein-cholesterol (LDL-C) are high risks for the development of cardiovascular disease (CVD). Phytosterols are well-known for their cholesterol-lowering activity. In recent years, consumers in the United States, Australia, and many European countries have been introduced to phytosterols-rich foods, which may be effective methods to reduce total and LDL-C levels. In 2000, Food Standards Australia and New Zealand (FSANZ) approved the use of phytosterol esters in table spreads and margarines at 13.7% (w/w), and extended the use for breakfast cereals and low fat milk and yogurt in 2006 (Food Standards Australia and New Zealand, 2000, 2006). In 2000, the FDA published the interim final rule (IFR) authorizing the use of a health claim for plant sterol/stanol esters and reduced risk for CVD on food labels. It is said that foods containing at least 0.65 g per serving of vegetable oil sterol esters (or 0.4 g sterols), eaten twice a day with meals for a daily total intake of at least 1.3 g (or 0.8 g sterols), as part of a diet low in saturated fat and cholesterol, may reduce the risk of heart disease (Federal Register 65 FR). In 2004, the EU Commission published Regulation 608/2004/EC concerning the labeling of foods and food ingredients with added phytosterols and/or phytosterol esters (EU-Commission Regulation, 2004).

The health-promoting effects of phytosterols on lowering LDL-C are based on numerous experimental and epidemiological studies. The relationship between phytosterol supplements and CVD is well-reviewed by Poli & Marangoni (2010) and John et al., (2007). In summary, although the exact mechanism of cholesterol-lowering activity of phytosterols is not very clear, it may (1) suppress the uptake of both dietary and endogenously-produced (biliary) cholesterol from the intensine, (2) replace the position of cholesterol in the "mixed micella," (3) phytosterols themselves are very poorly absorbed, and the small amount that is absorbed is actively re-excreted in bile, (4) mediate intestinal membrane-transport proteins such as ABCG5, ABCG8, and NPC1L1 and hence, reduce circulating cholesterol levels (Calpe-Berdiel et al., 2009; Jones, 1999; Marinangeli et al., 2006).

Several studies have indicated that phytosterols may have anti-cancer activity against human breast cancer (Awad et al., 2000), colon cancer (Awad et al., 1998), ovary cancer (McCann et al., 2003), stomach cancer (De Stefani et al., 2000), and lung cancer (Mendilaharsu et al., 1998). Phytosterols seem to act through multiple mechanisms of action, including inhibition of carcinogen production, cancer-cell growth, angiogenesis, invasion and metastasis,and through the promotion of apoptosis of cancerous cells. Phytosterols could also inhibit cancer development by lowering blood cholesterol as high blood cholesterol level and hence the concentration of cholesterol in lipid rafts of cell membranes are associated with reduced apoptosis of cancer cells (Bradford & Awad, 2007; Woyengo et al., 2009).

Antioxidant activities of phytosterol have been reported. White and Armstrong (1986) demonstrated that Δ-5 avenasterol may have valuable antioxidant activity. Additional studies have confirmed that Δ-5 avenasterol and other phytosterols that contain an ethylidene group possess antioxidant and antipolymerization properties, especially valuable during frying (Kochar, 2000; Tian & White, 1994). Whether these antioxidant properties have any significance to human health remains to be seen.

Other Minor Constituents

Mono- and diacylglycerols (MAGs and DAGs) are partial glycerides, which do not occur naturally in significant amounts except in palm oil from damaged fruits. 1-MAGs are more stable than their β-isomers, while 1,3-DAGs are the prominent DAGs in palm oil. The content of MAGs in palm oil is usually below 1%, and the major fatty acids of MAGs are identified as palmtic and oleic acid (Ooi et al., 1991). The quantification of DAGs in palm oil has been reported. The DAGs content of palm oils extracted from different varieties of palm fruit varied from 3.8 to 7.6%, while the average levels of DAGs reported for refined palm oil, olein and stearin were 6, 6.1, and 4.4%, respectively (Sambanthamurthi et al., 2000). Three types of DAGs exist in palm oil, with C34 (pal-mitoyloleoyl glycerol or PO) being the major, followed by C36 (dioleoylglycerol or OO) and C32 (dipalmitoyl glycerol or PP). The composition of DAGs in palm oil was dependent on the degree of fruit ripeness and the extent of hydrolytic degradation. More DAGs were found in ripe and over-ripe fruits and in fruits which were not processed immediately (Siew & Ng, 1997).

Palm oil also contains polar lipids such as phospholipids and glycolipids, which contains both lipophilic and hydrophilic functions. Glycolipids are the major polar lipids (1000–3000 ppm) (Goh et al., 1985). Structurally, glycolipids comprise of glycerol, ceramides, or lipopoly saccharides. The major glycolipid in palm oil is monoglycosyl diglyceride (MGDG) and diglycosyl diglyceride (DGDG), which has one or two sugars linked glycosidically to DAG. MGDG accounted for 22% to 25% of total glycolipids in oil from two varieties of oil palm, *dura* and *pisifera*, and DGDG for 42% to 45%. The predominant fatty acids in MGDG and DGDG were palmitic, stearic, oleic, linoleic, and linolenic acid. Besides MGDG and DGDG, steryl glycoside (SG) and acylated steryl glycoside (ASG) were also presented (Kulkarni et al., 1991).

Inorganic phosphorus and phospholipids are two forms of phosphorus compounds in CPO, with inorganic phosphorus being the most predominant. Phospholipids in palm oil are relatively low (5–130 ppm or 0.2–5 ppm as P) since only 4% of the phospholipids of the palm fruit remains in the CPO during the wet milling process (Goh et al., 1985). The main phospholipids in Indian palm oil are phosphatidyl choline (PC, 34% to 35%), phosphatidyl ethanolamine (PE, 22% to 25%), phosphatidylinositol (PI, 21% to 25%), cardiolipin (CL, 7% to 8%), and phosphatidyl glycerol (PG, 5% to 7%). The fatty acids in all these components are palmitic, stearic, oleic, and linoleic acids (Kulkarni et al., 1991). Other minor phospholipids component such as phosphatidic acid, diphosphatidyl glycerol, lysophosphatidyl ethanolanime, and traces of lysophosphatidyl choline and phosphatidylserine were also found in palm oil (Goh et al., 1982). Phospholipids act as an antioxidant synergist by chelating traces of prooxidant metal or as a prooxidant by dispersing metal impurities responsible for reduced oil stability. Both glycolipids and phospholipids are almost completely removed in refining processes; therefore, their content in commercial palm oil is very low.

Ubiquinone is another interesting and valuable minor component present in palm oil. The oil-soluble, vitamin-like substance structurally resembles with vitamin K, which includes a 1,4-benzoquinone head and a side chain consisted of 6–10 isoprenyl subunits. Coenzyme Q_{10} (CoQ_{10}) is the most common ubiquinone in palm oil. The CoQ_{10} were report to be about 10–80 ppm in CPO with analyses carried out by preparative thin layer chromatography (pTLC) and HPLC simultaneously (Hamid et al., 1995). Bonnie & Choo (2000) determined the content of CoQ_{10} in a commercial

Carotino red palm olein by HPLC, it ranged from 10 to 25 ppm. Using SFC, Ng et al. (2006) also reported 10–80 ppm and 1000–1500 ppm CoQ_{10} were contained in CPO and PFO, respectively. The structure of CoQ_{10} was confirmed by ^{1}H- and ^{13}C-NMR. At different stages of the oil palm fruit's development, CoQ_{10} can only appear from 12 weeks after anthesis (WAA) and the content increased gradually to 80 ppm at 22 WAA (Ng et al., 2009). Although CoQ_{10} is presented at relatively low concentrations in CPO, it acts as an antioxidant, boosts energy, enhances the immune system, and therefore, provide benefits in the prevention or treatment of cardiovascular diseases such as elevated cholesterol levels, high blood pressure, congestive heart failure, cardiomyopathy, mitral valve prolapse, coronary artery bypass surgery, and angina, neurodegenerative diseases such as Parkinson's disease and Huntington's disease (Shinde et al., 2005).

Squalene is a triterpene with 30 carbon atoms, containing six isoprene units, and it is known as an intermediate in the biosynthesis of sterols (Kelly, 1999). Squalene is reported to have anti-carcinogenic activity against skin, colon, and lung cancer. In addition, squalene is a powerful antioxidant which can scavenge free radicals from the body and thus reduce free radical oxidative damage to the skin (Huang et al., 2009; Rao et al., 1998). Determined by GC-FID, HPLC-Vis-UV, or SFC, squalene in CPO, RBD, PFO, and PFAD are reported to be 200–600 ppm, 415 ppm, 1000–1800 ppm, and 12,000–18,000 ppm, respectively (Choo et al., 1996, 2005; Goh et al., 1985; Lau et al., 2005). While using SC-CO_2 for selective extraction, the content of squalene in PFO could be 9690 ppm (Lau et al., 2008). Squalene is usually concentrated along with the concentration of tocopherols and tocotrienols. A process including transesterification of CPO, washing, three-stage of short path distillation has been patented. The resultant phytonutrient residue was rich in squalene (6.56%), vitamin E (3.98%), and sterols (8.40%) (Choo et al., 2006). A 28% squalene concentrate could be obtained by neutralization of PFAD, followed by enzyme hydrolysis, neutralization again, and then column chromatography or solvent extraction (Chua et al., 2007, 2008).

Apart from the above oil-soluble minor compounds, other water-soluble components are also presented in the oil palm fruits, as demonstrated by Neo et al. (2008). Unlike virgin olive oil, in which 50 to 1000 ppm phenolics are contained (Boskou et al., 2006), phenolic compounds in commercial palm oil is relatively low since those water-soluble phytochemicals partition into the aqueous phase in the wet milling process. Therefore, various phenolics will end up and can be recovered from the palm oil mill effluent (POME) resulting from the heat inactivation of polyphenol oxidase enzyme (Wattanapenpaiboon & Wahlqvist, 2003). A process has been made to recovery these water-soluble antioxidants from the sludge. It includes removal of undisolved solids, oleaginous parts, colloids, and higher weight molecules, adjustment of pH, and solvent extraction (Sambanthamurthi et al., 2008). These water soluble antioxidants found applications in the cosmeceutical industry (Sambanthamurthi et al., 2010) and in foods and edible oils (Tan et al., 2010).

The total phenolics content of the crude and ethanol extracts of palm fruits were found to be 40.3 and 49.6 mg GAE/g extract (dry basis), respectively, as determined by the Folin-Ciocalteu method (Balasundram et al., 2005). Phenolics in those extracts may be include gallic, chlorogenic, protocatechuic, gentisic, coumaric, ferulic, and caffeic acids, as well as catechins, hesperidin, narirutin, and 4-hydroxybenzoate (Tan et al., 2001). Phenolics also exist in the pressed fiber of the oil palm fruits and oil palm leaves (Nang et al., 2007; Ng & Choo, 2010). Using SC-CO_2 as extraction solvent, the water-soluble compounds together with oils were extracted from the palm-pressed fiber. Their concentration in PFO was from 0.1% to 0.2%, and 12 phenolic compounds were identified by GC-MS, most of which are structurally similar to those in olive oil. The water-soluble compounds showed higher antioxidant activity with 10 times more effective than pure α-tocopherol (Nang et al., 2007).

Conclusions

In the past decades, palm oil has become well known as the safe and non-toxic, easily digestible, efficiently utilized, and well absorbed vegetable oil for foods, pharmaceuticals, and oleochemicals. A lot of studies on palm oil and its minor constituents have been conducted. β-Carotene in palm oil may be important to improve vitamin A deficiency in addition to its antioxidant activity and protective effects against several forms of cancer. Tocotrienols are more powerful than tocopherols in terms of antioxidant and anti-cancer activity. Crude palm oil is unique in that it contains high content of carotenoids and tocotrienols, while almost 100% of carotenoids and much of tocotrienols may be lost from the processed oil. To retain as much of carotenoids and tocopherols and tocotrienols, proper refining or processing technology is necessary to increase the shelf life and nutrition value of palm oil.

The production of palm oil is expected to increase continuously. Biodiesel from palm oil is a new regenerative source for energy and will also increase with the exhaustion of petroleum. The refining process of palm oil, and processing for palm biodiesel, gives rise to PFAD, and palm phytonutrient concentrate, respectively. These by-products are a good source for extraction of vitamin E and other bioactive components. The technology used for extraction is crucial for the quality of the final products. In industrial-scale, molecular distillation is still a key technology in separation or concentration of minor constituents from palm oil or its by-products. Supercritical fluid extraction and pressure liquid extraction are emerging as new separation technologies, but their uses in industrial-scale are rarely reported.

There is strong competition between natural-derived and synthetic products. With the awareness of the safety issues, natural-derived products gain more interest among researchers and customers. This is true for the situation of natural carotenoids and tocopherols, which exhibit more safety and more bioactivities than synthetics. The opportunities for palm minor constituents in incorporation into foods, beverages, cosmetics, nutraceuticals, and dietary supplements as functional ingredients are vast. It can be assumed that the production and usage of palm oil minor constituents will continue to competitively increase.

Reference

Abidi, S.L. Chromatographic analysis of plant sterols in foods and vegetable oils. J. Chromatogr. A, 2001, 935, 173–201.

Abidi, S.L. Tocol-derived minor constituents in selected plant seed oils. J. Am. Oil Chem. Soc. 2003, 80, 327–333.

Abidi, S.L.; List, G.R.; Rennick, K.A. Effect of genetic modification on the distribution of minor constituents in canola oils. J. Am. Oil Chem. Soc., 1999, 76, 463–467.

Aggarwal, B.B.; Sundaram, C.; Prasad, S.; Kannappan, R. Tocotrienols, the vitamin E of the 21st century: Its potential against cancer and other chronic diseases. Biochem. Pharmacol. 2010, 80, 1613–1631.

Agarwal, M.K.; Agarwal, M.L.; Athar, M.; Gupta, S. Tocotrienol-rich fraction of palm oil activates p53, modulates Bax/Bcl2 ratio and induces apoptosis independent of cell cycle association. Cell Cycle 2004, 3, 205–211.

Ames, G.R.; Raymond, W.D.; Ward, J.B. The bleachability of Nigerian palm oil. J. Sci. Food Agric. 1960, 11, 194–202.

AOCS Official Method. Chapter 6–91; American Oil Chemists' Society: Champaign, IL, 1997.

AOCS. Official methods and recommended practices of the American Oil Chemists' Society, 6th ed. American Oil Chemists' Society: Champaign, IL, 2009.

Awad, A.; Tan, C.P.; Tan, Y.A. Determination of oil palm fruit phenolic compounds and their antioxidant activities using spectrophotometric methods. Int. J. Food Sci. Technol. 2008, 43, 1832–1837.

Awad, A.B.; von Holtz, R.L.; Cone, J.P.; Fink, C.S.; Chen, Y.C. Beta-sitosterol inhibits growth of HT-29 human colon cancer cells by activating the sphingomyelin cycle. Anticancer Res. 1998, 18, 471–473.

Azadmard-Damirchi, S.; Dutta, P.C. Stability of minor lipid components with emphasis on phytosterols during chemical interesterification of a blend of refined olive oil and palm stearin. J. Am. Oil Chem. Soc. 2008, 85, 13–21.

Baharin, B.S.; Abdul Rahman, K.; Abdul Karim, M.I.; Oyaizu, T.; Tanka, K.; Tanaka, Y.; Takagi, S. Separation of palm carotene from crude palm oil by adsorption. J. Am. Oil Chem. Soc. 1998, 75, 399–404.

Baker, R.; Gunther, C. The role of carotenoids in consumer choice and the likely benefits from their inclusion into products for human consumption. Trends Food Sci. Technol. 2004, 15, 484–488.

Balasundram, N.; Tan,Y.A.; Sambanthamurthi, R.; Manickam, K.S.; Samman, S. Antioxidant properties of palm fruit extracts. Asia Pac. J. Clin. Nutr. 2005, 4, 319–324.

Bansal, G.; Zhou, W.; Barlow, P.J.; Lo, H.L.; Neo, F.L. Performance of palm olein in repeated deep frying and controlled heating processes. Food Chem. 2010, 121, 338–347.

Barnicki, S.D.; Sumner, Jr., C.E.; Williams, H.C. Process for the production of tocopherol concentrates. US Patent: US5512691, 1996.

Basiron, Y.; Jalani, B.S.; Chan, K.W. Advances in oil palm research. Volume II. Malaysian Palm Oil Board, Ministry of Primary Industries: Malaysia, 2000; pp. 1036–1054.

Batistella, C.B.; Maciel, M.R.W. Recovery of carotenoids from palm oil by molecular distillation. Comput. Chem. Eng. 1998, 22, S53–S60.

Blaizot, P.P. Carotene from palm oil. US patent 2652433, 1953.

Bonnie, T.Y.P.; Choo, Y.M. Valuable minor constituents of commercial red palm olein: carotenoids, vitamin E, ubiquinones and sterols. J. Oil Palm Res. 2000, 12, 14–24.

Boskou, D.; Blekas, G.; Tsimidou, M. Olive oil composition. In Olive Oil: Chemistry and Technology. Boskou, D., Ed.; AOCS Press, Champaign, 2006, 41–72.

Bradford, P.G.; Awad, A.B. Phytosterols as anticancer compounds. Mol. Nutr. Food Res. 2007, 51, 161–170.

Breinhölder, P.; Mosca, L.; Lindner, W. Concept of sequential analysis of free and conjugated phytosterols in different plant matrices. J. Chromatogr. B 2002, 777, 67–82.

Briviba, K.; Schnabele, K.; Rechkemmer, G.; Bub, A. Supplementation of a diet low in carotenoids with tomato or carrot juice does not affect lipid peroxidation in plasma and feces of healthy men. J. Nutr. 2004, 134, 1081–1083.

Bruii, B.J. Analysis of vitamin E by HPLC. In The Encyclopedia of Vitamin E; Preedy, V.R., Watson, R.R., Eds.; CABI Publishing: United Kingdom, 2007; pp. 172–183.

Brunner, G. Counter-current separations. J. Supercrit. Fluid. 2009, 47, 574-582.

Burton, G.W.; Ingold, K.U. Autoxidation of biological molecules, l. The antioxidant activity of vitamin E and related chain-breaking phenolic antioxidants in vitro. J. Am. Chem. Soc. 1985, 103, 6472–6477.

Calpe-Berdiel, L.; Escolá-Gil, J.C.; Blanco-Vaca, F. New insights into the molecular actions of plant sterols and stanols in cholesterol metabolism. Atherosclerosis. 2009, 203, 18–31.

Carotenature (The carotenoids page), 2000. http://www.carotenature.com.

Chandrasekaram, K.; Ng, M.H.; Choo, Y.M.; Chuah, C.H. Effect of storage temperature on the stability of phytonutrients in palm concentrates. Am. J. Appl. Sci. 2009, 6, 529–533.

Chase, Jr., G.W.; Akoh, C.C.; Eitenimiller, R.R. Analysis of tocopherols in vegetable oils by high-performance liquid chromatography: comparison of fluorescence and evaporative light-scattering detection. J. Am. Oil Chem. Soc. 1994, 71, 877–880.

Cheah, K.Y.; Toh, T.S.; Koh, P.M. Palm fatty acid distillate biodiesel: next-generation palm biodiesel. Inform. 2010, 21, 264–266.

Chiu, M.C.; Coutinho, C.M.; Goncalves. L.A.G. Carotenoids concentration of palm oil using membrane technology. Desalination 2009, 245, 783–786.

Choo, Y.M.; Harrison, L.L.N.; Ma, A.N.; Basiron, Y. Extraction of palm vitamin E, phytosterols, and squalene from palm oil. US 7575767B2, 2009.

Choo, Y.M.; Lau, H.L.N.; Ma, A.N.; Basiron, Y. Extraction of vitamin E, phytosterols and squalene from palm oil. EP 1394144 B1, 2006.

Choo, Y.M.; Ng, M.H.; Ma, A.N.; Chuah, C.H.; Hashim, M.A. Application of supercritical fluid Chromatography in the quantitative analysis of minor components (carotenes,vitamin E, sterols, and squalene) from palm oil. Lipids 2005, 40, 429–432.

Choo, Y.M.; Yap, S.C.; Ong, A.S.H; Ooi, C.K.; Goh, S.H. Extraction of carotenoids from palm oil. Proc. Of the AOCS World Conference of Edible Fats and Oils: Basic Principles and Modern Practices; Erikson, D.R., Ed.; American Oil Chemist's Society: Champaign, IL, 1989.

Choo, Y.M.; Yap, S.C.; Ooi, C.K.; MA, A.N.; Goh, S.H. Production of palm oil carotenoid concentrate and its potential application in nutrition. Lipid-soluble antioxidants: Biochemistry and clinical application. Ong, A.S.H, Packer, L., Eds.; Basel Birkhauser: Verlag, 1992; pp. 243–253.

Choo, Y. M.; Yap, S.C.; Ooi, C.K.; MA, A.N.; Goh, S.H.; Ong, A.S.H. Recovered oil from palm pressed fiber: a good resource of natural carotenoids, vitamin E and sterols. J. Am. Oil Chem. Soc. 1996, 73, 599–602.

Chu, B.S.; Baharin, B.S.; Quek, S.Y.; Cheman, Y.B. Separation of tocopherols and tocotrienols from palm fatty acid distillate using hydrolysis-neutralization-adsorption chromatography method. J. Food Lipids 2003b, 10, 141–152.

Chu, B.S.; Quek, S.Y.; Baharin, B.S. Optimization of enzimatic hydrolisis for concentration of vitamin E in palm fatty acid distillate. Food Chem. 2003a, 80, 295–302.

Chua, C.S.L.; Baharin, B.S.; Man, Y.B.C.; Tan, C.P. Separation of squalene from palm fatty acid-distillate using adsorption chromatography. Eur. J. Lipid Sci. Technol. 2007, 109, 1083–1087.

Chua, C.S.L.; Baharin, B.S.; Man, Y.B.C.; Tan, C.P. Optimisation of enzymatic hydrolysis for concentration of squalene in palm fatty acid distillate. J. Sci. Food Agric. 2008, 88, 1512–1517.

Chuang, M.H.; Brunner, G. Concentration of minor components in crude palm oil. J. Supercrit. Fluid. 2006, 37, 151–156.

Darnoko, D.; Cheryan, M. Carotenoids from red palm methyl esters by nanofiltration. J. Am. Oil Chem. Soc. 2006, 83, 4, 365–370.

Darnoko, D.; Cheryan, M.; Moros, E.; Jerrel, J.; Perkins, E.G. Simultaneous HPLC analysis of palm carotenoids and tocopherols using a C-30 column and photodiode array detector. J. Liq. Chrom. Rel. Technol. 2000, 23, 1873–1885.

Das, S.; Lekli, I.; Das, M.; Szabo, G.; Varadi, J.; Juhasz, B.; Bak, I.; Nesaretam, K.; Tosaki, A.; Powell, S.R.; et al. Cardioprotection with palm oil tocotrienols: comparision of different isomers. Am. J. Physiol. Heart Circ. Physiol. 2008, 294, H970–978.

Das, S.; Nesaretnam, K.; Das, D.K. Tocotrienols in cardioprotection. Vitam. Horm. 2007, 76, 419–433.

Das, S.; Powell, S.R.; Wang, P.; Divald, A.; Nesaretnam, K.; Tosaki, A.; Cordis, G.A.; Maulik, N.; Das, D.K. Cardioprotection with palm tocotrienol: antioxidant activity of tocotrienol is linked with its ability to stabilize proteasomes. Am. J. Physiol. Heart Circ. Physiol. 2005, 289, H361–H367.

De Stefani, E.; Boffetta, P.; Ronco, A.L.; Brennan, P.; Deneo-Pellegrini, H.; Carzoglio, J.C. Plant sterols and risk of stomach cancer: A case-control study in Uruguay. Nutr. Cancer 2000, 37, 140–144.

Delia, B.; Rodriguez-Amaya, M.K.; Godoy, H.T.; Amaya-Farfan, J. Updated Brazilian database on food carotenoids: Factors affecting carotenoid composition. J. Food Comp. Anal. 2008, 21, 445–463.

Drotleff, A.M.; Ternes, W. Cis/trans isomers of tocotrienols—occurrence and bioavailability. Eur. Food Res. Technol. 1999, 210, 1–8.

Dutta, P.C.; Normén, L. Capillary column gas-liquid chromatographic separation of Δ5-unsaturated and saturated phytosterols. J. Chromatogr. A, 1998, 816, 177–184.

Edem, D.O. Palm oil: biochemical, physiological, nutritional, hematological, and toxicological aspects: a review. Plant Foods Human Nutr. 2002, 57, 319–341.

Eitsuka, T.; Nakagawa, K.; Miyazawa, T. Down-regulation of telomerase activity in DLD-1 human colorectal adenocarcinoma cells by tocotrienol. Biochem. Biophys. Res. Commun. 2006, 348, 170–175.

EU-Commission Regulation (EC) No. 608/2004 of 31 March 2004 concerning the labelling of foods and food ingredients with added phytosterols, phytosterol ester, phytostanols and/or phytostanol esters,OJ L97 of 1.04; 2004; pp. 0044–0045.

Fanali, S. Capillary electrochromatography of vitamin E. In The Encyclopedia of Vitamin E. Preedy, V.R., Watson, R.R., Eds.; CABI Publishing: United Kingdom, 2007; pp. 183–189.

Federal Register 65 FR 54685-54739, September 8, 2000—Food Labeling: Health Claims; Plant Sterol/Stanol Esters and Coronary Heart Disease; Interim Final Rule.

Fernandes, P.; Cabral, J.M.S. Phytosterols: Applications and recovery methods. Bioresource Technol. 2007, 98, 2335–2350.

Ferruzzi, M.G.; Sander, L.C.; Rock, C.L.; Schwartz, S.J. Carotenoid determination in biological microsamples using liquid chromatography with a coulormetric electrochemical array detector. Anal. Biochem. 1998, 256, 74–81.

Food Standards Australia and New Zealand. Application A410: Phytosterol esters derived from vegetable oils. Canberra. Food Standards Australia and New Zealand, 2000.

Food Standards Australia and New Zealand. Second Review Report. Application A433: Phytosterol esters derived from vegetable oils in breakfast cereals. Application A434: Phytosterol esters derived from vegetable oils in low fat milk and yoghurt. Application A508: Phytosterol esters derived from tall oils as ingredients in low fat milk. Canberra, Food Standards Australia and New Zealand, 2006.

Gapor, A.B.; Top, M.D.; Kato, A.; Ong, A.S.H. Studies on vitamin E and other useful compounds in PFAD and oil palm leaflets. International Oil Palm/Palm Oil Conferences: Technical Progress and Prospects. Ma, A.N., Maycock, J.H., Sieh, L.M.L, Augustine, M.A., Eds.; Palm Oil Research Institute of Malaysia: Bangi, 1987; pp. 124–128.

Gapor, T.; Wan Leong, L.; Ong, A.; Kawada, T.; Watanabe, H.; Tsuchiya, N. Production of high concentration tocopherols and tocotrienols from palm oil by-product. Australia Patent: AU-B-31360 1989, 89, 1–23.

Ghosh, S.; Bhattacharyya, D.K. Isolation of tocopherol and sterol concentrate from sunflower oil deodorizer distillate. J. Am. Oil Chem. Soc. 1996, 73, 1271–1274.

Ghosh, S.P.; Kulkarni, S.; Hieber, K.; Toles, R.; Romanyukha, L.; Kao, T.C.; Hauer-Jensen, M.; Kumar, K.S. Gamma-tocotrienol, a tocol antioxidant as a potent radioprotector. Int. J. Radiat. Biol. 2009, 85, 598–606.

Gibon, V.; Greyt, W.D.; Kellens, M. Palm oil refining. Eur. J. Lipid Sci. Technol. 2007, 109, 315–335.

Goh, S.H.; Choo, Y.M.; Ong, S.H. Minor constituents of palm oil. J. Am. Oil Chem. Soc. 1985, 62, 237–240.

Goh, S.H.; Khor, H.T.; Gee, P.T. Phospholipids of palm oil (*Elaeis guineesis*). J. Am. Oil Chem. Soc. 1982, 59, 296–299.

Goodwin, T.W. Metabolism, nutrition and function of carotenoids. Ann. Rev. Nutr. 1986, 6, 273–297.

Gu, J.Y.; Wakizono, Y.; Sunada, Y.; Hung, P.; Nonaka, M.; Sugano, M.; Yamada, K. Dietary effect of tocopherols and tocotrienols on the immune function of spleen and mesenteric lymph node lymphocytes in Brown Norway rats. Biosci. Biotechnol. Biochem. 1999, 63, 1697–1702.

Gunstone, F.D. Production of vegetable oils—forecasts for 2010/11. Lipid Technol. 2010, 22, 192.

Guthrie, N.; Gapor, A.; Chambers, A.F.; Carroll, K.K. Inhibition of proliferation of estrogen receptor-negative MDA-MB-435 and positive MCF-7 human breast cancer cells by palm oil tocotrienol and tamoxifen, alone and in combination. J. Nutr. 1997, 127, 544S–548S.

Hama, I.; Tanaka, Y.; Yogo, Y.; Okabe, T. Carotene-containing concentration. Japan patent No. 61109764, 1986.

Hamid, H.A.; Choo, Y.M.; Goh, S.H.; Khor, H.T. The ubiquinones of palm oil. In Nutrition, Lipids, Health and Disease. Ong, A.S.H., Niki, E., Packer, L., Eds.; AOCS Press: Champaign, IL, 1995; pp. 122–128.

Hara, N.; Hama, I.; Izumimoto, H.A.; Nakamura. Treatment of natural fats and oil for carotene recovery. Japan patent No. 6305074, 1988.

Harold, E.P.; Purcell, A.E. Carotenoid pigments of peanut oil. J. Am. Oil Chem. Soc. 1967, 44, 328–330.

Hartley, C.W.S. The botany of the oil palm. The Oil Palm. Longman: London, 1977; pp. 37–76

Hendler, S.S.; Rorvik, D. PDR for nutritional supplements (2nd Editon). Thomson Reuters, Physicians' Desk Reference: Montvale, NJ, 2008; pp. 69–76.

Hermizi, H.; Faizah, O.; Ima-Nirwana, S.; Ahmad Nazrun, S.; Norazlina, M. Beneficial effects of tocotrienol and tocopherol on bone histomorphometric parameters in Sprague-Dawley male rats after nicotine cessation. Calcif. Tissue Int. 2009, 84, 65–74.

Hood, R.L. Tocotrienols in metabolism. Phytochemicals—A new paradigm. Bidlack, W.R., Ed.; Technomic Publishing Company: Lancaster, 1998, pp. 33–51.

Huang, Z.R.; Lin, Y.K.; Fang, J.Y. Biological and pharmacological activities of squalene and related compounds: potential uses in cosmetic dermatology. Molecules 2009, 14, 540–554.

Huck, C.W.; Popp, M.; Scherz, H.; Bonn, G.K. Development and evaluation of a new method for the determination of the carotenoid content in selected vegetables by HPLC and HPLCMS-MS. J. Chromatogr. Sci. 2000, 38, 441–449.

Hussein, D.; Mo, H. Delta-tocotrienol-mediated suppression of the proliferation of human PANC-1, MIA PaCa-2, and BxPC-3 pancreatic carcinoma cells. Pancreas. 2009, 38, e124–136.

International Union of Pure Applied Chemistry: Method 2.401. In Standard Methods for the Analysis of Oils, Fats and Derivatives, 7th ed.; Paquot, C., Hautfenne, A., Eds.; Blackwell Scientific Publication: Oxford, 1987.

International Union of Pure Applied Chemistry: Method 2.403. In Standard Methods for the Analysis of Oils, Fats and Derivatives, 7th ed.; Paquot, C., Hautfenne, A., Eds.; Blackwell Scientific Publication: Oxford, 1987.

ISO 6799. Animal and vegetable fats and oils. Determination of composition of the sterol fraction. Method using gas chromatography. International Organisation for Standardization, Geneve, Switzerland, 1991.

ISO 12228: International Organisation for Standardization, Geneve, Switzerland, 1999.

Jacobs, L. Process for the production of tocotrienols. US Patent: US 0142083 A1, 2002.

Jacobsberg, B. Proceedings of First MARDI Workshop on Oil Palm Technology, Kuala Lumpur, 1974.

JECFA (Joint FAO/WHO Expert Committee on Food Additives). Compendium of food additive specification. Volume 1, 52/1, pp. 381–387.

John, S.; Sorokin, A.V.; Thompson, P.D. Phytosterols and vascular disease. Curr. Opin. Lipidol. 2007, 18, 35–40.

Jones, P.J. Cholesterol-lowering action of plant sterols. Curr. Atheroscler. Rep. 1999, 1, 230–235.

Kamal-Eldin, A.; Appelqvist, L. The chemistry and antioxidant properties of topherols and tocotrienols. Lipids 1996, 31, 671–701.

Kamal-Eldin, A.; Görgen, S.; Pettersson, J.; Lampi, A. Normal-phase high-performance liquid chromatography of tocopherols and tocotrienols: comparison of different chromatographic columns. J. Chromat. A, 2000, 881, 217–227.

Kamat, J.P.; Sarma, H.D.; Devasagayam, T.P.A.; Nesaretman, K.; Basiron, Y. Tocotrienols from palm oil as effective inhibitors of protein oxidation and lipid peroxidation in rat liver microsomes. Mol. Cell Biochem. 1997, 170, 131–138.

Kelly, G.S. Squalene and its potential clinical use. Altern. Med. Rev. 1999, 4, 29–36.

Khachik, F.; Beecher, G.R.; Goli, M.B.; Lusby, W.R.; Smith, J.C. Separation and identification of carotenoids and their oxidation products in extracts of human plasma. Anal. Chem. 1992, 64, 2111–2122.

Khachik, F.; Beecher, G.R.; Whittaker, F. Separation, identification and quantification of the major carotenoid and chlorophyll constituents in extracts of several green vegetables by liquid chromatography. J. Agric. Food Chem. 1986, 34, 603–616.

Khanna, S.; Parinandi, N.L.; Kotha, S.R.; Roy, S.; Rink, C.; Bibus, D.; Sen, C.K. Nanomolar vitamin E alpha-tocotrienol inhibits glutamate-induced activation of phospholipase A2 and causes neuroprotection. J. Neurochem. 2010, 112, 1249–1260.

Khatoon, S.; Krishna, A.G.G. Changes in tocopherols, tocotrienols and β–carotene during processing of palm oil extracted from Indian grown palm fruit. J. Lipids Sci. Technol. 2007, 39, 63–70.

Klingberg, S; Ellegård, L; Johansson, I; Hallmans, G; Weinehall, L; Andersson, H. Inverse relation between dietary intake of naturally occurring plant sterols and serum cholesterol in northern Sweden. Am. J. Clin. Nutr. 2008, 87, 993–1001.

Kochar, S.P. Stable and healthful frying oil for the 21st century. INFORM. 2000, 11, 642–647.

Komiyama, K.; Iizuka, K.; Yamaoka, M.; Watanabe, H.; Tsuchiya, N.; Umezawa, I. Studies on the biological activity of tocotrienols. Chem. Pharm. Bull. 1989, 37, 1369–1371.

Krinsky, N.I. The antioxidant and biological properties of the carotenoids. Ann. New York Acad. Sci. 1998, 854, 443–447.

Krinsky, N.I. Carotenoids and oxidative stress. In Oxidative Stress and Aging: Advances in Basic Science, Diagnostics, and Intervention. Cutler, R.G., Rodriguez, H., Eds.; World Scientific Publishing: New York, 2003; pp. 598–611.

Krinsky, N.I.; Johnson, E.J. Carotenoid actions and their relation to health and disease. Mol. Aspects Med. 2005, 26, 459–516.

Kuhad, A.; Chopra, K. Attenuation of diabetic nephropathy by tocotrienol: involvement of NFkB signaling pathway. Life Sci. 2009, 84, 296–301.

Kulkarni, A.S.; Khotpal, R.R.; Bhakare, H.A. Phospholipids and glycolipids in the oil from some varieties of Elaeis guineensis in India. Elaeis 1991, 3, 363–368.

Laakso, P. Analysis of sterols from various food matrices. Eur. J. Lipid Sci. Technol. 2005, 107, 402–410.

Lagarda, M.J.; García-Llatas, G.; Farré, R. Analysis of phytosterols in foods. J. Pharm. Biomed. Anal. 2006, 41, 1486–1496.

Lampi, A.M.; Kamal-Eldin, A.; Piironen, V. Tocopherols and tocotrienols from oil and cereal grains. In Functional Foods: Biochemical and Processing Aspects. Mazza, G., Maguer, M.L., Shi, J., Eds.; CRC Press LLC: Boca Raton, Florida, 2002.

Lau, H.L.N.; Choo, Y.M., Ma, A.N.; Chuah, C.H. Characterization and supercritical carbon dioxide extraction of palm oil (*Elaeis guineensis*). J. Food Lipids 2006, 13, 210–221.

Lau, H.L.N.; Choo, Y.M.; Ma; A.N., Chuah, C.H. Selective extraction of palm carotene and vitamin E from fresh palm-pressed mesocarp fiber (*Elaeis guineensis*) using supercritical CO_2. J. Food Eng. 2008, 84, 289–296.

Lau, H.L.N.; Puah, C.W.; Choo, Y.M., Ma, A.N.; Chuah, C.H. Simultaneous quantification of free fatty acids, free sterols, squalene, and acylglycerol molecular species in palm oil by high-temperature gas chromatography–flame ionization detection. Lipids 2005, 40, 523–528.

Law, M. Plant sterol and stanol margarines and health. Br. Med. J. 2000, 320, 861–864.

Lee, C.M.; Boileau, A.C.; Boileau, T.W.; Williams, A.W.; Swanson, K.S.; Heintz, K.A.; Erdman J.W. Review of animal models in carotenoid research. J. Nutr. 1999, 129, 2271–2277.

Lee, S.P.; Mar, G.Y.; Ng; L.T. Effects of tocotrienol-rich fraction on exercise endurance capacity and oxidative stress in forced swimming rats. Eur. J. Appl. Physiol. 2009, 107, 587–595.

Lee, S.P.; Yang, S.C.; Cheng, Y.S.; Lien, W.J.; Ng, L.T. Hepatoprotection by palm tocotrienol-rich fraction. Eur. J. Lipid. Sci. Technol. 2010, 112, 712–719.

Lietz, G.; Henry, C.A. Modified method to minimize losses of carotenoids and tocopherols during HPLC analysis of red palm oil. Food Chem. 1997, 60,109–117.

Liu, H.K.; Wang, Q.; Li, Y.; Sun, W.G.; Liu, J.R.; Yang, Y.M. Inhibitory effects of gamma-tocotrienol on invasion and metastasis of human gastric adenocarcinoma SGC-7901 cells. J. Nutr. Biochem. 2010, 21, 206–213.

MacLellan, M. Palm oil. J. Am. Oil. Chem. Soc. 1983, 60, 368–378.

Mamuro, H.; Kubota, Y.; Shina, H. Carotene concentrates. Japan patent No. 61282357, 1986.

Mangels, A.R.; Holden, J.M.; Beecher, G.R.; Forman, M.E.; Lanza, E. Carotenoid content of fruits and vegetables: an evaluation of analytic data. J. Am. Diet. Assoc. 1993, 93, 284–296.

Maniam, S.; Mohamed, N.; Shuid A.N.; Soelaiman, I.N. Palm tocotrienol exerted better antioxidant activities in bone than α-tocopherol. Basic Clin. Pharmacol. Toxicol. 2008, 10, 55–60.

Marinangeli, C.P.; Varady, K.A.; Jones, P.J. Plant sterols combined with exercise for the treatment of hypercholesterolemia: overview of independent and synergistic mechanisms of action. J. Nutr. Biochem. 2006, 17, 217–224.

Matsumoto, A.; Takahashi, S.; Nakano, K.; Kijima, S. Identification of new vitamin E in plant oil. J. Jpn. Oil Chem. Soc. 1995, 44, 593–597.

Mayamol, P. N.; Balachandran, C.; Samuel, T.; Sundaresan, A.; Arumughan, C. Process technology for the production of micronutrient rich red palm olein. J. Am. Oil. Chem. Soc. 2007, 84,587–596.

McCann, S.E.; Freudenheim, J.L.; Marshall, J.R.; Graham, S. Risk of human ovarian cancer is related to dietary intake of selected nutrients, phytochemicals and food groups. J. Nutr. 2003, 133, 1937–1942.

McIntyre, B.S.; Briski, K.P.; Gapor, A.; Sylvester, P.W. Antiproliferative and apoptotic effects of tocopherols and tocotrienols on preneoplastic and neoplastic mouse mammary epithelial cells. Proc. Soc. Exp. Biol. Med. 2000, 224, 292–301.

Mendilaharsu, M; Stefani, E.D.; Deneo-Pellegrini, H.; Carzoglio, J.; Ronco, A. Phytosterols and risk of lung cancer: A case-control study in Uruguay. Lung Cancer 1998, 21, 37–45.

Moh, M.H.; Tang, T.S.; Tan, G.H. Simultaneous determination of free fatty acids, partial acylglycerols and tocos in palm oil products using high-performance liquid chromatography. J. Food Lip. 2001, 8, 179–190.

Monde, A.A; Michel, F.; Carbonneau, M.; Tiahou, G.; Vernet, M.; Eymard-Duvernay, S.;Badiou, S.; Adon, B.; Konan, E.; Sess, D.; Cristol, J. P. Comparative study of fatty acid composition, vitamin E and carotenoid contents of palm oils from four varieties of oil palm from Côte d'Ivoire. J. Sci. Food Agric. 2009, 89, 2535–2540.

Moreau, R.A.; Whitaker, B.D.; Hicks, K.B. Phytosterols, phytostanols, and their conjugates in foods: structural diversity, quantitative analysis, and health-promoting uses. Prog. Lipid. Res. 2002, 41, 457–500.

Mortensen, A.; Skibsted L.H. Relative stability of carotenoid radical cation and homologue tocopherolxyl radicals. A real time kinetic study of antioxidant hierarchy. FEBS Lett 1997, 417, 261–266.

Mutalib, M.S.A.; Khazai, H.; Wahle, K.W.J. Palm-tocotrienol rich fraction (TRF) is a more effective inhibitor of LDL oxidation and endothelial cell lipid peroxidation than a-tocopherol in vitro. Food Res. Int. 2003, 36, 405–413.

Nafeeza, M.I.; Fauzee, A.M.; Kamsiah, J.; Gapor, M.T. Comparative effects of α-tocotrienol-rich fraction and tocopherol in aspirin-induced gastric lesions in rats. Asia Pac. J. Clin. Nutr. 2002, 11, 309–313.

Nakashima, K.; Virgona, N.; Miyazawa, M.; Watanabe, T.; Yano, T. The tocotrienol rich fraction from rice bran enhances cisplatin-induced cytotoxicity in human mesothelioma H28 cells. Phytother. Res. 2010, 1317–1321.

Nang, H.L.L.; May, C.Y.; Nang, M.A.; Hock, C.C. Extraction and identification of water soluble compounds in palm pressed fiber by SC-CO_2 and GC-MS. Am. J. Environ. Sci. 2007, 3, 54–59.

Neo, Y.P.; Ariffin, A.; Tan, C.P; Tan, Y.A. Determination of oil palm fruit phenolic compounds and their antioxidant activities using spectrophotometric methods. Int. J. Food Sci. Technol. 2008, 43, 1832–1837.

Nesaretnam, K.; Guthrie, N.; Chambers, A.F.; Carroll, K.K. Effect of tocotrienol on the growth of a human breast cancer cell line in culture. Lipids 1995, 30, 1139–1143.

Nesaretnam, K.; Selvaduray, K.R.; Razak, G.A.; Veerasenan, S.D.; Gomez, P.A. Effectiveness of tocotrienol-rich fraction combined with tamoxifen in the management of women with early breast cancer: a pilot clinical trial. Breast Cancer Res. 2010, 12, R81–R88.

Ng, M.H.; Choo, Y.M. Determination of antioxidants in oil palm leaves (*Elaeis guineensis*). Am. J. Appl. Sci. 2010, 7, 1243–1247.

Ng, M.H.; Choo, Y.M.; Ma, A.N.; Chuah, C.H.; Hashim, M.A. Separation of vitamin E (tocopherol, tocotrienol, and tocomonoenol) in palm oil. Lipids 2004, 39, 1031–1035.

Ng, M.H.; Choo, Y.M.; Ma, A.N.; Chuah, C.H.; Mohd, A.H. Separation of coenzyme Q10 in palm oil by supercritical fluid chromatography. Am. J. Appl. Sci. 2006, 3, 1929–1932.

Ng, M.H.; Choo, Y.M.; Ma, A.N.; Chuah, C.H.; Mohd, A.H. Determination of coenzyme Q9 and Q10 in developing palm fruits. J. Am. Oil Chem. Soc. 2009, 86, 201–205.

Normén, L.; Ellegårda, L.; Brants, H.; Dutta, P.; Andersson, H. A phytosterol database: Fatty foods consumed in Sweden and the Netherlands. J. Food Compos. Anal. 2007, 15, 123–142.

Nourooz-Zadeh, J.; Appleqvist, L. Isolation and quantitative determination of sterol oxides in plant-based foods: soybean oil and wheat flour. J. Am. Oil Chem. Soc. 1992, 69, 288–293.

Ojima, F.; Sakamoto, H.; Ishiguro, Y.; Terao, J. Consumption of carotenoids in photosensitized oxidation of human plasma and plasma low-density lipoprotein. Free Radic. Biol. Med. 1993, 15, 377–384.

Olson, J.A. Biochemistry of vitamin A and carotenoids. In Vitamin A Deficiency: Health, Survival and Vision. Sommer, A., West, K.P., Eds.; Oxford Press: New York,1996; pp. 221–250.

Omenn, G.S.; Goodman, G.E.; Thornquist, M.D.; Balmes, J.; Cullen, M.R.; Glass, A.; Keogh, J.P.; Meyskens, F.L.; Valanis, B.; Williams, J.H.; et al. Effects of a combination of beta-carotene and vitamin A on lung cancer and cardiovascular disease. N. Eng. J. Med. 1996, 334, 1150–1155.

Ong, A.S.H.; Choo, Y.M. Carotenoids and tocols from palm oil. In Natural Antioxidants: Chemistry, Health Effect, and Applications. Shahidi, F., Ed.; AOCS Press: Champaign, IL, 1997; pp. 133–149.

Ong, A.S.H.; Tee, E.S. Natural sources of carotenoids from plants and oils. In Methods in Enzymology. Academic Press: San Diego, CA, 1992; pp. 142–167.

Ooi, C.K.; Choo, Y.M.; Yap, S.C.; Basiron, Y.; Ong. A.S.H. Recovery of carotenoids from palm oil. J. Am. Oil Chem. Soc. 1994, 71, 423–426.

Ooi, T.L.; Leong, W.L. Adsorption of palm oil monoglycerides using rice husk ash. Elaeis 1991, 3, 317–323.

Ooi, T.L.; Ong, A.S.H; Mamuro, Y. Extraction of carotenes from palm oil: molecular distillation method. Japan Oil Chem. Soc. 1986, 35, 543–548.

Paiva, S.A.; Russell, R.M. β-Carotene and other carotenoids as antioxidants. J. Am. Coll. Nutr. 1999, 18, 426–433.

Palozza, P.; Calviello, G.; Emilia L.M.; Serini, S.; Bartoli, G.M. Canthaxanthin supplementation alters antioxidant enzymes and iron concentration in liver of Balb/c mice. J. Nutr. 2000, 130, 1303–1308.

Parker, R.A.; Pearce, B.C.; Clark, R.W.; Gordon, D.A.; Wright, J.J. Tocotrienols regulate cholesterol production in mammalian cells by post-transcriptional suppression of 3-hydroxy-3-methylglutaryl coenzyme A reductase. J. Biol. Chem. 1993, 268, 11230–11238.

Parker, R.S.; Swanson, J.E.; You, C.S.; Edwards, A.J.; Huang, T. Bioavailability of carotenoids inhuman subjects. Proc. Nutr. Soc. 1999, 58, 155–162.

Pauh, C.W.; Choo, Y.M.; Ma, A.N.; Chuah, C.H. The effect of physical refining on palm vitamin E (tocopherol, tocotrienol and tocomonoenol). Am. J. Appl. Sci. 2007, 4, 374–377.

Pauh, C.W.; Choo, Y.M.; Ma, A.N.; Chuah, C.H. Production of carotenoid and tocols concentrate from palm oil using supercritical carbon dioxide. J. Oil Palm Res. 2008, Special Issue, 12–15.

Phillips, K.M.; Ruggio, D.M.; Toivo, J.I.; Swank, M.A.; Simpkins, A.H. Free and esterified sterol composition of edible oils and fats. J. Food Compos. Anal. 2002, 15, 123–142.

Piironen, V.; Lindsay, D.G.; Miettinen, T.A.; Toivo, J.; Lampi, A.M. Plant sterols: biosynthesis, biological function and their importance to human nutrition. J. Sci. Food Agric. 2000, 80, 939–966.

Poli, A.; Marangoni, F. Phytosterols and cardiovascular health. Pharmacol. Res. 2010, 61, 193–199.

Posada, L.R.; Shi, J.; Kakuda, Y.; Xue, S.J. Extraction of tocotrienols from palm fatty acid distillates using molecular distillation. Separ. Purif. Technol. 2007, 57, 220–229.

Quackenbush, F.W.; Smallidge, R.L. Nonaqueous reversed phase liquid chromatographic system for separation and quantitation of provitamins. J. Assoc. Off. Anal. Chem. 1986, 69, 767–772.

Quilez, J.; Garcia-Lorda, P.; Salas-Salvado, J. Potential uses and benefits of phytosterols in diet: present situation and future directions. Clin. Nutr. 2003, 22, 343–351.

Qureshi, A.A.; Burger, W.C.; Peterson, D.M.; Elson, C.E. The structure of an inhibitor of cholesterol biosynthesis isolated from barley. J. Biol. Chem. 1986, 261, 10544–10550.

Qureshi, A.A.; Qureshi, N.; Wright, J.J.; Shen, Z.; Kramer, G.; Gapor, A.; Chong, Y.H.; DeWitt, G.; Ong, A.; Peterson, D.M. Lowering of serum cholesterol in hypercholesterolemic humans by tocotrienols (palm vitee). Am. J. Clin. Nutr. 1991, 53, 1021S–1026S.

Rao, C.V.; Newmark, H.L.; Reddy, B.S. Chemopreventive effect of squalene on colon cancer. Carcinogenesis 1998, 19, 287–290.

Reshma, M.V.; Saritha, S.S.; Balachandran, C.; Arumughan, C. Lipase catalyzed interesterification of palm stearin and rice bran oil blends for preparation of zero trans shortening with bioactive phytochemicals. Bioresour. Technol. 2008, 99, 5011 –5019.

Rice-Evans, C.; Sampson, J.; Bramley, P.M.; Holloway, D.E. Why do we expect carotenoids to be antioxidants in vivo. Free Radic. Res. 1997, 26, 381–398.

Rock, C.L. Carotenoids: Biology and treatment. Phannncol. Ther. 1997, 75, 185–197.

Rodríguez-Bernaldo de Quirós, A.; Costa, H.S. Analysis of carotenoids in vegetable and plasma samples: A review. J. Food Compos. Anal. 2006, 19, 97–111.

Rohr, R.; Trujillo-Quijano, J.A. Process for separating unsaponifiable valuable products from raw materials. 2005, US6846941.

Rossi, M.; Gianazza, M.; Alamprese, C.; Stanga, F. The effect of bleaching and physical refining on color and minor components of palm oil. J. Am. Oil Chem. Soc. 2001, 78, 1051–1055.

Ryan, E.; Galvin, K.; O'Connor, T.P.; Maguire, A. R.; O'Brien, N. M. Phytosterol, squalene, tocopherol content and fatty acid profile of selected seeds, grains, and legumes. Plant Foods Hum. Nutr. 2007, 62, 85–91.

Sakai, M.; Okabe, M.; Yamasaki, M.; Tachibana, H.; Yamada, K. Induction of apoptosis by tocotrienol in rat hepatoma dRLh-84 cells. Anticancer Res. 2004, 24, 1683–1688.

Sambanthamurthi, R.; Sundram, K.; Tan, Y.A. Chemistry and biochemistry of palm oil. Progr. Lipid Res. 2000, 39, 507–558.

Sambanthamurthi, R.; Tan, Y.A.; Manickam, K.S. Treatment of vegetation liquors derived from oil-bearing fruits. US 7387802 B2, 2008.

Sambanthamurthi, R.; Tan, Y.A.; Manickam, K.S.; Basri Wahid, M. Botanical extracts from oil-palm vegetation liquor for cosmeceutical application. US 2010/0209544 A1, 2010.

Sampathkumar, S. Process for recovering tocopherols from deodorizer sludge. US Patent: 4594437, 1986.

Scita, G. The stability of β-carotene under different laboratory conditions. J. Nutr. Biochem. 1992, 3, 124–128.

Sen, C.K.; Khanna, S.; Roy S. Tocotrienol: the natural vitamin E to defend the nervous system? Ann. N.Y. Acad. Sci. 2004, 1031, 127–142.

Serbinova, E.; Kagan, V.; Han, D.; Packer, L. Free radical recycling and intramembrane mobility in the antioxidant properties of α-tocopherol and α-tocotrienol. Free Radic. Biol. Med. 1991, 10, 263–275.

Sheppard, A.J.; Pennington, J.A.T.; Weihrauch, J.L. Analysis and distribution of vitamin E in vegetable oils and foods. In Vitamin E in Health and Disease. Packer, L., Fuchs, J., Eds.; Marcel Dekker: New York, 1993; pp. 9–31.

Shibata, A.; Nakagawa, K.; Sookwong, P.; Tsuduki, T., Oikawa, S.; Miyazawa, T. delta-Tocotrienol suppresses VEGF induced angiogenesis whereas alpha-tocopherol does not. J. Agric. Food Chem. 2009, 57, 8696–8704.

Shinde, S.; Patil, N.; Tendolkar, A.; Coenzyme Q10: A review of essential functions. Internet J. Nutr. Wellness 2005, 1(2).

Shirode, A.B.; Sylvester, P.W. Synergistic anticancer effects of combined gamma-tocotrienol and celecoxib treatment are associated with suppression in Akt and NF kappa B signaling. Biomed. Pharmacother. 2010, 64, 327–332.

Siew, W.L. Palm oil sterols. Palm Oil Develop. 1990, 12, 18–24.

Siew, W.L.; Berger. K.G. Malaysian Palm Kernel Oil, Chemical and Physical Characteristics, PORIM Technology Series No. 6,1981, PORIM, Kuala Lumpur, Malaysia.

Siew, W.L.; Ng, W.L. In Proceedings of the National Seminar on Palm Oil Milling, Refining Technology and Quality, 7–8 August 1997, Kota Kinabalu, Sabah, Malaysia, 1997, 124–140.

Simonne, A.H.; Eitenmiller, R.R. Retention of vitamin E and added retinyl palmitate in selected vegetable oils during deep-fat frying and in fried breaded products. J. Agric. Food Chem. 1998, 46, 5273–527.

Slover, H.T.; Thompson, R.H.; Merola, G.V. Determination of tocopherols and sterols by capillary gas chromatography. J. Am. Oil Chem. Soc. 1983, 60, 1524–1528.

Sookwong, P.; Nakagawa, K.; Yamaguchi, Y.; Shunjikato, T.; Kimura, F.; Miyazawa, T. Tocotrienol distribution in foods: estimation of daily tocotrienol intake of Japanese population. J. Agric. Food Chem. 2010, 58, 3350–3355.

Srivastava, J.K.; Gupta, S. Tocotrienol-rich fraction of palm oil induces cell cycle arrest and apoptosis selectively in human prostate cancer cells. Biochem. Biophys. Res. Commun. 2006, 346, 447–453.

Stahl, W.; Sies, H. Antioxidant effects of carotenoids: implication in photoprotection in humans. In Handbook of Antioxidants. Cadenas, E., Packer, L., Eds.; Marcel Dekker: New York, 2002; pp. 223–233.

Strohschein, S.; Rentel, C.; Lacker, T.; Bayer, E.; Albert K. Separation and identification of tocotrienol isomers by HPLC-MS and HPLC-NMR coupling. Anal. Chem. 1999, 71, 1780–1785.

Subagio, A.; Morita, N. Instability of carotenoids is a reason for their promotion on lipid oxidation. Food Res. Int. 2001, 34, 183–188.

Sundram, K.; Gapor, A. Vitamin E from palm oil: its extraction and nutritional properties. Lipid Technol. 1996, 4, 137–141.

Sundram, K.; Khor, H.T.; Ong, A.S.; Pathmanathan, R. Effect of dietary palm oils on mammary carcinogenesis in female rats induced by 7,12-dimethylbenz(a)anthracene. Cancer Res. 1989, 49, 1447–1451.

Sundram, T.; Thiagarajan, T.; Gapor, A.; Basiron Y. Palm tocotrienols: new antioxidants for the new millennium. INFORM, 2002, 13, 634–640.

Suzuki, Y.J.; Tsuchiya, M.; Wassall, S.R.; Choo, Y.M.; Govil, G.; Kagan, V.E.; Packer, L. Structural and dynamic membrane properties of α-tocopherol and α-tocotrienol: Implication to the molecular mechanism of their antioxidant potency. Biochem. 1993, 32, 10692–10699.

Sylvester, P.W.; Russell, M.; Ip, M.M.; Ip, C. Comparative effects of different animal and vegetable fats fed before and during carcinogen administration on tumorigenesis, sexual maturation and endocrine function in rats. Cancer Res. 1986, 46, 757–762.

Szulczewska-Remi, A.; Nogala Kałucka, M.; Kwiatkowski, J.; Lampart-Szczapa, E.; Rudzińska, M. Preparation of tocochromanol concentrate from red palm oil. J. Food Lipids 2005, 12, 112–123.

Tan, B. Palm carotenoids, tocopherols and tocotrienols. J. Am. Oil Chem. Soc. 1989, 66, 770–776.

Tan, Y.A.; Manickam, K.S., Sambanthamurthi, R. Water-soluble phenolics from the palm oil industry. In Biologically-active Phytochemicals in Food—Analysis, Metabolism, Bioavailability and Function. Pfannhauser, W., Fenwick, G.R., Khokhar, S., Eds.; The Royal Society of Chemistry, Cambridge, 2001; pp. 548–551.

Tan, Y.A.; Sambanthamurthi, R.; Manickam, K.S.; Basri Wahid, M. Palm phenolics and flavonoids as potent biological and chemical antioxidants for applications in foods and edible oils. US 2010/0183716 A1, 2010.

Tee, E.S.; Lim, C.L. The analysis of carotenoids and retinoids: A review. Food Chem. 1991, 41, 147–193.

Theriault, A.; Chao, J.T.; Gapor, A. Tocotrienol is the most effective vitamin E for reducing endothelial expression of adhesion molecules and adhesion to monocytes. Atherosclerosis 2002, 160, 21–30.

Tian, L.L.; White, P.J. Antioxidant activity of oat extract in soybean and cottonseed oils. J. Am. Oil Chem. Soc. 1994, 71, 1087–1094.

Toivo, J.; Piironen, V.; Kalo, P.; Varo, P. Gas chromatographic determination of major sterols in edible oils and fats using solid-phase extraction in sample preparation. Chromatographia 1998, 48, 745–750.

Top, A.G.; Leong, W.L.; Ong, A.S.H.; Kawada, T.; Watanabe, H.; Tsuchiya, N. Production of high concentration tocopherols and tocotrienols from palm oil by-products. US patent: US005190618A, 1993.

Verleyen, T.; Forcades, M.; Verhe, R.; Dewettinck, K.; Huyghebaert, A.; Greyt, W.D. Analysis of free and esterified sterols in vegetable oils. J. Am. Oil Chem. Soc. 2002a, 79, 117–122.

Verleyen, T.; Sosinska, U.; Ioannidou, S.; Verhe, R.; Dewettinck, K.; Huyghebaert, A.; De Greyt, W. Influence of the vegetable oil refining process on free and esterified sterols. J. Am. Oil Chem. Soc. 2002b, 79, 947–953.

Wald, G. The molecular basis of visual excitation. Nature 1968, 219, 800–807.

Wan Nazaimoon, W.M.; Khalid, B.A. Tocotrienols-rich diet decreases advanced glycosylation end-products in non-diabetic rats and improves glycemic control in streptozotocin-induced diabetic rats. Malays. J. Pathol. 2002, 24, 77–82.

Wang, S.H.; Furuno T.; Cheng, Z. Study of extraction of phytosterol from mass on pine raw tall oil. J. Wood Sci. 2002, 48, 505–511.

Watkins, T.; Lenz, P.; Gapor, A.; Struck, M.; Tomeo, A.; Bierenbaum, M. gamma-Tocotrienol as a hypocholesterolemic and antioxidant agent in rats fed atherogenic diets. Lipids 1993, 28, 1113–1118.

Wattanapenpaiboon, N.; Wahlqvist, M.L. Phytonutrient deficiency: the place of palm fruit. Asia Pacific J. Clin. Nutr. 2003, 12, 363–368.

Weng-Yew, W.; Selvaduray, K.R.; Ming, C.H.; Nesaretnam, K. Suppression of tumor growth by palm tocotrienols via the attenuation of angiogenesis. Nutr. Cancer 2009, 61, 367–373.

White, P.J.; Armstrong, L.S. Effect of selected oat sterols in the deterioration of heated soybean oil. J. Am. Oil Chem. Soc. 1986, 63, 525–529.

Woyengo, T.A.; Ramprasath., V.R., Jones, P.J.H. Anticancer effects of phytosterols. Eur. J. Clin. Nutr. 2009, 63, 813–819.

Wu, S.J., Liu, P.L.; Ng, L.T. Tocotrienol-rich fraction of palm oil exhibits anti-inflammatory property by suppressing the expression of inflammatory mediators in human monocytic cells. Mol. Nutr. Food Res. 2008, 52, 921–929.

Wu, S.J.; Ng, L.T. Antioxidant and antihepatoma activities of palm oil extract. J. Food Lipids 2007, 14, 122–137.

Xu, Z. Analysis of tocopherols and tocotrienols. Curr. Prot. Food Anal. Chem. 2002, D1.5.1–D1.5.12.

Yamada, Y.;Obayashi, M.; Ishikawa, T.; Kiso, Y.; Ono, Y.; Yamashita, K. Dietary tocotrienol reduces UVB-induced skin damage and sesamin enhances tocotrienol effects in hairless mice. J. Nutr. Sci. Vitaminol. (Tokyo). 2008, 54,117–123.

Yang, Z.; Xiao, H.; Jin, H.; Koo, P.T.; Tsang, D.J.; Yang, C.S. Synergistic actions of atorvastatin with gamma-tocotrienol and celecoxib against human colon cancer HT29 and HCT116 cells. Int. J. Cancer 2010, 126, 852–863.

Yap, S.C.; Choo, Y.M.; Ooi, C.K.; Ong, A.S.H.; Goh, S.H. Quantitative analysis of carotenes in the oil from different palm species. Elaeis 1991, 3, 309–378.

Yi, J.; Andersen, M.L.; Skibste, L.H. Interactions between tocopherols, tocotrienols and carotenoids during autoxidation of mixed palm olein and fish oil. Food Chem. 2011, 127, 1792–1797.

You, L.L.; Bahar, B.S.; Quek, S.Y.; Abdullah, M.A. Recovery of palm carotene from palm oil and hydrolyzed palm oil by adsorption column chromatography. J. Food Lipids 2002, 9, 87–93.

You, L.L.; Baharin, B.S.; Che Man, Y.B.; Takagi, S. Effect of degumming process on chromatographic separation of carotenes from crude and degummed palm oil. J. Food Lipids 2001, 8, 27–35.

Yu, S.G.; Thomas, A.M.; Gapor, A.; Tan, B.; Qureshi, N.; Qureshi, A.A. Dose-response impact of various tocotrienols on serum lipid parameters in 5-week-old female chickens. Lipids 2006, 41, 453–461.

Zachman, R.D. Vitamin A. Nutrition during infancy. Hanley & Belfus: Philadelphia, PA; C.V. Mosby: St. Louis, MO, 1988; pp. 253–263.

17

Enzymatic and Chemical Modification of Palm Oil, Palm Kernel Oil, and Its Fractions

Oi-Ming Lai[1], Seong-Koon Lo[2] and Casimir C. Akoh[3]

[1]Institute of Bioscience, Universiti Putra Malaysia, 43400 UPM Serdang, Selangor, Malaysia and Dept. of Bioprocess Technology, Faculty of Biotechnology & Biomolecular Sciences, Universiti Putra Malaysia, 43400 UPM Serdang, Selangor, Malaysia; [2]Sime Darby Research Sdn. Bhd., R&D Centre-Downstream, Lot 2664, Jalan Pulau Carey, 42960 Carey Island, Selangor, Malaysia; and [3]Department of Food Science and Technology, University of Georgia, Athens, GA, USA.

Introduction

Scientists have played around with the structures of palm oil and its fractions for many years. With increased understanding about the effects of certain fatty acids, their chain length, degree of unsaturation and positional distribution on the metabolism of triacylglycerols (TAG), scientists today are able to construct palm-based products which are tailor-made to meet specific nutritional or health requirements. Even if the types or location of fatty acids found in the original palm oil may not be the desired one, catalysts such as enzymes and chemicals can be used to modify it to suit the consumers' needs. These modified or newly synthesized lipids are known as structured lipids. For the palm oil industry, these structured lipids may be the answer to solving the love/hate relationship consumers have towards it.

Structured Lipids

There are many terms used to describe structured lipids (SL) such as structured TAG/triglycerides, tailor-made fats, designer lipids, interesterified fats, modified fats, restructured fats, novel TAG, and so on. Whatever the terminology that is used, SL is basically any lipids, which includes TAGs as well as acylglycerols such as monoacylglycerols (MAG), diacylgylycerols (DAG), and phospholipids, that have been chemically or enzymatically modified to include new fatty acids or restructured to have different fatty acid profiles at different positions on the glycerol molecule than that of the native lipid.

Production of Structured Lipids

Simple Blending

SL can be obtained from the simple blending of specific amounts of medium-chain triacylglcerols (MCT) and long-chain triacylglycerols (LCT). In such cases, there is no rearrangement of the FA within the same glycerol molecule. In the unstructured form, the mixture will maintain the original absorption rates of the individual TAG; for example, the MCT will be metabolized at a higher rate than the LCT. Intestinal absorption of fat is affected by the TAG structure, and the FA in the *sn*-2 position are preferentially absorbed, thus lipid structuring which is attainable only through the use of lipases is the answer in designing a potent SL targeted for a specific disease condition or food use. However, blending is an economical and easy way of obtaining a physical mixture of MCT and LCT.

Enzymatic Synthesis

Lipases (triacylglycerol acylhydrolases, EC 3.1.1.3), derived from bacterial, fungal or yeast sources can catalyze reversal of hydrolysis in the presence or absence of organic solvents and near anhydrous conditions (Macrae, 1985). Unlike esterases, which act on water-soluble interfaces, lipases can act on the interfaces between immiscible aqueous and lipid phases (oil-water interface). Lipase interesterification utilizes water during the hydrolysis of TAG to produce DAG, MAG, and free fatty acids (FFA). The products of hydrolysis will continue to increase during the interesterification process until equilibrium is achieved. Interesterification and hydrolysis by lipases follows the Ping Pong Bi Bi reaction for multisubstrate reactions (Hermansyah et al., 2010), with glycerides, fatty acids, and water being the main substrates.

Lipases can function efficiently in organic conditions as they may exhibit enhanced thermal stability, increased rigidity, and modified chemo- and stereoselectivity in organic solvents (Eltaweel et al., 2005; Kanjanavas et al., 2010; Rodriguez et al., 2009) When placed in an organic environment, the nonpolar lipid substrates showed increased solubility. Lipases can be used in several ways depending on the TAG modification strategy. Under water-limiting conditions, lipase-catalyzed reactions undergo a shift in the thermodynamic equilibria to the right in favor of ester synthesis over hydrolysis, which is the dominant reaction in the presence of excess water. Products of the reaction can also be easily recovered when a solvent with a low boiling point is used.

Lipases from various sources display positional, fatty acid, and stereospecificity, and this is the main difference between enzymatic and chemical interesterification reactions. The positional specificity of lipases on *sn*-1 and *sn*-3 positions of the TAG is due to the steric hindrance conflict that prevents the FA located at *sn*-2 from binding to the active site (Kovac et al., 1996). It was shown that reactions with such 1,3-specific lipases produce a mixture of products which include mixed TAG, 1,2- and 2,3-DAG, 2-MAG and FFA, and products of acyl migration. Temperature and extended reaction time, the two thermodynamic parameters that influence acyl migration, will cause the migration of FA from *sn*-2 into *sn*-1,3 positions (Li et al., 2010; Xu, 2000a). Some examples of 1,3-specific lipases include those from *Rhizomucor miehei, Rhizopus delemar, R. niveus, R. arrhizus, R. oryzae. R. japonicus, Humicola lanuginosa, Penicillium roquefortii, P. camembertii, Aspergillus niger* and *A. flavus* (Xu, 2000a, 2000b). Certain lipases do not possess positional or FA selectivity. Such lipases are known as non-specific lipases and produce interesterified products that are similar to those obtained through chemical interesterification (Ahmadi et al., 2008). Examples of such lipases include those derived from *Corynebacterium acnes, Staphylococcus aureus,* and *Candida* (Gunstone, 1994). Besides non-specific lipases, there are some lipases such as *C. antarctica* A lipase (Maria et al., 2005) which show specificity towards the FA at the *sn*-2 positions. Lipases that are able to differentiate between the two FA at the *sn*-1 and *sn*-3 positions are said to be stereospecific. Depending on the acyl groups, the lipases will hydrolyze the FA at different rates (Jaeger et al., 1992). Lipases that show selectivity towards certain FA substrates are known as FA/substrate specific lipases. The most common example would be the lipase from *Geotrichum candidum,* which is known to be selective towards long-chain fatty acids (LCFA) with a *cis*-9 double bond (Loo et al., 2007; Stransky et al., 2007).

For many years, the high cost of enzymes have limited the industrial application of lipases. With progress in genetic engineering, recombinant or mutant lipases with higher activity and improved properties are now available at lower costs, and it is hoped that this will make the industrial enzymatic production of SL a viable reality.

Chemical Synthesis

Before the discovery of lipases for lipid modification, chemical catalysts have been used. Chemical synthesis between butyric acid and glycerol has been reported since 1844 by Pelouze and Gelis. The most common type of chemical modification of oils and fats is chemical interesterification. Sodium alkoxides, specifically sodium methoxide, are the most commonly used chemical catalyst used chemical interesterification; however, other types of bases, acids, and metals are also available (Coenen, 1974).

There are two proposed mechanisms for chemical interesterification reactions, namely (1) the carbonyl addition mechanism and (2) the enolate intermediate mechanism or Claisen condensation. In principle, these two mechanisms are actually the same. In interesterification using sodium methoxide, the catalyst being a strong nucleophilic reagent attacks the carbonyl carbon of the fatty acid-glycerol ester bond and forms a tetrahedral intermediate. The resultant fatty acid methyl ester is then released, leaving behind a glycerylate anion. Reaction kinetics of a base-catalyzed hydrolysis of fatty acid esters show that the reaction is dependent on both ester and base concentration (second-order kinetics). The newly formed glycerylate anion acts as a nucleophile for the subsequent nucleophilic attack on carbonyl carbons. This constitutes the carbonyl addition mechanism. Both intra- and interesterification are possible with this mechanism. The increase in entropy (i.e., the randomization of fatty acids in all possible combinations within the TAG molecule) will act as the driving force behind the chemical interesterification reaction. In the Claisen condensation mechanism, the sodium methoxide removes a proton from the alpha carbon, thereby forming an enolate structure, which then forms a powerful nucleophile called a carbanion. The carbanion then attacks carbonyl groups, forming a β-keto ester and a glycerylate. The glycerylate is free to attack other carbonyl carbons and exchange esters within a TAG molecule and with other TAG molecules.

Apart from interesterification, acidolysis reaction between a fatty acid and a TAG can occur under acidic conditions. The acid catalyst, such as sulfuric acid, catalyses the hydrolysis of fatty acids from the TAG molecule, forming DAG and MAG. Fatty acids can then react with the exposed hydroxyl group of the DAG and MAG to form new TAG molecules. Although chemical acidolysis has the advantage of being a single-step process, the reversibility of the reaction limits the yield compared to that achieved when the reaction attains equilibrium. However, the yield could theoretically be pushed higher by having an excessive amount of fatty acids in the reaction, with the disadvantage of not being practical and economical. Chemical acidolysis using a basic catalyst is not possible due to the immediate conversion of fatty acids to its corresponding salts.

In a base-catalyzed alcoholysis reaction, the basic salt of the particular alcohol would be used as reactant. Acid-catalyzed alcoholysis can also take place readily. However, chemically catalyzed alcoholysis reactions are not usually applied in industrial modification of edible oils and fats.

Applications of Enzymatically-Synthesized Palm-Based Structured Lipids

Margarines and Spreads

Since its first invention, margarine, a water-in-oil emulsion, has undergone rapid development and innovation. The product range now includes table margarines, baker's margarines, specialized puff pastry margarines, and various low-calorie spreads which contain much higher levels of water and lower levels of fat than those legally required in margarines. Currently, healthful margarines such as zero-*trans* margarines, margarines with omega-3 fatty acids, and margarines for reducing blood cholesterol are found on supermarket shelves. The first margarines were made from animal fat, but today, in response to the demand of health-conscious consumers, most margarines are made using vegetable oils. According to the Codex Alimentarius (1992), the main ingredients which will affect the production and properties of margarines are fats and oils (minimum 80%), water (maximum 16%), emulsifier (generally 0.1–1.0%) and salt (maximum 4%). Other ingredients which can also be added include preservatives, antioxidants, colorings, and vitamins.

Table margarines are divided into tub and packet forms. Tub margarine should have lower solids at low temperatures as the product should be spreadable when taken direct from the refrigerator while packet table margarines, which resemble butter in consistency, have higher solids at low temperatures. Packet margarines are spreadable at ambient temperatures. All palm oil fractions, namely, palm olein, palm stearin, and palm kernel olein, have been used in table margarine formulations (Nor Aini & Miskandar, 2005). Palm oil-based margarines have a tendency to crystallize in the β′ form, which is the desirable polymorphic form for margarines and impart good plasticity over a wide working temperature range. Enzymatically interesterified *trans*-free palm-based fat blends suitable for use as table margarines have been reported (Adhikari et al., 2010; Hossein Goli et al., 2009; Kim et al., 2008; Lai et al., 1998a, 1998b, 1999; Mohd. Alim et al., 2009; Nor Aini & Miskandar, 2005; Siew et al., 2007).

Industrial/baker's margarines tend to have higher solid fat content (SFC) than table margarines. For optimum baking performance, it should contain a minimum of 10% SFC at 20°C to prevent oil exudation or oiling off. It should also contain a minimum of 8% SFC at working temperature (25°C). Pastry margarines contain the highest solids content among the three types of margarines in order to provide flakiness to the end product. Palm stearin, the solid fraction of palm oil, is a fully natural hard component and is commonly used in these formulations (Adhikari et al., 2010; Hossein Goli et al., 2009; Kim et al., 2008; Mohd. Alim et al., 2009; Zhang et al., 2001, 2004). Different ratios of palm stearin, palm oil, and palm kernel oil blends have been used to produce pastry margarines that are similar to commercial pastry margarines (Noor Lida et al., 1996), while interesterification of canola oil blend with palm stearin at 50:50 and 60:40 was found to be suitable for margarine making (Siew et al., 2007).

Low-calorie spreads are usually formulated using liquid oils to avoid unstable spread when hard fats are used. As such, the more liquid fraction of palm oil, palm olein, is used. Interesterified palm and palm kernel oils which have steeper melting profile and improved mouthfeel have also been used in the formulations.

Shortenings

Shortening is a 100% fat product made from animal fats and/or vegetable oil. It was first made using lard. In 1903, Joseph Crossfield and Sons patented a liquid-phase hydrogenation of oil production of shortening. Since then, hydrogenation has been used by most producers of shortening. Hydrogenated fats consist in part of *trans* fatty acids which may lead to coronary heart disease. Thus, there has been great interest in producing healthier shortenings that will simulate hydrogenated fats products but contain zero *trans* fatty acids. Processes such as interesterification and physical blending have managed to fulfill the industry's ability to meet both the specific functional and nutritional needs above (Berger & Nor Aini, 2005; Jeyarani & Reddy, 2003; Lee et al., 2008; Reshma et al., 2008).

Generally, shortenings which function to "shorten" or tenderize baked products can be grouped into 2 main groups, namely the pourable form which includes both the liquid and fluid shortenings and the dry forms which includes both solid and powdered shortenings. Regular cooking oils and salad oils which are not normally emulsified fall under the liquid shortening category while fluid shortening are opaque due to the presence of suspended solids which may be hard fat or emulsifiers. Solid shortenings generally have consistent and smooth texture while in powdered shortenings, the fats are encapsulated in water-soluble materials.

In the United States, shortenings tend to made from hydrogenated soybean oil, beef fat, and cottonseed oil while in Europe (UK, Germany, and Netherlands), they are mostly made of palm oil, hydrogenated rapeseed oil, and hydrogenated fish oils. Palm oil is β′ tending and has higher melting points than most other vegetable oils. As such, it is suitable to be used for making shortening which usually has melting points between 34 and 44°C (Jin et al., 2008; Nor Aini & Miskandar, 2005). In 2001, Chu et al. (2001) reported on experimental shortenings made from lipase-catalyzed palm stearin/palm kernel olein blend at 1:1 (w/w ratio) that had physical properties similar to domestic commercial shortening samples. The lipase transesterified solid frying shortening was also found to be more β′ tending than domestic Malaysian commercial shortenings made from non-hydrogenated hardstock such as palm oil, palm olein, and palm stearin. Imported frying shortening compared in the study were formulated from soybean oil and cottonseed oil. These shortenings also had high levels of β′ crystals and had flatter SFC curves than the experimental and domestic samples. The domestic samples were found to have better workability or plasticity at higher temperatures than the imported ones, probably because they were formulated for a tropical climate like Malaysia.

Vanaspati

In countries like India and Pakistan, a form of vegetable ghee known as vanaspati is often used as replacement for butterfat or ghee in cooking. The texture of the vanaspati used are grainy, firm, dry, and crumbly. Single or mixed blends of hydrogenated vegetable oils with other animal fats are sometimes used as a cheap alternatives to vegetable fat in vanaspati formulations. In India, many of the vanaspati formulations contain about 5 to 20% palm oil while in Pakistan, levels as high as 50% have been reported (Kheiri & Oh, 1983). As hydrogenation produces the undesirable *trans* FA, interesterification process has been used to produce vegetable fat with granular texture (Farmani et al., 2006). A *trans*-free vanaspati formulation was produced by blending interesterified palm stearin with other liquid oils such as soybean, rapeseed, and sunflower oil (Berger & Nor Aini, 2005; Jeyarani & Reddy, 2003; Nor Aini et al.,1999; Siew et al., 2007). Interesterified palm stearin was found to have lower SMP and SFC at all temperatures (Nor Aini et al., 1997a, 1997b).

Cooking/Frying Oils

Among all the palm oil fractions, it is only palm olein that is suitable to be used as cooking oil. Palm olein has a maximum cloud point of 10°C, which makes it suitable to be used as cooking oil in Malaysia and hot tropical countries. In temperate countries, palm olein will crystallize and become cloudy upon storage at low temperature. Cloud formation tends to influence a consumer's perception that the oil is of low quality. Palm olein can also be blended with other types of soft or liquid oils (Abdulkarim et al., 2010; Debnath et al., 2011; Naghshineh et al., 2010; Tabee et al., 2009). It is a very stable oil used for cooking and sometimes frying as it has the longest induction time of 40–44h at 100°C compared to other soft oils, indicating great stability against oxidation.

In order for palm olein to be marketed as cooking oil in temperate countries, it has to be double fractionated to superolein with a minimum IV of 60 and low cloud point (maximum of 6°C). This will ensure that the oil remains clear even when the temperature drops. The presence of DAG in palm oil and its fractions was found to have several disadvantages (Kamariah et al., 2005). DAG was found to have deleterious effect on the quality of palm oil when used in situations when palm oil has to crystallize. DAG together with TAG will form a eutectic mixture, making separation of these components difficult, and thus fractionation becomes incomplete. In their work, Kamariah et al. (2005) reported that better and improved iodine values and cloud point superolein can be obtained when palm olein with low DAG content was used during dry fractionation.

Cocoa Butter

Cocoa butter is the most expensive ingredient used in the production of chocolates. Cocoa butter is composed of more than 70% symmetrical TAG with oleic acid in the *sn*-2 position, namely POP, POS, and SOS. There are no other natural fat that resembles cocoa butter's physical properties. It melts between 32–35°C, is able to recrystallize during processing to form stable crystals, it is brittle at room temperature, and melts completely in the mouth. The uncertainty in the supply of cocoa butter and the volatile cocoa butter prices have forced confection producers to seek alternative sources. These alternative sources are made by modifying or blending of fats (Abigor et al., 2003; Ciftci et al., 2009; Khumalo et al., 2002; Pinyapong & Phutrakul, 2009; Wang et al., 2006). There are three types of cocoa butter alternatives, namely cocoa butter equivalents (CBE), which are the closest to cocoa butter in terms of physical and chemical properties and are totally compatible with cocoa butter; cocoa butter replacers (CBR), which are non lauric fats with similar FA distribution as cocoa butter but with different TAG structure and are compatible to cocoa butter only in small quantities; and, lastly, cocoa butter substitutes (CBS), which are lauric fats that are chemically different from cocoa butter but has some physical similarities.

Currently, palm oil, palm mid fraction, illipe (*Shorea stenoptera*) fat, sal (*Shorea robusta*) fat, shea (*Butyrospernum parkii*) butter, mango (*Mangifera indica*), piqui (*Caryocar brasiliennse Camb*) oil, and kokum (*Garcinia indica*) are the most common plant fat sources used to produce CBS. The most significant interesterification process using 1,3-positionally specific lipases is the production of CBS, a process commercialized by Fuji Oil Co. Japan. The 1,3-positional specificity of lipase was exploited to incorporate either palmitic or stearic acid into the natural fats mentioned above while maintaining the *sn*-2 positions with oleic acid, to produce FA composition that is similar to cocoa butter.

Non-Dairy Ice Cream

Ice cream is a rigid foam which has been partially beaten and frozen. It is, in fact, a well-ordered oil-in-water emulsion that is made up of fat, sugar, water, flavoring, and other components such as milk or cream which makes the cost of dairy ice cream expensive. Due to the rising cost in milkfat, vegetable fats have replaced milkfat in many European countries. Ice cream made of different vegetable fats were found to produce softer ice cream that melts easier than those made of milk fat or milk protein (Nazarudin et al., 2008; Sutar et al., 2009). The SFC at 0°C for a good vegetable fat used for ice cream should be high with melting points of up to 25–35°C. Laurical types of fats such as coconut oil and palm kernel oil are suitable because of their steep melting points (Choo et al., 2010; Sung & Goff, 2010). However, due to cost considerations, palm oil or hydrogenated palm oil and mixtures of them are more commonly used. For an even lower cost ice cream, palm stearin are sometimes used together with the palm oil mixtures. A study by Lai et al. (1998c) has shown that transesterification of palm stearin and anhydrous milkfat (AMF) at 40:60 (w/w) ratio using different microbial lipases was able to improve the properties of palm stearin, making it suitable for products such as ice cream. Liew et al. (2001) have used mycelium-bound lipase to transesterify PKO and AMF mixtures at 70:30 (w/w) ratio to produce a fat suitable for use in ice cream formulation.

Infant Formula

Human milk consists of 20–30% palmitic acid, with 33% located at the *sn*-2 position of the TAG (Szabo et al., 2010). The location of the palmitic acid, which is a crucial source of nutrient for infants, at the *sn*-2 position is important as there are reports of increased palmitic acid absorption in the 2-monoacylglycerol form rather than in free acid form (Mu & Hoy, 2004). However, most vegetable oils tend to have unsaturated fatty acids located at the *sn*-2 positions. Because of this, low-melting palm olein, which has 10–15% of its palmitic acid in the *sn*-2 position of the glycerol, is suitable to be used in production of infant food formulations when blended or enzymatically interesterified with other vegetable oils that rich with linoleic, linolenic acids, and PUFAs.

Other Uses

Salad Oil

With the growing interest in "foods that heal," the composition of salad oil has seen a shift towards the "balanced" and "healthy" side. The difference between salad oils and cooking oils lie in their oxidative stability, with cooking oil requiring higher stability at frying temperatures. Both salad and cooking oil are purified by refining and deodorization processes to give clear, bland, stable, and light-colored oil. Soft oils such as sunflower, soybean, corn, and canola oils are the most common oils used in salad oils. Generally, palm oil and its more liquid fraction, palm olein, do not make good salad oils on their own due to high cloud point values. A highly fractionated palm olein (superolein) with IV of 60–67 and low cloud point can be used as salad oil used in the manufacture of mayonnaise and salad dressing. The basic recipe in making a palm-based salad dressing include palm olein (IV 60–67), egg yolk, vinegar, starch, sugar, salt, mustard, and water. There are very few reports on enzymatic production of superolein. One such work documents the use of dry fractionation to produce superolein and soft stearin from crude oleins with different DAG contents which have been treated with a selective lipase (Kamariah et al., 2005).

Emulsifiers

Emulsifiers or surface-active agents function by reducing the surface tension of products such as shortenings and margarines. Monoacylglycerols (MAG) and diacylglycerols (DAG) are examples of emulsifiers often used in palm-based products. High amounts (40–45%) of MAG and DAG, which are partial glycerides, can be produced from chemically-catalyzed glycerolysis process where fats with glycerols are reacted at high temperatures of about 250°C. Chemical catalysts such as sodium ethylate or sodium hydroxide are commonly used. The partial glycerides obtained can be molecularly distilled to obtain concentrations of MAG and DAG as high as 90% or above (Fregolente et al., 2010). Commercially, chemical catalysts are the manufacturers choice catalyst. However, lipases have also been used to produce these partial glycerides from palm oil (Kawahara, 1993; Krishnamurthy, 1992) but only at research level. The benefits of using enzymes are obvious-mild reaction conditions, "cleaner" product with no formation of dark colored compounds or off-flavor due to the polymerization of unsaturated FA.

Medium-chain Triacylglycerols (MCT)

MCT have been used to meet the nutritional needs of patients with special dietary requirements such as fat malabsorption, metabolism difficulties related to cystic fibrosis, and in parenteral formulations for severely malnourished patients (Calder et al., 2010; Le et al., 2010; Siener et al., 2011). MCT is also an ingredient in cosmetic formulations due to their small molecular size and better solubility in aqueous solutions. They are also a good delivery system for lipophilic compounds such as vitamin E and γ-oryzanol in cosmetic formulations. Palm deodorizer distillate is a by-product of a vegetable oil refining process and it contains valuable compounds such as free fatty acids (FFA), vitamin E, and phytosterols. In 2007, Low et al. (2007) reported on an industrially feasible method of MCT production *via* enzymatic esterification of palm kernel deodorizer distillate and glycerol. The optimum MCT yield obtained from his work was 73.3% when 2 wt-% enzyme, reaction temperature of 90°C, reaction time of 10 h and a molar substrate ratio of 4:1 (MCFA:glycerol) was used.

Diacylglycerols (DAG)

DAG are esters of the trihydric alcohol glycerol in which two of the hydroxyl groups are esterified with fatty acids. They can exist in 2 structural isomers, namely, 1,2 and 1,3-DAG. The 1,3-DAG isomer has beneficial effects on obesity and weight-related disorders. It has been claimed to show the effects of body fat reduction in animals and humans. The physiological effect is believed to be attributed to its metabolic pathway, which is different from TAG metabolism. There have been numerous routes reported, either in literature or patents on DAG production. These methods often involve a chemical or enzyme catalyst. Several of the patents filed utilized palm oil, its fractions, or distillates to produce DAG *via* chemical or enzymatic method. Examples of patents that utilized enzymes in the production of palm-based DAG are that of Lo & Baharin (2001) and Lai et al. (2006). Lo & Baharin (2001) reported on a solvent-free method for DAG production using deodorizer distillates which acted as a source of FFA. In their work, glycerol was added to palm deodorizer distillates and enzymatically reacted to produce DAG with a yield of 60% and purity of 82.5% after purification with silica column chromatography (Lo et al., 2004a, 2004b). This process has the advantage of using low cost raw material compared to refined FFA, which can lower the cost of production. In 2006, Lai et al. disclosed an enzymatic partial hydrolysis process of TAG to produce DAG. The advantage of this process lies in the single-step hydrolytic reaction of TAG without addition of substrates such as glycerol. When palm olein was partially hydrolyzed with Lipozyme RMIM lipase and purified using short path distillation, a yield of DAG-enriched palm olein with 60% DAG and 40% TAG which was suitable for margarine (Cheong et al., 2009), baker's fats (Cheong et al., 2011), or shortening (Cheong et al., 2010) applications was obtained (Cheong et al., 2007). Recently, a patent was filed in Malaysia (Razam et al., 2010) that fractionated enzymatically-catalyzed palm oil-based DAG to DAG olein and stearin fractions with properties desirable for edible fat/oil products.

Chemical Synthesis of Palm-Based Structured Lipids

Chemical interesterification (CIE) has been used to rearrange the fatty acids in TAG to alter the plasticity of fats for products such as margarine and shortening. The physicochemical properties of the modified fat largely depend on the type and proportion of oils used in the CIE process. Examples of chemical modification of palm oil or its fractions for the synthesis of SL are scarce.

Chemical interesterification between palm stearin-palm olein and palm stearin-olive oil blends were reported to show various degrees of plasticity, which were suitable for commercial margarines and shortenings (De Martini Soares et al., 2009, 2010). The chemically interesterified blends at ratios of 4:6 and 3:7 palm stearin and olive oil were shown to be suitable for shortening applications with good plasticity and spreadability. Chemical interesterification between palm stearin and palm kernel olein were also reported (Norizzah et al., 2004). The resultant chemically interesterified blends showed lower slip melting points and solid fat content, while having desirable very fine spherulitic β′ crystals. The interesterified blends exhibited smoothness, with good aeration and creaming properties well suited for the production of margarine, shortening, non-temper-type confectionery fats, and fats for whipped creams. Similar outcomes were also shown by Alpaslan & Karaali (1998) on chemical interesterification of partially hydrogenated palm oil and refined olive oil.

In addition to the production of SL, numerous examples of chemical modifications on palm oil, palm kernel oil, and its fractions were reported for the production of non-food products, particularly biofuels, which are covered in Chapter 22 of this book.

Enzymatic Versus Chemical Synthesis

With the many applications of lipase catalysis in oils and fats modifications, it is evident that the progressive evolution of the industry has gradually favored the use of natural enzymes over chemical catalysts. However, it should be noted that each type of catalysis has its own advantages and disadvantages.

The advantages of chemical catalysis over the use of enzymes are definitely (1) lower catalyst cost, (2) lower catalyst concentrations used, (3) simpler industrial equipments, and (4) shorter reaction time. On a production cost point-of-view, these advantages definitely outweigh the justifications of enzymatic catalysis. However, with increasing global awareness on health and environment, one may question which type of catalysis is preferable? Enzymatic catalysis has the advantage of (1) having less or no hazardous and undesirable by-products; (2) the possibility of region-, stereo-, and fatty acid specificity; (3) milder process conditions used; and (4) reusability of the enzymes. The milder enzymatic processing conditions would prevent deterioration of the oil. The specificities that are possible with lipase catalysis allow the unique restructuring of oils and fats that is, otherwise, unachievable by chemical means.

On the other hand, in order to ascertain which type of catalysis to select for the production of SL, the following brand positioning elements should also be considered: (1) price positioning, (2) target positioning, (3) benefit positioning, and (4) the "green" feel good factor. Pricing of a SL product is critical as it can put the brand either at the top or bottom of the category. Consumers most often perceive highly-priced products with a high premium level, and vice versa. In this context, SLs from enzymatic catalysis can command a higher price position due to the *au naturale* perception that can be communicated. An SL may be targeted for low-end versus high-end consumer food products or industrial customers versus end consumers. As such, identifying which target group the SL is intended for is also important to determine whether the cost of production *via* chemical or enzymatic catalysis is justified. Thirdly, the type of benefit, be it nutritional or operational, of the SL to be communicated should be identified and considered. SL with specific nutritional benefits such as those with essential fatty acids at specific positions on the TAG will require the use of enzymatic catalysis, whereas chemical catalysis may be more suited for the production of a completely randomized SL fat for use in certain margarines. Lastly, the "green" theme has always been a premium and trendy selling point for many consumer products including SL. Generally, the brand positioning of certain types of SL produced via enzymatic catalysis may command a higher premium and targeted towards niche high-end consumers.

Conclusions

As consumers today become more aware of issues such as "*trans* fat," "hydrogenation," and "saturated fats," palm-based SL becomes very appealing and is a very real possibility in meeting the challenges of the palm oil consumer market. Just as micronutrients were lauded to prevent disease, so too can SL be used for its therapeutic value. With SL, we can combine the different positive and nutritionally valuable FA in palm oil to produce a potent lipid with maximum benefits and minimum adverse effects. Traditional industrial companies need to be more receptive to this area of research, which, in the long run, will produce new market of product developments. More intensive research is required before we can explore the commercial scale production of palm-based SL, which may well be the "fats of the future."

References

Abigor, R.D.; Maemer, W.N.; Foglia, T.A.; Jones, K.C.; DiCiccio, R.J.; Ashby, R.; Uadia, O. Production of cocoa butter-like fats by lipase catalysed interesterification of palm oil and hydrogenated soybean oil. J. Am. Oil Chem. Soc. 2003, 80(12), 1193–1196.

Adhikari, P.; Zhu, X.; Gautam, A.; Shin, J.; Hu, J.; Lee, J.; Akoh, C.C.; Lee, K. Scaled up production of zero trans margarine fat using pine nut oil and palm stearin. Food Chem. 2010, 119, 1332–1338.

Ahmadi, L.; Wright, A.J.; Marangoni, G. Chemical and enzymatic interesterification of tristearin/triolein-rich blends: Chemical composition, solid fat content and thermal properties. Eur. J. Lipid Sci. Technol. 2008, 110, 1014–1024.

Alpaslan, M.; Karaali, A. The interesterification-induced changes in olive and palm oil blends. Food Chem. 1998, 61, 301–305.

Benjapornkulaphong, S.; Ngamcharussrivichai, C.; Bunyakiat, K. Al_2O_3-supported alkali and alkali earth metal oxides for transesterification of palm kernel oil and coconut oil. Chem. Eng. J. 2009, 145, 468–474.

Berger, K.G.; Nor Aini, I. Formulation of zero trans acid shortening and margarines and other food fats with products of the oil palm. J. Am. Oil Chem. Soc. 2005, 82(11), 775–782.

Calder, P.C.; Jensen, G.L.; Koletzko, B.V.; Singer, P.; Wanten, G.J.A. Lipid emulsions in parenteral nutrition of intensive care patients: current thinking and future directions. Inten. Care Med. 2010, 36, 735–749.

Cheong, L.Z.; Arifin, N.; Lo, S.K.; Tan, C.P.; Long, K.; Yusoff, M.S.A.; Lai, O.M. Physicochemical, textural and viscoelastic properties of palm diacylglycerol bakery margarine during storage. J. Am. Oil Chem. Soc. 2009, 86, 723–731.

Cheong, L.Z.; Arifin, N.; Lo, S.K.; Tan, C.P.; Long, K.; Yusoff, M.S.A.; Lai, O.M. Baking performance of palm diacylglycerol bakery fats and sensory evaluation of baked products. Eur. J. Lipid Sci. Technol. 2011, 113, 253–261.

Cheong, L.Z.; Tan, C.P.; Long, K.; Yusoff, M.S.A.; Lo, S.K.; Lai, O.M. Production of a diacylglycerol-enriched palm olein using lipase-catalyzed partial hydrolysis: Optimization using response surface methodology. Food Chem. 2007, 105, 1614–1622.

Chu, B.S.; Ghazali, H.M.; Lai, O.M.; Che Man, Y.B.; Yusof, S.; Tee, S.B.; Yusoff M.S.A. Comparison of lipase-transesterified blend with some commercial solid frying shortenings in Malaysia. J. Am. Oil Chem. Soc. 2001, 78, 1213–1219.

Choo, S.Y.; Leong, S.K.; Lu, F.S.H. Physicocochemical and sensory properties of ice cream formulated with virgin coconut oil. Food Sci. Technol. Int. 2010, 16(6), 531–541.

Ciftci, O.N.; Fadiloglu, S.; Gogus, F. Conversion of olive pomace oil to cocoa butter-like fat in packed bed enzyme reactor. Biores. Technol. 2009a, 100, 324–329.

Ciftci, O.N.; Fadiloglu, S.; Gogus, F. Utilization of olive–pomace oil for enzymatic production of cocoa butter like fat. J. Am. Oil Chem. Soc. 2009b, 86, 119–125.

Coenen, J.W.E. Rev. Fr. Corps Gras. 1974, 21, 403–413.

Debnath, S.; Prakash, M.; Lokesh, B.R. Lipase mediated interesterification of oils for improving rheological, heat transfer properties and stability during deep frying. Food Bioproc. Technol. 2011.

De Martini Soares, F.A.S.; da Silva, R.C.; da Silva, K.C.G.; Lourenco, M.B.; Soares, D.F.; Gioielli, L.A. Effects of chemical interesterification on physicochemical properties of blends of palm stearin and palm olein. Food Res. Intl. 2009, 42, 1287–1294.

Eltaweel, M.A.; Raja Abdul Rahman, R.N.Z.; Salleh, A.B.; Basri, M. An organic solvent stable lipase from *Bacillus* sp strain 42. Annal. Microbiol. 2005, 55(3), 187–192.

Farmani, J.; Safari, M.; Hamedi, M. Application of palm olein in the production of zero trans Iranian vanaspati through enzymatic intertesterification. Eur. J. Lipid Sci. Technol. 2006, 108, 636–643.

Fregolente, P.B.L.; Pinto, G.M.F.; Wolf-Maciel, M.R.; Filho, M.R. Monogyceride and diglyceride production through lipase catalysed glycerolysis and molecular distillation. App. Biochem. Biotechnol. 2010, 160, 1879–1887.

Gunstone, F.D. Chemical properties. In Lipid Handbook, 2nd ed; Gunstone, F.D., Harwood, J.L., Padley, F.D., Eds.; Chapman and Hall: London, 1994; pp. 594–595.

Hermansyah, H.; Wjanarko, A.; Kubo, M.; Shibasaki-Kitakawa, N.; Yonemoto, T. Rigorous kinetic model considering positional specificity of lipase for enzymatic stepwise hydrolysis of triolein in oil-water system. Bioproc. Biosystem Engineer. 2010, 33, 787–796.

Hossein Goli, S.A.; Mat Shari, M.; Kadivar, M.; Keramt, J. The production of an experimental table margarine enriched with conjugated linoleic acid (CLA): Physical properties. J. Am. Oil Chem. Soc. 2009, 86, 453–458.

Jaeger, K.E.; Ransac, S.; Dijkstra, B.W.; Colson, C.; van Heuval, M.; Misset, M. Bacterial lipases. FEMS Microbiol. Rev. 1994, 15, 29–63.

Jeyarani, T.; Reddy, Y. Preparation of plastics fats with zero trans FA from palm oil. J. Am. Oil Chem. Soc. 2003, 80(11), 1107–1113.

Jin, Q.; Zhang, T.; Shan, L.; Liu, Y.; Wang, X. Melting and solidification properties of palm kernel oil, tallow and palm olein blends in the preparation of shortening. J. Am. Oil Chem. Soc. 2008, 85, 23–28.

Kamariah Long, M.; Jamari, A.; Ishak, A.; Karrupan, J.; Yeok, L.J.; Ahmadilfitri, R.A.L.; Lai, O.M. Physico-chemical properties of palm olein fraction as a function of diglyceride content in the starting material. Eur. J. Lipid Sci. Technol. 2005, 107, 754–761.

Kanjanavas, P.; Khuchareontaworn, S.; Khawasak, P.; Pakpitchareon, A.; Pothivejkul, K.; Santiwatanakul, S.; Matsui, K.; Kajiwara, T.; Chansiri, K. Purification and characterization of organic solvent and detergent tolerant lipase from thermotolerant *Bacillius* sp. RN2. Int. J. Mol. Sci. 2010, 11, 3783–3792.

Kawahara, Y. Progress in fats and oils food technology, INFORM 1993, 4, 663–667.

Kheiri, M.S.A.; Oh, F.C.H. Formulation of vegetable ghee/vanaspati. In Palm Oil Product Technology in the Eighties; Pushparajah, E., Rajadurai, M., Eds.; Incorporated Society of Planters: Kuala Lumpur, 1983; pp. 449–474.

Khumalo, L.W.; Majoko, L.; Read, J.S.; Ncube, I. Characterisation of some underutilized vegetable oils and their evaluation as starting material for lipase catalysed production of cocoa butter equivalents. Ind. Crops Prod. 2002, 16, 237–244.

Kim, B.H.; Lumor, S.E.; Akoh. C.C. *Trans* free margarine prepared with canola oil/ palm stearin/ palm kernel oil structured lipids. J. Agr. Food Chem. 2008, 56(17): 8195–8205.

Kovac, A.; Stadler, P.; Haalck, L.; Spener, F.; Paltauf, F. Hydrolysis and esterification of acylglycerols and analogs in aqueous medium catalyzed by microbial lipases. Biochim. Biophy. Acta. 1996, 1301, 57–66.

Krishnamurthy, R.E. Biotechnology offers route to new fats and oils, INFORM 1992, 3, 457–458.

Lai, O.M.; Ghazali, H.M.; Chong, C.L. Physical properties of *Pseudomonas* and *Rhizomucor miehei* lipase-catalysed transesterified blends of palm stearin:palm kernel olein. J. Am. Oil Chem. Soc. 1998a, 75, 953–959.

Lai, O.M.; Ghazali, H.M.; Chong, C.L. Effect of enzymatic transesterification on the melting points of palm stearin-sunflower oil mixtures. J. Am. Oil Chem. Soc. 1998b, 75, 881–886.

Lai, O.M. Production, properties and application of lipase-catalyzed transesterified palm stearin blends with palm kernel olein, sunflower oil and anhydrous milkfat. Doctoral thesis, Universiti Putra Malaysia, Selangor, Malaysia, 1998c.

Lai O.M.; Ghazali, H.M.; Chiew, C. Let Use of enzymatic transesterified palm stearin-sunflower oil blends in the preparation of table margarine formulation. Food Chem. 1999, 64, 83–88.

Lai, O.M.; Yusoff, M.S.A.; Lo, S.K.; Long, K.; Tan, C.P.; Lim, J.Y.; et al. Process for the production of diaclyglycerol. PCT International Application No. PCT/MY2006/000034 (in English), 2006.

Le, H.D.; Fallon, E.M.; Meijer, V.E.; Malkan, A.D.; Puder, M.; Gura, K.M. Innovative parenteral and enteral nutrition therapy for intestinal failure. Sem. Ped. Surgery 2010, 19, 27–34.

Lee, J.H.; Akoh, C.C.; Lee, K. Physical properties of trans free bakery shortening produced by lipase catalysed interesterification. J. Am. Oil Chem. Soc. 2008, 85, 1–11.

Li, W.; Du, W.; Li, Q.; Sun, T.; Liu, D. Study on acyl migration kinetics of partial glycerides: Dependence on temperature and water activity. J. Mol. Cataly. B: Enzymatic 2010, 63, 17–22.

Liew, M.Y.B.; Ghazali, H.M.; Long, K.; Lai, O.M.; Yazid, A.M. Physical properties of palm kernel olein-anhydrous milkfat mixtures transesterified using mycelium-bound lipase from *Rhizomucor miehei*. Food Chem. 2001, 72, 447–454.

Lo, S.K.; Baharin, B.S. Process for producing phytonutrient-enriched diacylglycerols. Malaysian Patent Application No. PI20014817 (English), 2001.

Lo, S.K.; Baharin, B.S.; Tan, C.P.; Lai, O.M. Diacylglycerols from palm oil doedorizer distillate: Part 1. Synthesis by lipase-catalyzed esterification. Food Sci. Technol. Intern. 2004a, 10(3), 149–156.

Lo, S.K.; Baharin, B.S.; Tan, C.P.; Lai, O.M. Diacylglycerols from palm oil deodorizer distillate: Part 2. Physical and chemical characterization. Food Sci. Technol. Intern. 2004b, 10(3), 157–161.

Loo, J.L.; Lai, O.M.; Long, K.; Mohd. Ghazali, H. Fatty acid preference of mycelium-bound lipase from a locally isolated strain of *Geotrichum candidum*. World J. Microb. Biotechnol. 2007, 23, 1771–1778.

Low, C.T.; Lo, S.K.; Mohamad, R.; Tan C.P.; Long, K.; Ismail, R.; Lai, O.M. Medium-chain triacylglycerols from palm kernel oil distillate: Optimization by response surface methodology. Eur. J. Lipid Sci. Technol. 2007, 109, 107–119.

Maria, P.D.; Oerlemans, C.C.; Tuin, B.; Bargeman, G.; Meer, A.; Germert, R. Biotechnology applications of *Candida antratica* lipase A: State-of-the-art. J. Mol. Cataly. B: Enzymatic. 2005, 37, 36–46.

Md. Alim, A.; Lee, J.H.; Lee, K. Production of lipase catalysed solid fat from mustard oil and palm stearin with linoleic acid by response surface methodology. J. Sci. Food Agric. 2009, 89, 706–712.

Mu, H.; Hoy, C.E. The digestion of dietary triacylglycerols. Progr. Lipid Res. 2004, 43, 105–133.

Naghshineh, M.; Ariffin, A.A.; Mohd. Ghazali, H.; Mirhosseini, H.; Mohammad, A. Effect of saturated/unsaturated fatty acid ratio on physicochemical properties of palm olein-olive blend. J. Am. Oil Chem. Soc. 2010, 87, 255–262.

Nazaruddin, R.; Syaliza, A.; Rosnani, I.W. The effect of vegetable fat on the physicochemical characteristic of dates ice cream. Intern. J. Diary Technol. 2008, 61(3), 265–269.

Noor Lida, H.M.D.; Ali, A.R.; Mahadhir, I. Blending of palm oil, palm stearin and palm kernel oil in the preparation of table and pastry margarine. Intl. J. Food Sci. Nutr. 1996, 47, 71–74.

Nor Aini, I.; Che Maimon, C.H.; Hanirah, H.; Zaiwiah, S.; Che Man, Y.B. Trans free vanaspati containing ternary blends of palm oil-palm stearin-palm olein and Palm oil-palm stearin-palm kernel oil. J. Am. Oil Chem. Soc. 1999, 76(5), 643–648.

Nor Aini, I.; Lina, L.; Yusoff, M.S.A.; Arif Simeh. Palm oil-based trans free vanaspati, Proc. of the 1997 PORIM Technol. Trans. Sem. 1997a, 28–39.

Nor Aini , I.; Miskandar, M.S. Utilization of palm oil and palm products on shortening and margarines. Eur. J. Lipid Sci. 2005, 109, 422–432.

Nor Aini, I.; Yusoff, M.S.A.; Arif Simeh. Palm-based trans free vanaspati, PORIM Information Series No. 66, 1997b, 1–4.

Norizzah, A.R.; Chong, C.L.; Cheow, C.S.; Zaliha, O. Effects of chemical interesterification on physicochemical properties of palm stearin and palm kernel olein blends. Food Chem. 2004, 86, 229–235.

Pelouze, J.; Gelis, A. Ann. Chim. Phys. 1844, 10, 434–456.

Pinyaphong , P.; Phutrakul, S. Synthesis of cocoa butter equivalent from palm oil by *Carica papaya* lipase catalysed interesterification. Chiang Mai J. Sci. 2009, 36(30), 359–368.

Razam, A.L.; Ab Karim, N.A.; Seong Koon, L.; Affandi Yusoff, M.S.; Ming, L.O. Process for Fractional Crystallisation of Palm-based Diacylglycerol Fat. Malaysian Application no. PI201004803, 2010.

Reshma, M.V.; Saritha, S.S.; Balachandarn, C.; Arumughan, C. Lipase catalysed interesterification of palm stearin and rice bran oil blends for preparation of zero trans shortening with bioactive phytochemicals. Biores. Technol. 2008, 99, 5011–5019.

Rodriguez, B.H.; Cordova, J.; Barzana, E.; Favela-Torres, E. Effects of organic solvents on activity and stability of lipase produced by thermotolerant fungi in solid state fermentation. J. Mol. Catalysis B: enzymatic 2009, 61, 136–142.

Siener, R.; Ehrhardt, C.; Bitterlich, N.; Metzner, C. Effect of fat spread enriched with medium chain triacylglycerols and a special fatty acid micronutrient combination on cardiometabolic risk factors in overweight patients with diabetes. Nutr. Met. 2011, 8, 21.

Siew, W.L.; Cheah, K.Y.; Tang, W.L. Physical properties of lipase catalysed interesterification of palm stearin with canola oil blends. Eur. J. Lipid Sci. 2007, 109, 97–106.

Stransky, K.; Zarevucka, M.; Kejik, A.; Wimmer, M.M.; Demnerova, K. Substrate specificity, regioselectivity and hydrolytic activity of lipases activated from *Geotrichum* sp. Biochem. Eng. J. 2007, 34, 209–216.

Sung, K.K.; Goff, H.D. Effect of solid fat content in ice creams containing palm kernel oil and high oleic sunflower oil. J. Food Sci. 2010, 75(3), 274–279.

Sutar, N.; Sutar, P.; Singh, G. Evaluation of different soybean varieties for manufacture of soy ice cream. Inter. J. Dairy Technol. 2010, 63(1), 136–142.

Szabo, E.; Boehm, G.; Beermann, C.; Weyermann, M. Fatty acid profile comparison in human milk sampled from the same mothers at the sixth week and sixth month of lactation. J. Ped. Gastr. Nutr. 2010, 50, 3.

Tabee, E.; Jagerstad, M.; Dutta, P.C. Frying quality characteristics of French fries prepared in refined olive oil and palm olein. J. Am. Oil Chem. Soc. 2009, 86, 885–893.

Wang, H.; Wu, H.; Ho, C.; Weng, X. Cocoa butter equivalent from enzymatic interesterification of tea seed oil and fatty acid methyl ester. Food Chem. 2006, 97, 661–665.

Xu, X. Enzymatic production of structured lipids: Process reactions and acyl migration. INFORM 2000a, 11, 1121–1131.

Xu, X. Production of specific–structured triacylglycerols by lipase catalysed reactions: a review. Eur. J. Lipid Sci. 2000b, 287–303.

Zhang, H.; Pedersen, L.S.; Kristensen, D.; Adler-Niessen, J.; Holm, H.C. Modification of margarine fats by enzymatic interesterification; Evaluation of a solid fat content based exponential model with two groups of oil blends. J. Am. Oil Chem. Soc. 2004, 81(7), 653–658.

Zhang, H.; Xu, X.; Nilsson, J.; Mu, H.; Adler-Nissen, J.; Hoy, C. Production of margarine fats by enzymatic interesterification with silica granulated *Thermomyces lanuginose* lipase in a large scale study. J. Am. Oil Chem. Soc. 2001, 78(1), 57–64.

18

Health and Nutritional Properties of Palm Oil and Its Components

Zulfitri Azuan Mat Daud, Deepinder Kaur, and Pramod Khosla
Department of Nutrition and Food Science, Wayne State University, Detroit, MI 48202, USA

Introduction

Palm oil (*Elaeis guineensis*) is the world's largest edible oil in terms of production and trade. Palm oil's balanced fatty acid composition (~50% saturated and ~50% unsaturated) makes it one of the most versatile oils for food industry applications. Moreover, crude palm oil is also one of the richest dietary sources of pro-vitamin A and vitamin E (primarily tocotrienols). This chapter reviews the nutritional properties of palm oil, both in terms of fatty acid composition (in relation to disease and health) as well as its constituents (pro-vitamin A and vitamin E) that are being increasingly exploited in nutrition research for their potential health benefits.

Palm oil and palm kernel oil, extracted from the orange-red mesocarp and seed, respectively, of the oil palm tree accounted for ~32% of the global production of oils and fats in 2009. Palm oil was the leading edible oil produced, overtaking soybean oil which accounted for ~22% of global production (MPOC, 2009). In terms of exports, in 2009, palm oil and palm kernel oil accounted for almost 61% of global exports. Palm oil is competitively priced vis-a-vis groundnut oil, corn oil, and soybean oil. Furthermore, palm oil yield per hectare is about 10 times the yield of soybean oil (Khosla & Sundram, 2010). Therefore, global palm oil output is obtained from less than 5% of the total arable land. The widespread availability of palm oil is finding increasing uses in the food sector. Fractionation procedures allow isolation of two main products, palm olein (liquid fraction) and palm stearin (solid fraction), which expand palm oil's usage in diverse food formulations (Edem, 2002). In addition, crude palm oil contains minor components, including carotenoids (some with Vitamin A activity) and Vitamin E (including a high concentration of tocotrienols), making palm oil a major player in various global public health initiatives.

Nutritional Properties of Palm Oil

Nutrition research related to palm oil has explored primarily its fatty acid composition and micronutrient content in relation to health and disease. Palm kernel oil, palm oil, and its fractions differ in their fatty acid composition in terms of the proportion of saturated fatty acids (SFA) and unsaturated fatty acids (UFA) and also in their primary SFA composition. Palm kernel oil is distinct from palm oil and its fractions with approximately 82% SFA comprised of lauric (~50%), myristic (~16%), and palmitic acid (~8%)—a composition closer to that of coconut oil. The UFA composition includes oleic acid (15%) and linoleic acid (~3%). Palm oil on the other hand contains approximately equal amounts of SFA and UFA. The primary SFA being palmitic acid (45%) along with stearic acid (5%). The UFA in palm oil include the monounsaturated fatty acid (MUFA), oleic acid, and the polyunsaturated fatty acid (PUFA), linoleic acid at 40% and 10%, respectively. Palm olein, compared to palm oil, contains lower amounts of palmitic acid and a higher amount of oleic acid (Edem, 2002). The differences in the fatty acid composition of palm kernel, palm oil, and palm olein potentially result in differential effects on lipids, lipoprotein metabolism, and various aspects of health, although these are generally "lumped together" and attributed to the SFA content.

Dietary Fat and Health

Cardiovascular disease (CVD) remains the primary cause of death in developed countries and one of the leading causes of death in developing countries (WHO, 2011). Traditionally, a lipid panel has been used to assess an individual's risk for CVD. Elevated plasma or serum total cholesterol (TC) and low-density-lipoprotein cholesterol (LDL-C) are associated with increased risk, while elevated high-density-lipoprotein cholesterol (HDL-C) is protective. Together the TC:HDL-C ratio provides a useful and simple index of CVD risk (Lemieux et al., 2001; Lewington et al., 2007; Mensink et al., 2003), while the LDL-C:HDL-C ratio offers a better predictive value for a future cardiovascular outcome such as acute myocardial infarction (Fernandez & Webb, 2008; Yokokawa et al., 2011). There are many dietary factors that can influence this lipid panel including macronutrient distribution and quality, fiber intake, various antioxidants, and phytochemicals. However, dietary fat probably is one of the nutrients that has received the greater scrutiny for its effect on CVD.

Early evidence relating the role of dietary fat to the risk of CVD was reported by Ancel Keys in the early 1950s. Keys and colleagues conducted the Seven Countries Study that compared the death rates in relation to the amount of fat eaten along with lifestyle factors. Keys's group demonstrated that increased blood cholesterol levels was associated with increased cardiovascular risk. (Although Keys had data for more than 20 countries, the reason for not including many of these are now in the realm of folklore.) The fact that TC can be affected by dietary fat intake resulted in greater attention being paid to this dietary component.

Initial research focused on modification in the amount and quality of fat to reduce TC. Hegsted, Keys, and colleagues delineated that SFA were twice as effective in raising TC as PUFA were in lowering it while MUFA and stearic acid (18:0) were considered neutral (Hegsted et al., 1965; Keys et al., 1965). When considering individual fatty acids, Hegsted's regression equation showed that myristic acid (14:0) raised TC four times more than palmitic (16:0) while PUFA (mainly linoleic acid) lowered it (Hegsted et al., 1965). These findings became the basis for numerous public health message aimed at replacing SFA with UFA. The fact that palm oil contains substantial amount of SFA (~50%) dragged it to the center of the controversy with it being branded as a cholesterol-raising fat based on the results of some early studies. Some of these compared palm oil as a source of fat with other dietary oil/fat while the other studies investigated the effect of individual SFA.

In initial studies from Grundy's group (Bonanome & Grundy, 1988; Grundy & Vega, 1988; Mattson & Grundy, 1985) (using liquid formula diets) in a metabolic ward setting, palm oil was shown to be hypercholesterolemic relative to 18:0, 18:1, and 18:2-rich fats/oils. Increases were noted in both LDL-C and TC. It should be noted that in these studies, fatty acids were not equalized across diets, and inadequate PUFA and/or the use of hypercholesterolemic subjects could have magnified the observed response. Furthermore, these studies utilized levels of fat (~40% of calories) that were typically consumed in Western populations.

When regular food-based diets were used in normocholesterolemic subjects and fatty acid exchanges were more realistic (~5% of total calories), palm olein (the liquid fraction from palm oil), was found to give comparable cholesterol values with olive oil and/or canola based diets (Choudhury et al., 1995; Ng et al., 1992; Sundram et al., 1994; Sundram et al., 2007). Recently, Tholstrup et al. (2011) and Voon et al. (2011) reported contrasting results when healthy subjects were fed palm olein or olive-oil based diets. In young normocholesterolemic subjects, the failure of palm olein to raise cholesterol could be attributed to the lack of any effect on endogenous cholesterol synthesis ascertained using stable isotopes (Clandinin et al., 1999; Clandinin et al., 2000; Cook et al., 1997; French et al., 2002).

The neutral effect of palm olein described in some of the above studies often occurs when total fat is around ~30% of total calories. A possible explanation for this phenomenon has been described by Hayes and co-workers (Hayes & Khosla, 1992; Hayes et al., 1991; Hayes et al., 1993; Khosla & Hayes, 1992; Khosla et al., 1997; Lindsey et al., 1990). Based on their work, individual SFA are not equally hypercholesterolemic. Secondly, any deleterious effects that SFA may have on cholesterol can be negated, if "threshold" levels of dietary (18:2) are present (~5% of total calories). At sub-threshold levels, the relative potencies of the various SFA are apparent, while above threshold levels the SFA appear "neutral." Therefore, palm olein, basically devoid of 12:0 and 14:0, has sufficient 18:2 to counter any deleterious effects of the other SFA. In the above model, 18:1 has minimal impact. The above concept of using 18:2 to essentially balance SFA in the diet is ingrained into most current dietary recommendations, which essentially advocate equivalent amounts of SFA and PUFA, with the remaining fatty acids being derived from MUFA. One of the surprise consequences of balancing fatty acids and even having a minimum amount of SFA present, resulted in an elevation of HDL-C observed in the study from Lindsey et al. (1990). A more detailed discussion of some of the early studies can be found in Khosla and Sundram (2011).

The argument as to whether the intake of SFA is causally related to CVD has been recently re-visited (Astrup et al., 2011; Mensink, 2011). In a meta-analysis by Siri-Tarino et al. (2010), 21 well-designed prospective epidemiological studies were analyzed to investigate the effect of SFA on CVD risks. A pooled analysis of 5–23 years of follow up comprising more than 300,000 subjects indicate that SFA was not associated with the increased risk of CHD (Relative Risk, RR = 1.07), stroke (RR = 0.81), and CVD (RR = 1.00) after adjusting for age, gender, and study quality.

In another meta-analysis by Jakobsen et al. (2009), the authors reported an inverse relationship between SFA and CVD. Ironically, some of the studies used in the analysis were also used as part of the other meta-analysis by Siri-Tarino et al. (2010). The authors reported that risk for coronary events was reduced when SFA was replaced with PUFA. However, replacing SFA with MUFA did not confer any beneficial effects while the effects of replacing SFA with carbohydrate varies depending on carbohydrate types (Jakobsen et al., 2009). Interestingly, in a systematic review of randomized clinical trials (RCTs), Ramsden et al. found that increasing n-6 PUFA did not confer any potential benefit but could increase CHD risk unless accompanied by an increase in n-3 PUFA (Ramsden et al., 2010).

The fact that many patients who develop CHD present without lipoprotein abnormalities has sparked considerable interest in the role of inflammation as an important factors that is implicated in development of CVD. Circulating inflammatory mediators such as acute phase C-reactive protein and cytokines participate in various levels of vascular insult and artheromatous change (Libby, 2002). A review by Kennedy et al. (2009) indicate that SF mediated inflammatory process in adipose tissue via accumulation of diacylglycerol and ceramide, promote recruitement, activation of immune cells, and induction of inflammatory genes in adipose tissue, thus increasing inflammation. Voon et al. (2011) conducted a randomized-crossover study among healthy human volunteers that tested the effects of 3 types of fats: (1) Palmitic acid rich palm olein, (2) Lauric and myristic acid rich coconut oil, and (3) oleic acid rich virgin olive oil. This study was meant to test the effects of common commercial fats in which each test fat contributed to 20% of the total energy rather than the effects of individual FA. Their results showed that the 3 diets did not differ in term of their effects on TNF-α, IL-1β, IL-6, IL-8, and hs-CRP. However, it should be noted that their study used relatively young volunteers (30 years old) which could have potentially contributed to non-significant observation on the effects of test diets on acute inflammatory profiles.

Palm Oil as an Alternative to *trans*-fatty Acids

When the campaign against tropical oils (coconut, palm kernel oil, and palm oil) was launched in the early 1980s, the food industry responded promptly by removing palm oil from their products to taper the public out-cry. Consequently, this left the food industry with limited choices, including the use of partially hydrogenated vegetable oils (PHVO) to maintain functionality and taste, especially for baked and fried products. In fact, PHVO offered a superior advantage from the food industry perspective as it was highly customizable to meet various manufacturing needs (McNamara, 2010). As a result, tropical oils were eventually replaced by PHVO that contained *trans*-fatty acids (tFA) in most fast food restaurant chains and processed-food manufacturers in Europe and the USA. In the late 1980s, it was conservatively estimated that in the USA tFA provided about 3% of total dietary energy (Hunter & Applewhite, 1991).

However, in the early 1990s, results from several groups that investigated health impacts of tFA stirred another controversy. Mensink and Katan (1990) showed that consumption of ~11% of total energy from SFA or tFA resulted in similar increase in LDL-C compared to MUFA, but tFA caused a more detrimental effect because they also lowered HDL-C. As a consequence the ratio of TC:HDL-C or LDL-C:HDL-C was also lowered. Similar outcomes were also reported by Judd and colleagues (1994), as well as a spate of studies from around the world.

The results from several epidemiological studies were consistent with the results from the clinical trials. Data from a longitudinal cohort in The Nurses' Health study demonstrated a strong association for CHD risk with the consumption of margarine and cookies which predominantly used PHVO (Willett et al., 1993). Similarly, after following this cohort for 14 years, Hu et al. (1997) reported that tFA consumption was related to the risk for CHD as opposed to SFA and MUFA that were not significantly related to CHD risk, while PUFA was inversely related to CHD risk. The authors also revealed that with 1.6% increase in total energy from tFA, the relative risk for CHD increased by 53%. Further follow up (after 20 years) continued to support the detrimental effect of tFA with respect to CHD risk, particularly among younger women (Oh et al., 2005).

These studies were instrumental in the resulting mandatory labeling for tFA contents in food products sold in the United States. Over the last 5 years, the food industry has been looking for alternatives to PHVO that do not cause adverse health effects, without sacrificing the original taste and baking quality. Several options are currently available, including interesterified fat, using naturally occuring saturated fat from tropical oils, or genetically modified oils seeds.

Interesterified fat involves chemical or enzymatic interesterification (IE) process in which vegetable oils are fully hydrogenated to form a fully saturated hard stock (either 16:0 or 18:0), followed by an exchange of acyl groups among the triacylglycerols to achieve the required degree of plasticity. Soybean oil and palm oil can be used to generate 18.0:16.0 ratios of 85:15 and 55:45, respectively (Hayes & Pronczuk, 2010).

This modifies the melting point and creates an oil that is suitable for deep frying and baking. A potential problem created by IE is that it causes alteration in SFA/UFA occupancy of the sn-2 position of the TAG molecule. In natural vegetable oils, SFA (16:0 and 18:0) usually occupy the sn-1 and sn-3 positions, but in the case of IE, they also occupy the sn-2 position. It has been postulated that this could potentially cause detrimental effects on lipoprotein metabolism in the case of high intakes, and affects immune function, glucose, and insulin metabolism (Hayes & Pronczuk, 2010).

While most studies generally suggest that IE is not an issue, a recent study by Sundram et al. suggested otherwise (Sundram et al., 2007). In this study, three types of fat (unmodified palm olein, trans-rich partially hydrogenated soybean oil, and interesterified fats) were investigated for their effects on lipids, plasma glucose, and insulin indices. Subjects were fed with the test diets using a cross-over design. Both the tFA rich-blend and the IE fat resulted in higher LDL/HDL ratio, indicating a detrimental effect on CHD risks. Moreover, the IE also altered glucose metabolism that was linked to depressed insulin and C-peptide. It appears that we have entered a new debate on the role of IE and human health.

A recent study by Mozaffarian and Clark evaluated the effects on CHD of replacing PHVO (with varying percentage of *trans*-fat) with specific fats and oils. The authors also evaluated the specific contributions of TC:HDL-C ratio, Lp(a), C-reactive protein and ApoB:ApoA ratio to overall CHD risk. The study showed that replacing 7.5% of energy from PHVO (containing 45% tMUFA) with palm oil decreased CHD risk by ~30%, while the figure for highly unsaturated oils (soybean or canola) was closer to 40% (Mozaffarian & Clarke, 2009).

Taken together, un-modified fat such as palm oil appear to be a better alternative for *trans*-fat replacement, as far as long-term health effects are concerned.

Palm Minor Constituents and Health

Palm fruit contains a significant amount of carotenoids, along with tocopherols and tocotrienols, phytosterols, and glycolipids. In fact, crude palm oil is one of the richest sources of carotenoids (500–700 ppm) and vitamin E (600–1000 ppm) (Goh et al., 1985; Yap et al., 1991). However, these antioxidants are destroyed during the conventional refining process. Through advancement of the technologies and R&D, a refined red palm oil, characterized by dark red color and processed using cold-press, would be able to retain much of the carotenoids and vitamin E that were originally present in crude palm oil (Nagendran et al., 2000). This red palm oil has been extensively studied for its health benefits.

Palm Vitamin E

Vitamin E is a collective term referring to eight structurally related compounds, namely α-, β-, γ-, δ-tocopherol and α-, β-, γ-, δ-tocotrienol. Each of these compounds comprises a 6-chromanol ring and an isoprenoid side chain in which tocopherols have a saturated side chain while tocotrienols possess an unsaturated side chain. Vitamin E is present in numerous food sources such as nuts, seeds, and vegetable oils. Among these, tocopherols are predominantly present in olive, sunflower, corn, and soy bean oils while tocotrienols are primarily from barley and rice bran, with crude palm oil being the richest source. The tocotrienol content in the tocotrienol-rich-fraction (TRF) of palm oil is estimated between 70 and 80% (McIntyre et al., 2000), predominantly consisting of γ- (46%) and α-tocotrienols (22%) (Sundram & Gapor, 1992). Vitamin E is essentially absorbed in the same manner as fat, namely incorporation into chylomicrons and transportation in the lymphatic system. From the point of view of health, there is great interest in research pertaining to vitamin E properties as antioxidants in which a defective antioxidant system is implicated in development of CVD, cancer, and other chronic diseases.

In the last several decades, there have been a substantial number of publications on vitamin E, mostly α-tocopherol, while the other members of vitamin E group have remained poorly understood. Much effort has been concentrated on α-tocopherol because this isomer is the most abundant of the eight present in the human body. This can also be explained by the selective affinity of α-tocopherol transfer protein (α-TTP) produced in the liver, together with tocopherol-associated protein for the α-tocopherol isomer, which is responsible for its transportation and endogenous accumulation. Another conflicting argument to neglect tocotrienols in basic science and clinical research has been that they have relatively low bioavailability as compared to tocopherols. However, current evidences indicate that tocotrienols are transported as part of plasma lipoproteins although with a shorter half-life than tocopherols (Fairus et al., 2006; Khosla et al., 2006). Furthermore, it has been documented that modest tocotrienol supplementation elevates blood levels from 1 to 3 uM (Khosla et al., 2006; O'Byrne et al., 2000), a level that may be sufficient to prevent neurodegeneration (Sen et al., 2007). In addition, studies show that tocotrienols accumulate in various tissues at different percentages suggesting that the vitamin E composition may be regulated individually at the tissue level (Arita et al., 1995; Hayes et al., 1993; Packer et al., 2001). A study over five generations of rats revealed that orally supplemented tocotrienol was delivered to all vital organs and in some tissues exceeding α-tocopherol concentration, suggesting an efficient transport system in vivo (Khanna et al., 2005). Interestingly, Traber et al. (1998) reported that tocotrienols rapidly penetrate through skin, which opens an exciting area of research in delivering effective dose without relying on α-TTP.

The TRF extracted from palm oil has been exploited for its properties in nutrition research. One of the postulated benefits is its potential as a cardioprotective molecule. Several studies have shown that micromolar amounts of tocotrienol inhibit cholesterol biosynthesis by suppressing the 3-hydroxy-3-methyl-glutarylCoA reductase enzyme (HMGR), causing reduction in endogenous cholesterol production (Pearce et al., 1994; Qureshi et al., 2001). This hypocholestrolemic effect is due to the side chain's unique ability to increase cellular farnesol that signals proteolytic degradation of HMGR (Goldstein & Brown, 1990; Vasanthi et al., 2011). Furthermore, several studies have demonstrated that tocotrienols could also prevent CHD through its ability in reversing arterial blockage. Nafeeza et al. (2001) showed that, a 10-week feeding of TRF from palm oil in rabbits fed with 2% cholesterol resulted in a lesser intima thickness as compared to cholesterol-fed rabbit without TRF supplementation. Similarly, Black et al. showed that mice fed atherogenic diet with TRL supplementation exhibit an average of 92–98% smaller lesions as compared to their counterparts. These anti-atherogenic properties of tocotrienols may be attributed to their cholesterol lowering effect and/or tocotrienols' ability to alleviate foam cell formation. A recent study showed that TRF attenuated development of atherosclerosis in ApoE-/- mice by modulating activities of proxisome proliferator-activated receptors (PPARs), transcription factors that play essential preventive roles in the development of atherosclerosis by regulating energy metabolism and inflammation (Li et al., 2010). Moreover, Das and colleagues demonstrated that isolated hearts pre-treated with TRF were protected against ischemia-reperfusion injury by preventing c-Src activation and stabilization of the proteosome (Das et al., 2005). Recently, Budin et al. (2009) showed that the normal morphology of the thoracic aorta was preserved with TRF supplementation suggesting its protective effect on the vascular wall (Budin et al., 2009). These evidences suggest that tocotrienols' function is not limited to its antioxidant ability per se. In fact, independent hypocholestrolemic, anti-thrombotic, and anti-ischemic properties of tocotrienols may modulate mechanisms that are cardio-protective.

There is growing evidence for tocotrienols' role as an anti-cancer agent. The earliest evidences were documented in 1989, in which two groups showed that oral and injected tocotrienols prevented tumor growth in animal models (Komiyama et al., 1989; Sundram et al., 1989). This is because tocotrienol is thought to possess anti-proliferative effects and could induce apoptosis through a mitochondria-mediated pathway. Several other reported anticarcinogenesis mechanisms of tocotrienols include inhibition of angiogenesis, suppression of HMG CoA reductase activity, and modulation of cell cycle by suppressing cyclin D resulting in cell arrest (Wada, 2009). Furthermore, it is known that the caspase cascade plays an important role in triggering and executing apoptotic signals within the cells. Basically, there are two types of caspases: initiator caspases (caspase 8, 10, 9, 2) and effector caspases (caspase 3, 7, 6). Sylvester's group has delineated potential mechanisms for the signaling pathways involved in tocotrienol-induced apoptosis (Samant et al., 2010; Shah et al., 2003; Sylvester & Shah, 2002). When a highly malignant mouse epithelial mammary cell line was subjected to a cytotoxic dose of tocotrienols it was found that caspase-8 and caspase-3 activity were largely increased. However, combined treatment of tocotrienols with selective caspase-8 inhibitor resulted in complete blockage of tocotrienol-induced apoptosis indicating the apoptosis pathway was mediated through caspase-8 (Sylvester & Shah, 2002).

In a different study, it was found that the severity of hepatocarcinogenesis in rats induced with 2-acetylaminofluorene (AAF) was decreased when AAF was administered together with palm TRF supplementation (Ngah et al., 1991). Recently, studies in cell culture and in rodents have revealed potential therapeutic value of tocotrienols in protecting against progressive loss in structure and function of neurons. Tocotrienols confer neuroprotective properties possibly by inhibition of c-src activity, which is a key player in executing neurodegeneration (Khanna et al., 2002).

Recently, Park et al. delineated a novel mechanism on the effect of α-tocotrienol supplementation to prevent stroke in vivo. The authors found that multidrug resistance associated protein-1 (MRP-1), a key mediator of intracellular oxidized glutathione efflux from neural cells, plays an important role as neuroprotectant. MRP-1 expression was found elevated in cortical tissue of α-tocotrienol supplemented mice that suggest MRP-1 is a sensitive target for α-tocotrienol intervention (Park et al., 2011).

Although there are a substantial amount of publications that show potential therapeutic use of tocotrienols once they have been isolated from palm oil, it is not clear how the tocotrienols as present in the native palm oil matrix can be exploited for human health. Furthermore, randomized clinical trials using isolated tocotrienols now must also take into account the fact that in the USA, tocotrienols have been granted GRAS (Generally Recognized as Safe) status by the Food and Drug Administration. As such the distinction between tocotreinols as pharmaceuticals/nutraceuticals is not as clear-cut.

Palm Carotene

Red palm oil contains high concentrations of beta- and alpha-carotene (pro-vitamin A), ranging from 500 to 800 mg of pro-vitamin A which is 15 times higher than those in carrot when compared weight-by-weight basis (Sundram et al., 2003). Furthermore, it has been reported that red palm oil has higher bioavailability than other vegetable sources (You et al., 2002). Because of this, red palm oil has been studied extensively in a dietary intervention approach to combat vitamin A deficiency, which is widely prevalent worldwide. To date, a number of studies have shown that beta-carotene-rich red palm oil can be used effectively to prevent vitamin A deficiency without the need for resorting to synthetic Vitamin A. In these studies, beta-carotene-rich red palm oil has been used either as a dietary supplement, in-home fortificant, or to fortify food items produced for distribution in targeted supplementary programs (Rice & Burns, 2010; Roels et al., 1963; van Stuijvenberg et al., 2000).

One of the earliest community-based intervention trials was in the 1960s in which a group of young boys, exhibiting clinical signs for vitamin A deficiency/xeropthalmia, were supplemented with either 2 g sugar, 2 g skim milk powder, 1 g coconut oil, 1 g red palm oil, or 2000 IU vitamin A acetate per kg body weight for 22 days. After 3 weeks, serum carotene increased in those who were fed red palm oil and night blindness was cured (Roels et al., 1963). In the 1990s, a study in an Indian population showed that children fed with traditional Indian sweets made with red palm oil for 1–2 months had increased blood retinol levels (Manorama et al., 1996). Similarly, school children who were fed biscuits baked in a commercialized baking fat containing red palm oil showed a significant improvement in serum retinol levels (van Stuijvenberg et al., 2000, 2001). Interestingly, the serum retinol level was comparable to those who received biscuits with added synthetic β-carotene. The increase in serum retinol level following red palm oil intake demonstrated the efficacy of red palm oil in delivering pro-vitamin A.

Boeteng et al. (2006) showed that rats that were fed with two different levels of red palm oil (7% or 14%) for 13 weeks had a significantly reduced number of aberrant crypt foci (ACF) as compared to those fed with soy bean oil at the same levels, indicating the effects of bioactive components such as tocotrienols and carotenoids that are highly present in RPO. In a study by Yamanushi et al., mice were fed a diet containing either 0.02% palm carotene, 0.02% beta carotene, or no carotenoids at two different levels of fat (4% and 16%), and lung tumorogenesis was induced by diesel exhaust particles. Interestingly, no adenocarcinoma or adenoma were found in a group that were fat palm carotene at 4% of fat (Yamanushi et al., 2001). In another tumorigenesis study by Nesaretnam's group, mice were injected with MCF-7 human breast cancer cells and were fed with commercial mouse pellet diet in which the intervention group was supplemented with 1000 ppm of palm carotene emulsion as daily drinking water. Their results showed that mice fed with palm oil carotene demonstrate a lesser tumor incidence, a lower tumor surface area, and metastasis to the lung as compared to the control group. These results suggest that palm oil carotene could potentially confer anti-cancer and anti-metastatic properties in mice injected with MCF-7 human breast cancer cells. This could be attributed to an enhancement of the immune system due to an increase in natural killer cells and B-lymphocyte populations, which play important role in tumor suppression (Nesaretnam et al., 2002). However, it should be noted that some of these study used a non-physiological dose of palm carotene, thus further testing is required in human populations.

Conclusions

The nutritional properties of palm oil with regards to its fatty acid composition in relation to health and disease, as well as from a food application stand point, have been studied for decades. Current evidence linking *trans*-fat with metabolic derangement places palm oil at an advantage as an alternative substitute. While the SFA has frequently been used as a negative in palm oil usage, there is growing evidence that the SFA composition of palm oil is not a concern. There are several potential gaps that need to be addressed by future research in understanding the role of palm oil in health and diseases—many of the evidences from epidemiological research are not from major palm oil consuming populations. Such research is very much needed to support our current understanding, which is mostly based on a small number of human dietary trials.

Nutritional role of palm minor components has garnered much attention these days. Several studies provide strong and convincing evidence that red palm oil can be effectively used in regular food items to prevent vitamin A deficiency. The competitive price of palm oil and good bioavailability of the palm carotenoids give it at a greater advantage than synthetic vitamin A. In addition, palm vitamin E, specifically tocotrienols, have been investigated for their anti-cancer, anti-thrombotic, and antioxidant effects that are implicated in various disease processes. Although the results from pre-clinical studies show promising results, further studies are needed to provide a mechanistic explanation and guidance for tocotrienols' practical use in human populations.

References

Arita, M.; Sato, Y.; Miyata, A.; Tanabe, T.; Takahashi, E.; Kayden, H.J.; Arai, H.; Inoue K. Human alpha-tocopherol transfer protein: cDNA cloning, expression and chromosomal localization. Biochem. J. 1995, 306(Pt 2), 437–443.

Astrup, A.; Dyerberg, J.; Elwood, P.; Hermansen, K.; Hu, F.B.; Jakobsen, M.U.; Kok, F.J.; Krauss, R.M.; Lecerf, J.M.; LeGrand, P.; et al. The role of reducing intakes of saturated fat in the prevention of cardiovascular disease: where does the evidence stand in 2010? Am. J. Clin. Nutr. 2011, 93, 684–688.

Black, T.M.; Wang, P.; Maeda, N.; Coleman, R.A. Palm tocotrienols protect ApoE +/– mice from diet-induced atheroma formation. J. Nutr. 2000, 130, 2420–2426.

Boateng, J.; Verghese, M.; Chawan, C.B. Red palm oil suppresses the formation of azoxymethane (AOM) induced aberrant crypt foci (ACF) in Fisher 344 male rats. Food Chem. Tox. 2006, 44, 1667–1673.

Bonanome, A.; Grundy, S.M. Effect of dietary stearic acid on plasma cholesterol and lipoprotein levels. N. Eng. J. Med. 1988, 318, 1244–1248.

Budin, S.B.; Othman, F.; Louis, S.R.; Bakar, M.A.; Das, S.; Mohamed, J. The effects of palm oil tocotrienol-rich fraction supplementation on biochemical parameters, oxidative stress and the vascular wall of streptozotocin-induced diabetic rats. Clinics 2009, 64, 235–244.

Choudhury, N.; Tan, L.; Truswell, A.S. Comparison of palmolein and olive oil: effects on plasma lipids and vitamin E in young adults. Am. J. Clin. Nutr. 1995, 61, 1043–1051.

Clandinin, M.T.; Cook, S.L.; Konard, S.D.; French, M.A. The effect of palmitic acid on lipoprotein cholesterol levels. Int. J. Food Sci. Nutr. 2000, 51, S61–71.

Clandinin, M.T.; Cook, S.L.; Konrad, S.D.; Goh, Y.K.; French, M.A. The effect of palmitic acid on lipoprotein cholesterol levels and endogenous cholesterol synthesis in hyperlipidemic subjects. Lipids 1999, 34, S121–124.

Cook, S.L.; Konrad, S.D.; Goh, Y.K.; French, M.A.; Clandinin, M.T. Palmitic acid effect on lipoprotein profiles and endogenous cholesterol synthesis or clearance in humans. Asia Pacific J. Clin. Nutr. 1997, 6, 6–11.

Das, S.; Powell, S.R.; Wang, P.; Divald, A.; Nesaretnam, K.; Tosaki, A.; Cordis, G.A.; Maulik, N.; Das, D.K. Cardioprotection with palm tocotrienol: antioxidant activity of tocotrienol is linked with its ability to stabilize proteasomes. Am. J. Physiol. Heart Circ. Physiol. 2005, 289, H361–367.

Edem, D.O. Palm oil: biochemical, physiological, nutritional, hematological, and toxicological aspects: a review. Plant Foods Hum. Nutr. 2002, 57, 319–341.

Fairus, S.; Nor, R.M.; Cheng, H.M.; Sundram, K. Postprandial metabolic fate of tocotrienol-rich vitamin E differs significantly from that of alpha-tocopherol. Am. J. Clin. Nutr. 2006, 84, 835–842.

Fernandez, M.L.; Webb, D. The LDL to HDL cholesterol ratio as a valuable tool to evaluate coronary heart disease risk. J. Am. Coll. Nutr. 2008, 27, 1–5.

French, M.A.; Sundram, K.; Clandinin, M.T. Cholesterolaemic effect of palmitic acid in relation to other dietary fatty acids. Asia Pac. J. Clin. Nutr. 2002, 11, S401–407.

Goh, S.H.; Choo, Y.M.; Ong, A.S.H. Minor constituents of palm oil. J. Am. Oil Chem. Soc. 1985, 62, 237–240.

Goldstein, J.L.; Brown, M.S. Regulation of the mevalonate pathway. Nature 1990, 343, 425–430.

Grundy, S.M.; Vega, G.L. Plasma cholesterol responsiveness to saturated fatty acids. Am. J. Clin. Nutr. 1988, 47, 822–824.

Hayes, K.C.; Khosla, P. Dietary fatty acid thresholds and cholesterolemia. FASEB 1992, 6, 2600–2607.

Hayes, K.C.; Pronczuk, A. Replacing trans fat: the argument for palm oil with a cautionary note on interesterification. J. Am. Coll. Nutr. 2010, 29, 253S–284S.

Hayes, K.C.; Pronczuk, A.; Liang, J.S. Differences in the plasma transport and tissue concentrations of tocopherols and tocotrienols: observations in humans and hamsters. Proc. Soc. Exp. Biol. Med.1993, 202, 353–359.

Hayes, K.C.; Pronczuk, A.; Lindsey, S.; Diersen-Schade, D. Dietary saturated fatty acids (12:0, 14:0, 16:0) differ in their impact on plasma cholesterol and lipoproteins in nonhuman primates. Am. J. Clin. Nutr. 1991, 53, 491–498.

Hegsted, D.M.; McGandy, R.B.; Myers, M.L.; Stare, F.J. Quantitative effects of dietary fat on serum cholesterol in man. Am. J. Clin. Nutr. 1965, 17, 281–295.

Hu, F.B.; Stampfer, M.J.; Manson, J.E.; Rimm, E.; Colditz, G.A.; Rosner, B.A.; Hennekens, C.H.; Willett, W.C. Dietary fat intake and the risk of coronary heart disease in women. N. Engl. J. Med. 1997, 337, 1491–1499.

Hunter, J.E.; Applewhite, T.H. Reassessment of trans fatty acid availability in the US diet. Am. J. Clin. Nutr. Aug 1991, 54, 363–369.

Jakobsen, M.U.; O'Reilly, E.J.; Heitmann, B.L.; Pereira, M.A.; Balter, K.; Fraser, G.E.; Goldbourt, U.; Hallmans, G.; Knekt, P.; Liu, S. Major types of dietary fat and risk of coronary heart disease: a pooled analysis of 11 cohort studies. Am. J. Clin. Nutr. 2009, 89, 1425–1432.

Judd, J.T.; Clevidence, B.A.; Muesing, R.A.; Wittes, J.; Sunkin, M.E.; Podczasy, J.J. Dietary trans fatty acids: effects on plasma lipids and lipoproteins of healthy men and women. Am. J. Clin. Nutr. 1994, 59, 861–868.

Kennedy, A.; Martinez, K.; Chuang, C.C.; LaPoint, K.; McIntosh, M. Saturated fatty acid-mediated inflammation and insulin resistance in adipose tissue: mechanisms of action and implications. J. Nutr. 2009, 139, 1–4.

Keys, A.; Joseph, T.; Grande, F. Serum cholesterol response to changes in the diet. IV. Particular saturated fatty acids in the diet. Metabolism 1965, 14, 776–787.

Khanna, S.; Patel, V.; Rink, C.; Roy, S.; Sen, C.K. Delivery of orally supplemented alpha-tocotrienol to vital organs of rats and tocopherol-transport protein deficient mice. Free Radic. Biol. Med. 2005, 39, 1310–1319.

Khanna, S.; Venojarvi, M.; Roy, S.; Sen, C.K. Glutamate-induced c-Src activation in neuronal cells. Methods Enzymol. 2002, 352, 191–198.

Khosla, P.; Hajri, T.; Pronczuk, A.; Hayes, K.C. Decreasing dietary lauric and myristic acids improves plasma lipids more favorably than decreasing dietary palmitic acid in rhesus monkeys fed AHA step 1 type diets. J. Nutr. 1997, 127, 525S–530S.

Khosla, P.; Hayes, K.C. Comparison between the effects of dietary saturated (16:0), monounsaturated (18:1), and polyunsaturated (18:2) fatty acids on plasma lipoprotein metabolism in cebus and rhesus monkeys fed cholesterol-free diets. Am. J. Clin. Nutr. 1992, 55, 51–62.

Khosla, P.; Patel, V.; Whinter, J.M.; Khanna, S.; Rakhkovskaya, M.; Roy, S.; Sen, C.K. Postprandial levels of the natural vitamin E tocotrienol in human circulation. Antioxid. Redox. Signal 2006, 8, 1059–1068.

Khosla, P.; Sundram, K. A supplement on palm oil—why? J. Am. Coll. Nutr. 2010, 29, 237S–239S.

Khosla, P.; Sundram, K. Nutritional characteristics of palm oil. In Reducing Saturated Fats in Foods; Talbot, G., Ed.; Woodhead Publishing: UK, 2011; pp. 112–127.

Komiyama, K.; Iizuka, K.; Yamaoka, M.; Watanabe, H.; Tsuchiya, N.; Umezawa, I. Studies on the biological activity of tocotrienols. Chem. Pharm. Bull. (Tokyo) 1989, 37, 1369–1371.

Lemieux, I.; Lamarche, B.; Couillard, C.; Pascot, A.; Cantin, B.; Bergeron, J.; Dagenais, G.R.; Despres, J.P. Total cholesterol/HDL cholesterol ratio vs LDL cholesterol/HDL cholesterol ratio as indices of ischemic heart disease risk in men: the Quebec Cardiovascular Study. Arch. Intern. Med. 2001, 161, 2685–2692.

Lewington, S.; Whitlock, G.; Clarke, R.; Sherliker, P.; Emberson, J.; Halsey, J.; Qizilbash, N.; Peto, R.; Collins, R. Blood cholesterol and vascular mortality by age, sex, and blood pressure: a meta-analysis of individual data from 61 prospective studies with 55,000 vascular deaths. Lancet 2007, 370, 1829–1839.

Li, F.; Tan, W.; Kang, Z.; Wong, C.W. Tocotrienol enriched palm oil prevents atherosclerosis through modulating the activities of peroxisome proliferators-activated receptors. Atherosclerosis 2010, 211, 278–282.

Libby, P. Inflammation in atherosclerosis. Nature 2002, 420, 868–874.

Lindsey, S.; Benattar, J.; Pronczuk, A.; Hayes, K.C. Dietary palmitic acid (16:0) enhances high density lipoprotein cholesterol and low density lipoprotein receptor mRNA abundance in hamsters. Proc. Soc. Exp. Biol. Med. 1990, 195, 261–269.

Manorama, R.; Brahmam, G.N.; Rukmini, C. Red palm oil as a source of beta-carotene for combating vitamin A deficiency. Plant Foods Hum, Nutr. 1996, 49(1), 75–82.

Mattson, F.H.; Grundy, S.M. Comparison of effects of dietary saturated, monounsaturated, and polyunsaturated fatty acids on plasma lipids and lipoproteins in man. J. Lipid Res. 1985, 26, 194–202.

McIntyre, B.S.; Briski, K.P.; Tirmenstein, M.A.; Fariss, M.W.; Gapor, A.; Sylvester, P.W. Antiproliferative and apoptotic effects of tocopherols and tocotrienols on normal mouse mammary epithelial cells. Lipids 2000, 35, 171–180.

McNamara, D.J. Palm oil and health: a case of manipulated perception and misuse of science. J. Am. Coll. Nutr. 2010, 29, 240S–244S.

Mensink, R.P. Dietary fatty acids and cardiovascular health—an ongoing controversy. Ann. Nutr. Metab. 2011, 58, 66–67.

Mensink, R.P.; Katan, M.B. Effect of dietary trans fatty acids on high-density and low-density lipoprotein cholesterol levels in healthy subjects. N. Engl. J. Med. 1990, 323, 439–445.

Mensink, R.P.; Zock, P.L.; Kester, A.D.; Katan, M.B. Effects of dietary fatty acids and carbohydrates on the ratio of serum total to HDL cholesterol and on serum lipids and apolipoproteins: a meta-analysis of 60 controlled trials. Am. J. Clin. Nutr. 2003, 77, 1146–1155.

Mozaffarian, D.; Clarke, R. Quantitative effects on cardiovascular risk factors and coronary heart disease risk of replacing partially hydrogenated vegetable oils with other fats and oils. Eur. J. Clin. Nutr. 2009, 63, S22–33.

MPOC. Reaching the Pinnacle of Success. Malaysian Palm Oil Council: Kuala Lumpur, 2009.

Nafeeza, M.I.; Norzana, A.G.; Jalaluddin, H.L.; Gapor, M.T. The effects of a tocotrienol-rich fraction on experimentally induced atherosclerosis in the aorta of rabbits. Malays. J. Pathol. 2001, 23, 17–25.

Nagendran, B.; Unnithan, U.R.; Choo, Y.M. Characteristics of red palm oil, a-carotene- and vitamin E-rich refined oil for food uses. Food Nutr. Bull. 2000, 2, 189–194.

Nesaretnam, K.; Radhakrishnan, A.; Selvaduray, K.R. Effect of palm oil carotene on breast cancer tumorigenicity in nude mice. Lipids 2002, 37, 557–560.

Ng, T.K.; Hayes, K.C.; DeWitt, G.F. Dietary palmitic and oleic acids exert similar effects on serum cholesterol and lipoprotein profiles in normocholesterolemic men and women. J. Am. Coll. Nutr. Aug 1992, 11, 383–390.

Ngah, W.Z.; Jarien, Z.; San, M.M.; Marzuki, A.; Top, G.M.; Shamaan, N.A.; Kadir, K.A. Effect of tocotrienols on hepatocarcinogenesis induced by 2-acetylaminofluorene in rats. Am. J. Clin. Nutr. 1991, 53, 1076S–1081S.

O'Byrne, D.; Grundy, S.; Packer, L. Studies of LDL oxidation following alpha-, gamma-, or delta-tocotrienyl acetate supplementation of hypercholesterolemic humans. Free Radic. Biol. Med. 2000, 29, 834–845.

Oh, K.; Hu, F.B.; Manson, J.E.; Stampfer, M.J.; Willett, W.C. Dietary fat intake and risk of coronary heart disease in women: 20 years of follow-up of the nurses' health study. Am. J. Epidemiol. 2005, 161, 672–679.

Packer, L.; Weber, S.U.; Rimbach, G. Molecular aspects of alpha-tocotrienol antioxidant action and cell signalling. J. Nutr. 2001, 131, 369S–373S.

Park, H.A.; Kubicki, N.; Gnyawali, S. Natural vitamin E α-tocotrienol protects against ischemic stroke by induction of multidrug resistance-associated protein 1. Stroke 30, 2011.

Pearce, B.C.; Parker, R.A.; Deason, M.E.; Dischino, D.D.; Gillespie, E.; Qureshi, A.A.; Volk, K.; Wright, J.J. Inhibitors of cholesterol biosynthesis. 2. Hypocholesterolemic and antioxidant activities of benzopyran and tetrahydronaphthalene analogues of the tocotrienols. J. Med. Chem. 1994, 37, 526–541.

Qureshi, A.A.; Sami, S.A.; Salser, W.A.; Khan, F.A. Synergistic effect of tocotrienol-rich fraction (TRF(25)) of rice bran and lovastatin on lipid parameters in hypercholesterolemic humans. J. Nutr. Biochem. 2001, 12, 318–329.

Ramsden, C.E.; Hibbeln, J.R.; Majchrzak, S.F.; Davis, J.M. n-6 fatty acid-specific and mixed polyunsaturate dietary interventions have different effects on CHD risk: a meta-analysis of randomised controlled trials. Br. J. Nutr. 2010, 104(11), 1586–1600.

Rice, A.L.; Burns, J.B. Moving from efficacy to effectiveness: red palm oil's role in preventing vitamin A deficiency. J. Am. Coll. Nutr. 2010, 29, 302S–313S.

Roels, O.A.; Djaeni, S.; Trout, M.E. The effect of protein and fat supplements on vitamin A-deficient Indonesian children. Am. J. Clin. Nutr. 1963, 12, 380–387.

Samant, G.V.; Wali, V.B.; Sylvester, P.W. Anti-proliferative effects of gamma-tocotrienol on mammary tumour cells are associated with suppression of cell cycle progression. Cell Prolif. 2010, 43, 77–83.

Sen, C.K.; Khanna, S.; Rink, C.; Roy, S. Tocotrienols: the emerging face of natural vitamin E. Vitam. Horm. 2007, 76, 203–261.

Shah, S.; Gapor, A.; Sylvester, P.W. Role of caspase-8 activation in mediating vitamin E-induced apoptosis in murine mammary cancer cells. Nutr. Cancer 2003, 45, 236–246.

Siri-Tarino, P.W.; Sun, Q.; Hu, F.B.; Krauss, R.M. Meta-analysis of prospective cohort studies evaluating the association of saturated fat with cardiovascular disease. Am. J. Clin. Nutr. 2010, 91, 535–546.

Sundram, K.; Gapor, A. Vitamin E from palm oil: its extraction and nutritional properties. Lipid Technol. 1992, 4, 137–141.

Sundram, K.; Hayes, K.C.; Siru, O.H. Dietary palmitic acid results in lower serum cholesterol than does a lauric-myristic acid combination in normolipemic humans. Am. J. Clin. Nutr. 1994, 59, 841–846.

Sundram, K.; Karupaiah, T.; Hayes, K.C. Stearic acid-rich interesterified fat and trans-rich fat raise the LDL/HDL ratio and plasma glucose relative to palm olein in humans. Nutr. Metab. 2007, 4, 3.

Sundram, K.; Khor, H.T.; Ong, A.S.; Pathmanathan, R. Effect of dietary palm oils on mammary carcinogenesis in female rats induced by 7,12-dimethylbenz(a)anthracene. Cancer Res. Mar 15 1989, 49, 1447–1451.

Sundram, K.; Sambanthamurthi, R.; Tan, Y.A. Palm fruit chemistry and nutrition. Asia Pac. J. Clin, Nutr. 2003 2003, 12, 355–362.

Sylvester, P.W.; Shah, S. Antioxidants in dietary oils: their potential role in breast cancer prevention. Mal. J. Nutr. 2002, 8, 1–11.

Tholstrup, T., Hjerpsted, J., Raff, M. Palm olein increases plasma cholesterol moderately compared with olive oil in healthy individuals. Am. J. Clin. Nutr. 2011, 94, 1426–1432.

Traber, M.G.; Rallis, M.; Podda, M.; Weber, C.; Maibach, H.I.; Packer, L. Penetration and distribution of alpha-tocopherol, alpha- or gamma-tocotrienols applied individually onto murine skin. Lipids 1998, 33, 87–91.

van Stuijvenberg, M.E.; Dhansay, M.A.; Lombard, C.J.; Faber, M.; Benade, A.J. The effect of a biscuit with red palm oil as a source of beta-carotene on the vitamin A status of primary school children: a comparison with beta-carotene from a synthetic source in a randomised controlled trial. Eur. J. Clin. Nutr. 2001, 55(8), 657–662.

van Stuijvenberg, M.E.; Faber, M.; Dhansay, M.A.; Lombard, C.J.; Vorster, N.; Benadé, A.J. Red palm oil as a source of beta-carotene in a school biscuit used to address vitamin A deficiency in primary school children. Int. J. Food Sci. Nutr. 2000, 51, S43–S50.

Vasanthi, H.R.; Parameswari, R.P.; Das, D.K. Multifaceted role of tocotrienols in cardioprotection supports their structure: function relation. Genes Nutr. 2011. May 21 Epub.

Voon, P.T.; Ng, T.K.; Lee, V.K.; Nesaretnam, K. Diets high in palmitic acid (16:0), lauric and myristic acids (12:0 + 14:0), or oleic acid (18:1) do not alter postprandial or fasting plasma homocysteine and inflammatory markers in healthy Malaysian adults. Am. J. Clin. Nutr. 2011, 94, 1451–1457.

Wada, S. Chemoprevention of tocotrienols: the mechanism of antiproliferative effects. Forum Nutr. 2009, 61, 204–216.

World Health Organization (WHO) Global Infobase—international comparison for cardiovascular disease. 2011; https://apps.who.int/infobase/Comparisons.aspx?. Accessed Nov 10, 2011.

Willett, W.C.; Stampfer, M.J.; Manson, J.E.; Colditz, G.A.; Speizer, F.E.; Rosner, B.A.; Sampson, L.A.; Hennekens, C.H. Intake of trans fatty acids and risk of coronary heart disease among women. Lancet 1993, 341, 581–585.

Yamanushi, T.T.; Ichinose, T.; Seto, H.; Sagai, M.; Torii, M.I.; Igarashi, O. The effect of dietary carotenoids on lung tumorigenesis induced by intratracheally instillated diesel exhaust particles. J. Nutr. Sci. Vitamin. 2001, 47, 32–39.

Yap, S.C.; Choo, Y.M.; Ooi, C.K.; Ong, A.S.H.; Goh, S.H. Quantitative analysis of carotenes in the oil from different palm species. Elaeis 1991, 3, 309–318.

Yokokawa, H.; Yasumura, S.; Tanno, K.; Ohsawa, M.; Onoda, T.; Itai, K.; Sakata, K.; Kawamura, K.; Tanaka, F.; Yoshida, Y. Serum low-density lipoprotein to high-density lipoprotein ratio as a predictor of future acute myocardial infarction among men in a 2.7-year cohort study of a Japanese northern rural population. J. Atheroscler. Thromb. 2011, 18, 89–98.

You, C.S.; Parker, R.S.; Swanson, J.E. Bioavailability and vitamin A value of carotenes from red palm oil assessed by an extrinsic isotope reference method. Asia Pac. J. Clin. Nutr. 2002, 11, S438–442.

19

Food Uses of Palm Oil and Its Components

Garima Pande[1], Casimir C. Akoh[1], and Oi-Ming Lai[2]
[1]Department of Food Science and Technology, University of Georgia, Athens, GA, USA and [2]Department of Bioprocess Technology, Universiti Putra Malaysia, 43400 UPM Sedang, Selangor, Malaysia

Introduction

Palm oil is an edible oil derived from the fruits of the oil palm (*Elaeis guineensis*), belonging to Palmaceae family. Palm oil is extracted from the mesocarp (pulp) of the fruit which contain 45–55% oil (Edem, 2002), whereas palm kernel oil is derived from the kernel (seed) of the oil palm. Oil palm originated from South Africa and was introduced to Malaysia as an ornamental plant. Worldwide production of palm oil has increased from 15.2 million tonnes in 1995 to 46.7 million tonnes in 2010, produced mainly by Indonesia (47%) and Malaysia (38%) (MVO, 2010). Palm oil is the world's largest utilized vegetable oil, with an estimated global consumption of 46.8 million tonnes in 2010 (MVO, 2010). It is an important and versatile vegetable oil which is used as a raw material for both food and non-food industries. Most of the oil is used for food applications such as cooking, margarines, spreads, confectionary fats, ice cream, emulsifiers, and vanaspati.

Composition of Palm Oil

Being a vegetable oil, palm oil is cholesterol-free. It has a unique fatty acid and triacylglycerol profile. It is naturally semi-solid at room temperature and does not require hydrogenation for use as a food ingredient. Palm oil is a rich source of carotenoids and tocopherols, which confer natural stability against oxidative deterioration.

Fatty Acid Profile

All oils and fats have the triacylglycerol (TAG) structure containing the fatty acids. Their differences lie in the types of fatty acids (RCOOH) attached to the glycerol backbone in positions 1, 2, and 3. Palm oil has a balanced fatty acid composition with approximately 50% saturated and 50% unsaturated fatty acids. The fatty acid composition of palm oil, palm olein, and palm stearin are shown in Table 19-A. Palmitic and oleic acids are the major fatty acids, along with linoleic acid and only a trace

Table 19-A. Fatty Acid Composition (wt%) and TAG Composition (%) of Palm Oil and Its Fractions.

Fatty acids	Palm oil	Palm olein	Palm stearin
C12:0	0.2	0.2	0.2
C14:0	1.1	1.0	1.3
C16:0	44.0	39.8	54.0
C18:0	4.5	4.4	4.7
C18:1n-9	39.2	42.5	32.3
C18:2n-6	10.1	11.2	7.0
C18:3n-3	0.4	0.4	0.1
C20:0	0.4	0.4	0.2
Others	0.1	0.1	0.2
TAG species			
C46	1.2	0.8	3.1
C48	8.1	3.3	23.7
C50	39.9	39.5	40.3
C52	38.8	42.7	25.3
C54	11.4	12.8	6.9
Others	0.6	0.9	0.7

TAG species are grouped on the basis of the total carbon number.
Source: Aini & Miskandar (2007).

amount of linolenic acid. This composition is significantly different from palm kernel oil, which is almost 85% saturated (Mukherjee & Mitra, 2009). Predominant saturated fatty acids with negligible amount of linolenic acid make palm oil oxidatively stable, thus increasing its food applications. Among saturated fatty acids, 17–23% of the palmitic acid is present at *sn*-2 position, and most of myristic and stearic acids are at *sn*-1,3 positions (Renaud et al., 1995). The fatty acid composition, mainly palmitic and stearic acids, can vary from country to country. Zaire, Indonesia, and Malaysia palm oils have 41–43%, 46–50%, 46–51% palmitic acid, respectively, and 4.4–6.3%, 2.0–4.0%, 1.5–3.5% stearic acid, respectively (Clegg, 1973).

Triacylglycerol Species

Crude palm oil contains mainly TAG (>95%), diacylglycerols (DAG), free fatty acids (FFA), and minor components (approximately 1%). Among the TAGs, saturated TAGs account for 7–10% while fully unsaturated ones account for 6–12% (Karleskind & Wolff, 1996). There are more than 33 TAG molecular species present in

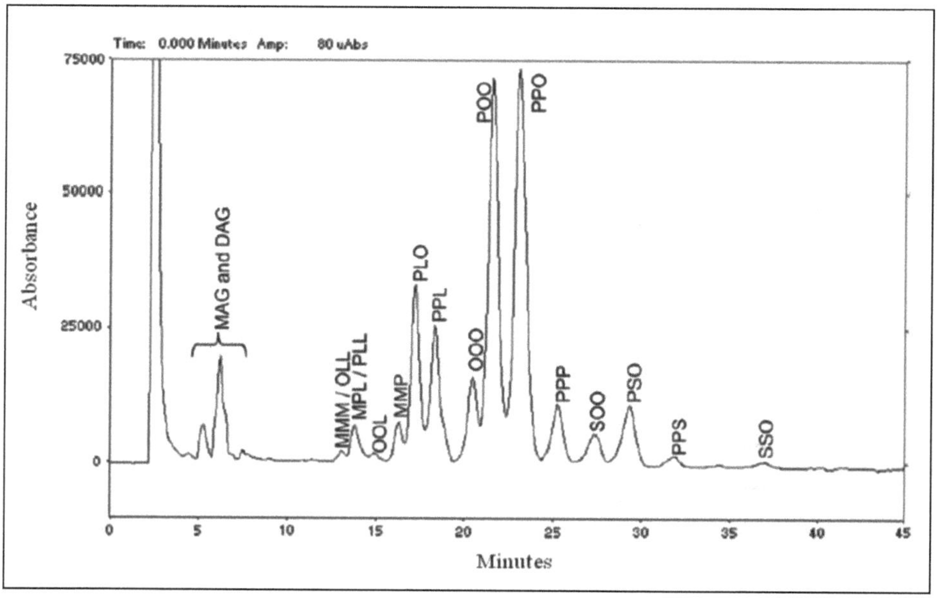

Fig. 19.1 TAG profile of palm oil. MAG, monoacylglycerol; DAG, diacylglycerol; M, myristic; P, palmitic, S, stearic; O, oleic; L, linoleic. Source: Reproduced with permission from Chen et al. (2007).

palm oil with symmetrical POP, POO, POL, and PLP being the major ones. TAG profile of palm oil is shown in Fig. 19.1. Based on total carbon number, TAGs are usually grouped into different classes, as shown in Table 19-A. More than 85% of the unsaturated fatty acids are located at the *sn*-2 position because of the specificity of the *sn*-2 position for unsaturated fatty acids (Karleskind & Wolff, 1996). The saturated fatty acids are better hydrolyzed and absorbed when present at *sn*-1, 3 positions, while unsaturated fatty acids are better absorbed at *sn*-2 position (Gebauer et al., 2005). Physical characteristics such as melting point and polymorphism of the palm oil are also influenced by its TAG molecular species to a great extent (Sambanthamurthi et al., 2000). The presence of the symmetrical TAGs, POP, and PLP makes it possible to fractionate palm oil to obtain fraction suitable for chocolates and confectioneries manufacture.

Minor Components

Tocopherols and tocotrienols are collectively known as Vitamin E. Vitamin E is the major lipid-soluble, membrane-localized antioxidant in humans. Their structure consists of a polar chromanol head and a hydrophobic prenyl tail which results in amphipathic nature of these compounds (Munne-Bosch & Alegre, 2002). Vegetable oils, especially the seed oils, are rich sources of tocopherols. Tocotrienols, on the other hand, are predominantly found in palm and rice bran oils (Sambanthamurthi et al., 2000). The total vitamin E content of crude palm oil ranges between 600 and 1000 parts per million (ppm), of which 18–22% are tocopherols and 78–82% are tocotrienols (Sambanthamurthi et al., 2000). Among the tocotrienols, gamma-tocotrienol is predominant (46%), followed by alpha- (22%), and delta-tocotrienols (12%) (Gapor & Ong, 1982). The vitamin E content of palm oil is partially lost during processing. It has been reported that refined, bleached, and deodorized (RBD) palm oil, palm olein, and palm stearin retain approximately 69, 72, and 76% of the original level of vitamin E in the crude oils, respectively (Sambanthamurthi et al., 2000). During fractionation of palm oil, vitamin E tends to partition preferentially into the olein fraction thereby making it more stable (Sambanthamurthi et al., 2000).

Carotenoids are a group of brightly colored compounds found in abundance in the plant kingdom. Carotenoids have long chains of alternating double and single bonds which allows extensive electron delocalization. The basic skeleton of carotenoids has 40 carbon atoms and can be modified by cyclization at one or both ends, by reducing certain double bonds, or by addition of oxygen containing functional groups (Halliwell & Gutteridge, 2007). Carotenoids are the precursor of vitamin A, which is essential for cell growth, differentiation, and vision. The dark orange-red color of crude palm oil is due to its high content of carotene (500–700 mg/L) (Edem, 2002). The major carotenoids in palm oil are beta- and alpha-carotenes (Edem, 2002). Carotenes are sensitive to oxygen and light. Carotenoids are removed during the refining and deodorization of the crude oil to produce a light yellow refined palm oil preferred by consumers (Cottrell, 1991).

Phytosterols are 28 or 29 carbon alcohols resembling cholesterol structurally. Most phytosterols contain 9–10 carbon atom side chain, compared to 8 carbon atom side chain in cholesterol (Lagarda et al., 2006). Phytosterols may decrease epithelial cell proliferation by suppressing bacterial metabolism of cholesterol and/or secondary bile acids in the colon and by increasing the excretion of cholesterol (Ling & Jones, 1995). Furthermore, phytosterols also show anti-cancer, anti-sclerotic, anti-inflammatory, and antioxidative effects (Kritchevsky & Chen, 2005). They are a major constituent of the unsaponifiable matter in oil. The total content of sterols in palm oil is about 0.03% (Sambanthamurthi et al., 2000). The major sterols in palm oil are beta-sitosterol (281.8 mg/L), campesterol (110.97 mg/L), and stigmasterol (55.97 mg/L) (Edem, 2002). Similar to other plant-derived edible oils, cholesterol content in palm oil is negligible and the phytosterol levels are further reduced during the refining process.

Palm oil also contains other minor constituents which affect its physical, chemical, and nutritional properties. Squalene is a 30-carbon hydrocarbon with six double bonds. It is a key intermediate in cholesterol synthesis and exhibits important biological activities. Crude palm oil contains about 200–350 ppm squalene (Sambanthamurthi et al., 2000). DAGs, MAGs, and FFAs are present in palm oil from damaged fruits. Palm oil contains mainly three types of DAGs, C34 (palmitoyloleoylglycerol, 54.4%), C36 (dioleoylglycerol, 33.0%), and C32 (dipalmitoylglycerol, 12.6%) (Siew, 1994). Presence of high amount of DAGs affects the crystallization and fractionation of palm oil (Kellens et al., 2007). The content of MAGs in palm oil is usually below 1% (Sambanthamurthi et al., 2000). Crude palm oil may contain 10% or more FFAs depending on handling procedures, but they are significantly reduced during refining.

The other important color pigment apart from carotenoid is chlorophyll. Chlorophylls have adverse effects on oxidative stability, hydrogenation, and refining of the oil (Sambanthamurthi et al., 2000). The visual effect of chlorophyll is nullified by the higher level of carotenoids in crude palm oil from ripe fruits (Sambanthamurthi et al., 2000). However, palm oil extracted from immature fruits contains more chlorophyll and less carotenoids. It was reported that the total chlorophyll content in Malaysian crude palm oil ranged from 897 to 4000 mg/kg (Tan et al., 2000). Palm oil also contains polar lipids such as glycolipids and phospholipids. Glycolipids consist of various types of long chain sugar derivatives which may contain a DAG, a ceramide backbone or a phosphorylated polysaccharide-lipid complex. The total glycolipids in palm oil range from 1000 to 3000 ppm (Sambanthamurthi et al., 2000). Phospholipids are asymmetrical phosphoric acid diesters comprised of C-C bonds, ester bonds, and phosphoester bonds. Phospholipids have a polar head group and a non-polar fatty acid region due to which they have an amphipathic nature (Erickson, 2008). Phospholipids are present in relatively small quantities (5–130 ppm) in palm oil as compared to other vegetable oils (Sambanthamurthi et al., 2000). The main phospholipids identified in Indian palm oil are phosphatidylcholine (34–35 %), phosphatidylethanolamine (22–26%), phosphatidylinositol (21–25%), cardiolipin (7–8%), and phosphatidylglycerol (5–7%). The predominant fatty acids in all these phospholipids were palmitic, stearic, oleic, and linoleic acids (Kulkarni et al., 1991).

Palm oil also contains low levels of phenolic compounds (<100 mg/L), which are responsible for the initial darkening of palm oil during frying (Berger, 1987). Other minor compounds in crude palm oil are tannins, trace flavonoids, terpenoids, aliphatic alcohols, hydrocarbons, ketones, wax esters, and methyl esters which are reduced during refining (Edem, 2002).

Fractionation of Palm Oil and Their Composition

Most of the oils and fats have only a limited application in their natural forms. Various physical and chemical modifications are required to diversify their food use. Fractionation is a selective physical and/or thermo-mechanical process to separate a mixture into two or more fractions with distinct physical and chemical properties (Kellens et al., 2007). It is a fully reversible modification process. The separation can be based on differences in solidification, solubility, or volatility of the different constituents (Kellens et al., 2007). Palm oil fractionation utilizes the differences in the crystallization properties of TAGs to separate a mixture into a low melting liquid fraction and a high melting liquid fraction. There are three different types of fractionation: dry fractionation, detergent fractionation, and solvent fractionation.

Palm oil is one of the most important fractionated oils, yielding a range of different fractions, as shown in Fig. 19.2. The TAGs of oils and fats can be classified into four types based on their composition, namely: trisaturated (SSS), disaturated (SSU/SUS), diunsaturated (SUU/USU), and triunsaturated (UUU) (Kellens et al., 2007). These

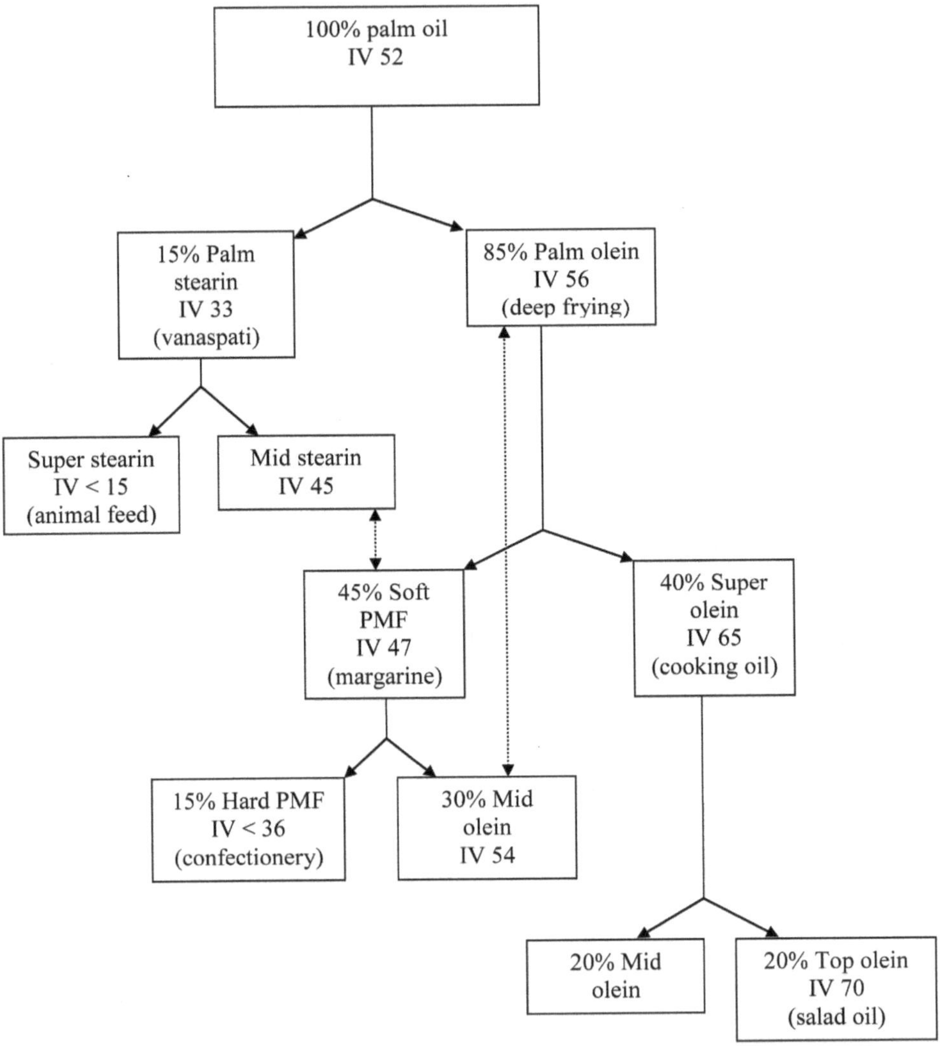

Fig. 19.2 Fractionation scheme of palm oil. IV, Iodine value. Source: Modified from Kellens et al. (2007); Gibon et al. (2009).

different TAGs exhibit different physico-chemical properties which determine their applications. Palm oil mainly contains 6–10% SSS (mainly PPP, tripalmitoylglycerol), 44–50% SSU-SUS (mainly POP, dipalmitoyl-oleylglycerol and PLP, dipalmitoyl-linoleylglycerol), 38–42% SUU-USU (mainly POO, dioleyl-palmitoylglycerol and PLO, palmitoyl-linoleyl-oleylglycerol), and 5–8% UUU (mainly OOO, trioleylglycerol and OOL, dioleyl-linoleylglycerol) (Kellens et al., 2007). These TAGs are separated from each other during different steps of fractionation, thereby yielding different fractions with distinct physical and chemical properties. Additionally, palm oil crystallizes in the beta prime form which helps in the fractionation and filtration process as large crystals are formed and easily filtered (Siew, 2002). Palm oil may be fractionated into two major fractions: a liquid oil (65–70%) olein and a solid fraction (30–35%) stearin (Kellens et al., 2007). In palm oil fractionation, palm olein is the main product and palm stearin is obtained as a co-product. The important physical properties of palm oil, palm olein, and palm stearin are given in Table 19-B.

Other fractions obtained are palm mid-fraction (PMF) and the double fractionated palm olein (super olein). PMF contains 60% palmitic and 40% oleic acid, high amount of SUS type TAG (POP), and is commonly used as a cocoa-butter equivalent (CBE) (Edem, 2002). Its physico-chemical properties lie in between palm olein and palm stearin. Palm olein and super olein are more unsaturated than palm oil or palm stearin and deteriorates less rapidly than many other vegetable oils. Another important product obtained from palm oil by chemical neutralization followed by deodorizing in mild conditions is red palm oil, which is rich in carotenes, tocopherols, and tocotrienols (Kellens et al., 2007). Red palm oil can also be further fractionated into olein and stearin, thereby facilitating its use in a wide range of food applications. Crude palm oil and refined palm olein have high oxidative and cold stability due to the high proportion of vitamin E, carotenoids, and minimal amount of polyunsaturated fatty acids (Edem, 2002; Kellens et al., 2007).

Table 19-B. Physical Properties of Palm Oil and Its Fractions.

Property	Palm oil	Palm olein	Palm stearin
Apparent density, g/mL at 50°C	0.8889–0.8896	0.8969–0.8977	0.8813–0.8844
Refractive index at 50°C	1.4521–1.4541	1.4589–1.4592	1.4482–1.4501
Saponification value, mg KOH/g oil	194–205	194–202	193–205
Unsaponifiable matter, wt%	0.19–0.44	0.30–1.30	0.30–0.90
Iodine value (g iodine/100 g fat)	50.4–53.7	56.0–59.1	27.8–45.1
Slip melting point, °C	33.8–39.2	19.2–23.6	46.6–53.8

Source: Sue (2009).

Palm Olein

Palm olein is the liquid fraction obtained during fractionation of palm oil, which involves crystallization under controlled temperature and removal of crystals by filtration. Palm olein contains higher amounts of oleic (39–45%) and linoleic acids (10–13%) compared to palm oil (Siew, 2002). It has a melting point of 18–20°C and therefore is a liquid at room temperature (25°C) (Siew, 2002). Various blends of palm olein with other vegetable oils such as soybean, sunflower, and canola oils increase the oxidative and cold stability of the mix (Siddique et al., 2010). There are different grades of palm olein based on the iodine value (IV) with potential for specific end-use applications. A more unsaturated fraction called super olein is obtained by further fractionation of olein. It contains 43–49% oleic and 10–15% linoleic acids (Sue, 2009). Super olein has an iodine value of 60 or above. Therefore, super olein has better clarity, stability, and lower tendency to turn cloudy compared to normal olein (Sue, 2009). The clarity of palm olein is significantly affected by its DAG content, which has a preference to separate into the olein phase during fractionation (Siew, 2002). Fig. 19.3 shows the melting and cooling thermograms of palm olein. Com-

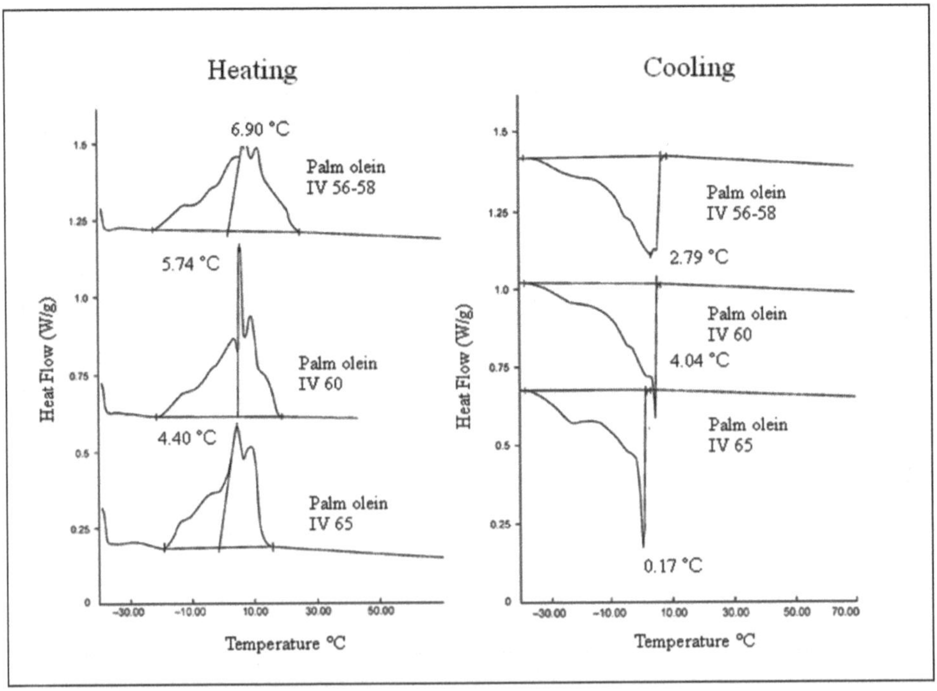

Fig. 19.3 Melting and cooling DSC curves of palm olein. IV, Iodine value. Source: Reproduced with permission from Siew (2002).

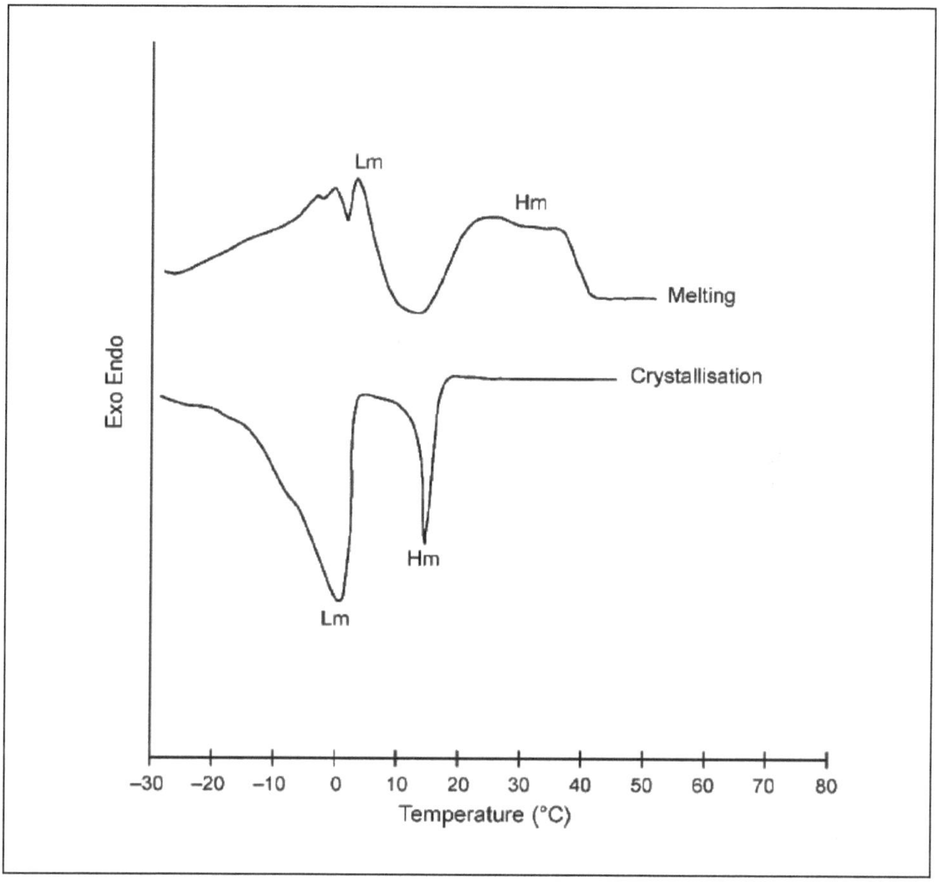

Fig. 19.4 Melting and cooling DSC curves of palm oil. Hm, high melting; Lm, low melting. Source: Reproduced with permission from Siew (2002).

pared to palm oil (Fig. 19.4), the crystallization thermogram shows a single broad peak (exotherm) with sharpness directly proportional to the iodine values. Similarly, a single broad endotherm is obtained during heating. Melting temperature decreases with increasing IV (Siew, 2002).

Palm Stearin

Palm stearin is the harder fraction of palm oil, containing a higher proportion of saturated fatty acids and TAGs with a higher melting point of 48–50°C. The palmitic acid content ranges from 49 to 68% and oleic content from 24 to 34% (Sue, 2009). A much harder stearin containing 79% palmitic acid and a tripalmitoylglycerol (PPP) content of 60% is also available. It is mainly used as hard stock for soft margarines and in infant fat formulas (Sue, 2009). Similar to super olein, a second fractionation of palm stearin yields another fraction called PMF, which is utilized in the manufacture of a cocoa butter equivalent (Siew, 2002). PMF can also be obtained by double fractionation of palm olein. This PMF is more suitable for CBE than the one obtained from stearin because of its higher concentration of disaturated TAG and a steep solid fat content (Kellens et al., 2007). Palm stearin has a wide range of composition, physical properties, solid fat content, and iodine values. These variations result in different types of palm stearin, which allows food manufacturers a wide choice of materials for their formulations. Palm stearin can provide the required solids in blends with unsaturated vegetable oils. It is a natural source of solid fat and can be used to provide increased stability for many types of foods that require solid fat functionality. Therefore, the use of hydrogenation can be avoided. Fig. 19.5 shows the heating and cooling profiles of different palm stearins.

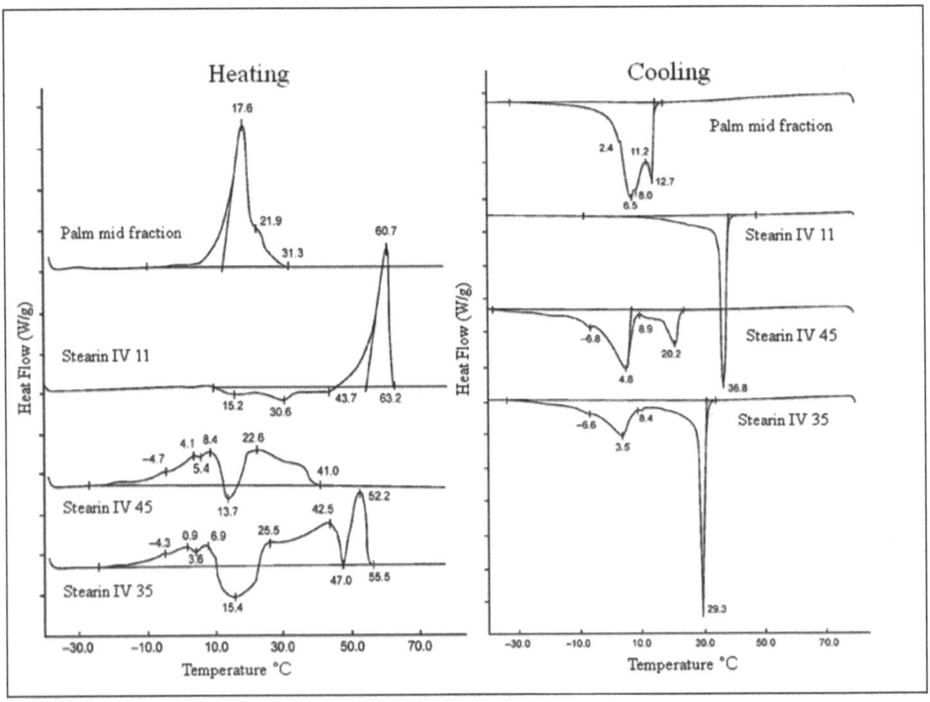

Fig. 19.5 Melting and cooling DSC curves of palm stearin. IV, Iodine value. Source: Reproduced with permission from Siew (2002).

Food Uses of Palm Oil, Its Fractions, and Components

Palm oil and its fractions are used in a number of food products ranging from margarine to salad dressings (Fig. 19.6). Certain characteristics of palm oil such as a high solid fat content (avoiding hydrogenation), high oxidative stability (long shelf life), high and low melting TAGs (wide plastic range), constant supply, competitive price, slow crystallization properties, structural hardness, and a tendency for recrystallization allow the use of the oil in diverse food applications (Edem, 2002).

Margarines and Shortenings

Margarine is a water-in-oil (w/o) emulsion comprised of at least 80% lipid phase (Aini & Miskandar, 2007). There are different types of margarines based on hardness and melting point. Hard and medium plastic are for bakery margarine and medium plastic and soft are for table margarine. Table margarine is further divided into tub and stick margarine (Weiss, 1983). Tub margarine should be spreadable at refrigeration temperature and therefore has low solids at low temperature. Stick margarine has relatively higher solid fats than tub margarine. It should have solid fat content of 28% at

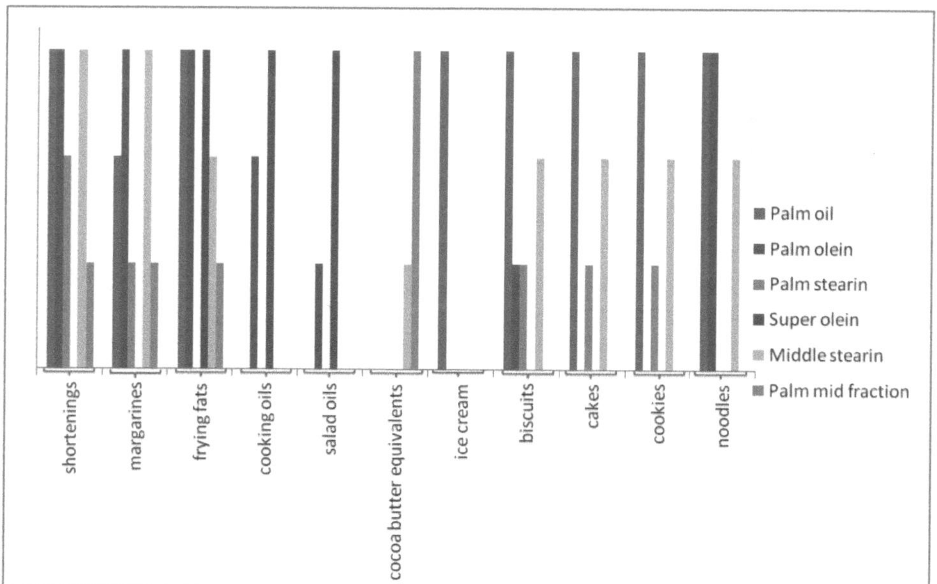

Fig. 19.6 Food applications of palm oil and its fractions. The height of the bars represents the magnitude of use. (Not drawn to scale.) Source: Modified from Kellens et al. (2007).

15°C to facilitate wrapping and between 12 and 15% at 20°C to avoid the separation of liquid and solid phases (Sahri & Idris, 2010). Also, the margarine produced should have a complete melting temperature of below 37°C to ensure the release of flavor and to impart smooth mouth-feel (Sahri & Idris, 2010). Many vegetable oils have been modified to produce margarine with required solid fat content and melting behavior. Hydrogenation is the main process to convert liquid oils such as soybean, canola, and sunflower oils to solid fats. Partial hydrogenation gives rise to *trans* fatty acids (TFA). Intake of high amounts of TFA is positively correlated with increased risk of coronary heart disease (Hu et al., 1997), inflammation (Mozaffarian et al., 2004), and cancer (Astorg, 2005). The natural solid-fat fraction of palm oil, palm stearin, is a suitable *trans*-free main ingredient in margarines, shortenings, and vanaspati.

Being semi-solid at ambient temperature makes palm oil very suitable for the production of margarines and shortenings. Palm oil contains an equal amount of unsaturated and saturated fatty acids, a unique property among vegetable oils and fats. Low levels of polyunsaturated fatty acids and high proportion of tocopherols and tocotrienols makes palm oil and its products resistant against oxidation (Sambanthamurthi et al., 2000). Palm oil and its fractions provide consistency, texture, and structure in margarines and shortenings without the need for hydrogenation. It has a melting point range of 32–40°C, which is very close to body temperature (Aini & Miskandar, 2007). Crude oils have slightly higher slip melting point than refined oils due to their higher content of DAGs and FFAs. An increase in DAG content by 10% has shown to reduce the solids content by 20% (Siew & Ng, 1999). The application of an oil is determined by its melting/crystallization behavior and solid fat content (SFC). Solid fat content of palm oil and its fractions are shown in Fig. 19.7. The melting and crystallization profiles of palm oil are shown in Fig. 19.4. Both the melting and cooling thermograms show two main endotherms and exotherms, respectively, indicating the presence of high and low melting fractions in the oil (Siew, 2002). TAG crystals exist in one of three major polymorphic forms known as alpha, beta, and beta prime. The alpha form is the least stable and the beta form the most stable. Fats containing predominantly beta prime TAG crystals impart firm texture, glossy appearance, and smooth mouth-feel to margarines, whereas those with predominantly beta TAG crystals results in softer fats with grainy texture/mouth-feel. Oils such as canola, soybean, sunflower, and olive crystallize in the beta form, whereas oils such as palm, palm kernel, and cottonseed crystallize in the beta prime form. Oils that crystallize in the beta prime form usually have increased TAG or fatty acid chain-length diversity. Palm oil consists of a diverse range of TAGs, which promotes beta prime crystal formation. Palm oil and palm olein contain mainly C50 and C52 which are beta prime tending. Palm olein has a higher content of the C52 and lower content of the C48 than palm stearin. The distribution and position of palmitic and stearic acids on the TAG molecules influence the crystallization behavior, melting properties, and polymorphic form of the oil or fat (Aini & Miskandar, 2007). In order for a TAG to exist in the beta prime crystal form, its carbon number should be kept as low as

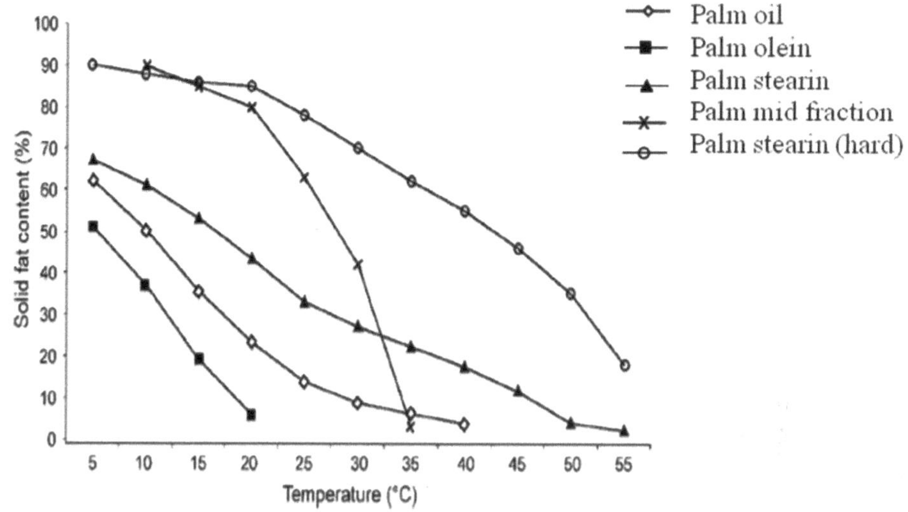

Fig. 19.7 Solid fat content of palm oil and its fractions. Source: Reproduced with permission from Siew (2002).

possible, preferably below C54. High content of palmitic acid (44%) promotes beta prime polymorph. Therefore, palm oil acts as beta prime crystal promoter in margarines and shortenings. Margarine made from canola oil exhibits beta crystal form whereas canola palm, soybean, and corn oil margarines show beta prime crystals (Aini & Miskandar, 2007). Crystal morphology is of great importance in the manufacture of margarine because it affects product consistency and acceptability. Palm stearin, the hard fraction of palm oil, is high in C48 TAG or tripalmitin (PPP). PPP is beta crystal tending. Palm stearin is a useful hard stock and contributes to the formulation of margarines and shortenings by providing structural support to the products. Palm stearin also helps maintain the products' structure and shape during storage and transportation temperature fluctuations (Yu et al., 2009). The main disadvantage with formulations containing high amounts of palm oil is its slow crystallization due to the presence of DAGs (5–8%). Furthermore, during storage there is an increase in SFC and this may lead to post-hardening (Siew, 2002).

The post-hardening problem in margarines and shortenings containing high amounts of palm oil could be overcome by using interesterification (Young, 1985). Interesterification can be carried out chemically or enzymatically by incorporating high melting fatty acids into oils or by reacting high melting fractions of natural oils or fully hydrogenated fats with liquid oils. This increases the solid fat content and oxidative stability of the product. Moreover, this process does not result in the formation of *trans* fatty acids, as is the case with partial hydrogenation, and therefore has no adverse effects on plasma lipoprotein profile. Structured lipids (SL) are defined as TAGs that have been structurally modified either by incorporation of new fatty acids or changing the position of existing fatty acids to yield novel TAGs of desired physical, chemical, and nutritional properties (Fomuso & Akoh, 2001). Various methods like hydrogenation, fractionation, blending, interesterification, esterification, and even bioengineering have been used for modification of natural TAGs. But SLs are produced by either chemical or enzymatic interesterification (Xu, 2000). Although chemical interesterification involves lower cost and easy scale-up process it lacks specificity. It results in the random redistribution of fatty acids on the TAG backbone to produce new acylglycerol species. Lipases are used for enzymatic interesterification and are preferred over chemical interesterification because they offer a better control over the final product. Selectivity could be regiospecific or preference for certain fatty acids as a result of chain length, saturation, or degree of unsaturation. Enzyme catalyzed reactions help to save energy and minimize thermal degradation (Osborn & Akoh, 2002).

Interesterified palm oil and/or palm stearin-based products are softer than one made from unmodified oil/fraction of oil blends. This is because more beta prime crystals are formed in the interesterified blend while beta crystals are predominant in direct blends of oils. Changes in consistency during storage and handling are also reduced in the interesterified products.

The percentage of palm oil and its fractions to be used in margarine formulations varies depending on the type of margarine required. Generally, 25% palm oil is used for margarine formulations. Higher amount of palm oil (up to 50%) can result in incomplete oral melt of the product. SLs containing up to 80% palm oil, and other oils can be prepared with desired texture and mouth-feel (Sue, 2009). If super olein is used, the percentage can be increased to 40% in premium tub margarine and 80% in stick margarine (Aini & Miskandar, 2007). High percentage of super olein can be used with low percentage of fully hydrogenated oil in a good table margarine formulation. Compared to palm oil and palm olein, lower percentage of palm stearin (10–15%) is used in tub margarine formulation by blending with liquid oil. With interesterification, this percentage can be increased to obtain a good margarine base stock (Aini & Miskandar, 2007). Palm olein and palm stearin can be mixed in different proportions and interesterified to obtain fats with varying plastic range, melting points, and solid fat content, increasing their possible food applications (Soares et al., 2009). Palm stearin is the most extensively used fraction in production of *trans*-free margarines and shortenings. A study by Lee et al. (2008) showed that interesterified blends of fully hydrogenated soybean oil, rapeseed oil, and palm stearin in 15:20:65, 15:40:45, and 15:50:35 ratios had desirable beta prime crystals, SFC, and a long plastic range. This study concluded that these interesterified fat blends had no *trans* fat and could be used as an alternative to hydrogenated bakery shortenings. *Trans*-free plastic shortenings have also been prepared with palm stearin and rice bran oil using enzymatic interesterification with Lipozyme RM IM as biocatalyst. This resulted in a healthier alternative to commercial shortenings with comparable SFC, melting and crystallization profiles, and polymorphism (Jennings & Akoh, 2010). Palm stearin has also been used with several other oils such as rice bran oil and coconut oil (Adhikari et al., 2010a), pine nut oil (Adhikari et al., 2010b), canola oil, and palm kernel oil (Kim et al., 2008) using enzymatic interesterification to produce *trans*-free margarine fat analogs. PMF has also been used to produce a *trans*-free spread by blending it with cottonseed oil and a SL synthesized by reacting canola oil with 40% stearic acid (w/w) (Lumor et al., 2010).

A pourable margarine is margarine in a fluid form. To maintain the fluidity and emulsion stability it should have a reduced level of SFC. Pourable margarine formulation differs from the conventional one by having a higher amount of liquid oil (Aini & Miskandar, 2007). Palm oil, palm olein, and super olein can be used for the formulation of pourable margarine with desired consistency (Aini & Miskandar, 2007). A suitable blend for pourable margarine would consist of palm oil, or palm olein, as the higher melting phase and soybean or sunflower oil as the liquid phase (Aini & Miskandar, 2007). On the other hand, industrial margarine requires high solids at 20°C. Palm oil contains 23% solid fat at 20°C, and therefore it is a valuable oil for industrial margarine (Siew, 2002). As high palmitic acid content facilitates good aeration of fat and sugar, 50–100% palm oil can be used in cake margarine (Sue, 2009). Pastry margarine requires relatively high solid fats and superior plasticity to withstand vigorous mechanical rolling. The function of the pastry margarine is to separate the dough layer and to produce a uniform flaky structure (Noor Lida et al., 1996). Various blends of palm oil, palm stearin, and palm kernel oil have been used successfully in the formulation of pastry margarine (Noor Lida et al., 1996).

Shortening was the term used to describe the function performed by naturally occurring solid fats, such as lard and butter, in baked foods. It now applies to fat products which affect the emulsification, texture, structure, aeration, and flavor of prepared foods. Shortenings comprised of 100% oil and fats with trace amounts of emulsifiers for improved stability (Siew, 2002). Palm oil, hydrogenated palm oil, and palm stearin are generally used for preparation of both soft and hard shortenings with desired beta prime polymorphic form, plasticity and spreadibility (Siew, 2002). The amount of palm oil and/or palm products in a cake shortening normally varies between 30 and 40%. Malaysian commercial shortenings are mostly based on 100% palm oil products (Sue, 2009). Shortenings entrap air bubbles during creaming which provides good aeration and structure to the product. In order to entrap air bubbles and stabilize the air pockets the shortening must be solid as well as plastic. This is best accomplished by a plastic shortening with a small beta prime crystal form (Aini & Miskandar, 2007). Shortenings of higher melting points provide better structural support to the cake batter but at the expense of mouth-feel. Shortenings prepared with 60% palm stearin (IV 36, m.p. 52°C) and 40% sunflower oil and with 20%

palm stearin (IV 36, m.p. 52°C) and 80% palm oil had better performance than the hydrogenated commercial shortening (Aini & Miskandar, 2007).

Vanaspati is an all-purpose hydrogenated fat widely used in the Indian subcontinent and also in some Middle Eastern countries (Sue, 2009). It is a substitute of ghee but unlike ghee which is made from butterfat, vanaspati is made from vegetable oils. Hydrogenation is the main process in the production of vanaspati. The texture of vanaspati varies from grainy crystals in liquid oil to dry and crumbly. These different textures can be obtained with different combinations of palm oil, palm olein, and palm stearin. The regulation of melting point and composition of vanaspati varies by country. In India and Pakistan, vanaspati should be hydrogenated to a slip melting point maximum of 41 and 39°C, respectively (Sue, 2009). Palm oil at ambient temperatures has a semi-solid texture like that of vanaspati, and the granular consistency can be obtained through hydrogenation or interesterification. Some common formulations for vanaspati are 20% palm stearin and 80% palm oil, 20% palm stearin (m.p. 48°C), 40% palm oil and 40% hydrogenated soybean oil (m.p. 32°C) (Sue, 2009). *Trans*-free vanaspati can be formulated using blends of palm stearin and other oils, interesterified blend of 70% palm stearin and 30% rapeseed oil (Sue, 2009), and also with ternary blends of palm oil/palm stearin/palm olein or palm oil/palm stearin/palm kernel olein (Kheiri & Oh, 1983). Using interesterification, a *trans*-free Iranian vanaspati fat was produced with palm olein, low-erucic acid rapeseed oil, and sunflower oil (Farmani et al., 2006).

Cooking and Frying

Deep-fat frying is one of the most popular methods for the preparation of food. It is an easy, fast, and a relatively cheap method that results in palatable foods preferred by the consumers. The food is immersed into hot oil, mostly at temperatures between 150 and 190°C, leading to heat and material transfer from the oil to the food, resulting in cooking of the food (Choe & Min, 2007). The oil not only acts as a heat transfer medium but also imparts flavor, smell, and taste to the fried product. But if the oil has deteriorated, undesirable off-flavors are produced. A number of chemical reactions such as oxidation, polymerization, and hydrolysis of the fatty acids produce volatile and non-volatile substances. Some of these compounds have adverse nutritional and health effects (Choe & Min, 2007). It also decreases the quality and stability of the fried product. Therefore, the choice and the oxidative and thermal stability of the oil are very important for the quality of the product as the oil becomes a significant part of the food, influencing the quality of the product. The fatty acid composition of the oil from the raw material is changed towards the fatty acid composition of the oil used for frying (Matthaus, 2007). Frying with olive oil results in a high amount of oleic acid in the product after frying, whereas saturated and monounsaturated fatty acids predominate in products fried in palm olein. Use of oils with a higher amount of polyunsaturated fatty acids significantly influences the oxidative stability of the product during storage (Matthaus, 2007).

Frying medium can be all-vegetable oil, all-animal fat, or a mixture of both. The main characteristic required of a frying shortening is good stability in terms of oxidation and polymerization. The content of unsaturated fatty acids, in particular polyunsaturated fatty acids, must be minimized in order to confer good oxidative stability to the frying shortening (Siew, 2002). Palm oil and olein contain no *trans* fatty acids, have a moderate amount of linoleic acid, and minimal amount of linolenic acid (Sue, 2009). This makes palm olein and palm oil good alternatives as a frying oil or fat. Also, palm oil is rich in tocopherols and tocotrienols, which impart excellent oxidative stability to it. Augustine et al. (1989) reported that palm olein was comparable with hydrogenated sunflower and cottonseed oils in terms of oxidative stability during frying. Palm oil is resistant to oxidation, polymerization, foaming, and maintains the quality of food. It extends the shelf life of the fried product, and does not produce any gummy or sticky residues in the equipment.

The choice of the frying oil should be based on economic, nutritional, and physical and chemical quality factors. Palm oil and its products, palm olein, double-fractionated palm olein, palm stearin, and red palm oil, are widely used in industrial frying (Siew, 2002). They have high oxidative stability and competitive price. The most popular commercial frying oil is palm olein which has a lower melting point (22–24°C), resulting in no waxy or greasy mouth-feel of the fried products (Siew, 2002). Zeddelmann & Wurzinger (1973) carried out heating experiments at laboratory scale without frying, comparing soybean, groundnut, hardened groundnut, hardened fish, and coconut oils with palm oil. Oxidized fatty acids, FFAs, and smoke point were the parameters assessed for the quality of fats and oils. They reported that palm oil and hardened groundnut oil had the greatest oxidative stability at high temperature, in comparison to animal fats or highly unsaturated oils (Zeddeimann & Wurziger, 1973). Xu et al. (1999) compared potato chips fried in high-oleic canola oils with different levels of linolenic acid, high oleic sunflower oil, and palm olein. The parameters analyzed were fatty acids, IV, color index, dielectric constant, FFAs, and the total polar compounds. They concluded that palm olein had comparable frying performance as high-oleic, low linolenic canola oil, and better performance than the other oils (Xu et al., 1999). In an another heating experiment, refined olive oil and high oleic rapeseed oil had similar oxidative stability, while palm olein had higher stability than both (Tabee et al., 2008).

Carotenoids, along with Vitamin E, protect the oil from thermal oxidation. During thermal oxidation carotene radicals are formed which are converted back to active carotene in the presence of tocotrienols. This synergistic relationship decreased the oxidation of oil during frying of potato slices at 163°C (Schroeder et al., 2006). Heating causes a significant reduction in Vitamin E, particularly, alpha-tocopherols, alpha, gamma, and delta-tocotrienols in palm oil (Adam et al., 2007).

Cocoa Butter-like Fats

Chocolate is one of the most popular foods all over the world. Chocolates are semi-solid suspension comprising approximately 70% sugar and cocoa in a fat phase (cocoa butter) (Afoakwa et al., 2007). Cocoa butter is responsible for the smooth texture, flavor release, gloss, brittleness, "snap," melting properties, and crystallization of chocolate (Lipp & Anklam, 1998). Cocoa butter is a unique fat having special characteristics like sharp melting point at body temperature and brittleness at room temperature. The predominant TAGs in cocoa butter are symmetrical with oleic acid at *sn*-2 position and palmitic acid and stearic acid at the other two positions, mainly POP, POS, and SOS (Xu, 2000). Shortage of supply, agricultural limitations, and some technological benefits have prompted the development of cocoa butter-like fats. The uncertainty of cocoa butter supply and fluctuations in cocoa butter prices have forced production of replacers for cocoa butter from other sources like vegetable oils.

Cocoa butter equivalents (CBE) are fats which behave like and are compatible with cocoa butter in any proportion without altering the melting, processing, and rheological properties of cocoa butter (Lipp & Anklam, 1998). CBE are fully compatible with cocoa butter and can be used in any proportion up to 100% for complete replacement; they also require the same tempering procedures as cocoa butter to prevent fat bloom (Gibon et al., 2009). CBE have similar chemical composition but its TAGs are derived from a source other than cocoa beans. The substrate which can be used for cocoa butter-like fats should preferably have oleic acid at *sn*-2 position. Some oils which contain abundant oleic acid at *sn*-2 position are high oleic sunflower oil, olive oil, PMF, high oleic canola oil, and Chinese vegetable tallow (Xu, 2000). PMF is the most commonly used starting substrate for lipase catalyzed CBE production due to its fatty acid profile, cost, and availability.

The hard PMF is a specialty fat widely used in the confectionery industry as an ingredient for CBE formulation. Hard PMF is rich in POP (>65%). This specific composition offers to the hard PMF a very steep melting profile similar to the one of cocoa butter for specialty fats applications. Common CBE formulations are comprised of hard PMF, shea butter stearin, illipe fat, and sal stearin in different proportions. CBE are also produced with soft PMF and stearic acid or stearic acid methyl ester by enzymatic interesterification using *sn*-1,3 specific lipases (Gibon et al., 2009). RBD palm oil and fully hydrogenated soybean oil were enzymatically interesterified at 1.6:1 weight ratios to produce cocoa butter-like fats (Abigor, 2003).

Cocoa butter replacers (CBR) are designed to replace cocoa butter typically in confectionery applications. Depending upon the compatibility with cocoa butter, CBR are sometimes classified as partial and complete. They have similar fatty acids distribution but a completely different structure of TAGs than cocoa butter and are compatible only in small ratios (Lipp & Anklam, 1998). CBRs are produced by hydrogenation of liquid oil, frequently followed by dry fractionation. Palm oil or palm olein are commonly used raw ingredients for CBRs. Due to the hydrogenation process these CBRs are also high in *trans* fat. The fractionated CBR have better melting profile and better eating quality than the non-fractionated ones but they are also more expensive (Gibon et al., 2009).

Other Uses

Red Palm Oil

Palm oil is a rich natural source of carotenoids and can be used for vitamin A fortification. Crude palm oil contains 500–700 ppm of carotenoids of which 56 and 35% are beta- and alpha-carotene, respectively. During the refining process, a major fraction of these carotenoids are lost. Using a two stage process involving pretreatment of the crude palm oil followed by bleaching and deodorization by molecular distillation, 80% of the original carotenoids and vitamin E can be retained (Mayamol et al., 2007). This oil is called red palm oil due to its high content of carotenoids. It is also a rich source of vitamin E (600–1000 ppm). Commercially, red palm oil is sold as Carotino® Classic which is 100% natural red palm oil and as Carotino® Premium which is a blend of vegetable oils. Red palm oil is *trans*- and cholesterol free. The oil also contains 18–25 mg/kg of Co Q10 and 325–365 mg/kg sterols (Tay & Choo, 2000). When used for frying, no significant difference was found between Carotino and canola oils for most of the food products (Hekmat & Haines, 2003). Red palm oil can be used as a functional food ingredient in baked goods, confectionaries, natural colorant in margarines and spreads, and also alone as cooking/frying and salad oils. Red palm olein has been used as a shortening in the formulation of novel functional biscuits by El-Hadad et al. (2010). The best formulation was obtained by replacing commercial shortening with 40 and 60% red palm olein. They reported that the formulated functional biscuits were acceptable based on sensory evaluation and contained 1.8 times more Vitamin E and 10.4–10.8 times more carotenoids than the control biscuits (El-Hadad et al., 2010). The high level of natural antioxidants also contributes to improved shelf life of the finished food products. Both carotenoids and vitamin E are also potent biological antioxidants (Edem, 2002). A number of studies have shown that red palm oil has potential to prevent vitamin A deficiencies in children (van Stuijvenberg & Benade, 2000), improve vitamin A status of lactating mothers and their infants (Canfield & Kaminsky, 2000), reduce atherosclerosis (Kritchevsky et al., 1999), and inhibit tumor growth (Kritchevsky, 2000).

Dietary Supplements

The bioactive minor components of palm oil can be isolated, concentrated, and used for fortification of foods and as dietary supplements in the form of capsules. These include carotenoids, vitamin E, phytosterols, and Co Q10. Palm fatty acid distillate, produced during the refining process, is also an important by-product that can be used as a functional ingredient. Among vegetable oils, palm oil is the richest source of carotenoids and tocotrienols. Carotenoids are important biological antioxidants, vitamin A precursors, and can also improve cardiovascular health (Mukherjee & Mitra, 2009). Tocotrienols have potent antioxidant and anticancer activities. Studies have shown that palm tocotrienols reduce the risk of stroke by preventing platelet aggregation and carotid artery blockage (Mukherjee & Mitra, 2009). Oral supplementation of alpha-tocotrienol-rich fraction of palm oil is capable of protecting the brain against oxidative damage and stroke-mediated neuropathy (Sen et al., 2010). Phytosterols inhibit intestinal absorption of cholesterol and lower serum LDL cholesterol. They can be extracted from fatty acid distillates where they are concentrated during refining. Co Q10 is a fat-soluble compound primarily synthesized by the body and also consumed in the diet. It is essential for electron transport and ATP production. Palm oil contains 10–80 mg/kg Co Q10 (Choo & Tay, 2000). Various methods have been used for the extraction of these bioactives such as selective solvent extraction, transesterification followed by molecular distillation (e.g., esters), adsorption chromatography, and membrane technology (Othman et al., 2010). New technologies such as supercritical fluid extraction and pressurized liquid extraction can also be used for extraction of heat labile bioactives such as carotenoids and vitamin E (Tan et al., 2007).

Baked Products

Fat is used in bread making to ensure a better texture and longer shelf life. An excellent shortening for bread is 100% RBD palm oil (Sue, 2009). Similarly, shortening is required in biscuit manufacturing for the dough, cream, and also for spraying. Palm oil is a suitable candidate for all three applications. The oil or fat used for biscuit dough should prevent fat bloom and improve the shelf-life. Generally, a blend of two or more oils is used. Palm oil rich in palmitic acid can be successfully used with other vegetable oils rich in oleic, linoleic, or linolenic acids (Sue, 2009). For use in the biscuit cream, the oil/fat should have a high SFC to maintain the texture and a melting temperature close to body temperature for complete oral melt. Hydrogenated palm kernel oil (PKO) or hydrogenated palm olein have been successfully used in the United Kingdom biscuit industry (Sue, 2009). Spraying the biscuits with oil after they are baked provides them a richer taste and glossy appearance. For this the oil should have relatively low SFC at ambient temperature and high oxidative stability as it is exposed to air and heat. Both palm oil and palm olein can be used as biscuit spraying oils (Sue, 2009).

Palm-based Emulsifiers

Most food products are in the form of emulsion consisting of two immiscible phases. Emulsions are stabilized with emulsifiers/surfactants. Emulsifiers influence the structure, texture, and quality of many foods. The type of fatty acid used in the lipid-based emulsifiers greatly influences its functionality. Vegetable fats are preferred over animal fats due to their healthful fatty acid profiles. Fully hydrogenated palm oil containing saturated fatty acids provides an alternative to *trans*-free emulsifiers (Chin et al., 2007). Palm based emulsifiers can be used in various food applications. The common emulsifiers used in the baking industry are the monoacylglycerols, lecithin, diacetyl tartaric acid esters of monoacylglycerols (DATEM), known as volume improver, and sodium stearoyl lactylates (SSL), which improves the texture, volume, and shelf life of bread (Chin et al., 2007). Chin et al. studied the effectiveness of fully hydrogenated palm oil based emulsifiers, namely, DATEM, SSL, and distilled monoacylglycerols (DMG), on bread quality. All three emulsifiers resulted in bread with larger volume, higher oven spring, lower weight and density, whiter color, finer crumb texture, and longer shelf life compared to control bread (Chin et al., 2007). Palm based emulsifiers (DMG) can also be used in low fat spreads (Young & Wassell, 2008). Some commercial palm based emulsifiers produced by Danisco are GRINDSTED® MONO-DI, GRINDSTED®LACTEM, GRINDSTED®CITREM, GRINDSTED®ACETEM, and DIMODAN®. These can be used to improve the texture and stability of various food products such as ice creams, margarines, chewing gums, desserts, and confectionaries.

Infant Formula

Human milk is considered the ideal source of nutrition for the growth and development of infants. Human milk fat (HMF) is the basis for the design of infant formulas. Vegetable oils are selected for production of infant formula based on their fatty acid profile. In human breast milk TAG, 60% or more palmitic acid is esterified at the *sn*-2 position, whereas most of the palmitic acid in vegetable oils such as palm oil and its olein fraction is present at the *sn*-1 and 3 positions (Martin et al., 1993). This positional difference renders palm oil or palm olein unsuitable for infant formula because during digestion, pancreatic lipase specifically hydrolyzes the fatty acids at the *sn*-1 and 3 positions, leaving saturated fatty acids (palmitic and stearic) unhydrolyzed. These saturated fatty acids are better absorbed as 2-MAGs (Kennedy et al., 1999). But if palmitic acid is present at *sn*-1 and *sn*-3 positions, as is the case for palm oil and palm olein, it is hydrolyzed to FFAs which form insoluble and unabsorbable calcium-fatty acid-soap complexes, thereby contributing to low calcium and fat absorption, low weight gain, and increased stool hardness (Kennedy et al., 1999; Yu et al., 2009). Enzymatic or chemical interesterification of palm oil or palm olein to enrich at *sn*-2 position with palmitic acid can be used to produce human milk fat analogs with desirable nutritional properties.

Other Uses

Ice cream is an oil-in-water (o/w) emulsion containing milk solids. Vegetable fats having a high SFC at 0°C and melting point of 35–37°C can be used to replace the milk fat. Palm oil, with a similar solid fat content profile to butterfat, is one such oil with suitable characteristics for ice-cream formulations (Siew, 2002). Palm oil is also used in the formulation of filled milk powders. Filled milk powder is produced by replacing milk fat with vegetable fat followed by spray drying of the product. This helps in reducing the cholesterol content of the product and also makes it suitable for lactose intolerant consumers. Generally, oils with high oxidative stability and bland taste such as palm oil and palm olein are suitable for use in filled milk powders (Sue, 2009). Powdered coffee whiteners are similar to filled milk powder except for its higher solid fats and addition of substances to make it free flowing and miscible. Popular choices for coffee whiteners' formulations are 100% palm oil and 100% palm olein (Sue, 2009). Whipped toppings are commonly used on coffee, ice cream, puddings, sodas, and cakes. Stability of the whipping cream is a critical aspect, particularly in hot climates. Palm oil based whipped toppings are more stable and functional than dairy whipping cream. This may be due to higher oleic acid content and unsaturation of palm oil which has a positive impact on the stability of foam (Shamsi et al., 2002). Palm olein is also used in the formulation of salad dressings. In addition to being cost effective, it is also oxidatively stable due to its lower iodine value and high content of Vitamin E (Sue, 2009). Palm oil is also used as an ingredient in dry soup mixes because of its high SFC and stability (Sue, 2009).

Conclusions

Palm oil and its various fractions are an important constituent of several food products. Palm oil has a unique fatty acid and TAG profile which makes it suitable for a number of food applications. Palm oil is also used as a source of fatty acids, vitamin E, and emulsifiers. Fractionation of palm oil results in various fractions with distinct physical and chemical properties. Customized blends of these fractions with other oils are used in different food products ranging from margarines to soup mixes. Palm oil and its fractions are able to replace certain traditional fats in these food products, mainly due to their competitive price and consistent supply. Palm stearin is helpful in providing the solid fat functionality without the use of hydrogenation, thus reducing *trans* fat intake in our diets. Interesterification of the oil also widens its scope of food applications. Interesterification can also be used to incorporate essential polyunsaturated fatty acids into palm oil to obtain a product rich in essential fatty acids with good antioxidant properties of palm oil (Fajardo et al., 2003).

References

Abigor, R.D.; William, N.M.; Foglia, T.A.; Jones, K.C.; DiCiccio, R.J.; Ashby, R.; Uadia, P.O. Production of cocoa butter like fats by the lipase catalyzed interestification of palm oil and hydrogenated soybean oil. J. Am. Oil Chem. Soc. 2003, 80, 1193–1196.

Adam, S.K.; Sulaiman, N.A.; Top, A.G.M.; Jaarin, K. Heating reduces vitamin E content in palm and soy oils. J. Biochem. Mol. Biol. 2007, 15, 76–79.

Adhikari, P.; Shin, J.; Lee, J.; Hu, J.; Zhu, X.; Akoh, C.C.; Lee, K. Production of *trans*-free margarine stock by enzymatic interesterification of rice bran oil, palm stearin and coconut oil. J. Sci. Food Agric. 2010a, 90, 703–711.

Adhikari, P.; Zhu, X.; Gautam, A.; Shin, J.; Hu, J.; Lee, J.; Akoh, C.C.; Lee, K. Scaled-up production of zero-*trans* margarine fat using pine nut oil and palm stearin. Food Chem. 2010b, 119, 1332–1338.

Afoakwa, E.O.; Paterson, A.; Fowler, M. Factors influencing rheological and textural qualities in chocolates—a review. Trends Food Sci. Technol. 2007, 18, 290–298.

Aini, I.D.; Miskandar, M.S. Utilization of palm oil and palm products in shortenings and margarines. Eur. J. Lipid Sci. Technol. 2007, 109, 422–432.

Astorg, P. Dietary fatty acids and colorectal and prostate cancer: epidemiological studies. B. Cancer 2005, 92, 670–684.

Augustine, M.A.; Heng, L.K.; Nor Aini, I. Frying experiments with hydrogenated vegetable shortenings. Pertanika 1989, 12, 167–172.

Berger, K.G. Palm oil. Chem. Ind. 1987, 24, 846–851.

Canfield, L.M.; Kaminsky, R.G. Red palm oil in the maternal diet improves the vitamin A status of lactating mothers and their infants. Food Nutr. Bull. 2000, 21, 144–148.

Chen, C.W.; Chong, C.L.; Ghazali, H.M.; Lai, O.M. Interpretation of triacylglycerol profiles of palm oil, palm kernel oil and their binary blends. Food Chem. 2007, 100, 178–191.

Chin, N.L.; Goh, S.K.; Rahman, R.A.; Hashim, D.M. Functional effect of fully hydrogenated palm oil-based emulsifiers on baking performance of white bread. Intl. J. Food Eng. 2007, 3, 1–17.

Choe, E.; Min, D.B. Chemistry of deep-fat frying oils. J. Food Sci. 2007, 72, R77–R86.

Choo, Y.M.; Tay, B.Y.P. Coenzymes Q in palm oil and palm oil products. Palm Oil Dev. Malays. Palm Oil Board 2000, 32, 18–19.

Clegg, A.J. Composition and related nutritional and organoleptic aspects of palm oil. J. Am. Oil Chem. Soc. 1973, 50, 321–324.

Cottrell, R.C. Nutritional aspects of palm oil. Am. J. Clin. Nutr. 1991, 53, 989S–1009S.

Edem, D.O. Palm oil: biochemical, physiological, nutritional, hematological, and toxicological aspects: a review. Plant Foods Human Nutr. 2002, 57, 319–341.

El-Hadad, N.; Abou-Gharbia, H.A.; El-Aal, M.H.A.; Youssef, M.M. Red palm olein: characterization and utilization in formulating novel functional biscuits. J. Am. Oil Chem. Soc. 2010, 87, 295–304.

Erickson, M.C. Chemistry and function of phopsholipids. In Food Lipids Chemistry, Nutrition, and Biochemistry, 3rd ed.; Akoh, C.C., Min, D.B., Eds.; CRC Press: Boca Raton, FL, 2008; pp. 39–61.

Fajardo, A.; Akoh, C.C.; Lai, O.M. Lipase-catalyzed incorporation of n–3 PUFA into palm oil. J. Am. Oil Chem. Soc. 2003, 80, 1197–1200.

Farmani, J.; Safari, M.; Hamedi, M. Application of palm olein in the production of zero-*trans* Iranian vanaspati through enzymatic interesterification. Eur. J. Lipid Sci. Technol. 2006, 108, 636–643.

Fomuso, L.B.; Akoh, C.C. Structured lipids: their applications and physical property testing methods. Food Sci. Biotechnol. 2001, 10, 690–698.

Gapor, A.B.; Ong, A.S.H. Protection of oils and fats against oxidative deterioration with special reference to palm oil. PORIM Bull. 1982, 5, 39–46.

Gebauer, S.; Harris, W.S.; Kris-Etherton, P.M.; Etherton, T.D. Dietary n-6:n-3 fatty acid ratio and health. In Healthful Lipids; Akoh, C.C., Lai, O.M., Eds.; AOCS Press: Champaign, IL, 2005; pp. 221–248.

Gibon, V.; Ayala, J.V.; Dijckmans, P.; Maes, J.; Greyt, W.D. Future prospects for palm oil refining and modification. OCL 2009, 16, 193–200.

Halliwell, B.; Gutteridge, J.M. Free radicals in biology and medicine; 4th ed.; Oxford University Press: New York, 2007.

Hekmata, S.; Haines, J. Comparison of the effects of Carotino and canola oils on the sensory properties of various foods. Nutr. Res. 2003, 23, 1211–1219.

Hu, F.B.; Stampfer, M.J.; Manson, J.E.; Rimm, E.; Colditz, G.A.; Rosner, B.A.; Hennekens, C.H.; Willett, W.C. Dietary fat intake and the risk of coronary heart diseases in women. N. Engl. J. Med. 1997, 337, 1491–1499.

Jennings, B.H.; Akoh, C.C. *Trans*-free plastic shortenings prepared with palm stearin and rice bran oil structured lipid. J. Am. Oil Chem. Soc. 2010, 87, 411–417.

Karleskind, A.; Wolff, J.P.; Eds.; Oils and fats manual: A comprehensive treatise—properties, production, application; Intercept: Andover, UK, 1996.

Kellens, M.; Gibon, V.; Hendrix, M.; Greyt, W.D. Palm oil fractionation. Eur. J. Lipid Sci. Technol. 2007, 109, 336–349.

Kennedy, K.; Fewtrell, M.S.; Morley, R.; Abbott, R.; Quinlan, P.T.; Bindels, J.G., Lucas, A. Double-blind, randomized trial of a synthetic triacylglycerol in formula-fed infants:effects on stool biochemistry, stool characteristics, and bone mineralization. Am. J. Clin. Nutr. 1999, 7, 920–927.

Kheiri, M.S.A.; Oh, F.C.H. Formulation of vegetable ghee/vanaspati. Proceedings of Palm Oil Product in the Eighties; Pusparajah, E., Rajanaidu, M., Eds.; Incorporated Society of Planters: Kuala Lumpur, Malaysia, 1983; pp. 449–474.

Kim, B.H.; Lumor, S.E.; Akoh, C.C. *trans*-Free margarines prepared with canola oil/palm stearin/palm kernel oil-based structured lipids. J. Agric. Food Chem. 2008, 56, 8195–8205.

Kritchevsky, D. Impact of red palm oil on human nutrition and health. Food Nutr. Bull. 2000, 21, 182–188.

Kritchevsky, D.; Chen, S.C. Phytosterols-health benefits and potential concerns. Nutr. Res. 2005, 25, 413–428.

Kritchevsky, D.; Tepper, S.A.; Kuksis, A. Effects of palm oil, randomized palm oil and red palm oil on experimental atherosclerosis. FASEB J. 1999, 13, A213.

Kulkarni, A.S.; Khotpal, R.R.; Bhakare, H.A. Phospholipids and glycolipids in the oil from some varieties of *Elaeis guineensis* in India. Elaeis 1991, 3, 363–368.

Lagarda, M.L.; Garcia-Llatas, G.; Farre, R. Analysis of phytosterols in foods. J. Pharm. Biomed. Anal. 2006, 41, 1486–1496.

Lee, J.H.; Akoh, C.C.; Lee, K. Physical properties of *trans*-free bakery shortening produced by lipase-catalyzed interesterification. J. Am. Oil Chem. Soc. 2008, 85, 1–11.

Ling, W.H.; Jones, P.J.H. Dietary phytosterols: a review of metabolism, benefits and side effects. Life Sci. 2008, 57, 195–206.

Lipp, M.; Anklam, E. Review of cocoa butter and alternative fats for use in chocolates—part A. compositional data. Food Chem. 1998, 62, 73–97.

Lumor, S.E.; Pina-Rodriguez, A.M.; Shewfelt, R.L.; Akoh, C.C. Physical and sensory attributes of a *trans*-free spread formulated with a blend containing a structured lipid, palm mid-fraction, and cottonseed oil. J. Am. Oil Chem. Soc. 2010, 87, 69–74.

Martin, J.C.; Bougnoux, P.; Antoine, J.M.; Lanson, M.; Couet, C. Triacylglycerol structure of human colostrums and mature milk. Lipids 1993, 28, 637–643.

Matthaus, B. Use of palm oil for frying in comparison with other high-stability oils. Eur. J. Lipid Sci. Technol. 2007, 109, 400–409.

Mayamol, P.N.; Balachandran, C.; Samuel, T.; Sundareson, A.; Arumughan, C. Process technology for the production of micronutrient rich red palm olein. J. Am. Oil Chem. Soc. 2007, 84, 587–596.

Mozaffarian, D.; Pischon, T.; Hankinson, S.E.; Rifai, N.; Joshipura, K.; Willett, W.C.; Rimm, E.B. Dietary intake of *trans* fatty acids and systemic inflammation in women. Am. J. Clin. Nutr. 2004, 79, 606–612.

Mukherjee, S.; Mitra, A. Health effects of palm oil. J. Hum. Ecol. 2009, 26, 197–203.

Munne-Bosch, S.; Alegre, L. The function of tocopherols and tocotrienols in plants. Crit. Rev. Plant Sci. 2002, 21, 31–57.

MVO Fact sheet palm oil, Product Board For Margarine, Fats and Oils: Rijswijk, Netherland. 2010; pp. 1–28.

Noor Lida, H.M.B.; Md. Ali, A.R.; Mahadir, I. Blending of palm oil, palm stearin and palm kernel oil in the preparation of table and pastry margarine. Int. J. Food Sci. Nutr. 1996, 47, 71–74.

Osborn, H.T.; Akoh, C.C. Structured lipid—novel fats with medical, nutraceutical, and food applications. Comp. Rev. Food Sci. Food Safety 2002, 3, 110–120.

Othman, N.; Manan, Z.A.; Wan Alwi, S.R.; Sarmidi, M.R.A review of extraction technology for carotenoids and vitamin E recovery from palm oil. J. Appl. Sci. 2010, 10, 1187–1191.

Renaud, S.C.; Ruf, J.C.; Petithory, D. The positional distribution of fatty acids in palm oil and lard influences their biologic effects in rats. J. Nutr. 1995, 125, 229–237.

Sahri, M.M.; Idris, N.A. Palm stearin as low *trans* hard stock for margarine. Sains Malaysiana 2010, 39, 821–827.

Sambanthamurthi, R.; Sundram, K.; Tan, Y. Chemistry and biochemistry of palm oil. Prog. Lipid Res. 2000, 39, 507–558.

Schroeder, M.T.; Becker, E.M.; Skibsted, L.H. Molecular mechanism of antioxidant synergism of tocotrienols and carotenoids in palm oil. J. Agric. Food Chem. 2006, 54, 3445–3453.

Sen, C.K.; Rink, C.; Khanna, S. Palm oil-derived natural vitamin E α-tocotrienol in brain health and disease. J. Am. Coll. Nutr. 2010, 29, 314S–323S.

Shamsi, K.; Man, Y.B.C.; Yusoff, M.S.A.; Jinap, S. A comparative study of dairy whipping cream and palm oil-based whipping cream in terms of FA composition and foam stability. J. Am. Oil Chem. Soc. 2002, 79, 583–588.

Siddique, B.M.; Ahmad, A.; Ibrahim, M.H.; Hena, S.; Rafatullah, M.; Omar, A.K.M. Physico-chemical properties of blends of palm olein with other vegetable oils. Grasa Y Aceites 2010, 61, 423–429.

Siew, W.L. Composition of palm glycerides and their effects on the properties of the oil, PhD Thesis, University of Malaya, Malaysia.1994

Siew, W.L. Palm oil. In Vegetable Oils In Food Technology: Composition, Properties and Uses, Gunstone, F.D., Ed.; CRC Press: Boca Raton, FL, 2002; pp. 59–97.

Siew, W.L.; Ng, W.L. Diglycerides in palm oil products: composition and effects in oil properties, In Physical Properties of Fats, Oils and Emulsifiers; Widlak, N., Ed.; AOCS Press: Champaign, IL, 1999; pp. 129–139.

Soares, F.A.S.D.M.; da Silva, R.C.; da Silva, K.C.G.; Lourenco, M.B.; Soares, D.F.; Gioielli, L.A. Effects of chemical interesterification on physicochemical properties of blends of palm stearin and palm olein. Food Res. Int. 2009, 42, 1287–1294.

Sue, T.T., Ed.; Pocketbook of palm oil uses. Malaysian Palm Oil Board: Kuala Lumpur, Malaysia, 2009.

Tabee, E.; Azadmard-Damirchi, S.; Jagerstad, M.; Dutta, P.C. Effect of a-tocopherol on oxidative stability and phytosterol oxidation during heating in some regular and high oleic vegetable oils. J. Am. Oil Chem. Soc. 2008, 85, 857–867.

Tan, Y.A.; Kuntom, A.; Siew, W.L.; Yusof, M.; Chong, C.L. PORIM Crude palm oil survey 97/98 quality and identity characteristics. PORIM Technol. 2002, 22.

Tan, Y.A.; Sambanthamurthia, R.; Sundramb, K.; Wahida, M.B. Valorisation of palm by products as functional components. Eur. J. Lipid Sci. Technol. 2007, 109, 380–393.

Tay, B.Y.P.; Choo, Y.M. Valuable minor components from commercial red palm olein. J. Oil Palm Res. 2000, 12, 14–24.

van Stuijvenberg, M.E.; Benade, A.J.S. South African experience with the use of red palm oil to improve the vitamin A status of primary schoolchildren. Food Nutr. Bull. 2000, 21, 212–214.

Xu, X. Production of specific structured triacyglycerols by lipase-catalyzed reactions. Eur. J. Lipid Sci. Technol. 2000, 102, 287–303.

Xu, X.O.; Tran, V.H.; Palmer, M.; White, K.; Salisbury, P. Chemical and physical analyses and sensory evaluation of sex deep-frying oils. J. Am. Oil Chem. Soc. 1999, 76, 1091–1099.

Young, F.V.K. Interchangeability of oils. J. Am. Oil Chem. Soc. 1985, 62, 373–375.

Young, N.; Wassell, P. Margarines and spreads. In Food Emulsifiers and Their Applications, 2nd ed.; Hasenhuettl, G.L., Hartel, R.W., Eds.; Springer: New York, 2008; pp. 307–325.

Yu, Z.B.; Han, S.P.; Zhu, C.; Sun, Q.; Guo, X.R. Effects of infant formula containing palm oil on the nutrient absorption and defecation in infants: a meta-analysis. Zhonghua Er Ke Za Zhi 2009, 47, 904–1010.

Zeddeimann, H.V.; Wurziger, J. Verhalten und beurteilung von fritierfetten in der praxis. Fette Seifen Anstrichm. 1973, 75, 18–24.

20

Palm Oleochemicals in Non-food Applications

Shoot Kian Yeong, Zainab Idris, and Hazimah Abu Hassan
Malaysian Palm Oil Board, Selangor Darul Ehsan, Malaysia

Introduction

The rapid increase in demand and unsustainable supply of fossil-based resources has resulted in the search for new technology that utilized agriculture, animal, forestry, and biomass as renewable feedstock, especially in most developed places like the United States and Europe. This new economic development presents a significant opportunity for palm oil and coconut oil producing countries such as Malaysia, Indonesia, Thailand, and Philippines.

Palm oil is currently one of the most sustainable and important key base stocks for what is known as "Modern" or "Green" industry due to its high productivity rate (Table 20-A) as compared to other crops. The world production of natural oil and fats started from 30 million tons in 1960 and had rapidly increased to a fourfold production volume of 120 million tonnes in 2002 (Höfer & Bigorra, 2007). Generally, in the field of natural oils and fats, a tremendous upward shift from animal to vegetable base stock was observed between 1960 and 2002, with Malaysia and Indonesia producing more than 80% of the total world production (Suyenty et al., 2007). The production volume of palm oil has relentlessly been increasing beginning in the year

Table 20-A. Productivity Rate of Different Oilseed Crops.

Oilseed crop	Productivity rate (MT/ha/year)
Palm/Palm kernel	4.4
Coconut	2.0
Rapeseed	1.2
Sunflower	0.7
Soya	0.4

Source: Rupilius & Salmiah, 2007.

2002 with total production volume of 25.034 million tonnes to 36.780 million tons in 2006, surpassing for the first time the production of soybean oil in 2006 which was 35 million tonnes (Rupilius & Salmiah, 2007).

The ever-increasing oil palm plantation acreage in Indonesia, taking over Malaysia as the major producer in 2006, has in some ways imposed challenges to the Malaysian palm oil Industry. However, high crude oil and ethylene prices, depleting fossil resources, and tremendous voluntary shift from animal to vegetable base stocks have created a new avenue in research and development, especially in catalysis and processes particularly focusing on oleochemical and their value added derivatives.

The chemical derivatives of natural oils and fats are technically known as oleochemicals. Derived from the chemically transformed triglycerides, major chemical components in most agricultural, forestry, marine, and animal resources can be further processed into their five basic constituents, namely fatty acids, fatty methyl ester, fatty alcohol, fatty amines, and glycerol.

The main feedstock for oleochemicals are coconut oil and tallow, resulting in oleochemicals having fatty acid moiety of even numbered saturated carbon chain lengths of 8 to 18 and that of the mono- and di-unsaturated constituent of 18:1 and 18:2 (Fig. 20.1). However, between 1985 to 1995 the significance of tallow was greatly reduced because of the BSE disease outbreak. Palm stearin (a palm oil fraction) has since replaced tallow in many applications and now becomes the major feed stock for oleochemicals.

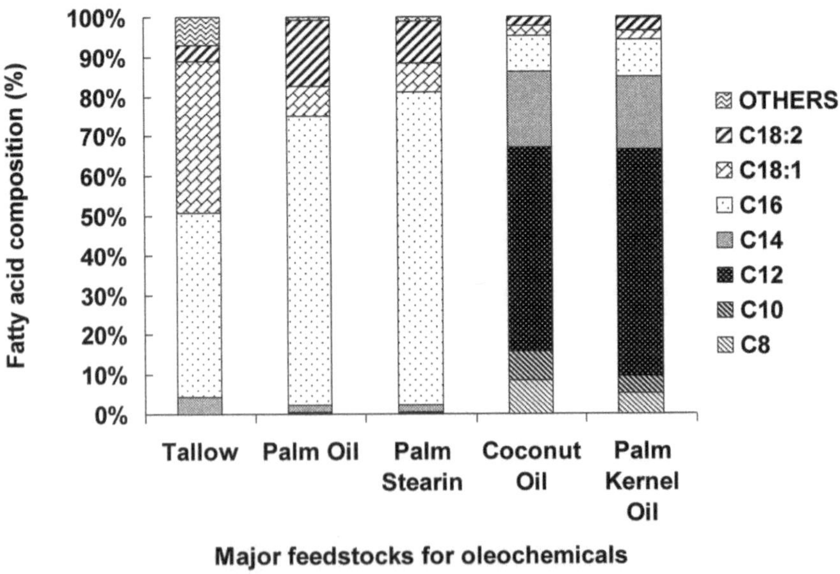

Fig. 20.1. Fatty acid composition of some major feedstock oils for oleochemicals.

In a similar fashion, prior to 1985 coconut oil was the most important source of fatty component with 12 and 14 carbons, commonly denoted as C12-C14 fatty acids, for the oleochemical industry. These C12-C14 carbon chain fatty acids play a very important role in the manufacture of surface active agents for the laundry and personal care products (Salmiah et al., 2000). During the period of 1985–1995, the global production of coconut oil could not meet the demand of these medium chain fatty acids. Coincidently, the production of palm oil was increasing rapidly and the oil from palm kernel has similar fatty acid composition to coconut oil. This has provided opportunity for palm kernel to be the major feedstock for C8-C14 carbon chain fatty acids.

Processing Routes for Basic Oleochemicals

The progress in oleochemical industries depends strongly on the development of three major research areas, namely process catalysis, process chemistry, and technology, which then lead to feasibility in economic, clean, and green processes and diversifying product functionality. Fig. 20.2 shows the basic processing route for production of the five basic oleochemicals from triglycerides of natural oil and fats.

In general, triglyceride molecules can be hydrolyzed or interchangeably split to form fatty acids and glycerol. When triglycerides are transesterified with a simple aliphatic alcohol, fatty acid alkyl ester and glycerol are produced. Alkyl ester could also

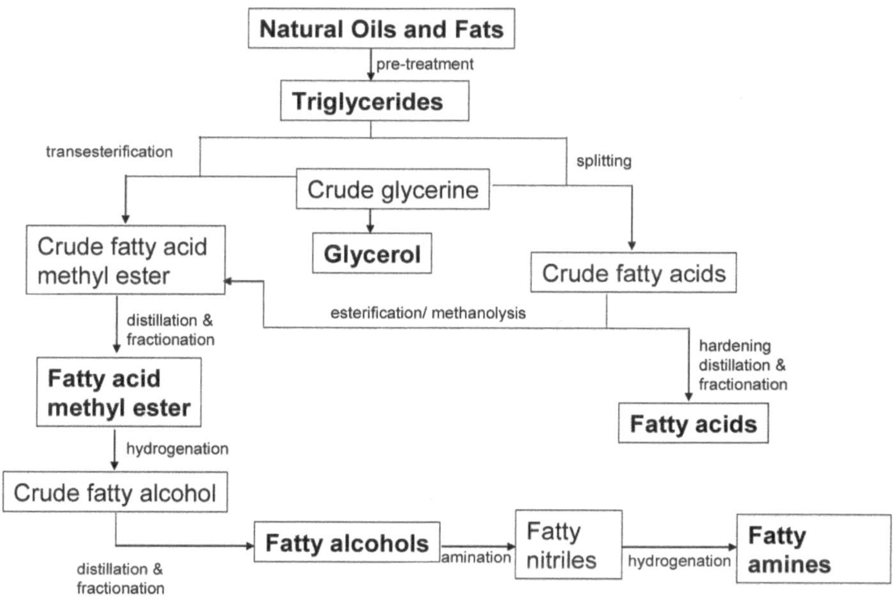

Fig. 20.2. Conversion of natural oils and fats into basic oleochemicals.

be produced through direct esterification of fatty acid and alcohol. Fatty acid methyl esters, one of the simplest form of fatty ester, produced either through transesterification of triglycerides and methanol or esterification of fatty acid and methanol, are normally used as feedstock for production of fatty alcohol through catalytic hydrogenation of fatty acid methyl ester. Fatty alcohol was known to be the most important oleochemical for the production of anionic surfactant for the detergent industries. Fatty alcohols can also be further processed into fatty amines through hydrogenation of fatty nitriles.

The production of fatty acids through hydrolysis or splitting of triglyceride molecules were considered the oldest and most important component in oleochemical industries. There are currently three major routes which are commonly adopted for the production of fatty acid and glycerol development, namely counter current fat splitting, alkaline hydrolysis followed by neutralization of the fatty acid to soap, and lipase-catalysed fat splitting. The most common process is the continuous counter current high pressure steam splitting at 55–70 bar and temperature of 250–260°C.

The splitting tower is normally integrated with a complex system or customized plants for the recovery of glycerine, fatty acid distillation and fractionation for specific mix cuts or pure cuts fatty acids, and hydrogenation for hardened fatty acids (Eierdanz, 1992; Lurgi Fatty Acid Technology).

The high splitting temperature of 245–255°C ensures adequate diffusion of water molecules into the counter flowing fats so that physical agitation can be avoided (Fig. 20.3). The crude triglycerides-based fat travels upward from bottom to the top of the tower while the heavy splitting water travels downward as a dispersed phase.

The second process for the production of fatty acids is through alkaline hydrolysis, which is not typically adopted even though the fatty acid soap can be eventually neutralized into fatty acids by adding an inorganic acid like HCl (Fig. 20.4).

The third option for the production of fatty acids is through immobilized lipase enzyme hydrolysis (Murty et al., 2002). Simulating lipase performance in bio-hydrolysis of fat in living system, the enzymatic hydrolysis of triglycerides may be carried out at body temperature of 35°C and atmospheric pressure. Unlike the chemical-hydrolysis

Fig. 20.3. Splitting process involving triglyceride and water for production of fatty acids.

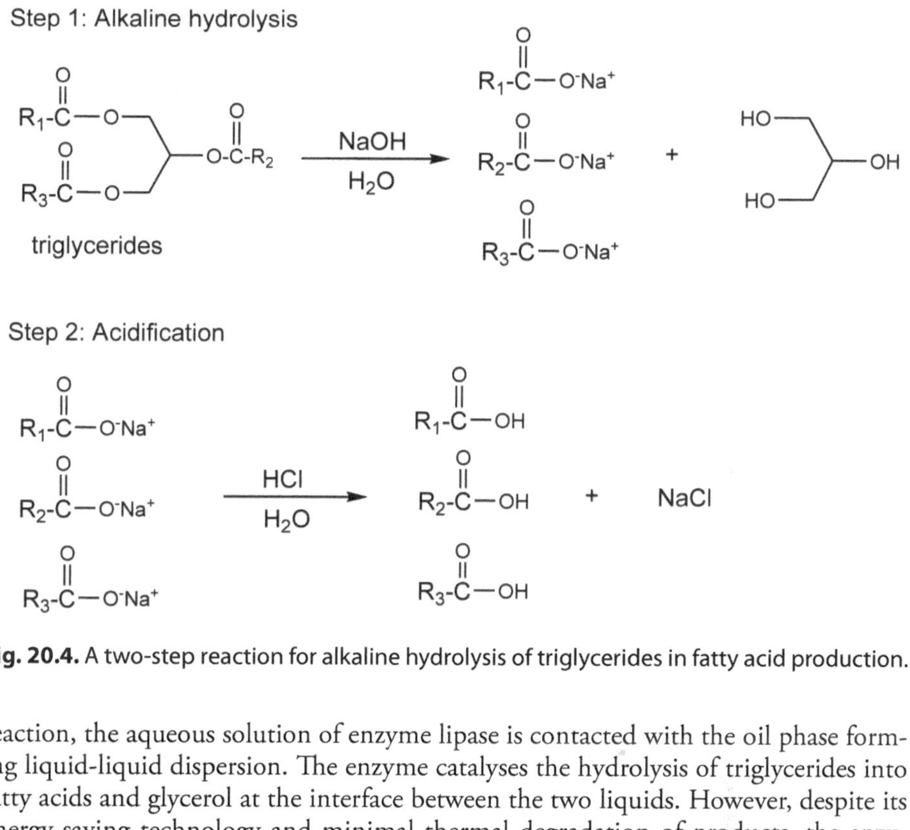

Fig. 20.4. A two-step reaction for alkaline hydrolysis of triglycerides in fatty acid production.

reaction, the aqueous solution of enzyme lipase is contacted with the oil phase forming liquid-liquid dispersion. The enzyme catalyses the hydrolysis of triglycerides into fatty acids and glycerol at the interface between the two liquids. However, despite its energy-saving technology and minimal thermal degradation of products, the enzymatic hydrolysis process is not widely used commercially due to several factors including long reaction period, limited technology for continuous processing, and high enzyme cost. However, only Japanese companies are actively producing premium products such as soap powder and high purity unsaturated fatty acids in continuous counter current hydrolysis reactors (Kosugu et al., 1990).

Production of Methyl Ester

Generally, fatty acid alkyl esters are produced either by direct esterification of fatty acids (Fig. 20.5) or by transesterification of triglycerides (Fig. 20.6) with the corresponding alcohols. Technically, both direct esterification and transesterification reactions are equilibrium reactions producing co-products such as water and glycerol, respectively (Hoydonckx et al., 2004).

Direct esterification can be conducted under catalytic or spontaneous condition, depending on the miscibility of the fatty acids and alcohols, temperature, and acidity of the fatty acids.

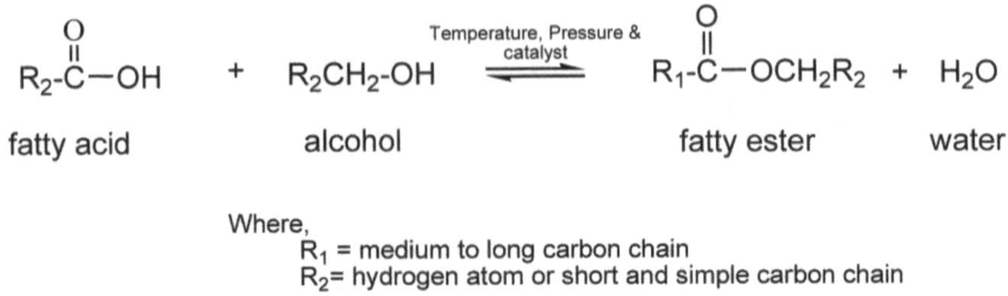

Fig. 20.5. Typical direct esterification reaction between fatty acid and alcohol for the production of fatty esters.

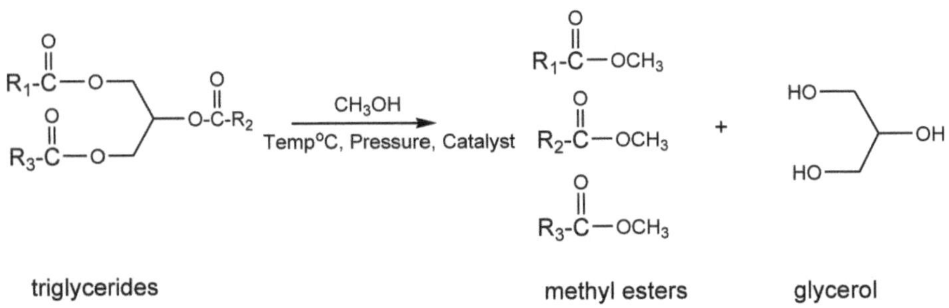

Fig. 20.6. Typical transesterification reaction of triglyceride for methyl ester production.

The equilibrium of the reaction may be displaced either through continuous removal of either one of the products and co-products or by feeding one of the reactants in excess. In catalytic esterification of fatty acids with alcohol, the choice of catalyst is crucial, depending on the stearic hindrance of both reactants. The most frequently used catalyst for direct esterification is the homogeneous Brønsted acids catalyst such as sulfuric acid, hydrochloric acid, arylsulfonic acid, phosphorus acid, or their mixtures. Under special circumstances, a few Brønsted acid catalyst such as diphenylammonium triflate (Wakasugi et al., 2000), triflurobenzeneboronic acid (Ishihara et al., 1996), and dodecylbenzenesulfonic acid (Manabe et al., 2002) have been proposed. However, when there is a need to avoid alcohol dehydration or racemization and to create acid-labile conditions, homogeneous Lewis acid catalyst may be preferred over Brønsted acid catalyst. Examples of Lewis acid compound are boron trifluoride; zinc oxalate; or complexes containing titanium, tin, hafnium, or zirconium (Hoydonckx et al., 2004). In the processing of renewable chemicals, especially those intended as intermediates or precursor for value added oleochemicals, research activities focused on solid, recoverable, and reusable heterogeneous catalysts. Examples of heterogeneous acid catalysts suitable for esterification reactions are hydrophobically modified zeolites, mesoporous ordered materials modified with sulfonic acid groups, heteropoly compounds such as carbon supported heteropolyacids, graphite bisulfate TiO_2 gel, and Lewis acid amorphous tin-silicon mixed oxides.

The transesterification reaction of triglycerides and methanol for the production of biodiesel methyl ester can be carried out, economically, with or without catalyst. Catalytic transesterification uses alkalis such as sodium hydroxide, potassium hydroxide, or their alkoxides as catalyst at temperatures of less than 100°C. Thermal non-catalytic transesterification may be conducted at 235°C and pressure of up to 6.2 MPa with methanol/triglyceride ratio of 21/1 giving conversion of 85% after 10 hr of reaction (Diosaku et al., 1998). However, under a supercritical condition, a triglyceride may be completely transesterified in 4 min (Saka & Kusdiana, 2001).

Production of Fatty Alcohol

The industrial importance of fatty alcohols lies on the diversified range of chemical reactions the primary hydroxyl group may undergo, resulting in a full range of multifunctional chemicals such as specialty esters, aldehydes, amines, and ether. Ethoxylation with ethylene oxide yields fatty alcohol ethoxylates, which are considered of high importance for detergent industries.

Production of natural fatty alcohols may either use triglyceride of plant and animal origin, wax esters from whale, or jojoba plant as their feed materials. The differences in the chemical structure of the feed materials will lead to some differential in the processing routes to produce these alcohols.

The key reaction for the production of fatty alcohols is hydrogenation of fatty acids or fatty acid methyl ester which involves reduction of carbonyl functionality by hydrogen gas (Fig. 20.7). Currently, there are 3 large scale hydrogenation processes adopted by the commercial producers:

- Suspension hydrogenation
- Trickle-bed hydrogenation
- Gas-phase hydrogenation

All these hydrogenation processes are carried out at 200–300°C and 20–30 bar using copper-containing, mixed-oxide catalysts. A milder condition which operates

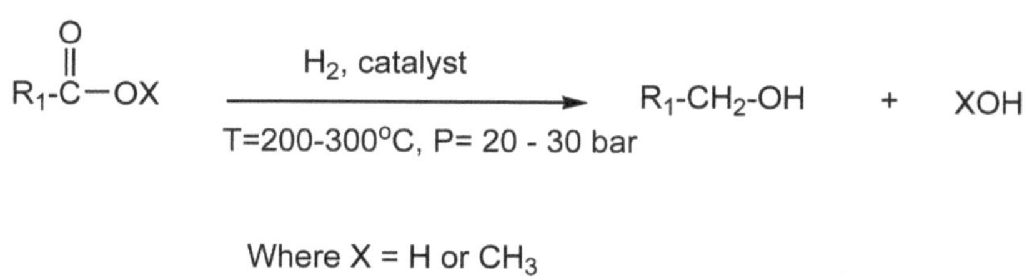

Fig. 20.7. Reduction of carbonyl group on feed material that produces fatty alcohol.

at 170–250°C and less than 10 bar has also been reported for Oleofina/B through the use of proprietary catalyst (Condea, 2000). Suspension hydrogenation uses powdered catalyst with specific particle size distribution, while trickle-bed and gas-phase hydrogenation use fixed-bed catalyst. Through fatty acid as feed materials, hydrogenation is carried out by feeding separately the hydrogen and recycled fatty alcohol catalyst slurry, which then forms the desired ester almost instantaneously. The ester is then hydrogenated in the same reactor at 260–300°C and 30 bars. The gas-phase hydrogenation process is suitable for a vaporized substrate or feed material such as methyl ester. The process requires extremely large excess of recycled gas (about 600 mole of H_2 per mole of ester) under high gas velocities with methanol used to aid evaporation. The product mixture consists of gas phase and liquid phase. The hydrogen in the form of gas is recycled while the methanol is stripped from the fatty alcohols. The crude fatty alcohol is then purified by distillation. On the other hand, in trickle-bed hydrogenation process, the substrates to be reduced or hydrogenated are used in liquid form and are normally of non-volatile properties such as wax esters and fatty acids. Further processing of the crude fatty alcohol is just the same as that from the gas-phase hydrogenation. Copper-chromite is the common catalyst used to produce saturated fatty alcohol. However, zinc is used if the double bond in the ester carbon chain needs to be preserved (Condea, 2000; Lurgi Fatty Alcohol Technology).

Production of Fatty Amine

Fatty amines are mostly produced for the production of quaternary ammonium compounds (QUATS) such as distearyldimethylammonium compounds.

Fatty amines can be produced from chemical transformation of fatty acids via fatty nitriles. The fatty acid is reacted with ammonia at high temperature in the presence of metal oxide catalyst. Fatty nitriles are hydrogenated to give primary, secondary, and tertiary amines (Fig. 20.8). Secondary and tertiary amines are the intermediates for QUATS. A quaternized triethanolamine esters, or commonly known as *di*-alkyl esterquats, is an example of which quats are used as cationic surfactant in fabric softener and hair conditioner. Fatty amines are converted to imidazolines, the intermediates for amphoteric and cationic surfactant.

Fatty nitriles are used as textile auxiliaries and also further derivatized for production of fatty amides and synthetic fibers, while fatty amides are used as water repellent agents, foam stabilizers, and additives in rubber compounding.

The Process to Produce Glycerol

Glycerol, or also commonly known as glycerine in the oleochemical industry, is obtained as a co-product of fat splitting or transesterification. It can also be produced synthetically from propylene via epichlorohydrin. This petrochemical pathway is slowly disappearing (Eierdanz, 1992). Today, almost all glycerine found in the market is derived from triglycerides.

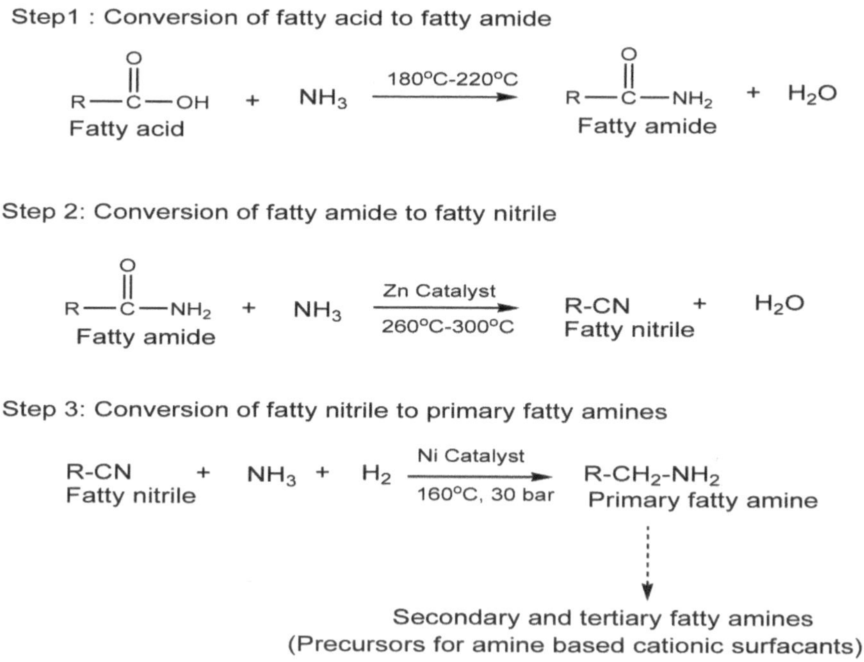

Fig. 20.8. Common processes involved in production of fatty amine.

Glycerol is a simple trihydric alcohol in the form of clear, odorless viscous liquid which has a sweet taste. It is also hygroscopic in nature. Although the terms glycerol and glycerine are often used interchangeably, subtle differences in their definitions do exist. Glycerol is the term used for the pure compound whereas glycerine refers to commercial grades, irrespective of their purity. Glycerol can be found in the triglyceride structure of oils/fats, and the content ranges from approximately 9 to 13.5%.

Natural glycerine is obtained primarily as a co-product from the production of fatty acid, fatty ester, or soap from oils and fats. Splitting or hydrolysis of oil, carried out under pressure and high temperature, produced fatty acids and sweetwater. The sweetwater contains 10–20% glycerol. Transesterification of oil with methanol, in the presence of a catalyst, yielded methyl esters and glycerine. Since the process does not use water, the glycerol concentration is higher. Saponification of an oil/fat with caustic soda formed soap and soap lye. The soap lye formed contains 4–20% glycerol and is also known as sweetwater or glycerine. Sweetwater or glycerine obtained as a co-product from the above three processes contains impurities and must be made to undergo further purification. Crude glycerine is seldom used directly (Salmiah et al., 2000). MPOB has recently filed a patent on the process of direct utilization of crude glycerol from biodiesel plant to produce polyglycerol (Hoong et al., 2009).

Oleochemical Applications

The largest portion of basic oleochemicals is used in the production of surfactants. Linear carbon chain oleochemicals of C12-C14 are the most important feedstock for surfactant industries especially in the household detergent application. These medium and long aliphatic chains provide the hydrophobic tail with different degrees of hydrophobicity. The hydrophilic head group is then provided through various chemical transformations, depending on the classification of surfactants required. Other application of oleochemicals is as substrates for polymerization that can produce polyalcohols (polyols) and can be subsequently formulated into polyurethane products. They can also be subjected to further derivatization into fine chemicals for industrial applications. High purity or pharmaceutical grade basic oleochemicals may also be used directly in high end products, for example as adjuvant in pharmaceutical ingredients, cosmetic and personal care products, and agrochemicals products.

Surfactant

Surfactants are amphiphilic molecules which are made up of a polar head group (ionic or non-ionic) and at least one hydrophobic tail (hydrocarbon chain). It is the hydrophobic chain that is normally derived from oleochemical-based feedstock. The polar head group normally exhibits a strong affinity for polar solvents, which is commonly water. It is often called hydrophilic head. In contrast, the apolar hydrocarbon chain, commonly called hydrophobic tail, exhibits strong affinity for apolar medium, mainly the oil phase. Basically, a surfactant molecule can be diagrammatically illustrated by the molecule shown in Fig. 20.9. In general, surfactants can be classified by their ionic properties (electrical charges on its hydrophilic head) in water such as anionic, non-ionic, cationic, and amphoterics.

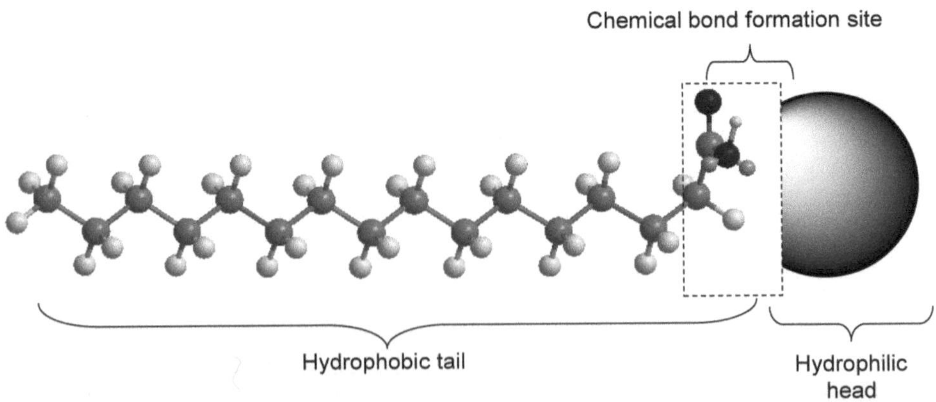

Fig. 20.9. Diagrammatic illustration of a surfactant molecule (reproduced from Zainab et al., 2010).

Due to the double affinity, a surfactant molecule exhibits a very strong tendency to migrate to interface whereby an interface is defined as the surface between two immiscible phases. Surfactants will orientate themselves as such; the polar group stays in water and the non-polar group displaced itself away from the water. These opposing forces that occur within the same molecule will change the properties at the interface. The phenomenon is termed as "adsorption" (Fig. 20.10). The largest end use market for surfactants is as household cleaning detergents (Fig. 20.11).

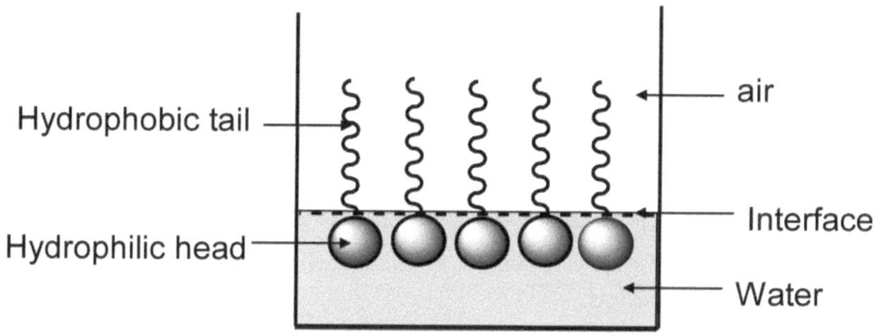

Fig. 20.10. Diagrammatic illustration of surfactant molecules alignment at air/water interface (reproduced from Zainab et al., 2010).

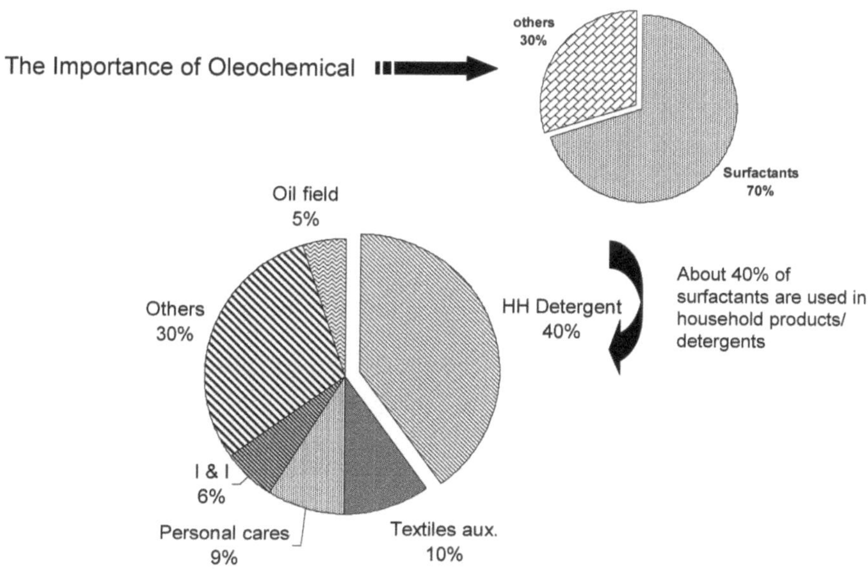

Fig. 20.11. Market share of oleochemical-based surfactants in household consumer products (reproduced from Hauthal, 2004).

Anionic Surfactant

Anionic surfactants are the major type of surfactants used in cleaning product formulations because of their excellent soil removal property. The soil removing property comes from the negatively charged head groups of the surfactant. The major types of anionic surfactants are linear alkyl benzene sulphonates (LABS), fatty alcohol ether sulphates (FAES), and fatty alcohol sulphates (FAS). Introduction of methyl ester sulphonates, which are derived from palm oil feedstock, into heavy duty laundry detergents is of interest as they perform better at lower water temperatures and in hard water than that of LABS. On the other hand, soap is considered the oldest and simplest form of anionic surfactant.

Soap. Soap making is likely one of the oldest chemical reactions known to mankind and in the cleaning technology based on oleochemical. In the 19th century, the animal rendering industry in United States provided the perfect feedstock, tallow, for soap making, which later the industry grew to what it is today. Most of the big names like Procter and Gamble (P & G), Kao, Unilever, and Henkel had also started their involvement during this period.

However, by the turn of the century, tallow-based soaps become less popular as a new invention on soap from palm-based oleochemicals by Unilever, then called Lever Brothers, who had their own oil palm plantations in 1911, emerged. Palmolive, a soap manufactured out of palm and olive oils was actually the world's largest selling soap at the turn of the century (Ellard, 2009). Today, Malaysia produces about 20% of the world's supply of soap chips or soap noodles using palm oil and palm kernel as feedstock.

The chemistry of soap making is very simple. Soaps are the salts of (mainly) saturated and unsaturated fatty acids having carbon number of C10 to C18. Soaps can be prepared through either saponification of the triglyceride or neutralization of the fatty acids by splitting triglycerides. Most soap producers in Malaysia preferred the neutralization process. It is a cleaner process where glycerine is not produced as a co-product. Therefore, soap manufacturer will not have to worry about the clean-up process for glycerine.

In order to make soap directly from oil, a blend of triglycerides is reacted with a strong sodium hydroxide solution that will produce soap and glycerine (Fig. 20.12).

For the neutralization process, the fatty acids obtained from the splitting process are first separated from glycerine. The fatty acids are then distilled, blended, and neutralized with NaOH (sodium hydroxide) solution to form soap (Fig. 20.13).

Almost all toilet soaps or laundry soaps are made with a blend of oils. The most common oils are coconut oil or palm kernel oil, which are generally called nut oils, whereas tallow or palm oils are generally referred to as non-nut oils. To get the best performance from soaps, both types of oils, the non-nut oils and nut oils, must be

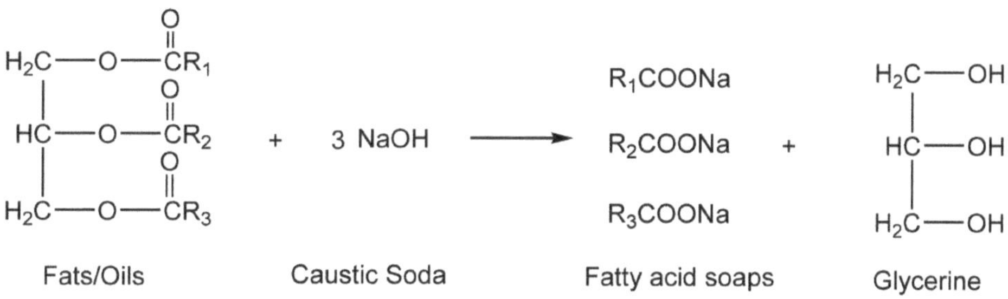

Fig. 20.12. Production of soap via saponification method.

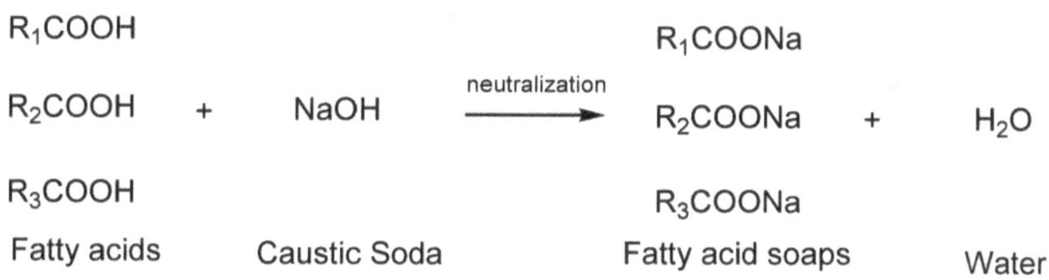

Fig. 20.13. Production of soap via neutralization method.

combined and blended. Table 20-B shows some of the distilled fatty acid blends for soap and their properties for soap making (Spitz, 2009).

The non-nut oils provide long-chained saturated fatty acids (C16/C18—palmitic and stearic acids) which make soaps almost insoluble at standard conditions and do not form lather. However, these almost insoluble soaps add to lather stability and hardness, which makes the soap a hard solid. Long chain fatty acids also gives soap better detergency property (cleaning performance).

On the other hand, the nut oils provide short-chained fatty acids (C12 fatty acid) that give soap the moderate stability but with high lather volume when they are dissolved. However, sodium laurate does not dissolve significantly until the temperature is over 40°C in order to give enough lather to the soap. However, when the non-nut and nut oils are blended together at the right proportion, it forms a eutectic mixture where the soap is soluble at low temperatures, producing both high lather and good detergency soap mixtures (Hall, 2009).

Other than color and fragrance, therapeutic additives are added to the soap to achieve a desired functional and/or marketing position. This has become the prime focus of soap formulators nowadays. The following is a list of several important categories of

Table 20-B. Specification of Distilled Fatty Acid Soap Blends.

Distilled fatty acid soap blends	Acid value (mg KOH/g sample)	Total fatty matter (%)	Moisture (%)	Iodine value (mg I_2/g sample)	Titre (°C)	Free caustic (%)	Sodium chloride (%)	Penetration value (mm)	Foaming power (ml/ml)*
90PO:10PKO	214	84	10	54.5	43.0	0.06	0.4	53	530/365
85PO:15PKO	217	81	11	52.2	42.2	0.06	0.4	52	520/355
80PO:20PKO	219	81	10	45.2	41.3	0.06	0.4	51	505/345
75PO:25PKO	221	82	11	44.5	38.4	0.04	0.4	50	513/343
70PO:30PKO	225	83	10	43.5	37.5	0.06	0.4	50	520/340
60PO:40PKO	229	80	11	38.2	37.5	0.03	0.3	50	510/350
40PO:40PS:20PKO	224	81	11	37.0	46.0	0.10	0.2	48	470/306
80PS:20PKO	219	83	12	27.0	48.0	0.10	0.5	48	510/345
70PS:30PKO	225	82	11	29.0	47.0	0.10	0.5	48	500/325
65PS:35PKO	228	82	10	25.2	45.0	0.10	0.5	48	525/355
60PS:40PKO	229	83	11	22.0	43.8	0.10	0.3	48	500/368

*Foaming power is indicated by foam volume immediately after 30 strokes/foam volume after 5 minutes
Abbreviations: PO = Palm oil, PKO = palm kernel oil, PS = palm stearin
Source: Spitz (2009).

additives in soaps. Table 20-C shows some of the oleochemical ingredients used as additives to soaps (Yeong et al., 2010a).

Methyl Ester Sulphonate (MES). Alpha-sulphonated methyl ester (α-SME), or known to the industry as methyl ester sulphonates (MES) (Fig. 20.14), is an oleochemical-based surfactant obtained by direct sulphonation of palm oil methyl esters (Fig. 20.15).

The MES can be differentiated from the conventional anionic surfactants derived from fatty alcohol ethoxylate through observation of its chemical structure specifically at the hydrophilic head group, as illustrated in Fig. 20.16.

The chemical structure of MES is slightly different from other common anionic surfactants derived from fatty alcohol because the sulphur group is attached to an alpha-carbon atom instead of to the oxygen atom. Palm methyl esters of C16/C18 carbon chain with low content of C18:1 and C18:2 are excellent feedstock for the production of MES. The quality of the feedstock is important in determining the final properties of the MES. Methyl ester of iodine (IV) < 0.5 is essential to produce a light color and low odor MES in order for the surfactant to be competitive with other surfactants.

Table 20-C. Oleochemical Ingredients Used in Soaps Making.

Desired property	Function	Ingredients
Emollients	To impart and maintain skin softness and pliability, and in general improve skin's overall appearance	Fatty esters (cetyl palmitate, isopropyl myristate, isopropyl palmitate, octyl palmitate) Alkoxylated ethers (PPG-20 cetyl ether, PPG-20 oleyl ether, PPG-20 myristyl ether) Alkoxylated alcohols (cetearyl alcohol, cetyl alcohol, stearyl alcohol)
Humectants/ moisturizers	Skin conditioning agent that increases the water level and moisture retaining ability of the skin	Alcohol (propylene glycol, octyl decanol, polyglycerine compounds) Polyol (glycerine, diglycerine)
Occlusive agents	To prevent moisture evaporation from the skin	Oil or wax (castor oil, coconut oil, jojoba oil, jojoba wax, cocoa butter, shea butter)
Dermabrasive/ exfoliating agents	To remove the outer layer of stratum cornea by scrubbing thus producing a smoother feel to the skin	Loofah, oatmeal, jojoba beads, various seeds
Drug components	Antimicrobial, anti-acne	Ingredients under FDA approval list

Fig. 20.14. Methyl ester sulphonate (MES) flakes from palm stearin methyl ester.

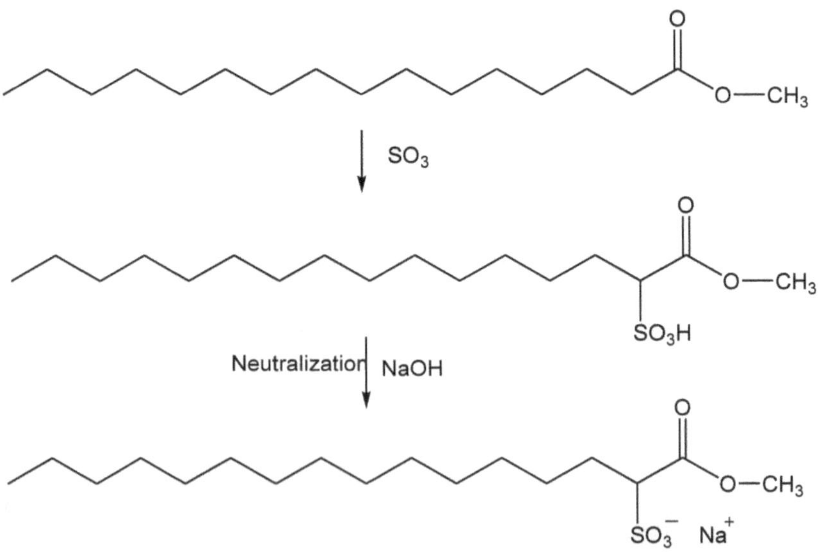

Fig. 20.15. Sulphonation reaction of palm-based methyl ester.

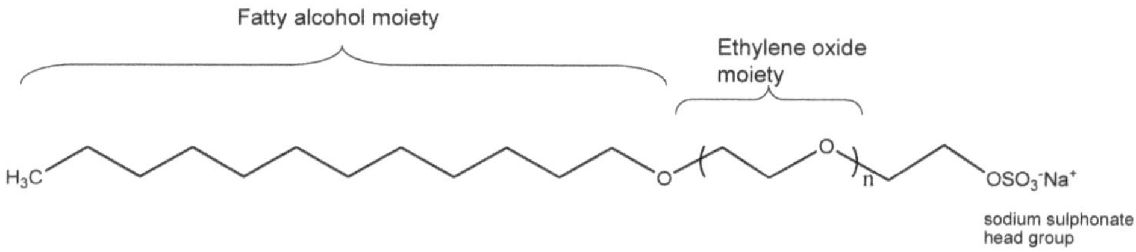

Sodium lauryl sulphate (SLS)

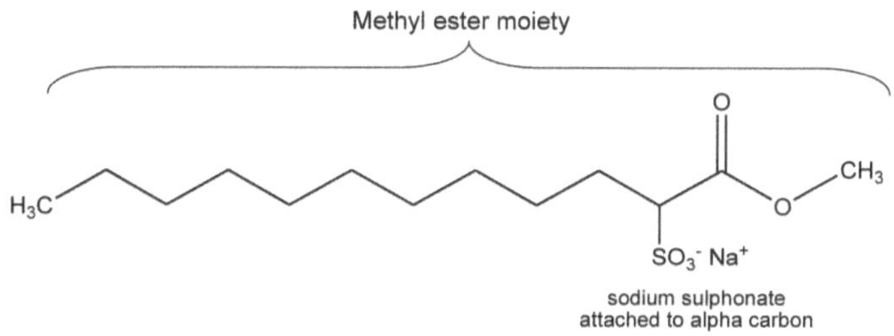

Methyl ester sulphonate (MES)

Fig. 20.16. Chemical structure of sodium lauryl sulphate (SLS) and methyl ester sulphonate (MES).

MES offers an environmentally friendly and viable alternative to the old workhorse anionic surfactant, LABS. MES can be produced at lower cost than LABS, which has been the driving force for its commercial production. As illustrated in Fig. 20.17, MES biodegrades faster and better than LABS (Razmah, 2002). In addition, Fig. 20.18 shows that MES has good calcium tolerance, thus greater detergency and less sensitivity to water hardness than other anionic surfactants. Detergency performance of MES with respect to the methyl ester of different carbon chain length (C12 to C18) is dependent on the chain length of methyl ester as the feedstock used. The detergency performance of MES with C16 and C18 carbon chain performed significantly well in both water hardnesses while that of LABS was only good in solution of low water hardness (Zulina et al., 2006).

Cationic Surfactant

Cationic surfactants are amphiphilic molecules, having a positively charged head group, dissociated in water to form an amphiphilic cation and an anion of halogen type or an alkyl sulphate. The positive charges on the molecule gave absorptive effect

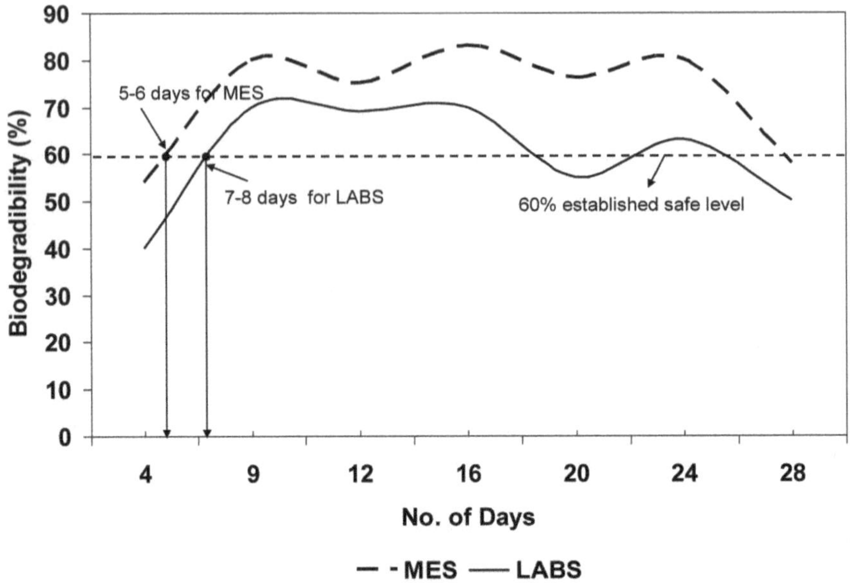

Fig. 20.17. Biodegradability of palm-based methyl ester sulphonates (C16-C18 MES) versus linear alkylbenzene sulphonates (LABS).

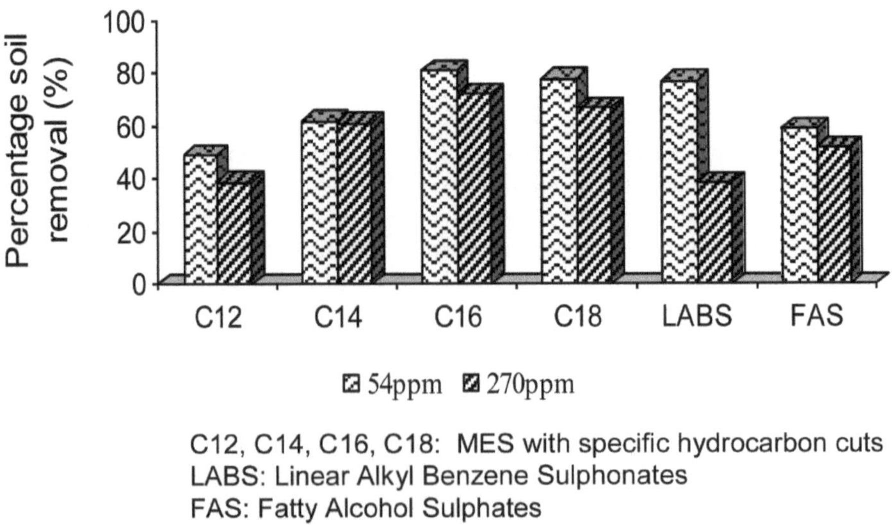

Fig. 20.18. Detergency of MES with respect to different methyl ester carbon chain length in comparison with LABS and FAS.

whenever they were exposed to negatively charged surfaces. Cationic surfactant found application as bactericide and active ingredients in fabric softening formula. They account for about 4% of the total world surfactant production. Quaternary ammonium compounds and quaternary triethanolamine ester are two types of cationic surfactant normally incorporated in fabric softener formulations, the largest product segment for cationic surfactants.

Quaternary Palm-Based Triethanolamine Ester (Esterquat). Esterquat types of cationic surfactants are gaining interest in household fabric conditioning products because of their good softening effect and excellent biodegradability with unique water repellency and reabsorbancy characteristic (Zainab et al., 2000). The major potential for palm-based esterquats, currently, is targeted for household fabric care industries (Haliza et al., 2007). The ester groups in palm-based esterquats provide a potential point for easy hydrolysis into fatty acids and smaller ammonium quats molecules when discharged into sewage systems so that our ecosystem is not being polluted with household discharge while maintaining quality in lifestyle (Haliza et al., 2007).

The palm-based esterquats project has further revealed the importance of saturated fraction of palm oil-based oleochemicals in household products, particularly surfactants. The process started with fatty acid as the source of alkyl group reacting with triethanolamine to give esteramine mixtures of different degree of esterified hydroxyl group (Zainab et al., 2009). The esteramine mixtures could then be subjected to quaternization with dimethyl sulphate in isopropanol. The two-step reaction is illustrated in Fig. 20.19.

Palm-based esterquats derived from palm stearin exhibit better softening performance than the commercial tallow esterquats due to its higher percentage of C16 carbon chain length, while that of tallow has significantly high amount of C18:1 and low in C16 carbons. (Haliza & Zainab, 2007; Zainab et al., 2009). Therefore, under normal practice tallow-based fatty acid were partially hydrogenated to obtain a certain degree of saturated hydrocarbon to give equivalent performance as palm stearin fatty acid-based esterquats (Fig. 20.20).

Emulsifier

Emulsifiers are also classified as surface-active agents due to the fact that they possess both hydrophilic and lipophilic characteristics which enable them to modify the interface property between air and water or water and oil. Emulsifiers are very useful in formulating products with ingredients of different polarity and solubility which would lead to a homogeneous and stable system.

Application of Surfactant/Emulsifier in Cosmetic and Personal Care Products. Cosmetics and personal care (CPC) products can be divided into four main categories: skin care, color cosmetic, hair care, and oral care products. The functions of skin care prod-

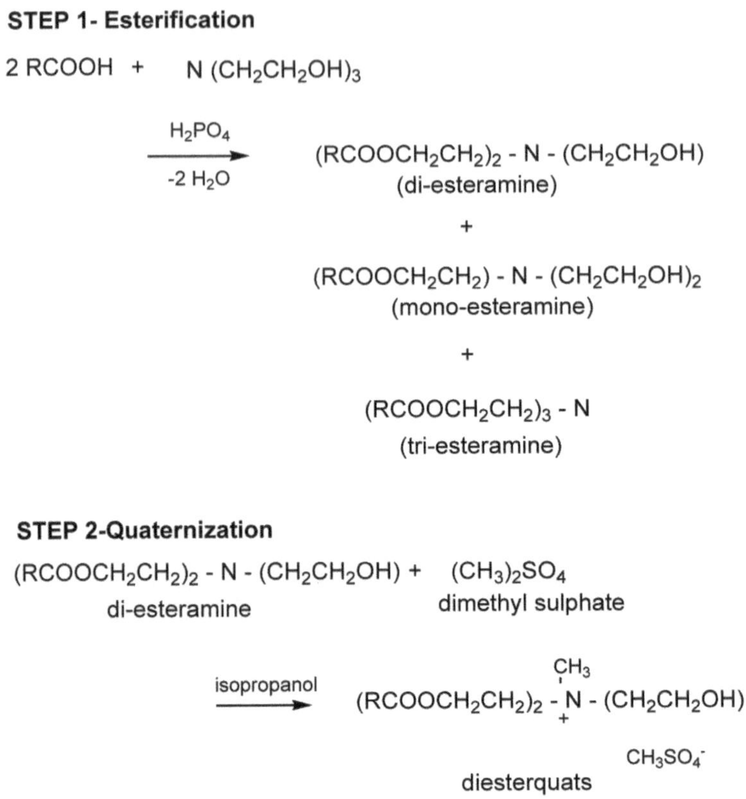

Fig. 20.19. A two-step reaction for production of esterquats derived from palm-based fatty acids.

ucts are to clean, nourish, moisturize, stimulate, and protect the largest and important organ of the human body which is the skin. Nowadays, cosmetic products are designed not only for beautifying but are also used as therapeutic products. This can be achieved by delivering the functional raw materials or active ingredients through skin barrier.

CPC products can be prepared in the form of lotion, cream, liquid, gel, and stick. Natural ingredients such as glycerine, fatty acids (lauric, myristic, palmitic, and stearic acids), fatty alcohols, and their esters can be used to replace synthetic ingredients. Table 20-D shows the list of synthetic ingredient and their natural based counterpart in a cream formulation. The total amount of natural based ingredients incorporated in formulations can range from 45 to 98%.

Almost all the cosmetic and personal care products are produced in the form of emulsion. In principle, a cosmetic or personal care product formulation must have an

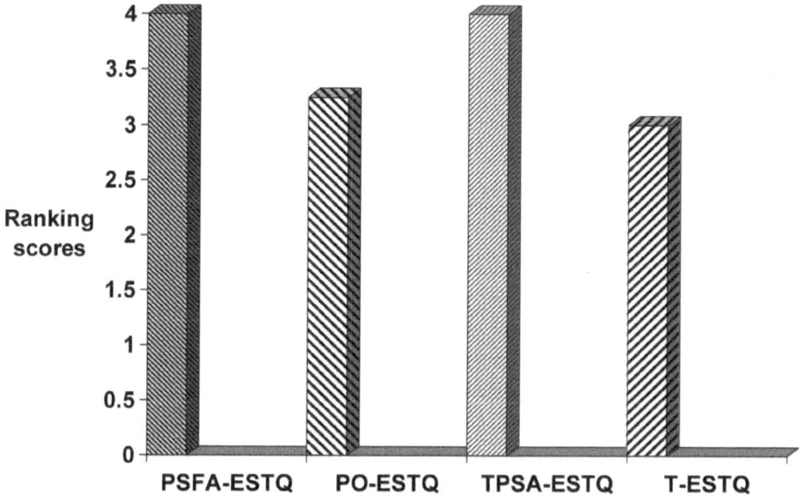

Note:
1.PSFA-ESTQ: palm stearin fatty acid-based esterquats.
2.PO-ESTQ: palm oil-based esterquats.
3.TPSA-ESTQ: triple pressed stearic acid-based esterquats.
4.T-ESTQ: tallow-based esterquats.
5.Ranking was calculated based on number of preference between a pair of test samples. The higher the ranking the better the softening power based on hand tactile evaluation.

Fig. 20.20. Hand tactile evaluation based on ranking score conducted on cotton towel treated with esterquats from various fatty acid feedstocks.

Table 20-D. Comparison of Synthetic and Natural Cosmetic Ingredients in Cream Formulation.

Functional constituents	Synthetic ingredients	Natural ingredients
Oil phase	Vaseline oil, silicones, esters of fatty acids	All vegetable oils (palm oil, palm kernel oil)
Water phase	Water	Water
Actives	Synthetic carotenes, humectants (glycerine)	Vegetables extracts, floral waters, natural vitamins, amino acids, alpha hydroxy acid, natural carotenes
Thickening agents	Chemical polymer: carbomer, carboxylic derivatives	Natural polymers: polysaccharides
Emulsifiers	Synthetic surfactants: ethoxylated alcohols, sorbitan derivatives	Natural polar molecules: proteins, lecithins
Fragrances	Synthetic fragrances	Essential oils extracted from plants/herbs
Antioxidants	BHA, BHT	Tocopherols (vitamin E)

oil phase and a water phase which is held together by an emulsifier system. An emulsion can be formulated as oil in water (O/W) or water in oil (W/O). From dermatological point of view, W/O emulsions are more preferable to O/W emulsions because of the natural lipid film on the skin is also a W/O emulsion. However, W/O formulation will leave a sticky and oily feel on the skin if used in tropical climates. Therefore, for tropical climates, most of the cosmetics are the O/W types (Yeong et al., 2010a).

In order to increase the topical therapeutic properties of cosmetic products with unique functional properties, bioactives are added into cosmetic formulation. Some of the common bioactive ingredients used are vitamins, plant extracts, and sunscreen agents. Vitamin E is one of the most important phytonutrients found in palm oil. It consists of eight naturally occurring isomers, a family of four tocopherols (alpha, beta, gamma, and delta) and four tocotrienols (alpha, beta, gamma, and delta) homologues. Vitamin E from palm oil is unique as it contains both tocopherols (20–30%) and tocotrienols (70–80%) (Ng et al., 2004). Tocopherols and tocotrienols have been shown to be excellent antioxidants (Choo et al., 2002; Nesaretnam et al., 2007). These palm-based bio-actives such as tocotrienols rich fraction (TRF) with and without carotenes are available commercially (Tocomin™, Gold-Tri E). Tocotrienols are a much more potent antioxidant than tocopherols (Nesaretnam et al., 2007). The chemical structure of tocotrienols and tocopherols are shown in Fig. 20.21.

Studies have shown that the administration of palm tocotrienol complex to animals results in a reduction of oxidative stress risks and, therefore, could be a potential anti-aging and oxidative stress preventing agent (Nesaretnam et al., 2007).

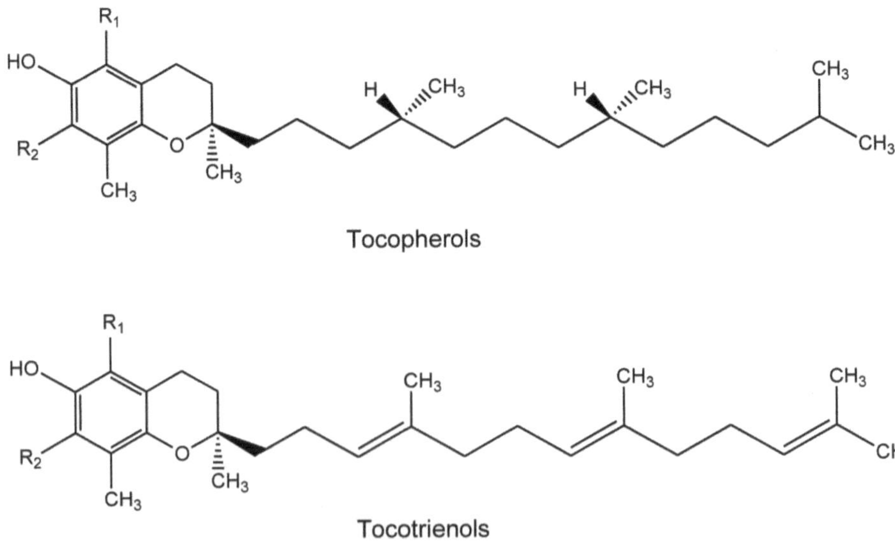

Fig. 20.21. Chemical structure of tocopherols and tocotrienols.

Palm-based cosmetic products formulated with palm tocotrienol rich fraction (TRF) have the ability to moisturize the skin by increasing skin hydration. A single dose application of the cream can maintain skin hydration for up to 3 hours, as shown in Fig. 20.22. Long term applications of tocotrienol creams have also been shown to enhance the skin hydration up to 14 days when compared to placebo (Fig. 20.23).

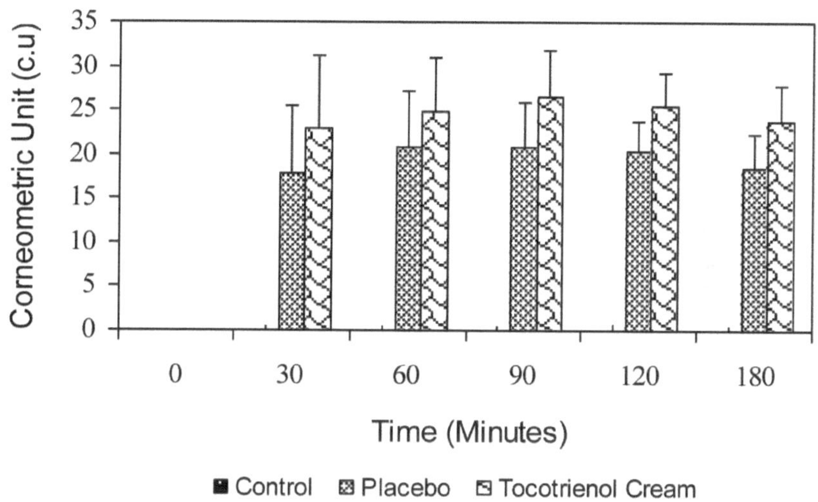

Fig. 20.22. Acute moisturizing effects of tocotrienol cream (reproduced from Zafarizal & Rosnah, 2009).

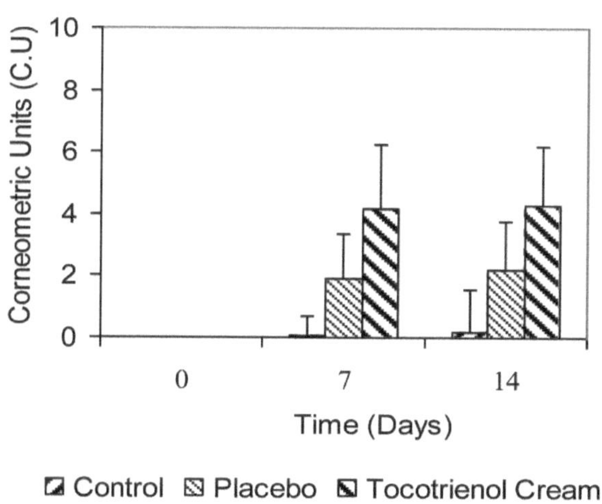

Fig. 20.23. Long term moisturizing effects of tocotrienol cream (Zafarizal & Rosnah, 2009).

The increase is skin hydration is further supported by reduction in transepidermal water loss (TEWL) level (Zafarizal & Rosnah, 2009).

Application of Surfactant/Emulsifier in Agrochemical. Agrochemical products are formulated to cater different types of application in agro industry. Examples of such formulations include aqueous concentrates (AC), emulsifiable concentrates (EC), wettable powders (WP), suspension concentrates (SC), emulsion in water concentrates (EW), suspo-emulsions, and water dispersible granules (WDG) (Ismail et al., 1996, 1998; Tadros, 1995).

There has been a shift recently in agrochemical formulations practices, for instance from emulsifiable concentrates (EC) to emulsion in water (EW), wettable powders (WP) to water dispersible granules (WDG), and powders or dusts to suspension concentrates (SC). This shift is said to provide products which are safer and more convenient to the end users. With this, the use of flammable or toxic solvents is avoided, and bulky and dusty products are being removed from the market. Other factors such as ease of handling, safer, and more convenient disposal of packing materials are becoming increasingly important.

There are two main components in agrochemical product formulations, namely the active and inert ingredients. The former is responsible for protecting targeted plants by destroying insects, fungi, and weeds or protecting them from attack by these pests. Meanwhile, the inert ingredients, which include wetting agents, dispersing agents, emulsifiers, spray adjuvant, solvents, carrier, and others, are added to the formulation to enhance the performance of the active ingredient (Bohmont, 1990).

In agrochemical industries, surfactants are used as additives primarily in pesticides (herbicides, insecticides, fungicides) and to a lesser degree in fertilizers, animal feeds, and soil treatments. When used as a pesticide adjuvant, surfactants provide emulsifying, dispersing, sticking, and foaming aid action. Emulsifying the active ingredients in liquid formulations create stable emulsions and extend shelf life. Dispersing agent is to aid application of the active ingredients. Sticking agent improves contact of the active ingredients onto the leave surface of the crop or weed, while a foaming aid controls spray drift during pesticide application.

The other inert ingredient where oleochemical can be used is in the area of solvent. Oleochemical, such as methyl esters, can be used to replace the petroleum-based solvent and this replacement have proven to be beneficial in reducing skin allergies among the farmers (Henkel, 1995; Ismail et al., 1998a, 2000, 2004, 2005).

Specialty Esters

Emollient

Emollients are substances that soften and soothe the skin. Emollients also commonly referred to as moisturizers. Emollient is a key component in the cosmetic products.

They are used to moisturize and prevent scaling of the skin. Emollients have three basic properties:

- Occlusion—providing a layer of oil on the surface of the skin to slow water loss and thus increase the moisture content of the stratum corneum.
- Humectant—increasing the water-holding capacity of the stratum corneum.
- Lubrication—adding a slip or glide across the skin.

Most emollients are in the forms of oil or grease, such as palm oil, mineral oil, squalene, lanolin, and fatty esters. Isopropyl myristate, or IPM, is the most common emollient used in CPC products. It has a very dry character and shows good spreading properties on the skin. A disadvantage of IPM is its high comedogenicity, which means that it exhibits clogging of skin pores thus causing various skin diseases such as acne. However, isopropyl palmitate (IPP) and isopropyl stearate were found to have lower comedogenic effect on skin and, thus, are increasingly used as skin emollient (James, 1989).

Lubricant

Lubricants are materials used to reduce friction between machinery parts in motion and minimize wear between interacting surfaces. Due to their vast applications in many industries such as industrial machineries, mining, metal working, fiber and textile manufacturing, agriculture, forestry, construction, railway system, and recreation sport, they tend to pollute the environment in small but widespread quantities. Most of these lubricants are non-aqueous liquids. In the past, almost all lubricants were mineral oil based products, causing the major part of environmental damage due to improper disposal, accidental spillage, and leakages and lost lubrication.

Environmental compatibility, high performance, and longer product life has been the driving forces for the shift from the traditional mineral oil lubricant to synthetic and quasi synthetic oils and biolubricant, at least in Western Europe. The selection of rapidly biodegradable lubricant, especially those classified as "not water pollutant" will offset the expenses of oil spillage or disposal. The majority of rapidly biodegradable lubricants are based on saturated and unsaturated ester oils. They are divided into five groups: monocarboxylic acid esters (monoesters), dicarboxylic acid esters (diesters), glycerol esters including natural oils and fats, polyol esters, and complex esters (Höfer & Bigorra, 2007). Table 20-E shows some of the earlier applications of natural oils and fats for lubrication purposes (Owei, 1988).

Using natural oils and fats as lubricants will give some limitations, as most of the time the esters are produced by nature and have fixed molecular structures. The limitations are poor cold flow behavior, low thermo-oxidative, and hydrolytic stability. Converting the natural oils and fats into synthetic esters will allow researchers to have better control of the desired properties of lubricants. Synthetic lubricant base

Table 20-E. Vegetable Oils and Animal Tallow and Their Usage.

Vegetable/animal oil	Common application
Olive oil	Automotive lubricants
Sperm whale oil	Spindle lubricant in textile mills, automotive transmission fluids, metal cutting fluids, instrument oils
Rapeseed oil	Metal forming processes
Castor oil	Gear lubricants
Coconut oil	Agents gas and petrol engine oils
Palm oil	Agents in steel industry for rolling thin gauge strip, railway wagon greases
Tallow	Agents in steam cylinder oils

oils offer better improved stability than refined petroleum oils but at a cost. More than 90% of chemical modifications are carried out at the carboxylic groups, while less than 10% have involved reactions at the hydrocarbon chain. Without sacrificing the favorable viscosity-temperature characteristics and lubricity, unsaturated vegetable oils can be converted into thermo-oxidatively stable products by saturation of the carbon-carbon double bonds. Chemical modification also includes transesterification and hydrolysis followed by esterification. The acid components normally having carbon 8 to 18 carbon chain length are obtained from natural oils and fats, while the alcohol component is derived from petroleum sources, as indicated in Table 20-F (Yeong, 1996). Most of the biodegradable synthetic oils are chemical esters which are known to offer superior thermal and oxidative stability. Prices for these niche products are higher than vegetable oils and significantly higher than petroleum-based lubricants (Erhan et al., 2006).

Diesters and polyol esters are probably the most commonly used esters for lubricant formulations. Diesters are prepared by combining linear diacid as the central component with generally a branched monofunctional alcohol. Diesters have good viscosity index (VI) and excellent pour points. The linear diacids contributes to the high VI because linear hydrocarbon chain is generally long. Branching in alcohol contributes to a low pour point property by maintaining fluidity and reducing its tendency to crystallize at lower temperatures. In general, commercially available diesters have kinematic viscosities that range from 2 to 8 cSt (100°C) and 5 to 50 cSt (40°C) and have excellent pour points (–40 to –70°C) (Rudnick, 2002).

Polyol esters consist of the reaction product of a multifunctional alcohol in combination with a monofunctional carboxylic acid. The carboxylic acid components are generally in the C7-C18 range. Yeong et al. (2010b) reported on a process to produce polyol esters using palm oleochemicals. Polyol esters generally have kinematic viscosities between 2 and 20 cSt (100°C) and 5 and 225 cSt (40°C), a wider range than for

Table 20-F. Raw Materials for Synthetic Esters for Lubricant.

Alcohol	Source
2-ethyhexyl alcohol	Petroleum
Isodecyl alcohol	Petroleum
Isotridecyl alcohol	Petroleum
Trimethylol propane	Petroleum
Neopentyl glycol	Petroleum
Pentaerythritol	Petroleum
Acids	Source
Adipic	Petroleum
Azelaic	Tallow, tall, palm
Sebacic	Castor
Hexanoic	Coconut, palm kernel
Octanoic	Coconut, palm kernel
Pelargonic	Petroleum, tallow, palm
Lauric	Coconut, palm kernel
Palmitic	Tallow, coconut, palm kernel
Oleic	Tallow, palm kernel

diesters. Polyol esters have pour points in the range of +10 to –40°C. Thermal and oxidative stabilities of synthetic esters are excellent and the pour points can also be very low. The structural variety of synthetic esters is the major strength of this technology (Rudnick, 2002).

Some of the successful applications of green and sustainable lubricants are environmentally friendly two stroke oils, especially for outboard engines, chain saw oils, tractor oils, synthetic wind turbine oils, hydraulic oils, and so on (Höfer & Bigorra, 2007).

Natural Oil Polyols (NOPs)

Polyol is one of the major components in polyurethane production. Polyurethanes are produced by reacting the polyols with isocyanates in the presence of suitable catalysts and additives. Methylene diphenyl diisocyanates (MDI) and toluene diisocyanates (TDI) are the two common isocyanates used in the production of polyurethanes.

Processing and properties of the resultant foam can be markedly influenced by the choice of starting polyol structure. Polyurethanes are linear polymers that have a molecular backbone containing carbamate groups ($-NHCO_2$). A urethane group is formed by the chemical reaction between an alcohol and an isocyanate. Polyurethane results from the reaction between alcohols with two or more reactive hydroxyl groups per molecule

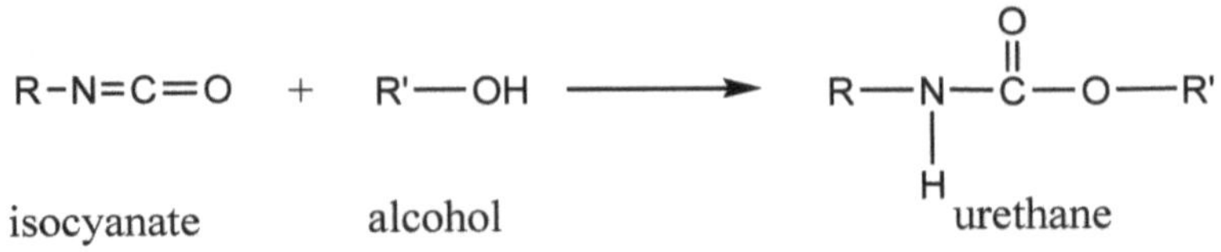

Fig. 20.24. Formation of urethane from isocyanate and alcohol.

(diols or polyols) and isocyanate that have more than one reactive isocyanate group per molecule (a diisocyanate or polyisocyanate). The polyurethane polymer-forming reaction occurs between an isocyanate and an alcohol as follows (Fig. 20.24).

Natural oil polyols, also known as NOPs or biopolyols, are polyols derived from vegetable oils by several different techniques. The primary use for these materials is in the production of polyurethanes. There are several routes to produce NOPs, which resulted in different properties even though the starting materials are triglyceride oils. Castor oil is the only commercially-available natural oil polyol that is produced directly from a plant. Other vegetable oils such as soybean, rapeseed, and palm oil would require some sort of modification because these vegetable oils do not have naturally occurring hydroxyl groups (http://en.wikipedia.org). The hydroxyl group has to be introduced via chemical reaction and the most common route is via epoxidation followed by hydrolysis process. Other routes include hydroformylation, ozonolysis, and esterification.

1. Polyols prepared by epoxidation followed by hydrolysis process
2. A chemical compound with a double bond is first reacted with an oxidizing agent such as peracid to form an epoxide ring. A ring opening reaction is then performed with an alcohol or water to form an ether or hydroxy linkage, respectively, in the presence of an acid catalyst (Fig. 20.25).

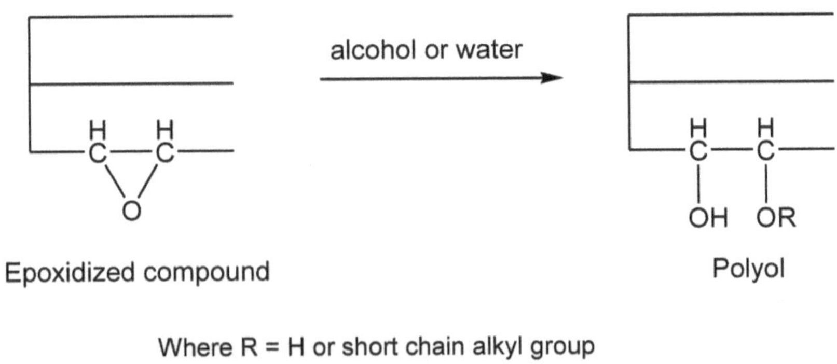

Fig. 20.25. Formation of polyols from epoxidized oil.

3. Polyols prepared by hydroformylation route. Polyols with primary hydroxyls can be obtained by hydroformylation. The process involved the reaction of carbon monoxide and hydrogen (syn-gas) with double bonds to introduce aldehyde groups which are subsequently converted to hydroxyls by hydrogenation (Fig. 20.26).

4. Polyols prepared via ozonolysis route. Polyols with terminal primary hydroxyls can be obtained by ozonolysis. Ozonolysis is also a very clean way of cutting double bonds. Since all major unsaturated fatty acids have one double bond at the 9,10 positions, the results of ozonoloysis are triglycerides with a

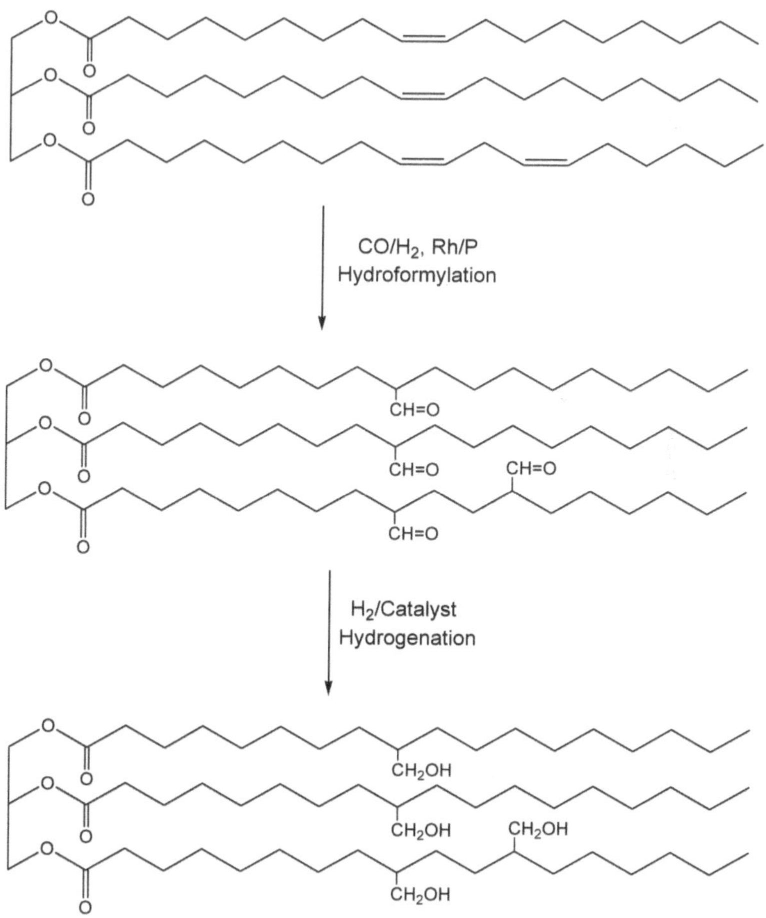

Fig. 20.26. Hydroformylation method for preparation of triglyceride based polyols (reproduced from Petrovic, 2008).

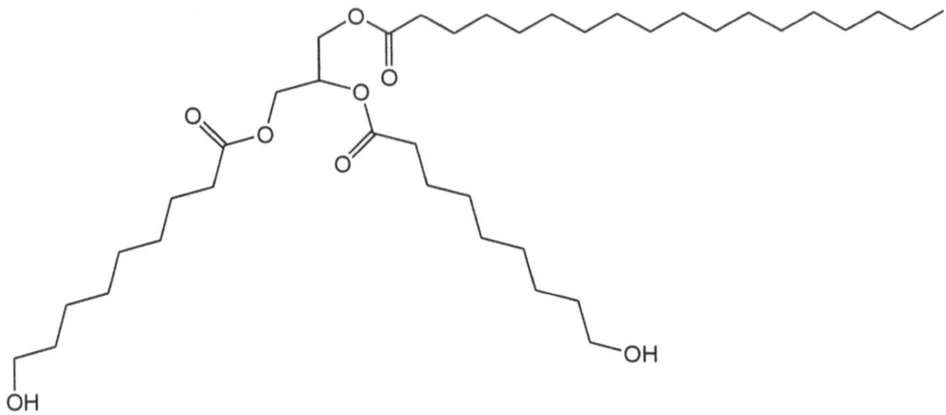

Fig. 20.27. A generalized diol component in the triglyceride-based polyol (reproduced from Petrovic, 2008).

maximum of three hydroxyl groups and a number of low molecular weight glycols or mono alcohols. Fig. 20.27 shows an example of diol obtained from a triglycerides molecule after ozonolysis.

5. Polyols prepared by esterification route. Transesterification of oils can be carried out by reacting oils with polyfunctional alcohols (e.g., glycerol) to obtain a mixture of monoglycerides and diglycerides (Fig. 20.28). Trans-esterified oils could be applied for the manufacture of coating resins, polyurethane adhesive systems for wood, alkyd resins, polyurethane prepolymers and higher polyol fatty acid polyesters (Petrovic, 2008; Yeong et al., 2010a).

Palm-Based Polyol

Being one of the major producers of palm oil, Malaysia through Malaysian Palm Oil Board has put a lot of efforts to develop palm oil-based polyols. The research and development of palm oil-based polyols (POP) and polyurethane products started in the early 1990s, and since then several types of palm-based polyols have been produced. They are named as POP Pioneer, POP Primer, POP Premier (which are triglyceride-based polyols), and PolyMO (which is fatty acid-based polyol). MPOB holds several patents on the technologies to produce these polyols (Hazimah et al., 2008; Hoong et al., 2009; Salmiah et al., 1995).

The triglyceride-based polyols are produced using RBD palm olein as the starting material. The first step is to convert the double bonds in the palm olein into epoxides. The reaction is catalysed by performic acid. The next step is to perform an alcholysis reaction onto the epoxy ring with polyhydric alcohol such as ethylene glycol, propylene glycol, and glycerol. The schematic reaction of the preparation of a triglyceride polyol is shown in Fig. 20.29. Table 20-G shows the properties of triglyceride-based

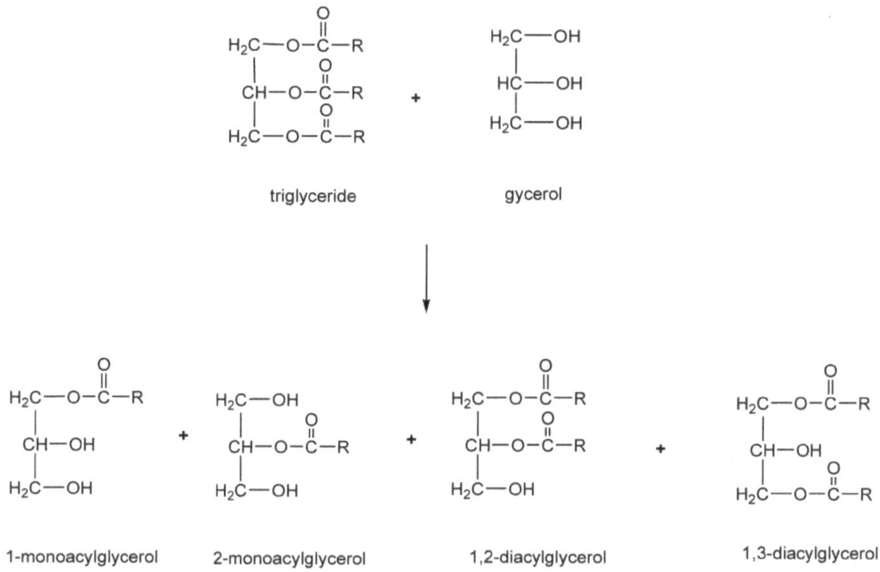

Fig. 20.28. Transesterification of an oil with glycerol to produce mono and di-glycerides.

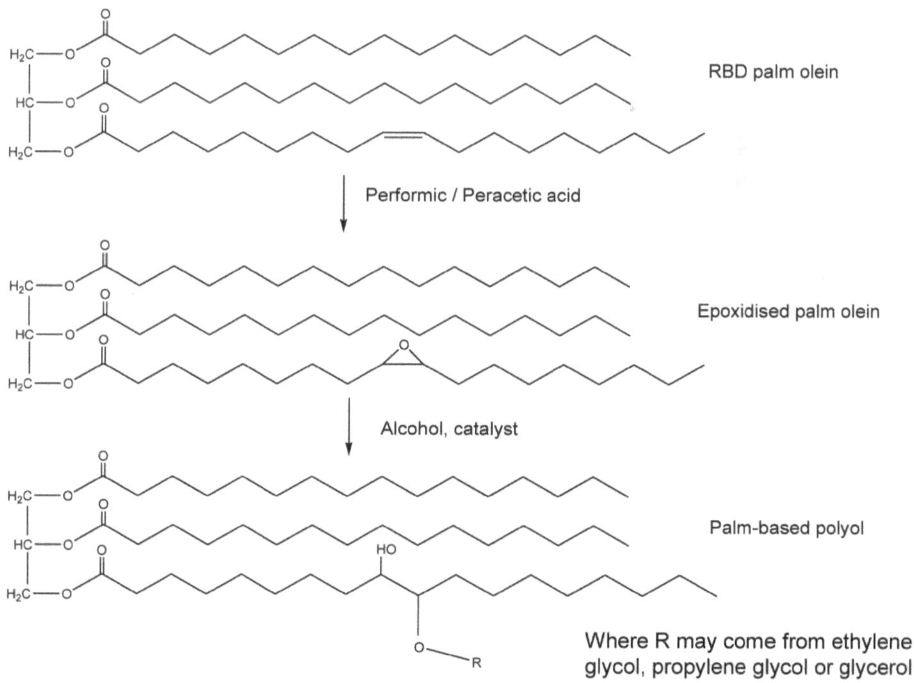

Fig. 20.29. Formation of polyols from crude palm oil or RBD palm olein.

Table 20-G. Properties of Palm Oil Based Polyol (POP Series).

	Palm oil based polyol		
Properties	POP PIONEER	POP PRIMER	POP PREMIER
OHV, mg KOH/g sample	90–150	100–130	60–85
AV, mg KOH/g sample	0.7	0.8	0.7
IV, g I_2/100g sample	6	5	6
Viscosity @ 25°C, cP	1800–9000	2000–9000	600–1000
pH	6	6	6
Moisture content, %	≤ 0.05	≤ 0.05	≤ 0.05

Note. OHV = hydroxyl value, AV = acid value, IV = iodine value.
Source: Maznee et al., 2008.

polyols from palm. These polyols are suitable for rigid, semi-rigid, and flexible polyurethanes (PUs) for various uses such as ceiling panels, flower foams, wall panels, cushion, flexible foams, automotive parts, etc. Several blends of palm oil products with other vegetable oils have also been used as feedstock for the preparation of these polyols.

Besides producing polyol directly from palm oil, MPOB has also invented a new type of polyol made from fatty acid and the properties of this new polyol are shown in Table 20-H. This polyol is referred to as PolyMO, which has higher functionality than the existing triglyceride-based polyols, and this has made it more suitable for making two components polyurethane (2K PU) coating formulations and decorative rigid PU foam as building material.

Table 20-H. Properties of PolyMO.

Properties	PolyMO
OHV, mg KOH/g sample	220–240
AV, mg KOH/g sample	10–14
IV, g I2/100 g Sample	45–50
Viscosity@ 25°C, cP	650–1500
pH	5–6
Moisture Content, %	0.1–0.3
Lovibond tintometer (L,a,b)	0.3L, 0.2a,1.5b

Note. OHV = hydroxyl value, AV = acid value, IV = iodine value.
Source: Hoong et al., 2008.

Palm-Based Polyurethane

Polyurethane (PU) can be classified into two categories (i.e., foam and non-foam). The properties of polyurethane such as compressive, density, flowability, thermal stability, tensile, and shear strength can be adjusted through modifications of the raw materials and formulations to meet requirements of different applications. PU products can be made with properties ranging from very soft and flexible, to elastomeric, and rigid. PU foams are generally divided into flexible, semi-rigid, and rigid, whereas PU non foams are divided into coating, adhesive, sealant, and elastomer (CASE) and binders.

Palm-based polyols have been used to formulate various types of polyurethane products by using suitable types of isocyanates. Among the PU products developed are flexible foam for pillow; mattresses; and automotive parts such as pad dash panel, carpet underlay, and car seat, rigid foam for decorative panel, ceiling panel, wall panel, and flower foam. Under non-foam categories, several products have been developed (i.e., binder for medium density fiber board, wood coating, and elastomer).

Palm-based polyols have a lot of potential in polyurethane applications that need to be explored aggressively both in foam and non-foam products. Several polyurethane based finished products such as decorative PU panels (Fig. 20.30), ceiling panel (Fig. 20.31), pillow (Fig. 20.32), and PU sandwiched panels (Fig. 20.33) have been successfully produced using palm-based polyols.

Fig. 20.30. Polyurethane decorative panels produced using palm-based polyols.

Fig. 20.31. Polyurethane ceiling panel from palm-based polyols.

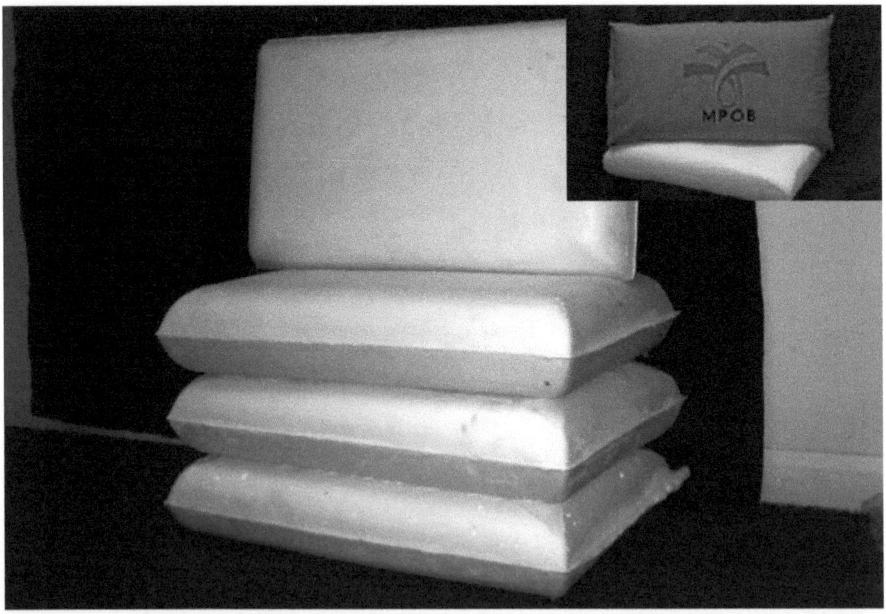

Fig. 20.32. Polyurethane pillow using palm-based polyol.

Fig. 20.33. Palm-based polyurethane sandwiched panel.

Conclusions

Traditionally palm oil is used for food. About 85 to 90% of palm oil is used in food application such as cooking oil, margarine, shortening, confectionery fats, etc. Only about 10–15% of palm oil is used for non-food application. Out of this, about 70% is used as surfactant either in food or non-food application.

Palm oil is first converted into basic oleochemicals such as fatty acid, glycerol, methyl esters, fatty alcohols, and fatty amines. These basic oleochemicals will serve as building blocks for many other chemicals. The basic oleochemicals industry started in the West, United States, and Western Europe because of the availability of tallow and soybean as the feedstock, and this is why most of the technology to produce these basic oleochemicals established in Malaysian oleochemical industry is brought in by multinational investors. About a century later, we saw many of the industries relocated in the Asia Pacific region because of the availability of cheap raw material such as palm oil and palm kernel oil, cheaper operating cost, and a growing consumer market. Increased demand in consumer product markets promotes demand in oleochemical derivatives and thus opens up a challenge in research and development for these oleochemicals derivatives.

References

Bohmont, B.L. Glossary for pesticide users. The Standard Pesticide User's Guide. Revised and Enlarged. Prentice Hall: Princeton, NJ, 1990; pp. 367 & 392.

Choo, Y.M.; Lau, H.L.N.; Puah, C.W.; Ng, M.H.; Bong, S.C.; Ma, A.N.; Yusof, B. Production of phytonutrients (carotenes, vitamin E, sterols, squalene, co-enzyme Q and phospholipids) from palm methyl esters. MPOB Information Series 168, 2002, TT No. 151.

Condea. Dr. Z presents all about fatty alcohols. 2000. http://www.zenitech.com.

Diasakou, M.; Louloudi, A.; Papayannakos, N. Kinetics of the non-catalytic transesterification of soybean oil. Fuel 1998, 77(12), 1297–1302.

Eierdanz, H. Oleochemistry: processes and products, Proceedings of the World Conference on Oilseed Technology and Utilization; Applewhite, T.H., Ed.; Champaign, IL, 1992; pp. 217–220.

Ellard, N. Global Overview and Opportunity for Oleochemicals in the Next Decades, Plenary Lecture 4, Proceedings of Oleo and Specialty Chemicals Conference, Kuala Lumpur, Malaysia, 2009; pp. 3–17.

Erhan, S.Z.; Adhvaryu, A.; Sharma, B.K. Chemically functionalized vegetable oils in synthetics. In Mineral Oils, and Bio-Based Lubricants Chemistry and Technology; Rudnick, L.R., Ed.; Taylor and Francis: USA, 2006; pp. 361–387.

Haliza, A.A.; Zainab, I. Fatty acid composition of methyl ester for esterquats production suitable for fabric softener. 2007, Patent Application No. PI 20072344.

Haliza, A.A.; Zainab, I.; Parthiban, S.; Salmiah, A. Palm based esterquats as an active ingredient in fabric softeners. Palm Oil Dev. 2007, 46, 1–7.

Hall, N. Implications of soap structure formulation and user properties. In Soap Manufacturing Technology; Spitz, L., Ed.; AOCS Press: Champaign, IL, 2009; pp. 83–87.

Hazimah, A.H.; Tuan Noor Maznee, T.I.; Mohd Norhisham, S.; Hoong, S.S.; Ooi, T.L.; Salmiah, A.; Kosheela Devi, P.P.; Cheong, M.Y. Process to produce polyols. United States patent application US 2008/0293913 A1. 2008.

Henkel. Inert ingredients and uses. Agricultural Specialties, A Newsletter for the Agricultural Industry, Henkel Corporation. Cincinnati, OH, 1995.

Höfer, R.; Bigorra, J. Green chemistry—a sustainable solution for industrial specialities applications. Green Chem. 2007, 9, 203–212.

Hoong, S.S.; Kosheela Devi, P.P.; Tuan Noor Maznee, T.I.; Mohd. Norhisham, S.; Nurul Ain, H.; Hazimah, A.H.; Salmiah, A. Low viscosity fatty acid based polyol, MPOB Information Series 420, 2008, TT No. 395.

Hoong, S.S.; Nur Musyahadah, M.; Mohd Norhisham, S.; Tuan Noor Maznee, T.I.; Hazimah, A.H.; Ooi, T.L.; Salmiah, A.H.; Mohd Basri, W.; Yusof, B. Process for the production of a polyol monomer. 2009, United States patent US 7629478.

Hoong, S.S.; Zailan, A.B.; Nik Siti Mariam, N.M.D.; Zainab, I.; Yeong, S.K.; Hazimah, A.H.; Salmiah, A. A process of producing polyglycerol from crude glycerol. 2009, Malaysian patent application PI 20091872 & PI 20092692, European patent application 10174216.

Hoydonckx, H.E.; De Vos, D.E.; Chavan, S.A.; Jacobs, P.A. Esterification and transesterification of renewable chemicals. Topics in Catalysis 2004, 27(1–4), 83–95.

Ishihara, K.; Ohara, S.; Yamamoto, H. 3,4,5-Trifluorobenzeneboronic acid as an extremely active amidation catalyst. J. Org. Chem. 1996, 61(13), 4196–4197.

Ismail, A.R. Application of vegetable oils and their oleochemicals in pesticide formulations. Palm Oil Dev. 1996, No. 24, 17–21 and 27–28.

Ismail, A.R.; Aishah, A.; Rita, M.; Ooi, T.L.; Salmiah, A. Palm-based insecticides formulations. MPOB Information Series 111, 2000, TT No. 90.

Ismail, A.R.; Dzolkifli, O.; Ooi, T.L.; Salmiah, A. Pesticide formulations based on oleochemicals derived from palm and palm kernel oils. The Planter, Kuala Lumpur 1998, 74(862), 9–16.

Ismail, A.R.; Ooi, T.L.; Salmiah, A. Environment friendly palm-based inert ingredient for EW-insecticide formulations. MPOB Information Series 243, 2004, TT No. 236.

Ismail, A.R.; Suhaimi, H.; Tiddy, G.J.T. Microemulsions with palm oil derivatives stabilised by a non-ionic surfactant: Effects of temperature and co-surfactant. J. Dispers. Sci. Technol. 2005, 26, 355–364.

James, E.F.J. Comedogenicity and irritancy of commonly used ingredients in skin care products. 1989. J. Soc. Cosmet. Chem. 40, 321–333.

Kosugu, Y.; Tanaka, H.; Tomizuka, N. Continuous hydrolysis of oil by immobilized lipase in a counter current reactor. Biotechnology Bioengineering, 1990, 36, 617–622.

Lurgi GmbH. Fatty Acid Technology. http://www.lurgi.com.

Lurgi GmbH. Fatty Alcohol Technology. http://www.lurgi.com.

Manabe, K.; Limura, S.; Sun, X.M.; Kobayashi, S. Dehydration reactions in water. Bronsted acid-surfactant-combined catalyst for ester, ether, thioether and dithioacetal formation in water. J. Am. Chem. Soc. 2002, 124(40), 11971–11978.

Murty, V.R.; Bhat, J.; Muniswaran, P.K.A. Hydrolysis of oils by using immobilized lipase enzyme: A review. Biotechnol. Bioprocess Eng. 2002, 7, 57–66.

Nesaretnam, K.; Wong, W.Y.; Mohd. Basri, W. Tocotrienols and cancer: beyond antioxidant activity, Eur. J. Lipid Sci. Technol. 2007, 109, 445–452.

Ng, M.H.; Choo, Y.M.; Ma, A.N; Chuah, C.H.; Mohd Ali, H. Separation of vitamin E (tocopherol, tocotrienol and tocomonoenol) in palm oil, Lipids 2004, 39, 1031–1035.

Odi-Owei, S. Tribological properties of some vegetable oils and fats, Lubrication Engineers, 1988, 45, 685–690.

Petrovic, Z.S. Polyurethanes from vegetable oils. Polymer Rev. 2008, 48, 109–155.

Razmah, G. The effect of disalt on the biodegradation of methyl ester sulphonates (MES). J. Oil Palm Res. 2002, 14, 45–50.

Rudnick, L.R. A comparison of synthetic and vegetable oil esters for use in environmentally friendly fluids. In Biobased Industrial Fluids and Lubricants. Erhan, S.Z., Perez, J.M., Eds.; AOCS Press: Champaign, IL, 2002; pp. 46–58.

Rupilius, W.; Salmiah, A. Palm oil and palm kernel oil as raw materials for basic oleochemicals and biodiesel. Eur. J. Lipid Sci. Technol. 2007, 109, 433–439.

Saka, S.; Kusdiana, D. Biodiesel fuel from rapeseed oil as prepared in supercritical methanol. Fuel 2001, 80(2), 225–231.

Salmiah, A.; Kang, Y.B.; Mohtar, Y. Basic oleochemicals in advances. In Oil Palm Research, Yusof, B., Jalani, B.S., Chan K.W., Eds.; Malaysian Palm Oil Board: Malaysia, 2000; Chapter 29; pp. 1141–1194.

Salmiah, A.; Parthiban, S.; Dieter, W. Polyurethane foams from palm oil and palm oil products. 1995. Singapore patent 55223, Malaysian patent MY-114189-A and Indonesia patent application P9622884.

Spitz, L. Soap manufacturing technology. Spitz, L., Ed.; AOCS Press: IL, 2009; p. 452.

Suyenty, E.; Sentosa, H.; Agustine, M.; Anwar, S.; Lie, A.; Sutanto, E. Catalyst in basic oleochemicals. Bull. Chem. React. Eng. Cat. 2007, 2(2–3), 22–31.

Tadros, F.T. Emulsions. Surfactant Agrochem. 1995, 54, 93–132.

Tuan Noor Maznee, T.I.; Nurul 'Ain, H.; Kosheela Devi, P.P.; Mohd. Norhisham, S.; Hoong, S.S.; Yeong, S.K.; Hazimah, A.H. Improved Palm-based Polyols: POP Pioneer, POP Primer and Premier, MPOB Information Series 491, 2008, TT No. 394.

Wakasugi, K.; Misaki, T.; Yamada, K.; Tanabe, Y. Diphenylammonium triflate (DPAT): efficient catalyst for esterification of carboxylic acids and for transesterification of carboxylic esters with nearly equimolar amounts of alcohols. Tetrahedron Letters 2000, 41(27), 5249–5252.

Wikipedia. http://en.wikipedia.org/wiki/Natural _oil_polyols, accessed on 13 Oct 2011.

Yeong, S.K. Vegetable oil and its derivatives for lubricants. Palm Oil Dev. 1996, 24, 6–10.

Yeong, S.K.; Norashikin, A.; Ismail, A.R.; Srihanum, A. Applications of oleochemicals and their derivatives. Lecture Notes of Introductory Course on Oleochemicals (ICOC), MPOB, 2010a, 137–184.

Yeong, S.K.; Ooi, T.L.; Salmiah, A. Lubricant base from palm oil and its by-products, 2010b, United States patent US7, 781,384 B2 and Malaysian patent MY-141241-A.

Zafarizal Aldrin, A.H.; Rosnah, I. Palm tocotrienols rich fraction in topical application. Proceeding of Palm International Nutra-Cosmeceutical Conference (PINC), Malaysia, 2009.

Zainab, I.; Haliza, A.A. Determination of reaction parameters for preparation of lauryl esteramine. Proceedings of Oleo & Specialty Chemicals Conference (Unedited). Kuala Lumpur, Malaysia, 2009; pp. 138–145.

Zainab, I.; Salmiah, A. Performance of palm-based vs. tallow based esterquats in fabric softener. OFIC, Putra World Trade Centre, Kuala Lumpur, 2000.

Zainab, I.; Zulina, A.M.; Arniza, M.Z. Oleochemical based surfactant. Lecture notes in Introductory Course on Oleochemicals (ICOC), MPOB, 2010, 105–136.

Zulina Abdul, M.; Razmah, G.; Parthiban, S.; Zahariah, I.; Salmiah A. Alpha-Sulphonated Methyl Ester (SME) as an active ingredient in palm-based powder detergents. J. Surf. Detergents 2006, 9, No. 2, 161–167.

21

Oil Palm Biomass for Various Wood-based Products

Anis Mokhtar, Kamarudin Hassan, Astimar Abdul Aziz, and Choo Yuen May
Malaysian Palm Oil Board, P.O. Box 10620, 50720 Kuala Lumpur

Introduction

The Malaysian wood-based sector generated more than RM20 billion (USD 6.7 billion) in 2009, but declined to RM18 billion (USD 6 billion) in 2010. The raw materials for the sector are derived mainly from the natural forests and rubberwood, with a small addition from forest plantation timber. The timber resource scenario, particularly for the production of plywood, sawn lumber, and fiber-based biocomposite products, in Peninsular Malaysia has changed in many respects during the last 20 years and continues to change. Available timber is now generally smaller in diameter and lower in quality than before. Many species that were deemed undesirable in the past are now being used because of a shortage of more desirable species of timber. During this period, demand for timber from the country's forests increased significantly.

As a result of high demand and short supply of timber, the prices of timber have increased rapidly. One response to this situation of shortage of high quality timber, increasing demand for wood and fiber-based biocomposite products, and rising timber costs would be to look for alternative raw materials for some of the products that have traditionally been made from wood. One such material is felled oil palm stems (OPT), a sustainable non-wood source which is available in abundance at replanting operations. Based on the replanting regime of 25 years, it is estimated that about 0.51% of the total area under oil palm, or 23,500 hectares, are due for replanting every year in Malaysia. This would involve the felling of approximately 3.19 million palms generating more than 5.62 million cubic meters of oil palm trunks (Kamarudin et al., 1997). Thus, the quantities at hand could make a very substantial contribution to the production of plywood, sawn lumber, and fiber-based biocomposites without depleting the nation's wood resources from the natural forest and forest plantations.

In spite of their availability, the commercial use of OPT to act as raw materials for the manufacture of plywood, sawn lumber, and fiber-based biocomposites has yet to reach the desired level. The commercial exploitation of oil palm trunks for a variety of products depends on a number of factors. These include a long security of sustainable supply, the economic system for handling them, and a reliable design of processing and manufacturing equipment to cater for their physical forms and sizes. In addition, patterns of variation within stems vary from butt to crown (Lim & Gan, 2005) and may further complicate by site location and growth characteristics.

For veneer and sawn lumber, wood density is widely used as an estimator of product quality because of its good relationship with various wood properties. For example, wood density plays a prominent role for wood strength, workability and decay resistance (Barbosa et al., 2004; Zobel et al., 1989). However, relationships between wood density and other wood properties are often indirect, making them difficult to interpret (Barrett & Kelogg, 1989; Bendtsen, 1978: MacPeak et al., 1990). This is mainly due to the lack of information about its cellular structure or organization such as the relative proportions of the different cell types, the dimensions of these cells, and the thicknesses of the cell walls. For the same wood density, differences at the cellular level may occur, causing variations in wood properties (Zhang et al., 1992).

An enhanced use of oil palm stems from the manufacture of industrial commodities or niche products with minimal environmental impact requires knowledge of the performance of these materials.

Sources and Availability of Oil Palm Biomass in Malaysia

Oil Palm Plantation Area

Malaysia's palm oil plantations are spread out across both Peninsular and East Malaysia, with the states of Sabah and Sarawak showing the largest proportion of planted area. The total oil palm planted area in Malaysia rose by 3.5% to 4.85 million hectares in 2010 compared to 4.69 million hectares in 2009 (Table 21-A). The area expansion occurred mainly in Sabah and Sarawak, with combined growth of 5.8%, or 127,748 hectares, compared to 1.4%, or 34,858 hectares, in Peninsular Malaysia. Sabah remained the largest oil palm planted state with 1.41 million hectares, or 29% of the total planted area, followed by Sarawak with 919,418 hectares (18.9%), Johor with 717,398 hectares (14.8%), and Pahang with 688,866 hectares (14.2%).

The total oil palm matured area in 2010 increased by 3.1% to 4.20 million hectares and accounted for 86.6% of the total planted area. Sarawak recorded a significant increase in oil palm matured area by 10.9%, or 70,584 hectares, while in Sabah the matured area rose 2.3%, or 28,089 hectares. Peninsular Malaysia registered an increase of 27,829 hectares, or 1.3%, over the previous years. Sabah has the largest oil palm matured area accounting for 30.0% of the total matured area, followed by Sarawak (17.1%), Johor (15.1%), and Pahang (14.3%) (Table 21-B). Historically, an increase in planted area for oil palm was the main driver of biomass volume growth.

Table 21-A. Oil Palm Planted Area in Malaysia, 2009/2010 (Hectares).

State	2009	2010	%
Johor	712,448	717,398	0.7
Kedah	78,384	78,539	0.2
Kelantan	113,185	132,715	17.3
Melaka	51,193	51,350	0.3
Negeri Sembilan	166,501	164,362	–1.3
Pahang	675,667	688,866	2.0
Perak	373,854	382,527	2.3
Perlis	234	234	0
P. Pinang	13,588	13,588	0.1
Selangor	139,544	129,000	–7.6
Terengganu	165,216	166,086	0.5
Peninsular Malaysia	2,489,814	2,524,672	1.4
Sabah	1,361,598	1,409,676	3.5
Sarawak	839,748	916,418	9.5
Sabah&Sarawak	2,201,346	2,329,094	5.8
MALAYSIA	4,691,160	4,853,766	3.5

Source: MPOB Statistic (2010).

Table 21-B. Oil Palm Matured Area, 2009/2010 (Hectares).

State	2009	2010	%
Johor	632,510	635,934	0.5
Kedah	74,702	73,968	–1.0
Kelantan	84,851	100,394	18.3
Melaka	48,451	48,888	0.9
Negeri Sembilan	150,969	149,405	–1.0
Pahang	587,698	600,398	2.2
Perak	338,737	345,182	1.9
Perlis	234	234	0.0
P. Pinang	13,319	13,312	–0.1
Selangor	128,312	118,512	–7.6
Terengganu	136,861	138,246	1.0
Peninsular Malaysia	2,196,644	2,224,473	1.3
Sabah	1,233,056	1,261,154	2.3
Sarawak	646,002	716,586	10.9
Sabah&Sarawak	1,879,058	1,977,740	5.3
MALAYSIA	4,075,702	4,202,213	3.1

Source: MPOB (2010).

Oil Palm Biomass

In 2001, Malaysia's palm oil industry produced almost 80 million dry tons of solid biomass per annum. This volume is projected to increase to 95–110 million dry tons until 2020. However, the oil production represents only about 10% of the total biomass produced by the industry. The remaining 90%, mainly in the form of lignocellulosic material, need to be exploited commercially to benefit the industry and the country as a whole (Husin et al., 1995; Jalani et al., 1999). In line with the crude palm oil production, the oil palm industry also produces by-products from two sources, namely from oil palm mills and oil palm plantation. The by-products from oil palm mills consist of shell, empty fruit bunch (EFB), pressed fruit fibers, palm oil mill, and effluent (POME), while the by-products from plantation comprises oil palm trunk and oil palm fronds (OPF). Gurmit (1994) had reported a scheme for estimation of oil palm biomass (both from the palm oil mill and plantation) based on per hectare of the oil palm plantation, as summarized in Table 21-C. It is estimated around 21–22% empty fruit bunch, 12–16% mesocarp fibers, and 5–7% of shells are being generated from the mills. Although EFB are available for industrial applications, the palm oil industry believes that the fronds should be recycled in the plantation, not only because of the nutrient content but also for other purposes such as organic matter, soil moisture conservation, and prevention of surface water run off (Chan et al., 1981). It should be noted that manufacturing wood-based products from OPF requires only the vascular bundles from the petiole of the OPF (Fig. 21.4.B). The petioles are less than half of the total weight of the fronds and generally contain the least amount of nutrients compared to the leaflets.

Availability of Oil Palm Biomass

Malaysia has 4.83 million hectares of oil palm, producing 16.99 million tonnes of crude palm oil and 4.29 million tonnes of crude palm kernel oil in 2010. It is estimated that more than 70,000 hectares of oil palm area are due for replanting every

Table 21-C. Estimated Oil Palm Biomass Distribution in One Hectare of Oil Palm Plantation.

Oil palm biomass	Dry weight
Felled oil palm trunk (during replanting)	74.48 kg
Oil palm fronds	
During replanting	14.47 kg
Annual pruning	10.40 kg
Empty fruit bunch	1.646 tonnes
Mesocarp fibre	1.626 tonnes
Palm kernel shells	0.938 tonnes

Source: Gurmit (1994).

year. This would involve the felling of approximately 9 million palms. Apart from the felled oil palm trunks, the oil palm industry also generates massive quantities of oil palm fronds and empty fruit bunches available from replanting and through routine field and mill operations.

The availability of OPT, OPF, and EFB were evaluated by several researchers including Husin et al. (1985) and Sing (1994). Kamarudin et al. (1997) and Norini et al. (1997) projected the availability up to year 2020 (Table 21-D) based on the total oil palm area 2.3 million hectare. Number of felled oil palm tree was estimated from 10 to 15 million palms from the year 2011 to 2020. However, for the fronds, about 6 to 7 million tonnes (dry weight) per year is available from the pruning exercise.

Empty Fruit Bunches

In oil milling, after the fresh fruit bunches (FFB) are processed, the end economic products are crude palm oil, oil, and palm kernel, while the kernel on crushing produces kernel oil and kernel meal. The biomass co-products besides those of oil and kernel from FFB processing are empty fruit bunches (EFB), fiber, shell, and effluent. A diagram rep-

Table 21-D. Estimated Availability of Oil Palm Biomass (from year 1996 to 2020).

	1996	2011–2013	2014–2016	2017–2020
Oil palm trunk				
Total oil palm				
Are due for replanting (%)	2.4	4.9	4.1	3.4
Total replanting (ha)	56,157	114,654	95,935	79,556
Number of felled palms	7 637,381	15,592,987	13,047,193	10,819,624
Amount of fibre bundles	1,130,180	1,742,360	2,307,450	2,307,450
Amount of parenchyma (1/yr, dw)	664,605	1,356,902	1,356,367	941,524
Amount of bark (t/yr)	303,051	618,730	517,713	429,323
Total of whole trunk (t/yr)	2,097,836	4,283,082	3,583,803	2,971,924
Pruned fronds				
Ha of palms > 7 yrs (%)	79	70.4	72.9	73.9
Total ha of palms	1,848,508	1,647,278	1,705,775	1,729,174
Amount of leaf stalks (t/yr, dw)	14,584,731	12,997,026	13,458,568	13,643,183
Amount of petiole (t/yr, dw)	7,634,340	6,803,260	7,044,853	7,141,490
Empty fruit bunches				
Hectarage of palms > 3 yrs (%)	92.7	85.3	87.6	86.3
Total ha of palms	2,169,072	1,995,921	2,049,738	2,019,320
Mean FFB production (t/hr/yr)	20.12			
Total FFB production (t/yr)	43,641,738	40,157,932	4,124,073	40,628,716
Amount of bunch stalk (t/yr, dw)	673,392	619,637	636,345	626,901
Amount of spikelet (t/yr, dw)	2,402,478	2,210,694	2,270,303	2,236,611
Amount of whole EFB (t/yr, dw)	3,075,870	2,830,331	2,906,647	2,863,512

Note. t = tonne; dw = dry weight.
Source: adapted from Kamarudin et al. (1997).

resentative of input and output from a palm oil mill is shown in Fig. 21.1. At a typical palm oil mill, the sterilized FFB will go through a threshing process to separate the fruits and nuts. After threshing, the co-products are the stalks with empty spikelets, termed simply EFB (Fig. 21.2.A and 21.2.B). The freshly sterilized empty fruit bunches contain 70% water, 30% dry matter, and 2% of crude palm oil. Kamarudin et al. (1997) reported that EFB consists of around 20–25% of stalk and 75–80% of spikelets.

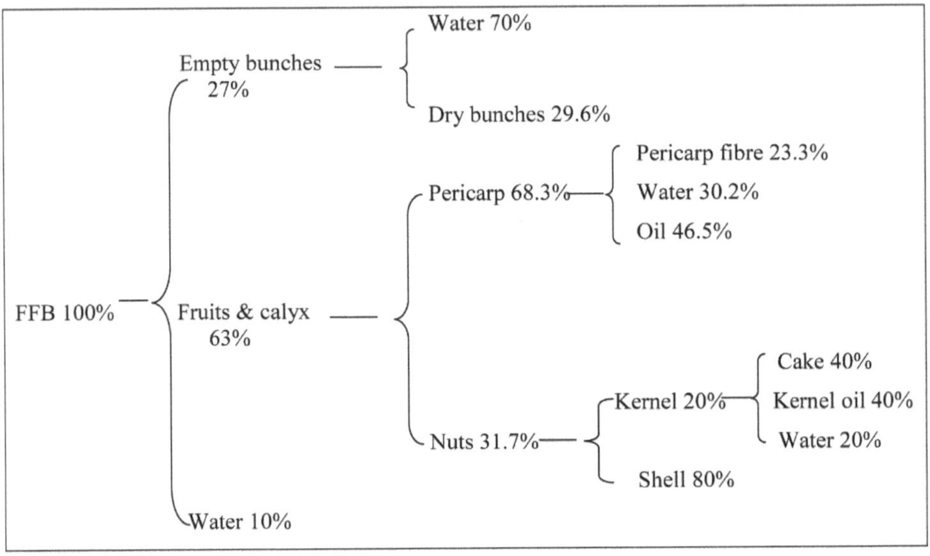

Fig. 21.1. A diagrammatic representation of input and output of a palm oil mill. Source: Kirkaldy & Sutanto, 1976.

(a) Empty fruit bunch

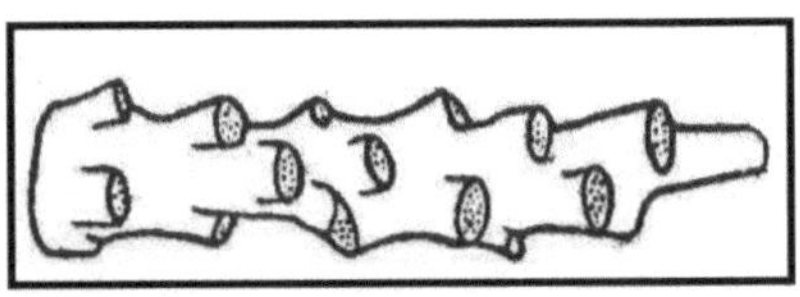

(b) Empty fruit bunch stalk

Fig. 21.2. Empty fruit bunch.

Oil Palm Trunks

Oil palm is a monocotyledonous species, and therefore OPT does not have cambium, secondary growth, growth rings, ray cells, sapwood, heartwood, branches, and knots (Fig. 21.3.A). The increase in diameter of OPT results from the overall cell division and enlargement of cells in vascular bundles. In the cross-sectional direction, OPT can be divided into three main parts which are easily distinguishable, namely the cortex, the peripheral region, and the central zone (Fig. 21.3.B). The cortex consists of a narrow tissue of approximately 1.5 cm to 3.5 cm in width. It makes up the outer part of the stem just underneath the bark. It is largely composed of ground parenchyma with numerous longitudinal fibrous strands and vascular bundles. The peripheral region with narrow layers of parenchyma and congested vascular bundles gives rise to a sclerotic zone, which provides the main mechanical support for the OPT. The central zone, which makes up about 80% of the total area, is composed of slightly larger and widely scattered vascular bundles embedded in the thin-walled parenchymatous ground tissue. The bundles increase in size and are more widely scattered towards the core of the OPT (Killmann & Lim, 1985). Choon et al. (1991) also mentioned that the vascular bundles increase from the butt end to the top of the palm, and with regards to physical properties, there is great variation of density at different parts of the trunk. Bakar et al. (2001) stated that the moisture content (MC) at the butt end is higher than the other parts and it decreases from the butt end to the top. Killmann & Choon (1985) mentioned that a gradual increase in MC is indicated along the trunk height from the butt end to the top and the density decreased linearly with trunk height and towards the center of the trunk.

Relating the mechanical properties, the above physical conditions influence the strength properties of the trunk. Choon et al. (1991) reported that bending strength, compression strength parallel to grain, and hardness of the trunk is generally poor

(a) Oil palm trunk

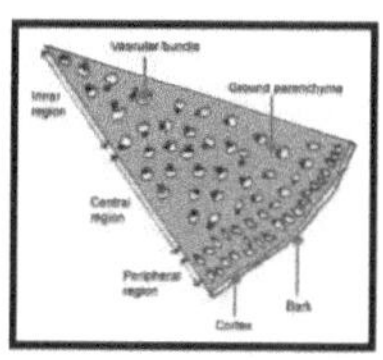

(b) Cross section of oil palm trunk

Fig. 21.3. Oil palm trunk.

compared to other timber species including rubberwood and coconut tree. In addition, Ratanawilai et al. (2006) stated that the mechanical properties of oil palm trunk were approximately two times lower than those of teak and rubberwood.

Physical Properties of Oil Palm Trunk

a. Moisture Content

Killmann & Lim (1985) stated that initial moisture content of the oil palm trunk varies between 100 and 500%. Lim & Khoo (1986) reported that a gradual increase in moisture content is indicated along the trunk height and towards the central region, with the outer and lower zone having far lower values than the other two zones. The trend in moisture content increment can be explained by the distribution of the parenchymatous cells which retain more moisture than vascular bundles. Kamarudin et al. (1997) reported that from the total trunk volume, the amount of parenchyma tissues, vascular bundle, and bark are approximately 32%, 54%, and 14%, respectively. Parenchyma tissues are more abundant towards the apex of the palm trunk as well as towards the center region, while Bakar et al. (1998) stated that based on depth of the trunk, the highest moisture content was reached at the center of trunk and a gradual decrease to the outer part of trunk. These values were between 258% and 575%. An increase in the number of vascular bundles was caused of a decreasing in percentage of parenchyma cells which have a high capacity in water absorption. Bakar et al. (1998) also stated that based on the trunk height, there was a tendency for the moisture content to decrease from the bottom to the top of the oil palm tree.

b. Density

Due to its monocotyledonous nature, there is a great variation of density values at different parts of the oil palm trunk. The density or specific gravity is considered the most important characteristic in the utilization of OPT. It is defined as the amount of the material per unit volume. Generally, the density is highest at the peripheral region of the bottom end and lowest at the central core region of the top end of the OPT. Density values range from 200 to 600 kg m^{-3} with an average density of 370 kg m^{-3} (Lim & Khoo, 1986). According to Lim & Khoo (1986), the density of oil palm trunk decreases linearly with the trunk height and towards the center of the trunk. The outer region throughout the trunk shows density values over twice those of the inner regions. These variations are due to several factors. For example, across the trunk, the density is influenced largely by the number of vascular bundles per square unit, which decreases towards the center. However, variations in density along trunk height are due to the vascular bundles being younger at the top end of the palm. For solid-wood product such as sawn lumber and veneer, density is widely used as an indicator of product quality because of its good relationship with various properties such as strength, workability, and decay resistance (Burdon et al., 2004).

Chemical Properties of Oil Palm Trunk. In OPT, there are variations in chemical composition along the longitudinal and radial directions of the stem. At 1.8 m above ground level, there is no definite trend for the extractives which was extracted using alcohol benzene, alkali solubility, holocellulose content, lignin content, and pentosan content (Yusof et al., 1984). However, the alpha-cellulose content was observed to increase from the pith towards the periphery while the hot water solubility and ash content decreased in the direction. In general, the central portion of the stem yielded higher hot water and alkali soluble compositions than the outer zone. The components in water and alkali soluble generally indicate greater quantity of starch and short chained polysaccharides present in the inner portion of OPT, where parenchymatous tissue is present in abundance. Husin et al. (1985) has studied the chemical properties of oil palm trunk and found that the oil palm trunk has lignin and lignocelluloses contents markedly lower but shows higher content of extractives, as well as water and alkali soluble components, than coconut and rubberwood. The comparison of the proximate chemical composition of oil palm trunk with coconut and rubberwood are reported in Table 21-E.

In another study, Mansor & Ahmad (1990) observed that the bottom part contained high amounts of lignin while that for the central core region in the upper portion was very low. This is due to the quantity of fibrous vascular bundles, which increase in number towards the peripheral region. These vascular bundles contain

Table 21-E. Chemical Composition of Oil Palm Trunk, Coconut, and Rubberwood.

Chemical composition	Oil palm trunk	Coconut palm	Rubberwood
Alcohol-benzene	9.8	2.6	1.5
Hot water soluble	14.2	2.8	4.8
Alkali soluble (1% NaOH)	40.2	22.9	19.2
Holocellulose	45.7	66.7	67.0
Alpha-cellulose	29.2	n.a	41.5
Acid-insoluble lignin	18.8	25.1	26.0
Pentosan	18.8	22.9	19.4
Ash	2.3	2.8	1.5

Note. n.a = not available.
Source: Husin et al. (1985).

Table 21-F. Mechanical Properties of Oil Palm Trunk at Different Height.

Axis	MOE (MPa)			MOR (MPa)			Compression (MPa)			Hardness (side) (N)		
height (m)	A	B	C	A	B	C	A	B	C	A	B	C
12	3207	1555	743	17	9	8	6.3	4.4	–	1036	748	713
11	5112	1681	1081	31	10	–	16.1	–	–	1252	779	645
10	4185	858	743	26	12	–	15.3	7.1	–	1534	1521	–
9	279	1809	–	20	12	–	11.4	5.6	5.2	1581	1613	–
8	4724	1904	–	30	14	–	11.3	7.0	–	1718	1496	–
7	5088	2418	–	31	12	–	17.9	10.0	–	1630	1803	–
6	6694	3945	–	39	15	–	15.1	8.5	–	2021	1949	–
5	6600	3807	2641	40	27	20	23.9	11.5	5.5	2222	1949	–
4	7232	3965	1552	40	24	12	24.5	12.7	6.2	2140	1440	345
3	7165	4228	2035	47	27	17	24.1	15.1	6.2	2013	1220	412
2	7958	4540	1484	44	20	10	26.7	14.6	5.0	2461	1436	573
1	5305	4219	1482	30	23	10	21.4	15.2	7.0	2419	1356	600
Mean	5505	2911	1470	32.9	17.8	12.8	17.8	10.2	5.9	1836	1443	548

Notes. A = peripheral region; B = middle region; C = central region.
Source: Killmann & Lim (1985).

thicker and older cells, which give rise to the higher lignin content of the lower end of OPT. The lignin content range varies between 15% and 21.7%. The ash content, also observed to be similar throughout the trunk, has a range between 3.0% and 3.3%. Free sugars amounting to over 10% throughout the stem height were also observed. The core regions contain higher proportions of free sugars, as shown by the methanol extracts, while at the peripheral zones the free sugar content is at the lowest. Analysis by High Performance Liquid Chromatography (HPLC) revealed that sucrose, glucose, and fructose are the three main free sugars of the OPT. The level of sucrose, the major sugar component, remains fairly constant throughout the stem height. On the other hand, glucose level decreases with increasing stem height while that of fructose increases.

Mechanical Properties of Oil Palm Trunk. Mechanical properties are directly proportional to density of the timber. The means for various mechanical properties at different axial direction of OPT are given in Table 21-B (Killmann & Lim, 1985). The modulus of elastic (MOE), which is a measure of the stiffness or rigidity of the material, varies in the OPT between 740 MPa and 7960 MPa, with the highest value found in the lower peripheral region of the OPT. The lowest MOE is in the core of the central region near the top end of the trunk. The modulus of rupture (MOR) indicates the maximum bending strength or maximum fiber stress before failure. Thus, it is measure of the resistance for the material against breakage. The MOR, varying just as the MOE, is very low compared to conventional timber species, as indicated in Table 21-F.

Killmann & Lim (1985) have investigated the mechanical properties of oil palm trunk (30 years old) and compared it to the other species, such as coconut and rubberwood (Table 21-G). Mechanical properties of oil palm trunk reflect the density variation observed in the trunk both on radial and in the vertical direction. Bending

Table 21-G. Comparison of Mechanical Properties of Oil Palm Trunk with Other Wood.

Property	Oil palm	Coconut palm	Date palm	Rubberwood	Poplar
Density O.D. (kg m^{-3})	220–550	250–850	410	530	360–560
MOE (MPa)	800–8000	5300	1719–2745	8800	8300
MOR (MPa)	8–45	36	11.3–22.6	58	76
Compression (MPa)	5–25	24	6.3–10.1	26	36
Hardness (N)	350–2450	4230	2000	4320	2500

Note. MOR: Modulus of rupture; MOE: Modulus of elasticity.
Source: Killmann & Lim (1985).

strength values are obtained from the peripheral lower portion of the trunk, and the central core of the top portion of the trunk gives the lowest strength. Bending strength of oil palm trunk is comparable to coconut wood but lower compared to rubberwood. Variation of the compression strength parallel to grain also follows the same trend as the bending strength. The compression strength value is comparable to rubberwood at similar density value. The hardness value of oil palm trunk is lower than rubberwood, as well as coconut palm. Bakar et al. (1998) reported on the mechanical properties on the *Tenera* species and concluded that all the properties tested, including MOE, MOR, and hardness, were decreased from the outer to the center and from the bottom to the top of the trunk, where the influence of the trunk depth was higher than the trunk height. Bakar et al. (1998) also reported that the average MOE values at various positions indicated a gradual decrease in MOE along the trunk height and depth. Variation of the MOR also follows the same trend as the MOE.

Oil Palm Fronds

The availability of fronds during replanting and pruning had been reported and calculated using an estimate of 14.47 t ha^{-1} and 10.4 t ha^{-1}, respectively (Chan et al., 1981; Gurmit, 1994). The availability of frond from the crown at felling during replanting in year 2010 is summarized in Table 21-H. The dry weight of fronds available for biomass utilization from 5% of the mature palm during replanting is about 3.03 million tonnes.

The total dry weights of fronds obtained from annual pruning and during normal routine harvesting from the mature areas in Malaysia are shown in Table 21-I. Taking into calculation that the pruning is being carried out for the palm that are over 7 years old (mature palm) and with 5% replanting rate, the estimated availability of the oil palm fronds for 2010 was calculated to be about 43.7 million tonnes. Kamarudin et al. (1997) has reported on the projection of oil palm fronds in Peninsular Malaysia between 2007 up to 2015 from the pruning and during replanting and the amount is about 45.49 million tonnes.

Table 21-H. Availability of Fronds During Felling at Replanting in Year 2010 (Million Tonnes).

Location	Oil palm (mature ha)	Frond, dry wt (million tonnes)
Peninsular Malaysia	2,224,473	1.60
Sabah	1,261,154	0.91
Sarawak	716,586	0.52
Total	4,202,213	3.03

Note. Based on 5% replanting of mature areas at 14.47 t/ha of dry weight of fronds at felling; from Chan, 1999.

Table 21-I. Total Availability of Fronds from Annual Pruning and Routine Harvesting in 2010 (Million Tonnes).

Location	Oil palm (mature ha)	Frond, dry wt (million tonnes)
Peninsular Malaysia	2,224,473	23.1
Sabah	1,261,154	13.1
Sarawak	716,586	7.5
Total	4,202,213	43.7

Note. Based on 10.47 t dry weight/ha from pruning; from Chan, 1999.

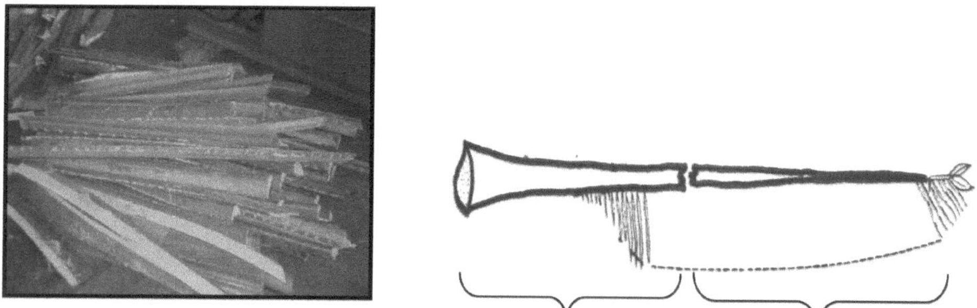

(a) Oil palm fronds (b) Petiole Rachis

Fig. 21.4. Oil palm fronds

Wood-based Products

Oil Palm Plywood

Plywood is one of the most commonly used wood products available today. It is extremely versatile, strong and durable, and comes in various grades and thickness. Plywood consists of layers of veneer laid crossways to each other, glued and pressed together to form a sheet. The world demand for plywood is expected to increase at a rate of 1% per annum, while demand for round wood will also be on the increase at a rate of 1.6% up to a year 2030. Malaysia is one of the major producers of tropical plywood and ranked second in the world for export of this panel product. Oil palm trunk has been successfully turned into plywood by peeling into veneers and layering with adhesives. The success of producing OPT plywood principally depends mainly on the proper way of peeling the oil palm trunk and drying of oil palm veneers. OPT based plywood could potentially be produced on a large scale with acceptable quality and perceived market acceptance. At the plantation, normally OPT logs are felled and

(a) Felling of oil palm tree

(b) Cutting of OPT

Fig. 21.5. Felling and cutting of OPT during replanting.

cross-cut to desired lengths that are 2.7, 5.5, or even 8.2 m (9, 18, or 27 feet) for easy transportation, as can be seen in Fig 21.5.

The process flows for producing OPT plywood are shown in Figs. 21.6.A–21.6.I. At the storage yard, Figs. 21.6.A and 21.6.B, OPT logs are grouped together (bottom and upper portion) in the same group for easy peeling to increase the efficiency and recovery of the veneer.

First, the bark of the OPT must be removed before proceeding to peeling to get the veneers. The basic equipment for rotary cutting of veneer is similar throughout the world and the only difference is on the control parameters for peeling oil palm trunks. A rotary lathe machine was used to peel the oil palm logs into veneers. For OPT, two types of peeling processes were applied. First rotary lathe machine is used to debark the OPT, as shown in Fig. 21.6.C. The purpose of this process is to remove the outer bark without substantially damaging the logs.

The debarked OPT are shown in Fig. 21.6.D. After that, the second peeling process takes place using either a rotary or spindle-less lathe machine, as shown in Fig. 21.6.E. Positioning the veneer knife correctly plays an important role in producing the quality veneer. The angle of the lathe knife also determines the uniformity of the veneer thickness as the log was peeled. The thickness of veneer varies between the outer and inner portions of the OPT logs. The peeled veneer is cut to 0.9 m or 1.2 m (3 or 4 ft) for easy handling.

Veneers are dried using a mechanical conventional dryer or roller dryer (Fig. 21.6.F) and the temperature used ranges between 130°C and 150°C. Steam was supplied from the boiler and the temperature was controlled for drying the veneers up to about 3–5% moisture content. After drying, the veneers are conditioned to ambient

Fig. 21.6.A. OPT at the storage yard.

Fig. 21.6.B. Grouping of OPT logs.

Fig. 21.6.C. Debarking of OPT.

Fig. 21.6.D. OPT without bark.

Fig. 21.6.E. Peeling of OPT.

Fig. 21.6.F. Drying of OPT veneer.

Fig. 21.6.G. Gluing of OPT veneer.

Fig. 21.6.H. Pressing of OPT plywood.

Fig. 21.6.I. Finished products: OPT Plywood.

temperature before spreading the glue using a glue spreader, as shown in Fig. 21.6.G. The amount of glue used ranges from 268 g m^{-2} to 376 g m^{-2} (25 to 35 g ft^{-2}). After gluing, the veneers were pressed using cold and hot-pressing machines to produce plywood. The temperature used for hot-pressing ranges from 120°C to 150°C, while the pressure applied ranges from 6 to 15 kg cm^{-3} (Fig. 21.6.I). The temperature, pressure, and time used for different thickness of plywood are shown in Table 21-J, while the finished product is shown in Fig. 21.6.H.

The plywood after pressing is then taken to a finishing process where edges are trimmed and the face and back may or may not be sanded to get the smooth surface using the sand paper grit P-60.

The recovery rates for making 1 cubic meter of plywood from oil palm trunk are shown in Table 21-K. The amount of wastage seemed to be high in the peeling section due to debarking, clipping, and also trimming processes. For 1 cubic meter of oil palm trunk consumed, approximately 47.4% could be used for making plywood.

The strength properties of oil palm trunk plywood are shown in Table 21-L. All the tests have been carried out according to the method as stipulated in Japanese Standard

Table 21-J. Hot-pressing Parameters for Different Thickness of OPT Plywood.

Base board thickness(mm)	Temperature (°C)	Pressure (kgm^{-2})	Time
9	105	40 125	30 sec 9 min
12	105	40 125	30 sec 10 min
18	105	40 125	30 sec 14 min

Source: Anis et al. (2006).

Table 21-K. Recovery Rate Per Cubic Meter of Oil Palm Trunks.

Processing stage	Amount of wastage, % (based on volume)	Recovery rate, % (based on volume)
Storage yard		100.00
Peeling	39.00	61.00
Drying	1.51	59.49
Gluing	6.00	53.49
Finishing	6.09	47.43

Source: Anis et al. (2006).

Table 21-L. Strength Properties of Oil Palm Plywood.

	OPT Plywood	JAS 233:2003
Test direction: Perpendicular to grain		
Bending strength, MPa		
3 mm thick	30.8	8.0
9 mm thick	33.5	12.0
12 mm thick	37.0	20.0
Test direction: Parallel to grain		
Bending strength, MPa		
3 mm thick	10.6	8.0
9 mm thick	20.8	12.0
12 mm thick	25.2	20.0

Method JAS 233:2003. From the results, the strength properties of oil palm trunk plywood exceeded those standard values and this indicates that oil palm trunk is suitable for making plywood.

Oil Palm Lumber

Oil palm trunk also could be turned into lumber through air drying and kiln drying processes but need some adjustment and modification at the process parameters. This is due to the different properties from other wood species, as shown in Table 21-M. OPT is a highly anisotropic material, and its strength and stiffness properties perpendicular to grain lumber also shrink and swell with varying moisture contents (Mohd Fahmi et al., 2008). Of the total stem volume, the amount of parenchyma tissue, vascular bundles, and bark (cortex) are approximately 32%, 54%, and 14%, respectively (Kamarudin et al., 1997), which make it susceptible to numerous types of drying defects (Ho et al., 1985).

Palm lumber has its own idiosyncrasies, including varying density and a high fiber content. Due to its monocotyledonous nature, there is a great variation of density values at different parts of the oil palm tree. Density values ranges from 200 to 600 kg m^{-3} with an average density of 370 kg m^{-3} (Lim & Khoo, 1986). Kiln drying of palm lumber is completely different than kiln drying other timber, particularly due to high moisture variation (100–350%). The oil palm trunk logs collected during replanting were cut into 9 or 18 ft before sawn into lumber. Normally, the sawing machines were used to saw the OPT logs (Figs. 21.7.A and 21.7.B). The lumber from OPT does not have good machining properties. It is “slightly difficult to very difficult” to work with, depending on the machining process used, and gives very rough machined surfaces (Ho et al., 1985). The rough surfaces are either due to raised vascular bundle or tearing of the vascular bundles. This makes the finishing process difficult. Generally, the initial recovery of sawn lumber from OPT ranges from 51% to 70% for the butt logs and 65% to 70% for the top logs (Table 21-N). After air-drying, the average recovery is 24% for the butt logs and only 6% for the top logs. This reduction is due to the seasoning defects. The main defects are cupping, twisting, collapse (Fig. 21.8.A), and checks (splits) between vascular bundles and parenchymatous tissues (Fig. 21.8.B).

Table 21-M. Properties of Oil Palm Lumber in Comparison with Other Wood Species.

Property	Oil palm	Coconut palm	Rubberwood
Density	220–550	250–850	530
MOE (MPa)	800–8000	5300	8800
MOR (MPa)	8–45	36	50
Compression (MPa)	8–25	24	25
Hardness (N)	350–2450	4230	4320

Note. MOE = modulus of elasticity; MOR = modulus of rupture.
Source: Anis et al. (2006).

(a) Sawing of oil palm trunk (b) Oil palm lumber

Figs. 21.7.A and 21.7.B. Sawing of oil palm trunk.

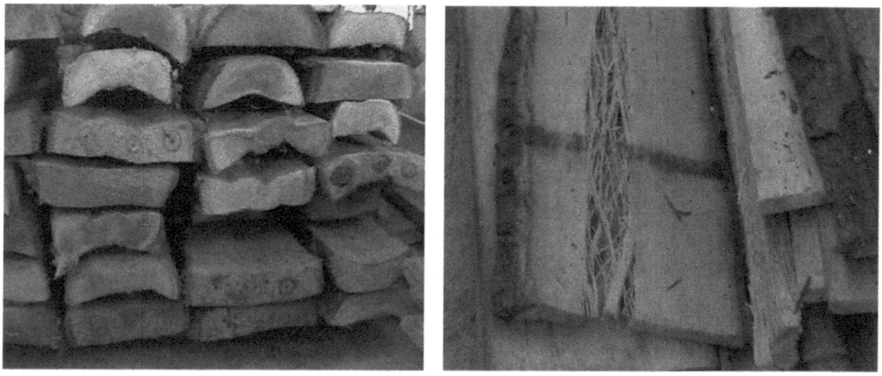

(a) Collapse (b) Cracking

Fig. 21.8. Drying defects of sawn lumber.

Table 21-N. Recovery of Lumber from Oil Palm Trunk.

No of trunk	Portion	Log diameter (cm)	Log volume (cm^3)	Recovery (%)	
				Off saw	After air-drying
1	Butt	42	0.39	60	38
	Top	36	0.28	75	24
2	Butt	39	0.34	70	21
	Top	33	0.25	68	5
3	Butt	36	0.31	57	24
	Top	30	0.20	66	3
4	Butt	38	0.33	53	28
	Top	34	0.25	65	0
5	Butt	38	0.32	51	9
	Top	29	0.19	72	0

Source: Haslett (1999).

Fig. 21.9. Kiln dryer for drying sawn OPT.

The effect of seasoning of OPT by both air-drying and kiln drying (Fig. 21.9) on the quality of boards are summarized in Table 21-O. OPT lumbers suffer from various drying defects, including raised grain, warping and collapse. Most of the defects are from the central zone. Therefore, in seasoning of oil palm lumbers, boards from the peripheral zones of the stem have to be separated from the central region, and a

Table 21-O. Drying Properties of Oil Palm Lumber.

Drying method	Type of board	Initial m.c (%)	Final m.c (%)	Time to dry (days)	Drying defects
Air drying	Boards from peripheral and central zone, 25 mm thick	128.0	15.7	60	Little staining. Boards from center suffer from collapse.
Kiln drying, Schedule B*	Mixture of boards, 25 mm thick	128.0	8.5	15	About 40% of the boards suffer from drying degrades.
	Mixture of boards, 50 mm thick	205.0	10.2	24	About 40% of the boards suffer from drying degrades.
Kiln drying, Schedule E*	Mixture of boards, 25 mm thick	528.1	9.6	9	About 40% of the boards suffer from serious raised grain, warping and collapse.
	Boards from the peripheral zone	130.0	10.0	4	About 10% of the boards suffer from drying degrades.

Source: Ho et al. (1985).

fast kiln drying method is recommended. Drying boards from the central region is not recommended because the recovery is too low.

Medium Density Fiberboard

The global demand for MDF is increasing rapidly. Currently, there are eight MDF producers in Malaysia, mostly using rubberwood as their raw material. Projections show that in the near future the supply of rubberwood will not meet the demand for producing MDF (Husin et al., 1985), and the possible alternative raw material that can be used for making MDF is from oil palm biomass.

Medium density fiberboard (MDF) is a nonstructural fiber-based panel that is composed of wood fibers bonded together with resin under heat and pressure. The fibers (either from oil palm biomass or other woody materials) were refined using defibrator or refiner before mixing it with resin and wax. A bonding agent or resin is used to coat the surface of the fibers (Hochstrate & Weber, 1998). The resonated fibers are dried and then formed into mats, which are finally hot pressed to form homogeneous board. The density range is usually between 660 and 850 kgm^{-3}. The process flow for producing MDF from oil palm biomass is presented in Fig. 21.10. In contrast to

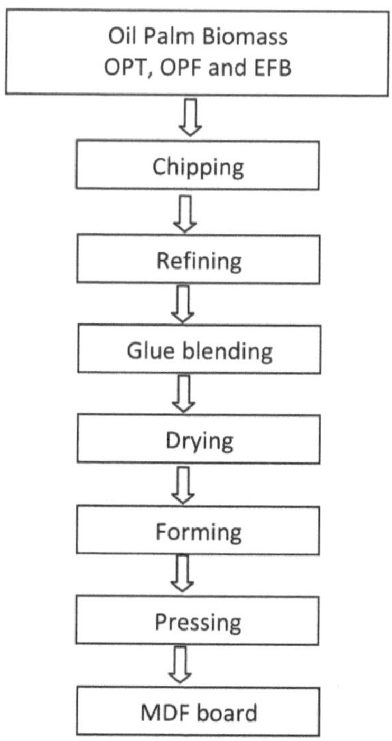

Fig. 21.10. Flow chart for the production of medium density fiberboard.

particleboard, MDF has more uniform density throughout the board and has smoother, tighter edges than can be machined. It can be finished to a smooth surface and grain printed, eliminating the need for veneers and laminates. Most of the thicker MDF panels (1.27 to 1.91 cm) are used as core material in furniture panels. MDF panels thinner than 1.27 cm are typically used for siding.

Studies by Malaysian Palm Oil Board (MPOB) showed that EFB and OPT is a potential material to be blended with tropical and rubberwood for producing MDF. The board showed excellent strength, a light color (for MDF made from OPT), and has a good surface appearance. Boards produced using OPT blended with rubberwood and tropical wood has been successfully commercialized with Dongwha Malaysia. On the other hand, the MDF produced from EFB appears darker in color compared to MDF produced from rubberwood. The oil content of the EFB fibers is believed to interfere in the bonding of the resin with the fibers. Therefore, a study was carried out to apply chemical pre-treatment to remove the residual oil (Ridzuan et al., 2002). The study found that removing the oil improved the MDF properties. It was found that boiling fiber in water is a better way than boiling in 2% sodium hydroxide (NaOH). The average value for bulk density, pH, oil content, and length for the fiber before and after the pre-treatments are shown in Table 21-P.

The lower bulk density from the water fiber was due to the fineness of the fiber. In the NaOH pre-treatment, the fiber remained clumpier or less broken up. Nevertheless, both pre-treatments reduce the bulk density from that of the original fiber,

Table 21-P. Properties of EFB Fibers after Pre-treatment.

		Pretreatment	
Property	EFB	Hot water	2% NaOH
Bulk density (kg m^{-3})[a]	106A	42B	57C
pH[a]	5.48A	5.45A	6.56B
Oil content (%)[a]	1.1A	0.14B	0.07B
Single fibre length (mm)[b]	0.25		
Fibre length class[c]			
>28 mesh (>0.595 mm)		71.76	87.42
28–48 mesh (0.595–0.297 mm)		10.07	4.20
48–100 mesh (0.297–0.149 mm)		5.24	3.93
100–200 mesh (0.149–0.074 mm)		3.49	1.44
<200 mesh (<0.074 mm)		9.44	3.01

Notes. [a]Values with the same following letters are not significantly different at $\alpha = 0.01$; [b]Arithmetic mean of 30,000 fibers; [c]Average of two determinations.

although all three were above the recommended minimum of 32 kg m^{-3} (Suchland & Woodson, 1985). The treated fiber with NaOH was significantly less acidic, contained less residual oil (0.07%), and was coarser than the fiber boiled in water.

The physical and mechanical properties of MDF board produced from the treated fiber are presented in Table 21-Q. Overall, the MDF boards made from EFB boiled in water performed better than EFB treated with NaOH. The MDF board produced using EFB boiled in water had lower water absorption (WA) and thickness swelling (TS) than the board using EFB treated with NaOH.

There are also studies conducted on the blending of the oil palm biomass with rubberwood and also tropical wood. There are also few findings on the different blending ratios used, but the differences may be due to the preparation of the materials. Khalil & Wan (1991) found that the MDF prepared from the blending of 20:80 (EFB:rubberwood) showed better mechanical and physical properties than that of the 50:50 blending.

Particleboard

The suitability of oil palm biomass for manufacturing particleboard is obvious from the research done since the 1980s. The collaboration work with a company called Muar Hardboard in Malaysia on manufacturing school furniture has shown the superior strength properties of oil palm biomass board, even when compared to the component steel structure of the furniture. Particleboards made from 100% OPT and in combination with particles from other timber species at a ratio of 50:50 were also examined. Results showed that homogeneous particleboard manufactured with 100% oil palm fiber with a density higher than 700 kgm^{-3} and 8% adhesive content passed the minimum standard requirement in the BS 5669:1979 Type 1 (Table 21-R). The board made from mixture with other species had better properties than those with 100% oil palm particles and easily passed the standard (BS 5669:1979 Type 1).

The use of EFB for making particleboard was attempted based on the British Standard BS 5669:1989. A study by Yamani et al. (1994) concluded that bending

Table 21-Q. Properties of Medium Density Fiberboards Produced Using Treated EFB.

	Mechanical Properties			Physical Properties				Density
Pre-treatment	MOR MPa*	MOE GPa*	IB KPa*	WA2 %*	WA24 %*	TS2 %*	TS24 %*	D avg kg m^{-3}
Water	37A	2.7A	806A	48B	58B	19B	22B	1041A
2% NaOH	22.8B	1.9B	696B	71A	78A	37A	40A	1057A

Notes. *Values with the same following letters are not significantly difference at $\alpha = 0.05$; MOE = modulus of elasticity; MOR = modulus of rupture; WA2, WA24 = water absorption at 2 and 24 hr; TS2, TS24 = thickness swelling at 2 and 24 hr; Davg = average density.

Table 21-R. Properties of Homogeneous Particleboard from Oil Palm Trunk and Mixture with Other Wood Species.

Sample	Resin (%)	Density (kgm^{-3})	MOR (Mpa)	Internal bond (Mpa)	Screw hold (N)	Water absorption (%)
100% OPT	8	569	10.2	0.23	383	44.5
100% OPT	8	751	16.7	0.55	568	56.9
100% OPT	10	612	11.9	0.40	414	40.9
OPT:rubber wood 50:50	8	707	27.1	0.97	891	39.8
OPT:waste wood 50:50	8	731	22.1	0.79	930	16.3
BS 5691 Type 1		610–640	14.0	0.35	360	6.0

strength can be improved by increasing the resin content and the dimensional stability improved by the addition of wax. It could be an important source of material for particleboard manufacture.

Molded Particleboard

The process of manufacturing molded particleboard is generally the same as normal particleboard except that the particle mat is placed in mold with desired shape and design before pressing (Fig. 21.11). OPT and OPF have also been used as raw materials in molded particleboard manufacturing (Hassan & Jalani, 1993). Several items, such as tabletops of various sizes and shapes have been successfully manufactured in industrial scale plant. Board manufactured using 18% resin content and density of 700% kgm^{-3} had sufficient strength and pass the British Standard BS 5669:1979 Type 1 (Table 21-S).

The advantage of producing molded particleboard is that its minimizes or eliminates the successive processing steps, as normally done using conventional process producing similar products (Hassan et al., 1997). Molded particleboards from oil palm biomass are easily manufactured without much alteration to the existing machinery in commercial plant. The strength properties of tabletops from OPF are lower than those made from OPT. This is probably due to the presence of high proportion of parenchyma tissue in the chip, which allowed compaction far exceeding the normal compaction ratio for particleboard which is around 1:1.3. It is also possible that the presence of parenchyma tissues and cells hinder or limit the contact between adhesive particles and the surface of the OPF particles reducing the total glued area, which lead to lower strength properties.

Table 21-S. Strength Properties of Moulded Tabletop Particleboard Made from Oil Palm Biomass.

Properties	OPT	Fronds	BS 5669:1979 Type 1
Density (kgm-3)	679	689	700
MOR (Mpa)	14.90	12.79	13.8
MOE (Mpa)	2,577	1,813	2,000
Internal bond (Mpa)	1.71	1.62	0.32

Source: Hassan & Jalani (1993).

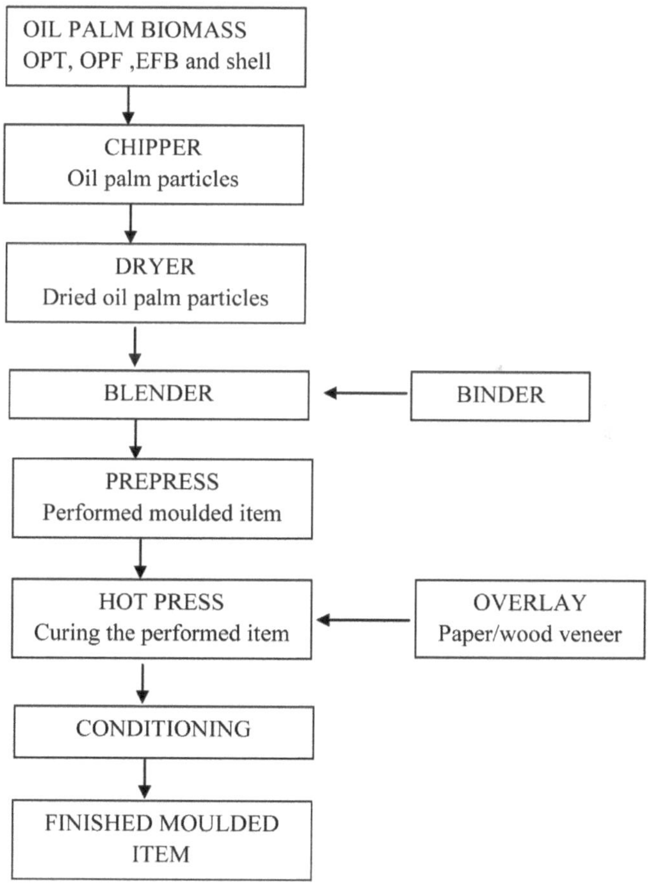

Fig. 21.11. Process flow for manufacturing oil palm molded particleboard. Source: Hassan & Jalani (1993).

Conclusions

Based on the research and development findings by MPOB and various agencies, oil palm biomass has a great potential as a raw material for producing products for the wood-based industry. The product quality showed that it is comparable with the products made from tropical wood. However, in the process line and parameters, it needs to be adjusted accordingly based on the oil palm biomass characteristic as compared with normal wood.

References

Anis, M.; Kamarudin, H.; Ismadi, I.; Choo, Y.M.; Mohd, B.W. Challenges in making plywood from oil palm trunk. Proc of the 8th Pacific Rim Bio-based Composites Symp. Kuala Lumpur, 2006; pp. 211–127.

Bakar, E.S.; Hadi, Y.S.; Sumardi, I. Quality improvement of oil palm wood: Impregnation with pf. Jurnal Teknologi Hasil Hutan 2001, 14(2), 26–31.

Bakar, E.S.; Rachman, O.; Hermawan, D.; Karlinasari, L.; Rosdiana, N. Utilization of oil palm tree as building and furniture materials (1). Physical and chemical properties and durability of oil palm wood. Jurnal Teknologi Hasil Hutan 1998, 11(1), 1–12.

Barbosa, R.I.; Fearnside, P.M. Wood density of trees in open savannas of the Brazilian Amazon. Forest Ecol. Manage. 2004, 199, 115–123.

Barrett, J.D.; Kelogg, R.M. Bending strength and stiffness of second-growth Douglas-fir dimension lumber. Forest Products J. 1989, 41(10), 35–43.

Burdon, R.D.; Kibblewhite, R.P.; Walker, J.; Megraw, R.A. Juvenile versus mature wood. A new concept, orthogonal to corewood versus outerwood, with special reference to *Pinus radiate* and *Pinus taeda.* Forest Sci. 2004, 50(4), 399–415.

Chan, K.W. Biomass production in the oil palm industry. In Oil Palm and the Environment—A Malaysian Perspective; Gurmit, S., Lim, K.H., Teo, L., Lee, K.D., Eds.; Malaysian Oil Palm Grower's Council, November 1999; pp.41–53.

Chan, K.W.; Watson, I.; Lim, K.L. Use of oil palm waste for increased production. In Soil Science and Agricultural Development in Malaysia; Pushparajah, E., Chin, S.L., Eds.; Malaysian Soil Science Society: Kuala Lumpur, 1981; pp. 213–242.

Choon, K.K.; Kilmann,W.; Choon, L.S.; Mansor, H. Characteristic of oil palm stem. In Oil Palm Stem Utilization; Shaari, K., Khoo, K.C., Ali, A.R.M., Eds.; Forest Research Institute: Malaysia, 1991; pp. 15–28.

Gurmit, S. Management and utilization of oil palm by-products. Proc of 3rd National Seminar on Utilisation of Oil Palm Tree and Other Palms; Kuala Lumpur, 1994; pp. 12–18.

Haslet. Suitability of oil palm trunk for timber uses. J. Tropical Forest Sci. 1990, 2(3), 243–251.

Hassan, K.; Jalani, S. Industrial moulding of oil palm particles. 1. Suitability of oil palm trunk and frond for moulded table-tops. PORIM Bull. 1993, 27, 1–7.

Hassan, K.; Salam, A.; Jalani, S.; Darus, A. The economic feasibility of oil palm biomass moulded particleboard. Proc. of the Fourth National Seminar on Oil Palm Tree Utilization. Kuala Lumpur, 1997; pp. 41–53.

Ho, K.S.; Choo, K.T.; Hong, L.T. Processing, seasoning and protection of oil palm limber. Proc. of the National Symposium of Oil Palm By-products for Agro-based Industries. Kuala Lumpur, 1985; pp. 43–51.

Husin, M.; Sukaimi, J.; Darus, A. Oil palm biomass utilization in wood-based industries. Proc. of the 1995 PORIM National Oil Palm Conference Technologies in Plantation 'The Way Forward'. Kuala Lumpur, 1995; pp. 171–187.

Husin, M.; Zakaria, Z.Z.; Hasan, A.H. Potential of palm by-products as raw material for agro-based industries. PORIM Bull. 1985, 11, 7–15.

Jalani, B.S.; Chan, K.W.; Ariffin, D. System approach to potential products from biomass in oil palm. In Proc. 1999 PORIM International Palm Oil Congress 'Emerging Technologies and Opportunities in the Next Millennium. Palm Oil Research Institute of Malaysia: Bangi, 1999; pp. 402–413.

Kamarudin, H.; Mohd, H.; Ariffin, D.; Jalani, S. An estimated availability of oil palm biomass in Malaysia. PORIM Occasional Paper 37. Palm Oil Research Institute of Malaysia: Kuala Lumpur, 1997.

Khalil, A.H.; Wan, R.W.M. Availability and distribution of oil palm stem. In Oil Palm Stem Utilization—A Review of Research; Khozirah et al., Eds. Research Pamphlet No. 107. Forest Research Institute Malaysia, Kepong, 1991; pp. 11–14.

Killmann, W.; Lim, S.C. Anatomy and properties of oil palm stem. Proc. of the National Symposium on Oil Palm By-products for Agro-based Industries, Kuala Lumpur, 1985, 18–42.

Kirkaldy, J.L.R.; Sutanto, J.B. Possible utilisation of by-products from oil palm industry. Planter 1976, 72(838).

Mansor, H.; Ahmad, A.R. Carbohydrate in the oil palm stem and their potential use. J. Tropical Forest Sci. 1990, 2(3), 220–226.

Mohd Fahmi, A.; Othman, S.; Rokiah, H.; Razak, W. Preliminary study on shrinkage and swelling properties of oil palm trunk in organic liquid. Proc. of the National Seminar on Utilisation of Oil Palm Tree: Strategies for Commercial Exploitation, 2008, 199–204.

Lim, S.C.; Gan, K.S. Characteristic and utilization of oil palm stem. Timber Technol. Bull. 2005, 35.

Lim, S.C.; Khoo, K. Characteristic of oil palm trunk and its potential utilization. Malaysian Forester 1986, 49(1), 3–22.

Macpeak, M.D.; Burkhart, L.F.; Weldon, D. Comparison of grade, yield and mechanical properties of lumber produced from young fast-grown and older slower-grown planted slash pine. Forest Prod. J. 1990, 40(1), 11–14.

Norini, H.I.; Asma,W.; Azmi, M.I.M. Availability of oil palm residue in Peninsular Malaysia. Proc. of the Fourth National Seminar on the Oil Palm Tree Utilization. Kuala Lumpur, 1997; pp. 16–24.

Ratanawilai, T.; Chumthong, T.; Kirdkong, S. An investigation on the mechanical properties of trunk of palm oil trees for the furniture industry. J. Oil Palm Res. Special Issue. 2006, 114–121.

Ridzuan, R.; Shaler, S.; Mohd, A.J. Properties of medium density fiberboard from oil palm empty fruit bunch fibre. J. Oil Palm Res. 2002, 14, 2, 34–40.

Sing, G. Management and utilization of oil palm by-products. In Proc. of the Third National Seminar on Utilization of Oil Palm Tree and Other Products. Kuala Lumpur, 1994; pp. 138–144.

Yamani, S.A.K.; Ahmad, A.J.; Kassim, J. Particleboard manufacture from EFB of oil palm. Proc. of the National Seminar on the Utilization of Oil Palm Tree and Other Palm. Kuala Lumpur, 1994; pp. 130–144.

Yusoff, M.N.M.; Khoo, K.C.; Lee, T.W. Preliminary characteristic of oil palm stem as raw material for pulp and paper. Malaysian Forester 1984, 47(1), 28–42.

Zhang, S.Y.; Zhong, Y. Structure-property relationship of wood in East-Lioning oak. Wood Sci. Technol. 1992, 26, 139–149.

Zobel, B.J.; Van Buijtenen, J.P. Wood variations—its causes and control. Springer-Verlag: Berlin, 1989.

22

Oil Palm as Bioenergy Feedstock

Robiah Yunus[1], Rozita Omar[1], Zurina Zainal Abidin[2], and Dayang Radiah Awang Biak[3]

[1]Green Engineering and Sustainable Technology Laboratory, Institute of Advance Technology, Universiti Putra Malaysia, 43400 UPM Serdang, Selangor, Malaysia; [2]Biosensor Laboratory, Institute of Advance Technology, Universiti Putra Malaysia, 43400 UPM Serdang, Selangor, Malaysia; and [3]Department of Chemical & Environmental Engineering, Faculty of Engineering, Universiti Putra Malaysia, 43400 UPM Serdang, Selangor, Malaysia.

Introduction

Increasing concern on energy security and environmental issues such as emission of greenhouse gasses has raised the interest towards the development of renewable bioenergy as an alternative energy to fossil-based fuel. Bioenergy or biofuel is the renewable energy contained in the materials derived from biological sources known as biomass. As one of the most productive biomass (crops) sources, oil palm offers great potential as bioenergy feedstock. Palm oil, a major product from oil palm, is gaining widespread acceptance across the world as a source for biodiesel. Its utilization as bioenergy feedstock also brings other environmental benefits such as reduced levels of CO_2, black smoke of carbon particulates, carbon monoxide, and sulfur dioxide, as it is a cleaner energy.

The production of palm oil inevitably generates oil palm waste coming from the plantation or the effluent itself which needs to be addressed in order to avoid additional environmental issues. The feasibility of using these oil palm biomass and palm oil mill effluent (POME) as a feedstock for bioenergy production has received overwhelming interests worldwide in the effort to confront the energy crisis. Intensive utilization of these oil palm wastes through various methods such as thermochemical, chemical, biological, and physical routes has made it viable to harness renewable energy, bioethanol, biomethanol, bio-oil, biobriquettes, biogas, biodiesel, and biogasoline.

This chapter will highlight the availability, demand, and potential and future direction of oil palm resources from oil to biomass waste as the feedstock for bioenergy conversion.

Bioenergy Potential from Oil Palm

Oil palm and its products are major sources of bioenergy. Worldwide, palm oil industry is sizeable with Malaysia and Indonesia as the main producers. The world production has tremendously increased from about 9.22 million tons per year in the 1990s to 42.3 million tons (average of 5 years) in 2010 (Halsall, 2011; May, 2011; MPOB,

2011a; Nellens, 2009). Palm oil is consumed mostly as food or used to produce soaps, detergents, and cosmetics. With a good calorific value (39 MJ/kg), palm oil can also be directly combusted as fuel that produces 1.64 exajoules (1 exa = 10^{18}) of energy annually (Halsall, 2011). However, the future of palm oil as an energy source is conflicted with its role as a food source.

Table 22-A shows the actual and forecasted world consumption of palm oil for the period of 1985–2020. During 1990–2009, the share of palm oil in the major oil and fat consumption had increased by twofolds from 13.9% to 27.5% (Nellens, 2009). The use of palm oil as biofuel was virtually nonexistent in the early 1990s. The use of palm oil as a biofuel resource began only in the late 1990s with only 40,000 tons of palm oil converted into biofuel (Nellens, 2009). This amount was predicted to increase by 139% by 2020. In a biodiesel plant with an annual capacity of 220 million liters of biodiesel, the cost of producing 1 liter of biodiesel from palm oil is estimated to be US$0.86. This cost is far below that of soybean-based (USA) biodiesel (~1.62 US$/liter). However, the biodiesel processed from tallow (UK) costs the least with the estimated value of only US$0.5 /liter (Bauen et al., 2009). Increase in fossil fuel prices and excess supply of oils and fats for food promise a bright future for palm oil as one of the sustainable energy sources.

Table 22-A. Palm Oil—World Consumption (Million Tons).

5-year Average						
Total Usage	2016–2020F*	2011–2015F	2006–2010F	2001–2005	1996–2000	1991–1995
EU-27	7.2	6.22	5.02	3.65	2.17	1.64
US	1.41	1.08	0.78	0.24	0.13	0.12
China	9.30	7.89	6.19	3.22	1.47	1.14
India	6.00	5.10	4.14	3.61	2.20	0.37
Indonesia	8.95	6.50	4.47	3.19	2.82	1.80
Malaysia	4.30	3.56	2.40	1.66	1.23	0.88
Pakistan	2.30	2.08	1.71	1.37	1.11	1.06
Others	30.04	24.07	17.45	11.21	7.41	6.24
TOTAL	69.50	56.50	42.16	28.15	18.54	13.25
Palm oil consumption						
As biofuel	5.60	3.70	1.71	0.19	0.04	0.00
As food	48.30	39.50	29.90	20.36	13.32	9.41
Others[a]	15.60	13.30	10.55	7.60	5.18	3.84

[a]Consumption for feeds and in oleochemical industry
*F indicates forecasted values
Source: Nellens (2009).

Palm biodiesel has been used as a diesel substitute since 1996 (Sumathi et al., 2008) in which crude palm oil (CPO), crude palm stearin, and crude palm kernel oil (CPKO) were converted into their methyl ester products. In Malaysia, Envodiesel, a blend of 5% palm olein with 95% petrodiesel, was launched in 2006 (Bernama, 2011). The current statistics show that there are more than 15 main biodiesel producers (MPOB, 2011b) in Malaysia with a total annual biodiesel production in 2009 approximated at 236,564 tons (MPOB, 2011c). In the European Union alone the demand for biodiesel is increasing, particularly due to the various incentives (Ong et al., 2011). Indonesia and Malaysia are the two main producers of biodiesel with a total production of 22,135 million liters per year.

For every ton of palm oil produced, approximately 9 tons of non-oil biomass will be generated. This includes the oil palm trunk (OPT), the oil palm fronds (OPF), the empty fruit bunches (EFB), the palm kernel shell (PKS), and the mesocarp fibers (MF). Each bunch of fresh fruit has about 21–23% palm oil, 6–7% palm kernel shell, 14–15% mesocarp fiber, and 23% EFB (Hambali et al., 2011; Ng et al., 2011; Yusoff, 2006). For each ton of FFB processed, approximately 0.67 m^3 of palm oil mill effluent (POME) is produced (Ng et al., 2011; Sridhar & AdeOluwa, 2009). Table 22-B summarizes the amount of biomass generated by palm oil industries recently.

Although oil palm biomass can be converted to various value-added products such as fertilizer, animal food, adsorbent (Shuit et al., 2009), pulp and paper (Chew & Bhatia, 2008), medium density fiberboard (Husin et al., 2005), molded oil palm products to be used in furniture, building, packaging and automobile industries (Malaysian Palm Oil Council, 2006), its potential as a source of renewable energy is more promising, as shown in Table 22-B. Apart from capturing the CO_2, oil palm plantation and subsequent processing relieve the dependence on the industry to fossil fuel, also. There are five types of biofuel that can be produced from oil palm biomass, namely bioethanol, biomethanol, biobriquettes, synthesis gas, and pyrolysis oil (bio-oil) (Shuit et al., 2009).

Table 22-B. Oil Palm Biomass (Dry Basis) Collected in 2009 in Malaysia and Indonesia.

Oil Palm Wastes	Collected Amount (million tons)	Higher Heating Value (MJ/kg)	Energy Potentials
EFB	12.74	18.8	493.4 MW[a]
Palm kernel shell	8.54	20.1	
Mesocarpfiber	14.13	19.0	
Oil palm trunk and frond	123.5	16.6	
POME	106.56	20 MJ/m^3	483 MW[b]

Source: Chin et al. (2008); Hambali et al. (2010); Ng et al. (2011).
[a]Value is estimated based on 25% thermal efficiency.
[b]Value is estimated based on 21% heat input for 7200 hours.

EFB and MF are the main biomass generated by the mill. Currently, EFB is being used as fuel to generate steam at the mill, as mulch in the plantation, or as pulp and paper or mould pulp and paper for food packaging (Ong et al., 2011). EFB has tremendous potential to produce renewable energy due to its high lignocellulosic (hemicelluloses and cellulose) contents. The cellulose and hemicellulose contents in EFB can be up to 45–50% and 25–35%, respectively (Sridhar & AdeOluwa, 2009). Various methods can be applied to convert lignocellulosic materials into reducing sugars [xylose (C5) or glucose (C6)], and subsequently the reduced sugars are fermented to produce ethanol. The typical glucose yield ranges between 50 and 70% for dilute acid hydrolysis process and 75 and 95% for enzymatic hydrolysis technique (Wu et al., 2010). Different types of microorganisms, such as *Saccharomyces cerevisae, Escherichia coli,* and *Zymomonas mobilis* are needed to convert these sugars into ethanol. The overall production is currently quite low due to sugar consumption selectivity and low ethanol tolerance of the microorganisms. Production of 1 kg of ethanol requires 4.653 kg of EFB after considering the conversion of EFB to sugar via dilute hydrolysis process (Tan et al., 2010). Bioethanol is mainly used as fuel additive to reduce the emission of carbon monoxide and other smog-causing emissions. Depending on the capability of the vehicle, it is possible to mix bioethanol and gasoline up to a ratio of 85:15 (Shuit et al., 2009).

At the palm oil mill, dry EFB, dry MF, and PKSs are shredded or pulverized before being burned to produce steam or generate electricity. A total of 204 tons of EFB is used to produce 114 tons of palm oil fiber fuel (POFF) that has an HHV of 8.80 GJ/ton to produce 1 TJ of energy (Evald, 2005). The price of POFF is highly sensitive to the transportation price due to its low density (Chiew et al., 2011). A high volume of POFF is required to supply a specific amount of energy required.

EFB and MFs are pressed or densified into pellets and bales as an alternative to drying, which makes the transfer, delivery, and feeding much easier. Even though the moisture removal from this process is slightly low (i.e., 45% compared to 67% when the drying process is applied), moist pellets after shredding are much easier to compact and handle. The bales can be stored up to 5 days only because they will either be naturally degraded or affected by spores or fungi (Evald, 2005). The bales are much easier to transport and thus suitable for household use or for efficient co-firing plant combustions. The price of EFB bales fuel is approximately 172 RM/ton for a transport distance of 30 km. Assuming that the moisture content in the bales is 17%, the price is equal to 11.8 RM/GJ with an LH value of 14.5 GJ/ton (Evald, 2005). In 2009, more than 300 mills self-generated electricity for their internal uses as well as supplied electricity for the surrounding areas (Shuit et al., 2009). To date, approximately 25 oil palm-based projects have licenses to generate and supply electricity with a total connection to grid approximated at 194.4 MW. Three of these projects are biogas based, while the remaining are using oil palm wastes as the biomass supplies (MPOB, 2010).

Palm kernel shells are the fibrous shell fractions left after nut crashing and removal in the palm oil mill. They are easily handled in bulk directly from the product line to the end use. The shells are a mixture of broken fractions, roughly 5–8 mm in diameter up to about half a shell. The moisture content in the shell is very low (i.e., between 11 and 13%) compared to other biomass from oil palm industries. Furthermore, it also contains some palm oil residues, making its calorific value slightly higher than the average lignocellulosic samples, as presented in Table 22-B. PKS comparably possesses the best quality biomass fuel in that it has a uniform size distribution and is easy to crush and handle with limited biological activity because of low or no moisture content. Therefore, it is easy to transport and typically suitable to be used in grate combustion of dust firing (Evald, 2005). PKS and EFB can be mixed to form briquettes where they are typically heated at a high temperature and pressure is applied using the screw press technology. Sawdust is typically blended into the mixture to enhance burning. The pellets or briquettes are available throughout the year, have a high calorific value, and longer burning duration. With a low (or no) moisture content, the briquettes do not emit smog, thus, making it the most potential renewable energy source in the future (Shuit et al., 2009).

Pruning and replanting are some seasonal activities that are done at oil palm plantation. The pruning process of the fronds ensures that an optimum number of fronds are left on the palm tree for good growth of FFB. In general, the productivity of oil palm tree declines after 20–25 years of age. The old trees are cut down to make way for the replanting process of new seedlings. In 2006–2010, it was estimated that 120,000 ha of oil palm is replanted (Basiron & Chan, 2006). Currently, almost all pruned fronds are discarded between the rows of oil palm tree for soil conversation, erosion control, and nutrient recycling. The trunks produced from replanting process are mostly discarded or burned at the plantation site. The cut trunks, even though they have a very high moisture content (~76%), are one of the enormous biomass resources in the oil palm plantation. The average glucose content in the sap of the fresh cut trunks is approximately 85.2 g/L. Upon storage (30 days), the glucose concentration in the sap increases to approximately 153 mg/mL (Kosugi et al., 2010; Yamada et al., 2010).

EFB, PKS, and MF are also used in gasifier or pyrolysis chamber to produce hydrogen or synthesis gas (syngas). Biomethanol, a less popular product compared to bioethanol, is one of the products obtained from the gasification process. It is mainly used in spark ignition engines due to its high octane ratings and is typically produced via the catalyst-aided gasification process. Other fuel gases (H_2 and methane) are released during the gasification process; liquid fuel (methanol) is obtained when the products enter the catalytic reactor. At present, the demand for liquid fuel is not as high, so the production of hydrogen or other syngas via gasification and pyrolysis is more preferred.

Sterilizer condensates, hydrocyclone wastes, and separator sludges are the main types of wastewater generated in palm oil mill. The amount of POME generated for each ton of crude palm oil (CPO) produced is approximately 0.9–1.5 m^3 (Ma & Ong, 1985). POME is rich in organic carbon with a very high biological oxygen demand (BOD), chemical oxygen demand (COD), and nitrogen contents. In the previous years, POME was treated in an open pond and was converted to fertilizers and used in nearby farms or vegetation areas. At the same time, greenhouse gases called biogas (35% carbon dioxide and 65% methane) are released in the atmosphere making this system no longer preferred. The gas is lighter than air and has a higher calorific value. In the recent years, the anaerobic digestion of POME has been done in an anaerobic digester whereby the biogas formed is captured and used as energy source. For each ton of FFB processed, approximately 0.7 ton of effluent is produced (Ee, 2009). The potential amount of energy that can be harnessed from an anaerobic digestion of POME of a mill with FFB processing capacity of 530 thousand ton per year is approximately 292 TJ (Ee, 2009).

The oil palm industry not only provides nutritional food to the world but also produces renewable and sustainable energy, secures the energy supplies, and reduces the emission of greenhouse gases and smog.

Oil Palm as a Bioenergy Resource

Palm oil industry produces palm oil and by-products, namely oil palm waste both from the milling process and plantation. Palm oil that is traditionally used as food and for manufacturing products such as soap and cosmetics is currently getting acceptance as biofuel. Oil palm wastes, especially from the milling process generally, are used in-house for heat and energy production (Yusoff, 2006). More recently, oil palm waste is finding its way into the market as value added products. However, there is always a surplus of oil palm wastes; therefore, their use as bioenergy resources is an attractive alternative for the current high energy demand and more environmentally concerned world.

Palm Oil

Palm oil is extracted from the mesocarp (pulp) of the fruit of the oil palm, and the palm kernel oil is derived from the palm kernel (seed) of the oil palm. Bioenergy or biofuel that is contained in the palm oil can be produced as either biodiesel or advanced biodiesel. In 2010, the global biofuel production grew from 16 billion liters in 2000 to more than 100 billion liters, converting millions of tons of vegetable oils such as tallow, grains, and sugar cane to biofuel. Most conventional biofuel technologies need to improve conversion efficiency, cost, and overall sustainability to cater to the growing market. In addition, advanced biofuels should be given a priority which requires substantial investment in research, development and demonstration units and government support for commercial-scale advanced biofuel plants.

Biodiesel

Biodiesel is the bioenergy derived mainly from plant oils or animal fats and has shown great potential as a substitute to petroleum-derived diesel for compression ignition engine. Palm oil, like other vegetable oils, can be used to produce biodiesel, as either a simply mixed processed palm oil with petrodiesel, or converted through transesterification to produce a palm oil methyl ester blend, which meets the international EN 14214 specification. Biodiesel can also be manufactured from used or recycled cooking oils or greases or animal fats. Since palm oil biodiesel is plant based, it is an environmentally friendly, non-toxic, renewable, and biodegradable biofuel, which can potentially reduce greenhouse gas emissions and mitigate climate changes. The palm biodiesel has received extensive attention due to these advantages and has become an important alternative to conventional diesel fuel.

Biodiesel can be used pure or in a blend (commonly B5 or B20, which contains 5% or 20%, respectively, biodiesel mixed with fossil diesel). More than 50 countries, including both OECD (Organization for Economic Cooperation and Development) and non-OECD countries, have adopted blending targets or mandates and several more have announced biofuel quotas for future years (Table 22-C). These policies developed across the past decade aimed at promoting the use of biofuels locally but had tremendous effects on global market development (Lamers et al., 2011).

Biodiesel can be used to substitute fossil diesel. Over 50 plant species produce oils that have potential for use as fuel, but most are very expensive. Besides palm oil (Abdullah et al., 2009; Boey et al., 2009; Kansedo et al., 2009), rapeseed oil (Liu & Wang, 2009; Rashid & Anwar, 2008; Shi & Bao, 2008), canola (Dizge et al., 2009; Thanh et al., 2010), soybean oil (Georgogianni et al., 2009; Qi et al., 2009, Wang et al., 2011), and corn oil (Bi et al., 2010) have been used for biodiesel production and have good potential as diesel substitutes. Non-edible vegetable oils, such as one extracted from *Jatropha curcas,* have also been found to be suitable for biodiesel production (Azhari et al., 2011; Lu et al., 2009; Vyas et al., 2009). Another possible source of lipids for biodiesel is oil-rich microalgae feedstock, but it is still at the research and demonstration phase.

Despite biodiesel being uneconomic due to the rise in feedstock price, many countries have supported the development of a domestic biodiesel industry for social and environmental reasons. The presumed environmental benefit of biodiesel is most notably in terms of reducing greenhouse gas (GHG) emissions and energy security. However, over the last few years, there has been vigorous debate about the extent to which biofuels cause GHG reductions (Edwards et al., 2010). The expansion of the vegetable oil industry has been associated with deforestation, release of carbon from vegetation and soil, forest fires, soil erosion, water pollution, and biodiversity loss. These sustainability issues may limit the viability of biodiesel exports, but policy changes in the future may improve the access to the US and EU subsidies (Pio Lopez

Table 22-C. Overview of Biofuel Blending Targets and Mandates.

Country / Region	Current mandate/ target	Future mandate/ target	Current status (mandate [M]/ target [T])
Argentina	B7	NA	M
Australia	NSW: B2	NSW: B5 (2012)	M
Bolivia	B2.5	B20 (2015)	T
Brazil	B5	NA	M
Canada	B2–B3 (in 3 provinces)	B2 (nationwide) (2012)	M
Chile	B5	NA	T
China (9 provinces)	E10 (9 provinces)	NA	M
Colombia	B10	B20 (2012)	M
Costa Rica	B20	NA	M
Dominican Republic	NA	B2 (2015)	NA
European Union	5.75% biofuels	10% renewable energy in transport	T
India		(2017)	M
Indonesia	B2.5	B5 (2015); B20 (2025)	M
Japan	500 Ml/y	800 Ml/y (2018)	T
Korea	B2	B2.5 (2011); B3 (2012)	M
Malaysia	B5	NA	M
Mozambique	NA	B5 (2015)	NA
Norway	3.5% biofuels	5% (2011)	M
Paraguay	B1	NA	M
Paraguay	B1	NA	M
Peru	B2	B5 (2011)	M
Philippines	B2	B5 (2011)	M
South Africa	NA	2% (2013)	NA
Taiwan	B2	NA	M
Thailand	B3	B5 (2011)	M
Uruguay	B2	B5 (2012)	M
United States	0.8 billion gallons	136 billion litres (2022)	M
Vietnam	NA	50 Ml biodiesel (2020)	NA
Zambia	NA	B10 (2011)	NA

B – biodiesel (B2 = 2% biodiesel blend); NA – not available.
Source: International Energy Agency (2011).

& Laan, 2008). However, at current domestic production rates of biodiesel in most countries, it is unlikely to be driving deforestation, due to low production levels.

Palm Oil Biodiesel

There are three methods of producing biodiesel from palm oil, as shown in Fig. 22.1. The first method is the direct blending of straight vegetable oil (SVO) with petroleum diesel. In Malaysia, the B5 biodiesel is a blend of 5% refined palm olein (processed palm oil) and 95% petroleum diesel. However, the use of SVO as biodiesel has yet to gain widespread acceptance and is not generally recommended due to the risks of engine damage and gelling of the lubricating oil (IEA, 2011). The second method is by converting the palm oil to methyl esters through transesterification reaction. The biodiesel is usually blended with petroleum diesel prior to use in compression ignition engines (diesel engines) without any engine modification. Co-products of biodiesel production, mainly protein meal and glycerin, are important factors in the overall economics of the process. The third method is by converting the palm oil via

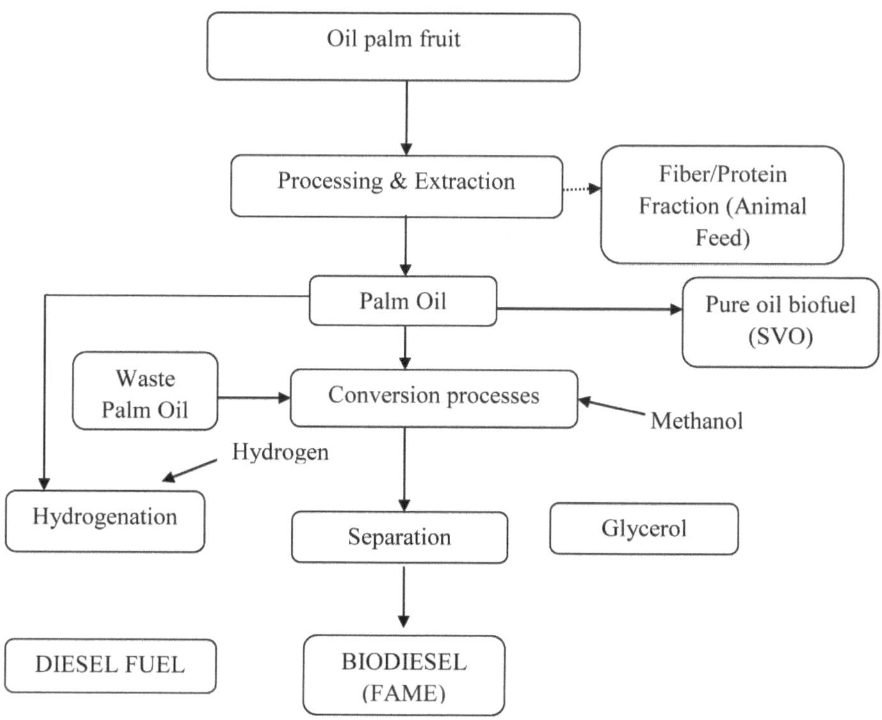

Fig. 22.1. Production route for biodiesel from palm oil.

hydrogenation to advanced biodiesel which has similar properties as diesel fuel. The fuel can be used directly in diesel engines.

The Renewable Fuels Standard (RFS2) estimates biodiesel production will be at least 1 billion gallons (3.78 billion liter) or 2.8% of the total demands for biofuel of 36 billion gallons (136 billion liter) by 2022 (Lamers et al., 2011). Conventional biofuels (>20% GHG savings) are allowed to contribute 15 billion gallons and the advanced biofuels (>50% GHG savings) shall cover the remaining 21 billion gallons. IEA (2011) estimated that by 2050, biofuels would provide 27% of total transport fuel and contribute in particular to the replacement of diesel, kerosene and jet fuel. Table 22-D shows the top 10 countries in terms of biodiesel production capacity (Johnson & Holloway, 2007) where Malaysia tops the list. The feedstock available for development of biodiesel is 22% for palm oil, 28% for soybean oil, 20% for animal fats, and 11% for coconut oil, while rapeseed, sunflower, and olive oils constitute only 5% each (Sharma & Singh, 2009). Globally, 600,000 tons of crude palm oil were used for biodiesel production in 2005, and this was estimated to grow to one million tons in 2007 (Chowdhury, 2007).

In a scenario in which production costs are strongly coupled with vegetable oil prices, they would remain slightly more expensive than fossil fuels. In addition, biodiesel produced by the oilseed-producing countries would also be more competitive as the cost of feedstock represents 80% of the production cost (Yusuf et al., 2011). The total cost of biodiesel in EU produced from palm oil imported from Malaysia is estimated at US$754/ton, as compared to the biodiesel from the local rapeseed oil priced

Table 22-D. Major Biodiesel Producers (Johnson & Holloway, 2007).

Country	Production volume (million liters)	Production cost[a] (US$/liter)
Malaysia	14,540	0.53
Indonesia	7,595	0.49
Argentina	5,255	0.62
USA	3,212	0.70
Brazil	2,567	0.62
The Netherlands	2,496	0.75
Germany	2,024	0.79
Philippines	1,234	0.53
Belgium	1,213	0.78
Spain	1,073	1.71

[a]Average production cost per liter is calculated from all available lipid feedstock prices, increased by a US$0.12 refining cost and decreased by US$0.04 for the sale of byproducts.

at US$996/ton (Table 22-E). If the biodiesel is produced from soybean imported from USA, the total cost would be US$801/ton. The average retail biodiesel price in Germany in 2007 was at US$1,332/ton.

Several processes are under development that aim to produce fuels with properties very similar to those of diesel and kerosene. Advanced biodiesel and biokerosene will become increasingly important since demand for low carbon fuels with high energy density is expected to increase significantly in the long term. Advanced biodiesel is not widely available at present but could become fully commercialized in the near future, since a number of producers have pilot and demonstration projects underway (USDOE, 2009).

Palm Oil Biodiesel Quality

A conventional palm oil biodiesel is essentially fatty acid methyl ester (FAME) of a palm oil derived from a reaction between palm oil and methanol. Although, other alcohols can also be used (e.g., propanol, butanol, isopropanol, tert-butanol, branched

Table 22-E. Estimated Production Costs for Biodiesel from Palm Rapeseed and Soybean Oils.

Cost comparison (US$/ton)	Palm oil	Rapeseed	Soybean
Feedstock (FOB at producing country)	547	800	601
Biodiesel production cost:			
Solvents, acids and chemicals	47	–	–
Other costs	35	–	–
Adjustment for energy parity with petroleum diesel (based on 90% of kJ/kg of energy of petrol–diesel)	55	–	–
Total	**137**	**196**	**150**
Cost of biodiesel	684	996	751
Estimated freight and insurance cost to Rotterdam	70	—	50
Total cost in EU	**754**	**996**	**801**
Local distribution (approximation)	30–50	30–50	30–50
[a]Total cost at petrol kiosks in EU	784–804	1,029–1,046	831–851
[b]Price of retail biodiesel (Germany)	**1,322**		

[a]Assuming production plant with capacity > 100,000 ton/annum; other figures based on pricing as of March 2007.
[b]FOLicht based on UFOP Marktinformation (three-month average retail prices from November 2006 to January 2007).
Source: Yusuf et al. (2011).

alcohols, and octanol), the cost of production using these alcohols is much higher and the properties are less notable. The palm oil contains 32–45% palmitic acid (C16:0), 2–7% stearic acid (C18:0), 38–52% oleic acid (C18:1), and 5–11% linoleic acid (C18:2). Naturally the fatty acid compositions of palm biodiesel follow its mother tree, which predominantly comprises of C18 and C16 fatty acids.

The energy content of biodiesel varies between 88% and 99% of the energy content of diesel, depending on the feedstock and esterification process used (Love & Cuevas-Cubria, 2007). The cetane number of palm biodiesel from various sources has been estimated to vary from 48 to 61 (Bala, 2005). The physical properties of palm biodiesel are compared against the petroleum diesel and the EN 14214 standard, as shown in Table 22-F. The appearance of palm biodiesel is a clear amber-yellow liquid with a viscosity similar to that of petroleum diesel. It is non-flammable with a flash point of 423 K, as compared to 337 K for petroleum diesel. Since it is plant based, palm biodiesel is biodegradable and non-toxic, and it significantly reduces toxic and other emissions when burned as a fuel.

Since palm biodiesel contains mainly saturated fatty acids, it has a high cloud point and pour point because they are precursors for crystallization (Sadrolhosseini et al., 2011). Cloud point and pour point are two significant parameters used to evaluate the biodiesel fuel. The cloud point is the temperature at which dissolved solids appear in diesel or biodiesel fuel while the pour point is the lowest temperature at which the fuel can flow. These parameters depend on the concentration of C16:0, C18:0, and

Table 22-F. Physical Properties of Palm Biodiesel.

Parameters	Standard Test Method	Limit (EN 14214) Min	Max	Diesel	Palm biodiesel
Density (kg/m^3)	ASTM D 40	–	–	836–850	835
Kinematic Viscosity (40°C) (mm^2/s)	ASTM D 445	3.5	5.0	4–8	4.50
Pour Point (°C)	ASTM D 97	By customer		–10	–20
Cloud Point (°C)	ASTM D 2500	By customer		–6	NA
Flash Point (°C)	ASTM D 93	120	–	45–60	71.9
Calorific Value (MJ/kg)	ASTM D 240	–	–	42–46	39.89
Cetane Index	ASTM D 976	51	–	40–55	48
Carbon Residue (%)	ASTM D 189	–	0.30	0.17	0.05
Total Acid Number (mg KOH/g)	ASTM D 974	–	0.50	NA	1.054

Source: Azhari et al. 2011.

C18:1 in the product. These are the index of the biodiesel quality and show the ability of the biodiesel for application at low temperatures (Van Gerpen et al., 2004). The low-temperature properties can be improved by blending with biodiesel from unsaturated feedstock (Sharma et al., 2009).

Other major disadvantages of biodiesel are its higher viscosity, lower energy content, higher nitrogen-oxides (NO_x) emissions, lower engine speed and power, injector coking, engine compatibility, higher engine wear, and high price. Due to its high flash point, palm biodiesel exhibits cold-start problems and fuel-pumping difficulty due to the higher viscosity. Its fatty acid origins also contribute to higher acid value, hence higher copper-strip corrosion. This increases fuel consumption when biodiesel or its blends are used in direct proportion to the share of the biodiesel content, in comparison with neat petroleum diesel.

Biodiesel has become more attractive recently because of its environmental benefits. Inherent characteristics of the palm oil in the palm biodiesel reduce the level of several diesel pollutants, including sulfur dioxide, carbon monoxide, carbon dioxide, polycyclic aromatic hydrocarbon (PAH), and nitrated PAH emissions. Combustion of biodiesel alone provides over a 90% reduction in total unburned hydrocarbons (HC) and a 75–90% reduction in PAHs (Canakci et al., 2009). The use of biodiesel decreases the solid-carbon fraction of particulate matter (PM) and reduces the sulfate fraction. The biodiesel content in the fuel contributes to the emission of nitrogen oxides, and some biodiesels produce more nitrogen oxides than others. However, some additives have shown promise in moderating the increase in NO_x emissions in biodiesel (Yusuf et al., 2011).

In general, biodiesel contains 11% oxygen by weight and contains no sulfur. The structural oxygen content of a fuel improves its combustion efficiency due to the increase of the homogeneity of oxygen with the fuel during combustion (Dermibas, 2007a). Thus, the use of biodiesel can extend the life of diesel engines because it lubricates more than petroleum diesel fuel (Canakci et al., 2009). The higher heating values (HHVs) of biodiesels are relatively high. The HHVs of biodiesels (39–41 MJ/kg) are slightly lower than that of gasoline (46 MJ/kg), petroleum diesel (43 MJ/kg), or petroleum (42 MJ/kg), but higher than coal (32–37 MJ/kg) (West et al., 2008).

Industrial and Plantation Oil Palm Wastes

Generally 90% of the oil palm fresh fruit bunches becomes waste consisting of MF, PKS, and EFB. In addition, the wastewater known as palm oil mill effluent is also produced, as the palm oil extraction process utilizes a large amount of steam. Both solid and liquid wastes from the mill can generate surplus energy for the mill and nearby estate communities. Suitability of technology and economy of biomass waste to energy conversion largely depend on its characteristics, availability, and logistics. Maintenance of oil palm plantation for high crop yield results in plantation wastes such as fronds from pruning and oil palm trunk from replanting.

Fuel and Physical Characteristics of Oil Palm Wastes

Waste-to-energy conversion processes include thermal, chemical, biochemical, and physical treatments. As different characteristics of the waste require different types of treatment, it is imperative to separate their characteristics according to the product of interest. Fuel and physical characteristics are crucial in thermal treatment for production of heat, gas fuel, and solid fuel. Table 22-G shows these characteristics for oil palm industrial wastes. The values are compared with coal, the current solid fuel commonly used for heat and electricity generation worldwide. Variability of biomass cannot be controlled; therefore, the utilization of biomass such as oil palm waste as feedstock must be able to be monitored during the conversion processes. The most practical method for monitoring is through the chemical composition of the feed material, although the final products and residues must also be evaluated (Demirbas, 2009a).

The physical properties such as density, particle size, and shape distribution affect the fuel preparation of the intended raw material. High solid density of EFB and PKS were reported compared to fiber, as shown in Table 22-G. However, this value could not be used to evaluate the choice of reactor configuration, as knowledge of bulk density of the fuel is better suited for this purpose. Generally biomass has low bulk density around 0.5 g/cm^3 as compared to the density of coal at 1.3 g/cm^3 (Demirbas, 2004a), which would require densification such as briquetting or palletizing, especially for modern technology thermal conversion such as fluidized bed. Briquetting will not only facilitate transportation and storage (McKendry, 2002) but will also increase the combustion efficiency of thermal treatment as more homogeneous feed is achieved (Husain et al., 2002; Nasrin et al., 2008) but at a lower cost. Low bulk density of oil palm wastes for biochemical, chemical, and physical treatments would mean that a larger size of reactor may be needed.

The high moisture content like in oil palm fiber, EFB, trunk, and frond would generally decrease the calorific value of raw material (Kataki & Konwer, 2002) thus incurring additional costs for drying and could cause ignition and combustion problems (Demirbas, 2004b) for thermal treatment. A decrease in devolatilization rate because of the need to dry off the water first would result in a longer reaction time (de Diego et al., 2003). Nevertheless, this high moisture biomass can be an advantage in biochemical conversion methods such as hydrolysis and enzymatic treatment. Furthermore, high moisture biomass is not very practical and energy efficient in conventional thermal treatment such as combustion, gasification, and pyrolysis. A new alternative treatment named gasification that uses supercritical water technology utilizing water properties at critical points offers the advantage of high moisture feed of above 50% for hydrogen gas production (Kelly-Yong et al., 2007).

Generically, the agricultural residues had high volatile matter above 60% (Wan Azlina et al., 2007), giving high volatility and reactivity advantages (Demirbas, 2004a) as compared to coal. The volatile matter of fiber, PKS, and EFB are high (i.e.,

Table 22-G. Fuel and Physical Characterization of Oil Palm Industrial Wastes.

Properties	Fiber	PKS	EFB	Frond	Trunk	Coal (Mukah Sarawak)
Solid density (g/cm^3)	0.75	1.42	1.39	NA	NA	NA
Initial Moisture Content (%)	37.2–42.0	11.0–21.4	57.2–67.0	60	67–78	NA
Proximate Analysis (wt %, db)						
Ash	6.6–10.2	2.24–10.5	3.02–7.3	2.2–4.3	2.4–10.3	5.8
Volatile Matter	68.8–73.7	69.2–86.5	70.5–83.9	NA	61.3–75.2	42.2
Fixed Carbon	12.6–18.8	6.55–20.5	8.65–17	NA	10.0–28.4	48.8
Ultimate Analysis (wt %, daf)						
C	43.2–46.6	41.3–67.5	40.9–49.1	NA	41.9	54.4
H	5.24–5.66	4.57–8.48	5.42–7.33	NA	5.98	5.29
N	1.59–1.73	0.08–1.61	0.00–1.56	NA	3.76	1.75
S	0.10–0.19	0.09–0.35	0.1–0.68	NA	0.35	0.25
O (by diff.)	39.5–49.8	22.1–53.0	40.2–51.78	NA	43.24	38.3
HHV (MJ/kg)	19–19.6	16.3–20.1	16.8–19.4	NA	14.4–17.52	24.6

db – dry basis; daf– dry ash–free basis; NA – not available.
Sources: Abnisa et al. (2011); Damodaran (2005); Goh et al. (2010); Husain et al. (2002); Idris et al. (2010); Khor et al. (2010); Lim & Lim (1992); Luangkiattikhun et al. (2008); Ma et al. (2005); Omar et al. (2011); Razuan et al. (2010); Uemura et al. (2011); Yamada et al. (2010); Yang et al. (2006).

above 68%) with EFB volatile matter falling in the higher range, between 70 and 84%, (i.e., almost half the content of coal). High volatile matter content indicates better suitability for liquid fuel production (Demirbas, 2005a) such as from pyrolysis process. On the other hand, the ash content is slightly higher than coal, which will inversely affect the HHV (Demirbas, 2002). Furthermore, ash from biomass is denser as compared to coal ash deposits, therefore making it harder to remove and reduce the heat transfer efficiency (Demirbas, 2005a). In addition, high ash content in biomass might affect liquid yields in fast pyrolysis negatively (Chiaramonti et al., 2007). High ash content may result in low quality feed material for bioethanol production such as in some variety of oil palm EFB, fiber, and PKS. The amount of oxides in the ash such as silica oxide also has a negative impact on the bioethanol production processes (Binod et al., 2010).

The heat content in a biomass is related to the oxidation state of the fuel in natural state whereby the carbon is more important compared to hydrogen. As shown in Table 22-G, carbon content of oil palm biomass is lower than that of a low rank coal. Typical value of carbon content in coal varies from 65 to 85%, giving a high calorific value (Demirbas, 2004a). In addition, high oxygen content in oil palm biomass will have a negative effect both on energy content and products produced in thermal conversion such as pyrolysis and gasification. Nonetheless, the energy content of oil palm industrial waste is at least 65% of the low rank coal; therefore, it is suitable for use as a solid fuel alone or mixed with coal, especially PKS. Conversion to solid biofuel via carbonization is also attractive to increase the energy content. Renewability, availability, and negative values of oil palm waste partially outweigh the coal advantages as fuel. Although the sulfur content in low rank coal compared here is similar to that of oil palm biomass, especially the EFB, other types of coal can contain sulfur as high as 7.5% (Demirbas, 2004a). Low sulfur content might also be beneficial, as it reacts with calcium to form calcium sulfate which was favorable to sustain furnace operation (Baxter et al., 1998). Also, the low amount of sulfur would decrease the tendency of SO_x production if combustion method is used for energy conversion (Harimi et al., 2005).

Chemical Composition of Oil Palm Biomass

Table 22-H shows the chemical components of oil palm biomass. Cellulose content is higher in all parts of oil palm tree except for PKS, which is high in lignin content. PKS lignin content is similar to that of nut shells (Sun & Chen, 2002) which is more suitable for thermal treatment. Coupled with high density, it becomes the fuel of choice in the mill for heat generation. Recently, the price of PKS increased due to its usage as renewable activated carbon for gas and water treatment resulting in a higher fiber mixture in boilers in the mills. Generally, for high cellulosic biomass, the liquid product from pyrolysis reaction contains acids, alcohols, aldehydes, ketones, esters, heterocyclic derivatives, and phenolic compounds (Klass, 1998). High lignin content

Table 22-H. Chemical Components of Oil Palm Industrial Biomass (Moisture-free wt%).

Types of oil palm residue	Cellulose	Hemicellulose	Lignin	Extractives	Ash
Empty fruit bunch	38.3–59.7	22.1–35.3	17.8–22.1	0.1–3.21	1.6–5.36
Palm kernel shells	20.8–27.7	21.6–22.7	44.0–50.7	4.8	1.0
Fibers	33.9–34.5	26.1–31.8	25.7–27.7	6.9	3.5
Trunks	34.5–41.02	31.8–32.4	24.1–25.7	3.7–5.35	2.2–4.3
Fronds	30.4–56.03	27.51–40.4	21.7–24.51	1.7–4.4	2.4–5.8

Sources: Abdul Khalil et al. (2008); Abnisa et al. (2011); Kelly–Yong et al. (2007); Koba et al. (1990); Sulaiman & Abdullah (2011).

could result in high liquid yields during pyrolysis (Demirbas, 2007b) but could contain high unwanted polycyclic aromatic compounds, as well (Tsai et al., 2007).

Generally, all lignocellulosic biomass consists of three basic polymers: cellulose, hemicelluloses, and lignin. Cellulose is basically a long-chain polymerized glucose of 5,000–10,000 units. Hemicellulose is a mixture of various polymerized monosaccharides such as glucose, mannose, galactose, xylose, arabinose, and galacturonic acid (Mohan et al., 2006). High total cellulose and hemicellulose content (i.e., above 60%), especially in EFB, trunks, and fronds, are potential sources for bioethanol production. Nonetheless, unlike sugars such as starch, lignocellulosic material has a more complicated structure which makes it more difficult to be converted to sugar (Liu & Shen, 2008). The hemicellulose will first need to be hydrolyzed into their corresponding monomers prior to enzymatic process to produce bioethanol. However, before the chemicals or enzymes can access the hemicellulose, disruption of the naturally resistant carbohydrate-lignin shield is a prerequisite (Yunus et al., 2010). Lignin is the heterogeneous biopolymer responsible as a support to a plant by strengthening of xylem cells (wood) in trees. The higher the lignin content is, lesser the accessibility of the chemical or enzyme to reach the holocellulose. PKS contains the highest lignin among the oil palm industrial wastes. Therefore, EFB and fibers are the better candidates for bioethanol production from oil palm wastes, although the existence of lignin (17–28%) will still require pretreatment for more rapid and high yields (Wyman, 1999).The conversion and recovery efficiency of cellulose to glucose for low cellulose (~40%), high lignin (>20%) biomass such as oil palm biomass is reported to be around 0.76 when using the acid hydrolysis method (Demirbas, 2005b).

Cellulose degradation after hydrolysis mainly consists of glucose, while hemicellulose degradation liberated mainly xylose, mannose, galactose, glucose, and acetic acid. High temperature and pressure during the hydrolysis of cellulose could produce high acid, which is inhibitory to microorganisms during fermentation to produce bioethanol (Palmqvist et al., 1997). Lignin content in oil palm biomass is partially broken down to produce phenolic compounds which are also inhibitory to fermentation of bioethanol (Heipieper et al., 1994).

Conversion Processes

A conversion process is used to increase the calorific value of material to be used as fuel. The process is also used to convert material physically and/or chemically for more efficient and suitable usage in the currently available technologies. Biofuel from oil palm can be divided into two parts: the palm oil and the oil palm wastes. The palm oil can be used as fuel without modification in low concentration, due to its inherent problems. Two conversion methods are currently available, namely chemical and thermal. Specifically, they are transestered to produce biodiesel and catalytic cracking to produce mainly gasoline, kerosene, and diesel. Oil palm wastes are generally used in-house directly for heat and electricity to save on fuel, where thermal conversion is most suited for the purpose. Conversion methods of the oil palm biomass for liquid and gas biofuels are as outlined in Fig. 22.2. Three common routes have been used: thermal, chemical, and biochemical.

Thermochemical Process

Thermochemical process is one of the main conversion processes in producing biofuel. The high temperature is capable of cracking long chain oxygenated molecules such as palm oil into smaller straight chain hydrocarbons similar to petroleum products such as gasoline and diesel. Thermochemical methods are used to produce heat, gas, liquid, or solid fuel. Three routes are mainly used, namely combustion with excess

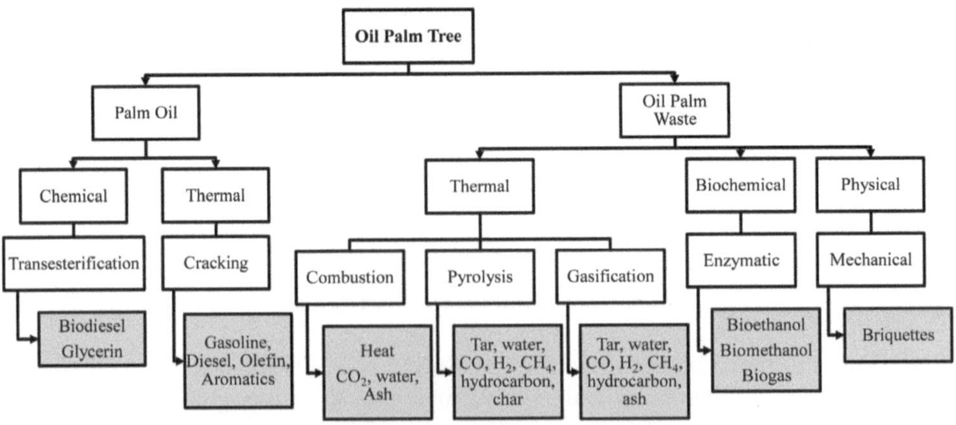

Fig. 22.2. Conversion process of palm oil and oil palm waste into biofuels.

air producing mainly heat; gasification using partial air or oxygen to produce low to medium fuel gas; and heating in the absence of air (pyrolysis), producing gas, liquid, and solid fuel. The choice of route will generally depend on technical and economic evaluations. Combustion of biomass is the oldest thermochemical conversion method, therefore the most established and responsible for over 97% of the world's bioenergy production (Demirbas, 2004a). However, gasification and pyrolysis are gaining popularity as the energy produced from these methods can be utilized both in energy and transportation sectors as biofuel.

Combustion of Oil Palm Wastes

The energy priority of oil palm mills is to produce heat and electricity at minimal cost and in an environmentally friendly way. Therefore, many oil palm mills utilize either fiber and PKS or a mixture of fiber, PKS, and EFB as feed fuel for the boilers where they are converted to mainly water, CO_2, ash, and heat. This type of energy production is commonly called co-generative system, where both steam and electricity are produced from the heat produced, depending on the mill requirements. As of 2003, over 360 palm oil mills in Malaysia produced their own energy totaling up to 338 MW_e (Ma et al., 2004). The limitation of oil palm mill energy generation is it is inefficient due to variability of the feed material. Improvement can be done by briquetting oil palm solid wastes (Husain et al., 2002). Most of the millers favor fiber and PKS as the feed into their boilers because they are readily used. As the demand of PKS for value added product increases, oil palm millers are beginning to utilize EFB, although some pretreatments such as shredding and cutting are required. Thereafter, the feeding process is similar to that of PKS and fiber although the maximum temperature for combustion is limited to around 900°C as the high organochloride content in EFB poses corrosive problems at higher temperatures (Chua et al., 2009).

Fiber and EFB, with their inherent bulky characteristic and low calorific value similar to other lignocellulosic biomass, are not suitable for fluidized bed combustion unless they are co-fired with coal. Co-firing of coal-fiber and coal-PKS could give 69.4–92% efficiency in a 10 kW pilot scale fluidized bed by increasing the coal portion in the mixture (Wan Azlina, 2005). The use of boiler for combustion generally faces heavy ash problems, particularly when EFB is used due to the high potassium content and low combustibility in a boiler at high flame temperature on grate at full load. Fluidized bed boilers offer several advantages such as high boiler efficiency, long continuous operation, higher ash purity, and lower stoichiometric air ratio, which in turn result in low ash production and NO_x emission compared to using a stoker boiler. A fluidized bed combustion test of EFB by Malaysia Palm Oil Board (MPOB)-Japan researchers in a 0.2 MW fluidized bed at a bed temperature of 700°C for up to 100-hour operation indicated good combustibility and fluidity. There was some potassium accumulated in the bed; however, the rate usually decreases with time and the fluctuation of moisture between 43 and 53% is not a problem (Ino et al., 2003).

Pyrolysis of Palm Oil

Pyrolysis of fats and oil is the direct conversion of the substance into smaller chain compounds by thermal cracking. The process is done usually with the aid of mesoporous catalysts such as HZSM-5 and zeolite (Twaiq et al., 2004) by reducing the viscosity (Ali & Hanna, 1994). In the catalytic pyrolysis process, the cleavage of chemical bonds will occur and yield smaller molecules (Jain & Sharma, 2009). The mechanism of reaction varies among researches but generally consists of the decomposition of the glycerides forming fatty acids and acrolein and then further cracking into short chain hydrocarbons, as illustrated in Equations 22.1 to 22.4 (Maher & Bressler, 2007).

1. Decomposition of glycerides:

$$\text{Glycerides} \rightarrow \text{RCOOH} + \text{R}'\text{COOH} + \text{R}'''\text{COOH} + \text{R}''\text{CH} = \text{CO} \quad (22.1)$$

2. Decomposition of fatty acids:

$$\text{RCOOH} \rightarrow \text{CO}_2 + \text{RH} + \text{H}_2\text{O} + \text{RCOR} \quad (22.2)$$

3. Decomposition of ketenes and acrolein:

$$\begin{aligned} &\text{R}''\text{CH} = \text{CO} + \text{CH}_2 = \text{CHCHO} + \text{RCOCH}_2\text{R} + \text{RCOCH}_2\text{R} \rightarrow \\ &\quad \text{CO} + \text{C}_2\text{H}_4 + \text{RHC} = \text{HCR} + \text{R} - \text{R} + \text{R}_2 + \text{CH}_2\text{CHO} \end{aligned} \quad (22.3)$$

4. Decomposition into elements; dehydrogenation, decomposition, alkylation, isomerization, aromatization and polymerization of paraffin:

$$\text{C}_n\text{H}_{2n+2} \rightarrow \text{C} + \text{H}_2 + \text{C}_n\text{H}_{2n} + \text{C}_m\text{H}_{2m} \quad (22.4)$$

The compounds produced from the pyrolysis or thermal cracking of fats and oils are generally paraffin and olefin, similar to those present in petroleum sources. The yield of gasoline range product during the catalytic pyrolysis of vegetable oil depends greatly on the composition of the oil (Bielansky et al., 2010), reaction temperature (Demirbas, 2009b), residence time (Ong & Bhatia, 2010), and type of catalyst used (Maher & Bresslar, 2007). However, oxygen removal from the process decreases the products' benefits of being an oxygenated fuel. This decreases their environmental benefits and generally produces more fuel similar in properties of gasoline than diesel, with the addition of some low value materials (Ma & Hanna, 1999).

Pyrolysis of Oil Palm Wastes

Pyrolysis of biomass has become increasingly used because the process condition can be optimized to produce high energy density oils, as well as charcoal and intermediate energy gas (Demirbas & Arin, 2002). Two common types of pyrolysis are carbonization and fast pyrolysis. The former targets at the solid product and the latter emphasizes on the liquid (called pyrolysis oil) by optimizing the temperature, heating rate, and fuel size. Pyrolysis of biomass is a series of complex reactions but can be summarized as in Equation 22.5. Carbonization of EFB under hydrothermal condition produced charcoal with a higher calorific value of 25.8 MJ/kg as compared to that under nitrogen purging (22.0 MJ/kg) for substitute of coal as solid fuel (Inoue, 2010). A higher calorific value of carbonized oil palm frond at 28.9 MJ/kg can be obtained when pyrolyzed at 750°C (Khor & Lim, 2006), whereas oil palm trunks are not suitable for charcoal production as the calorific value of the product is low and ash content is high (Lim & Lim, 1992).

$$\text{Lignocellulosic biomass} \rightarrow \text{Volatiles} + CO + CO_2 + H_2 + H_2O + C_nH_{n+2} + \text{tar} \quad (22.5)$$

PKS pyrolysis using laboratory scale fluidized bed could produce up to 58% liquid fraction, including 10% water at a moderate temperature of 500°C (Islam et al., 1999). Oil product fraction contained a very low concentration of paraffin and no polycyclic aromatic hydrocarbon (PAH), with phenol being the most abundant compound. The calorific value is comparable to that of wood oil but only half that of diesel. However, the tar from biomass waste is a mixture of highly oxygenated complex compounds (Abnisa et al., 2011) and is chemically unstable (Ani & Islam, 1998). Therefore, the liquid products need further upgrading. Pyrolysis of oil palm trunk yielded a high condensate, but at high temperatures the condensate yield is comparable to gas. The char produced has a higher calorific value compared to the feed, and the oil produced has good properties when compared to No. 6 fuel oil except for its acidic nature and lower calorific value (Khor et al., 2010). Bio-oil produced from EFB pyrolysis shows a higher calorific value compared to wood-derived bio-oil although inferior to both light and heavy fuel oil (Sulaiman & Abdullah, 2011).

Pretreated and catalytically pyrolyzed EFB catalytically for the production of bio-oil shows high phenolic fraction in the bio-oil (Misson et al., 2009). Ground MF, PKS, and EFB were pyrolyzed in counter current-fixed bed reactor by Yang et al. (2006) at a high temperature around 900°C to maximize the gas production obtaining 70 wt% gas with around 33 and 41 vol% of H_2 and CO, respectively, with a medium heating value of 14.7 MJ/m^3. Addition of nickel as a catalyst to the pyrolysis process increased the H_2 yield by threefolds. The long reaction time in conventionally heated pyrolysis has attracted the use of microwave as a heating source. The use of a microwave absorber shortens the reaction time as well as omits the need to dry the oil palm biomass (Omar et al., 2011), which produces high gas products (Omar, 2010; Salema & Ani, 2011).

Gasification of Oil Palm Wastes

Gasification technology has attracted a lot of attention as it has the ability to produce mechanical and electrical power from biomass in a small scale at an affordable price. This method does not directly produce heat or electricity, but the fuel to drive internal gas engines can be used as chemical feedstock, as shown in a series of reactions in Equations 22.6 to 22.11. The carbon is produced from the pyrolysis reaction of the biomass before proceeding with the gasification reaction. As opposed to combustion process, the products—syngas—can be used *in situ* or distributed according to demand. The composition of syngas is generally about 20–30% carbon monoxide, 10–15% hydrogen, 1–5% methane, less than 1% other hydrocarbons, 5–10% carbon dioxide, and nitrogen in the range of 50%.

$$C + O_2 \rightarrow CO_2 \tag{22.6}$$

$$C + {}^1/_2O_2 \leftrightarrow CO \tag{22.7}$$

$$C + CO_2 \leftrightarrow 2CO \tag{22.8}$$

$$C + H_2O \rightarrow CO + H_2 \tag{22.9}$$

$$CH_4 + H_2O \leftrightarrow CO + 3H_2 \tag{22.10}$$

$$CH_4 + 2H_2O \leftrightarrow CO_2 + 4H_2 \tag{22.11}$$

EFB downdraft gasification resulted in a higher hydrogen content compared to wood and bagasse, even though a lower calorific value was obtained (Erlich & Fransson, 2011). The estimated cost of hydrogen production via air gasification is only higher compared to gasification with CO-shift but significantly lower compared to electrolyzed hydrogen and high pressure pyrolysis (Mohammed et al., 2011). Catalytic gasification of oil palm biomass such as MF, EFB, and PKS could enhance the syngas production and hydrogen content, as shown by several researchers (Ismail et al., 2011; Li et al., 2009).

Industrial-scale Thermochemical Conversion of Oil Palm Biomass in Malaysia

There are only a few industrial scale oil palm biomass power plants in Malaysia. A 14 MW_e power plant using a feed mixture of fiber, PKS, and EFB in boiler and steam turbine for electricity generation (10 MW) was already operational since 2004 in Sabah, East Malaysia, by TSH Bio-Energy Sdn. Bhd. (Asian Institute of Technology, 2004). A full-scale model power plant of 10 MW electricity generation had been constructed in 2008 in Negeri Sembilan, West Malaysia, under the Small Renewable Energy Power (SREP) program based on the TSH Bio-Energy plant. There are other smaller co-generation projects such as the one at Sungai Dingin Palm Oil Mill by Kumpulan Guthrie Bhd. (2 MW_e) and Sahabat Palm Oil Mill by FELDA Palm Industries (7.5 MW_e)(CDM Executive Board, 2006).

Ab. Rahman et al. (2004) studied a pilot-scale fixed-bed gasification of oil palm biomass. The process achieved 22–37% thermal conversion efficiency, and this was slightly higher as compared to combustion which was at 15–25%. The requirement for flue gas treatment is reduced without black smoke emission and with a less operational problem while ash could be reused as a fertilizer. The project investment is estimated at RM 1.5 million per MW with a payback period of two years. Genting Bio-Oil plant in Ayer Hitam, Malaysia, produces 1.2 tons per hour of pyrolysis oil using comminuted EFB containing 5–10% moist as feedstock using fast pyrolysis technology technique (Venderbosch et al., 2006). The plant produces high quality "green oil" for local energy and chemical markets (terHarmsel & Bottenberg, 2006). This pyrolysis oil can be used in a wide variety of applications, including direct co-combustion at power plants and boilers for heat and electricity generation.

Chemical Conversion

There are several methods which are generally acceptable to produce biodiesel. Among those highlighted in this section are transesterification, microemulsion, biocatalyst, and supercritical methanol. While transesterification, bioconversion, and supercritical methanol are reactive processes, microemulsion is a non-reactive process. In general, microemulsion fuels are made by direct blending of conventional diesel fuel (DF) and/or vegetable oil, a simple alcohol, an amphiphilic compound such as a surfactant, and a cetane improver (Demirbaş, 2003). It is designed to tackle the problem of high viscosity vegetable oils by reducing the viscosity of oils with solvents such as simple alcohols. Microemulsions (1–150 nm) are formed spontaneously from two normally immiscible liquids and one or more ionic or non-ionic amphophiles (Scwab et al., 1987). All microemulsions with alcohols such as butanol, hexanol, and octanol must meet the maximum viscosity requirement for No. 2 diesel fuel.

One of the cheapest processes selected for producing biodiesel from vegetable oils or animal fats is the transesterification reaction (Schuchardt et al., 1998). Other processes usually require a high capital investment in producing biodiesel and viscosity of triglycerides to a range close to that of conventional diesel fuel (Knothe & Dunn, 2001). Barnwal & Sharma (2005) reported that transesterification of vegetable triglycerides with alcohol is the preferred technique for producing biodiesel. In general, there are two methods of transesterification. One method uses a catalyst and the other is without a catalyst. The method with a catalyst has a long history of development for biodiesel production. The process usually takes place under a moderate reaction temperature in order to minimize the vaporization of the reactant (alcohol). This process has widely been used to reduce the viscosity of triglycerides to a range close to that of conventional diesel fuel (Knothe & Dunn, 2001). The overall transesterification reaction is given in Equation 22.15. The reverse reaction of methanolysis of fats is the major route for fatty acid methyl ester glycerolysis (Ma & Hanna, 1999).

$$\text{Palm oil Triglycerides} + \text{Methanol} \leftrightarrow \text{Palm Oil Methyl Ester} + \text{Palm Oil Di-glycerides} \qquad (22.12)$$

$$\text{Palm Oil Diglycerides} + \text{Methanol} \leftrightarrow \text{Palm Oil Methyl Ester} + \text{Palm Oil Monoglycerides} \qquad (22.13)$$

$$\text{Palm Oil Monoglycerides} + \text{Methanol} \leftrightarrow \text{Palm oil Methyl Ester} + \text{Glycerol} \qquad (22.14)$$

Overall reaction:

Palm Oil Triglycerides + 3Methanol ↔ 3Palm Oil Methyl Ester + Glycerol (22.15)

Shay (1993) reported that commercially, alkaline-catalyzed reactions are used more often than acid catalysts, as the reaction is faster and less corrosive. Nonetheless, the oils sometimes contain a significant amount of free fatty acids that cannot be converted into biodiesel by the alkaline catalyst; rather, it produces alkaline soaps via saponification. Thus, an alternative method is to use an acid catalyst (Furuta et al., 2004). Acid catalysts require longer reaction time compared to alkaline catalysts, thus it is not economical to use them. Enzyme (lipase) catalyzed production of biodiesel with primary or secondary alcohol has also been reported (Ana et al., 2003). No complex operation is needed (i.e., for the recovery of glycerol and removal of catalyst and salt). However, the reaction yield and reaction time are still unfavorable compared to the alkaline-catalyzed reaction systems (Haas, 2005).

Transesterification without a catalyst can also be achieved using supercritical methanol method. Since the reaction is non-catalytic and extremely fast, it has encouraged numerous investigations to explore the synthesis of biodiesel by transesterification of vegetable oils in supercritical methanol. Supercritical methanol method without using any catalyst is more suitable for the production of biodiesel from high free fatty acid feedstock, as saponification reaction is greatly reduced. The reactivity of rapeseed oil and its fatty acids and other vegetable oils in supercritical alcohol had also been investigated (Demirbas, 2003).

Advanced biofuel includes hydrotreated vegetable oil (HVO) and biomass-to-liquid (BtL) diesel, also referred to as Fischer-Tropsch (FT) diesel (IEA, 2011). HVO is produced by hydrogenating vegetable oils or animal fats (Bacovsky et al., 2010). The first large-scale plants are already in operation in Finland and Singapore, but the process has not yet been fully commercialised (Bacovsky et al., 2010). Fischer-Tropsch diesel is produced by a two-step process in which first the oil is converted to a syngas rich in hydrogen and carbon monoxide, and then the syngas is catalytically converted through FT synthesis into a broad range hydrocarbon liquids, including synthetic diesel and biokerosene. Advanced biodiesels exhibit properties very similar to those of petroleum diesel and kerosene and are blendable with fossil fuels in any proportion and should be fully compatible with the existing engines (IEA, 2011).

Biochemical Conversion

Conversion via biological route refers to the use of a living organism, usually microorganisms, to accomplish certain chemical reactions. Often, the organism converts a substance into its chemically modified form for further use. Possible energy products obtained from oil palm waste through bioconversion can be bioethanol, biomethanol, biogas such as methane, and also biohydrogen. Bioconversion is essential for products which are not feasible to be produced non-biologically and favored as an option for a green and low cost technology.

Oil palm wastes are all of lignocellulosic in nature and rich in sugar. These biomasses can be fermented to produce a second generation of bioethanol and biomethanol. Both fuels can be used in vehicles. Bioethanol production from biomass is one good example of industrial process for renewable energy production. In comparison to other biomasses such as starch, the production of ethanol from cellulose and lignocellulose from oil palm biomass is slightly laborious and very expensive due to its nature.

Unlike other biomasses with different nature, the bioconversion for a lignocellulosic biomass such as EFB and palm press cakefiber (PPF) entails pretreatment, hydrolysis, fermentation, and separation processes (Balat, 2011; Demirbas, 2009a). Fig. 22.3 illustrates a typical sequence of processes associated with biological conversion of lignocellulosic biomass. Pretreatment is vital to alter structural, compositional, and physicochemical barriers for the bioconversion to occur (Kumar et al., 2009). It aims to disrupt cellulose crystalline structure and remove lignin for the ease of the subsequent hydrolysis process (Fig. 22.4). Various techniques such as ammonia fiber

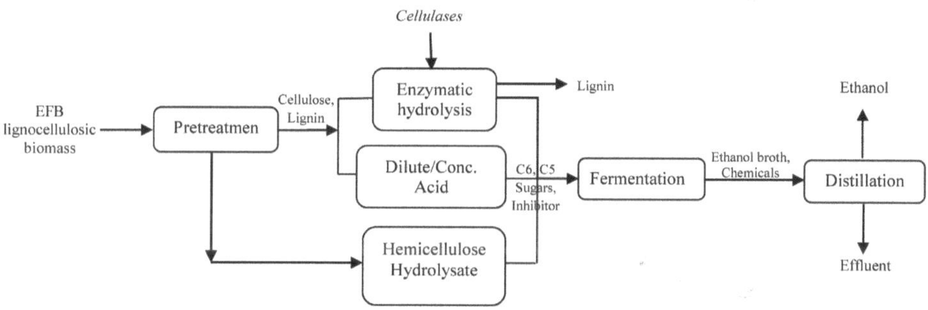

Fig. 22.3. Schematic process route of biochemical degradation of EFB lignocellulosic biomass for bioethanol production.

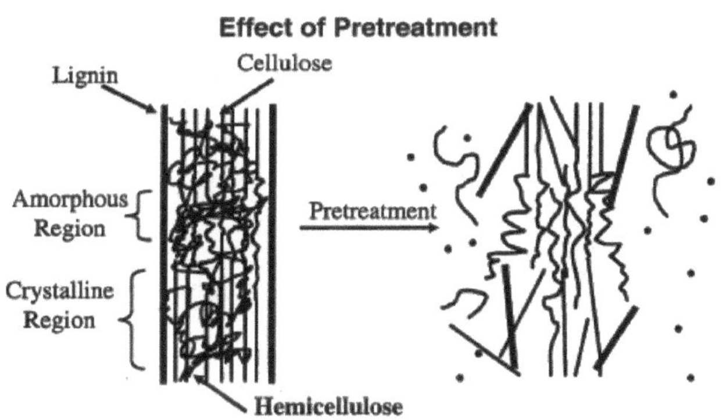

Fig. 22.4. Schematic representation of the effect of pre-treatment process on the lignocellulosic EFB and PPF biomass. Adapted from Mosier et al. (2005) and as Published in Hsu et al. (1980).

explosion, chemical and biological treatment, thermomechanical process, pyrolysis, and autohydrolysis are used for the pretreatment step. An inefficient pretreatment process can result in a low yield of bioethanol, especially when the process is solely based on one method only (Piarpuzan et al., 2011). Recent work (Hamzah et al., 2011) employed a high pressure of 10 MPa and temperature of 121°C combined with alkaline pretreatment process and resulted in a great disintegration of EFB and also removed hemicelluloses, lignin, and silica barriers. The use of ultrasound pretreatment (Yunus et al., 2010) on EFB at 90% amplitude for 45 minutes also had shown remarkable increment of xylose yield. Although this pretreatment step can prove costly in biomass to biofuel conversion (Alvira et al., 2009), improvement on the efficiency of the process through intensive research can help lower the cost (Chandel et al., 2007; Mosier et al., 2005).

Pretreatment of lignocellulosic biomass may degrade products with an inhibitory effect on the fermentation process. These inhibitors have toxic effects on the microorganism, thus, reducing the ethanol yield. The level of toxicity depends on the fermentation conditions such as cell physiology, dissolved oxygen content, and pH of medium. The major types of inhibitors are furfural, phenols, levulinic acid, formic acid, acetic acid, 5-hydromethylfurfural, and aldehydes. Inhibitor formation should be controlled through manipulation of pretreatment process conditions and other measures (Lynd et al., 2002).

Hydrolysis is extremely important in breaking down the cellulose and hemicellulose into their corresponding sugar components (Taherzadeh & Karimi, 2007). Typically, three main processes are employed for hydrolysis by using dilute acid, concentrated acid, or enzymes (Sun & Cheng, 2002). In dilute acid hydrolysis, a high temperature of 160–230°C (Balat, 2011), pressure around 10 atm (Iranmahboob et al., 2002), and acid concentration of 2–5% are used (Broder et al., 1995). Concentrated acid hydrolysis meanwhile is achieved with acid concentration in the range of 10–30% (Balat, 2011; Iranmahboob et al., 2002), at a much lower temperature of <50°C, and at atmospheric pressure. Low yield of ethanol is generally obtained at dilute acid conditions due to shorter retention time; sugar degradation also occurs. This leads to low conversion of cellulose to glucose.

Enzymatic hydrolysis breaks down the bonds in the cellulose and hemicellulose to their sugar components which are C-6 (hexose:galactose and mannose) and C-5 sugars (pentose: xylose and arabinose), respectively. Enzyme hydrolysis is slow in comparison to chemical methods but more favorable due to its selectivity for cellulose conversion to glucose. Different kinds of cellulases such as endoglucanases, exoglucanases, β-glucosidases and cellobiohydrolases may be utilized to cleave the cellulose and hemicellulose bonds (Balat, 2011). The cellulase can be produced by bacteria that belong to *Clostridium, Cellulomonas, Bacillus, Thermomonospora, Ruminococcus, Bacteriodes, Erwinia, Acetovibrio, Microbispora,* and *Streptomyces* species (Sun & Cheng, 2002) and from fungi such as *Sclerotium rolfsii, P. chrysosporium, Trichoderma, Aspergillus, Schizophyllum* and *Penicilium,* in particular from *T. reesei* (Balat, 2011). The endoglucanases attack the cellulose chain randomly to produce polysaccharides of shorter length while exoglucanases attach to the non-reducing ends of these shorter chains and remove cellobiose. β-glucosidase hydrolyzes cellobiose and other oligosaccharides to produce glucose. Several factors that may affect enzymatic hydrolysis are substrates, cellulose activity, reaction conditions, and inhibition. Optimization of the hydrolysis process and enhancement of the cellulase activity can improve the yield and rate of hydrolysis (Sun & Cheng, 2002). Enzymatic hydrolysis of alkaline-treated EFB at a high temperature and pressure showed enhancement of glucose production until pH 4.8 and 50°C by adding both cellulase and β,1-4 glucosidase at a ratio of 5:1 during the hydrolysis process (Hamzah et al., 2011). This suggests that the yield can be improved through enzyme manipulation.

Fermentation also involves the use of microorganisms to ferment sugars in EFB for food while at the same time produces ethanol and byproducts. Fungi, bacteria (*Z. mobilis, E. coli,* and *Klebsiella oxytoca*), and yeast can be used for fermentation of glucose. Commercially, robust yeast like bakers' yeast or *S. cerevisiae* is often used. Firstly the invertase enzyme in the yeast catalyzes the hydrolysis of sucrose and converts it into glucose and fructose. Later, the zymase enzyme present in the yeast as well converts glucose and fructose into ethanol. The yeast can only convert the hexose to ethanol but not the pentose. Only a few microorganism strains from bacteria and yeast can ferment the pentoses, (i.e., *Pichia stipitis, Candida shehatae,* and *C. parapsilosis*) (Lynd et al., 2005). Several thermophilic anaerobic bacteria, for instance *Thermoanaerobacter ethanolicus, Clostridium thermohydrosulfuricum, T. mathranii, T. brockii* and *C. thermosaccharolyticum* (Avci et al., 2006; Cook & Morgan, 1994; Lamed & Zeikus, 1980; Larsen et al., 1997) also have been examined for bioethanol production. However, these thermophilic anaerobic bacteria have low (<2% v/v) tolerance of ethanol (Georgiva et al., 2007).

Normally, the actual yield is expected to be lower than the theoretical value since the microorganism requires part of the substrate for cell growth and maintenance. Fermentation is a function of hydraulic retention time, temperature, solid recycle ratio, sludge age, and mixing (Balat, 2011; Dermirbas, 2009a). It can be performed as a batch, fed-batch, or continuous process. The most suitable process not only depends on the kinetics of the microorganism and the lignocellulosic hydrolysates but also on the associated cost (Balat, 2011). Many approaches have been considered to improve bioethanol yield. Using compatible mixed culture of fungi and yeast (Kabbashi et al., 2007) or adopting strategies in performing hydrolysis step are among ways to increase ethanol yield from EFB conversion (Millati et al., 2011).

Research has been conducted to use genetically modified microorganisms (GMO) for improving the efficiency and selectivity of the fermentation process. This includes fermenting 5-carbon sugars into ethanol with high efficiency and degrading both glucose and xylose to useful chemicals such as lactic acids (Patel et al., 2005; Wyman, 1999). In addition, work on combining hydrolysis and fermentation steps or simultaneous saccharification and fermentation (SSF) has been intensified (Ballesteros et al., 2004; Sun & Cheng, 2002). In SSF, production of both biofuels and chemicals happens in the same reactor and hence reduces cost. Fungal cellulases for SSF are most active at 50–55°C while fermentation microorganisms are effective below 35°C. Normally, the SSF process is carried out at a compromise temperature that favors fermentation over cellulase activity. Much research now is focusing in combining pretreatment and SSF to maximize the productivity in the most cost effective way. Direct microbial conversion, a combination of cellulase production, cellulose hydrolysis, and glucose fermentation in a single step, is another approach to bioethanol production (Balat, 2011). Although, this strategy seems to be attractive in terms of cost, it may suffer from low bioethanol yield, limited growth in hydrolysate, and low tolerance of microorganisms to ethanol (Zaldivar et al., 2001). Despite the technological improvement, many plants are still currently in testing, demonstration, or at the pilot plant stage (Meng Hon, 2010).

Another potential bioenergy from oil palm comes from biogas like methane and biohydrogen. Biogas can be compressed, similar to natural gas, and used to power vehicles too. Anaerobic digestion of biowastes like POME and sludge in the absence of oxygen produces biogas which consists of methane and carbon dioxide and also contains impurities like hydrogen sulfide, moisture, and particulate matter. Anaerobic digestion generally occurs in the temperature range of 10–71°C (Lam & Lee, 2011). The anaerobic reaction of the lignocellulosic waste like EFB involves four stages, namely hydrolysis, acidogenesis, acetogenesis, and methanogensis. Similarly, the hydrolysis steps aims to disintegrate waste to assist the release of cell components and organic matter. The complex organic compounds such as protein, carbohydrates, and lipids are degraded to form smaller molecules of sugars, fatty acids, and amino acids. Thermal, mechanical, and chemical treatments are possible methods for accelerating hydrolysis. Hydrolysis is followed by acidogenesis, acetogenesis, and methanogenesis processes (Demiral & Scherer, 2008), which can be summarized as follows:

1. Acidogenesis Reaction

$$\text{Simple monomers} \xrightarrow{\text{fermentation}} \text{Volatile fatty acids} + H_2 + CO_2 + \text{acetic acids} \quad (22.16)$$

2. Acetogenesis Reaction

$$\text{Volatile fatty acids} \xrightarrow{\text{acid forming bacteria}} H_2 + CO_2 + \text{acetic acid} \quad (22.17)$$

3. Methanogenesis Reaction

$$\text{Hydrogen, acetate} \xrightarrow{\text{acetrophic/hydrogenotrophic}} CH_4 + CO_2 \quad (22.18)$$

Waste stabilization is accomplished when methane gas and carbon dioxide are produced. This biogas can then be harnessed to produce energy and fuel such as electricity, heat, and natural gas.

Typical systems that are used to treat POME consist of conventional lagoon system and open and closed digesting tanks (Yacob et al., 2009). New technologies using fluidized bed reactors (Idris et al., 2003), membrane technology (Ahmad et al., 2009; Wu et al., 2007), up-flow anaerobic sludge blanket (UASB) reactor (Borja et al., 1996), and up-flow anaerobic sludge fixed-film reactor (Zinatizadeh et al., 2007) are also explored to enhance the anaerobic treatment.

Biohydrogen can be produced through various methods including dark fermentation and photofermentation of lignocellulosic biomass (Chong et al., 2009a) and POME (Ismail et al., 2010; Yusoff et al., 2009). In photofermentation, *Rhodobacter sphaeroides* and *Rhodopseudomonas palustris* (Jamil et al., 2009) are used to convert carbohydrates and organic acids in POME to form CO_2 and hydrogen. Factors such as low light conversion and high energy demand in photofermentation limit its practicality. Alternative dark fermentation of POME utilizes acidogenic bacteria such as *Enterobacter, Bacillus, Clostridium* (Chong et al., 2009b), *Thermoanaerobacterium* (O-Thong et al., 2008), and even a mixed culture (Yusoff et al., 2009) to produce higher hydrogen yield compared to photofermentation and thus is favored. Both of these methods still have many limitations to overcome, and to date, the production of biohydrogen is still not ready for commercial scale.

Physical Processes

Various types of physical processes are used to generate energy from oil palm. For instance, bunch sterilization process with a high pressure steam followed by stripping, digestion, and pressing is conducted in the mill to extract palm oil from the MF. The sterilization process is usually conducted at 130°C for 1 hour whereby the pressure built up reaches 3 kg/cm^2 (Corley & Tinker, 2003). Pressing is the most important stage that determines the throughput of a palm oil mill. The pressing and separation technology has shifted from centrifuge used in many earlier mills to hydraulic press that was introduced in the early 1920s. In the late 1950s the screw press was introduced, and the throughput of the screw press unit can reach as high as 20 tons of FFB/h (Corley & Tinker, 2003). The extracted palm oil can be physically fractionated into palm olein and stearin, directly used as biofuel, or chemically transesterified to produce palm oil methyl esters.

Palm kernels that are separated from the MFs are crushed to obtain the kernel and remove shells. The kernel is pressed; the palm kernel oil is collected and refined while the cake is processed for animal feed. The dried shells are then pulverized and mixed with MF, the mixture is compacted using screw extrusion or piston press to form briquettes (Hussain et al., 2002; Nasrin et al., 2008; Shuit et al., 2009). Starch (Hussain et al., 2002) and sawdust (Shuit et al., 2009) are typically used as binders and also to increase the HHV of ground kernels. The temperature and pressure applied during the extrusion or pressing process are approximately 150–250°C and 7MPa, respectively (Nasrin et al., 2008). The products are shaped into logs of 490 mm (length) by 50 mm (diameter). For easy ignition, the outer surfaces of the logs are carbonized. These briquettes can be directly used as household fuel or in co-firing system.

Conclusions

Renewable energy is a precious commodity similar to any other energy. Presently, renewable energy represents 5% of all prime energy use, but by the year 2060, it is strongly predicted that it will reach 70%. In this chapter we discussed the energy potential of various parts of the oil palm resources that include its EFB, PKS, fronds biomass from harvesting, palm oil, and also its effluent. Bioenergy from oil palm biomass can be produced through various methods and can be further improved through intensified research and development in this field. Great utilization of oil palm resources in producing renewable energy indicates its capability in meeting global energy demand. Technology improvement and advancement in areas such as enzyme and microorganism manipulations, process integrations, and operating condition optimization can lead to better product yield and cost reduction. For effective long term commitment to the renewable energy from palm oil and its biomasses, adequate policies, sound regulatory framework, and attractive incentives are vital. A sustainable and careful approach is also a prerequisite towards the development of the bioenergy from oil palm to secure its benefits rather than imposing more harm. Potential risks include increasing the emission of climate change gases instead of controlling them, damaging food security, damaging ecosystems and biodiversity, and exacerbating social conflict.

References

Ab Rahman, Z.; Mamat, R.; Menon, R. Gasification technology using palm oil mill biomass for syngas production and renewable energy. MPOB Technology Transfer Seminar 2004, Bangi, Malaysia, 24 June 2004.

Abdul Khalil, H.P.S.; SitiAlwani, M.; Ridzuan, R.; Kamarudin, H.; Khairul, A. Chemical composition, morphological characteristics, and cell wall structure of Malaysian oil palm fibers. Polymer-Plastics Technol. Eng. 2008, 47, 273–280.

Abdullah, A.Z.; Salamatinia, B.; Mootabadi, H.; Bhatia, S. Current status and policies on biodiesel industry in Malaysia as the world's leading producer of palm oil. Energy Policy 2009, 37, 5440–5448.

Abnisa, F.; Wan Daud, W.M.A; Husin, W.N.W.; Sahu, J.N. Utilization possibilities of palm shell as a source of biomass energy in Malaysia by producing bio-oil in pyrolysis process. Biomass Bioenergy 2011, 35, 1863–1872.

Ahmad, A.L.; Chong, M.F.; Bhatia, S. A comparative study on the membrane based palm oil mill effluent (POME) treatment plant. J. Hazard Material 2009, 171, 166–174.

Ali, Y.; Hanna, M.A. Alternative diesel fuels from vegetable oils. Biores. Technol. 1994, 50, 153–163.

Alvira, P.; Tomás-Pejó, E.; Ballesteros, M.; Negro, M.J. Pretreatment technologies for an efficient bioethanol production process based on enzymatic hydrolysis: A review. Biores. Technol. 2010, 101, 4851–4861.

Ana, V.; Pizarro, L.; Park, E.Y. Lipase-catalyzed production of biodiesel fuel from vegetable oils contained in waste activated bleaching earth. Process Biochem. 2003, 38, 1077–1082.

Ani, F.N.; Islam, M.N. Pyrolytic recycling of agro-industrial solid wastes in Malaysia. J. Inst. Energy 1998, 71, 55–58.

Asian Institute of Technology; Successful conversion of palm oil wastes into useful energy in COGEN 3 Information Sheet FSDP special issue.www.cogen3.net/staff.html, 2004 (Accessed on 21/2/2006).

Avci, A.; Donmez, S. Effect of zinc on ethanol production by two *Thermoanaerobacter* strains. Process Biochem. 2006, 41, 984–989.

Azhari, M.S.; Yunus, R.; Mohd Ghazi, T.I.; Choong, T.S.Y. Alcoholysis of *Jatropha curcas* triglycerides using sodium methoxide as catalyst, Pertanika. J. Sci. Technol. 2011, 9, 3161–3165.

Bacovsky, D.; Dallos, M.; Wörgetter, M. Status of 2nd Generation Biofuels Demonstration Facilities in June 2010, A report to IEA Bioenergy Task 39. http://www.bioenergy2020.eu/files/publications/pdf/2010-bericht-demoplants.pdf (Accessed on 26/05/2011).

Bala, B.K. Studies on biodiesels from transformation of vegetable oils for diesel engines. Energy Educ. Sci. Technol. 2005, 15, 1–43.

Balat, M. Production of bioethanol from lignocellulosic materials via the biochemical pathway: a review. Energy Convers. Manage. 2011, 52, 858–875.

Ballesteros, M.; Oliva, J.M.; Negro, M.J.; Manzanares, P.; Ballesteros, I. Ethanol from lignocellulosic materials by a simultaneous saccharification and fermentation process (SFS) with *Kluyveromyces marxianus.* CECT 10875, Proc. Biochem. 2004, 39, 1843–1848.

Barnwal, B.K.; Sharma, M.P. Prospects of biodiesel production from vegetable oils in India. Renew Sustainable Energy Rev. 2005, 9, 363–378.

Basiron, Y.; Chan, K.W. Oil palm: the agricultural producer of food, fiber and fuel for global economy. Oil Palm Ind. Econ. J. 2006, 8(1), 1–17.

Bauen, A.; Berndes, G; Junginger, M.; Londo, M; Vuille, F. Bioenergy—a sustainable and reliable energy source: A review of status and prospects. IEA Bioenergy, 2009, 92.

Baxter, L.L.; Miles, T.R.; Miles Jr., T.R.; Jenkins, B.M.; Milne, T.; Dayton, D.; Bryers, R.W.; Oden, L.L. The behavior of inorganic material in biomass-fired power boilers: field and laboratory experiences. Fuel Process Technol. 1998, 54, 47–78.

Bernama, P.M. launches "Envo Diesel" biofuel. March 21, 2006. http://www.bernama.com/bernama/v3/news (Accessed on 1st June 2011)

Bi, Y.; Ding, D.; Wang, D. Low-melting-point biodiesel derived from corn oil via urea complexation. Biores. Technol. 2010, 101, 1220–1226.

Bielansky, P.; Reichhold, A.; Schonberger, C. Catalytic cracking of rapeseed oil to high octane and olefins. Chem Eng and Processing: Process Intensification 2010, 49, 873–880.

Binod, P.; Sindhu, R.; Singhania, R.R.; Vikram, S.; Devi, L.; Nagalakshmi, S.; Kurien, N.; Sukumaran, R.K.; Pandey, A. Bioethanol production from rice straw: An overview. Biores. Technol. 2010, 101, 4767–4774.

Boey, P.L.; Maniam, G.P.; Abd Hamid S. Biodiesel production via transesterification of palm olein using waste mud crab (*Scylla serrata*) shell as a heterogeneous catalyst. Biores. Technol. 2009, 100, 6362–6368.

Borja, R.; Banks,C.J.; Sánchez, E. Anaerobic treatment of palm oil mill effluent in a two-stage up-flow anaerobic sludge blanket (UASB) reactor. J. Biotechnol. 1996, 45, 125–135.

Broder, J.D.; Barrier, J.W.; Lee, K.P.; Bulls, M.M. Biofuels system economics. World Res. Rev. 1995, 7, 560–569.

Canakci, M.; Ozsezen, A.N.; Arcaklioglu, E.; Erdil, A. Prediction of performance and exhaust emissions of a diesel engine fueled with biodiesel produced from waste frying palm oil. Expert. Syst. Appl. 2009, 36, 9268–9280.

CDM Executive Board; Clean Development Mechanism simplified project design document for small-scale project activities. http://cdm.unfccc.int/Projects/DB/SGS-UKL1141127479.95/view .html, (Accessed on 09/08/2006).

Chandel, A.K.; Es, C.; Rudravaram, R.; Narasu, M.L.; Rao, L.V.; Ravindra, P. Economics and Environmental impact of bioethanol production technologies: an appraisal. Biotechnol. Mol. Biol. Rev. 2007, 2, 14–32.

Chew, T.L.; Bhatia, B. Catalytic processes towards the production of biofuels in a palm oil and oil palm biomass-based refinery. Biores. Technol. 2008, 99, 7911–7922.

Chiaramonti, D.; Oasmaa, A.; Solantausa, Y. Power generation using fast pyrolysis liquids from biomass. Renew. Sustainable Energy Rev. 2007, 11, 1056–1086.

Chiew, L.Y.; Iwata, T.; Shimada, S. System analysis for effective use of palm oil waste as energy resources. Biomass Bioenergy 2011, doi:10.1016/j.biombioe.2011.03.027

Chin, C.M.; Wahid, M.B.; Weng, C.K. Availability and potential of biomass resources from the Malaysian palm oil industry for generating renewable energy. Oil Palm Bull. 2008, 56, 23–28.

Chong, M.L.; Rahim, R.A.; Shirai, Y.; Hassan, M.A. Biohydrogen production by *Clostridium butyricum* EB6 from palm oil mill effluent. Int. J. Hydrogen Energy 2009b, 34, 764–771.

Chong, M.L.; Sabaratnam,V.; Shirai, Y.; Hassan, M.A. Biohydrogen production from biomass and industrial wastes by dark fermentation. Int. J. Hydrogen Energy 2009a, 34, 3277–3287.

Chua, N.S.; Lee, K.H.; Tong, S.L. Personal communication. KekSeng (M) Bhd, Masai, Johor, Malaysia, 2009.

Cook, G.M.; Morgan, H.W. Hyperbolic growth of *Thermoanaerobacter thermohydrosulfuricus* (*Clostridium thermohydrosulfuricum*) increases ethanol production in pH-controlled batch culture, Appl.Microbiol.Biotechnol. 1994, 41, 84–89.

Corley, R.H.V.; Tinker, P.B. The Oil Palm; 4th ed.; Blackwell Science: Kent, 2003.

Damodaran, S. Biomass fired boilers for power generation design consideration. BioGen. Workshop. Seri Kembangan, Selangor, Malaysia; 25–26 July 2005.

de Diego, L.F.; Garcia-Labiano, F.; Abad, A.; Gayan, P.; Adanez, J. Effect of moisture content on devolatilization times of pine wood particles in a fluidized bed. Energy Fuels 2003, 17, 285–290.

Demirbas, A. Combustion characteristics of different biomass fuels. Prog. Energy Comb. Sci. 2004a, 30, 219–230.

Demirbas, A. Effect of initial moisture content on the yields of oily products from pyrolysis of biomass. J. Anal. Appl. Pyrolysis. 2004b, 71, 803–815.

Demirbas, A. Relationship between initial moisture content and the liquid yield from pyrolysis of sawdust. Energy Sources 2005a, 27, 823–830.

Demirbas, A. Bioethanol from cellulosic materials: a renewable motor fuel from biomass. Energy Sources Part A 2005b, 27, 327–337.

Demirbas, A. Importance of biodiesel as transportation fuel. Energy Policy 2007a, 35, 4661–4670.

Demirbas, A. The influence of temperature on the yields of compounds existing in bio-oils obtained from biomass samples via pyrolysis. Fuel Proc. Technol. 2007b, 88, 591–597.

Demirbas, A. Biofuels—securing the planet's future energy needs; Springer-Verlag: London, 2009a.

Demirbas, A. Recovery of gasoline range fuels from vegetable oil. Energy Sources, Part A 2009b, 31, 420–426.

Demirbas, A. Biodiesel fuels from vegetable oils via catalytic and non-catalytic supercritical alcohol transesterifications and other methods: a survey. Energy Convers. Manag. 2003, 44, 2093–2109.

Demirbas, A. Gaseous products from biomass by pyrolysis and gasification: effects of catalyst on hydrogen yield. Energy Conv. Manag. 2002, 43, 897–909.

Demirbas, A.; Arin, G. An overview of biomass pyrolysis. Energy Sources 2002, 24, 471–482.

Demirel, B.; Scherer, P. The roles of acetotrophic and hydrogenotrophic methanogens during anaerobic conversion of biomass to methane: a review. Rev. Env. Sci. Biotechnol. 2008, 7, 173–190.

Dizge, N.; Keskinler, K.; Tanriseven, A. Biodiesel production from canola oil by using lipase immobilized onto hydrophobic microporous styrene–divinylbenzene copolymer. Biochem. Eng. J. 2009, 44, 220–225.

Edwards, R.; Mulligan, D.; Marelli, L. Indirect land use change from increased biofuels demand—Comparison of models and results for marginal biofuels production from different feedstocks. EUR—Scientific and Technical Research Series; Publication Office of the European Union: Luxemberg, 2010.

Ee, L.M. Effluent treatment system in palm oil mills: Part II—Biogas from anaerobic digestion. Palm Oil Eng. Bull. 2009, 93, 23–26.

Erlich, C.; Fransson, T.H. Downdraft gasification of pellets made of wood, palm-oil residues respective bagasse: Experimental study. Appl. Energy 2011, 88, 899–908.

Evald, A. Renewable energy resources. Integrated Resource Planning Report, 2005, 1–39.

Furuta, S.; Matsuhashi, H.; Arata, K. Biodiesel fuel production with solid superacid catalysis in fixed bed reactor under atmospheric pressure. Cataly. Commun. 2004, 5, 721–723.

Georgieva, T.I.; Skiadas, I.V.; Ahring, B.K. Effect of temperature on ethanol tolerance of a thermophilic anaerobic ethanol producer *Thermoanaerobacter* A10: modeling and simulation. Biotechnol. Bioeng. 2007, 98, 1161–1170.

Georgogianni, K.G.; Katsoulidis, A.P.; Pomonis, P.J.; Kontominas, M.G.Transesterification of soybean frying oil to biodiesel using heterogeneous catalysts. Fuel Process. Technol. 2009, 90, 671–676.

Goh, S.H.; Lee, K.T.; Bhatia, S. Hot compressed water pretreatment of oil palm fronds to enhance glucose recovery from production of second generation bioethanol. Biores. Technol. 2010, 101, 7362–7867.

Haas, M.J. Improving the economics of biodiesel production through the use of low value lipids as feedstocks: Vegetable oil soapstock. Fuel Process. Technol. 2005, 86, 1087–1096.

Halsall, L.I. Towards a sustainable energy economy—The role of palm. J. Oil Palm Env. 2011, 2, 8–14.

Hambali, E; Thahar, A.; Komarudin, A. The potential of oil palm and rice biomass as bioenergy feedstock. 7th Biomass Asia Workshop. Jakarta, Indonesia; 29 November–1 December, 2010.

Hamzah, A.; Idris, A.; Shuan, T.K. Preliminary study on enzymatic hydrolysis of treated oil palm (*Elaeis*) empty fruit bunches fibre (EFB) by using combination of cellulase and β-1-4 glucosidase. Biomass. Bioenerg. 2011, 35, 1055–1059.

Harimi, M.; Megat Ahmad, M.M.H.; Sapuan, S.M.; Idris, A. Numerical analysis of emission component from incineration of palm oil wastes. Biomass. Bioenergy 2005, 28, 339–345.

Heipieper, H.J.; Weber, F.J.; Sikkema, J.; Kewelo, H.; de Bont, J.A.M. Mechanism of resistance of while cells to toxic organic solvents. TIBTECH 1994, 12, 409–415.

Hsu, T.A.; Ladisch, M.R.; Tsao, G.T. Alcohol from cellulose. Chem. Technol. 1980, 10(5), 315–319.

Husain, Z.; Zainac, Z.; Abdullah, Z. Biquetting of palm fiber and shell from the processing of palm nuts to palm oil. Biomass Bioenergy 2002, 22, 505–509.

Husin, M.; Ridzuan, R.; Anis, M.; Wan Hasamudin, W.H.; Kamaruddin, H.; et al. Research and development of oil palm biomass in wood based industries. Malaysian Palm Oil Board Bull. 2005.

Idris, A.B.; Noor, M.J.M.M.; Al-Mamun, A. Modelling of anaerobic fluidized bed bioreactor in the treatment of palm oil mill effluent. Asian J. Microbiol. Biotechnol. Environ. Exp. Sci. 2003, 5, 137–144.

Idris, S.S.; Norazah, A.R.; Ismail, K.; Azil, B.A.; Abd Rashid, Z.; Aris, M.J. Investigation on thermochemical behavior of low rank Malaysian coal, oil palm biomass and their blends during pyrolysis via thermogravimetric analysis (TGA). Biores. Technol. 2010, 101, 4584–4592.

Ino, T.; Tanigucgi, K.; Ohmura, Y.; Aoki, T.; Takemoto, M.; Muraoka, T.; et al. Fluidized bed combustion test results of empty fruits bunches. 2003 MPOB International Palm Oil Congress (Chemistry and Technology). Putrajaya, Malaysia, 24–28 August 2003, 43–51.

Inoue, S. Hydrothermal carbonization of empty fruit bunches. J. Chem. Eng. Japan 2010, 11, 972–976.

International Energy Agency (IEA); Technology Roadmap: Biofuels for Transport, OECD/IEA, Paris, www.iea.org/papers/2011/biofuels_roadmap.pdf (Accessed on 26/05/2011).

Iranmahboob, J.; Nadim, F.; Monemi, S. Optimizing acid hydrolysis: A critical step for production of ethanol from mixed wood chips. Biomass Bioenergy 2002, 22, 401–404.

Islam, M.N.; Zailani, R.; Ani, F.N. Pyrolytic oil from fluidized bed pyrolysis of oil palm shell and its characterization. Renew. Energy 1999, 17, 73–84.

Ismail, I.; Hassan M.A.; Abdul Rahman, N.A.; Soon C.S. Thermophilicbiohydrogen production from palm oil mill effluent (POME) using suspended mixed culture. Biomass Bioenergy 2010, 34, 42–47.

Ismail, K.; Yarmo, M.A.; Taufiq-Yap, Y.H.; Ahmad, A. Calcium oxide as potential catalyst for gasification of palm oil empty fruit bunch to produce syngas. Adv. Mat. Res. 2011, 173, 178–183.

Jain, S.; Sharma, M.P. Prospects of biodiesel from Jatropha in India: A review. J. Renew. Sust. Energy Rev. 2009, 14, 763–771.

Jamil, Z.; MohamadAnnuar, M.S.; Ibrahim, S.; Vikineswary, S. Optimization of phototrophic hydrogen production by *Rhodopseudomonas palustris* PBUM001 via statistical experimental design. Int. J. Hydrogen Energy 2009, 34, 7502–7512.

Johnston, M.; Holloway, T.A. Global comparison of national biodiesel production potentials. Env. Sci. Technol. 2007, 41, 7967–7973.

Kabbashi, A. N.; Alam, M.Z; Faharurrazi Tompang, M. Direct bioconversion of oil palm empty fruit bunches for bioethanol production by solid state bioconversion. IIUM Eng. J. 2007, 8(2), 37–45.

Kansedo J.; Lee K.T.; Bhatia S. Biodiesel production from palm oil via heterogeneous transesterificatio. Biomass Bioenergy 2009, 2, 271–276.

Kataki, R.; Konwer, D. Fuelwood characteristics of indigenous tree species of north-east India. Biomass Bioenergy 2002, 22, 433–437.

Kelly-Yong, T.L.; Lee, K.T.; Abdul Rahman, M.; Bhatia, S. Potential of hydrogen from oil palm biomass as a source of renewable energy worldwide. Energy Policy 2007, 35, 5692–5701.

Khor, K.H.; Lim, K.O.; ZainalAlimuddin, Z.A. Laboratory-scale pyrolysis of oil palm trunk. Energy Sources, Part A2010, 32, 518–531.

Khor, K.H.; Lim, K.O. Carbonisation of oil palm frond. Int. Energy J. 2006, 7, 239–243.

Klass, D. Biomass for renewable energy, fuel and chemicals. Academic Press: San Diego, 1998.

Knothe, G.; Dunn, R.O. Biofuels derived from vegetable oils and fats. In Oleochemical Manufacture and Applications; Gunstone, F.D., Hamilton, R.J., Eds.; Acad. Press Publishers England: Sheffield, 2001; pp. 106–163.

Koba, Y.; Ishizaaki, A. Chemical composition of palm oil fiber and its feasibility as cellulose raw material for sugar production. Agric. Biol. Chem. 1990, 54, 1183–1187.

Kosugi, A.; Tanaka, R.; Sulaiman, O.; Hashim, R.; Abdul Hamid, Z.A.; MohdYusof, M.N.; Ibrahim, W.A.; Takamitsu, A.; Murata, Y.; Nirasawa, S.; et al. Bioethanol and bioplastic production using sap from old oil palm trunks: A promising source of sugars for biomass refinery. 7th Biomass Asia Workshop. Jakarta, Indonesia; 29 November–1 December 2010.

Kumar, P.; Barret, D.M.; Delwiche, M.J.; Stroeve, P. Methods for pretreatment of lignocellulosic biomass for efficient hydrolysis and biofuel production, Ind. Eng. Chem. Res. 2009, 48, 3713–3729.

Lam, M.K.; Lee, K.T. Renewable and sustainable bioenergies production from palm oil mill effluent (POME): win-win strategies toward better environmental protection, Biotechnol. Adv. 2011, 29, 124–141.

Lamed, R.; Zeikus, J.G. Glucose fermentation pathway of *Thermoanaerobium brockii*. J. Bacteriol. 1980, 14, 1251–1257.

Lamers P.; Hamelinck C.; Martin Junginger M.; Faaij A. International bioenergy trade—Are view of past developments in the liquid biofuel market. Renew. Sustainable Energy Revs. 2011, 15, 2655–2676.

Larsen, L.; Nielsen, P.; Ahring, B.K. *Thermoanaerobacter mathranii* sp. Nov., an ethanol-producing, extremely thermophilic anaerobic bacterium from a hot spring in Iceland. Arch. Microbiol. 1997, 168, 114–119.

Li, J.; Yin, Y.; Zhang, Z.; Liu, J.; Yan, R. Hydrogen-rich gas production by steam gasification of palm oil wastes over supported tri-metallic catalyst. Int. J. Hydrogen Energy 2009, 9108–9115.

Lim, K.O.; Lim, K.S.; Carbonization of oil palm trunks at moderate temperatures. Biores. Technol. 1992, 40, 215–219.

Liu, R.; Shen, F. Impacts of main factors on bioethanol fermentation from stalk juice of sweet sorghum by immobilized *Saccharomyces cerevisiae* (CICC 1308). Biores. Technol. 2008, 99, 847–854.

Liu, Y.; Wang, L. Biodiesel production from rapeseed deodorizer distillate in a packed column reactor. Chem. Eng. Process: Process Intensive 2009, 48, 1152–1156.

Love G.; Cuevas-Cubria C. Outlook for biofuels in Australia: the challenges ahead. Australian Commodities, Australian Bureau of Agricultural and Resource Economics. www.abareconomics.com/interactive/ac_mar07/htm/a5.htm (Assessed on 30/5/2011).

Lu H.; Liu Y.; Zhou H.; Yang Y.; Chen M.; Liang B. Production of biodiesel from *Jatropha curcas* L. Oil. Comput. Chem. Eng. 2009, 33, 1091–1096.

Luangkiattikhun, P.; Tangsathitkulchai, C.; Tangsathitkulchai, M. Non-isothermal thermogravimetric analysis of oil-palm solid wastes. Biores. Technol. 2008, 99, 986–997.

Lynd, L.R.; Weimer, P.J.; Van Zyl, W.H.; Pretorius, I.S. Microbial cellulose utilization: fundamentals and biotechnology. Microb. Molec. Biol. Rev. 2002, 66, 506–577.

Lynd, L.R.; Van Zyl, W.H.; McBride, J.E.; Laser, M. Consolidated bioprocessing of cellulosic biomass: an update. Curr. Opin. Biotechnol. 2005, 16, 577–583.

Ma, A.N.; Choo, Y.M.; Cheah, K.Y. Development of renewable energy in Malaysia. ICS-Unido International Conference on Exploitation of Renewable Resources and Renewable Energy, Trieste, Italy, 10–12 June 2004.

Ma, A.N.; Ong, A.S.H. Pollution control in palm oil mills in Malaysia. JAOCS 1985, 62, 261–266.

Ma F.; Hanna M.A. Biodiesel production: A review. Biores. Technol. 1999, 70, 1–15.

Maher, K.D.; Bressler, D.C. Pyrolysis of triglyceride materials for the production of renewable fuels and chemicals. Biores. Technol. 2007, 98, 2351–2368.

Malaysian Palm Oil Board (MPOB); Statistics 2008. http://econ.mpob.gov.my/economy/annual/stat2008/World6.2.pdf (Accessed on 27/05/2011a).

Malaysian Palm Oil Board (MPOB); Directory of the Malaysian palm oil industry—biodiesel. http://econ.mpob.gov.my/economy/biodiesel/Directory%20of%20Biodiesel%20Producers%209.2.11.pdf (Accessed on 27/05/2011b).

Malaysian Palm Oil Board (MPOB); Summary of the Malaysian oil palm industry, 2009. http://econ.mpob.gov.my/stat/web_report1.php?val=200984 (Accessed on 27/05/2011c).

Malaysian Palm Oil Board (MPOB); Datasheet. Palm Oil Engineering Bulletin, 2010, 94, 53.

Malaysian Palm Oil Council. Palm oil industry—a learning experience, Poster A. http://www.mpoc.org.my/ (Accessed 01/11/2006).

May, Y.C. Overview of the Malaysian Oil Palm Industry, 2010. http://econ.mpob.gov.my/economy/Overview%202010_final.pdf (Accessed on 27/05/ 2011).

McKendry, P. Energy production from biomass (part 1): overview of biomass. Biores. Technol. 2002, 83, 37–46.

Meng Hon, J.L. A case study of palm empty fruit bunch as energy feedstock. Segi. Rev. 2010, 3, 3–15.

Millati, R.; Wikandari, R.; Trihandayani, E.T.; Cahyanto, M.N.; Taherzadeh, M.J.; Niklasson, C. Ethanol from oil palm empty fruit bunch via dilute-acid hydrolysis and fermentation by *Mucor indicus* and *Saccharomyces cerevisiae.* Agric. J. 2011, 6, 54–59.

Misson, M.; Haron, R.; Ahmad Kamaroddin, M.F.; Saidina Amin, N.A. Pretreatment of empty fruit bunch for the production of chemicals via catalytic pyrolysis. Biores. Technol. 2009, 100, 2867–2873.

Mohammed, M.A.A.; Salmiaton, A.; Wan Azlina, W.A.K.G.; Mohammed Amran, M.S.; Fakhru'l-Razi, A. Air gasification of empty fruit bunch for hydrogen-rich gas production in a fluidized-bed reactor. Energy Conv. Manag. 2011, 52, 1555–1561.

Mohan, D.; Pittman, C.U.; Steele, P.H. Pyrolysis of wood/biomass for bio-oil: a critical review. Energy Fuels 2006, 20, 848–889.

Mosier, N.; Wyman, C.; Dale, B.; Elander, R.; Lee, Y.Y.; et al. Features of promising technologies for pretreatment of lignocellulosic biomass. Biores. Technol. 2005, 96, 673–686.

Nasrin, A.B.; Ma, A.N.; Choo, Y.M.; Mohamad, S.; Rohaya, M.H.; Azali, A.; Zainal, Z. Oil palm biomass as potential substitution raw materials for commercial biomass briquettes production. Am. J. Appl. Sci. 2008, 5, 179–183.

Nellens, P. Production push: Oil palm rises to challenge. Global Oils Fats Bus. Mag. 2009, 6(4), 10–14.

Ng, F.Y.; Yew, F.K.; Basiron, Y.; Sundram, K. A renewable future driven with Malaysian palm oil-based green technology. J. Oil Palm Env. 2011, 2, 1–7.

Omar, R. Evaluation of microwave pyrolysis of oil palm empty fruit bunches. PhD Thesis, Universiti Putra Malaysia, 2010.

Omar, R.; Idris, A.; Yunus, R.; Khalid, K.; Aida Isma, M.I. Characterization of empty fruit bunch for microwave-assisted pyrolysis. Fuel 2011, 90, 1536–1544.

Ong, H.C.; Mahlia, T.M.I.; Masjuki, H.H. A review on energy scenario and sustainable energy in Malaysia. Renewable and Sustainable Energy Reviews 2011, 15, 639–647.

Ong, Y.K.; Bhatia, S. The current status and perspectives of biofuel production via catalytic cracking of edible and non-edible oil. Energy 2010, 35, 111–130.

O-Thong, S.; Prasertsan, P.; Intrasungkha, N.; Dhamwichukorn, S.; Birkeland, N.K. Optimization of simultaneous thermophilic fermentative hydrogen production and COD reduction from palm oil mill effluent by *Thermoanaerobacterium*-rich sludge. Int. J. Hydrogen Energy 2008, 33, 1221–1231.

Palmqvist, E.; Hahn-Hagerdal, B.; Szengyel, Z.; Zacchi, G.; Reczey, K. Simultaneous detoxification and enzyme production of hemicelluloses hydrolysates obtained after steam pretreatment. Enzyme Microbial. Technol. 1997, 20, 286–293.

Patel, M.A.; Ou, M.S.; Ingram, L.O.; Shanmugam, K.T. Simultaneous saccharification and co-fermentation of crystalline cellulose and sugar cane bagasse hemicellulose hyrolysate to lactate by a thermotolerant acidophilic *Bacillus* sp. Biotechnol. Prog. 2005, 21, 1453–1460.

Piarpuzan, D.; Quintero, J.A.; Cardona, C.A. Empty fruit bunches from oil palm as a potential raw material for fuel ethanol production. Biomass Bioenergy 2011, 35, 1130–1137.

Pio Lopez, G.; Laan, T. Biofuels, At What Cost? Government support for biodiesel in Malaysia, 2008, International Institute for Sustainable Development (IISD), www.iisd.org/pdf/2008/biofuels_subsidies_malaysia.pdf. (Accessed on 26/05/2011).

Qi, D.H.; Geng, L.; Chen, H.; Bianh, Y.Z.H.; Liu, J.; Ren, C.H. Combustion and performance evaluation of a diesel engine fueled with biodiesel produced from soya bean crude oil. Renew. Energy 2009, 34, 2706–2713.

Rashid U.; Anwar F. Production of biodiesel through optimized alkaline catalyzed transesterification of rapeseed oil. Fuel 2008, 87, 265–273.

Razuan, R.; Chen, Q.; Zhang, X.; Sharifi, V.; Swithenbank, J. Pyrolysis and combustion of oil palm stone and palm kernel cake in fixed-bed reactors. Biores. Technol. 2010, 101, 4622–4629.

Sadrolhosseini, A.R.; Moksin, M.M; , Harrison Lau, L.K.; Norozi, M.; Mat Yunus, W.M.; Zakaria, A. Physical properties of normal grade biodiesel and winter grade biodiesel. Int. J. Mol. Sci. 2011, 11, 2100–2111.

Salema, A.A.; Ani, F.N. Microwave induced pyrolysis of oil palm biomass. Biores. Technol. 2011, 102, 3388–3395.

Schuchardt, U.; Sercheli, R.; Vargas, R.M. Transesterification of vegetable oils: a review. J. Brazilian Chem. Soc. 1998, 9, 199–210.

Schwab, A.W.; Bagby, M.O.; Freedman, B. Preparation and properties of diesel fuels from vegetable oils. Fuel 1987, 66, 1372–1378.

Sharma, Y.C.; Singh, B. Development of biodiesel: current scenario. Renew. Sustainable Energy Rev. 2009, 13, 1646–1651.

Shay, E.G. Diesel fuel from vegetable oils: status and opportunities. Biomass Bioenergy, 1993, 4, 227–242.

Shi, H.; Bao, Z. Direct preparation of biodiesel from rapeseed oil leached by two-phase solvent extraction. Biores. Technol. 2008, 99, 9025–9028.

Shuit, S.H.; Tan, K.T.; Lee, K.T.; Kamaruddin, A.H. Oil palm biomass as a sustainable energy source: A Malaysian case study. Energy 2009, 34, 1225–1235.

Sridhar, M.K.C.; AdeOluwa, O.O. Palm oil industry residues. In Biotechnology for Agro-Industrial Residues Utilisation. Singh-nee' Nigam, P., Pandey, A., Eds.; Springer: New York, 2009; pp. 341–355.

Sulaiman, F.; Abdullah, N. Optimum conditions for maximizing pyrolysis liquids of oil palm empty fruit bunches. Energy 2011, doi: 10.1016/j.energy.2010.12.067.

Sumathi, S.; Chai, S.P.; Mohamed, A.R. Utilization of oil palm as a source of renewable energy in Malaysia. Renew. Sustainable Energy Rev. 2008, 12, 2404–2421.

Sun, Y.; Cheng, J.Y. Hydrolysis of lignocellulosic materials for ethanol production: as review. Biores. Technol. 2002, 83, 1–11.

Taherzadeh, M.J.; Karimi, K. Acid-based hydrolysis processes for ethanol from lignocellulosic materials: a review. Biores. 2007, 2, 472–499.

Tan, H.T.; Lee, K.T., Mohamed, A.R. Second-generation bioethanol (SGB) from Malaysian palm empty fruit bunch: Energy and exergy analyses. Biores. Technol. 2010, 101, 5719–5727.

terHarmsel, J.; Bottenberg, H. Small scale commercial bio-oil plant commissioned in South-east Asia. Technology Horizons, Winter 2006.

Thanh, L.T.; Okitsu, K.; Sadanaga, Y.; Takenaka, N.; Maeda, Y.; Bandow, H.; Ultrasound-assisted production of biodiesel fuel from vegetable oils in a small scale circulation process. Biores. Technol. 2010, 101, 639–645.

Tsai, W.T.; Mi, H.H.; Chang, Y.M.; Yang, S.Y.; Chang, J.H. Polycyclic aromatic hydrocarbons (PAHs) in bio-crudes from induction-heating pyrolysis of biomass wastes. Biores. Technol. 2007, 98, 1133–1137.

Twaiq, F.A.; Mohamed, A.R.; Bhatia, S. Liquid hydrocarbon fuels from palm oil by catalytic cracking over aluminosilicate mesoporous catalysts with various Si/Al ratios. Microporous Mesoporous Materials 2004, 64, 95–107.

Uemura, Y.; Omar, W.N.; Tsutsui, T.; Yusup, S. Torrefaction of oil palm wastes. Fuel 2011, Doi:10.1016/j.fuel.2011.03.021.

USDOE (US Department of Energy); Alternative Fuels & Advanced Vehicles Data Center. Energy Efficiency and Renewable Energy, www.afdc.energy.gov/afdc (Accessed on 26/05/2011).

Van Gerpen, J.; Shanks, B.; Pruszko, R.; Clements, D.; Knothe, G. Biodiesel Analytical Methods, August 2002–January 2004; Technical Report for U.S. Department of Energy Office of Energy Efficiency and Renewable Energy by Midwest Research Institute, NREL/SR-510-3624; National Renewable Energy Laboratory: Golden, CO, USA, 2004, 31–33.

Venderbosch, R.H.; Gansekoele, E.; Florijn, J.F.; Assink, D.; Ng, H.Y. Pyrolysis of palm oil residue in Malaysia. Thermal. Net. News. 2006, 1, 2–3.

Vyas, A.; Subrahmanyam, N.; Patel, P.A. Production of biodiesel through transesterification of Jatropha oil usin KNO3/Al2O3 solid catalyst. Fuel 2009, 88, 625–628.

Wan Azlina, W.A.K.G. Co-combustion of biomass fuels with coal in a fluidised bed combustor. PhD Thesis, University of Sheffield, Sheffield, United Kingdom, 2005.

Wan Azlina, W.A.K.G.; Norhamizah, A.G.; AzilBahari, A.; MohdSalleh M.A.; Omar, R.; Ali, S.; Idris, A. Characterization of agricultural waste as biofuels: Significant and impact. Symposium of Malaysian Chemical Engineers, Serdang, Malaysia; 12–14 December 2007.

Wang, X.; Liu, X.; Zhao, C; Ding, Y.; Xu, P. Biodiesel production in packed-bed reactors using lipase–nanoparticle biocomposite. Biores. Technol. 2011, 102, 6352–6355.

West A.H.; Posarac, D.; Ellis, N. Assessment of four biodiesel production processes using Hysys. Biores. Technol. 2008, 99, 6587–6601.

Wu, T.Y.; Mohammad, A.W.; Md. Jahim, J.; Anuar, N. Palm oil mill effluent (POME) treatment and bioresources recovery using ultrafiltration membrane: effect of pressure on membrane fouling. Biochem. Eng. J. 2007, 35, 309–317.

Wu, X.; McLaren, J. Madl, R.; Wang, D. Biofuels from lignocellulosic biomass. In Sustainable Biotechnology. Singh, O.V., Harvey, S.P., Eds.; Springer: Heidelberg, 2010; pp. 19–41.

Wyman, C.E. Biomass ethanol: technical progress, opportunities, and commercial challenges. Ann. Rev. Energy Env. 1999, 24, 189–226.

Yacob, S.; Shirai, Y.; Hassan, M.A. Treatment of palm oil wastewaters. In Waste Treatment in the Food Processing Industry. Wang, L.K., Hung, Y-T., Lo, H.H., Yapijakis, C., Eds.; CRC Press: Boca Raton, 2009; pp. 101–118.

Yamada, H.; Tanaka, R.; Sulaiman, O.; Hashim, R.; Hamid, Z.A.A.; Yahya, M.K.A., Kosugi, A.; Arai, T.; Murata, Y.; Nirasawa, S.; et al. Old oil palm trunk: A promising source of sugars for bioethanol production. Biomass Bioenergy, 2010, 34, 1608–1613.

Yang, H.; Yan, R.; Chen, H.; Lee, D.H.; Liang, D.T.; Zheng, C. Pyrolysis of palm oil wastes for enhanced production of hydrogen rich gases. Fuel Proc. Technol. 2006, 87, 935–942.

Yunus, R.; Salleh, S.F.; Abdullah, N.; Awang Biak, D.R. Effect of ultrasonic pre-treatment on low temperature acid hydrolysis of oil palm empty fruit bunch. Biores. Technol. 2010, 101, 9792–9796.

Yusoff, M.Z.M.; Hassan, M.A.; Abd-Aziz, S.; Rahman, N.A.A. Start-up of biohydrogen production from palm oil mill effluent under non-sterile condition in 50 L continuous stirred tank reactor. Int. J. Agric. Res. 2009, 4, 163–168.

Yusoff, S. Renewable energy from palm oil–innovation on effective utilization of waste. J. Cleaner Prod. 2006, 14, 87–93.

Yusuf, N.N.A.N.; Kamarudin, S.K.; Yaakub, Z. Overview on the current trends in biodiesel production. Energy Convers. Manag. 2011, 52, 2741–2745.

Zaldivar, J.; Nielsen, J.; Olsson, L. Fuel ethanol production from lignocelluloses: a challenge for metabolic engineering and process integration. Appl. Microbiol. Biotechnol. 2001, 56, 17–34.

Zinatizadeh, A.A.L.; Mohamed, A.R.; Mashitah, M.D.; Abdullah, A.Z.; Isa, M.H. Optimization of pre-treated palm oil mill effluent digestion in an up-flow anaerobic sludge fixed film bioreactor: a comparative study. Biochem. Eng. J. 2007, 35, 226–237.

23

Waste and Environmental Management in the Malaysian Palm Oil Industry

Mohd Ali Hassan and Suraini Abd-Aziz
Department of Bioprocess Technology, Faculty of Biotechnology and Biomolecular Sciences, Universiti Putra Malaysia, 43400 Serdang, Selangor, Malaysia

Introduction

In pursuing its objective to attain a developed country status, Malaysia has embarked on rapid industrialization programs. However, with industrialization comes the problem of pollution. In this regard, the effluent discharges from palm oil industry contributed to the pollution problem (DOE, 2011).

The palm oil industry plays an important role in the economic development of Malaysia in enhancing the economic welfare of the population. Despite obvious benefits of this industrial development, it also contributes to environmental pollution, both at the input and the output sides of its activities. On the input side, palm oil mills use large quantities of water and energy in the production process. On the output side, manufacturing processes generate large quantities of wastewater, solid waste/by-products, and air pollution. Waste and waste management is one of the palm oil sustainability indicators besides capital, soil conservation practices, nutrients, pest management, biodiversity, energy, product value, and finally contribution to society.

In order to ensure that the industrialization process does not progress at the expense of the environment, Malaysia has put in place a fee structure to control the level of the palm oil industries effluent discharge and to enhance the development of environmentally sound technology in the disposal of the effluents discharge of the industry. The Environmental Quality Act, 1974, is the principal regulation in Malaysia to prevent, abate, and control pollution. Malaysia is aware from the onset that the command and control mechanism alone is insufficient to address pollution problems in the country. Hence, the Act incorporates provisions containing punitive as well as economic measures for the control of effluents discharge into the environment (DOE, 2011).

The imposition of the effluent fee has, in general, reduced the discharge of effluents from the palm oil industry into the water system due to factors as follows:

a. Progressive phasing out of standards has allowed the industries to adapt to the more stringent standards

b. The reduction of effluent fees over time due to the more stringent standards has allowed the industries to invest into more technologically improved treatment facility
c. Despite the increase in the number of palm oil mills, the level of pollution in the river system remain low
d. The tremendous reduction in the amount of fees collected indicates compliance on the part of the industry
e. The manner in which the effluent fee was structured serves as an incentive for the industry to improve their technological capability to reduce pollution arising from their production activities.

Waste Generation by Palm Oil Industry

The palm oil industry wastes were generated, first, at the oil palm plantations during pruning (fronds), harvesting (fronds), and replanting (trunks). It is generated, second, at the palm oil mills, which includes biogas and biomass (mesocarp fiber, shell, empty fruit bunch, and decanter cake). Fig. 23.1 summarizes the wastes generation of the palm oil industry in general.

Solid Waste

The oil palm plantations generate huge amounts of waste such as trunks, fronds, empty fruit bunches, shells, and fibers. These wastes comprise biomass in the form of lignocelluloses, which have potential for generating energy. According to Goh et al. (2010), the total area of oil palm cultivation in the year 2007 in Malaysia was 4,304,914 hectares.

Oil palm frond is one of the most abundant agricultural by-products in Malaysia. Almost all pruned fronds are discarded in the plantation, mainly for nutrient recycling and soil conservation. Oil palm frond has great potential for use as a roughage source or as a component in compound feed for ruminants.

Oil palm empty fruit bunches, besides being rich in plant nutrients, also improve soil physical and chemical properties in the following manner when used as mulch:

a. Increases soil organic matter content
b. Improves soil structure
c. Increase infiltration and aeration
d. Reduces run-off
e. Improves soil water retention
f. Increases soil fauna micro activity
g. Increases cation exchange capacity
h. Reduces soil temperature fluctuation

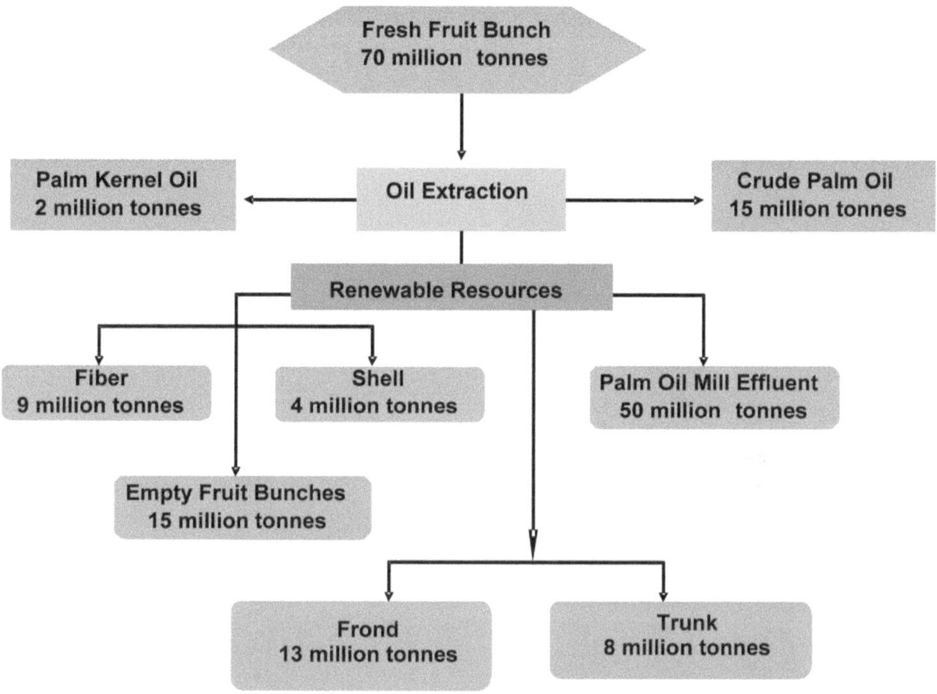

Fig. 23.1. Overall palm oil wastes generation (based on approximate material balance on wet basis).

Table 23-A shows the quantities of oil palm residues produced in Malaysia in 2007. Other by-products such as palm kernel cake and decanter cake are also produced in the palm oil extraction. In 2005, the palm kernel cake generated from the extraction palm oil process was 2.11 million tonnes (Sumathi et al., 2008).

The evacuation of empty fruit bunches from the palm oil mill to the field is by tractor-trailer where the empty fruit bunches is side-tipped to the "station" between the palms along the harvester's paths. In estates where rail transportation is available, the loco/wagons/tractor system is the most suitable for transporting and spreading of empty fruit bunches. Empty fruit bunches mulching is carried out once a year. The empty fruit bunches may be incinerated to produce potash which is applied in the plantation as fertilizer by mulching. The fiber and shell materials are used as boiler fuel. The palm kernel is usually sold to palm kernel oil producers for the extraction of the palm kernel oil (Thani et al., 1999).

Oil palm trunks are another component of the oil palm tree that has great use after they have been replaced during the process of replanting. Previously, the trunks were burnt before replanting, and this has caused severe environmental problems. With the

Table 23-A. Types of Biomass and Quantity Produced (Goh et al., 2010).

Types		Quantity (thousand tonnes)	Moisture Content (wt%)	Dry Weight (thousand tonnes)
Oil palm fronds		46,837	60.0	18,735
Oil palm empty fruit bunches		18,022	65.0	6,308
Oil palm fibers		11,059	42.0	6,414
Oil palm shells	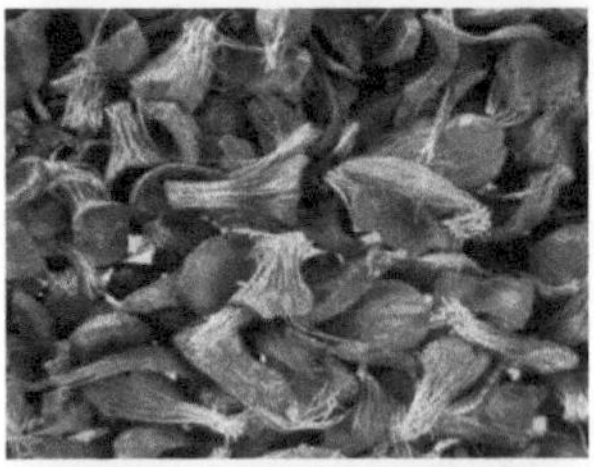	4,506	7.0	4,190
Oil palm trunks		10,827	75.9	2,609

zero burning policies being strictly implemented by palm oil plantation companies in Malaysia, the trunks have to be eliminated using a different approach. The trunks are felled first and then chipped in sizes not more than 10 cm and then transported to trenches which are dug at the fourth palm row. If the trunks are shredded, they are normally stacked at the second palm row in the field. In the case of contour planting, the shredded palms are spread in the inter rows and left to rot. Trunks of mature palms that have been chipped or shredded can be recycled more effectively. They contain up to 1,000 kilograms per hectare of potassium. Exposure to the sun and rain gives a slow breakdown of the material which subsequently turn into natural fertilizer. Chipping or shredding releases all the nutrients within 2–3 years and are fully absorbed by the plants and palms planted in the rows and inter rows. The trunks of palm trees can also be used as lumber or made into fiberboards. There are companies that have ventured into manufacturing veneer and medium density fiberboards (MDF) using the oil palm trunks (Deforestation Watch, 2011).

Palm kernel shell can be considered a pellet form because of its natural form, low moisture content, low ash, and low sulfur content. Palm kernel shell has been used as a burner for power plants due to its high calorific value. One of the most potential uses of palm kernel shell is to process it to become charcoal or biochar.

Oil palm fibers consist of several solid wastes produced by palm oil mill. The empty fruit bunch (EFB) fiber is the most easily available because of logistics reasons. The EFB has the highest fiber yield and is the only material commercially utilized for fiber extraction, but there are good potentials for the exploitation of the other two materials (i.e., oil palm fronds and trunks). The standard for EFB is unique in that it is an indigenous standard and is the first in the world that directly relates to the oil palm residues. Extensive study and research had been conducted on the critical parameters affecting the quality of the end product. The critical characteristics of oil palm fiber include the fiber length, moisture content, oil content, and impurities. Details on the acceptable limits and methods of determination are given in the Malaysian Standard, MS 1408:1997 (P)—Specification for oil palm empty fruit bunch fiber (SIRIM Berhad, 2011).

Oil palm decanter cake or palm oil sludge is the material that remains after decanting the palm oil effluent. Decanter cake produced after the oil mesh from digester passes through the centrifugal decanter to purify the oil before sending to a storage tank. Decanter cake can be filter-pressed before drying to produce dehydrated palm oil mill effluent. It can be centrifuged in the wet state after undergoing anaerobic, thermophilic and acidophilic fermentation to produce fresh centrifuged sludge solid. Currently this sludge solid is used as fertilizers since it contains amino acids, crude protein, and fiber. Basically, this waste still contains 30–40% of oil and the solid wastes are currently disposed directly in landfills without treatment, causing severe water and air pollution problems. Recovery of oil and the reuse of spent bleaching clay and decanter cake is the areas where great opportunity exists for cost saving in the oil processing industry (Gopal, 2009).

Waste Water

Palm oil mill effluent, or POME, is the effluent generated from the final stages of palm oil production in the mill. For every tonne of crude palm oil extracted from milling, about 2.5 tonnes of POME is generated (Sulaiman et al., 2009), and in 2005, about 66.8 million tonnes of POME were produced (Vairappan & Yen, 2008). It is colloidal suspension containing 95–96% water, 0.6–0.7% oil, and 4–5% total solid including 2–4% suspended solid. Raw POME is low in pH content, large amount of total solid (40,500 mg/L), oil and grease (4000 mg/L), chemical oxygen demand (COD) (50,000 mg/L), and biological oxygen demand (BOD) (25,000 mg/L). If it is discharged directly into receiving waterways, it has the potential to cause adverse environmental consequences. In addition, palm oil mill effluent has a foul smell and can cause odor pollution.

Gaseous Emissions

The two major problems associated with air emission are biogas released by POME in the ponds during anaerobic digestion and boiler ash. At the milling stage of palm oil production, the boiler is the most significant contributor of air pollutants (Yusoff & Hansen, 2007). The composition of boiler ash is a mix of clinkers and ash. Typically, the Malaysian palm oil mills burn some of the wastes to produce electricity and steam required for sterilization of the fresh fruit bunches (Yusoff, 2006). This is economically efficient but the combustion process of the boiler releases emissions such as particulate matters, CO, NO_x, and SO_2 (Ahmad et al., 2004). In order to produce 1 ton of crude palm oil, the boiler stack is estimated to release 0.64 kg of NO_x and 5.64 kg C to the air (Yusoff & Hansen, 2007). The biogas is a mixture of mainly methane and carbon dioxide. Methane, a greenhouse gas, is 20 times more harmful than carbon dioxide on climate change. The non-recovered biomethane emission from POME contributed the highest impact towards the environment and, therefore, makes the overall processes not environmentally friendly.

Palm oil mills are generally self-sufficient in terms of energy requirements due to the availability of adequate quantities of fiber and shell materials that are used as solid fuel in the stream boiler. There are two principle sources of air pollution in the mills (i.e., the boiler and incinerator) that are caused by incomplete combustion of the solid waste materials (waste fiber, shell materials, and empty fruit bunches) (Thani et al., 1999). The main practice of treating POME is by using ponding and/or open digesting tank systems (Ma et al., 1999). The emission of green house gases (CH_4 and CO_2) from these systems to the atmosphere has been recently reported as a source of air pollution from the palm oil mills (Yacob et al., 2005).

Current Waste Treatment Process in Palm Oil Industry

POME is undoubtedly the largest waste generated from the oil extraction process. For every tonne of oil palm fresh fruit bunch, it is estimated that 0.5–0.75 tonnes of POME is discharged from the mill (Hassan et al., 2004). In 2004, more than 40 million tonnes of POME was generated from 372 mills in Malaysia. One of the significant drawbacks in anaerobic treatment of wastewater is the difficulty of start-up procedures, which can be time-consuming and unproductive. Major problems lie in the inconsistency of effluent and the establishment of the most suitable microbial population for the waste to be treated. Adequate seeding of the bioreactor is critical, as the growth rates of anaerobic micro-organisms tend to be slow, especially methanogens. A few reports have shown promising application of bioreactors and process designs in the treatment of POME (Borja & Banks, 1995; Faizal & Unno, 2001; Mustapha et al., 2003; Najafpour et al., 2005).

The current POME treatment system in palm oil mills involves a series of biological processes, namely anaerobic, facultative, and aerobic treatments, and it is at the anaerobic stage that a large amount of methane is released to the atmosphere and the major reduction of POME polluting strength occurs (Hassan et al., 2004). The idea of anaerobic digestion offers several advantages and an ideal solution for organic waste treatment for the production of useful methane gas as a valuable product, low volume of sludge generation which can be used as fertilizer, and low energy requirement process (Poh & Chong, 2009). The anaerobic degradation process of organic matter occurs in four metabolic stages, namely hydrolysis, acidogenesis, acetogenesis, and methanogenesis (Bouallagui et al., 2005). POME is a colloidal slurry consisting of water, oil, and fine cellulosic materials. In the first stage the insoluble organics will undergo liquefaction by extracellular enzymes before being taken up by the acidogenic bacteria. The soluble organic components are then converted into organic acid alcohols, hydrogen, and carbon dioxide by acidogens. In the third stage acetic acid, hydrogen, and carbon dioxide are produced, and finally methane is produced by methanogenic bacteria from acetic acid, hydrogen, and carbon dioxide. The subject of anaerobic digestion of organic waste for biogas generation has been a subject of interest since decades ago.

The technology of anaerobic digestion for the bioconversion of organic materials for biogas has been reported for different types of organic waste sources such as dairy cattle waste (Wohlt et al., 1990), semisynthetic waste waters (Converti et al., 1993), particulate industrial waste (Adeel & Levine, 1994), banana peel and pineapple waste (Bardiya et al., 1996), animal waste (Karim et al., 1996), cattle-manure slurry (Ong et al., 2000), domestic wastes (Gallert et al., 2003; Krzystek et al., 2001), swine and piggery waste (Hill & Bolte, 2000; Sanchez et al., 2005), fruit and vegetable wastes (Bouallagui et al., 2005; Kalia et al., 2000; Parawira et al., 2005), and olive mill wastewater (Boubaker & Ridha, 2006).

Waste water treatment process in palm oil industry usually uses conventional biological treatment. Generally, two types of treatment systems in palm oil industry were implemented (i.e., biological treatment with land application and biological treatment without land application).

In biological treatment with land application, the waste water or POME was treated until anaerobic digestion, after which it can be spread to the plantation as an irrigation water and liquid fertilizer. Nowadays, biological treatment with land application is a common waste water treatment system in the palm oil industry. The biological treatment without land application system is the old waste water treatment system in palm oil industry. In this system the target of waste water treatment process is to reach the effluent standard set by the Department of Environment. After anaerobic digestion, the POME was treated in facultative pond, aerobic pond, and sometimes use sand bad filter. This system needs a lot of energy for aeration and using this system a lot of organic materials were lost (di-agroindustry, 2011).

Palm Oil Mill Effluent Treatment Systems

Anaerobic Digestion System

Generally, the treatment system for POME (Fig. 23.2) are operated on two–phase anaerobic digestion process followed by natural aeration process. This two-phase anaerobic process gives excellent pollutant destruction efficiency of above 95%, while natural aeration ensures that the final pollutant levels in the effluent are within the limits set by DOE. In the anaerobic digestion process, the raw palm oil mill effluent is first converted into volatile fatty acids by acid forming bacteria. The volatile acids are then converted into methane and carbon dioxide (APOC, 2011).

The advantages of anaerobic digestion system are

a. the two phase system allows greater control of digester environmental conditions
b. long solid retention times allow better biodegradation efficiencies
c. additional settling of liquor ensures minimum loading to the aerobic process
d. there is capability to cope with full effluent load, regardless of fluctuation

Fig. 23.2. Ponding system for treatment of POME.

Open Ponding and Open Tank Systems

The raw effluent is treated using a ponding system (Fig. 23.2) comprising three phases (i.e., anaerobic, facultative, and algae processes). Although the system takes a longer retention time of 90 days, it is less sensitive to environmental changes, stable, efficient, and could guarantee excellent pollutant biodegradation efficiency of above 95% (APOC, 2011).

Ponding is a general term which includes waste stabilization lagoons (ponds) and oxidation ponds. The term oxidation pond has also been loosely used and can mean aerobic, facultative, maturation, or sometimes it may even be used for anaerobic ponds. Ponding essentially employs a biological method of treatment for wastewaters. It is also used for settling of sludge or suspended solids (sedimentation ponds). Ponding can be used where land space is available. It can achieve a reasonable degree of treatment, is low in construction and operating costs and is easily maintained, as the technology required is relatively unsophisticated. Ponds have been used extensively in several other countries for the treatment of industrial wastewaters amenable to biological treatment (Wong, 1980).

From the baseline study (Yacob et al., 2006a) of methane emission from anaerobic ponds of POME treatment from two anaerobic ponds in Felda Serting Palm Oil Mill, Negeri Sembilan, Malaysia for 52 weeks, the methane content was between 35.0% and 70.0% and biogas flow rate ranged between 0.5 and 2.4 L/min/m^2. The total methane emission per anaerobic pond was 1043 kg/day. The total methane emission calculated from the two equations derived from relationships between methane emission and total carbon removal and POME discharged were comparable with field measurement. This study also revealed that anaerobic pond system is more efficient than open digesting tank system for POME treatment. Two main factors affecting the methane emission were mill activities and oil palm seasonal cropping. This is supported by its application to almost 90% of the palm oil mills in Malaysia. A linear equation established from the study could be useful in quick estimation of the total methane emission from the palm oil industry.

The baseline study of Yacob et al. (2005) of methane emission from open digesting tanks of POME treatment in Felda Serting Palm Oil Mill, Negeri Sembilan, Malaysia (Fig. 23.3) for 52 weeks from 3600 m^3 open digesting tanks indicated that the methane content was between 13.5% and 49.0%, which was lower than the value of 65% reported earlier. The biogas flow rate ranged between 0.8 L/min/m^2 and 9.8 L/min/m^2. Total methane emission per open digesting tank was 518.9 kg/day. A correlation was also established between methane emission and COD where 0.109 kg of methane for every kilogram of COD removed. For every tonne of palm oil mill effluent discharged, an average of 5.5 kg of methane is emitted from the anaerobic treatment.

Data showed that a long-term observation is crucial to determine the methane emission, as it is severely governed by the seasonal cropping of oil palm. Mills activities will also influence the quality and quantity of POME discharged, which in turn affect the anaerobic process.

Fig. 23.3. Anaerobic digestion system using open tanks.

Improved Closed Digester System

POME generated through oil extraction processes has a great impact to the industry. Owing to its chemical properties and volume of discharge, a large wastewater treatment is required to reduce the polluting strength of POME before safe discharge. Thus, the selection and performance of the treatment system determine the quality of wastewater discharged. A simple and innovative bioreactor process that is capable of treating POME efficiently is superior to the conventional system, as it operates with very short hydraulic retention times, takes high organic loading, requires less space, and is more environmentally friendly.

An improved 500 m^3 closed digester (Fig. 23.4) was constructed to evaluate the POME treatment efficiency for a comparison study with the open digester system at Felda Serting Hilir, Negeri Sembilan, Malaysia. Prior to actual treatment, the closed digester was subjected to a start-up operation, which is crucial to the overall POME treatment. During the start-up operation, the system demonstrated a remarkable performance of high COD removal efficiency (up to 97%) and satisfactory ratio of volatile fatty acids:alkalinity (VFA:Alk) between 0.1 and 0.3. The lowest hydraulic retention time (HRT) at 17 days was achieved in less than 3 months. Initial biogas production rate was high, however it declined during higher organic loading rates (OLR). This was attributed to sudden variations of POME chemical properties that affect the system stability. The start-up strategy used for this process has achieved its objectives by creating an active microbial population which was expressed in terms of key performance parameters such as % COD removal efficiency, pH, VFA:Alk, and HRT (Yacob et al., 2006b).

Fig. 23.4. Biogas pilot plant systems in palm oil industry at Felda Serting Hilir, Negeri Sembilan, Malaysia.

Composting System

The composting system offers an effective solution to the oil palm industry's perennial problem of waste disposal. The composting system utilizes 100% palm oil mill effluent and empty fruit bunches and uses a technically advanced method to convert these waste materials into compost.

Co-composting of empty fruit bunches with partially treated POME was investigated to be one of the options for the waste utilization and could offer many environmental and economic benefits. Compost is safe to be used in agriculture, however it depends on the production of good quality compost, matured and low in heavy metals and salt content. These materials could be composted and used for oil palm plantation purposes and partial substitute of inorganic fertilizers. Successful compost stabilization process, however, depends on maintaining a suitable environment for process control including moisture content, oxygen level, carbon-nitrogen ratio, nutrients, and temperature (Ekinci et al., 2006 and Liang et al., 2006).

Baharuddin et al. (2008) studied the co-composting of empty fruit bunches and partially treated POME in order to investigate the performance of windrow composting piles of these materials in the pilot scale. Data showed that it can be used as an alternative method for converting these materials using a simple and traditional windrow composting pile. The temperature obtained in the pilot-scale composting met the sanitary requirement for pathogen-killing. The compost obtained in this study has a

C/N ratio of 12 with a considerable amount of calcium, magnesium, phosphorus, potassium, and other micronutrients. Besides that, the compost product contained very low levels of heavy metals. Thus, the compost might be suitable to be used as fertilizer for the oil palm plantation or as a soil amendment.

Future Trends: Palm Oil Industry and Clean Development Mechanism

Among the palm oil mill industry wastes, POME is being produced at a huge volume and is highly polluting. The major source of POME is the usage of steam and water in the extracting processes and washing activities in the mill (Agamuthu, 1995). Various treatments have been applied to treat POME in order to meet the BOD and COD discharge standards of 100 and 150 mg/L, respectively (Md Noor, 1999). Conventional systems such as anaerobic ponds or lagoon and open tank digesters are widely used. It requires long residence times in excess of more than 20 days for satisfactory treatment (Ma, 1999). During the treatment of POME, the organic matters are consumed by micro-organisms leading to the release of a biogas mixture as the end products. In the context of environmental issue, these gases are considered greenhouse gases (GHG) because the mixture consists of methane, carbon dioxide, and traces of hydrogen sulfide. Unfortunately, these gases, especially methane (good potential for renewable energy), is not utilized but released into the atmosphere, thus contributing to global GHG emission. Until now, there is no reliable data to indicate the quantity of these GHG being emitted from the wastewater treatment facility of the palm oil industry.

In line with the abundance of renewable resources in Malaysia, the government has encouraged the energy producers such as Tenaga Nasional Malaysia Berhad (TNB), Sarawak Electricity Supply Corporation (SESCO), and Sabah Electricity Board (SEB) to utilize these renewable resources for power generation. In doing so, they have a target to provide 5% of the whole electric power supply from new energy technologies including biomass power generation. Based on this policy, TNB, the largest electric power company, has set a price to purchase electricity up to $0.10/kWh. One of the most promising resources for biomass power generation is wastes from palm oil industry. Briefly, electric power supply to Malaysia is contributed by TNB, SESCO, and SEB. TNB was initiated in 1990 from the National Electric Board by the government policy as a special private institute whose shares are completely owned by the Malaysian Government. SEB also followed the same procedure as TNB in 1998. Electric power demand in 1999 in Malaysia was 54,225 GWh and 92% came from the demand in Peninsular (Malaysian Electricity Supply Industry Trust Account, 2000).

Clean Development Mechanism (CDM) project concerning effective usage of methane emitted in the POME treatment for the electric power generation in Malaysia has been in place for several years. The research work includes field work to quantify methane emission from palm oil mill and bench study to evaluate information gathered from palm oil industry. This task was carried out by the combined team of Japanese and Malaysian researchers (Shirai et al., 2003).

Globally, there is a great concern for reducing the emission of the GHG such as nitrous oxides, methane, and carbon dioxide into the atmosphere particularly by the developed countries through CDM. The mechanism which was established under Kyoto Protocol in 1997 has a dual purpose of assisting the non-Annex I Parties in achieving sustainable development and also assisting the Annex I Parties (developed countries) in achieving compliance with their quantified GHG emission commitments (Ellis et al., 2007). Being a developed country, Japan has committed to reduce the emission of GHG by 6% (about 60 million tonnes of CO_2 equivalent) from year 2008 until 2012, based on 1990 emission level. Through the CDM project, developed countries such as Japan could achieve its GHG reduction target and also promote the environmental-friendly and sustainable development projects in the host countries like Malaysia (Shirai et al., 2003). In addition the CDM project may also provide substantial local economic growth (by allowing the host countries to receive transfer of financial, technological, and human resources) and social sustainability (by human capital formation, job creation, and building of local capacity) to the host countries (Del Rio, 2007). From the bioenergy perspective, CDM could also be used as a platform to demonstrate and disseminate new and modern bioenergy technology with lower investment risks and enhanced project's cost-efficiency (Silveira, 2005).

From now until 2012, the window for CDM business is open and could be explored from various biomass resources, especially in the palm oil industry where a huge amount of biomass is available and ready to be exploited for commercialization. Developed countries such as Japan should urgently take this opportunity to actively involve in the CDM business. In return, Malaysia will benefit from sustainable development in the palm oil industry from environmental, economic, and social perspectives.

Sustainable Palm Biomass Refinery in Malaysia Through Clean Development Mechanism (CDM) Project

In developing the sustainable palm oil industry, a paradigm shift towards palm biomass is needed. In the paradigm shift, palm biomass should not be regarded as waste but as a renewable and sustainable resource. Due to the uncertainties of biomass utilization, there is a need to develop technologically proven methods and economically feasibilities for such technologies. The value chain in the sustainability of palm

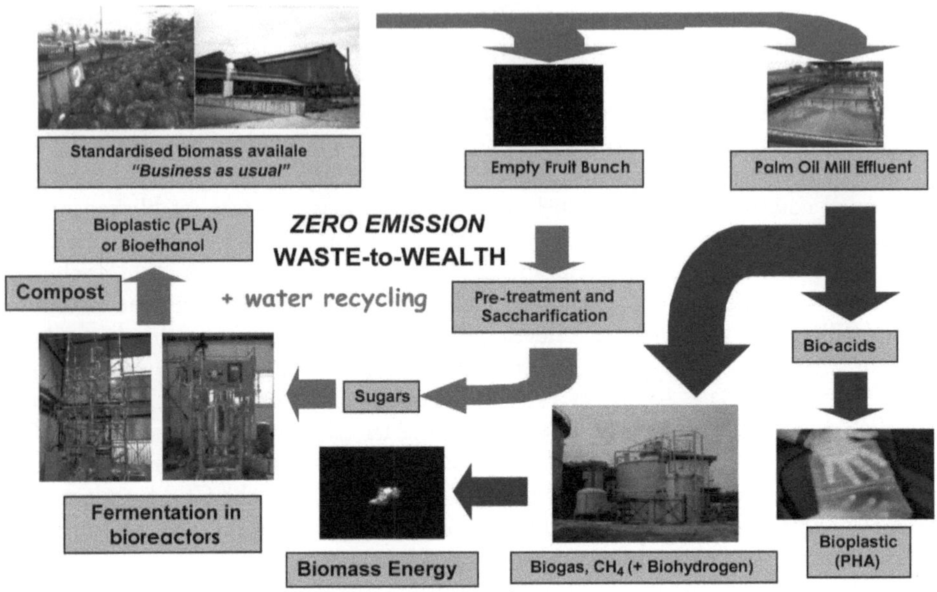

Fig. 23.5. Sustainable palm biomass refinery concept.

biomass refinery starts from production of fuel, feed, fiber, food, and finally fine chemicals. Fig. 23.5 summarizes the sustainable palm biomass refinery concept.

For empty fruit bunches and other biomass from palm oil tree, quite extensive research has been developed to produce higher value-added products such as fuel briquette, pulp and paper, medium density fiberboard, automotive components, fiber strands, plywoods, blockboards, particleboards, glucose for further conversion to ethanol, citric acid, butanol, and other single cell protein from hemicellulose and lignin (Basiron & Simeh, 2005). In a recent study, Nasrin et al. (2008) demonstrated that empty fruit bunches, palm kernel expeller, and sawdust could be compacted to produce briquettes fuel at high temperature and pressure using a screw extrusion technology with good combustion properties. This briquette fuel is suitable for boiler and incinerator to produce steam and electricity. The future second-generation biofuel will be based on the conversion of cellulosic fiber or biomass into liquid fuel (Basiron, 2007). This is a more promising future direction for the biomass utilization, and future research should be directed into this area.

For POME, currently there are two methods of utilization. The first is the anaerobic treatment to produce biogas or organic acids (mixture of acetic, butyric, and propionic acids) and eventually bioplastics such as polyhydroxyalkanoates. The second would be composting with empty fruit bunches. Both approaches still produce effluent, and further polishing is required before discharge as to comply with the Department of

Table 23-B. Regulation for Watercourse Discharge for Palm Oil Mill Effluent.

Parameters	Discharge level (units in mg/L except for pH)
Biochemical oxygen demand (BOD)	100
Suspended solids	400
Oil and grease	50
Ammoniacal nitrogen	150
Total nitrogen	200
pH	5–9

Source: Hassan et al. (2004).

Environment (DOE) Malaysia regulations, as shown in Table 23-B. In the future, the regulations for the effluent discharge level from the palm oil mills in peninsular Malaysia will become more stringent (BOD level of less than 20 mg/L). At this level many of the mills may face problems to comply, and advanced technologies may be required to handle it. The technology for water recycling (zero-discharged) may be possible, and research should be focused for cost-effective treatment technologies, and this will open up new opportunities for technology transfer. Among other challenges in future sustainable palm oil production is the need to implement a scheme to enable sustainably produced palm oil to be certified with full traceability through the Round Table on Sustainable Palm Oil (RSPO), land conservation for forest, and wildlife conservation issues (Basiron, 2007). Further information on the trade and traceability of palm oil and RSPO can be found in Chapters 24 and 25, respectively.

Future Outlook

The future for the renewable energy is bright, especially in the developing countries where the trend of renewable energy market is shifting from the fringe to the mainstream for sustainable development. The support for renewable energy has been building in government, multilateral organizations, industry, and non-governmental organization (NGOs) at local, national, and global levels (Martinot et al., 2002). This is due to recognition of renewable energy contribution to the rural development, energy independence, and effort for climate change mitigation. With the increase in the world's population, higher energy demand, and intensity, the market share of the renewable energy is expected to increase from 16.5% in 2004 to 25% (high scenario in 2020) and further increase to 40–50% (high scenario in 2050) (Martinot et al., 2007).

Malaysia is committed to continue its focus on sustainable development to support the economic growth, enhanced competitiveness, and contribute towards achieving a balanced development. The strategies outlined include reducing high dependence on petroleum products, increasing use of alternative fuels, promoting greater use of renewable energy for power generation, fostering a more conducive environment to support the implementation of Small Renewable Energy Program (SREP) by reviewing the terms and conditions of the Renewable Energy Power Purchase Agreement (REPPA), and issues related to project viability (EPU, 2008).

One of the tangible approaches is to actively and carefully explore other potential CDM projects. Currently, most of the registered CDM projects in Malaysia are utilizing oil palm empty fruit bunches for renewable energy generation. This is because empty fruit bunches are readily available with consistent characteristic. Moreover, presently, mulching requires additional cost and incineration causes air pollution. With CDM, empty fruit bunches are combusted for energy and certified emission reduction (CER) is claimed through the displacement of more fossil fuel and methane avoidance from biomass decay. In addition, the investors could further benefit from the sales of electricity to TNB at RM0.30/kWh.

Malaysia has already submitted inventory for GHG emission (carbon dioxide, methane, nitrous oxide) in 1994 (Chong & Philip, 2001). They focused on 5 sectors: energy (coal, oil, and natural gas), industry (cement production), agriculture (livestock and paddy field), wastes (dumping and wastewater treatment), and land use (forestry management). In the wastes sector methane is emitted from the anaerobic biological treatment. Palm oil industry is classified into the wastes sector because treatment facilities emit a large amount of methane mainly from anaerobic digesting tank and lagoons. The study also indicates that the quantity and quality of the methane emitted from the wastes sector is largely influenced by the characteristic of the wastes, amount of wastewater, and the treatment methods (aerobic or anaerobic). In this inventory, methane emission from the palm oil industry and rubber industries was estimated in the wastes sector, as both industries are the largest GHG source in Malaysia.

It is estimated that 2.23 million tonnes of methane was emitted yearly from Malaysia, while methane emission from palm oil industry is approximately 0.214 million tonnes. This was the fourth largest value, following that from dumping field, oil and natural gas industry, and paddy field, and occupies around 10% in the whole methane emission from Malaysia. Since the greenhouse effect of methane is around 20 times higher than carbon dioxide, the greenhouse effect from palm oil industry is equivalent to 4.28 million tonnes CO_2 emission annually (Sulaiman et al., 2005). It is expected that the mitigation methods identified would help to address this issue.

References

Agamuthu, P. Palm oil mill effluent-treatment and utilization. In Waste Treatment Plant; Sastry, C.A., Hashim, M.A., Agamuthu, P., Eds.; Narosa Publishing House: New Delhi, 1995; pp. 338–360.

Ahmad, A.L.; Azid, A.R.; Yusof, A.R.; Seetharamu. Emission control in palm oil mills using artificial neural network and genetic algorithm. Computers Chem. Eng. 2004, 28, 2709–2715.

APOC. 2011. http://www.americanpalmoil.com/sustainable-pome.html. Accessed on 1 Nov 2011.

Baharuddin, A.S.; Abd-Aziz, S.; Rahman, N.A.R.; Hassan, M.A.; Wakisaka, M.; Shirai, Y. Co-composting of empty fruit bunches and partially treated palm oil mill effluents in pilot scale. Int. J. Agric. Res. 2008, 4(2), 69–78.

Bardiya, N.; Somayaji, D.; Khanna, S. Biomethanation of banana peel and pineapple waste. Biores. Technol. 1996, 58, 73–76.

Basiron, Y. Palm oil production through sustainable plantations. Eur. J. Lipid Sci. Technol. 2007, 109, 289–295.

Basiron, Y.; Simeh, M.A. Vision 2020—The palm oil phenomenon. Oil Palm Ind. Econ. J. 2005, 5(2), 1–10.

Borja, R.; Banks, C.J. Comparison of an anaerobic filter and an anaerobic fluidized bed reactor treating palm oil mill effluent. Process Biochem. 1995, 30, 511–521.

Bouallagui, H.; Touhami, Y.; Ben Cheikh, R.; Hamdi; M. Bioreactor performance in anaerobic digestion of fruit and vegetable wastes. Process Biochem. 2005, 40, 989–995.

Chong, A.L.; Philip, M. Malaysia National Response Strategies to Climate Change, Ministry of Science, Technology and Environment Malaysia Report, 2001.

Converti, A.; Borghi, M.D.; Ferraiolo, G. Influence of the organic loading rate on the anaerobic treatment of high strength semisynthetic waste waters in a biological fluidized bed. Chem. Eng. J. 1993, 52, B21–B28.

Deforestation Watch. Palm Oil: An Environmentally Friendly Tree Crop. http://deforestation watch.org/index.php/Key-Papers/Palm-Oil-An-Environmentally-Friendly-Tree-Crop.html. Accessed on 1 November 2011.

Del Rio, P. Encouraging the implementation of small renewable electricity cdm projects: an economic analysis of different options. Renew. Sustain. Ener. Rev. 2007, 11, 1361–1387.

Department of Environment Malaysia. Effluent Related Fee in the Palm Oil Industry of Malaysia. http://www.unescap.org/drpad/vc/conference/ex_my_14_erf.htm. Accessed on 1 Nov 2011.

di-agroindustry 2011. http://di-agroindustry.blogspot.com/2010/03/waste-water-treatment process-in-palm.html. Accessed on 1 Nov 2011.

Economic Planning Unit, Prime Ministers Department, Ninth Malaysia Plan 2006–2010 (http://www.epu.jpm.my/rm9/htm/english.htm (accessed on 15 Sept 2008).

Ekinchi, K.; Keener, H.M.; Akbolat, D. Effect of feedstock, airflow rate, and recirculation ratio on performance of composting system with air recirculation. Bioresour. Technol. 2006, 97, 922–932.

Ellis, J.; Winkler, H.; Corfee-Morlot, J.; Gagnon-Lebrun, F. CDM: taking stock and looking forward. Ener. Policy 2007, 35, 15–28.

Faisal, M.; Unno, H. Kinetic analysis of palm oil mill wastewater treatment by a modified anaerobic baffled reactor. Biochem. Eng. J. 2001, 9, 25–31.

Gallert, C.; Henning, A.; Winter, J. Scale-up of anaerobic digestion of the biowaste fraction from domestic wastes. Water Res. 2003, 37, 1433–1441.

Goh, C.S.; Tan, K.T.; Lee, K.T.; Bhatia, S. Bio-ethanol from lignocellulose: status, perspectives and challenges in Malaysia. Bioresour. Technol. 2010, 101(13), 4834–4841.

Gopal, D. Oil recovery from palm oil solid wastes. Bac Eng Thesis, University Malaysia Pahang, 2009.

Hassan, M.A.; Yacob, S.; Shirai, Y. Treatment of palm oil wastewaters. In Handbook of industrial and hazardous wastes treatment. Wang, L.K., Hung, Y., Lo, H.H., Yapijakis, C., Eds.; Marcel Dekker: New York, 2004; pp. 719–736.

Hill, D.T.; Bolte, J.P. Methane production from low solid concentration liquid swine waste using conventional anaerobic fermentation. Bioresour. Technol. 2000, 74, 241–247.

Kalia, V.C.; Sonakya, V.; Raizada, N. Anaerobic digestion of banana stem waste. Bioresour. Technol. 2000, 73, 191–193.

Karim, K.; Hoffmann, R.; Klasson, T.; Al-Dahhan, M.H. Anaerobic digestion of animal waste: Waste strength versus impact of mixing. Bioresour. Technol. 2005, 96, 1771–1781.

Krzystek, L.; Ledakowicz, S.; Kahle, H.J.; Kaczorek, K. Degradation of household biowaste in reactors. J. Biotechnol. 2001, 92, 103–112.

Liang, C.; Das Mcclendon, K.C. The influence of temperature and moisture content regimes on the aerobic microbial activity of a biosolids composting blend. Bioresour. Technol. 2003, 86, 131–137.

Ma, A.N. Treatment of palm oil mill effluent. In Oil Palm and the Environment—A Malaysian Perspective; Singh, G., Lim, K.H., Leng, T., David, L.K., Eds.; Malaysian Oil Palm Growers Council, 1999; pp.113–126.

Malaysian Electricity Supply Industry Trust Account. Feasibility study on grid connected power generation using biomass cogeneration technology, 2000; pp. 26–32.

Martinot, E.; Chaurey, A.; Lew, D.; Moreira, J.R.; Wamukonya, N. Renewable energy market in developing countries. Ann. Rev. Energy Environ. 2002, 27, 309–348.

Martinot, E.; Dienst, C.; Weiliang, L.; Qimin, C. Renewable energy futures: targets scenarios and pathways. Ann. Rev. Environ. Resour. 2007, 32, 205–239.

Md. Noor, M. Environmental legislation: Environmental Quality Act, 1974. In Oil Palm and the Environment—A Malaysian Perspective; Singh, G., Lim, K.H., Leng, T., David, L.K., Eds.; Malaysian Oil Palm Growers Council, 1999; pp. 261–264.

Mustapha, S.; Ashhuby, B.; Rashid, M.; Azni, I. Start-up strategy of a thermophilic upflow anaerobic filter for treating palm oil mill effluent. Trans. IChemE 2003, 81, 262–266.

Najafpour, G.; Yieng, H.A.; Younesi, H.; Zinatizadeh, A. Effect of organic loading on performance of rotating biological contactors using palm oil mill effluents. Process Biochem. 2005, 40, 2879–2884.

Nasrin, A.B.; Ma, A.N.; Choo, Y.M.; Mohamad, S.; Rohaya, M.H.; Azali, A.; Zainal, Z. Oil palm biomass as potential substitution raw materials for commercial biomass briquettes production. Am. J. Appl. Sci. 2008, 5(3), 179–183.

Ong, H.K.; Greenfield, P.F.; Pullammanappallil, P.C. An operational strategy for improved biomethanation of cattle-manure slurry in an unmixed single stage digester. Bioresour. Technol. 2000, 73, 87–89.

Parawira, W.; Murto, M.; Read, J.S.; Mattisson, B. Profile of hydrolases and biogas production during two-stage mesophilic anaerobic digestion of solid potato waste. Process Biochem. 2005, 40, 2945–2952.

Sanchez, E.; Borja, R.; Travieso, L.; Martin, A.; Colmenarejo, M.F. Effect of organic loading rate on the stability, operational parameters and performance of a secondary upflow anaerobic sludge bed reactor treating piggery waste. Bioresour. Technol. 2005, 96, 335–344.

Shirai, Y.; Wakisaka, M.; Yacob, S.; Hassan, M.A.; Suzuki, S. Reduction of methane released from palm oil mill lagoon in Malaysia and its countermeasures. Mitigation Adapt. Strat. Global Change 2003, 8, 237–252.

SIRIM Berhad, Malaysia. Standards for Oil Palm Fiber. http://www.aseansec.org/7011.htm. Accessed on 1 November 2011.

Sulaiman, A.; Busu, Z.; Hassan, M.A.; Wakisaka, M.; Shirai, Y. Clean development mechanism project for methane capture from palm oil mill effluent treatment in Malaysia. Biomass Asia Workshop 2005. www.biomass-asia-workshop.jp/biomassws/05workshop/poster/P-20.pdf. Accessed on 1 November 2011.

Sulaiman, A.; Zakaria, M.R.; Hassan, M.A.; Shirai, Y.; Busu, Z. Co-digestion of palm oil mill effluent and refined glycerin wash water for chemical oxygen demand removal and methane production. Am. J. Environ. Sci. 2009, 5, 639–646.

Sumathi, S.; Chai, S.P.; Mohamed, A.R. Reviews; Utilization of oil palm as a source of renewable energy in Malaysia. Renew. Sustainable Ener. 2008, 2(9), 2404–2421.

Vairappan, C.S.; Yen, A.M. Palm oil mill effluent (POME) cultured marine microalgae as supplementary diet for rotifer culture. J. Appl. Physics 2008, 20(5), 603–608.

Wohlt, J.E.; Frobbish, R.A.; Davis, C.L.; Bryant, M.P.; Mackie, R.I. Thermophilic methane production from dairy cattle waste. Bio. Wastes 1990, 32, 193–207.

Wong, K.K. Application of ponding systems in the treatment of palm oil mill and rubber mill effluents. Pertanika 1980, 3(2), 133–141.

Yacob, S.; Hassan, M.A.; Shirai, Y.; Wakisaka, M.; Subash, S. Baseline study of methane emission from open digesting tanks of palm oil mill effluent treatment. Chemosphere 2005, 59, 1575–1581.

Yacob, S.; Hassan, M.A.; Shirai, Y.; Wakisaka, M.; Subash, S. Baseline study of methane emission from anaerobic ponds of palm oil mill effluent treatment. Sci. Total Environ. 2006a, 366, 187–196.

Yacob, S.; Shirai, Y.; Hassan, M.A.; Wakisaka, M.; Subash, S. Start-up operation of semi-commercial closed anaerobic digester for palm oil mill effluent treatment. Process Biochem. 2006b, 41, 962–964.

Yusoff, S. Renewable energy from palm oil: Innovation on effective utilization of waste. J. Cleaner Production 2006, 14(1), 87–93.

Yusoff, S.; Hansen, S.B. Feasibility study of performing an life cycle assessment on crude palm oil production in Malaysia. Int. J. Life Cycle Assess. 2007, 12(1), 50–58.

24

Traceability of RSPO-certified Sustainable Palm Oil

Marieke Leegwater[1] and Gerrit van Duijn[2]
[1]Manager of Sustainable Production at Product Board for Margarine Fats and Oils and Former Chair of RSPO Standing Committee Trade and Traceability and [2]Oil Supply Chain Consultant and Former Oil Supply Technology Director, Unilever R&D Vlaardingen

Introduction

Traceability

Different people understand different things when speaking about traceability. Wikipedia gives as a definition of traceability: the ability to chronologically interrelate uniquely identifiable entities in a way that is verifiable. The Codex Alimentarius, the worldwide food safety standard setting organization, defines traceability as the ability to follow the movement of a food through specified stage(s) of production, processing and distribution (Codex Alimentarius Commission, 2011).

Traceability for foods boils down to this: there should be uniquely identifiable consignments that can be followed through a production chain. This is the way the concept is usually used in food safety contexts. A distinction is often made between upward traceability and downward traceability, where upward refers to back to origin (one step back) and downward to the downstream path (one step forward). Another distinction is that of continuous traceability (the ability to generate information on the origin of a product at any given moment) versus step-wise traceability that is generated by accumulating data of different operators.

Until recently the concept of traceability was mainly used by companies to manage food safety risks related to possible contaminations of consignments. However, nowadays sustainable production is becoming more and more important and traceability is used to secure this as well. This can sometimes create confusion; it appears that when traceability is referred to in an ethical or environmental context the concept is often used in a broader sense. For example, the Roundtable for Sustainable Palm Oil (RSPO) uses the traceability concept to define a link between producers of sustainable products and end users of these products. Within RSPO context, this is logical, as the goal of RSPO is to increase sustainable production of palm oil products. The mechanisms for supply chain organization are designed to contribute to that. In this chapter we use the concept of traceability in the strict sense, or maybe better said, the food safety sense.

Traceability in the Palm Oil Chain

Palm oil is usually traded, processed, and stored as a commodity. This means there is no qualitative differentiation between different consignments, and it is usually traded in large volumes. This is done to maximize efficiency of transport and storage and to reduce costs. It makes it easy to imagine that palm oil and traceability do not necessarily come hand in hand. On the contrary, due to the fact that palm oil can easily be exchanged with other palm oil, the possibility of the unique identification of consignments is often lost quickly.

The introduction of the Roundtable on Sustainable Palm Oil in 2003 meant that a qualitative distinction was created between RSPO-certified Sustainable Palm Oil and non-RSPO-certified Sustainable Palm Oil. This required RSPO to develop systems in order to create value for the RSPO-certified Sustainable Palm Oil and ways to link end users of palm oil to producers thereof.

This chapter gives a detailed overview of the supply chain of palm oil and the way traceability is organized during the different steps. Subsequently, it shows the systems that have been designed by RSPO to connect producers and end-users of RSPO-certified Sustainable Palm products and evaluates whether these systems provide additional traceability of palm products.

Traceability in the Current Palm Oil Chain

Traceability Requirements that Apply to Palm Oil

Traceability of palm oil and palm products is secured through legal and contractual requirements. The main ones are the European General Food Law and the FOSFA (Association for International Trading in Oils, Fats and Oilseeds. It developed standard contracts for the trade in vegetable oils and fats.) standard contracts #54, #80 and #81. These requirements were developed to secure food safety.

The European General Food Law (Regulation EC/178/2002) requires all European operators producing food and feed to ensure the foods they produce are traceable. It defines traceability as the ability to trace and follow food, feed, and ingredients through all stages of production, processing, and distribution. According to the definitions, "stages of production, processing, and distribution" means any stage, including import, from and including the primary production of a food, up to and including its storage, transport, sale, or supply to the final consumer.

The European Commission has issued a Guidance document indicating that to comply with the above legal requirements, operators need to be able to identify the immediate supplier of their products and their immediate receivers and the steps the food has followed (one step back–one step forward). Operators do not have to be able to demonstrate this information at any given moment in time (continuous traceability).

The trade in crude palm oil is usually done based on FOSFA contracts:

- Contract # 80 is for Crude Unbleached Palm Oil–Cost Insurance and Fright (CIF) (The shipper has to pay the cost of shipment up to the ship, the insurance cost of the cargo, and the freight cost up to destination port.)
- Contract # 81 is for Refined Palm and Palm Kernel Oil Products–CIF
- Contract # 54 is for Crude Palm Kernel oil (Contract for vegetable and marine oil in bulk)–CIF
- Contract # 53 is for Vegetable and Marine Oil in Bulk–Free On Board (FOB) (The shipper/trader pays the costs up to the ship and insurance cost, but freight charges is paid by the Buyer/Consignee.)

All contracts require that at payment of the contracted sum, the buyer receives the Bill of Lading and all shipping documents, including certification documents from the superintendent. The Bill of Lading is the legal document transmission which ensures the transmission of ownership. It contains information on

- Port of Loading,
- Port of Delivery,
- the quantity,
- the time and date of loading,
- the amount,
- the original lot, and
- the ship tanks in which it was loaded.

For an example of a Bill of Lading, see Fig. 24.1.

The Certificates of the superintendent contain information on

- The shipped weight ascertained at the port of loading
- Particulars of the time and place of loading, including from which shore tanks the consignment was loaded

This means that if palm oil or palm kernel oil, refined or crude, is traded in accordance with FOSFA contracts #80, #81, and #54, the buyer receives the documents that indicate which ship was used, where it was loaded, and which shore tank(s) were used to fill the ship tankers.

CHEMICAL BUSINESS

Shipper

BILL OF LADING | B/L NO.

Consignee (if 'Order' state Notify Party)
TO ORDER

Notify Party (only if not stated above: otherwise leave blank)

• Pre-carriage by	• From		
Ocean Vessel Voy. No.	**Port of loading** SANDAKAN, MALAYSIA		
Port of discharge ROTTERDAM	**• Final destination (if on-carriage)**	**Freight payable at**	**Number of Original B s/L.** 3 (THREE)

Description of goods PALM OIL (MALAYSIAN CRUDE UNBLEACHED PALM OIL IN BULK) Quantity said to be. 250.000 Metric Tons

Loaded in Tank Nos. 2P,2S & 3S

'CLEAN ON BOARD'
'FREIGHT PREPAID'

ABOVE PARTICULARS DECLARED BY SHIPPER

THIS SHIPMENT was loaded on board the vessel as part of one original lot of .... stowed into .... with no segregation as to parcels.

The shipment of .... kilos of Oil is part of a shipment of .... kilos of Oil in bulk made in accordance with a Contract of Affreightment dated .... for which .... sets of Bills of Lading have been issued and for which the vessel is relieved from all responsibilities to the extent it would be if one set only would have been issued. The vessel under takes to deliver to the holders of the respective Bills of Lading as near as practicable that proportion of the Oil actually loaded in the designated tank (s) which is represented by the percentage of the quantity in the Bill of Lading to the total co-mingled quantity.

Shipped in apparent good order and condition on board the above vessel at the port of loading for carriage to the port of discharge or so near thereto as the vessel may safely get always afloat, and for delivery thereat to the consignee or his assigns on payment of freight. This shipment is carried subject to the clauses overleaf and to the terms, clauses, conditions and exceptions of the Contract of Affreightment/Charter Party dated 29/10/2008 between .... and PROCUREMENT COMPANY SDN BHD and all terms whatsoever of the Contract of Affreightment/Charter Party apply to and govern the rights of the parties concerned in this shipment. If there should be any conflict between the conditions set out overleaf and the terms, clauses, conditions and exceptions of the Contract of Affreightment/Charter Party, the latter shall prevail.

FREIGHT PARTICULARS

In witness whereof the Master has signed 3 (THREE) Originals and 4 (FOUR) copies of Bills of Lading of this tenor and date, one of which being accomplished the others will be void.

Dated at ....
FEBRUARY .... 20....

COPY NOT NEGOTIABLE

• Applicable only to pre-carriage/on-carriage in accordance with clauses 6 and 8
"Weight and Quantity Unknown"

Fig. 24.1. Example of Bill of Lading. Source: Private communication with anonymous company representative, August 12, 2011.

Two types of shipping contracts are used to divide transaction costs and responsibilities between buyer and seller:

- **FOB** (Free On Board at named port of loading). The seller must load the goods on board the ship nominated by the buyer. The costs and risk transfers from seller to buyer when the goods pass the ship's rail.
- **CIF** (Cost, Insurance, and Freight to named destination port). The seller pays the costs and freight to bring the goods to the port of destination. However, the risk is transferred from seller to buyer once the goods have crossed the ship's rail at the port of loading. The seller must produce and pay the insurance for the buyer.

The shipping contract form has, in principle, no relationship with traceability. However, FOB is more often used for special deliveries under control of the buyer, while most standard deliveries are CIF. This has the indirect effect that FOB deliveries are often perceived to have a better traceability.

Traceability of Palm Oil in Practice

A schematic overview of the palm oil supply chain between harvesting in the plantation and overseas shipping is shown in Fig. 24.2. The schematic overview between

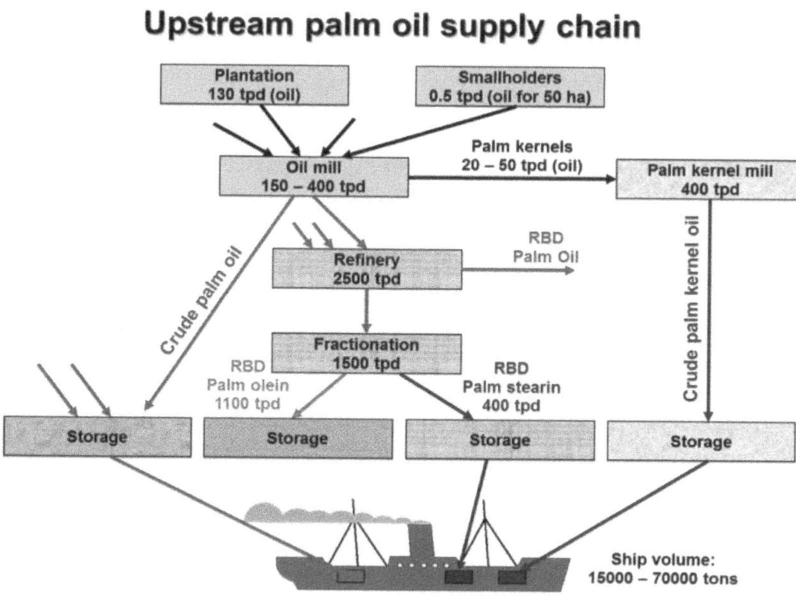

Fig. 24.2. The palm products supply chain from plantation until overseas shipping. As example, some indicative data are included. Source: Gerrit van Duijn, August 20, 2011.

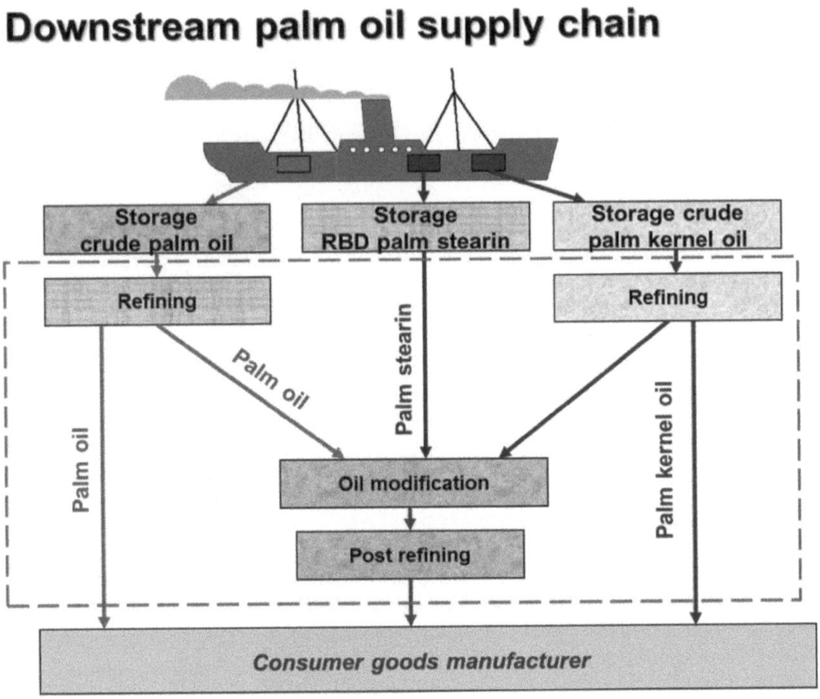

Fig. 24.3. The palm products supply chain from overseas shipping to the use in consumer goods manufacturing. Source: Gerrit van Duijn, 20 August 20, 2011.

discharge after shipping and use in consumer goods manufacturing is shown in Fig. 24.3. The main steps and traceability systems in this supply chain are given below.

Harvesting in the Plantation

Ripe fruit bunches are harvested manually and collected for transport to a nearby oil mill. The degree of ripeness of a fruit bunch is detected by counting the number of loose fruit under the palm tree. For quality reasons, fruit bunches have to be processed within 3 days after harvesting, and for high quality oils even within 24 hours. Therefore, large plantations process their fruit bunches in their own oil mill, which is located in, or close to, the plantation. Part of the fruit bunches processed in an oil mill can be bought (directly or via a middle man) from neighboring smallholders. A smallholder runs a plantation smaller than 50 ha. In most cases it is a family operated business; some smallholders get technical support from a nearby large plantation and others are fully independent.

Traceability in the Plantation

A plantation harvester will mark the fruit bunches after harvesting by writing a unique code on the cut-off surface of the bunch. This allows plantation management to allocate poor quality fruit bunches (over- or under-ripe, rotten) to a harvester and his harvested area. Harvester payment may also be connected to the number and quality of harvested bunches. Fruit bunches collected from smallholders by a middle man are mixed with bunches from other smallholders and have a limited traceability because they are usually not marked.

Oil Extraction and Kernel Separation in the Oil Mill

Fruit bunches are transported to the oil mill by trucks. The trucks discharge on a ramp from which small rail wagons (around $2m^3$) are filled. The bunches in the rail wagons are sterilized with low pressure steam in long horizontal cylinders. The fruitlets are separated from the sterilized bunches by a rotating sieve. A screw press separates the liquid (oil + sludge) from the solids (fibers + kernels). After decanting, the oil is stored in crude palm oil tanks. The kernels are separated from the fibers, washed, dried, and sold to palm kernel mills.

Traceability in the Oil Mill

The delivered fruit bunches can be traced to plantation/harvester by the code on the bunch or to the middle man/ smallholder by the delivering truck. Samples of delivered bunches can be graded to check the quality. Oil mills normally start to operate at the end of the morning and continue until all deliveries of that day have been processed. It is not possible to link produced oil to delivered fruit bunches within this one-day production window. Several days of crude oil and kernel production will be stored in large tanks or silos. Hence, crude oil and kernels delivered from storage can be traced to the production period in which the storage tank or silo has been filled. Palm kernel oil mills process mixtures of kernels from various palm oil mills. Crude palm kernel oil can only be traced to the group of kernel suppliers that delivered in the silo(s) used during the production period.

Refining and Fractionation in the Region of Origin

The largest market for refined palm oil and palm oil fractions is in the region of origin (e.g., Asia for Malaysian and Indonesian palm products). High capacity refining and fractionation plants deliver the following products:

- RBD palm oil. Bleached and deodorized palm oil used for many local applications in food and non-food. RBD palm oil is also shipped to other regions depending on export and import duties and refining availability in the receiving region.

- RBD palm olein. The liquid fraction of palm oil after bleaching, deodorization, and fractionation. The palm olein yield is 75–80% of the fractionated palm oil. Palm olein is largely used in Asia and Africa as cooking oil; shipping to other regions is less important.
- RBD palm stearin. The solid fraction of palm oil after bleaching, deodorization and fractionation. The palm stearin yield is 20–25% of the fractionated palm oil. Palm stearin has a local use as vegetable fat and for soap making. It is also exported to Europe as vegetable fat and feedstock for oleo chemicals manufacturing.

Traceability in Refining and Fractionation

Crude palm oil is transferred from the storage tank(s) in the oil mill(s) to the crude oil storage tank(s) in the refinery by road tankers or ships (mainly barges and coasters) or by pipeline in case of a combined extraction/refining operation. The crude oil transport is in sealed tanks and the operation is well documented and fully traceable. Good practice is to receive crude oils in one tank and to feed the refinery from a different tank. The feedstock of the refinery can be traced to the suppliers that delivered in the feeding tank. Poor practice is to receive crude oil in the same tank that feeds the refinery at the same time. In that case traceability is including all suppliers that delivered in the crude oil tank during a long period before the checked production run. The refining process can be batch, continuous, or semi-continuous (sequence of small batches). Batch and semi-continuous processes can be fully traceable via batch tracking. The output of a continuous process traces back to the crude oil feed tank with a correction for the oil content of the refining equipment, which after change-over is still from the previously used feed tank. Fractionation is a batch process and can be fully traceable.

Overseas Shipping

Palm products are transported from oil mills or fractionation plants to large shore tanks in the port of departure by sealed road tankers or barges. The shore tanks for crude palm oil normally contain deliveries of more oil mills since the shipped parcel size is large compared to the oil mill production capacity. Palm fractions and palm kernel oil are shipped in smaller volumes.

Three types of sea-going vessels are used for the overseas transport of palm products:

- Bulk tankers: These range in capacity from 15,000 to 75,000 tonnes and have a varying number of different sized tanks usually with three separated storage compartments and with interconnecting valves between the tanks. These vessels are best suited for the carriage of large volumes of crude palm oil. They can be loaded and discharged with interconnecting valves open. Loading and discharge may occur in several ports or port locations because, for example,different suppliers can book the same ship or the fully loaded ship is too deep for certain ports or locations in the port.

- Parcel tankers: These are more sophisticated ships in the 15,000 to 50,000 tonnes range. They have a varying number of different sized tanks and are designed to carry a variety of different bulk liquids which are kept fully segregated. Each tank or group of tanks has its own dedicated pipelines and pumps. These ships are suited to transport smaller parcels of palm fractions and palm kernel oil. For a picture of a parcel tanker see Fig. 24.4.
- Coasters: These are smaller ships (750–6500 tonnes) suited for short sea voyages. They are frequently used for further regional sea transport after transshipment from ocean going vessels.

Capacity of Cargo Tanks

Compartment	Volume 98%	St.St/ Coating	Tank Type	Pump Capacity	Heating Capacity
10CP	1802	Stainless Steel	Cargo		$0,00
10CS	1803	Stainless Steel	Cargo		$0,00
11CP	888	Stainless Steel	Cargo		$0,00
11CS	889	Stainless Steel	Cargo		$0,00
12CP	1775	Stainless Steel	Cargo		$0,00
12CS	1775	Stainless Steel	Cargo		$0,00
13CP	1025	Stainless Steel	Cargo		$0,00
13CS	1043	Stainless Steel	Cargo		$0,00
14CP	702	Stainless Steel	Cargo		$0,00
14CS	691	Stainless Steel	Cargo		$0,00
1CP	1030	Stainless Steel	Cargo		$0,00
1CS	1014	Stainless Steel	Cargo		$0,00
2CP	1622	Stainless Steel	Cargo		$0,00
2CS	1622	Stainless Steel	Cargo		$0,00
3CP	876	Stainless Steel	Cargo		$0,00
3CS	876	Stainless Steel	Cargo		$0,00
4CP	1800	Stainless Steel	Cargo		$0,00
4CS	1800	Stainless Steel	Cargo		$0,00
5CP	888	Stainless Steel	Cargo		$0,00
5CS	889	Stainless Steel	Cargo		$0,00
6CP	1799	Stainless Steel	Cargo		$0,00
6CS	1799	Stainless Steel	Cargo		$0,00
7CP	889	Stainless Steel	Cargo		$0,00
7CS	889	Stainless Steel	Cargo		$0,00
8CP	1801	Stainless Steel	Cargo		$0,00
8CS	1802	Stainless Steel	Cargo		$0,00
9CP	889	Stainless Steel	Cargo		$0,00
9CS	889	Stainless Steel	Cargo		$0,00
Total Center Tanks:	35563				
Total Wing Tanks:	0				
Grand Total:	35563				

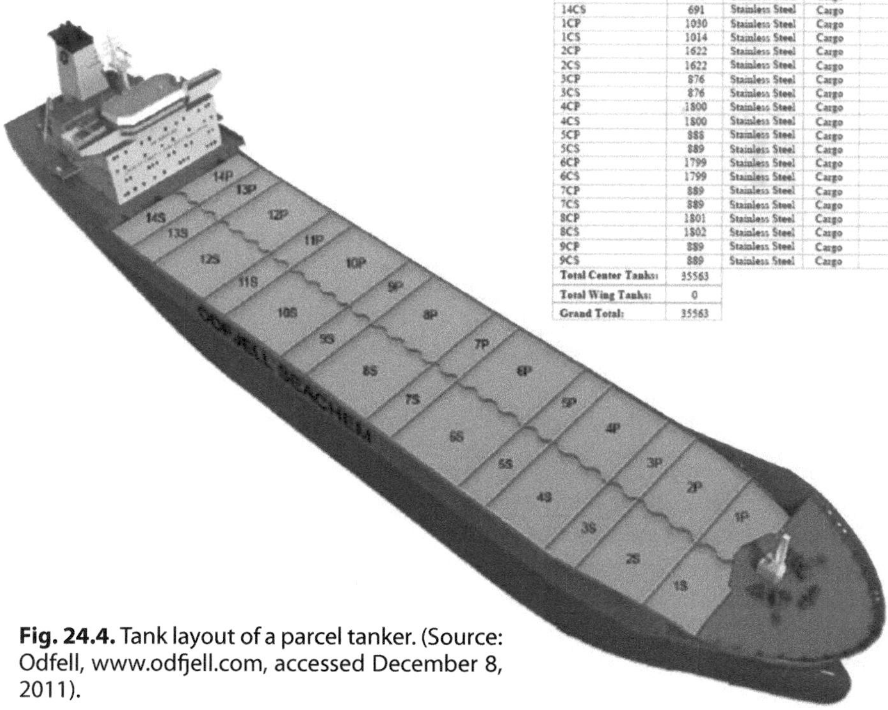

Fig. 24.4. Tank layout of a parcel tanker. (Source: Odfell, www.odfjell.com, accessed December 8, 2011).

The operations at the ports of loading and discharge are supervised by independent superintendents. Superintendents can represent the interest of both the seller(s) and the buyer(s). The activities supervised at the port of departure are

- Sampling of the shore tanks
- Analyses of the contractual parameters by an approved laboratory
- Check of ship conditions, previous cargoes, and coating
- Loading pump plan agreed by captain and superintendent. The pump plan includes the detailed routing from shore tank to receiving ship tank or from barge/coaster to ship tank in case of trans-shipment
- Loading, sampling of ship tanks, and sealing of tanks if required
- Heating and pumping instructions for the sea voyage

The activities supervised at the port of arrival are

- Check of ship condition and previous cargoes
- Check registered conditions during the voyage (like temperature, pumping actions, etc.)
- Check the volume by measuring the height and temperature in the ship tank
- Sampling of ship tanks
- Positive release decision in agreement with buyer(s)
- Analyses of contractual parameters at an approved laboratory
- Discharge pump plan agreed by captain and superintendent. The pump plan includes the detailed routing from ship tank to receiving shore tank or to barge/coaster in case of trans-shipment
- Measuring the height and temperature in the shore tank (before and after discharge) to determine delivered volume

Traceability of Overseas Shipping

The shore tanks in the port of loading can be traced to the oil mills or refineries who have delivered in these tanks. In case of loading in bulk tankers with open interconnecting valves, the full content of the ship will be a mixture of the contents of all tanks or ships (in case of trans-shipment) that have loaded into that bulk tanker. At arrival, each delivery will be a mixture of the full content of that bulk tanker. In case of loading into dedicated tanks of a parcel tanker, the content of each ship tank can be traced to the shore tanks that have discharged in that ship tank by the loading pump plan. Receiving shore tanks at the port of discharge can be traced to the tanks of the parcel tanker by the discharge pump plan. Trans-shipments to barges and coasters have similar traceability as discharge into land tanks.

Refining and Modification after Overseas Shipping

The further processing of oils and fats, intended for human consumption, after overseas shipping is given by Table 24-A.

Further processing will normally include bleaching and deodorization as a minimum treatment. This refined palm oil, palm oil fractions, or palm kernel oil can be used as such or after modification of their chemical/physical properties. The modification technologies are

1. Fractionation; crystallization followed by separation in a high and low melting fraction.
2. Hydrogenation; saturation of the double bonds to increase the melting point.
3. Interesterification of one, two, or more components to modify the molecular composition of the product.

Palm products not intended for human consumption (mainly oleo chemicals, feed, and fuel) may be used after sea transport without further refining.

Table 24-A. Rules for Processing after Shipment.

Type of oil/fat	Type of conveyance	Rule to apply
All (crude, RBD, fully refined)	Other than deep sea vessel	Ship tanks must be foodstuff-dedicated. Crude and RBD oils can be further processed to improve quality.
Fully refined	Deep sea vessel	No further processing is required if the ship tanks are of stainless steel or epoxy coated and if the three previous cargoes transported in the tank have been foodstuffs.
Crude or RBD	Deep sea vessel	The oils can be used for human consumption after further processing if the ship tanks are stainless steel or epoxy coated and the direct previous cargo is foodstuff or from the list of allowed previous cargoes (EU or FOSFA). For tanks of other material, the three previous cargoes must have been foodstuffs or from the list of allowed previous cargoes.

Traceability of Refining and Modification

Traceability of refining and fractionation is similar to that of refining and fractionation in the region of origin (see above). Hydrogenation and chemical interesterification are normally batch processes and can be fully traceable by batch tracking. Enzymatic interesterification is a continuous process with possible quality change overs; the output traces back to the oil feed tank(s) with a correction for the oil content of the enzyme reactors which after change-over are still filled by oil from the previously used feed tank(s).

Consumer Goods Manufacturing

Only a very limited number of consumer products contain palm oil or fractions as the majority ingredient (e.g., deep frying fats, soaps). However, it is a minority ingredient in many, many food and non-food consumer products like deep fried chips and snacks, spreads, bakery products, confectionary products, personnel care products, detergents, etc. The palm products transported to the consumer goods manufacturing may be in bulk by road tanker or tank container or in small packed units like drums, bag in box, flakes, etc. On the production site, the oil may be stored in tanks or in the packaging in which it is delivered. Transfer to the production line can be by pumping from the storage tank or directly from the delivered packaging.

Traceability at the Consumer Goods Manufacturer

Consumer products have a product code on pack specifying the date and time of production, the production shift, and a unique batch number. This information allows specification of the used ingredients and tracing of the tanks or packaging from which these ingredients were delivered. Tanks can be fully linked to road tanker or barge delivery, if the oil is received in a different tank than the tank feeding the production line. In case of simultaneous filling and emptying in the same tank, the oil can only be traced to all loads that were delivered in this tank during a long period before the checked production run.

Summary

The European General Food law requires identification of the immediate supplier and identification of the subsequent receiver for each operation in the supply chain. Table 24-B shows that this definition of traceability applies throughout the complete supply chain of palm products. This example shows how a product bought by a consumer can be traced back via the consumer goods manufacturer and the refinery to the ship(s) that have delivered the oil. The seller of the ship load can trace the oil in the bulking station to the delivering oil mills. The oil mills can further trace it to plantations and these to the harvesters and harvested areas in the plantations. Every upstream step in the supply chain contains a larger total volume. The small quantity

Table 24-B. Traceability of Palm Products Throughout the Supply Chain.

Supply chain operation	Traceability to immediate supplier	Traceability to subsequent recipient
Consumer	Retailer where the product was bought and consumer goods manufacturer via brand name and product code	
Consumer goods manufacturer	Refined oil tanks in supplying refinery	Products delivered to retailer via product code
Refinery/oil modification plant	Shore tanks, often customer dedicated	Consumer goods manufacturers
Tanks in sea going vessel	Shore tanks in port(s) of loading via pumping plan(s)	Shore tanks in port(s) of discharge via pumping plan(s)
Shore tanks in bulking station	Oil mills delivering to bulking station	Tanks in sea going vessel
Oil mill	Fruit bunches from plantation and smallholders (via middle man)	Shore tanks in bulking station (oil)
Plantation (fruit bunches)	Harvester and harvested area	Oil mill to which bunches are delivered

of palm product in a consumer unit may originate from a mixture of palm oil from a large shipment, produced by a few dozen palm oil mills, supplied by a few hundred individual harvesters.

Traceability of RSPO-certified Sustainable Palm Oil

RSPO Supply Chain Organization

RSPO Supply Chain Models

RSPO-certified Sustainable Palm Oil is palm oil that is produced by plantations that have been certified against the Principles and Criteria for Sustainable Palm Oil. At present, this concerns roughly 8% (4 million ton) of the global palm oil production (Oil world predicts the global production of palm oil in 2011 at 50 million tonnes, OIL WORLD Statistics Update, September 16, 2011). The production of RSPO-certified Sustainable Palm Oil is concentrated in Indonesia (40%) and Malaysia (48%). The Figs. 24.5 and 24.6 give an overview of the development of the annual production capacity of RSPO-certified Sustainable Palm Oil and RSPO-certified Sustainable Palm Kernel Oil in general and per country.

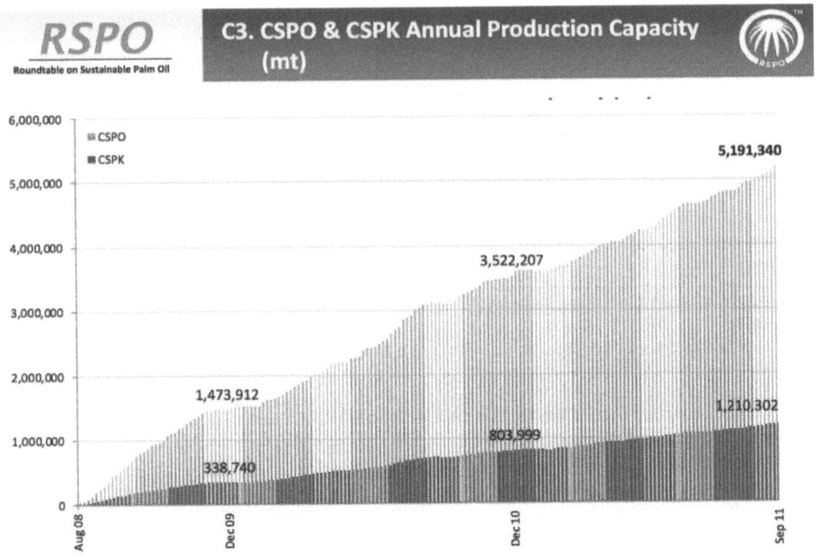

Fig. 24.5. Overview development annual production capacity RSPO-certified sustainable palm oil and RSPO-certified sustainable palm kernel oil. Source: RSPO, www.rspo.org, accessed December 10, 2011.

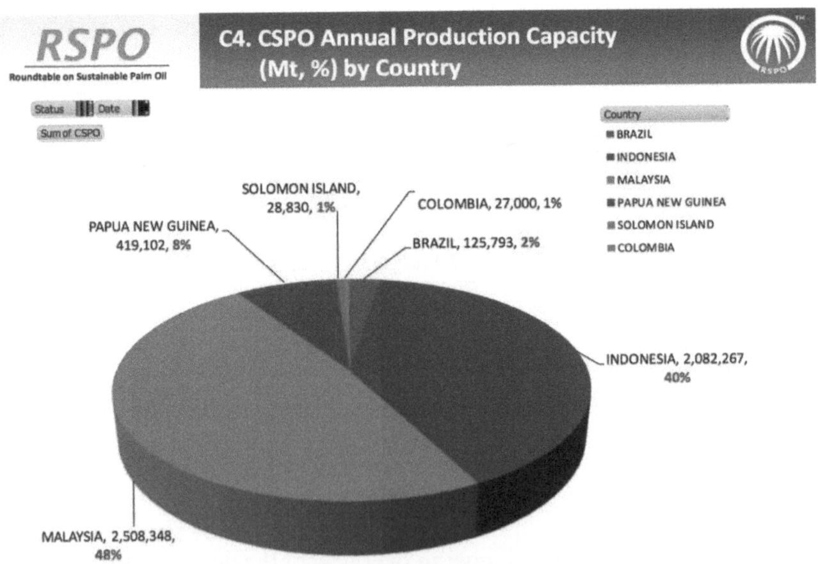

Fig. 24.6. Overview production capacity RSPO-certified sustainable palm oil and RSPO-certified sustainable palm kernel oil per country. Source: RSPO Web site, October 2011.

To make sure the Certified Sustainable Palm Products can be bought and used or claimed in consumer products, the RSPO has designed four trading systems that link the Certified Oil to end users in the market: Identity Preserved (IP), Segregated (SG), Mass Balance (MB), and Book & Claim (B&C) certificate trading. These systems are also called "supply chain models."

When the Identity Preserved, the Segregated, or the Mass Balance model is used, the sustainable palm products are followed through the chain in a different degree of detail. In case the Identity Preserved model is used, the palm products that are traded are identifiable back to the individual oil mill and physically isolated from all other palm products at every stage of the chain. This means that operators in all steps of the chain who purchase or sell the Identity Preserved RSPO-certified Sustainable Palm products know from which plantation the palm oil came. For a schematic overview of this model, see Fig. 24.7.

When the Segregated model is used, operators need to make sure the RSPO-certified Sustainable Palm products are kept separate from non-certified material. They do not have to know the origin. This model thus allows for mixing of sustainable palm products from various sources. Operators are not required to follow the material within their facility. For a schematic overview of this model, see Fig. 24.8.

When the Mass Balance model is used, operators buy a certain quantity of RSPO-certified Sustainable Palm products (Mass Balance or Segregated), but they can mix certified palm products with non-certified products oil as long as they make sure they do not ship more than they have received. When the operator buys in accordance with the Mass Balance model, it is thus never sure if the palm oil product that was received actually contains RSPO-certified Sustainable Palm products. For a schematic overview of this model, see Fig. 24.9.

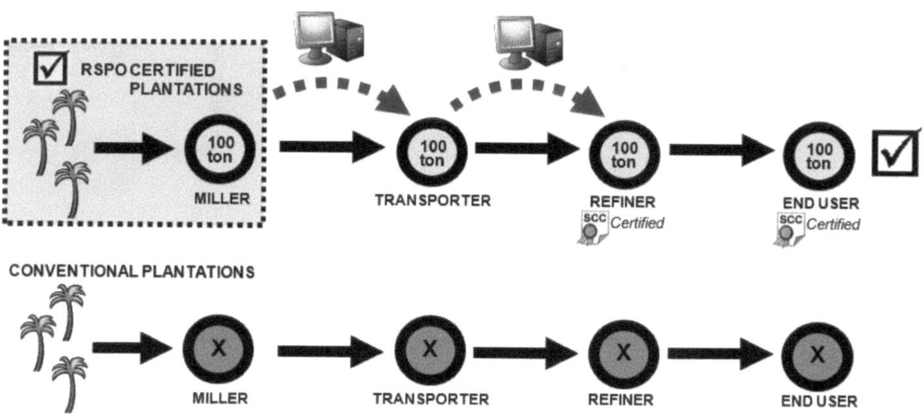

Fig. 24.7. RSPO identity preserved model. Source: RSPO Supply Chain Certification Systems, November 2009.

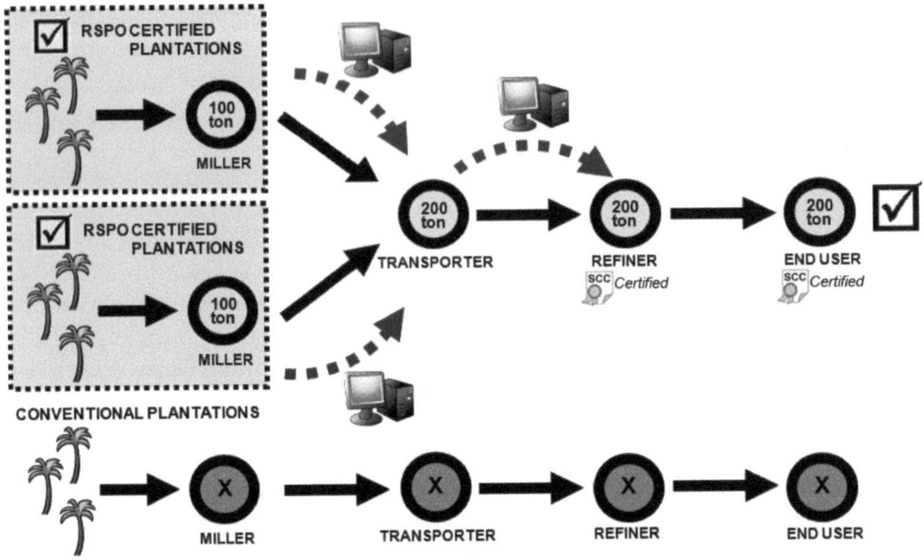

Fig. 24.8. RSPO segregated model. Source: RSPO Supply Chain Certification Systems, November 2009.

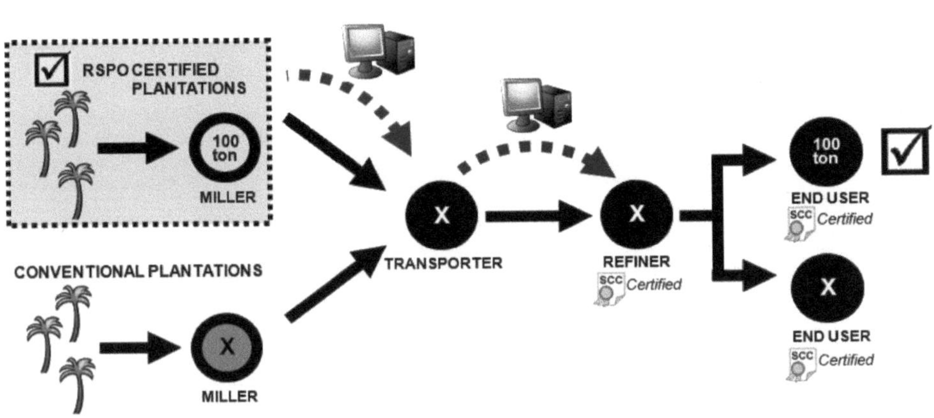

Fig. 24.9. RSPO mass balance model. Source: RSPO Supply Chain Certification Systems, November 2009.

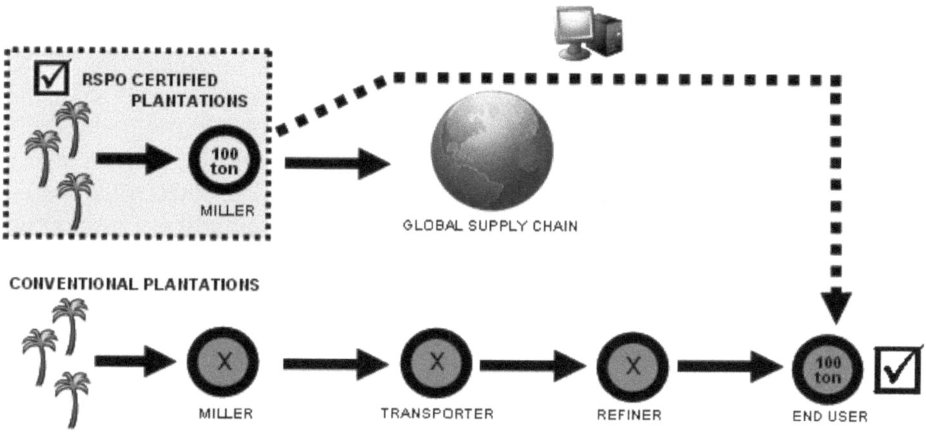

Fig. 24.10. RSPO Book and Claim Model. Source: RSPO Supply Chain Certification Systems, November 2009.

The Book and Claim system allows users of palm oil to buy certificates from a platform to offset their use of palm products. End-users buy certificates directly from producers and the steps in the chain are not followed. When this option is used, there is no link between the physical palm products used and the certificates that are traded. This system thus does not bring additional traceability of the palm oil that is traded. For a schematic overview of this model, see Fig. 24.10.

All rules regarding the different systems are laid down in the RSPO Supply Chain Certification Requirements (RSPO SCCS), which were approved by the RSPO Executive Board on 5 November 2009. They require all operators, that legally own and physically handle RSPO-certified material and who want to claim they purchased or sell RSPO-certified Palm Oil, to be certified. There are three exceptions to the rule that all operators who want to buy and sell RSPO-certified sustainable palm oil need to be RSPO Supply Chain Certified (SCCS Certified):

- operators who do not, at any moment in time, have physical control over certified palm oil;
- retail companies that do not manufacture, modify, (re)package or (re)label consumer goods in any way; and
- Palm Oil Mills. This is a missing link in the system. RSPO is addressing this by introducing SCCS requirements for palm oil mills. The RSPO "Voluntary Add On" that is developed to demonstrate compliance with the EU Renewable Energy Directive requires palm oil mills to be RSPO SCCS certified.

Transaction Registration System—UTZ System

To allow the four different RSPO-approved systems to work in parallel, the RSPO has set up a Web-based Transaction Registration System. The main purpose of this system is to prevent overselling of certified sustainable palm products. Besides this, it ensures RSPO can collect a volume-based levy over the sustainable palm oil that is traded. This system is currently managed by UTZ Certified (www.utzcertified.org).

The Transaction Registration System works by keeping accounts of stocks of operators producing, shipping, and receiving RSPO-certified sustainable palm products. It registers production, shipments, and delivery of three types of operators trading in RSPO-certified Sustainable Palm products: palm oil mills, intermediate supply chain partners (traders and refiners), and end buyers. The system works up and until refinery level, because after that palm products are usually dispersed over many different products in which it is usually not a major component.

When an oil mill is certified it is assigned a "production stock" by the Certification Body based on the production of the previous year. The Certification Body enters this data in the system and the mill can subsequently sell this volume, in accordance with the supply chain models as designed by RSPO. Once the system registers the shipment, the stock of the mill is reduced with the volume that was shipped.

When the Book and Claim system is used, the mill owner can offer a corresponding number of certificates at his offering price at the RSPO Trading Platform for certificates. This platform is currently run by GreenPalm (see www.greenpalm.org). When the palm oil mill uses the trading models with a link to the physical chain (Mass Balance, Segregation, or Identity Preserved) it makes a Shipping Announcement to the system. This way the system knows the volume has to be deducted from the stock of the mill. The buyer receiving the oil needs to make a Shipping Confirmation determining the volume that is added to his administrative stock.

By keeping account of the stocks of the different players in the chain, the system can verify if an operator, who envisages delivering a certain volume of certified sustainable palm oil, can actually deliver this. To guarantee that operators use the System, the RSPO Supply Chain Certification System requires operators who produce or trade RSPO-certified palm oil products, including the final palm oil refinery, to report their receptions and shipments.

At present, the system also functions as a levy instrument for RSPO. Via the system RSPO charges a levy of 1 USD per ton of Sustainable Palm Oil that is traded for RSPO projects. The levy is charged to the first buyer of the mill. This is usually an intermediate supply chain player.

It appears that the Transaction Registration System does not deliver all functions it was foreseen to deliver. One of the reasons for this is that Book and Claim certificates are traded and registered on group level, usually encompassing several mills, and the Transaction Registration System administrates at individual mill level. RSPO is considering how IT support services can be adapted to better fit the needs of RSPO members.

Requirements of End-Users—Food and Fuel

End-users of RSPO-certified Sustainable Palm products can make a claim on the sustainability of the palm oil in their product when they apply one of the supply chain models, as defined by RSPO. The claim they can make depends on the supply chain model that is followed. If they use the MB, the S, or the IP system, they can use the RSPO Trademark. To make the claim, operators need to check the supply chain certification of their supplier and have a license to use the RSPO Trademark.

Some end-users have specific requirements regarding the traceability of palm oil. This concerns for example producers of biofuels, who want to contribute to the targets for selling renewable energy. According to the European Renewable Energy Directive (2009/28/EC), they need to be able to demonstrate that the vegetable oil they used meets sustainability requirements.

RSPO has designed a "Voluntary Add On" to the generic RSPO systems that enable producers and operators in the chain to comply with the requirements in the Renewable Energy Directive. This voluntary add on contains additional requirements; for example, according to this system purchase, orders for RSPO-certified palm oil or palm oil products shall specify the country of origin. The European Commission is evaluating whether this voluntary add on can be considered as sufficient to meet their requirements.

Presently, on average about 50% of the RSPO-certified Palm Oil that is produced is sold. This number is expected to increase; many users of palm products have committed to only source sustainable palm oil as of 2015. This concerns among others General Mills (http://www.generalmills.com/Responsibility/Sourcing/palm_oil_statement.aspx), Unilever (http://www.unilever.com/sustainability/environment/agriculture/palmoil/commitment/index.aspx), Tesco (http://cr2010.tescoplc.com/buying-and-selling-our-products-responsibly/sourcing-our-raw-materials-responsibly/palm-oil.aspx), and Procter and Gamble (http://www.pg.com/en_US/sustainability/point_of_view/palm_oil.shtml). The UK-based retailer Sainsbury's (http://www.j-sainsbury.com/cr/index.asp?pageid=48) has even committed to use only sustainable palm oil in 2014 for their own brand products.

When RSPO-certified Sustainable Palm Oil was just on the market in 2009, most of the trade was done via the Book and Claim certificates. Since then the use of the physical supply chain models "Segregated" and "Mass Balance" steadily increases. At present, 60% of the trade is done via Book and Claim certificates and 40% via the physical supply chain models. For an overview of the development of the market uptake of RSPO-certified Sustainable Palm Oil and the use of the different RSPO supply chain models, see Figs. 24.11 and 24.12.

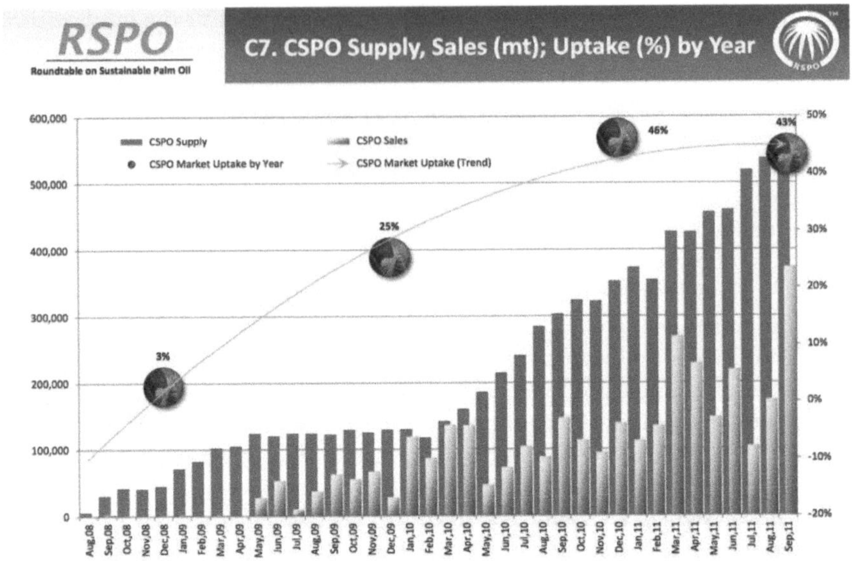

Fig. 24.11. Uptake of RSPO-certified sustainable palm oil since inception RSPO. Source: RSPO Web site, October 2011.

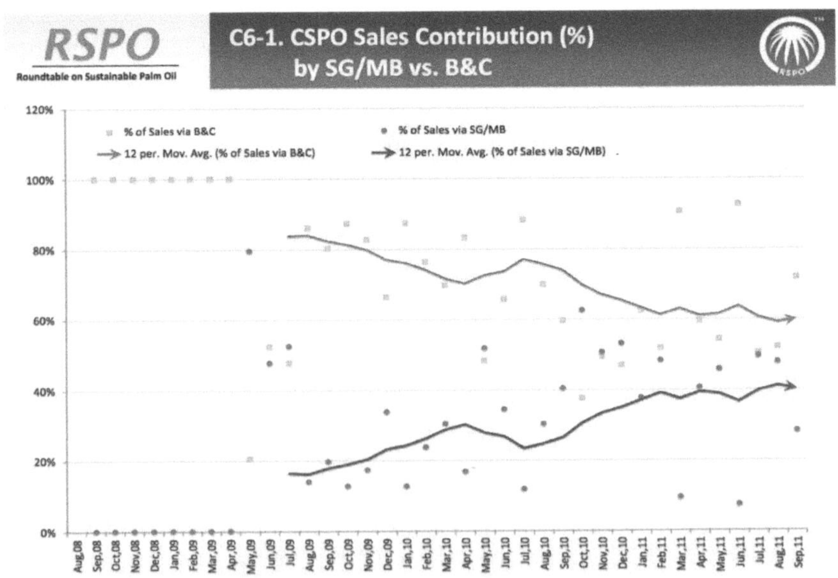

Fig. 24.12. Use of RSPO supply chain models B & C and SG/MB in practice since inception RSPO. Source: RSPO, www.rspo.org, accessed December 10, 2011.

Costs and New Option to Use the RSPO Mass Balance Model

The additional costs of palm products that is traded in accordance with the different models differ depending on availability and supplier. In general, it can be said that buying certificates is the cheapest option to buy RSPO-certified Sustainable Palm Oil, or better said, offset the used palm product. This is logic, as in this case the palm oil does not need to be physically followed in the chain; then follow in order of increasing price: the Mass Balance Model, the Segregation model, and the Identity Preserved Model.

In general it can be said that the more complex the product (fractions or derivatives), the higher the costs of Identity Preserved or Segregation. This is due to the accumulation of costs to keep the material segregated and identifiable through different steps in the chain. The costs are further increased when not all components of a product can be sold as sustainable with a premium. For example, fractionation of 5 tons of palm oil delivers 4 tons of palm olein and 1 ton of palm stearin. A considerable part of the olein market is currently not prepared to pay the additional sustainability premium. Full allocation of the premium to stearin results in a 5 times higher premium compared to the straight palm oil.

When the supply of RSPO-certified Segregated Sustainable Palm products is increasing, the additional premium for the segregated products will decrease. In November 2010, RSPO introduced the new option to use the Mass Balance model to help building the flow of segregated palm oil to Europe. This model allows fractionation plants to buy segregated material and sell the equivalent volume of fractionated product with the Mass Balance claim. This model helps operators who can only sell part of their products with a premium to reduce the premium, as they do not need to buy the full volume needed to produce that material.

Situation in Practice

Does RSPO Supply Chain Organization Bring Additional Traceability?

The above description shows the RSPO Supply chain models are designed to stimulate and increase the sustainable production of palm oil by connecting producers of sustainable palm oil with the end-users thereof. They are not set up to create additional traceability. This is logical, considering that the goal of RSPO is to stimulate and increase sustainable production and not to create traceability.

Nevertheless, when the RSPO Identity Preserved model is used to trade palm products additional traceability is generated. In this case, palm products need to be traceable back to a uniquely identifiable mill. This ensures that the traceability is "continuous" instead of "step wise" and makes the traded palm "more" traceable. However, in practice this model is usually only used for relatively small deliveries for niche products (like sustainable organic oils).

When the "Segregated" model is used to trade palm products, the RSPO systems ensure that the RSPO-certified Palm Oil is not mixed with other palm oil but no additional requirements with regard to traceability are set. In this case, the RSPO systems do not bring additional traceability. If the Book and Claim or the Mass Balance model is used, the physical palm products are not followed through the chain, so it is clear that these RSPO systems do not deliver additional traceability.

Despite this, the RSPO factsheets describing the supply chain options in details rate the four supply options for "the degree of traceability" they deliver:

- The Identity Preserved model has five drops, indicating delivered palm oil is fully traceable to a uniquely identifiable RSPO-certified point of origin;
- The Segregated model has 4 drops indicating delivered palm products only contain RSPO-certified sustainable palm oil, and it is guaranteed that it is a mix of RSPO-certified palm oil from a variety of RSPO-certified points of origin, but not traceable to one unique plantation;
- The Mass Balance option has 2 drops. It indicates delivered physical palm oil is likely not directly linked with the RSPO-certified palm oil at the RSPO-certified point of origin. However, the amount of palm oil with the Mass Balance claim at the end of the supply chain is equal to the amount of certified palm oil that has entered the physical supply chain;
- The GreenPalm option has only one drop, indicating the delivered physical palm oil is not linked to the RSPO-certified sustainable certificate. In other words, this option directly connects the RSPO-certified sustainable producers to end-users of palm oil, without the involvement of the entire chain.

It is clear this analysis is not aligned with the definition of traceability we use in this chapter. The RSPO Supply chain models create additional transparency in the supply chain of RSPO-certified Palm products, but only when the Identity Preserved Model is used is additional traceability generated. The Book and Claim system allows end-users to buy certificates directly of plantation owners. The Segregated and the Mass Balance model ensure that all operators a chain of RSPO-certified Products are certified. The certificates demonstrating the supply chain certifications can be found on the RSPO Web site. In practice, it appears this increased transparency contributes to ensuring better production practices of palm oil, which is the main goal of RSPO.

Nevertheless, the origin of palm oil entering Europe can always be traced back as it has to comply with the European Food Safety Legislation and the FOSFA contract clauses. These demands make sure that it has to be known to the importer where individual shipments arriving in Europe came from and where they were loaded.

Additional Traceability Brings Additional Costs

From the above analysis it becomes clear that there is clearly a trade-off between costs and traceability. If palm oil is shipped in small containments, in separated tanks, it is possible to identify exactly what the origin is. However, this comes at extra costs of separate storage, transport, and refining. Because palm oil is usually a small ingredient in products, the benefits of the traceable palm oil cannot be clearly advertised on products; this means that the added cost of this traceability is often not valued.

Conclusions

Both RSPO-certified and non-certified palm products are traceable because the European food safety legislation and the FOSFA contract clauses require operators to know their suppliers and their customers (one step back and one step forward). Both RSPO-certified and regular palm oil is usually not traceable in the sense that at any given moment in time, the origins of a consignment can be identified (continuous traceability). There is one exception: If RSPO-certified Sustainable Palm Oil is traded in accordance with the RSPO Identity Preserved system, it is always known where a consignment came from. In practice this system is not often used.

The RSPO supply chain models are designed to stimulate and increase the production of sustainable palm oil by connecting sustainably produced palm oil with their end-users. This is what they deliver. The use of the models generates additional transparency in the supply chain of RSPO-certified Sustainable Palm products and stimulates sustainable production practices. The RSPO systems that are currently regularly used to trade RSPO-certified Sustainable Palm Oil—Segregated, Mass Balance, and Book and Claim—do not generate additional traceability of palm oil. The RSPO Segregated model guarantees sustainable palm oil is kept segregated from non-certified palm oil and allows end-users of these products to make claims on that. The Mass Balance model guarantees that operators in a chain are certified and have a volume in volume out accounting system. The Book and Claim model ensures end-users can buy certificates of certified plantations to offset their palm oil use.

With continuing announcements of companies to be fully sustainable in 2015, it becomes clearer and clearer that the European market at a certain point will only use sustainable palm oil. This will eliminate additional segregation costs but not necessarily lead to increased traceability of palm oil. As the chain is organized to generate the maximum level of efficiency and minimize costs, increased traceability leads to a strong increase of supply chain complexity, and thus increase in the costs of palm oil and palm oil products. This will only create additional consumer value if also additional claims can be made on end-products (like organic, special quality, specific origin, etc).

Acknowledgments

The authors thank Mr. Sietse Buisman, Product Line Manager at Cargill Refined Oils Europe, and Mr. Martijn Schneider, Manager Production Planning at Loders Croklaan Europe, for their review and constructive suggestions for this chapter.

References

Association for International Trading in Oils, Fats and Oilseeds. FOSFA Contract # 53 for Vegetable and Marine Oil in Bulk.

Association for International Trading in Oils, Fats and Oilseeds. FOSFA Contract # 54 for Crude Palm Kernel Oil.

Association for International Trading in Oils, Fats and Oilseeds. FOSFA Contract # 80 for Crude Unbleached Palm Oil.

Association for International Trading in Oils, Fats and Oilseeds. FOSFA Contract # 81 for Refined Palm and Palm Kernel Oil Products.

Codex Alimentarius, Codex Alimentarius Commission. Codex Alimentarius Commission Procedural Manual, 20th edition, FAO and WHO, 2011.

European General Food Law, Regulation EC/178/2002.

European Renewable Energy Directive 2009/28/EC.

General Mills. http://www.generalmills.com/Responsibility/Sourcing/palm_oil_statement.aspx (accessed 12/20/2010).

GreenPalm. http://www.greenpalm.org (accessed 12/10/2011).

ISTA Mielke, OIL WORLD Weekly, September 2011

Odfell, http://www.odfjell.com, (accessed 12/08/2011)

Procter and Gamble. http://www.pg.com/en_US/sustainability/point_of_view/palm_oil.shtml (accessed 12/10/2011).

Roundtable on Sustainable Palm Oil. http://www.rspo.org (accessed 12/10/2011).

Roundtable on Sustainable Palm Oil. RSPO Supply Chain Certification Systems. Approved by the RSPO Executive Board on 5 November 2009.

Sainsbury's. http://www.j-sainsbury.com/cr/index.asp?pageid=48 (accessed 12/10/2011).

Tesco. http://www.tescoplc.com/media/60113/tesco-cr-report-2011.pdf (accessed 12/10/2011).

Unilever. http://www.unilever.com/sustainability/environment/agriculture/palmoil/commitment/index.aspx (accessed 12/10/2011).

UTZ Certified. http://www.utzcertified.org (accessed 12/10/2011).

25

Sustainable Development of Palm Oil Industry

Jan Kees Vis[1], Cheng Hai Teoh[2], Mavath R. Chandran[3], Matthias Diemer[4], Simon Lord[5], and Ian McIntosh[6]

[1]Unilever R&D Vlaardingen, P.O. Box 114, 3130AC Vlaardingen, The Netherlands; [2]CSR Asia, Suite 15-02A, Plaza See Hoy Chan, Jalan Raja Chulan, 50200 Kuala Lumpur, Malaysia; [3]Platinum Energy Sdn Bhd, Unit A-11-06, Menara UOA Bangsar, No.5, Jalan Bangsar Utama 1,59000, Kuala Lumpur, Malaysia; [4]WWF Schweiz, Hohlstrasse 110, Postfach, 8010 Zürich, Switzerland; [5]New Britain Palm Oil, 18 Duxton Hill, Singapore, 08960; and [6]53A East End , Walkington, E Yorks, HU17 8RX, United Kingdom

Introduction

Scope of the Chapter

The United Nations Conference on Environment and Development (UNCED) or the Earth Summit in Rio de Janeiro in 1992 laid the foundation for a global shift towards sustainable development. Agenda 21, which was the key outcome of Summit, provided the blueprint for governments, business, industry, and other stakeholders to take action towards a sustainable path. UNCED produced two legally binding agreements, the Convention on Biological Diversity (CBD) and the UN Framework on Climate Change (UNFCCC), and also adopted the Forestry Principles to guide governments and stakeholders on the conservation and sustainable management of all types of forests. The commitments made at the UNCED were further strengthened by UN member states in September 2000 when they pledged to achieve eight tangible and quantifiable Millennium Development Goals (MDGs) by 2015.

Business and industry covering all sectors are expected to promote responsible entrepreneurship and contribute towards the attainment of the UNCED objectives and the MDGs. Being a natural resource-based industry, provisions in the CBD and UNFCCC as well as the Forestry Principles are of direct relevance to the palm oil industry (Teoh, 2003); the MDGs are also significant, particularly with regard to reduction of extreme poverty and environmental sustainability.

Against this backdrop, the past decade has seen an on-going debate over the sustainability of the palm oil industry. While palm oil has been a key driver for economic development in the tropics, through creation of rural employment and reduction of poverty, the rapid growth of the industry to meet increasing demand for palm oil, for both food and non-food uses, has met with criticisms over practices deemed unsustainable and detrimental to nature and local/indigenous communities. This chapter discusses the significant sustainability issues confronting the sector and how the entire supply chain is addressing the problem through the establishment of a multi-stakeholder platform, the Roundtable on Sustainable Palm Oil (RSPO), to promote the production and use of sustainable palm oil. While sustainable production is critical, it must also make economic sense to the producers; therefore, the business case for supporting sustainable palm oil is also discussed, using case studies of the companies who pioneered the production of certified sustainable palm oil.

Overview of the Palm Oil Industry

Among the 17 oils and fats in the global market, palm oil has emerged as the leader, accounting for about 45.9 million tonnes in 2010, which is 33% of total world's production. Its nearest rivals are soybean oil (39.3 million tonnes) and rapeseed oil (23.5 million tonnes). Production of palm oil has increased ten-fold in the past 40 years, mainly as a result of its lower price and higher productivity when compared to other vegetable oils. Today, palm oil is produced in 43 countries, and the world's largest producers are Indonesia and Malaysia which jointly account for about 85% of the global production. Palm oil is also the most widely traded edible oil and the major importers are India, China, and the EU, which collectively account for about 40% (18.3 million tonnes) of the total production.

In view of the growing world population and rising standard of living, coupled with increasing demand for non-food uses and renewable energy, it is projected that the global demand for palm oil by 2015 would be about 62 to 63 million tonnes (an increase of more than 36% over 2010 production) and could increase to 76 to 77 million tonnes by 2020 (Mielke, 2011). Even assuming that part of the projected demands could be met by increases in productivity, it is estimated that a further 12 million hectares of new land would have to be developed to satisfy this increase by 2020.

Given the relatively limited land available for expansion in Southeast Asia, especially in Malaysia, and the push for foreign investment in the sector in other tropical countries, it is likely that future production of palm oil could shift towards countries in Africa and South America where governments, such as Liberia and Nigeria, Cameroon, Gabon in Africa, and Brazil and Colombia in Latin America, have been proactive in supporting the sector. However, such expansion may be limited by government policy. A case in point is Brazil where some 29 million hectares in the Amazonian Basin may be suitable for oil palm development, but the Government has announced future development would be limited to 5 million hectares mainly on non-forested land.

The Palm Oil Sustainability Debate

Setting the Context

March 2011 marks 100 years of commercial production of palm oil in Indonesia where the first plantations were established Pulau Raja and Tanah Itam Ulu estates in Sumatra in 1911 (GAPKI, 2011). That the industry has expanded and flourished over a century is clear evidence of its economic sustainability and relevance to society at large. Yet, in the broader context of sustainability, which includes the environmental and social dimensions, the industry has attracted much controversy and debate in recent times. Has the rapid expansion in production in the past few decades to meet the world's demand for food and non-food applications contributed to the deterioration of the world's state of the environment and its finite natural resources? Headlines and campaigns such as "How the palm oil industry is cooking the climate" (Greenpeace, 2007), "The guilty secrets of palm oil: Are you unwittingly contributing to the devastation of the rain forest?" (Hickman, 2009), and "Ghosts in our own land" (Colchester & Jiwan, 2006) have portrayed the negative social and environmental impacts of the industry. On the other hand, the industry and its supporters have maintained that "palm oil is the only product able to meet a large portion of the world's increasing demand for oil-based consumer goods, foodstuff and biofuel" (MPOC, 2009).

This section discusses the pros and cons of the palm oil industry with regard to its sustainability performance and how the debate has evolved and paved the direction for a multi-stakeholder approach to address the problem.

Positive Impacts—Contribution of the Palm Oil Industry

Economic Development and Growth

Palm oil has been a key driver for economic development in producing countries in Southeast Asia, Papua New Guinea, Central and West Africa, and tropical countries in South America. In particular, large-scale development of the sector was seen in Malaysia and Indonesia as part of the strategy to diversify away from rubber production and for rural development and poverty alleviation. The sector is a major contributor to the export earnings of these producing countries. In Malaysia, for example, the export value of palm oil and its derivatives was equivalent to USD 16.5 billion, or about 7.5% of the Gross National Product (GNP), and was the 4th largest contributor to the national economy in 2009 (Khoo, 2010.) The palm oil industry is one of the 12 National Key Economic Areas (NKEA) that drives the Economic Transformation Programme (ETP) for Malaysia. In Indonesia, palm contributed about USD 7.9 billion to GNP in 2007 (World Bank, 2009), and the national revenues of other producers in Southeast Asia are expected to continue to increase as more land comes into production in a generally bullish global palm oil market.

While palm oil has been traded mainly as a raw material in the form of crude palm oil (CPO), this commodity has been a catalyst for the development of downstream industries, from the refining of CPO initially to the emergence of the specialty fats and oleochemical industries (soaps, household detergents, etc.) and more recently in the production of palm-based biofuels. In addition, palm oil and palm kernel oil and its by-products are important ingredients in cosmetics and nutraceutical products.

Social Development and Job Creation

Being labor intensive, particularly for the harvesting of the ripe fruit bunches, the palm oil industry is a major employer and provides ample job opportunities for the rural population. The number of hectares per worker for field and harvesting operations in oil palm estates is about 8 hectares, whereas in the annual oilseeds crops such as soybean oil, one worker is able to cultivate more than 200 hectares through highly mechanized field operations. In Malaysia, the sector provided direct employment for about 570,000 workers in 2009 and another 290,000 workers were employed in the downstream industries. As Malaysia has a shortage of workers, the palm oil sector is heavily dependent on migrant workers, especially from Indonesia and Bangladesh. In Indonesia, more than 3 million people are involved in the sector, and in Papua New Guinea (PNG) the plantations are the second largest employer next to the civil service.

Although the development and expansion of the palm oil industry has been largely driven by the private sector, the smallholder sector is a major player, and in some countries it is the main producer, especially in Nigeria where about 70% of the oil palm area is under smallholders. In Thailand, the smallholder sector accounts for 76% of the area under oil palm but is responsible for only 32% of the total palm oil production. Smallholders in Indonesia, Malaysia, and PNG cultivate more than 40% for the national area under oil palm and account for 30 to 35% of the national production (Bethe, 2010).

Globally, more than 3 million heads of smallholder families are involved in the industry (www.rspo.org), either as independent smallholders or part of land development schemes supported by national governments. As early as 1956 and with support from the World Bank, the Federal Land Development Authority (Felda) spearheaded development of the smallholder sector in Malaysia to address the dual objectives of resettling the poor and the landless and diversification from rubber. Today, Felda is the largest producer of palm oil in Malaysia with 720,000 ha of oil palm, and 112,635 landless families have been resettled (Ahmad Tarmizi, 2009). The commercial arm of Felda (Felda Holdings Berhad) is 51% controlled by Felda settlers, staff, and their families through their membership in the co-operative known as *Koperasi Permodalan Felda* (Hanim Adnan, 2011).

In Indonesia, smallholder development started with the nucleus estates scheme (*Perkebunan Inti Rayat* or PIR) that was introduced by the government and supported financially by the World Bank from 1978. The PIR scheme was an integral part of the transmigration (*transmigrasi*) program to resettle the poor and landless from Java, Bali, and Sumatra to less densely populated islands, particularly Borneo (Kalimantan). Under this scheme, a government-linked plantation company develops plasma areas of 2 hectares each for the settlers around the company's nucleus estate (Vermeulen & Goad, 2006). At its peak between 1979 and 1984, 535,000 families, or almost 2.5 million people, were moved under the transmigration program. The nucleus estate schemes were eventually replaced with the cooperatives scheme known as KKPA (*Koperasi Kredit Primer Anggota*) that was introduced by the Indonesian government from 1995 (Suharto, 2009).

Nutritional Aspects—Palm Oil in Human Nutrition

Oils and fats form an integral part of our diets, and today palm oil is among the major oils consumed globally. Choice of the oil or fat has traditionally been based on quality characteristic, product functionality, sensory properties, oxidative stability, and resistance to development of rancidity in foods. Although palm oil easily fits these criteria, its path towards universal acceptance has been fraught with many challenges due to its relatively high content of saturated fatty acids. Through a systematic nutritional research program spanning nearly four decades, much of misgivings about palm oil have been overcome.

Current knowledge suggests that not all saturated fatty acids have adverse effects on health, especially cardiovascular health, and their role in the etiology of heart health is being re-evaluated (Siri-Tarino et al., 2010). Quality and quantity both impact nutritional outcomes, signaling the possibility of making healthy choices from most oils and fats including from palm oil. The realization that monounsaturated oleic acid is not just a neutral dietary component has spurred significant interest in monounsaturated-rich oils including olive and palm olein (Khosla & Sundram, 2011).

Palm oil has been the subject of nutritional scrutiny for the last 40 years. It contains the highest number of polyunsaturated fatty acids of all the saturated fats (e.g., about five times more than coconut oil or butter fat). With the right balance, saturated fatty acids help maintain high-density lipoprotein (HDL) while polyunsaturated fatty acids decrease low-density lipoprotein (LDL). Balanced dietary fatty acid composition has long been advocated by the American Heart Association and such balanced compositions are readily apparent when palm oil is included as a major component of such a recommended American Heart Association (AHA) dietary fat blends (Sundram et al., 2003). The emerging realization that triacylglycerol structure and placement of fatty acids in the sn-2 position could impact nutrition and physiology of fats adds a new dimension to these issues (Karupaiah & Sundram, 2007). Each of these areas has been addressed for palm oil and has contributed to the positioning of palm as a healthy edible oil choice (Khosla & Sundram, 2011).

The food industry's choices are not always nutritionally sound, and the previous preference for *trans*-rich partially hydrogenated fats in food are testimony to this. The adverse health effects of trans fats has been proven conclusively and has led to global legislations and increasing demands from consumers for healthy alternatives with zero *trans* content (Mozaffarian & Stampfer, 2010). An expert committee, comprising leading nutrition researchers, concluded that palm oil is a healthier alternative to replace *trans*-fat in food manufacturing (Hayes, 2010). Whereas *trans*-fatty acids increase LDL-cholesterol and decrease HDL-cholesterol, palm oil is an attractive alternative that may actually moderately increase healthy HDL levels. Palm oil has found favorable niches in *trans* fat replacements, and several human clinical trials have clearly demonstrated the beneficial health outcomes (Khosla & Sundram, 2011). Palm oil's rich content of bioactives (carotenes, vitamin E tocotrienols, and phenolic acids) are also nutritionally attractive. Red palm oil, rich in pro-Vitamin A carotenes including beta-carotene, helps prevent Vitamin A deficiency. The estimated 190 million preschool-age children and 19 million pregnant women at risk in developing countries from blindness, anemia, and even death due to Vitamin A deficiency could benefit from red palm oil inclusion in their diets since several clinical interventions have proven the effectiveness of red palm oil in preventing vitamin A deficiency in such at-risk populations (Rice, 2010). This food based approach is also proving more attractive since there is a historical track record of palm oil consumption dating back nearly 5000 years.

Palm Vitamin E tocotrienols, especially alpha tocotrienol, may help protect the brain against neurological damage. Tocotrienols may help protect the brain's sensitive nerve cells, possibly limiting the damage due to stroke, and helping to accelerate recovery of functional loss (Sen et al., 2010). Tocotrienols in palm oil display potent anti-cancer activity in several model systems. Although tocopherols have shown no ability to fight cancer cells (Sylvester et al., 2010), the tocotrienols in palm oil have the potential to block the activation of NF-KB and may help the body better regulate its inflammatory and immune responses. Despite such findings (gained from scientific efforts spanning nearly 160 studies), palm oil researchers continue to explore palm oils nutritional potential.

Oil Palm as the Most Productive and Affordable Vegetable Oil Crop

The world average productivity of oil palm in 2009 was about 3.6 tonnes oil per hectare, while in well managed plantations more than 7.0 tonnes of oil per ha has been achieved (IOI Corporation, 2009). Compared with the other vegetable oils, land productivity of oil palm was more than 6–10 times higher than the competing edible oils. In terms of total production, oil palm contributed about 38% of the 8 major vegetable oils requiring about 5.3% of the world's total area under vegetable oil crops. (Kholsa & Sundram, 2010). Palm oil is the cheapest source of vegetable oil and has historically been marketed at a substantial discount to soybean oil. Although fluctuations in the price of palm oil have been broadly in tandem with the price of fossil fuel, it remains the most affordable oil, especially for consumers in the developing world. The combination of high productivity and competitive price is possibly one of the key factors for the continued world demand for this commodity. In addition to this, oil palm can produce commercially viable yields on degraded land, which may be considered less suitable for other crops.

Negative Impacts of the Palm Oil Boom

Key Sustainability Challenges Associated with Palm Oil Industry

The key sustainability challenges confronting the palm oil sector were reviewed in a discussion paper (Teoh, 2010) that was commissioned by the World Bank Group (WBG) to facilitate a series of global consultations in 2010 as part of the WBG's effort on development of a new strategy for their involvement in the palm oil sector. A summary of the diverse range of issues and concerns identified during the consultations is and the proceedings for each consultation session are available from www.ifc.org/palmoilstrategy/.

Although there were differences in areas of concern at the detail level which reflected the regional or local realities, there was a high degree of consensus on key sustainability issues relating to deforestation; loss of biodiversity; climate change; and social issues with regard to land tenure, rights of indigenous peoples and local communities, and the well-being of smallholders, who form the backbone of the sector in several producing countries. These issues have also been the focus of Non-Government Organizations (NGOs) and media campaigns against the industry. A brief overview is given in this section.

Deforestation. The conservation of global forest resources is a major concern of many stakeholder groups including NGOs, national governments, and international organizations such as the UN Food and Agriculture Organization (FAO). Serious losses of forests with high conservation values have resulted from logging for timber, infrastructure development, and conversion to agricultural crops. Although producers, especially in Malaysia have maintained that oil palm has been developed through conversion from other perennial tree crops such as rubber and cocoa or on logged-over forests by the timber industry, the expansion of the sector, from the perspective of NGOs and other stakeholders, has often been associated with deforestation. Clay (2004) stated that there is a direct relationship between the growth of the industry and deforestation in Malaysia and Indonesia. An evaluation of land cover by the FAO in 2005 had indicated that between 1990 and 2005, about 55 to 59% of oil palm expansion in Malaysia and Indonesia had been at the expense of forest cover (Koh & Wilcove, 2008). The linkage between oil palm development and forest conversion has been demonstrated in some studies in Indonesia. For example, an assessment of deforestation and degradation in the Province of Riau, Sumatra, showed a 65% loss of forest cover between 1982 and 2007, or about a loss of about 4.2 million hectares over the 25-year period. It was estimated that the cultivation of oil palm contributed to about 24% of the forest loss in Riau (Uryu et al., 2008).

Deforestation has serious impacts on the whole ecosystem through loss of biodiversity, degradation of watershed resulting in hydrological changes, diminished carbon storage, and increased greenhouse gas emissions that contribute to climate change. Local communities and indigenous peoples can also be deprived of the valuable ecosystem services provided by forests. The economic and social values of these services need to be considered, as an analysis by Koh & Wilcove (2007) estimated that the value of an intact tropical forest to society is over USD 6,000 per ha compared to the returns of about USD 2,100 per ha per year from mature oil palm plantations.

The discussion on the linkage between oil palm and deforestation is often complicated by the lack of a common agreement on definitions of what types of forests are being destroyed. It is likely that original tropical rainforests have already been logged by the timber industry and the resulting secondary forests are exploited further before its conversion for oil palm plantations. The industry tends to hold the view that plantations have been developed on logged over forests. However, NGOs and other stakeholders have maintained that such areas may still have high conservation values and thus, need to be conserved. Irrespective, there is ample evidence that valuable forest areas were and are being cleared to make place for plantations. In fact, as recently as 2010 encroachment by oil palm developments within Tesso Nilo National Park has been acknowledged at the ministerial level (Jakarta Post, 2010). Likewise, despite Zero Burning policies, forest fires still originate in the vicinity of oil palm plantations, and there is evidence that they have been set purposefully to expand areas (http://eyesontheforest.or.id/).

In principle, governments of palm oil-producing countries acknowledge the need for forest conservation, and relevant laws have been enacted to protect the forests. In 2007, the governments of Indonesia, Malaysia, and Brunei signed the historic "Heart of Borneo" agreement to conserve about 220,000 sq. km of tropical forests in the 3 countries. These efforts were undermined by the so-called Mega Oil Palm Plantation scheme along the Kalimantan border, which was proposed to encompass 1.8 million hectares. As a result of massive interventions by NGOs and poor suitability, this project was more or less abandoned. Recently, Indonesia has introduced a 2-year moratorium on the clearing of primary forests and peatlands as part of the national effort to combat climate change under the scheme known as Reducing Emissions from Deforestation and Forest Degradation (REDD). While these commitments augur well for forest conservation, the threat of further forest conversion could continue if commitments are not matched with good governance and effective enforcement of the relevant laws and regulations. This concern was reflected in the WBG stakeholder consultations in 2010 when weak governance framework was consistently identified as a major challenge towards sustainable development of the palm oil sector. Among the wide range of governance issues raised by stakeholders, the need for clear and effective spatial planning frameworks, laws and enforcement mechanisms were considered crucial (World Bank Group, 2010).

Given the critical role forests have in providing and maintaining ecosystems services, conversion of high conservation value forests for agricultural or other development is clearly not a sustainable option. New business models that use less valuable land for oil palm expansion have to be considered. Shifting plantations to idle lands could be a win-win solution (WWF HOB NI, 2011) if agreements can be reached with existing users of the land and if defaulted concessions are revoked. The expanse of these areas is considerable: estimates for degraded land or *alang alang* (*Imperata cylindrica*) grasslands range from 8 to 20 million hectares in Indonesia. In this context, it is noted that the Government of Brazil (2010) announced the program for Development of Sustainable Palm Oil in Brazil, which would limit the expansion of oil palm cultivation to 5 million hectares of degraded land.

Biodiversity Loss. The loss of biodiversity is closely associated with deforestation and forest degradation. The major oil palm producing countries in Southeast Asia as well as emerging producers in Africa and South America are also mega-biodiversity centers, and numerous endangered species are under threat from development. Being a monoculture, it is clear that an oil palm plantation will have lower biodiversity than natural forests. The plight of large flagship species such the Asian elephant, Asian rhino, Sumatran tiger, and the orangutan have received considerable attention from NGOs and the media. The survival of these species is threatened by forest clearing; fragmentation of their forest habitats has led to frequent human–wildlife conflicts, particularly in the case of the Sumatran tiger and elephants. As a consequence, elephants continue to be poisoned when they raid plantations in search for food. In poorly managed plantations, poaching or illegal wildlife trade, for example with immature orangutans, still occurs. The occurrence of forest fires and the use of fire for land clearing for agriculture have also caused serious losses to wildlife as well as flora.

The survival of the orangutans become the focal point of numerous NGO campaigns and the subject of media attention against the palm oil industry. The orangutan has become symbolic of the problems related to deforestation and forest degradation and is used as barometer of the health of forests (Iko Hari et al., 2006). The Bornean orangutan has been classified by the World Conservation Union (IUCN) as "endangered," while the Sumatran orangutan is listed as "critically endangered." Estimates of the wild orangutan population in Borneo vary from 45,000 to 60,000 individuals, but in Sumatra there are about only 7,300 orangutan (Nellamann et al., 2007).

A major factor that contributes to biodiversity loss is the inadequate attention given by project proponents during the planning phase to ascertain if the areas earmarked for conversion has significant biodiversity value. Although national regulations usually require an Environmental Impact Assessment (EIA) or an *Amdal* in Indonesia, the assessment of biodiversity in the proposed project site may not be sufficiently comprehensive. In addition, where they are comprehensive, the biodiversity assessment often limits itself to an inventory of species found in the area and may not incorporate appropriate management plans for the conservation of any rare, endangered, or threatened species identified.

Climate Change. Deforestation and conversion of forest land for large scale monoculture has also been linked to greenhouse gas (GHG) emissions and the impact on climate change. The awareness of this linkage had been accentuated since the publication of the Stern Review, which stated that about 18% of global greenhouse gas (GHG) emissions can be attributed to deforestation (Stern, 2006). Land use change, particularly in the conversion of carbon rich lowland tropical peat areas, could contribute significantly to GHG emissions, and this has become a major concern.

As land on mineral soil becomes less readily available in Southeast Asia, the expansion of palm oil has extended into lowland peat forests. It has been estimated that Indonesia has 22.5 million hectares of peat soil which is about 12% of the total land area. Drainage of this massive store of carbon and conversion to palm oil could contribute to enormous greenhouse gas emissions. Fires from degraded peatland cause further CO_2 emissions and contribute to the regional haze problem in Southeast Asia. A Delft Hydraulics study showed that CO_2 emissions arising from decomposition of drained peatlands in Indonesia are likely to be 632 Mt/yr, with the range of 355 to 874 Mt/yr (Hooijer et al., 2006). Further, an estimated average emission of 1400 Mt/yr was caused by peatland fires over 1997–2006.

Use of fire for land clearing is another source of GHG emissions. Although the ASEAN countries signed an agreement to address trans-boundary haze pollution in 2002 and had adopted a regional policy on zero burning (ASEAN Secretariat, 2003), the incidence of haze resulting from open burning continues annually, albeit at a much lower intensity than the great fire of 1997/98. While oil palm plantations companies have generally adopted the zero burning practice, use of fire for land clearing remains a traditional practice among farmers and smallholders, as heavy machinery required for implementing the "zero burn" approach is often not available to them.

Release of methane from palm oil effluent treatment plants is a significant concern as it is more potent than CO_2 in terms of global warming potential. The most commonly used system for treatment of palm oil mill effluents (POME) before they are allowed to be discharged into waterways is the anaerobic digestion of POME through a series of ponds. As most of the effluent treatment plants have open ponds, the methane generated is released directly into the atmosphere. More than 80% of the 423 POME treatment plants in Malaysia use the open pond system. Progressive companies have taken measures to capture the methane and use it for generation of electricity, often under the auspices of the Clean Development Mechanism (CDM) under the Kyoto Protocol. However, the sector as a whole faces the major challenge to implement methane capture in all treatment plants as the capital expenditure required could not be justified unless it is offset by revenue from credits for carbon emissions reductions. In some countries, for example PNG, oil palm mills are linking with national power authorities to implant such methane capture systems into the national electrical grid as renewable energy-driven power stations.

The production of palm oil-based biofuels as a renewable energy will depend on compliance to the European Union (EU) Directive for Renewable Energy, which requires biofuels to achieve minimum reductions in GHG emissions compared to fossil fuels. The target for 2010 was a 30% reduction, and this would be raised to 60% by 2017. However, currently there is much contention over the assignment of emissions savings for palm oil, and the industry has claimed that the potential emissions savings from palm oil has been understated (MPOC, 2008). Unless this discrepancy is addressed, the potential use of palm oil as a feedstock for renewable fuels from a carbon life cycle perspective is curtailed.

Social Issues. Although development of the palm oil sector has contributed to socio-economic development and an effective vehicle for poverty reduction, the 2010 WBG stakeholder consultations identified a wide range of social issues concerning indigenous peoples and local communities, smallholders, and the plantation workers in all producing countries. The problems are more acute in Indonesia, and these have been well documented in numerous publications such as *Losing Ground: The human rights impacts of oil palm expansion in Indonesia* (Marti, 2008) and *Land is Life: Land rights and oil palm development in Sarawak* (Colchester et al., 2007). For many indigenous peoples who have used their land for generations, the development of large scale agriculture is a cultural transformation that can disrupt their way of life and denies them access to environmental services provided by forests when they are replaced by plantations.

Many of the social issues and conflicts relate to questions over land rights and ownership, land use, and land acquisition without fair compensation. For example, more than 570 conflicts between companies and the government with indigenous peoples and local communities in the palm oil sector have been recorded in Indonesia in 2008 while in Sarawak, 40 of the total of 150 cases of land conflicts involved oil palm companies. An underlying cause of land conflicts are inconsistencies of the laws relating to the recognition of and respect for native customary rights. For example, although the 1945 Constitution of Indonesia recognizes the existence of traditional political entities based on the heritage of indigenous peoples, some laws enacted subsequently, such as the Basic Forest Law No5, have been discriminatory towards indigenous peoples with respect to their access to forest ecosystem goods and services (Jiwan, 2009). In Malaysia, although the 1957 Sarawak Land Code recognizes the Dayak's right to land, the Forest Ordinance of 1953 restricted or controlled their activities in Permanent Forests areas and was intended to control the practice of shifting agriculture (Raunter et al., 2005).

For economy of scale, development of oil palm plantations requires large tracts of land, and it may involve the acquisition of land belonging to individuals and local communities. Conflicts may arise if the local land owners do not wish to be involved in the development or, in other cases, where there is lack of clarity over the legal ownership of the land which could result in multiple claims over the same piece of land. An example is the land dispute between IOI Corporation Berhad and Kayan and Kenyah natives in Sarawak. Although the High Court has ruled in 2010 that IOI could continue to occupy the land, the natives must be given compensation that is established by the Registrar. However, in the process of seeking the final resolution, a major challenge encountered was "the absence of a clear identity or identification of the persons who constitute the affected parties" (IOI Corporation, 2011). Conflicts between plantation companies and local communities or indigenous peoples also arise from a lack of adequate consultation and agreement among the parties especially with regard to the terms of acquisition and compensation. Such disputes can be minimized if the acquiring company has undertaken effective due diligence following the principle of Free Prior Informed Consent (FPIC). This means that the land owners, through consultation, receive full disclosure of information from the developer and fully understand the implications of the transaction before freely agreeing to sell or participate in a project.

The social issues surrounding smallholders are those factors which affect their livelihood. These can be summarized as follows:

- agronomic constraints (knowledge of and access to extension services pertaining to best management practices for sustainable production (e.g., land quality, input quality such as planting materials, and fertilizers);
- involvement in the supply chain (access to credit to meet development and operating expenses, consistent access to mills and reliable infrastructure to ensure timely delivery of oil palm bunches from the smallholder farms to the processing centers, and equitable prices for smallholders' fresh fruit bunches); and
- access to enabling environment (organizing structures of smallholders, legal structures including land tenure, market dynamics including impact of price volatility, and social dynamics such as health and education) (FSG, 2010).

The impact of these factors varies with the type of smallholders. In general, "supported" or "schemed," which are tied to specific mills in out grower schemes or land development such as nucleus estate system, are better placed than "independent" smallholders in terms of access to credit, infrastructure and milling facilities, and the availability of advice on sound agricultural practices from the associated plantation companies or land development agencies such as Felda. Although independent smallholders have the freedom to sell their crop to any dealer or mill, the monopoly held by the mill may make it difficult for smallholders to obtain fair prices for their produce and so ensure they maintain a passive role in this rural economy.

Raising productivity is perhaps the biggest challenge faced by the smallholder sector as there is a wide productivity gap between smallholders and plantations. For example, the average yield in the smallholder sector in Indonesia was 35% and 40% lower than the production from private and government–owned plantations (Suharto, 2009). Actual productivity varied considerably, depending on the type of smallholders. Even among the supported smallholders in nucleus estate schemes, there was up to 50% variability around the mean production level (Zen et al., 2005). Low productivity, particularly among the independent smallholders is often ascribed to the lack of capacity of the smallholders in various aspects of cultivation and production, from use of inferior planting materials, poor soil and pest management, and harvesting of under-ripe or unripe fruit bunches (Ayat Rahman et al., 2008). Poor quality of the harvested crop leads to lower oil extraction and thus less income for the farmers.

Issues related to workers in the palm oil sector include the need for a fair and safe working environment with regard to health and occupational safety, freedom of association, fair wages, and various forms of discrimination, including sexual harassment in the workplace. National laws in producing countries such as Indonesia and Malaysia have provisions to meet the basic rights of workers such as terms of employment and occupational health and safety, but their effective and consistent enforcement is often an area of concern. Although terms for minimum wages and benefits may be set through collective agreements between employers and workers' unions or by relevant Government agencies, these may vary with states or provinces/districts. For example, while collective wage agreements between the employer association, Malaysian Agricultural Producers Association, and the respective staff and workers' unions have been practiced in Peninsular Malaysia since the 1960s, collective agreements are not done in Sabah and Sarawak which account for more than 40% of the palm oil production in Malaysia.

The more vulnerable groups are women workers, children, and immigrant labor. In general, women workers are not fully aware of gender related issues; sexual harassment in the workplace and treatment of gender issues have often not been given adequate attention by male dominated plantation management. Women are a significant component of the plantation workforce and they are employed mainly in field operations such as maintenance of oil palm nurseries, planting, weeding and application of pesticides. Spraying of hazardous pesticides such as paraquat has been reported to cause serious health problems to women and they are at greater risks when working during pregnancy. The risks posed by handling organophosphate pesticides are even greater. These chemicals absorbed through the skin or inhaled can cause damage to the central nervous system and ultimately death.

The issue of child labor can be complex because of the nature of work in the plantation, particularly during crop harvesting. While companies may comply with minimum legal age requirements for engagement of young workers, they may not have effective control over children working voluntarily alongside their parents in the fields. Children often assist their parents in the collection of loose palm fruitlets from the ground. Loose fruitlets have the highest oil content and efficient recovery will enhance the earnings of the harvester. While it may be difficult to prevent children from contributing to the family income, the important consideration is that these activities must not adversely affect their access to formal education.

The palm oil industry in Malaysia is highly dependent on foreign workers; more than 400,000 workers, mainly from Indonesia, help to keep the industry going. Although employers are required by Malaysian laws to provide the same employment benefits as Malaysian workers to the guest workers, in reality, they may face discrimination and are often paid less than the minimum wage (Marti, 2008).

Food Security

Access to sufficient quantities of food will be a critical issue this century. Global food security, particularly for those living closest to the poverty line, will be heavily reliant on low-cost supply of food. In addition, the world population is expected to increase by 2 billion to 9 billion by 2050. The United Nations (UN) Food and Agricultural Organization (FAO) has forecast a 70% increase in food demand to feed the additional mouths (FAO 2009). Food security exists when all people, at all times, have physical and economic access to sufficient, safe, and nutritious food to meet their dietary needs and food preferences for an active and healthy life (FAO 2003). In September 2010, the UN expounded eight Millennium Development Goals. At the top of the list is the goal to eradicate extreme poverty and hunger.

Palm oil and palm kernel oil supply the bulk of the world's trade in oils and fats and contributed nearly 60% of the market share in 2010. Palm oil can continue to play a significant role in ensuring food security. Through research and development efforts and investments in advanced technology, the industry can balance the need to meet increased demand with the need to achieve higher productivity per unit area with the lowest possible environmental impact.

There are, however, major hurdles to overcome in the pursuit of food security. Of major concern is the diminishing availability of already scarce arable land. A FAO analysis showed arable land to have decreased from 0.75 hectare per capita in 1922 to 0.23 hectare in 2005 and predicted this to further decrease to 0.19 hectare in 2030. Thus, there is a crucial need to optimize the use of land. Oil palm has demonstrated high land productivity compared to the competing edible oils. There is currently a shortage of oils and fats in some parts of the developing world due to weather induced disasters, political unrest, infrastructure problems, and trade distortions, among other factors, thus leading to food inflation. But abundant amounts of palm oil remain available all year around at an affordable price, lending a reliable hand in efforts to curb hunger.

Evolution of the Palm Oil Sustainability Debate

What is the controversy over the sustainability of palm oil about? Why are NGO and media campaigns directed at the palm oil sector and not the other vegetable oils such as soybean oil and rapeseed oil? Is it just another technical barrier to trade? These sorts of questions were raised in the past decade or so, as palm oil production continues to expand to meet ever increasing global demand. This section gives an overview on how the debate over the impact of the oil palm boom on natural resources and people and climate change has developed since the late 1990s.

The Year the World Caught Fire

Although forest conservation is a core objective of the major international NGOs, it was the "year the world caught fire" (WWF, 1997) that accentuated their response to the massive expansion of the palm oil industry in past decades in Southeast Asia and its potential impact on forest resources. During 1997 and 1998, large scale forest fires raged in Southeast Asia, Australia and Papua New Guinea, Central and South America, Europe, Russia, China, and USA. In Southeast Asia, the fires in Indonesia were most extensive and about 9.5 million hectares of land were affected, of which about 49% was under forests. The economic cost was estimated to be USD 5 to 10 billion and the health of up to 70 million people in the region was adversely affected (Rowell & Moore, 2000).

In 1997, WWF Nederland initiated studies on the underlying causes for the forest fires, and the resulting report had provided a linkage of several market players, including multi-national corporations (MNCs) like Unilever and financial institutions to the oil boom in Indonesia. This led WWF Germany to mount their campaign in 1998 on "Burning Forests for Margarine"; although the campaign was directed at German companies, it also raised the awareness of consumers in other European countries (AidEnvironment, 2007). In a review of the global forest fires commissioned by WWF and IUCN, Rowell & Moore (2000) stated "up to 80% of the fires in Sumatra and Kalimantan were started by plantation companies, using fire as a tool to clear land." The great fires became the trigger for the "conservation change mechanism" (AidEnvironment, 2007) and catalyzed international NGOs to intensify their campaigns on oil palm's impact on deforestation.

The WWF Forest Conversion Initiative

To address the increasing threats posed by the expansion of oil palms in Southeast Asia as well as soybean in South America to tropical forests, WWF initiated its campaign on "strategic action on palm oil and soy" in 2000. This was renamed as the "Forest Conversion Initiative" (FCI) in 2001. The early goal of the FCI was "by 2005, ensure that High Conservation Value Forests, fresh water ecosystems and habitats of key species in focal ecosystems are no longer threatened by expansion of palm oil and soy" (WWF, 2002a). A number of strategies focused on engaging the main players in the palm oil supply chain were to be employed to achieve the FCI objective.

A major FCI strategy was to transform the market through the establishment of roundtables for palm oil and soy to bring about collective actions by the key actors in the two industries (WWF, 2002a). Of the two commodities, more rapid progress had been achieved in the development of the roundtable for palm oil.

Shifting Focus from Deforestation and Social Impacts to Climate Change

During the early phase of the sustainability debate, concern was focused on deforestation, loss of biodiversity, and the social impacts caused by the expansion of the sector. In the WWF position paper on oil palm in February 2002 (WWF, 2002b), the key elements of sustainability within the oil palm industry should include "maintenance of high conservation value forests" and "positive social impacts" to ensure the long term well-being of plantation workers and local communities. The report, "Greasy palms—palm oil, the environment and big business" by Friends of the Earth (2005) focused on "rainforest destruction in your shopping basket." The main impacts of oil palm highlighted in the report were deforestation, forest fires, pollution, and social conflicts over land rights. Absent in this and other publications of the time were mention of the direct impacts of the oil palm expansion on greenhouse gas (GHG) emissions and climate change. The negative impacts of oil palm development on local communities and indigenous peoples in Indonesia has been a focus of local NGOs, Sawit Watch, and Walhi. In 1998, Sawit Watch was set up as a network of members and local contacts working with local communities in Sumatra, Kalimantan, and Sulawesi. The objective was to "support local communities who have lost their forests and livelihoods because of large scale oil palm expansion and for those in the forestlands who continue to resist the development" (Sawit Watch, 2002). They worked together with the Forest Peoples Programmes and other international NGOs to champion the cause of the local communities and indigenous peoples (Colchester et al., 2006; Marti, 2008).

At a meeting in Kuala Lumpur organized by the voluntary oil palm Roundtable Organizing Committee in August 2003, the focus was on establishing the impact between oil palm development and deforestation and seeking a clear definition of sustainability for the sector. The deliberations were supported by 2 discussion papers prepared by ProForest:

- Palm oil, forests and sustainability
- Defining sustainability in palm oil production: An analysis of existing sustainable agriculture and oil palm initiatives

This roundtable meeting led to the formation of the Roundtable on Sustainable Palm Oil (RSPO) in 2004, one of the first multi stakeholder commodity fora and the subsequent establishment of a Standards and system for certifying sustainable palm oil production. Concurrently, supply chain mechanisms were developed to facilitate trade in certified sustainable palm oil. Despite these efforts, the sustainability debate continued. Social and Environmental NGOs continue to identify and bring forth further cases of deforestation and the plight of endangered species, especially the orang-utan, involving oil palm. Through the work of Sawit Watch, Forest Peoples Programme and Oxfam, social issues conflicts have been brought to a wider audience. From 2007 onwards, attention has shifted to serious concerns over the growing demand for palm oil as a feedstock for production of renewable energy particularly after the announcement of EU's ambitious and binding 10% target for biofuels in road transport by 2020. As rapeseed oil production would not be able to meet the projected demands for biofuel in Europe, palm oil was considered to be the suitable alternative feedstock. If the EU's demand for biofuel were to be met by palm oil alone, it has been estimated that another 4 million hectares of oil palm would have to be planted (Sheil et al., 2009). Given the limited availability of mineral soils in the major producing countries, future expansion is likely to be on high carbon stock land, especially peat swamps.

Focus on Climate Change

While the Stern Report (Stern 2006) identified deforestation as a significant source of global GHG, it was the Delft Hydraulics assessment of CO_2 emissions from drained peatlands in Southeast Asia that had added significantly to the sustainability debate from the climate change perspective. The concern is drainage of tropical peatland for oil palm cultivation would expose the peat to oxidation, resulting in release of CO_2 over extended periods. The study concluded that emissions arising from decomposition of peat combined with emissions from peatland fires that occurred during 1997 to 2006 have made Indonesia the third largest emitter of greenhouse gases, after China and the USA (Hooijer et al., 2006). However, this assertion had been challenged by many in the industry (Paramananthan, 2008; Suharto, R., cited by Hanin, 2009; Melling, cited by Wong, 2010). While the contribution of peatland to GHG emissions and climate change remains a contentious issue, the Delft Hydraulics findings have been used to support NGOs and media campaigns. Irrespective of the relative contributions of peat, emissions from fires, land use changes, and other sources have contributed to the fact that Indonesia ranks among the top emitters of GHG world-wide. According to Wicke et al. (2011) roughly 7 million hectares of forests in Indonesia were converted to palm oil by 2008, which have resulted in a substantial release of carbon from biomass removal and soil disturbance.

The Multi-Stakeholder Response—The Roundtable on Sustainable Palm Oil (RSPO)

From RT1 (2003) to RT8 (2010)

While the different actors in the palm oil supply are taking their own actions to address the sustainability challenges from their respective perspectives, real and meaningful results will only be achieved through constructive engagement and participation of all relevant players. The idea of a multi-stakeholder roundtable for palm oil was proposed by WWF Switzerland and supported by other NGOs at a Workshop on "Addressing Investors and Consumers of Palm Oil and Soy" in Zurich in November 2002. The workshop agreed "to launch a multi-stakeholder roundtable for palm oil including producers, traders, retailers, and NGOs in mid-2003 in a South East Asian venue" (WWF 2002a).

WWF engaged with Unilever and other players in Europe and the MPOA in Malaysia to form an informal organizing committee (RTOC) to organize the proposed roundtable meeting in Kuala Lumpur. The roundtable meeting (now called RT1) was a watershed as it was the first time that the players in the oil palm industry met with civil society members to discuss sustainability issues. Attended by more than 200 delegates from 16 countries, the meeting was a contentious one with no agreement over what constitutes sustainable palm oil. However, agreement was reached to further develop the concept of a roundtable and the RTOC was mandated to look into the establishment of a multi-stakeholder Roundtable on Sustainable Palm Oil (RSPO). In addition, the meeting adopted of a Statement of Intent (SOI) in which signatories pledged their support to promote the development of sustainable palm oil. The SOI was conceived as an interim measure to garner support from stakeholders for the initiative and more than 45 parties signed the SOI.

In the subsequent months, the RTOC worked on the governance, statues, by-laws, and roadmap for the proposed roundtable. The challenge of determining a fair and equitable composition of the Executive Board was preeminent. After intense debate, the RTOC recommended a Board with oil palm grower and NGOs (environmental and social) members holding 25% each of the total seats while the rest of the supply chain (representing 5 sectors) hold the remaining 50% of seats. The statues and by-laws and administrative organization of the RSPO were finalized at the RTOC meeting in April 2004 in Zurich and the RSPO was established as a not-for-profit association under the Swiss code with the objective of promoting the growth and use of sustainable palm oil products through credible global standards and engagement of stakeholders.

The pioneer signatories to the formation of the RSPO were

- Aarhus United UK Ltd
- Karlshamns AB
- Malaysian Palm Oil Association
- Migros Genossenschafts Bund
- Unilever NV
- WWF Switzerland

Following the establishment of RSPO, the RTOC was retired and an interim Executive Board, led by Unilever, and a Secretary General were appointed to manage the affairs of RSPO until the 1st General Assembly later in that year.

The RSPO has made unprecedented progress since its formation and is today the most advanced global multi-stakeholder initiative that is dedicated to a single commodity. The global nature of the RSPO is expressed through the organization of Roundtable meetings (RT) and general assembly on an annual basis and the production of an international standard defining sustainable palm oil. (The development of the RSPO Principles & Criteria was supported by HSBC and Doen Foundation.) Developed through a multi-stakeholder process (within the initial 2 years), the standard is supported by a comprehensive certification system and trade and traceability mechanisms for the whole supply chain. The first plantation for sustainable palm production was certified in late 2008, and to-date, about 4.2 million tonnes of sustainably produced palm oil has been certified. The progress made by RSPO from RT1 in 2003 to RT8 is highlighted in Table 25-A.

Table 25-A. Milestones in the RSPO Journey.

Year	RT Meeting	Milestones
2003	RT1 (Kuala Lumpur)	1st ever meeting of key stakeholders in the palm oil supply chain in August. Discussion on definition of "sustainable palm oil" initiated. Adoption of the Statement of Intent (SOI) to support the Roundtable process.
2004	RT2 (Jakarta)	RSPO formed in Zurich, Switzerland, on 8th April. RSPO Secretariat established in Kuala Lumpur. Criteria Working Group (CWG) appointed; work commenced on Principles & Criteria (P&C).
2005	RT3 (Singapore)	P&C adopted; 2-year pilot testing of P&C by 14 companies commenced. Study on supply chain traceability options initiated. Verification Working Group (VWG) formed to develop RSPO certification systems.
2006	RT4 (Singapore)	Members' Code of Conduct adopted. RSPO Indonesian Liaison Office (RILO) established in Jakarta. Task Force on Smallholders appointed. Development of National Interpretations (NIs) of the generic P&C commenced.
2007	RT5 (Kuala Lumpur)	RSPO Certification System adopted and launched at RT5. Generic P&C was revised and approved by General Assembly in November. Board-level Standing Committees on Certification & Standards, Trade & Traceability and Communications & Claims established.
2008	RT6 (Bali)	Approval of the National Interpretations (NIs) of the generic P&C for Indonesia, Malaysia and PNG. 1st RSPO certification issued in August and maiden shipment of CSPO arrived in Rotterdam in November. 1st Latin America RSPO meeting held in Colombia in October. Greenhouse Gas Working Group (GHGWG1) appointed.
2009	RT7 (Kuala Lumpur)	RSPO Supply Chain Certification System launched. RSPO P&C—Guidance for schemed smallholders adopted. Biodiversity Technical Committee (BTC) appointed. 2nd Greenhouse Gas Working Group (GHGWG2) appointed.
2010	RT8 (Jakarta)	Organizational review of RSPO by PricewaterhouseCoopers; adoption of a new organizational structure. Procedure for new planting of oil palms (NPP) came into force in January. RSPO P&C—Guidance for independent smallholders for group certification adopted. Launch of the RSPO logo for CSPO at RT8. 25,000 oil palm smallholders associated with 5 companies received RSPO certification. RSPO additional guidelines for compliance with EU Renewable Energy Directive (RED) requirements adopted.

Overview of RSPO

Vision and Mission (http://www.rspo.org/?q=page/16)

The vision of RSPO is "RSPO assures palm oil contributes to a better world" and its mission is "to advance the production, procurement and use of sustainable oil palm products through, the development, implementation and verification of credible global standards and the engagement of stakeholders along the supply chain."

In support of its mission, RSPO works on a range of projects and activities through working groups and task forces and has developed the necessary standing committees to empower and oversee these. The initial effort was focused on the development of the RSPO Principles and Criteria (P&C) for Sustainable Palm Oil Production, National Interpretations of the P&C for major palm oil producer countries, and RSPO certification systems and trade mechanisms. Communication of RSPO's work and activities to members, stakeholders, and the broader public is mainly through its websites (www.rspo.org and www.rspo.eu), annual roundtable meetings, and outreach events.

Membership

RSPO has three categories of membership; Ordinary Membership (OM) consists of the main players in the palm oil supply chain, namely the oil palm growers, processors and traders, manufacturers of consumer goods, retailers, banks and investors, environmental NGOs, and social or developmental NGOs. Affiliate Membership is open to any individual or organizations that are not included in the 7 Ordinary Membership sectors and are interested in supporting the objectives and activities of RSPO. The third category, which was introduced more recently, enables the organizations that are active in the supply chain for certified sustainable palm oil but do not purchase more than 500 MT palm oil products per year to be Supply Chain Associates. While Affiliate Members and Supply Chain Associates can participate in RSPO activities and attend the General Assembly meetings, they have no voting rights.

The growth of RSPO membership has been impressive, growing from 50 Ordinary Members in the 1st year of its establishment to currently (as of 15 June 2011) having 471 Ordinary Members, 90 Associate Members, and 52 Supply Chain Associates. Among the Ordinary Members (OM) the dominant players are Processors & Traders (182 members), Consumer Goods Manufacturers (145), and Growers (87), representing 36.8, 30.8, and 18.5% of the total OM membership, respectively (Fig. 25.1); most of the members are from the biggest palm oil producing countries, with Malaysia and Indonesia accounting for 36.4% of the total OM membership. Among the consuming countries, UK has the largest number of Ordinary Members (16.8%), followed by the Netherlands (9.5%), Germany (9.3%), and France (8.2%).

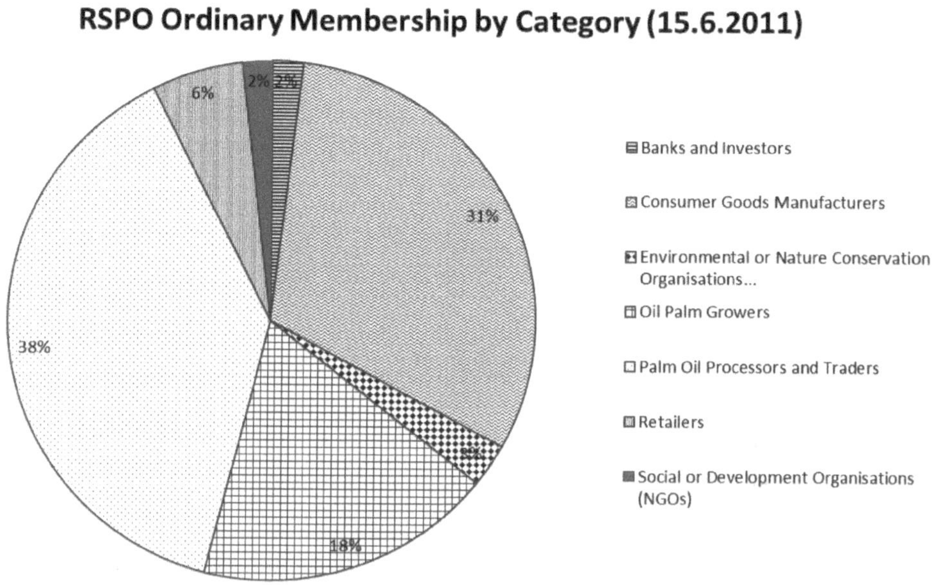

Fig. 25.1. RSPO ordinary membership, by sectors (15 June 2011).

The profile of the membership sectors has changed over time; while the growers dominated during the formative years, membership of processors and traders grew steadily since 2005 when the RSPO P&C were adopted and has since remained the largest sector in RSPO. Membership among consumers goods increased significantly since trade in CSPO commenced, but the number of oil palm growers appear to have reached a plateau in recent years. The number of ordinary members among NGOs, banks, and retailers has remained static in the past 5 years (Fig. 25.2).

Governance

The highest authority in RSPO is the annual General Assembly (GA) of members. The assembly elects members to the Executive Board, vote on resolutions tabled by members, and approves the audited accounts and financial budgets. RSPO is managed by an Executive Board comprising 16 seats representing the 7 membership sectors. Four seats are allocated to grower members representing Malaysia, Indonesia, the Rest of the World, and the smallholder sector, and four seats are for civil society organizations with two seats each assigned to environmental NGOs and social/developmental NGOs. Two seats each are allocated to the five supply chain sectors. On the decision making process, the GA decides by majority votes while the Executive Board makes decisions by consensus.

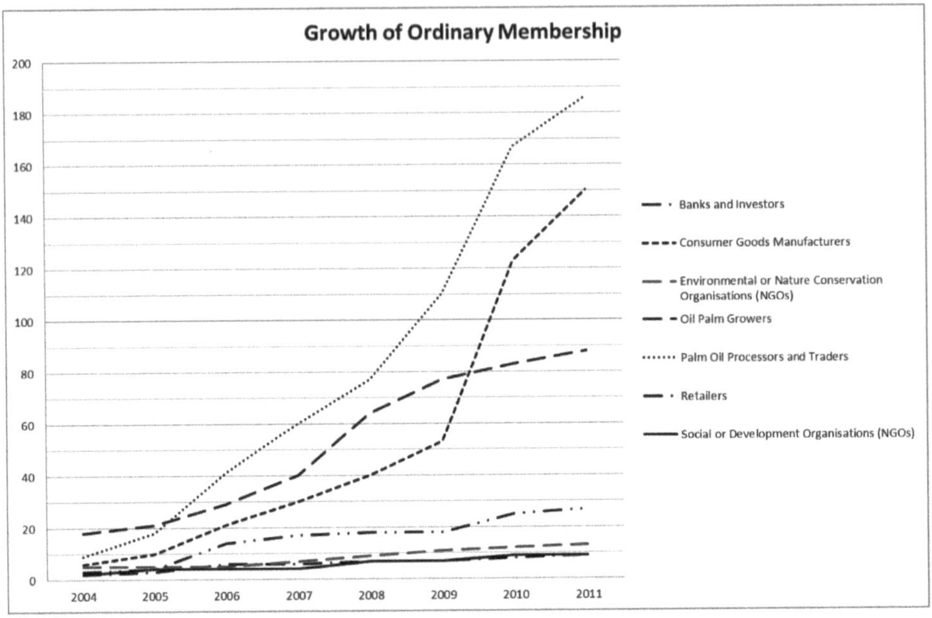

Fig. 25.2. Profiles of RSPO ordinary membership by sectors over time.

The day-to-day management of RSPO is handled by a Secretary General who is supported by a Secretariat located in Kuala Lumpur and a satellite office, the RSPO Indonesian Liaison Office (RILO) in Jakarta.

All RSPO members are governed by its statutes and by-laws, Code of Conduct, and anti-trust guidelines. The Code of Conduct stipulates that "it is fundamental to the integrity, credibility and continued progress of the RSPO that every member supports, promotes and works towards the production, procurement and use of sustainable palm oil" (www.rspo.org). The RSPO has put in place a grievance process to address complaints against members for alleged non-compliances of RSPO's statues and by-laws, Code of Conduct, and other regulations.

Development of Standards and Certification Systems and Supply Chain Mechanisms for Certified Sustainable Palm Oil (CSPO)

RSPO Certification System for Production

A robust and credible certification system based on measurable and verifiable performance indicators is essential for assuring the production of sustainable palm oil.

The RSPO certification system that was adopted in June 2007 consists of three basic elements (RSPO, 2007):

- **Certification Standard** which sets out the requirements which must be met and against which certification assessments are made. The RSPO Standard is the RSPO Principles and Criteria for Sustainable Palm Oil Production (RSPO P&C) consisting of 8 principles and 39 criteria with a range of indicators and guidelines to cover the social, environmental, economic, and legal aspects of production (RSPO, 2005). RSPO allows the development of National Interpretations of the generic indicators and guidance; when approved, these National Interpretations will be used as the basis for certification in the respective producing countries. To-date, National Interpretations of the P&C have been approved for Indonesia, Malaysia and Papua New Guinea, Solomon Islands, and Ghana.
- **Accreditation Process.** Before Certification Bodies (CBs) are allowed to undertake independent assessments on producers against the RSPO standard, they must be accredited by national or international accreditation bodies to ensure they are credible and competent to undertake the assessments. CBs are assessed for their conformance to ISO Guide 65 and/or ISO Guide 66 and approved by RSPO.
- **Certification Process** establishes whether or not the RSPO standard, the Principles and Criteria has been met and is carried out by an accredited Certification Body. The unit of certification is the palm oil mill and its supply base, including oil palm fresh fruit bunches supplied by associated smallholders and out growers. Organizations with more than one operating unit or subsidiary companies are required to submit a time-bound plan for certification of the remaining units or subsidiary companies. Transparency is of paramount importance in the certification process, starting with a public announcement of the intention to conduct the audit by the CB to public availability of documents related to the assessments, including the results of the certification and a summary of pertinent findings.

RSPO Supply Chain Certification

Users and processors wishing to claim that certified sustainable palm oil (CSPO) is used or supported in their products are required to comply with the requirements of the RSPO *Supply Chain Certification Systems* (RSPO, 2009). As in the case for CSPO production, the certification is undertaken independently by accredited Certification Bodies. Any batch of CSPO can be traded through any of the following supply chain mechanisms:

- **Identity Preserved (IP).** The Identity Preserved (IP) supply chain model, as defined by RSPO (2009), "assures that the RSPO certified palm oil and its derivatives delivered to the end user is uniquely identifiable to the mill and its supply base and is kept physically isolated from all other oil palm sources throughout the supply chain (including other segregated RSPO-CSPO sources)." The IP model requires all players along the supply chain to maintain full segregation and traceability of the product, from the producing plantation to the end-user. Among the 4 models, the IP route provides the highest level of confidence on traceability and allows the supply chain to make the highest level of claim. However, being a more complex process, it is also the most costly to implement.

- **Segregation.** As defined by RSPO, "the Segregation supply chain model assures that RSPO certified palm oil and its derivatives delivered to the end user comes only from RSPO certified sources. It permits the mixing of RSPO certified palm oil from a variety of certified sources." While the Segregation model assures that the palm oil originates from RSPO certificated mills and plantations, the product cannot be traced to the specific production units, unlike the IP model. Consequently, the level of traceability and claim will be comparatively lower, but the cost is also expected to be lower as well.
- **Mass Balance.** "The Mass Balance supply chain model administratively monitors the trade of RSPO certified palm oil and its derivatives throughout the entire supply chain, as a driver for mainstream trade in sustainable palm oil." As the model allows mixing of RSPO certified oil with non-RSPO palm oil, the level of traceability is low, but being considerably less complex that the preceding two models, it is conceived that this practical approach could stimulate the trade in certified sustainable palm oil. The Mass Balance model was designed in manner to ensure that there is strict reconciliation between the volume of certified sustainable palm oil bought and the quantity sold.
- **Book and Claim.** "The Book and Claim supply chain model provides tradable certificates for RSPO certified palm oil to the palm oil supply base. The supply base may then offer these certificates on a web based transaction system to end users who choose to support specific volumes of RSPO certified palm oil and or their derivatives." Under the Book and Claim model which was conceived by Aarhus United UK (now AarhusKarlshamns UK, or AAK), the certificate for sustainable palm oil is traded separately from the physical oil. This is based on the premise that there are no biophysical differences in the properties of conventional and certified palm oil expect in the manner in which they are produced to comply with the stringent social and environmental requirements specified by RSPO.

Progress and Challenges in the Production and Trade in CSPO

Following the adoption of the RSPO Principles & Criteria as the standard for sustainable palm oil in November 2005, pilot testing was undertaken by several producer companies in Malaysia, Indonesia, and PNG over a 2-year period. The first RSPO certification for production of sustainable palm oil was awarded to United Plantations Berhad (Malaysia) on 26 August 2008; this was followed by New Britain Palm Oil Ltd (PNG) in the same month and PT Musim Mas (Indonesia) on 6 January 2009. Trade of CSPO commenced in November 2008 when the first shipment of 500 tonnes of CSPO produced by United Plantations reached Rotterdam.

The volume of CSPO increased steadily as more producers received certification. As of 15 June 2011, there are 102 certified palm oil mills which are capable of producing more than 4.24 million tonnes of CSPO. The largest producer is Malaysia where 60 mills can produce about 2.30 million tonnes or 54% of total production. Indonesia is the 2nd largest producer accounting for about 35% (Table 25-B).

Although sufficient quantities of CSPO have become available since early 2009, the initial market uptake was disappointing. By October 2009 when an estimated one million tonnes of CSPO was available, only 195,000 tonnes was traded, which was less than 5% of the EU's annual consumption of palm oil. In order to stimulate the update of CSPO, WWF introduced the "Palm Oil Buyers' Scorecard" to assess the progress of sourcing of CSPO by European companies during March to September 2009 (WWF, 2009). The impact of this campaign, along with actions taken by companies to fulfill their time-bound targets for CSPO sourcing, saw a marked improvement in the trade in CSPO during 2010 when about 1.28 million tonnes or 56 percent of the CSPO produced had been purchased (www.rspo.org/eu). Currently, the market uptake of CSPO is about 50%.

Among the four supply chain mechanisms for trade in CSPO, the Book and Claim model which is operated under the GreenPalm Ltd platform handled 67% of the total volume of CSPO sold [to-date, as of 1 June 2011 (www.rspo.org/eu/)]. Under the Book and Claim system, any user of palm oil can purchase certificates to match their purchases of physical palm oil on the conventional market, thus allowing them

Table 25-B. Production of Certified Sustainable Palm Oil (as at 15 June 2011).

Country	Major Producers	Mills	Palm Oil (tonnes)	Palm Kernel (tonnes)
Malaysia	Sime Darby Bhd, IOI Group, Wilmar International, United Plantations, KLK Bhd, Felda	60	2,289,818	568,915
Indonesia	Musim Mas Group, Cargill, PT London Sumatra, Wilmar International	33	1,493,322	310,246
PNG & Solomon	New Britain Palm Oil Ltd, SIPEF	8	434,403	96,756
Colombia	Daabon–CI Tequendama	1	27,000	2,000
	TOTAL	102	4,244,543	977,917

to claim support of production and use of sustainable palm oil. The Book and Claim is widely accepted as it is a practical and the lowest cost option among the 4 supply chain options but it has the lowest level of traceability since the source of the palm oil—certified or from outside RSPO is unknown.

For a commodity that is an ingredient in more than half the packed supermarket products, a sustainability trademark that is shown on the product label is an effective and practical way to assure the consumer that it contains palm oil that is sustainably produced. Towards this end, the RSPO unveiled the trademark for CSPO (Fig. 25.3) at the 8th Roundtable Meeting (RT8) in November 2010 in Jakarta. According to the RSPO President, Jan-Kees Vis, "the RSPO trademark will assure consumers that products they buy contribute to sustainable cultivation of palm oil and other palm products. This trademark will mean that palm cultivation does not contribute to the sustained destruction of valuable tropical forests or damage the interests of people in the regions where the palms are grown" (RSPO, 2010b). The CSPO trademark was formally launched on 1 June 2011, and consumer goods manufacturers and retailers can use the logo on their products, according to the RSPO guidelines on claims and communication (RSPO, 2011).

Fig. 25.3. RSPO trademark for certified sustainable palm oil.

SWOT Analysis on RSPO

Table 25-C provides a snapshot of the strengths and weaknesses of RSPO and the threats to and opportunities for the organization.

Strengths

The key strength of RSPO is its multi-stakeholder nature that has involved both ends of the palm oil value chain and its stakeholders, especially civil society to find a credible approach to address the sustainability challenges confronting the sector. According to Pye (2009), it "has become the most successful stakeholder initiative of all

Table 25-C. SWOT Analysis on RSPO.

STRENGTHS	WEAKNESSES
Most advanced global multi-stakeholder sustainability initiative dedicated to a single commodity. Comprehensive Principles & Criteria/ standards with objective performance indicators and certification systems developed and implemented. Supply chain mechanisms for trade and traceability of CSPO developed and implemented. Active participation from main players at both ends of the value chain. Large and comprehensive membership	Perceived imbalanced governance structure from perspectives of growers and NGOs. "Lacks teeth" and capacity to monitor behavior of members. Weak communications with internal and external stakeholders. Lack of an effective Secretariat. Inadequate attention to reduction of GHG emissions in P&C. Failure to engage governments. Supply outstrips demand Failure to deal with grievances rapidly
OPPORTUNITIES	**THREATS**
RSPO can make a real difference towards forest conservation, mitigation of climate change and social justice. Transformation of the global palm oil market; CSPO being the mainstream commodity. Engagement of governments of producer countries and emerging economies (China and India). Development of new inclusive sustainable business models palm oil production and use. RSPO to become the benchmark for other commodity roundtable initiatives.	Risk of being superseded by alternative certification schemes (e.g., SAN, ISPO) Loss of credibility through poor management of members' behavior. RSPO standards and certification system seen as affront to sovereignty of producer countries. Perceived as a technical barrier to trade. Perceived as overly biased towards environmental and social aspects at the expense of economic development and poverty reduction

times." While "RSPO is not perfect, it represents a tested and constantly improving multi-stakeholder opportunity to avert continued habitat loss for much of Southeast Asia's most threatened terrestrial biodiversity" (Paoli et al., 2010). Through the multi-stakeholder approach, RSPO has developed a robust set of standards with objective and verifiable performance indicators that is the corner stone of the independent third party certification system for the production of sustainable palm oil.

Weaknesses

While the establishment of RSPO was welcomed by stakeholders, continuing instances and allegations against its members over deforestation and social conflicts has drawn severe criticisms from stakeholders, especially NGOs. One of the strongest critics is Greenpeace, which accused RSPO of "greenwashing" in not taking action against errant members (Greenpeace, 2009). Laurance et al. (2010) stated that RSPO "lacks teeth" and organizational capacity to monitor the behavior of its members.

The governance structure has been seen as imbalanced, albeit from differing perspectives of the stakeholder groups. From the NGO point of view RSPO is dominated by industry members representing growers, processors, manufacturers, and other corporate sectors (Laurance et al., 2010). A similar view was offered by Paoli et al. (2010) but they stressed that, by virtue of their membership numbers, processors and traders and manufactures are the most dominant voice as they jointly have about 65% of the voting membership. On the other hand, the palm oil producers who maintained that while they carry the most responsibility to deliver sustainable oil to market, they do not have sufficient say in the decision making process. In their view, the producers' seats should be 50% and equal those of all non-producers on the Executive Board. They proposed a resolution at the General Assembly in 2010 for the review of the structure of the Executive Board and this was adopted by the meeting.

Another major weakness is inadequate attention given to address climate change. Although the RSPO P&C has a general requirement under Criterion 5.6 on reduction of pollution and emissions, there were no specific indicators or guidance to address greenhouse gas emissions. This is a concern for NGOS and a number of stakeholders and has been a source of criticism and campaigns.

Threats

Among the threats to RSPO, perhaps the most critical is the potential loss of organizational credibility among its members and stakeholders for shortcomings in delivering results against the stated objectives and mission of the organization and/or not meeting their expectations, be it perceived or real. Loss of confidence could spur others to look for alternatives to achieve sustainable palm oil production. This is already happening as the largest producing countries have announced their intention to implement national certification schemes that could compete with RSPO.

Recently, the Government of Indonesia announced its intention to implement the Indonesian Sustainable Palm Oil (ISPO) system. The ISPO standard will be mandatory for all producers and it will be based on the classification of plantations according to the Decree of Ministry of Agriculture No 7, 2009 (Suharto, 2010). Pilot implementation commenced in March 2011 with 20 companies participating in trial certification and all oil palm companies are expected to undergo certification in 2014 (Jakarta Post, 2011). In Malaysia, the Malaysian Palm Oil Board (MPOB) launched in August 2007, a set of voluntary Codes of Practice on good agricultural practices for oil palm estates and smallholders, good processing practices for milling and refining of palm oil and other processes (Kuntom, 2010). However, in the longer term, Malaysia would develop a set of certification standards; an added benefit of the national scheme is the "creation of a global brand for Malaysian palm oil" (Basiron, 2011).

Although RSPO has developed modified versions of the RSPO P&C for certification of palm oil produced by smallholders, the Sustainable Agriculture Network (SAN) is also offering this service for compliance with its Sustainable Agriculture Standard and Group Certification Standard that have been applied to more than 20 crops, mainly in small farms and co-operatives, in over 25 countries. Growers which comply with these standards are able to use the Rainforest Alliance trademark for their products. Amendments to the Sustainable Agriculture Standard now include requirements for best practices for oil palm plantations (SAN, 2010).

Opportunities

RSPO is the only organization that can make a real difference to the sustainable production of palm oil globally. It can assure the integrity of tropical forest, respects the rights of communities and indigenous peoples while providing the need for economic development and the ever increasing global demand for the commodity. Being the most advanced multi-stakeholder initiative that is dedicated to a single commodity, RSPO could leverage its strength to influence others towards sustainable production.

While RSPO is essentially a voluntary business-to-business initiative that has so far not formally involved the participation of national governments, it is timely for RSPO to engage with relevant governments of producing as well as consuming countries. RSPO could support governments of producing countries to strengthen the policy and statutory environment to ensure sustainable production. With the consuming countries, particularly the emerging economies, RSPO could support the development of public sector polices that stimulate uptake of CSPO.

Endeavours to Improve the Effectiveness of RSPO

Following a review of the organizational structure and performance by PricewaterhouseCoopers (PwC) in mid-2010, the Executive Board adopted a new management model for the Secretariat and a new way of working by the Executive Board itself. The role and responsibilities of the Secretary-General have been broadened and will be supported by Directors responsible for Finance & Administration, Communications, and Technical, and an Indonesia Director to head the RSPO Indonesian Liaison Office. At the Executive Board level, 4 Standing Committees focusing on Certification & Standards, Trade & Traceability, Communications & Claims, and Finance. Each Standing Committee is headed by an Executive Board member and supported by a dedicated Manager from the Secretariat. Working Groups on topical areas will report to the respective Standing Committees, while Task forces report directly to the Working Groups (RSPO, 2010).

Building on the recommendations from the PwC review and acting on the 2010 GA resolution, which called for a review of the RSPO governance structure, the RSPO Executive Board undertook a comprehensive strategic planning exercise in March 2011 to re-examine the purpose of its existence and develop an appropriate strategy for the next 5 years. The key output from this effort was the agreement on a new vision and mission for RSPO, as follows:

RSPO Vision

- *RSPO will transform the market to make sustainable palm oil the norm.*

RSPO Mission

- *To advance the production, procurement, finance and use of sustainable palm oil products;*
- *To develop, implement, verify, assure and periodically review credible global standards for the entire supply chain of sustainable palm oil;*
- *To monitor and evaluate the economic, environmental and social impacts of the uptake of sustainable palm oil in the market;*
- *To engage and commit all stakeholders throughout the supply chain, including governments and consumers.*

On the need to strengthen the RSPO requirements on greenhouse gases (GHGs) emissions, a Greenhouse Gas Working Group (GHGWG1) was established in November 2008 with the mandate to review all stages of the palm oil supply chain, with emphasis on understanding GHG emissions from land use changes in development of new plantations and emissions from operations in existing plantations. While GHGWG1 agreed on the need for voluntary actions to reduce emissions, there was no consensus on the dimensions of GHG emissions. A second Greenhouse Gas Working Group (GHGWG2) was convened in late 2009 to review the earlier findings and to establish a process to reduce GHG emissions through a voluntary mechanism. GHGWG2 is organized into six work streams, which includes emissions from peatlands. Meanwhile, RSPO has issued additional guidance to the RSPO P&C to allow members to comply with the requirements in the EU Directive 2009/28/EC on the promotion of the use of energy from renewable sources (EU RED).

The Business Case for Sustainable Palm Oil Production—Going Beyond Certification

Drivers for Sustainable Palm Oil Supply Chain

In reviewing the work of the RSPO, it is clear that pressure from the supply chain, particularly from civil society organizations, has been a major driver for the production and use of sustainable palm oil. Transnational NGO campaigns have essentially produced the "boomerang effect" (Pye, 2010) where drawing from greater consumer awareness (e.g., through WWF's *Check Your Oil* campaign in 2002 to 2004) and using the threat of negative branding, NGOs have put pressure on the major manufacturers of consumer goods and retailers and on financial institutions to push their upstream producers to adopt sustainable production practices that are consistent with the RSPO P&C and other relevant standards. For example, the Anglo-Dutch multinational, Unilever, took a public stance against deforestation by suspending future purchases of palm oil from its Indonesian supplier, PT SMART (part of Sinar Mas Group) on the grounds of their alleged contribution to destruction and development of peatland (Unilever, 2009). Nestlé followed this approach in March 2010 by discontinuing sourcing of palm oil from Sinar Mas following a Greenpeace campaign against "KitKat" (Hornby, 2010).

Besides the campaigns, some companies in developed economies have also taken their own initiatives for sustainable and ethical sourcing in support of the corporations' long term commitment to sustainability. For example, Unilever has the Sustainable Agriculture Initiative to support the sustainable production of the key natural raw materials it requires for its products. This included the sustainable production of palm oil, and Unilever has produced "Sustainable Agriculture Code" which is a generic standard for sustainable agriculture applicable across crops and raw materials (Unilever, 2010). MIGROS, the largest supermarket chain in Switzerland, became the first European retailer to commit to sourcing of sustainable palm oil in 2001 (Financial Times, 2002). In collaboration with WWF Switzerland, they developed a set of criteria for sustainable production with regard to compliance with environmental and social standards and legal requirements, and their suppliers are required to undergo independent verification of performance against these criteria. Both of these initiatives had preceded the formation of the RSPO, and the Unilever guidelines as well as the MIGROS' criteria contributed towards the development of RSPO's P&C.

Banks and financial institutions play a vital role in promoting production and trade in sustainable palm oil. In general, banks which subscribe to the Equator Principles [The Equator Principles is a voluntary set of standards for determining, assessing environmental and social risks in a project. (http://www.equator-principles.com/)] have put in place policies and procedures for sustainable investments in the sector. The World Bank Group, which was the catalyst for early development of the palm oil industry in many developing countries, especially in Indonesia and Malaysia, could be a major driver for sustainable palm oil. In its new strategy for engagement of the sector, the WBG has stipulated that future financing for oil palm development by the public and private sectors must be consistent with sustainability principles such as the RSPO P&C and full compliance of its Performance Standards for social and environmental sustainability (World Bank, 2011).

Investors as well as shareholders are increasingly demanding higher levels of transparency amongst oil palm companies and are seeking to improve their reporting of Environmental, Social, and Governance issues. Some companies are responding by producing sustainability reports using the RSPO standards as part of the assurances mechanisms within the Global Reporting Initiative (GRI) standards that govern such reports. In 2009, Kulim Malaysia Berhad and New Britain Palm Oil Ltd (PNG) were amongst the first to produce sustainability reports to GRI (G3 guidelines), and in 2010, Daabon and Wilmar International also produced sustainability reports.

To date, most of the CSPO traded has been consumed by companies in Europe. The share of CSPO going to Europe is expected to increase rapidly in coming years, as many companies have made public commitments to time-bound targets for procurement of sustainable palm oil. Companies such as Unilever and Nestle have committed to source 100% of requirements as CSPO by 2015. The world's largest retailer, Wal-Mart, has also made the requirement for sustainably sourced palm oil for all products under its brand around the globe, including China.

A major development that could accelerate the demand for CSPO is the manifesto by the Dutch Task Force on Sustainable Palm Oil, in which palm oil consuming companies in the Netherlands have collectively pledged to source only CSPO by 2015. When implemented, this effectively means that virtually all palm oil used in the Netherlands would have been sustainably produced. In the UK, the government is considering introduction of policy interventions to stimulate sustainable palm oil sourcing and consumption. An analysis of various policy options indicates that the highest positive impacts would be achieved by a combination of a public procurement policy that incorporates a time-bound goal, together with targeted support and awareness rising to drive action across UK supply chains (ProForest, 2011).

From the perspective of the upstream producers, a price premium for CSPO could stimulate production, and they expect incremental revenue for the additional effort required to meet the stringent RSPO standards. However, trading in CSPO to-date has shown that price premium has not been a significant contributing factor to expanding volumes of CSPO and is unlikely to be so in the future.

Risks of Unsustainable Palm Oil Production and Procurement

Maintaining or improving corporate reputation was identified by about 36% of the survey respondents as the most important reason for addressing sustainability issues in the 2010 McKinsey Global Survey (McKinsey, 2010) on how companies manage for sustainability. This could also apply to companies in the palm oil supply chain, from upstream producer companies to manufacturers and retailers and banks. The corporate brand and reputation is an easy target of transnational NGO campaigns if the companies have been found or even perceived to be responsible for unsustainable practices that contribute to environmental degradation or cause of conflicts with local communities or indigenous peoples. While it may be difficult to assess the financial impact of campaigns on brands, the negative publicity generated could change the public perception of the company. Regaining their trust and confidence in the brand may require much effort.

Other risks related to unsustainable production or procurement of palm oil include:

- Market access could be limited.
- Access to finance could be more difficult as banks and financial institutions could place stringent sustainability criteria on loans.
- "License to operate" may be compromised for non-compliance of legal requirements and failure to effectively engage with local communities and indigenous peoples.
- Loss of competitiveness as non-compliant companies would be treated as laggards.

Benefits of Sustainable Palm Oil Production

While the risks of unsustainable palm oil production have been well defined, the business case for undergoing RSPO certification may be less convincing. Considering that much effort has to be expanded to meet the stringent requirements of the RSPO standards for sustainable production, there has been an expectation of attractive price premiums for certified sustainable palm oil (CSPO). Although reasonable premiums were paid after trade of CSPO commenced in late 2008, the incremental revenue had declined with time, and currently, based on trade in GreenPalm certificates, the premium is in the region of USD 2 per tonne CSPO. It is clear that price premium alone will not be a motivating factor for producer companies and a more convincing business case that considers financial and non-financial benefits would be necessary.

A survey conducted by WWF (WWF HOB NI, 2011) on 46 respondents in countries covered under the Heart of Borneo initiative has identified several benefits from good environmental and socially responsible practices. More than 60% of respondents have identified good environmental and social risk management as the main benefit from sustainability activities. Enhancement of corporate public image and improved relationships with stakeholders (Government, NGOs and communities) were next most important. About 35% of respondents felt that long term profitability could be enhanced but less than 10% stated no business benefit.

For this chapter, a survey on experiences and benefits realized by companies that had implemented the RSPO Principles & Criteria and achieved RSPO certification for sustainable palm oil production was conducted in May 2011. Unlike questionnaire surveys, which tend to elicit responses to closed-ended questions, this survey solicited actual experiences and results through a number of open-ended questions. All companies which have achieved RSPO certification were invited to participate and nine companies (Box 1) shared their experiences. These companies represented about 40% of the total number of certified palm oil mills, accounting for about 45% (or 1.90 million tonnes) of the current production of CSPO.

The main benefits or improvements with regard to the company's economic, social, and environmental performance are summarised in Table 25-D, and salient points are discussed in the following sections. As these the results are based on responses to open-ended questions, it is not prudent to provide an overall ranking of the benefits reported.

Corporate Reputation and Management

As in the WWF survey, improved corporate image and reputation from their commitment towards sustainable production is a major benefit, as reported by 4 companies. As the first Sarawak-owned company to receive RSPO certification, Keresa Plantations & Mill (Keresa) has reported improvement of its public image within the State and has become a model for plantation best practices for neighboring companies as well as educational institutes. Cargill's certified company, PT Indoli in Indonesia, has also been regarded as a benchmark to others in the industry. For PT Bakrie Sumatera Tbk in Indonesia, the enhanced public image reduces the risk of conflict with NGOs and the public.

Box 1. Respondents to the Business Case Survey

1. PT Bakrie Sumatera (B)
2. Cargill Tropical Oils Asia Unit (C)
3. Felda (F)
4. Keresa Plantations & Mill (Ke)
5. Kuala Lumpur Kepong Berhad (Kl)
6. Kulim (Malaysia) Berhad (Ku)
7. Musim Mas Group (M)
8. New Britain Palm Oil Ltd (N)
9. Wilmar International Ltd (W)

Table 25-D. Benefits from Implementation of RSPO P&C and Certification for Sustainable Palm Oil Production.

Benefits	Respondents
Corporate reputation & management	
Improved corporate image and reputation	B, C, Ke, Kl
P&C as a driver for corporate strategy and management system	Ke, M, N
Economic	
Better access to international markets, especially Europe	Kl, M, N, W
Price premium for CSPO	B, C, Ke, Kl
Enhanced overall profitability	Ke, Kl, N
Improved operational efficiency and/or productivity	B, C, Ke, Kl, W
Creation and use of best management practices (BMPs)	B, C, Ku
Better relationship with financers for RSPO certified companies	B, W
Social	
Change in mindset and attitude of staff and workers; better team work	B, Ke, Ku, M
Better OHS risk management and reduction in workplace accidents	C, Ke, Kl, Ku, N,W
Improved communication and relationship with workers	Ke, Ku
More stable or loyal workforce; lower turnover rate	Ke, W
Improved communication and engagement with smallholders	C, Ke, M, N,W
Enhanced productivity and well-being of associated smallholders	M, N
Obtaining social license to operate through consultation and engagement of communities	W
Environmental	
Reduction in GHG emissions through methane capture in effluent treatment plants	Kl, N, W
Greater awareness of and stewardship for High Conservation Areas	C, K, N, W
Better assessment, management and monitoring of wastes and pollution (including domestic wastes)	C, Ke, Ku, N,W
Judicious use of pesticides and implementation of Integrated Pest Management (IPM)	B, C, Ku, N
Governance	
More systematic documentation and access to information on operational and non-operational aspects, especially legal requirements	F, C
Improved compliance of regulatory requirements	Ku
Better understanding of legal and regulatory requirements	F, W
Improved relationship with local government authorities and regulators	C, Ke
Better understanding of stakeholder engagement	W, N

While the RSPO P&C serve primarily as the standard for the certification system for sustainable palm oil, some companies have used them as the basis for driving corporate strategies aligned to sustainability. New Britain Palm Oil Ltd (NBPOL), which operates oil plantations in Papua New Guinea and the Solomon Islands, has implemented the RSPO P&C as the core foundation for their long term business strategy and is crucial to their license to operate in future. It has redefined its new vision as "to demonstrate that palm oil can be produced and consumed responsibly and sustainably—creating livelihoods and opportunities for tens of thousands in the provinces where we grow palm oil and nutrition for millions where we sell palm oil."

Similarly, Bakrie Sumatera has used their experience with the RSPO P&C to refocus and redefine their corporate mission in 2010 as to "Nurture and sustain the wealth of our community by extracting the optimum value creation through environmentally friendly operation and leveraging core expertise in multi crops and global operations."

The Musim Mas Group in Indonesia has developed and implemented an in-house Management System for Production of Sustainable Palm Oil that is based on the RSPO P&C. The key drivers of the system are Principle 3 (Commitment to long term economic viability) and Principle 8 (Commitment to continuous improvement). Priority is making continuous improvement corporate culture through the implementation of a systematic problem solving process called the Musim Mas Improvement Process (MMIP). Based on nominations from operational management, a Steering Committee selects improvement projects to be undertaken by cross-functional teams supported by in-house facilitators. After the recommendation of the project teams have been successfully implemented, the improvements are institutionalized by incorporating them in the Group's Standard Operating Procedures. To date, 3 series of improvement projects to address problems such as social conflict, pollution control, and management of peat soils have been implemented.

Keresa's top management has set up a dedicated Total Quality Management (TQM) team to initiate sustainable practices, ensure adherence to quality standards, and drive continual improvement within the context of the RSPO P&C. Through the TQM process, the company has developed new policies and procedures for production of sustainable palm oil. The improved management system covers a range of important areas including labor standards, occupational safety and health, environmental management, and supply chain management. According to Keresa, the robust system in place now ensures all their operations are assessed and made compliant to their overall sustainability goal.

Economic

Four companies have reported that RSPO certification has enabled better access to international markets, especially to Europe, but in the case of Kuala Lumpur Kepong Berhad (KLK), it has found niche markets for CSPO in Japan. Wilmar International

Ltd realized that RSPO certification has given them a competitive advantage in opening new markets, especially in the biofuels market in Europe. NBPOL sees their commitment to sustainability as a means to securing the long term access to the European market and this has been enhanced by the opening of the world's first palm oil refinery in Liverpool, UK, to process 100% certified and traceable sustainable palm oil.

Since implementation of the RSPO P&C, Keresa, KLK and NBPOL have seen an improvement in their overall profitability, but it is difficult to ascribe the amount attributable to sustainability performance as profitability is determined by many factors, especially commodity prices. However, four respondents have received some premium for their certified palm oil while five companies have seen improvements in operational efficiencies and/or productivity. On the other hand, Felda, which has the mandate to uplift the socio-economic status of the settlers stated that the settlers in their schemes which have been certified did not get the expected monetary benefits from certification. There was no price premium, while considerable cost and effort had been expended on documentation and preparation for certification.

Through the implementation of better management practices (BMPs), Bakrie Sumatera has reported about 5% increase in palm oil yield per hectare while Keresa achieved 13% increase in Fresh Fruit Bunch (FFB) production per hectare from 2007 to 2009. Productivity at PT Indoli (Cargill) had increased from 20–21 tonnes FFB per ha to 24–25 tonnes FFB per ha. Assuming an oil extraction rate of 22% and the palm oil price of USD 1100 per tonne, the gross incremental revenue could be about USD 1000 per hectare per year. Besides productivity enhancement, Keresa reported that improvement in operational efficiencies through the application of appropriate BMPs has resulted in a 30% reduction in chemical usage.

According to Bakrie Sumatera, financial institutions give preference to RSPO certified companies while Wilmar continued to enjoy good relations with their financiers to support future growth because of their commitment to production of certified sustainable palm oil.

Social

Among the non-tangible benefits, perhaps the most significant could be the motivational effect the implementation of the RSPO P&C has on proactive companies and their people. Respondents have seen a positive change in the mind-set and attitude of their staff and workers. According to Bakrie Sumatera, the RSPO P&C has instilled a greater sense of commitment and pride among its management and employees and has achieved better team work and uniformity of approach to all operations. In Wilmar, their plantation and mill management now appreciate the importance of obtaining the social license from communities where they operate.

Better occupational safety and health management (OHS) has been identified as a major benefit by six respondents. Through provision of appropriate training on safe work practices such as proper handling of pesticides and use of personal protective equipment (PPE), respondents have reported a decline in the incidence of workplace accidents. Although no data was provided to support this, KLK stated that their OHS performance had earned the company the first placing in the National Occupational Safety and Health Award for 2010.

Several companies have implemented practices that are consistent with the requirements of RSPO P&C, such as decent pay and working conditions, housing for workers and their dependents, access to medical facilities, and schooling for workers' children. Oil palm plantations in Malaysia are heavily dependent on migrant workers for field mill operations, particularly for crop harvesting. While legal requirements would restrict the workers from bringing along their families, many migrant workers have raised children in the plantations. As they are non-citizens or considered "stateless", the children would not be eligible to attend Malaysian schools. To address this problem, plantation companies have collaborated with the Borneo Child Aid Society to set up primary schools in the plantations. (See Box 2.)

Relationship between workers and stakeholders and management has also improved through better communications and the provision of grievance procedures. Kulim (Malaysia) Berhad has given priority to addressing gender related issues and has implemented the WOW (Women Onwards) initiative to provide a platform to understand their rights and to be heard. With improvements in the work conditions and environment, Keresa has been able to maintain a very low rate of turnover of workers (average rate being less than 2%), while Wilmar stated that the improvements have contributed to a more stable workforce.

Under the RSPO certification system, the unit of certification consists of the palm oil mill and its supply base, including associated smallholders "the mill will develop and implement a plan to ensure that 100% of associated smallholders and out grow-

Box 2. Giving the children of migrant workers a future

Oil palm plantations are highly dependent on migrant workers from neighboring countries and many of them have started to have families on the plantations, but their children do not have access to education and are not eligible for Malaysian public schools, as many of them are considered "stateless" children. The Borneo Child Aid Society (Humana) has worked collaboratively with RSPO members such as Wilmar International Ltd and Kuala Lumpur Kepong Berhad to set up Humana schools in Sabah and Sarawak. Since the launch of its first school in 2007, Wilmar has to-date established 14 schools for 1009 pupils. The Group will build 2 more schools in 2011. KLK set up a Humana school for 256 stateless children aged 16 and below in palm oil mill in Tawau, Sabah, in 2011 and plans to build another 4 schools in Sabah in 2012 to educate about 1200 children. The provision of free education to the workers' children supports the UN Millennium Development Goal on primary education for children.

ers are of certifiable standard within 3 years" (RSPO, 2007). As agronomic standards and productivity of smallholdings are usually poor when compared to plantations, considerable effort would be required to bring them to comply with the RSPO P&C. Examples of actions taken by some companies to improve the performance of smallholders and prepare them for certification are

- NBPOL has improved the communications and engagement of their 7500 smallholder growers in West New Britain through provision of training, extension services and agronomic training, and assisting the growers to improve their financial management. The efforts of the smallholders were rewarded by NBPOL in January 2011 with the payment of a "sustainability bonus" based on fresh fruit bunch productivity achieved by each grower in 2010.
- Musim Mas has been able to train and mentor their associated smallholders in cooperative (KKPA) scheme to implement sustainable practices in oil cultivation and production. This was done through dedicated managers and staff who support the smallholders in administrative and technical matters such as formulation of fertilizer recommendations. Being better managed, FFB productivity and quality from the KKPA schemes have improved. As a reward for achieving RSPO certification, Musim Mas pays the smallholders an incentive premium.
- PT Hindoli (Cargill) has focused on achieving good environmental stewardship by its associated smallholders and has given them appropriate training to assume this responsibility. Safe use of pesticides is a priority and each smallholder co-operative (KUD) has a centralized chemical/fertilizer store that is built by the co-operatives. Smallholders are trained on proper use of personal protective equipment (PPE) and safe handling of agrochemicals. Herbicide spraying is done only in harvesters' paths and palm circles. Manual weeding or mechanical grass cutting is encouraged wherever possible.

Environmental

Compliance of RSPO P&C has encouraged some companies to re-examine how they use resources and manage wastes and pollution resulting from their operations. There has been a greater awareness on environmental impacts and better management practices (BMPs) on various operations such as zero burning, integrated pest management (IPM), segregation of domestic wastes, recycling, effluent treatment, and production of biofertilizers have been implemented by several respondents. On recycling, Kulim Malaysia worked with the Malaysian Croplife and Public Health Association (MCPA) and Department of Agriculture to recycle triple-rinsed HPDE pesticides containers into pellets for making products such as planter pots and seedling trays. Through recycling, Kulim realized an annual savings of about USD 40,000 which, would have been the normal cost of disposal of HDPE containers from all their plantations.

Risk assessment and monitoring of operational performance are critical aspects of environmental management. NBPOL stated that "regular sampling and independent analysis coupled with quick feedback to the operating units have meant that resource management has become a powerful tool and as a result has been included in our managers' key performance indicators."

Compliance of the RSPO criterion on the need to undertake assessments of rare, threatened, and endangered species, and high conservation value (HCV) habitats has been a challenge to many companies undergoing the certification process. However, progressive companies have recognized the importance of this requirement for forest and biodiversity conservation. NBPOL and Kulim have emphasised the need to undertake rigorous assessments. Some activities undertaken by Kulim on HCV conservation are given in Box 3.

Palm oil mill effluent (POME) is normally treated in large open ponds; facultative digestion of the organic matter results in the release of methane gas with a global warming potential of 21 times that of carbon dioxide. Companies such as NBPOL, KLK, and Wilmar have implemented methane capture plants and reduced the GHG emissions from the palm oil milling operations. The trapped methane would be used for generation of electricity and reduce the dependence on fossil fuel. In NBPOL, the surplus electricity generated is fed into the local electricity grid.

Governance

Through implementation of the RSPO P&C, plantation management have gained a better understanding of the legal and regulatory requirements in respect to running

Box 3. Biodiversity and HCV Stewardship

Under RSPO Principle 5, plantations are required to undertake assessments of rare, threatened, or endangered species, and HCV habitats and their conservation must be incorporated in management plans. Kulim (Malaysia) Berhad has given serious attention to biodiversity and HCV conservation. Some of the activities undertaken are

1. Conduct of biodiversity assessments with Malaysian Nature Society in their plantations in Tg Lenai, Johor.
2. Briefing sessions during daily muster to raise the awareness of workers' role in HCV conservation.
3. Participation in the Johor Wildlife Conservation Project on tiger conservation.
4. A Biodiversity Bulletin is issued regularly to employees at all operational levels to advise them on sightings of animals in plantations and recommended actions

their operations. This has led to a better level of compliance of applicable laws and regulations and reduction in penalties and fines. According to Kulim, the penalties imposed for non-compliances in the past could vary from RM 5,000 (USD 1,700) to RM 200,000 (USD 66,700) per palm oil mill.

Improvement in legal compliance has been partly attributed to better documentation systems that have been implemented. Cargill (PT Indoli) and Felda have identified better and structured documentation and improved access to information as the most significant improvement they have obtained from the RSPO P&C. Prior to implementation of the criteria, documentation in Felda was focused on operational activities to monitor progress of work and consumption of material inputs. While other related documents may be available, they are not readily retrieved and consequently classified as "missing documents." With a structured documentation process, the availability of relevant and complete documents has made Felda employees more aware of requirements of regulatory bodies and it has also assisted management in operational planning, for example, training programs for employees.

Overall Benefits from Implementation of RSPO P&C and Certification

While the price premium as a key driver for the business case for certified sustainable palm remains rather elusive, experiences in this survey from nine palm oil companies which collectively contribute to about 45% of the world's CSPO production currently have demonstrated clear tangible and intangible benefits from implementation of the RSPO P&C and certification. Improvements in operational efficiencies and enhancement of crop productivity would contribute positively to the economic bottom-line of the company. From the social perspective, the motivational impact of the RSPO P&C cannot be underestimated, as some companies have reported positive changes in mind sets and attitudes of their employees and workers and associated smallholders. For the natural environment, there has been better awareness and effort on resource and waste management and conservation of habitats and biodiversity.

> "The framework of the RSPO P&C allows us to maximise production and as such is now an integral part of encouraging best demonstrated practices in all that we do" (New Britain Palm Oil Ltd).

> "Plantations and mills begin to understand stakeholder engagement better than before. They now understand better that their operations can bring about not only positive impacts but negative impacts as well" (Wilmar International Ltd).

> "In conclusion, RSPO compliance has generally improved the standard of living in the operating units. The company's assets are better managed and there is more concern for the well-being of employees and the natural environment" (Kulim (Malaysia) Berhad).

Outlook

"RSPO will transform the market to make sustainable palm oil the norm" is a compelling and challenging vision that sets a clear direction for the RSPO specifically and for the palm oil industry as a whole. Market transformation will mean that sustainably produced palm oil will become the mainstream commodity, from the perspective of production and consumption.

On the demand side, critical volumes or "tipping points" for CSPO uptake need to be achieved in order to drive the market transformation. Uptake in the near future from consumers in developed countries, especially in Europe, is optimistic as major companies and MNCs have made public commitments to source only sustainably produced palm oil by 2015. However, CSPO consumption by India and China, the world's largest users of palm oil, will be a major challenge unless a clear business case at the enterprise level could be demonstrated. A recent scenario study by Oxfam International suggested that for market breakthrough, 20% of global demand for palm oil should be met by certified production by 2013, and 30 and 40% as CSPO by 2016 and 2019, respectively (Verburg, 2010). In developing this scenario, Oxfam International had assumed that three emerging markets (China, India, and Indonesia) would make the most significant contributions to the future demand for CSPO. The most ambitious scenario was for China, and it was projected that the proportion of CSPO in relation to total demand could be 10% in 2013, 20% in 2015, 30% in 2017, and 40% in 2019. What strategies and actions should RSPO and players adopt to meet these very challenging targets?

On the supply side, meeting the market transformation goal would mean that the new production areas would have to comply with established principles and practices for production. The business-as-usual model that causes deforestation or gives rise to social conflicts is no longer an option. The challenge for RSPO and the industry is to develop new business models that support economic growth without negative impacts to people and the natural environment. New inclusive models should focus on realizing the tremendous potential in raising the productivity of the smallholder sector, particularly independent smallholders, in meeting the rising demand for palm oil in the future. While the projected global demand for palm oil would result in opening land for plantations, future development should be directed at utilization of degraded or idle land.

Given the unprecedented progress achieved since its formation in 2004 and the recent improvements in the organizational structure and management, RSPO is well placed to take the lead in transforming the global market for sustainable palm oil across the supply chain, from production to consumption. The RSPO standards and certification systems and supply chain mechanisms are now well accepted in the market place and they provide the basis for production and trade in certified sustainable palm oil. The business case for production of CSPO has also been established. Experiences from several companies, which account for about 45% of the current production CSPO, have derived economic benefits as well as non-financial benefits such as improvements in corporate image, operational efficiencies, and the well-being of its workers and associated smallholders.

The single most enduring strength of the RSPO is the multi-stakeholder nature of the association. To meet the challenges of the future, each stakeholder must rise to the challenge. The Executive Board and Secretariat must demonstrate good governance and diligently enforce the principles of sustainability. Taking a strong stance on those failing to uphold the standard will send a powerful message to a world that still distrusts palm oil. The producers themselves must continue to push with their time-bound plans to certify all their estates and mills and urge others to join and commit to the RSPO process. The environmental NGOs, with their high level of trust amongst the general public need to react strongly not only to the negatives within the industry but also to those players journeying towards sustainability. For this to happen there must be a complete de-coupling of palm oil with forest destruction. The social NGOs also need to place their weight behind those trying to bring about change but will only do so once they see real change within the social landscape and a commitment by producers to bring about poverty reduction and prosperity to communities who are freely engaged with the oil palm sector. Investors need to recognize and use their strength to devise financial strategies that reward excellence in performance while penalizing recalcitrant players. Processors and retailers need the courage to implement the various supply chain options and to listen to what the consumers want and then to move more towards the producers in creating traceable and sustainable supply chains.

References

Ahmad Tarmizi, A. Felda—A success story. Global Oils Fats Bus. Mag. 2008, 5, 1, 7–11.

AidEvironment. Towards effective conservation strategies: The application of strategic principles to increase the impact and sustainability of WWF conservation efforts. Report commissioned by WWF Nederland, on behalf of WWF US and WWF UK, 2007.

ASEAN. Guidelines for the Implementation of the ASEAN Policy on Zero Burning. The ASEAN Secretariat, Jakarta, 2003.

Ayat, R.; Abdullah, R.; Mohd Shariff, F.; Arif Simeh, M. The Malaysian palm oil supply chain: The role of the independent smallholder. Oil Palm Ind. Econ. J. 2008, 8(2), 17–27.

Basiron, Y. Preparing the palm oil industry for the opportunities and challenges of tomorrow. Paper presented at MPOC Reach & Teach, January 2011.

Bethe, E. Increasing benefits for oil palm smallholders. Presentation at the 8th Roundtable Meeting on Sustainable Palm Oil, Jakarta, Indonesia, November, 2010.

Clay, J.W. World agriculture and the environment: a commodity-by-commodity guide to impacts and practices. World Wildlife Fund, USA, 2004.

Colchester, M.; Jiwan, N. Ghosts on our own land: Indonesian oil palm smallholders and the Roundtable on Sustainable Palm Oil. Forest Peoples Programme & Sawit Watch, 2006.

Colchester, M.; Jiwan, N.; Andiko Sirait, M.; Firdaus, A.Y.; Surambo, A.; Pane, H. Promised Land: Palm oil and land acquisition in Indonesian—implications for local communities and indigenous peoples. Forest Peoples Programme & SawitWatch, 2006.

Colchester, M.; Wee, A.P.; Wong, M.C.; Jalong, T. Land is life: Land rights and oil palm development in Sarawak. Forest Peoples Programme & SawitWatch, 2007.

Food and Agriculture Organisation. Trade Reforms and Food Security: Conceptualizing the Linkages, 2003.

Food and Agriculture Organisation. How to Feed the World in 2050, October 2009.

Financial Times, "Migros commits to buying green palm oil". http://ft.com. 23 January, 2002.

Friends of the Earth. "Greasy palms—palm oil, the environment and big business". 2005.

FSG. Improving livelihoods of palm oil sector smallholders: The role of the private sector. Commissioned by the IFC of the development of the WBG Framework for engagement of the palm oil sector, 2010.

GAPKI, Celebrating 100 years—Commercial Oil Palm Industry in Indonesia, 2011. http://oilpalm100.com/index.html (accessed 22 May 2011).

Government of Brazil website. Brazil Launches National Program for Environmentally Sustainable Palm Oil Production, 6 May 2010. http://www.brasil.gov.br/para/press/press-releases/may/brazil-launches-national-program-for-environmentally-sustainable-palm-oil-production/br_model1?set_language=en.

Greenpeace. How the palm oil industry is cooking the climate. Greenpeace International, The Netherlands, 2007.

Hanim, A. Threat to market access looms. StarBiz 12 November 2009.

Hanim, A. Transforming Felda. Starbiz 21 May 2011.

Hayes, K.C. Fatty acid expert roundtable: Key statements about fatty acids. J. Am. Coll. Nutr. 2010, 3, 285S–288S.

Hickman, M. The guilty secrets of palm oil: Are you unwittingly contributing to the devastation of the rain forests? 2009. http://www.independent.co.uk/environment/the-guilty-secrets-of-palm-oil-are-you-unwittingly-contributing-to-the-devastation-of-the-rain-forests-1676218.html (Accessed 19th March 2010).

Hooijer, A.; Silvius, M.; Wosten, H.; Page, S. Peat-CO2: Assessment of CO2 emissions from drained peatlands in SE Asia. Delft Hydraulics report Q3943, 2006.

Hornby, C. Nestle drops palm oil supplier after report. 2010. http://www.reuters.com/article/idUSTRE62G3PM20100317 (Accessed 16 April 2010)

Iko Hari, Y.; Susanto, P.; SAleh, C.; Andayani, N.; Prasetyo, D.; Atmoko, S.S.U. Guidelines for better management practices on avoidance, mitigation and management of human-orangutan conflict in and around oil palm plantations. WWF Indonesia, 2006.

Industry Week. Unilever backs moratorium on palm oil deforestation in Indonesia, 2008. http://www.industryweek.com/articles/unilever_backs_moratorium_on_palm_oil_deforestation_in_indonesia_16245.aspx (Accessed 16 April 2010).

IOI Corporation Berhad. Giving Back: Annual Report, IOI Corporation, 2008.

IOI Corporation Berhad. Re: Breach of RSPO Code of Conduct 2.3 and Certification Systems 4.2.4 (c). Letter of 26 April to RSPO, 2011.

Jakarta Post. Encroachement in Tesso Nilo National Park still going on: minister, 2010. (http://www.thejakartapost.com/news/2010/09/24/encroachment-tesso-nilo-national-park-still-going-minister.html

Jakarta Post. Oil palm plantations to take part in ISPO pilot projects, 2011. http://www.thejakartapost.com/news/2011/02/07/20-oil-palm-plantations-take-part-ispo-pilot-projects.html

Jiwan, N. Political economy of the Indonesian palm oil industry: A critical analysis. Presentation at ISEAS Workshop on the Oil Palm Controversy in Transnational Perspective. Singapore: Institute of Southeast Asian Studies, March 2009.

Karupaiah, T.; Sundram, K. Effects of stereospecific positioning of fatty acids in triacylglycerol structures in native and randomized fats: a review of their nutritional implications. Nutr. Metab. 2007, 4, 16, doi:10.1186/1743-7075-4-16.

Khoo, K.M. Some key productivity and efficiency issues facing the oil palm industry from a management perspective. Paper presented at OFIC 2010 Congress-Oils and Fats Industry: Challenges and Innovative Solutions. October, 2010.

Koh, L.P.; Wilcove, D.S. Cashing in palm oil for conservation. Nature August 2007, 448, 30, http://dl.dropbox.com/u/939465/Koh&Wilcove2007_Nature_448_993.pdf.

Koh, L.P.; Wilcove, D.S. Is oil palm agriculture really destroying tropical biodiversity. Conservation Letters 1, 2008, 60–64.

Khosla, P.; Sundram, K. Introduction. J. Am. Coll. Nutr. 2010, 29, 3(S).

Khosla, P.; Sundram, K. Nutritional characteristics of palm oil. In Reducing Saturated Fats in Foods. Talbot, G., Ed. Woodhead Publishing: PA, 2011; pp. 112–124.

Kuntom, A. MPOB Codes of Practice. Presentation from www.mpob.gov.my, 2010.

Laurance, W.F.; Koh, L.P.; Butler, R.; Sodhi, N.S.; Bradshaw, C.J.A.; Neidel, J.D.; Consunji, H.; Vega, J.M. Improving the Performance of the Roundtable on Sustainable Palm Oil for Nature Conservation. Society for Conservation Biology. Conserv. Biol., 2010.

Malaysian Palm Oil Council. Four steps to rationality for sustainable biofuels. Global Oils Fats Bus. Mag. 2008, 5, 3.

Malaysian Palm Oil Council. Palm Oil—The Green Answer, 2009. <http://www.mpoc.org.my/Palm_Oil_-_The_Green_Answer.aspx> (Accessed 10 August 2009).

Marti, S. Losing Ground: The human rights impacts of oil palm expansions in Indonesia. Friends of the Earth, LifeMosaic & SawitWatch, 2008.

McKinsey. How companies manage sustainability, 2010. http://www.mckinseyquarterly.com/How_companies_manage_sustainability_McKinsey_Global_Survey_results__2558.

Mielke, T. Global supply and demand outlook for palm and lauric oils—Trends and future prospects. Presentation at MPOC POINTERS 2011 internet conference, 2011. http://www.pointers.org.my/report_details.php?id=46

Mozaffarian, D.; Stampfer, M.J. Removing industrial trans fat from foods. BMJ 2010, 340, c1826.

Nellemann, C.; Miles, L.; Kaltenborn, B.P.; Virtue, M.; Ahlenius, H. The Last Stand of the Orangutan. State of Emergency: Illegal Logging, Fire and Palm Oil in Indonesia's National Parks. UNEP 2007.

Paoli, G.D.; Yaap, B.; Wells, P.L.; Sileuw, A. CSR, oil palm and the RSPO: Translating boardroom philosophy into conservation action on the ground. Tropical Conserv. Sci. 2010, 3(4), 438–446.

Paramanathan, S. Tussle over peatlands. Global Oils Fats Bus. Mag. 2008, 5, 3, Pullout.

ProForest, Review of policy options relating to sustainable palm oil procurement. DEFRA, April 2011.

Pye, O. An analysis of transnational environmental campaigning around palm oil. Presentation at ISEAS Workshop on the Oil Palm Controversy in Transnational Perspective. Institute of Southeast Asian Studies: Singapore, March 2009.

Rautner, M.; Hardiono, M; Alfred, R.J. Borneo: Treasure island at risk—Status of forest, wildlife and related threats on the island of Borneo. WWF Germany, Frankfurt, 2005.

Rice, A.L. Moving from efficacy to effectiveness: Red palm oil's role in preventing vitamin A deficiency. J. Am. Coll. Nutr. 2010, 3, 302S–313S.

Roundtable on Sustainable Palm Oil. RSPO Principles and Criteria for Sustainable Palm Oil Production. Public release version. RSPO, Kuala Lumpur, 17 October, 2005.

Roundtable on Sustainable Palm Oil. RSPO Certification Systems. RSPO, Kuala Lumpur, 2007.

Roundtable on Sustainable Palm Oil. RSPO Supply Chain Certification Systems. RSPO, Kuala Lumpur, 2009.

Roundtable on Sustainable Palm Oil. RSPO Newsletter Issue #20, October–December, 2010a.

Roundtable on Sustainable Palm Oil. Roundtable unveils 'sustainable palm oil' trademark. Press release, November, 2010b.

Rowell, A.; Moore, P.F. Global review of forest fires. WWF International & The World Conservation Union (IUCN), 2000.

SAN Sustainable Agriculture Standard, July 2010.pdf

Sawit Watch. Letter of 7th October 2002 to the NGO Workshop in November, 2002 in Zurich.

Sen, C.; Rink, C.; Khanna, S. Palm oil derived natural vitamin E alpha-tocotrienol in brain health and disease. J. Am. Coll. Nutr. 2010, 3, 314S–323S.

Sheil, D.; Casson, A.; Meijaard, E.; van Nordwijk, M.; Gaskell, J.; Sunderland-Groves, J.; Siri-Tarino, P.W.; Sun, Q.; Hu, F.B.; Krauss, R.M._Meta-analysis of prospective cohort studies evaluating the association of saturated fat with cardiovascular disease. Am. J. Clin. Nutr. 2010, 91, 535–546.

Sundram, K.; Hayes, K.C.; Siru, O.H. Both dietary 18:2 and 16:0 may be required to improve the serum LDL/HDL cholesterol ratio in normocholesterolemic men. J. Nutr. Biochem. 1965, 4, 179–187.

Sylvester, P.W.; Kaddoumi, A.; Nazzal, S.; El Sayed, K.A. The value of tocotrienols in the prevention and treatment of cancer. J. Am. Coll. Nutr. 2010, 3, 324S–333S.

Stern, N. Stern Review: The Economics of Climate Change. Cambridge University Press: Cambridge, UK, 2006.

Suharto, R. Sustainable production in Indonesia. Presentation at the China International Oil and Oilseeds Summit 2009, 8–10 July 2009, Beijing.

Suharto, R. Indonesian sustainable palm oil system (ISPO). Presentation at the 8th Roundtable Meeting on Sustainable Palm Oil, November, 2010, Jakarta, Indonesia.

Teoh, C.H. Global directions in sustainable development from Rio to Johannesburg: Challenges and opportunities for the oil palm industry in Malaysia. 4th International Planters Conference 2003: Globalisation and its Impact on the Palm Oil Industry 2003.

Teoh, C.H. Key Sustainability Issues in the Palm Oil Sector: A discussion paper for multi-stakeholder consultations, 2010. (http://www.ifc.org/ifcext/agriconsultation.nsf/Content/Key Documents).

Unilever. Unilever takes a stance against deforestation. Press Release, 11 December 2009.

Unilever, Sustainable Agriculture Code. 2010. http://www.unilever.com/images/sd_Unilever_Sustainable_Agriculture_Code_2010_tcm13-216557.pdf.

Uryu, Y. And 15 co-authors. Deforestation, forest degradation, biodiversity loss and CO2 emissions in Riau, Sumatra, Indonesia. WWF Indonesia Technical Report, 2008.

Verburg, J. An international NGO view on the market breakthrough for sustainable palm oil. Presentation at the 8th Roundtable Meeting (RT8) on Sustainable Palm Oil, Jakarta, 2010.

Vermeulen, S.; Goad, N. Towards better practice in smallholder palm oil production. Natural Resource Issues Series No. 5. International Institute for Environment and Development, London, UK, 2006.

Wertz, K.; Kanninen, M. The impacts and opportunities of oil palm in Southeast Asia: What do we know and what do we need to know? Occasional paper no. 51, CIFOR, Bogor, Indonesia, 2009.

Wicke, B.; Sikkema, R.; Dornburg, V.; Faaij, A. Exploring land use changes and the role of palm oil production in Indonesia and Malaysia. Land Use Policy 2011, 28, 193–206.

Wong, J. Study: CO_2 emission lower from peatland. StarBizWeek 17 July 2010, SBW12.

World Bank Group, The WBG's Framework for Engagement in the Palm Oil Sector (Consultation Draft), 2010.

World Bank Group. The World Bank Group Framework and IFC Strategy for Engagement in the Palm Oil Sector, March 31, 2011.

WWF. Summary NGO Workshop on WWF Forest Conversion Initiative, Zurich, November 20022002a.

WWF. Oil Palm Position Paper, February 2002b.

WWF International. The year the world caught fire—Discussion Paper. WWF International, 1997.

WWF HOB NI, Business solutions: Delivering the Heart of Borneo Declaration 2011.

WWF International. WWF palm oil buyers' scorecard 2009, WWF International/Forest Conversion Programme, 7 pp. www.panda.org/palmoilscorecard.

Zen, Z.; McCarthy, J.F.; Gillespie, P. Linking pro-poor policy and oil palm cultivation. Australia Indonesia Governance Research. Policy Brief 5, 2008.

Editors and Contributors

Editors

Oi-Ming Lai is a Professor in Enzyme Technology from the Department of Bioprocess Technology, Faculty of Biotechnology and Biomolecular Sciences, Universiti Putra Malaysia. She obtained her Ph.D. in Biotechnology from Universiti Putra Malaysia in 1999. She is the coeditor of *Healthful Lipids* (AOCS Press), the author or co-author of more than 100 referenced SCI publications, more than 8 book chapters and the holder of 8 patents. She has taught undergraduate and graduate courses in enzyme technology, food biotechnology, industrial biotechnology, and food biocatalysis. Oi-Ming's research program includes the development and improvement of structured lipids, upscaling and bioreactor design and kinetics, and product diversification of various oilseed components and its by-products. She has won numerous professional awards including the WIPO's Best Invention by Woman Award in 2008 in Brussels, Belgium. She also sits on the Editorial Board of the Food and Bioprocess Technology Journal published by Springer USA.

Chin-Ping Tan is an associate professor in the Department of Food Technology, Faculty of Food Science and Technology, Universiti Putra Malaysia. He obtained his B.Sc. in Food Science and Technology and Ph.D. in Food Processing from Universiti Putra Malaysia in 1998 and 2001, respectively. From 2002 to 2004, he served as a JSPS Postdoctoral Fellow at National Food Research Institute, Japan. He has a strong interest in physicochemical properties and changes in quality of palm oil products. His current research interests center on lipid science and technology, food nanotechnology, food emulsion, food processing, flavor sciences, and thermal analysis. He has published 5 book chapters, over 120 refereed papers in international refereed journals and presented more than 130 papers in various national and international conferences. In June 2011, he was appointed Visiting Professor at King Saud University, Riyadh, where he is involved in the research related to extraction and modification of plant oil from underutilized oilseeds. To date, he has received more than 40 research awards by various national and international bodies. Most recently, he was awarded the ProSPER.Net-Scopus Young Researcher Award 2010, TWAS Young Affiliate Fellow (2010–2014) and Universiti Putra Malaysia's Excellent Researcher in Science and Technology in 2011.

Casimir C. Akoh received the Ph.D. degree (1988) in food science from Washington State University, Pullman. He holds B.S. and M.S. degrees in Biochemistry from Washington State University and the University of Nigeria, Nsukka, respectively. He is a Distinguished Research Professor of Food Science and Technology and an Adjunct Professor of Foods and Nutrition at The University of Georgia, Athens. He is the coeditor of the book *Carbohydrates as Fat Substitutes* (Marcel Dekker, Inc.), coeditor of *Healthful Lipids* (AOCS Press), editor of *Handbook of Functional Lipids* (CRC Press), the author or coauthor of over 210 referenced SCI publications, and the holder of three U.S. patents. He is an ISI Highly Cited Researcher in Agricultural Science. Springer Publisher listed his paper as the top 5 most cited articles in 2008/2009 in the *J. Am. Oil Chem. Soc.* He is a Fellow of the Institute of Food Technologists (2005), American Oil Chemists' Society (2006), and the American Chemical Society (2006). He serves on the editorial boards of six journals and is a member of the Institute of Food Technologists, the American Oil Chemists' Society, and the American Chemical Society. He has received numerous international professional awards for his work on lipids including 1998 IFT Samuel Cate Prescott Award, 2003 D. W. Brooks Award, 2004 AOCS Stephen S. Chang Award, 2008 IFT Stephen S. Chang Award, 2008 IFT Research and Development Award, 2009 AOCS Biotechnology Division Lifetime Achievement Award, and 2012 Supelco/Nicholas Pelick-AOCS Research Award. His research interest is on improving human health through healthful lipids produced by enzymatic reactions.

Contributors

Suraini Abd-Aziz received her B.Sc. (Hons) from National University of Malaysia, M.Sc. degree from University of Wales, Swansea, United Kingdom, and completed her Ph.D. in Biochemical Engineering at the University of Wales, Swansea, United Kingdom, in 1997. She is a Professor and now Head, Department of Bioprocess Technology, Faculty of Biotechnology and Biomolecular Sciences, Universiti Putra Malaysia. Her greatest achievement was that she was selected to be the first Malaysian and South-east Asian to win the prestigious UNESCO-L'Oreal Co-Sponsored Fellowship for Young Women in Life Sciences 2001 for Asia and The Pacific region. Following that in 2002, she was awarded the Young Asian Biotechnology Prize by the Bioscience and Bioengineering Society, Japan. Her research interests are in the field of Industrial and Environmental Biotechnology focused on the utilization of agro wastes for the production of value-added products such as enzymes, flavor, and fine chemicals. She also served as an internal and external examiner for MS and PhD Thesis Examination Boards, reviewer for some journal papers and books, expert panel for Malaysian government research grants and scholarship applications. Currently, she has published more than 80 cited journals internationally and more than 100 proceedings internationally and nationally.

Meilina Ong Abdullah obtained a B.Sc. degree in Applied Science from Universiti Sains Malaysia (Penang, Malaysia), M.Sc. in Industrial Microbiology from University of Malaya (Kuala Lumpur, Malaysia), and Ph.D. in Plant Biotechnology from Universiti Putra Malaysia (Selangor, Malaysia). Meilina has worked for the last 17 years with the Malaysian Palm Oil Board (MPOB) and is currently a Principal Research Officer and Head of the Breeding and Tissue Culture Unit in MPOB. Her research interest includes biomarker discovery and development, epigenenomics and systems biology of the oil palm. During her career at MPOB, Meilina has established MPOB's Microscopy Services, developed and patented diagnostic markers for tissue culture amenity and conformity.

Zurina Zainal Abidin A graduate of the University of Manchester, Dr. Zurina Zainal Abidin developed her career as a process engineer in Elektrisola (M) Sdn Bhd before joining Universiti Putra Malaysia (UPM) as a Tutor in 1999. She is now a senior lecturer in the Department of Chemical and Environmental Engineering. She is also an associate researcher for the Biosensor Group at the Institute of Advance Technology, Universiti Putra Malaysia. She is a member of several international and national professional bodies such as AIChE, Institution of Engineers Malaysia, and Board of Engineers Malaysia. During her service at the university, she has authored and co-author 16 papers in related fields. She received her B.Sc. in Chemical Engineering from Loughborough University (UK) and did her Master in Biotechnology at the University of Manchester Institute of Science and Technology (UMIST, UK). She completed her Ph.D. in 2006 from University of Manchester (UK) in Biochemical Engineering. Her research is mainly on biochemical engineering, which includes separation technology either in chemical and biochemical processes; AC electrokinetics, microfluidics, and wastewater treatment.

Roland Doye Abigor is a Research Biochemist, an assistant Director and the Head of Biochemistry Division in the Nigerian Institute For Oil Research (NIFOR), Benin City, Edo State, Nigeria. He studied Biochemistry in the University of Benin, Benin City, Nigeria, where he received his B.Sc., M.Sc., and Ph.D. (2000). His research interests include lipase extraction from a variety of plant and animal sources, their purification methods and applications in niche value addition processes and product development. He is the lead scientist of the team that developed a 10 L biodiesel reactor for the production of biodiesel. He has served and is currently serving on some important National committees. He is also a member of the American Oil Chemists' Society (AOCS), Chemical Society of Nigeria, Biochemistry and Molecular Biology Society of Nigeria, and Biofuels Society of Nigeria.

Dr. Omorefe Asemota is a Director of Research at the Nigerian Institute for Oil Palm Research, Benin City, Edo State, Nigeria, where he heads the Technology and Product Development Department. He received his B.Sc. Biology degree from the University of Ife and a Ph.D. in plant physiology (Molecular biology) from the Université Paris 7 in 1986. He is active in drought tolerant research, gene isolation and characterization, plant development, host pathogen relationship, as well as mechanisms of fruit ripening. Dr. Asemota is a recipient of numerous fellowships from the Institut des Sciences Vegetale, Gif-sur-Yvette cedex—France; Plant Biotechnology Institute, Saskatoon, Canada; The National University, Malaysia; The International Institute for Tropical Agriculture, Nigeria; etc.

Dr. Dayang Radiah Awang Biak joined Universiti Putra Malaysia in 1996 as a tutor. She received her B.Sc. (Chemical Engineering) from Case Western Reserve University, Cleveland, USA, in 1996 and pursued her Ph.D. in Chemical Engineering at University of Birmingham, United Kingdom, until 2000. Her research focus was on food processing, specifically crystallization of palm kernel oil. The coverage of her research interests now expands to the synthesis of advanced material and bio-energy from food waste and fermentable sugar, as well as applications of nanotechnology in food and advanced material synthesis.

Jane Mara Block received her B.Sc. in Pharmacy and Biochemistry from Federal University of Santa Catarina (Brazil) in 1988. She completed her Masters in Food Science in 1992 at the same university and her Doctorate degree at Campinas State University (Campinas, Brazil). Currently, she is an Associate Professor at Federal University of Santa Catarina. She has over 20 years of experience in Food Science with an emphasis on fats and oils processing and technology. Her areas of research interest include fats and oils oxidation, extraction and antioxidant activity of bioactive compounds of natural origin, formulation of zero-*trans* fats using neural networks and crystallization and performance of interesterified fats based on soybean oil blends. She completed a Post-Doctorate at the

University of Guelph in 2009–2010. She was the President of the Brazilian Fats and Oils Society from 2004–2006 and 2006–2008.

Mavath R. Chandran was educated at King's College, South Australia, and the University of Adelaide, graduating in Agricultural Economics and Technology in 1960. He is a 50+ year veteran of the agro-commodities industry with professional experience in the executive management of business operations within the tree crop plantation sector. He spent 36 years in the plantation industry then headed the Malaysian Palm Oil Association. He was involved in the formation of Roundtable on Sustainable Palm Oil (RSPO), elected as its first Honorary Member and currently the Advisor to the Executive Board. He holds directorships in various commodity related business ventures and continues to serve the industry in an advisory and consultancy capacity at both regional and international levels.

Gait Fee Chung received his B.Agr.Sc. degree from University of Malaya (UM) in 1975 and M.Sc. degree from University of Science, Malaysia (USM) in 1977. He has been a crop protectionist and independent consultant since 2007, advising in the management of pests, diseases, weeds of plantation crops, and training personnel from plantations/agriculture sectors. From 1976 to 1981, he was employed in a research station of an international agrochemical company, where he conducted research and development of pesticides for the control of agricultural and public health pests in Malaysia. He joined the research department of a large plantation company in Selangor from 1981 to 2007 as crop protectionist, responsible for research in the management of pests, diseases and weeds of several crops (oil palm, rubber, cocoa, durian, mango, and greenhouse vegetable) and the advisory work in many estates.

Zulfitri Daud received his B.Sc. in Dietetics from the Universiti Sains Malaysia, his M.H.Sc. from Universiti Kebangsaan Malaysia, and his M.Sc. in Nutrition and Food Science from Wayne State University, USA. A tutor at Universiti Putra Malaysia, he is currently pursuing his Ph.D. and R.D. at Wayne State University. His research interest is in renal nutrition with a special interest in lipid metabolism pertaining to renal disease.

Matthias Diemer is an ecologist by training and received his Ph.D. at the University of Innsbruck in 1990. After working as a researcher at various universities and a stint in experimental education, he joined WWF Switzerland in 2003. He has served as Head of the Forest Unit, and currently leads the International Unit of WWF Switzerland. From 2003 on he represented the WWF network as Head of the Forest Conversion Initiative and served between 2004 and 2008 as board member and vice president of the RSPO. He is a member of various professional organizations (BES, GfÖ, etc.) and a lecturer for environmental studies at the University of Zürich.

Ir. Gerrit van Duijn has a Ph.D. in Applied Physics of the Delft University of Technology (1982). He started his professional career as researcher at Unilever Research, specializing in the technology of edible oil hydrogenation and became subsequently oil processing advisor for Unilever refineries. In 1990 he was transferred to Astra Calvé (Unilever France), as oil processing production manager responsible for a refinery with around 150,000 tons per annum production of sunflower oil, rapeseed oil, hydrogenated, and interesterified fats. After his return to Unilever Netherlands in 1996, he was involved in the restructuring of Unilever's oil processing activities.

From 1998 until November 2011, he directed the technological support of oils and fats supply to Unilever (from 3rd parties and own refineries). He is now a freelance consultant specialized in oil processing know how transfer. He is vice president of Euro Fed Lipid (the European federation for the science and technology of Lipids).

Dr. Véronique Gibon graduated from the University of Namur in Belgium, with a Ph.D. degree in physical chemistry and a specialty in crystallography (1984).

After her thesis, she worked as Research Assistant at the University of Namur, her major field of research being crystallization of edible fats; she is author of several scientific publications in this area. In 1995, she joined the industry and is now R&D Project Manager at the company Desmet Ballestra Group in Belgium.

Isona Leonard Gold is an oils and fats Chemist at the Nigerian Institute for Oil Palm Research, Benin City, Edo State, Nigeria. He holds a Ph.D. in Food Chemistry from the University of Benin (BNI, Nigeria). Isona L. Gold's research programs cover structural modification of fats and oils for enhanced nutritional performance and development of processing technologies for food and non-food value addition to products of oil palm and date palms in Nigeria. He is a member of the Chemical Society of Nigeria and the Institute of Chartered Chemists of Nigeria.

Teoh Cheng Hai was trained as an agronomist and has more than 40 years of experience in the plantation crops industry. He was Director, R&D (1989–1995), and Director, Quality and Environment (1995–2000), for Golden Hope Plantations Berhad (now part of Sime Darby Berhad) He was the Honorary Advisor, Plantation Agriculture, for WWF-Malaysia (2001–2004); in this capacity he was involved in the engagement of the Malaysian palm oil industry leading to the formation of the RSPO. He was appointed the 1st Secretary-General of RSPO in 2004/05, Advisor to the RSPO Executive Board in 2005/06, and Advisor to the RSPO Indonesia Liaison Office in 2006/07. Cheng Hai has served as a resource person/facilitator on assignments for various organizations including RSPO, WWF, Asian Productivity Organization (APO), ASEAN Secretariat, Department of Environment Malaysia, China Chamber of Commerce for Import & Export of Foodstuffs, Native Products and Animal By-Products (CFNA). Currently, he is a Non-Executive Director of CSR Asia and ProForest South East Asia.

Khairudin Hashim received his B.Sc. and M.Sc. from the Universiti Putra Malaysia (formerly known as Universiti Pertanian Malaysia) in 1979 and 1990, respectively. Subsequently, he obtained his Master in Management from the Asian Institute of Management in the Philippines. He has been Head, Group Sustainability and Quality Management of Sime Darby Berhad since 2011. Previously, he was the Head, R&D Centre of Sime Darby Plantation for eight years after working as scientist, senior scientist, and chief scientist. His experience covers all aspects of the oil palm industry from producing seeds, agronomy of oil palm, processing of oil palm fruits to refining and product developments from palm oil.

Hazimah Abu Hassan obtained her B.A. in Chemistry from California State University at Chico in 1981. She received her M.Sc. and Ph.D. degrees from Universiti Putra Malaysia (UPM) in 1990 and 1998, respectively. She started her service at Malaysian Palm Oil Board (MPOB) as a research officer in July 1990. She has a strong interest in organic chemistry and her research fields involve developing palm-based polyols and polyurethane products, synthesis of new oleochemicals, and recovery of valuable chemicals from oleochemical wastes. Under the Polymer and Composite Group, where she was a group Leader, they have developed processes to produce palm-based polyols for polyurethane products, and as a team, the group holds several patents on these processes. She is currently the Director of Advanced Oleochemical Technology Division of MPOB since April 2007.

Mohd Ali Hassan received his B.Sc. and M.Sc. degrees from Leeds University and completed his Ph.D. in Environmental Biotechnology at Okayama University Japan in 1997. He is currently a Professor of Environmental Biotechnology at the Department of Bioprocess Technology, Faculty of Biotechnology and Biomolecular Sciences, University Putra Malaysia (UPM). He has been actively pursuing research and development in oil palm biomass and effluent treatment and utilization with Professor Yoshihito Shirai from Kyushu Institute and Technology Japan, in collaboration with Felda Palm Industries Sdn. Bhd. (for more details, visit: www.ebgroup.upm.edu.my). The joint research had led to several pilot plants, and the biogas project has now been commercialized under the Clean Development Mechanism. Mohd Ali Hassan is currently the Dean of the faculty, Chief Editor of UPM's Pertanika Journal of Science and Technology and Vice President of the Asian Federation of Biotechnology (AFOB).

Ian Henson initially trained as a horticulturist, being awarded two National Diplomas in Horticulture by the UK Royal Horticultural Society. He then went on to obtain a B.Sc. in Botany at the University of Durham followed by a Ph.D., specialising in plant physiology, at University College of Wales, Aberystwyth. After a period of post-doctoral research in the Botany Department of the University of Glasgow, Henson moved to the Plant Breeding Institute at

Cambridge where he worked for six years on drought resistance of tropical cereals in association with the International Crops Research Institute for the Semi-arid Tropics in India and the International Rice Research Institute in the Philippines. A continued interest in drought research then took him to Western Australia to work further on crop production under limited rainfall. In 1985 he joined the Palm Oil Research Institute of Malaysia as a Senior Research Fellow where he led the oil palm physiology program, conducting comprehensive measurements of plantation carbon balance. He later undertook consultancies in oil palm research in Colombia before returning to Malaysia and the newly formed Malaysian Palm Oil Board (MPOB), initiating work on oil palm crop modelling, an interest in which he is still engaged. Henson has published extensively in the areas of plant and crop physiology, crop micrometeorology and modelling, with over 90 papers and reviews on oil palm and a similar number on other topics, and has developed computer software programs for simulating oil palm flowering, growth and yield and, most recently, for assessing the carbon footprint of oil palm plantations.

Peng Hu received his B.Sc. and M.Sc. degrees in Food Science and Engineering from Wuhan Polytechnic University. Currently, he is the professional director of Specialty Oils & Fats at Wilmar (Shanghai) Biotechnology R&D Center Co., Ltd, where his work focuses on structural, functional, and low-trans, low-saturated Specialty Oils & Fats using innovative technologies such as enzymatic modification and microbial fermentations.

Zainab Idris received her B.Sc. at the University of Laurentian in Ontario, Canada in 1988. She then pursued her M.Sc. at the Universiti Putra Malaysia in 1998 and completed her Ph.D. at the Universiti Kebangsaan Malaysia in 2010. She started her career in chemistry as a production chemist involving rubber examination gloves and rubber thread. She pursued her research interest in Malaysian Palm oil Board (MPOB) with a special interest in surfactant synthesis from palm oil based oleochemicals, particularly on fatty amine based surfactants for fabric softeners. She currently holds a position as the Head of Process Engineering and Design Unit, specializing in developing scale-up protocols and prototypes reactors for oleochemicals derivatives and products. She also holds a position as the group leader for the Specialty Chemical Group, focusing on reaction optimization, as well as characterization of palm based oleo derivatives in Advanced Oleochemical Technology Division of MPOB.

Celestine E. Ikuenobe was educated at the University of Ibadan, Ibadan, Nigeria where he received B.Sc., M.Sc., and Ph.D. from the Department of Agronomy. His research interest covers weed biology and management, and soil fertility in the oil palm, coconut, raphia, and date palms. He is currently a Director of Research at the Nigerian Institute for Oil Palm Research, Benin City, Edo State, Nigeria. He has taught at the University of Benin as Associate Lecturer and Ambrose Alli University as a visiting Lecturer. Dr. Ikuenobe is currently a National Expert on Oil Palm with the United Nations Industrial Development Organization. He is a member of many learned Societies and currently the Editor-in-Chief of the Nigerian Journal of Weed Sciences.

Maizura Ithnin completed her B.Sc. and Ph.D. degrees in Genetics at the Universiti Kebangsaan Malaysia in 1992 and 1999, respectively. She began her career at the Malaysian Palm Oil Board (MPOB) in 1992 where she was given a task to apply molecular marker technique to evaluate the genetic diversity of both the oil palm germplasm collection as well as its advanced breeding populations. When the MPOB Oil Palm Genebank Centre was established, Maizura took charge in establishing the *in vitro* cryopreservation facilities that houses approximately 15,000 accessions of oil palm genetic materials. Her experience in molecular markers was then applied in the genetic stability studies of the cryopreserved materials. Her current interest lies in associating molecular markers with economically important traits of oil palm.

Yuanrong Jiang received her Ph.D. degree in Food Science from Jiangnan University, where she lectured for six years. Since 2006, she has worked as senior engineer and R&D director in Wilmar International Limited. She has strong academic and industrial experience in oils and fats. Her research interests involve nutrition and health of fatty acids, oxidative stability and antioxidants of oils and fats, and their processing and flavor. Her outstanding research brought her more than 10 scientific achievement awards from the Chinese government, over 40 publications, and several patents. She is a standing director of Chinese Nutrition Society (CNS) and Food Sub-association and Oils & Fats Sub-association of Chinese Cereals and Oils Association (CCOA).

Deepinder Kaur received her B.Sc. and Postgraduate Degree in Dietetics from University of Delhi, India. To pursue her interest further, she completed her M.Sc. and Ph.D. degrees in Nutrition and Food Sciences at Wayne State University (Michigan, U.S.A) in 2011. During the course of her studies, she was involved in projects studying the effects of dietary fat on lipoprotein metabolism involving pathways affecting Alzheimer's disease and Cardiovascular disease in animal models and in humans. Her research focus is on the effects of various dietary macronutrient manipulations on weight management and lipoprotein metabolism in humans.

Dr. Pramod Khosla has been on the faculty at Wayne State University since 1996. He has been working on diet and lipoprotein metabolism for almost 25 years. Additionally, he has been doing palm oil-related research since 1989. Dr. Khosla is the author of 70+ original articles and book chapters and has consulted extensively for the oils and fats industry for some fifteen years. His current research interests include the interaction of dietary fat/carbohydrate on lipoprotein metabolism, life-style risk factors in various ethnic groups and oral supplements for decreasing inflammation. Recently, he was Guest Editor for a Special Supplement on Palm Oil that was published in the J. Am. Coll. Nutr. in June 2010.

Marieke Leegwater works as sustainability manager at the Dutch Product Board for Margarine, Fats and Oils: an Industry association for the Dutch Vegetable Oils and Fats industry. Within the Board she is responsible for improving the sustainability of Dutch soy and palm oil imports. In that role she supports the work of the Roundtable Sustainable Palm Oil (RSPO) and the Round Table on Responsible Soy (RTRS) and ensures communication about these initiatives in The Netherlands. For both palm oil and soy Marieke has contributed to the creation of international standards for sustainable production (Principles and Criteria) and the development of certification systems therefore. From 2009 to 2011 Marieke chaired the RSPO Standing Committee Trade and Traceability. She currently is a member of the RSPO Standing Committee Communication and Claims, and she supports the work of the Dutch Task Force Sustainable Palm oil. Furthermore she runs the secretariat of the Dutch Task Force Sustainable Soy. Marieke studied law at Leiden University. After graduating she started working at the Product Board, first as a food safety expert, later as a sustainability manager.

Chong Chiew Let is a Senior Research Fellow at the Malaysian Palm Oil Board (MPOB). He received his B.Sc. Joint Honours degree in Chemistry and Physics Studies from the University of Hull and his Ph.D. from Imperial College, University of London.

Chong has been actively involved in research on various aspects of palm oil, palm kernel oil, and their products. His main areas of interest are in quality of palm oil, palm oil chemical and physical properties, oil crystallization, oil transportation, storage and handling and environment management. His current research focus is on the development of processes for enhanced fractionation of palm oil and palm kernel oil where the crystallization of triacylglycerols is an important part of the process.

While at MPOB he has been awarded the Gold Medal Award and various excellent service awards. He has published papers in various peer reviewed journals and contributed to some book chapters.

Dr. Seong-Koon Lo received a Ph.D. in enzyme technology from Universiti Putra Malaysia in 2008. He is now a research scientist with Sime Darby Research Sdn. Bhd., where his interest in the field of edible oils and fats is explored. He began his research interest by working on modification of oils and fats using lipases. Immobilized lipases were applied as a green alternative to conventional chemical catalysts for processing edible oils and fats. Dr. Lo has co-authored several book chapters and patents on edible oils and fats technology.

Simon Lord has 25 years management experience in the oil palm sector. He began his career with NBPOL in 1995 after 10 years with Unilever. As Head of Research he developed and expanded the Groups research facilities. Simon currently is the Director of Sustainability New Britain Palm Oil (NBPOL). He is also a Director of the NBPOL foundation, a registered charity in PNG. He has been involved with the RSPO since 2004. He was a Vice President of RSPO (2006–2008) and Chair of the Standards and Certification sub group (2008). Simon has been a member of the numerous working Groups including the Criteria, Verification, New Planting, GHG and Smallholder think tanks from 2004 to 2010. He has facilitated in the National Interpretation of the RSPO standards in numerous countries and is currently on the RSPO Executive Board as alternate member for the Rest of the World producer sector. He is Director of Global Sustainability Associates (the consultancy wing of NBPOL) which provides advice on matters relating to production, efficiency and sustainability of palm oil. Simon holds a Ph.D. in Environmental Micro-biology from Bath University (UK).

Dr. Leslie Low is a Research Officer at the Malaysian Palm Oil Board, where he has been conducting research on oil palm genomics and genome informatics since 1999. He is also the Head of the Bioinformatics Unit and Group Leader of the Data Mining & Analysis Group. His research interests include data mining and annotation of oil palm genome data, sequence analysis, genome comparisons and transcriptomics. He received his B.Sc. in Biotechnology and Ph.D. in Plant Molecular Biology from University Putra Malaysia (Selangor, Malaysia).

Maria Madon is a Principal Research Officer (Group Leader Karyobiology) in the Malaysian Palm Oil Board (MPOB). She received her B.Sc. degree from Carleton University, Ontario, Canada, in Biology and completed her MS.c. and Ph.D. in Genetics at the National University of Malaysia (Bandar Baru Bangi, Selangor, Malaysia) in 1995 and 2000, respectively. Her areas of research are mainly on the oil palm cytogenetics, molecular cytogenetics, genetics, cellular biology, molecular biology, breeding and tissue culture in the efforts to generate elite oil palm planting material. MPOB offers Graduate Student Assistantship fund to postgraduate students from local universities to pursue their M.Sc. and Ph.D. programs and she has supervised nine postgraduate students (one Ph.D. and 8 M.Sc.) to date. Currently, she is also a member of the Technical Evaluation Panel for Agribiotechnology research projects funded under the Ministry of Science, Technology and Innovation (MOSTI). Throughout her service in MPOB, she has won several awards including MPOB's best publication award.

Ian McIntosh has over 30 years' experience in the food industry, which included posts with Mars Incorporated, Grand Met Foods, Northern Foods and Aarhus Karlshamn (AAK). Between 1998 and 2006, he was the Managing Director of AAK UK (formerly Aarhus United and the UK's largest palm oil refinery) and from 2006 he assumed responsibility for all the Groups businesses in the USA, Mexico and the UK. Mr. McIntosh served on the RSPO's organizing committee in 2003 and then served on the Executive board from 2004 to 2011. He was also the founder and Managing Director of the Green Palm program which operates the "Book & Claim" program for trading CSPO certificates. He holds a Masters Degree in Engineering Design and an MBA degree.

Jude Uche Obibuzor, received his Doctor of Philosophy (Ph.D.) Analytical Chemistry (2009), Master of Science (M.Sc.), Chemistry (1994) and Bachelor of Science (B.Sc) Industrial Chemistry (1988), respectively from the Federal University of Technology, Owerri, University of Lagos, Akoka-Yaba, Lagos and University of Nigeria, Nsukka. He is currently a research chemist and Assistant Head, Date Palm/Shea Tree/Jojoba Research and Development Department in the Nigerian Institute For Oil Research (NIFOR), Benin City, Nigeria. His current research interest is centered on the value addition along the value chain of palms and shea tree with bias to second and third generation biofuels from palms wastes vegetable oil and other emerging sources and nutraceuticals. He is a member of a team that developed a 10 L biodiesel reactor for the production of biodiesel from palm oil and other vegetable oils and less tedious method of producing high quality shea butter. He is a member of the Expert/working committee on standards for Edible Vegetable Oil/Allied Products of the Standard Organization of Nigeria (SON) Lagos, 1997 (the committee produced the current SON Edible Vegetable Oil/Allied Products standards). He is also a member of the American Oil Chemists' Society (AOCS), Chemical Society of Nigeria, Institute of Chartered Chemists of Nigeria (ICCON), and Biofuels Society of Nigeria.

Dr. Dere Okiy graduated from the University of Ibadan, Nigeria with a B.Sc. in Biochemistry in 1969. He had his M.Sc. and M.Phil. in Food Science from the University of Reading and London, respectively. His Ph.D. was in applied Chemistry from the University of Ife, Nigeria in 1983. His area of research interest includes fats and Oils chemistry with emphasis on the crystallization behavior of Palm oil and the reactions of crude palm oil at elevated temperatures. He has published numerous articles on fats and oils in the tropics. He recently retired as the Executive Director of the Nigerian Institute for Oil Palm Research (N.I.F.O.R.).

Emmanuel Okogbenin received his B.Sc. and M.Sc. from the University of Benin and currently on his Ph.D. program in Biochemistry. He joined the Nigerian Institute for Oil Palm Research April 2008 as research officer in training. His current position is research officer 1. He is working on improving oxidative stability of stored palm oil. His other areas of interests include Biofuel production, product and process improvement.

Dr. Rozita Omar is a senior lecturer at Chemical & Environmental Engineering Department in Universiti Putra Malaysia. She is also an associate researcher at the Institute of Advanced Technology, Universiti Putra Malaysia. She is currently a member of several international and national professional bodies such as AIChE, ISWA and Board of Engineers Malaysia. Over a short period of 8 years of her service at the university, she has managed to author and co-author 10 papers in her field. She received her B.Sc. in Chemical Engineering from Colorado School of Mines, USA, and her M.Sc. and Ph.D. from Universiti Putra Malaysia. Her research interest is in the area of wastes treatment and renewable energy production particularly in using microwave heating.

Garima Pande is currently a Ph.D. student at the University of Georgia, Athens, GA, USA. She received a B.Sc. degree in Agriculture from G. B. Pant University of Agriculture and Technology, India and an M.S. degree in Food Science from the University of Georgia. Her research foci are on phytochemicals and lipid biotechnology. She is student member of IFT and AOCS. She received first place award at the 2011 IFT Biotechnology Division Graduate Student Paper Competition.

Ana Cristina Pinheiro do Prado received her B.Sc. in Pharmacy and Biochemistry from the Federal University of Santa Maria (Brazil) in 2005. She completed her Masters in Food Science in 2008 at the Federal University of Santa Catarina (Brazil) and currently is completing her Ph.D. in Food Science there. She has experience in Pharmacy and Food Science with emphasis in Chemistry and Biochemistry. She works as a researcher investigating the following topics: antioxidant activity of bioactive compounds of natural origin and oxidation of oils and fats. She has published research papers in the following journals: Grasas y Aceites, Brazilian Journal of Food Technology, Experimental and Toxicologic Pathology, Ecotoxicology and Environmental Safety and International Journal of Food Sciences and Nutrition.

Rajinder Singh obtained B.Sc. degree in Applied Science from Universiti Sains Malaysia (Penang, Malaysia), M.Sc. in Biotechnology from University of Malaya (Kuala Lumpur, Malaysia) and Ph.D. in Plant Genetics from Universiti Putra Malaysia (Selangor, Malaysia). Rajinder has worked for the last 17 years with the Malaysian Palm Oil Board (MPOB) and is currently a Principal Research Officer and Head of the Genomics Research Unit in MPOB. His research interest includes marker development, genome sequencing, DNA Microarray, and marker assisted breeding. During his career at MPOB, Rajinder has developed and patented diagnostic DNA markers for oil palm breeding.

Aik Chin Soh obtained his BAgrSc (Hon) and MAgrSc degrees from University of Malaya, Malaysia and his Ph.D. degree from Oregon State University, USA. He is a Special Professor at the University of Nottingham Malaysia, a private consultant and an advisor to Applied Agricultural Resources Sdn Bhd where he was the Director of Research and Principal Plant Breeder. Dr. Soh has been involved in oil palm breeding and tissue culture for the past 30 years and was responsible for the development of the high oil yielding semi-dwarf hybrid and tissue culture clonal varieties. He has contributed more than 85 research papers and book chapters in leading international publications and was a reviewer of the breeding and genetics programs of MPOB and CIRAD, the two leading oil palm research institutes. He received the Malaysia Toray Science Foundation Outstanding Scientist Award in 2005 and was appointed as Fellow of the Academy of Sciences Malaysia in 2006.

Jan Kees Vis was trained as a chemist and received a Ph.D. in Heterogeneous Catalysis in 1984. He joined Unilever in 1985, in Research & Development. For more than 10 years he worked on Environmental Life Cycle Analysis, Environmental Management Systems, environmental auditing, standard setting, training, reporting and preparation of policy proposals on environmental issues. From 2001 to 2010, he was Global Supply Chain Director Sustainable Agriculture for Unilever. In 2010, his job title changed to Global Director Sustainable Sourcing Development, making him responsible for developing sustainability standards for all of Unilever's raw and packaging materials purchases. Jan Kees has been involved in, and holds or has held board positions in, Roundtable on Sustainable Palm Oil, Sustainable Agriculture Initiative Platform, Sustainable Food Laboratory and the Roundtable for Responsible Soy.

Xuebing Xu received his Ph.D. degree in Chemical Engineering from the Technical University of Denmark. His research focus is in enzyme technology and lipid product/process development. The research is in enzymatic technology, physical characterization of lipids and products, as well as process monitoring technology, and green solvent technology. Dr. Xuebing Xu has published more than 200 papers and chapters.

Shoot Kian Yeong received her Bachelor of Science in Chemistry from the University of Science Malaysia, Master of Science degree from University of Malaya, and Ph.D. Degree from Imperial College London. She joined Malaysian Palm Oil Board (MPOB) as a researcher in 1992, involving biotransformation of palm oil. She later worked on the synthesis and formulation of palm-based lubricant and grease. She holds the position as the Head of Synthesis and Product Development in the Advanced Oleochemical Technology Division of MPOB since 2007. She specializes in non-food application of palm oil which includes development of oleochemical derivatives for various applications such as cleaning products, polyurethanes, lubricants and industrial chemicals.

Tiankui Yang received his B.Sc. and M.Sc. degrees from Henan University of Technology, and his Ph.D. degree in Food Science and Technology from China Agricultural University, associated with Technical University of Denmark, in 2002. He was an assistant professor at BioCentrum-DTU, Technical University of Denmark from 2002 to 2004. He is currently a professor at the Department of Bioscience & Biotechnology of Dalian University of Technology, and senior consultant of Wilmar (Shanghai) Biotechnology R&D Center Co., Ltd. He is an executive member of the council of China Oils and Fats Association and China Proteins Association. His research interests focus on enzyme technology, biofuels, and functional lipids.

Dr Robiah Yunus is a Professor in Chemical Engineering from the Department of Chemical and Environmental Engineering, Faculty of Engineering, Universiti Putra Malaysia. She is a chartered chemical engineer, and a fellow at the Institution of Chemical Engineers, UK. She obtained her first degree in Chemical Engineering from the University of Alabama, USA, followed by the M.Sc. degree in Integrated Design of Chemical Plant from the University of Leeds, UK. In 2004, she completed her Ph.D. and was appointed as the Head of the Department of Chemical and Environmental Engineering, UPM in January 2004. Prior to joining Universiti Putra Malaysia, she served as a lecturer at the Universiti Technology Malaysia. She has taught undergraduate and graduate courses in various Chemical Engineering courses such as chemical process principles, thermodynamics, heat transfer, mass transfer as well as teaching process safety courses in other professional postgraduate programs. Besides teaching, she is also active in research primarily in the area of bio-based products synthesis, which encompasses chemical and energy transformation of bio-renewable resources. She has led numerous research projects worth more than RM9.5 million and has authored around 70 publications in international journals. She has also filed 8 patents, of which one has been granted in 2010. Currently, she is the deputy director for Institute of Advanced Technology, Universiti Putra Malaysia.

Yanping Zou received his B.Sc. and M.Sc degrees in Food Science and Engineering from Wuhan Polytechnic University and his Ph.D. degree in Biochemical Engineering from East China University of Science and Technology. He carried out his postdoctoral research with Dr. Dayuan Zhu (Shanghai Institute of Material Medica, Chinese Academy of Science, Shanghai, China), Dr. Se-Kwon Kim (Pukyong National University, Pusan, South Korea), and Dr. Sam Chang (North Dakota State University, Fargo, U.S.A.). His research interests have included valued products from Agri-by-products, bio-active compounds from Chinese Traditional Medicine, fruits, and vegetables, and biological evaluation of active constituents.

Index

D

G

P

FSC
www.fsc.org
MIX
Papier | Fördert gute Waldnutzung
FSC® C083411

Kolibri 360 GmbH
Ferdinand-Jühlke-Straße 7
99095 Erfurt, Deutschland
produktsicherheit@kolibri360.de